MICROBIAL EXOPOLYSACCHARIDES

Developments in Applied Microbiology and Biotechnology

MICROBIAL EXOPOLYSACCHARIDES

An Integrated Approach to Green Applications

Edited by

ONKAR NATH TIWARI

Principal Scientist, Centre for Conservation and Utilisation of Blue Green Algae (CCUBGA), Division of Microbiology, ICAR-Indian Agricultural Research Institute (PUSA), New Delhi, India

PARTHA PALIT

Assistant Professor, Department of Pharmaceutical Sciences, Assam University (A Central University), Silchar, Assam, India

PANNAGA PAVAN JUTUR

Group Leader, Omics of Algae Group, Industrial Biotechnology, International Centre for Genetic Engineering and Biotechnology (ICGEB), New Delhi, India

ELSEVIER

ACADEMIC PRESS

An imprint of Elsevier

elsevier.com/books-and-journals

Academic Press is an imprint of Elsevier

125 London Wall, London EC2Y 5AS, United Kingdom
50 Hampshire Street, 5th Floor, Cambridge, MA 02139, United States

ISBN: 978-0-443-36619-2

For Information on all Academic Press publications visit our website at https://www.elsevier.com/books-and-journals

Publisher: Andre Wolff
Acquisitions Editor: Kattie Washington
Editorial Project Manager: Namrata Aggarwal
Production Project Manager: Jenela Priscy
Cover Designer: Gopalakrishnan Venkatraman

Typeset by Aptara, New Delhi, India

Contents

4. Green extraction and production of high-yielded exopolysaccharides: Optimization of regulatory process parameters for industrial scale-up

Jothieswari Dhamotharan, Radha Rayi, Shaik Naveen Taj, Mopuri Deepa Reddy, and Sibbala Subramanyam

5. Role of exopolysaccharides as prebiotics in health-promoting nutraceutical formulation development

Sahanaz Khatun, Sohini Chatterjee, Mousumi Tudu, Suparna Garai, Pijush Kanti Shit, Abhishek Mohanta, Amalesh Samanta, and Arnab De

6. Probiotics potential of novel natural exopolysaccharides: Prospects and scope

Kaushiki Priya, Suparna Garai, Mousumi Tudu, Sohini Chatterjee, Abhishek Mohanta, Md. Nahid Zaman, and Amalesh Samanta

7. Exopolysaccharide-producing probiotics and their applications in translational medicine

Aiswarya S.V., Roopam Kumawat, Merugu Shashank Goud, Kesavan Bhaskar Reddy, Avijeet Singh Oinam, and Onkar Nath Tiwari

8. Microbial exopolysaccharides: A potential source for bio-fertilizers in eco-friendly green farming for crop improvement and productivity

Kuppam Chandrasekhar, Md. Nazneen Bobby, Davuluri Syam Babu, Alugunulla Venkata Narayana, Ambati Ranga Rao, and Prasun Kumar

9. Exopolysaccharides as raw nanobiomaterials for 3D printing technology

Blessy Maruthukunnel Mani, Sherin Shaji, Blessy Babukutty, Martin George Thomas, and Sabu Thomas

10. Potential application of EPS in nano-biosensor development toward biomedical diagnosis

Bedanta Bhattacharjee, Nayan Ranjan Ghose Bishwas, Hrishav Das Purkayastha, Ananya Bhattacharjee, Partha Palit, and Abhishek Bhattacharjee

11. Potential application of EPS as a cleansing agent

Amresh Kumar, Gaurav Gupta, Partha Palit, and Manabendra Dutta Choudhury

12. Recent progress and outlook of natural EPS for biomedical application

Onkar Nath Tiwari, Dipankar Ghosh, and Kesavan Bhaskar Reddy

13. Potential application of natural EPS in pest control in organic farming

Dipankar Ghosh, Indira Mikkili, and Onkar Nath Tiwari

14. Microbial exopolysaccharides (EPS) as a potential nano-theragnostic agent: Current perspectives and scope

Nilimanka Das, Rishabh Rana, Subodh Kumar, and Deepak Kumar

15. EPS: A promising antioxidant and antibiofilm agent

Laxmipriya Koijam, Avijeet Singh Oinam, Bijaya Thongataba, Tabitha Langhu, Md. Nazneen Bobby, and Onkar Nath Tiwari

16. EPS: A potential anti-inflammatory agent

Paramita Paul, Sucheta Sarkar, Mrinmoy Pratim Chetia, Tapan Kumar Shaw, Tarun Kumar Dua, Ranabir Sahu, and Gouranga Nandi

17. EPS: A potential application in bone tissue engineering for repairing and regeneration

Subhajit Mandal, Tanmay Jit, Ranabir Sahu, Gouranga Nandi, and Partha Palit

18. EPS: A potential agent as wound healer

Amlan Bishal, Tapan Kumar Shaw, Bratati Bandyopadhyay, Mohini Mondal, and Paramita Paul

19. The significant role of exopolysaccharides in biodegradable, renewable, and sustainable bioplastic production technology

Vadde Venkatesh, Gundu Shravani, Prasanna Kumar Desu, and Vanitha Kondi

20. Chemico-structural features and respective biomedical utility co-relationships of different categorized exopolysaccharides

Akshayaa Sankar, Ramachandran Venkataraman, Ramya Devi Durai, Sivanjineyulu Veluri, Marek Brzeziński, and Vedha Hari B Narayanan

21. Current status and advancements of natural exopolysaccharides as potential excipients and adjuvants in the pharmaceutical field

Sanjukta Sen, Piu Jana, Esha Das, and Gouranga Nandi

22. Latest status and advancements of EPS in the food industry and sustainable eco-friendly packaging

Dilip Kumar Roy, Partha Palit, Aparna Beena Unni, and Rajesh Kumar Mukherjee

23. Integrated application of exopolysacharides as pesticide and herbicide for green cultivation

Prasanna Sathiadevan Prathibha, Vinayaka Hegde, and Hiba Kanhirala

Contributors

Tijo Antony Department of Chemistry, Pavanatma College, Murickassery, Kerala, India; School of Energy Materials, Mahatma Gandhi University, Kottayam, Kerala, India

Ajmi Ayoob International and Inter University Centre for Nanoscience and Nanotechnology, Mahatma Gandhi University, Kottayam, Kerala, India

Davuluri Syam Babu Department of Biotechnology, Vignan's Foundation for Science, Technology and Research, Vadlamudi, Guntur, Andhra Pradesh, India

Blessy Babukkutty Federation University, Gippsland Campus, Melbourne, VIC, Australia

Blessy Babukutty School of Chemical Sciences, Mahatma Gandhi University, Kottayam, Kerala, India

Bratati Bandyopadhyay Bharat Technology, Uluberia, Howrah, West Bengal, India; Department of Pharmaceutical Technology, JIS University, Agarpara, Kolkata, West Bengal, India

Abhishek Bhattacharjee Department of Pharmaceutical Sciences, Assam University (A Central University), Silchar, Assam, India

Ananya Bhattacharjee Symbiosis Institute of Technology, Symbiosis International (Deemed University), Lavale, Pune, Maharashtra, India

Bedanta Bhattacharjee School of Pharmaceutical Sciences, Girijananda Chowdhury University, Tezpur Campus, Guwahati, Assam, India

Amlan Bishal Bharat Technology, Uluberia, Howrah, West Bengal, India

Nayan Ranjan Ghose Bishwas Department of Pharmaceutical Sciences, Dibrugarh University, Dibrugarh, Assam, India

Md. Nazneen Bobby Department of Biotechnology, Vignan's Foundation for Science, Technology and Research, Vignan University, Vadlamudi, Guntur, Andhra Pradesh, India

Marek Brzeziński Centre of Molecular and Macromolecular Studies, Polish Academy of Sciences, Lodz, Poland

Kuppam Chandrasekhar Department of Biotechnology, Vignan's Foundation for Science, Technology and Research, Vadlamudi, Guntur, Andhra Pradesh, India

Sohini Chatterjee Division of Microbiology and Pharmaceutical Biotechnology, Department of Pharmaceutical Technology, Jadavpur University, Kolkata, West Bengal, India

Mrinmoy Pratim Chetia School of Pharmacy, Arunachal University of Studies, Namsai, Arunachal Pradesh, India

Cintil Jose Chirayil Department of Chemistry, Newman College, Thodupuzha, Kerala, India

Manabendra Dutta Choudhury Department of Life Sciences and Bioinformatics, Assam University & Ravindranath Tagore University, Hojai, Assam, India

Esha Das Division of Pharmaceutics, Department of Pharmaceutical Technology, University of North Bengal, Darjeeling, West Bengal, India

Nilimanka Das Regional Institute of Pharmaceutical Science and Technology (RIPSAT), Agartala, Tripura, India

Nirupam Das Department of Pharmaceutical Sciences, Assam University (A Central University), Silchar, Assam, India

Arnab De Department of Pharmaceutics, Bharat Technology, Uluberia, Howrah, West Bengal, India

Debasmita Deb Department of Pharmaceutical Sciences, Assam University (A Central University), Silchar, Assam, India

Prasanna Kumar Desu Principal & Professor, Dr. Samuel George Institute of Pharmaceutical Sciences, Markapur, Andhra Pradesh, India

Jothieswari Dhamotharan Department of Pharmaceutical Analysis, Sri Venkateswara College of Pharmacy (Autonomous), Chittoor, Andhra Pradesh, India

Tarun Kumar Dua Department of Pharmaceutical Technology, University of North Bengal, Darjeeling, West Bengal, India

Ramya Devi Durai Pharmaceutical Technology Laboratory, School of Chemical & Biotechnology, SASTRA Deemed to be University, Thanjavur, Tamil Nadu, India

Suparna Garai Division of Microbiology and Pharmaceutical Biotechnology, Department of Pharmaceutical Technology, Jadavpur University, Kolkata, West Bengal, India

Dipankar Ghosh Microbial Engineering and Algal Biotechnology Laboratory, Department of Biosciences, JIS University, Agarpara, Kolkata, West Bengal, India

Merugu Shashank Goud Centre for Conservation and Utilization of Blue Green Algae (CCUBGA), Division of Microbiology, ICAR-Indian Agricultural Research Institute (PUSA), New Delhi, India

Gaurav Gupta Department of Immunology, University of Manitoba, Winnipeg, MB, Canada

Santosh Kumar Guru Department of Pharmacology, National Institute of Pharmaceutical Education and Research, Hyderabad, Telangana, India

Vinayaka Hegde Head, Crop Protection Division, ICAR-Central Plantation Crops Research Institute, Kasaragod, Kerala, India

Abu Md Ashif Ikbal Department of Pharmaceutical Sciences, Assam University (A Central University), Silchar, Assam, India

Piu Jana Division of Pharmaceutics, Department of Pharmaceutical Technology, University of North Bengal, Darjeeling, West Bengal, India

Tanmay Jit Department of Pharmaceutical Sciences, Assam University (A Central University), Silchar, Assam, India

Tomy Muringayil Joseph Department of Polymer Technology, Faculty of Chemistry, Gdańsk University of Technology, Gdańsk, Poland

Hiba Kanhirala Young Professional, Entomology Section, ICAR-Central Plantation Crops Research Institute, Kasaragod, Kerala, India

Hanieh Kargarzadeh Centre of Molecular and Macromolecular Studies, Polish Academy of Sciences, Lodz, Poland

Rajeev Kaushik Division of Microbiology, ICAR-Indian Agricultural Research Institute, New Delhi, India

Sahanaz Khatun Division of Microbiology and Pharmaceutical Biotechnology, Department of Pharmaceutical Technology, Jadavpur University, Kolkata, West Bengal, India

Laxmipriya Koijam Department of Biotechnology, T.S. Paul Manipur Women's College, Mongsangei, Manipur, India

Vanitha Kondi Department of Pharmaceutics, Vishnu Institute of Pharmaceutical Education and Research, Narsapur, Medak, Telangana, India

Amresh Kumar Department of Zoology, M.L. Arya College, Purnia, Bihar, India

Deepak Kumar Department of Pharmaceutical Chemistry, School of Pharmaceutical Sciences, Shoolini University, Bajhol, Himachal Pradesh, India

Prasun Kumar Department of Biotechnology, School of Engineering and Technology, Sharda University, Greater Noida, Uttar Pradesh, India

Subodh Kumar Department of Pharmaceutics, School of Pharmaceutical Sciences, Shoolini University, Bajhol, Himachal Pradesh, India; Department of Pharmaceutical Chemistry, School of Pharmaceutical Sciences, Shoolini University, Bajhol, Himachal Pradesh, India

Roopam Kumawat Centre for Conservation and Utilization of Blue Green Algae (CCUBGA), Division of Microbiology, ICAR-Indian Agricultural Research Institute (PUSA), New Delhi, India

Tabitha Langhu Department of Biotechnology, T.S. Paul Manipur Women's College, Mongsangei, Manipur, India

Debarshi Kar Mahapatra Chitkara College of Pharmacy, Chitkara University, Rajpura, Punjab, India

Subhajit Mandal Gitanjali College of Pharmacy, Birbhum, West Bengal, India; Department of Pharmaceutical Technology, University of North Bengal, Darjeeling, West Bengal, India

Blessy Maruthukunnel Mani International and Inter University Centre for Nanoscience and Nanotechnology, Mahatma Gandhi University, Kottayam, Kerala, India

Indira Mikkili Department of Biotechnology, Vignan's Foundation for Science, Technology and Research, Vignan University, Vadlamudi, Guntur, Andrha Pradesh, India

Abhishek Mohanta Division of Microbiology and Pharmaceutical Biotechnology, Department of Pharmaceutical Technology, Jadavpur University, Kolkata, West Bengal, India

Mohini Mondal Bharat Technology, Uluberia, Howrah, West Bengal, India; Department of Pharmaceutical Technology, JIS University, Agarpara, Kolkata, West Bengal, India

Rajesh Kumar Mukherjee Department of Pharmaceutical Technology, Brainware University, Kolkata, West Bengal, India

Gouranga Nandi Department of Pharmaceutical Technology, University of North Bengal, Darjeeling, West Bengal, India

Alugunulla Venkata Narayana Department of Biotechnology, Vignan's Foundation for Science, Technology and Research, Vadlamudi, Guntur, Andhra Pradesh, India

Vedha Hari B Narayanan Pharmaceutical Technology Laboratory, School of Chemical & Biotechnology, SASTRA Deemed to be University, Thanjavur, Tamil Nadu, India

Avijeet Singh Oinam Department of Botany, Regional College, Imphal, Manipur, India

Partha Palit Department of Pharmaceutical Sciences, Assam University (A Central University), Silchar, Assam, India

Paramita Paul Department of Pharmaceutical Technology, University of North Bengal, Darjeeling, West Bengal, India; Institute of Pharmacy, Bankura, West Bengal, India

Arpita Avinash Prabhu Department of Biotechnology, National Institute of Pharmaceutical Education and Research, Hyderabad, Telangana, India

Prasanna Sathiadevan Prathibha Senior Scientist, Entomology, Crop Protection Division, ICAR-Central Plantation Crops Research Institute, Kasaragod, Kerala, India

Kaushiki Priya Division of Microbiology and Pharmaceutical Biotechnology, Department of Pharmaceutical Technology, Jadavpur University, Kolkata, West Bengal, India

Hrishav Das Purkayastha Department of Pharmaceutical Sciences, Assam University (A Central University), Silchar, Assam, India

Rishabh Rana Faculty of Biotechnology and Applied Sciences, Shoolini University, Bajhol, Himachal Pradesh, India; Department of Pharmaceutical Chemistry, School of Pharmaceutical Sciences, Shoolini University, Bajhol, Himachal Pradesh, India

Ambati Ranga Rao Department of Biotechnology, Vignan's Foundation for Science, Technology and Research, Vadlamudi, Guntur, Andhra Pradesh, India

Meena Rathore Division of Microbiology, ICAR-Indian Agricultural Research Institute, New Delhi, India

Neeraj Kumar Ravi Department of Pharmaceutics, National Institute of Pharmaceutical Education and Research, Hyderabad, Telangana, India

Radha Rayi Department of Pharmaceutical Chemistry, Mohan Babu University, Tirupati, Andhra Pradesh, India

Kesavan Bhaskar Reddy Apollo Institute of Pharmaceutical Sciences, The Apollo University, Chittoor, Andhra Pradesh, India

Mopuri Deepa Reddy Department of Pharmaceutical Chemistry, Annamacharya College of Pharmacy (Autonomous), Kadapa, Andhra Pradesh, India

Dilip Kumar Roy Department of Pharmaceutical Technology, JIS University, Agarpara, Kolkata, West Bengal, India

Aiswarya S.V. Centre for Conservation and Utilization of Blue Green Algae (CCUBGA), Division of Microbiology, ICAR-Indian Agricultural Research Institute (PUSA), New Delhi, India

Ranabir Sahu Department of Pharmaceutical Technology, University of North Bengal, Darjeeling, West Bengal, India

Amalesh Samanta Division of Microbiology and Pharmaceutical Biotechnology, Department of Pharmaceutical Technology, Jadavpur University, Kolkata, West Bengal, India

Akshayaa Sankar Pharmaceutical Technology Laboratory, School of Chemical & Biotechnology, SASTRA Deemed to be University, Thanjavur, Tamil Nadu, India

Sucheta Sarkar Guru Nanak Institute of Pharmaceutical Science & Technology, Panihati, Kolkata, West Bengal, India

Sanjukta Sen Division of Pharmaceutics, Department of Pharmaceutical Technology, University of North Bengal, Darjeeling, West Bengal, India; Gitanjali College of Pharmacy, Birbhum, West Bengal, India

Sherin Shaji International and Inter University Centre for Nanoscience and Nanotechnology, Mahatma Gandhi University, Kottayam, Kerala, India

Pushpendra Sharma Division of Microbiology, ICAR-Indian Agricultural Research Institute, New Delhi, India

Tapan Kumar Shaw Department of Pharmaceutical Technology, JIS University, Agarpara, Kolkata, West Bengal, India

Pijush Kanti Shit Division of Microbiology and Pharmaceutical Biotechnology, Department of Pharmaceutical Technology, Jadavpur University, Kolkata, West Bengal, India

Gundu Shravani Department of Pharmaceutical Chemistry, Vishnu Institute of Pharmaceutical Education and Research, Narsapur, Medak, Telangana, India

Pankaj Kumar Singh Department of Pharmaceutics, National Institute of Pharmaceutical Education and Research, Hyderabad, Telangana, India

Vaibhav Soni Department of Pharmaceutics, National Institute of Pharmaceutical Education and Research, Hyderabad, Telangana, India

Sibbala Subramanyam Department of Pharmaceutical Chemistry, Vignan Foundation for Science Technology and Research, Guntur, Andhra Pradesh, India

Kirti Suman Division of Microbiology, ICAR-Indian Agricultural Research Institute, New Delhi, India

Shaik Naveen Taj Department of Pharmaceutics, Sri Padmavathi Mahila Visvavidyalayam, Tirupati, Andhra Pradesh, India

Martin George Thomas Institute Charles Gerhardt, ICGM UMR, Montpellierm, France; Department of Physics & Electronics, CHRIST (Deemed to be University), Bengaluru, Karnataka, India

Sabu Thomas Department of Physics & Electronics, CHRIST (Deemed to be University), Bengaluru, Karnataka, India; International and Inter University Centre for Nanoscience and Nanotechnology, Mahatma Gandhi University, Kottayam, Kerala, India; School of Chemical Sciences, Mahatma Gandhi University, Kottayam, Kerala, India; Department of Chemical Sciences, University of Johannesburg, Johannesburg, Gauteng, South Africa; Trivandrum Engineering Science & Technology Research Park, Sreekariyam, Trivandrum, Kerala, India; School of Energy Materials, Mahatma Gandhi University, Kottayam, Kerala, India

Bijaya Thongataba Department of Zoology, T.S. Paul Manipur Women's College, Mongsangei, Manipur, India

Onkar Nath Tiwari Centre for Conservation and Utilisation of Blue Green Algae (CCUBGA), Division of Microbiology, ICAR-Indian Agricultural Research Institute (PUSA), New Delhi, India

Prashant Tiwari College of Pharmaceutical Sciences, Dayanand Sagar University, Ramanagara, Karnataka, India

Mousumi Tudu Division of Microbiology and Pharmaceutical Biotechnology, Department of Pharmaceutical Technology, Jadavpur University, Kolkata, West Bengal, India

Aparna Beena Unni Institute of Physics, University of Silesia, Chorzów, Poland

Rini Thresia Varghese Department of Chemistry, Newman College, Thodupuzha, Kerala, India

Sivanjineyulu Veluri Centre of Molecular and Macromolecular Studies, Polish Academy of Sciences, Lodz, Poland

Ramachandran Venkataraman Pharmaceutical Technology Laboratory, School of Chemical & Biotechnology, SASTRA Deemed to be University, Thanjavur, Tamil Nadu, India

Vadde Venkatesh Department of Pharmaceutics, Vishnu Institute of Pharmaceutical Education and Research, Narsapur, Medak, Telangana, India

Md. Nahid Zaman Division of Microbiology and Pharmaceutical Biotechnology, Department of Pharmaceutical Technology, Jadavpur University, Kolkata, West Bengal, India

CHAPTER

1

Natural exopolysaccharides: Introduction, structural classification, and importance as a nanobiomaterial

Rini Thresia Varghese[a], *Tijo Antony*[b,c], *Hanieh Kargarzadeh*[d] and *Cintil Jose Chirayil*[a]

[a]Department of Chemistry, Newman College, Thodupuzha, Kerala, India [b]Department of Chemistry, Pavanatma College, Murickassery, Kerala, India [c]School of Energy Materials, Mahatma Gandhi University, Kottayam, Kerala, India [d]Centre of Molecular and Macromolecular Studies, Polish Academy of Sciences, Lodz, Poland

1.1 Introduction to natural exopolysaccharides

1.1.1 Definition and origin of exopolysaccharides

The term, natural exopolysaccharides (EPSs) has been obtained from the mixture of two terms, respectively, "exo" meaning external, and "polysaccharides" are the different monomers of sugar-containing carbohydrate moieties. EPSs are a type of biopolymer that exist in nature in the form of structural components of plants, animals, and microorganisms. They possess a high molecular weight (MW) in the order of 10–10,000 kDa. These exopolymers are typically made up of sugars, proteins, or other biopolymers, and noncarbohydrates like proteins, nucleic acids, enzymes, etc., and they are often referred to as EPSs, exoproteins, or biofilms, depending on their chemical composition. Another definition of EPSs is that they are produced by marine microorganisms and consist of a complex mixture of biopolymers, mainly consisting of polysaccharides, in addition to nucleic acids, lipids, proteins, and humic substances. EPSs are a broad group of compounds, consisting of repeating sugar units, and they differ significantly in terms of structure, composition, and function depending on the producing organism. They are biopolymers and are involved in many biological processes, including the formation of biofilms, resistance to environmental stress, and cell-cell communication. They possess several

DOI: https://doi.org/10.1016/B978-0-443-36619-2.00001-2

characteristics like biocompatibility, biodegradability, efficacy, edible, and nontoxicity. The EPSs are high MW natural metabolic biopolymeric matrices secreted by a variety of microorganisms and comprise repeating units of glucose, galactose, and rhamnose. The number of sugar units involved and the type of sugar determine the physicochemical characteristics and functional integrity of the polymer (Tiwari et al., 2024).

EPSs are primarily made up of carbohydrate fractions such as d-glucose, d-mannose, d-galactose, l-rhamnose, d-galacturonic acid, N-acetyl-d-glucosamine, and N-acetyl-d-galactosamine, as well as noncarbohydrate components that are abundant in carboxyl, phosphate, and pyruvate fractions.

1.1.2 Historical background and discovery

Sutherland coined the term "exopolysaccharide" in 1972 (Osemwegie et al., 2020) to describe high MW carbohydrate biopolymers made by microbes. Many distinct types of bacteria secrete bacterial EPSs, which have a variety of structures. In the middle of the 19th century, an EPS known as dextran was discovered in wine. *Leuconostoc mesenteriodes* was recognized as the prokaryote that produced it, marking the beginning of the history of bacterial EPSs. Other EPSs have been found over time, including xanthan, alginate, and cellulose. Exocellular polysaccharides, another name for EPSs or EPSs, are frequently released into the environment from the membrane level, and since they appear to be free, renewable, and unattached. Specialized fermentation procedures can readily extract them from organic biomasses. The term "exopolysaccharide" was first described as high MW polymers of carbohydrates produced by marine bacteria. EPSs composed of repeated sugar moieties may be associated with other molecules like proteins, organic, and inorganic compounds, lipids, DNA, and metal ions. The proper actions of EPS and its exact function depend on the ecological niches and structural units of the concerned microorganisms (Yadav et al., 2024).

1.1.3 Significance in modern biomaterials science

EPSs, due to their exceptional combination of properties like biocompatibility, nontoxicity, antimicrobial and antitumor activities, are of use in biomedical fields. They are widely used for tissue engineering applications, drug delivery systems, and viscosupplements.

EPS, which is a polymer with a wide range of uses, is formed by certain bacteria in their late logarithmic and early stationary phases. It possesses the properties of being hygroscopic, swelling, and viscous. Furthermore, bacterial EPSs contain certain chemical groups that enable them to form three-dimensional carrier molecules that help to retain a prolonged regulated release while delivering medications to precise locations in animal models. Their noncytotoxicity, biocompatibility, and antimicrobial, anti-inflammatory, and antitumor properties can make them attractive tools for creating three-dimensional carrier molecules, such as hydrogels, aerogels, liposomes, micro- and nanoparticles, scaffolds, films, and sponges, which can be used to deliver the active pharmaceutical ingredients to particular target cells and tissues. In addition, they provide an ambient environment that promotes the growth of many types of cells, which makes them suitable biomaterials for regenerative medicine. The microbial-derived EPSs are being explored in multiple biomedical dimensions with extensive studies on their structural, physio-chemical, and health-promoting properties that are bioactive, such as antiulcer, antigastritis, immunomodulatory, anti-inflammatory, and cholesterol-lowering effects, and gut-related diseases. Also, they have been used for wound-healing applications extensively.

Herein, we discuss the different applications of EPSs in different fields, especially in the field of biomedicine. Different forms of EPSs exist, which can be obtained from different forms of biomass, which are readily available in nature and is a renewable source.

The several types, synthesis mechanisms, composition, and different application possibilities have been discussed. The literature study also highlights how these novel materials can be utilized in the field of emerging biomaterial applications.

1.2 Structural classification of exopolysaccharides

1.2.1 Structural features

The main types of carbohydrates found in nature are biopolymers called polysaccharides. Long chains of various simple sugars (mono-, di-, and oligosaccharides) linked by glycosidic linkages polymerize to produce a carbonic chain that determines their shapes. The structural feature description of EPS generally includes the monosaccharide composition and conformation, glycosidic bond conformation, repeating units, MW range, linkage sites, and spatial structural arrangement. EPSs are a different class of biopolymers that are common and widespread as they are produced by a wide variety of micro-organisms, bacteria of all taxa, and microbes, and are excreted into the external environment. EPSs are polysaccharides with a heterogeneous structure that is made up of repeating sugar monomers and noncarbohydrate substituents. The vast array of EPS origins is reflected in the wide range of molecule sizes, compositions, and sequences. They are made up of several types of monosaccharides such as glucose, galactose, fructose, rhamnose, xylose, fructose, and mannose. EPS have two types of extracellular secretion states: Capsular polysaccharides that adhere to the microbial cell wall to form a capsule, or slime polysaccharides that are very loosely attached or sometimes completely released into the surrounding environment. Another classification of EPS includes EPS with the straight chain of monosaccharide units, known as linear EPSs, for example, pullulan (PUL), and EPS containing arms and bent monomer structures, such as EPS-W1 extracted from *Lactobacillus plantarum* W1, are called branched EPSs (Wang et al., 2023).

1.2.2 Conformational arrangement

EPSs show different properties under different conditions, such as solvent, pH, and temperature, which directly influence their potential application and the solid-state conformation. In the case of a majority of EPSs, a conformational transition is observed during the experimental conditions, which can lead to a more stable conformation. In general, the constitution of EPS regarding the nature and the regularity of the repetition of the monomers, the type of bonds, and the existence and also the length of side chains, is responsible for its conformational features. There are presently four types of EPS conformation proposed, which include: (1) Extended and ribbon-like chain, (2) flexible helical chain, (3) rigid and crumpled chain, and (4) very flexible chain. When the EPS is present in solution, the properties depend on the interactions that occur, which are polymer–polymer, polymer–solvent, and the competition between them. The capacity to create suitable arrangements between the chains and to adopt stable conformations is directly influenced by the constitution of the EPS and also the external conditions (especially the hydration degree). Hence, the understanding of the EPS composition and conformation is a major step before applying the materials to different systems and applications (Carvalho et al., 2021).

1.2.3 Classification based on monomers

EPSs are categorized based on their chemical structure, basically into homopolysaccharides and heteropolysaccharides. They can be hetero-polymeric with a range of high MWs (10–2000 kDa) and made up of repeating units of ten or fewer monosaccharide residues (Birch et al., 2019). Based on the chemical composition, EPS are categorized into homopolysaccharides, containing the same monomer units, and heteropolysaccharides, containing multiple or different monomer units. The EPS are generally categorized into two groups according to their chemical structure: Homopolysaccharides, which have only one type of sugar unit, and heteropolysaccharides, which have several different sugar units within their repeating units. Heteropolysaccharides consist of common sugar units like glucose, galactose, and fructose, but in some cases, sugars like rhamnose, fucose, xylose, and mannose are also part of the monosaccharide unit. This distinction is frequently crucial in determining the functional characteristics and possible uses of EPS (Padmanabhan and Shah, 2020). It is evident to note that the structure and composition of EPS, such as the glycosidic bond, monosaccharide composition, MW, and branching patterns determines their properties and also affect the biological activity of EPS. They can possess different structures like chain structure, triple helical structure, random coil formation, and rod-like structure, which affect and enhance their different biological properties.

1.2.4 Classification based on microbial sources

EPSs are polymers produced by living organisms like bacteria, fungi, and algae to protect themselves from environmental factors. They are complex biopolymers that are produced by various microorganisms, and they serve a variety of important roles, such as providing structural support, protecting the microbes from environmental stress, and aiding in adhesion to surfaces. Among the natural sources for obtaining EPS, the microbial EPSs such as alginate, chitosan, PUL, scleroglucan (Scl), xanthan gum, bacterial cellulose, etc., have favorable biological and mechanical properties. The EPS is classified into several types based on the microorganism that produces them, which causes changes in their structural features and properties. Microbial EPS yield varies based on the type of strain and factors such as the conditions of the fermentation parameters, such as pH, temperature, etc. Hyaluronan, kefiran, and gellan are examples of microbial polysaccharides that have been tested to produce biodegradable films. Based on microbial sources, they can be classified into:

1.2.4.1 *Bacterial exopolysaccharides*

These types of EPS are produced by bacteria like *Pseudomonas, Lactobacillus, Acetobacter*, and *Streptococcus*. Bacterial EPS are often composed of repeating sugar units and may be homopolysaccharides or heteropolysaccharides. Certain bacterial EPS may also have functional groups such as acetyl, phosphate, or pyruvate attached to the sugar units. Bacterial EPSs are high molecular-weight carbohydrate biopolymers excreted by bacteria into the extracellular environment. The most widely used EPS source is bacteria due to their quick reproduction, ease of renewal, and compatibility with the majority of EPS isolation techniques. Bacterial EPSs are easily detached from cells and are loosely linked slime layers. The number of EPS produced by bacteria is determined by the strains, the medium's composition, and the culture's parameters, including temperature, pH, and the carbon/nitrogen ratio. There are several applications for bacteria-derived EPS, such as *Streptococcus* EPS, which are important in pharmaceuticals (Netrusov et al., 2023), as they can help with drug delivery systems and have been studied for their immunomodulatory properties, and *Lactobacillus* EPS, which are used in the food industry for texture enhancement in dairy products, as well as in probiotics for gut health.

1.2.4.2 *Fungal exopolysaccharides*

These EPS are produced by Fungi like *Aspergillus, Neurospora*, and *Saccharomyces*. Fungal EPSs are secreted into the extracellular medium in the form of biofilm or capsules. Fungal EPSs are classified as acidic or neutral (i.e., β-glucans [βGs]) based on the presence or absence of uronic acid. Fungal EPS can be linear or branched and often consists of mannose, glucose, and galactose units. Hyperbranched fungal EPSs, for example, lentinan from Lentinus edodes and schizophyllan (SPG) from *Schizophyllum* commune, have numerous functional groups with high density and low viscosity. Their structures can vary greatly depending on the species and growing conditions (Hamidi et al., 2022). *Saccharomyces cerevisiae* (baker's yeast) EPS are utilized in the food and beverage industries, particularly for improving texture. Similarly, *Aspergillus* EPS are used in the food industry for the production of bioactive compounds.

1.2.4.3 *Algal exopolysaccharides*

EPSs obtained from different algae such as *Spirulina, Chlorella*, and *Porphyridium*, are referred to as algal EPSs. Algal EPS is typically composed of sulfated polysaccharides with galactose, sulfate groups, and uronic acids. The structures of these EPSs are highly diverse depending on the type of algae and the environmental factors during the growth conditions. Algal EPS represent a wide variety of chemical structures with high-molecular-weight polysaccharides (10–30 kDa) with homopolymeric and heteropolymeric compositions. Polysaccharides in algae are important producers of agar, alginate, carrageenan, dextran, gellan, PUL, and xanthan gum. These types of polysaccharides have properties such as gelling, stabilizing, or thickening, which are important agents as food additives. For instance, *Spirulina* and *Chlorella* EPS are used in the food industry for their potential prebiotic effects and ability to enhance gut health. Algal EPS are also utilized in the cosmetics industry for their moisturizing, antiageing, and anti-inflammatory properties.

1.3 Functional properties of exopolysaccharides as nanobiomaterials

1.3.1 Biodegradability and biocompatibility

EPSs are high-molecular-weight polymers that are secreted by a microorganism in the surroundings and are composed of sugar residues. EPS generally possesses inherent properties of nontoxicity, biodegradability, and biocompatibility. EPSs are naturally occurring biopolymers with outstanding biodegradability and biocompatibility, making them highly valuable for a range of applications. They possess excellent immunogenicity, cytocompatibility, and biocompatibility, which are ideal properties for use in biomedical devices, especially for the controlled delivery of active ingredients into the targeted cells (Carvalho et al., 2021). EPSs are biocompatible and biodegradable, and their degradation metabolites are mostly nontoxic, which can be attractive properties for further applications in medical devices. For example, an EPS biopolymer, PUL, is a nontoxic, biodegradable, biocompatible, nonimmunogenic, noncarcinogenic, nonmutagenic polymer with fantastic water absorptivity and impermeability to oxygen, and is used in pharmaceutical and biomedical industries and food packaging. The biodegradation of EPS involves extracellular enzymes, including EPS hydrolases and glycoside hydrolases, which cleave the polysaccharide chains into smaller oligosaccharides or monomeric sugars. The rate of biodegradation depends on factors such as the chemical composition of the EPS, the structure of the polysaccharide chains, and environmental conditions like temperature, oxygen levels, pH, etc. Films prepared from EPS showed good biodegradation for 28-day experiments with inoculum prepared from soil samples. EPS has been shown to support cell adhesion, proliferation, and differentiation, especially in the context of tissue engineering. They can provide a hydrophilic and biocompatible surface for cell attachment, which is essential for the formation of tissue-like structures.

1.3.2 Viscosity, gelation, and stability

EPS are capable of strong viscosity-enhancing properties when dissolved in water or other solvents, which makes them valuable agents as thickeners and stabilizers in industrial applications. Mostly, EPSs give bacteria a gel-like consistency in a solid medium and enhance the viscosity in a liquid medium. Some EPS exhibit **shear-thinning behavior**, meaning their viscosity decreases under applied shear stress (e.g., during mixing or pumping), making them easier to process. Additionally, viscosity can change with temperature; for example, **xanthan gum** maintains its viscosity across a wide temperature range, while other EPS, like **gellan** is more temperature sensitive.

1.4 Techniques for the extraction, purification, and characterization of exopolysaccharides

1.4.1 Isolation techniques

1.4.1.1 Precipitation

EPS are generally isolated through precipitation using solvents like methanol, followed by centrifugation and freeze-drying. Various solvents are used for precipitating EPSs. Most commonly, water and cold ethanol are used. Chloroform has been used in combination with butanol to remove lipid particles from EPSs. However, before isolation, EPS-producing microorganisms must be grown under optimal conditions because the quantity and distribution of EPSs are highly dependent on growth conditions. They can be grown by sugar nucleotide precursor synthesis. Several factors affect the production of EPSs, including the components of the culture medium (carbon source, nitrogen source, mineral salts), and also the culture conditions like pH (temperature, initial inoculum, incubation time, and the presence or lack of oxygen). The precipitation mechanism usually involves the use of 95% or 100% ethanol (v/v). Because ethanol is less polar than water, it breaks the hydration shell generated by water molecules interacting with the hydrophilic regions of EPS, lowering its solubility. EPS is further collected by centrifugation, dissolved in distilled water or Milli-Q water, and purified in several ways. Various other precipitating chemicals used to precipitate EPSs include propanol, acetone, and isopropanol. The precipitation method is widely used due to its simplicity, efficiency, minimal chemical contamination, and ability to handle large volumes, paving effortless ways for EPS extraction. However, some disadvantages of this method include incomplete precipitation, co-precipitation of impurities (proteins, nucleic acids, and other biopolymers), and the large volumes of nonsolvents, like ethanol, required (Yadav et al., 2024).

1.4.1.2 *Ultrafiltration*

Ultrafiltration is another method utilized for the extraction of EPS, which is simpler and more precise. Ultrafiltration operates on the concept of size exclusion and pressure-driven separation. Ultrafiltration effectively separates EPS from smaller contaminants, such as salt and tiny organic compounds, according to size. In this approach, the coagulated proteins and cell fractions from the liquid medium are removed by centrifugation, and the EPS by ultrafiltration. Different MW polysaccharides are separated in batches by gradual addition of speed in the differential centrifugation method. The high MW polysaccharides are separated at low speed, whereas the low MW polysaccharides are separated at higher speeds. Ultrafiltration is a membrane-based separation process that concentrates and purifies EPSs using semi permeable membranes. It is efficient, scalable, and chemical-free. However, obstacles such as membrane fouling, high initial costs, and potential limitations in MW selectivity must be addressed for successful application.

1.4.2 Purification

Ion-exchange chromatographic columns were used for the intensive purification of EPSs, e.g., Sephadex columns. A cation exchange resin can be used for EPS extraction. For instance, cation exchange resin separated the biofilm constituents that required positively charged ions for their attachment.

1.4.2.1 *Anion-exchange chromatography*

Anion-exchange column chromatography is a widely used analytical technique for polysaccharide isolation and purification. It is typically applied to eliminate negatively charged contaminants, such as proteins and nucleic acids, which bind to the positively charged resin. The underlying separation mechanism relies on ion exchange and adsorption-desorption. The most widely used anion exchangers so far are DEAE-cellulose, DEAE-sephadex, DEAE-sepharose, and phenyl-sepharose. In this method, the dialyzed and lyophilized crude EPS samples are dissolved in Tris-HCl (0.02 M, pH 7.4) buffer and loaded onto the anion exchange columns. The adsorption ability generally increases with an increase in acidic groups in the polysaccharide, usually, the adsorption capacity of straight-chain polysaccharides is greater than that of branched-chain polysaccharides. The challenges of this technique include its cost and careful optimization, and it is inappropriate for neutral or weakly charged EPS, and also extremely sensitive to ionic strength and pH.

1.4.2.2 Size-exclusion chromatography or gel-filtration chromatography

Here, the dialyzed and anion exchange chromatography eluted fractions of EPS are loaded in the GFC column, and the elution is performed using ultrapure water with an optimized flow rate. In this method, the dialyzed and AEC-eluted fractions of EPS are loaded in the GFC column, and the elution is performed using ultrapure water with an optimized flow rate. GFC uses size-based separation to purify EPSs without denaturing them, eliminating minute contaminants by desalting. It does this without the requirement for a specific buffer or sample loss from charge.

1.4.2.3 Dialysis

To further purify the EPS, dialysis can be performed against a buffer or distilled water to remove small molecules, salts, and residual solvents. Dialysis is performed using a dialysis membrane to remove tiny neutral sugars, media pollutants, small proteins, salts, and damaged EPS sections against distilled water/Milli-Q water or phosphate buffer saline (PBS, pH 7.4). Dialysis, which is carried out as the last step before drying, is a crucial process since it aids in the removal of carbohydrates with a low MW. Prior to dialysis, distilled water or Milli-Q water is used to dissolve the recovered EPS precipitates. Dialysis is particularly useful if a high degree of purity is required. Dialysis effectively eliminates minor impurities (salts, residual sugars, solvents, and tiny metabolites), does not require organic solvents, is easy to use, and does not require harsh chemicals or conditions, but requires more time.

1.4.3 Characterization

Once the EPS is isolated and purified, various analytical techniques can be used to analyze the structure, MW, and functional properties of the EPS. By combining different methods, researchers can efficiently isolate and purify EPSs for further study and potential applications in industries like food, pharmaceuticals, and biotechnology. The freeze-dried or lyophilized samples of EPSs have to be initially weighed and analyzed to determine the quantity and structural characteristics. The colorimetric method using a spectrophotometer analyzes the amount of sugar and its related components by the phenol-sulfuric acid method. The molar mass or MW of these biopolymers occurs in the range of 13×10^5 Da and can be determined by techniques such as high-performance liquid chromatography with gel permeation columns and also by GC-MS and NMR (^{13}C and ^{1}H) to analyze the structure, glycosidic linkages, and monomeric units.

1.4.3.1 Analysis of morphological properties

The morphological properties of purified EPS, such as the surface morphology and distribution of molecules, can be studied through different characterization techniques. For several purposes, the microstructure and surface morphology of EPSs are crucial. For example, the linear structure exhibits high viscosity and good water solubility, while the rough and irregular surface shows attraction to water molecules and pseudo-plastic behavior. Atomic force microscopy (AFM) and scanning electron microscope (SEM) are highly effective in determining the physical characteristics of EPSs and are employed to examine the microstructure and surface morphology of the modified EPS.

1.4.3.2 *Analysis of molecular weight*

The MW of EPS is determined using gel permeation chromatography (GPC) or SEC and high-performance size-exclusion chromatography (HPSEC) with RID and multiangle laser light scattering (MALLS) detectors. EPS properties and functions are greatly influenced by their MW. A lower MW is linked to the compound's bioactivity (antioxidant activities), while a greater MW makes aqueous solutions more viscous and demonstrates strong anticancer properties.

1.4.3.3 *Analysis of monosaccharide composition of exopolysaccharides*

The composition of monosaccharides depends on the media composition, type of strains, and culture conditions, like water activity, pH, and temperature. The composition of monosaccharides is an important feature affecting the biological activities of EPSs. Almost all monosaccharides occur in the D-configuration except for L-arabinose, L-fucose, and L-rhamnose, which become essential components that can influence the biological activity of EPSs, such as antitumor, anti-inflammatory response, and antioxidant. The purified EPSs are hydrolyzed using enzymes or acids, and the released monomers are determined using different chromatographic methods, such as thin-layer chromatography (TLC) or high-performance thin-layer chromatography (HPTLC), HPLC combined with UV or RID or diode array (DA) detectors, gas chromatography (GC), gas chromatography-mass spectrometry (GC-MS) (Yadav et al., 2024).

1.5 Exopolysaccharides as a nanobiomaterial: Key applications and case studies

The field of nanobiomaterials has witnessed remarkable growth in recent years, with EPSs emerging as a groundbreaking class of materials that bridge the gap between natural polymers and nanotechnology. These remarkable biomolecules, produced by various microorganisms as part of their survival mechanisms, have garnered significant attention due to their exceptional properties, including biocompatibility, biodegradability, and versatile chemical structures. The ability of EPS to form nanoscale structures and their potential for modification have opened up numerous possibilities in biomedical applications, tissue engineering, drug delivery systems, and beyond.

The exploration of EPS as nanobiomaterials has led to innovative developments across multiple sectors, supported by numerous case studies demonstrating their practical applications and effectiveness. EPS-based nanomaterials have shown promising results in both laboratory settings and clinical applications, from enhancing drug delivery efficiency to creating sophisticated tissue scaffolds. Their natural origin and eco-friendly nature align perfectly with the growing demand for sustainable materials in medical and industrial applications, while their unique chemical and physical properties enable the development of advanced therapeutic and diagnostic tools. This comprehensive analysis examines the key applications of EPS as nanobiomaterials, supported by relevant case studies that highlight their practical implementation and future potential.

1.5.1 Biomedical applications

EPSs have emerged as revolutionary biomaterials in the medical field, offering diverse applications across multiple therapeutic domains. Their inherent biocompatibility and structural versatility make them ideal candidates for various biomedical applications. In tissue engineering, EPS serve as excellent scaffolds that support cell growth and tissue regeneration. Their ability to mimic the natural extracellular matrix provides an optimal environment for cell adhesion, proliferation, and differentiation. Moreover, their mechanical properties can be fine-tuned to match specific tissue requirements, making them suitable for applications ranging from soft tissue repair to bone regeneration.

Drug delivery represents a significant application where EPS have demonstrated remarkable potential. These natural polymers can be engineered into various delivery systems, including nanoparticles, microspheres, and hydrogels (Charoenwongpaiboon et al., 2021; Dahiya and Nigam, 2023; Laubach and Sani, 2023). The controlled release properties of EPS-based delivery systems help maintain therapeutic drug concentrations over extended periods, thereby enhancing treatment efficacy while reducing side effects. Their ability to respond to specific physiological stimuli also enables the development of smart delivery systems that can release drugs in response to environmental changes such as pH, temperature, or enzymatic activity. The inherent biocompatibility and biodegradability of many EPS make them particularly attractive candidates for the development of advanced drug delivery systems (Marković et al., 2016; Pathak, 2019; Gezginç et al., 2022). Their ability to be formulated into various forms, including hydrogels, nanoparticles, and microspheres, offers significant versatility in designing drug delivery systems tailored to specific therapeutic needs (Marković et al., 2016; Pathak, 2019). EPS-based hydrogels, for example, can provide a controlled release of therapeutic agents over extended periods, a crucial feature for many drug therapies (Marković et al., 2016). Kalimuthu et al. explored the use of EPS from the endophytic bacterium *Pseudomonas otitidis* as a drug delivery vehicle for quercetin, an anticancer flavonoid. The isolated EPS demonstrated biocompatibility and biodegradability, making it suitable for drug delivery applications. Quercetin-loaded EPS nanoparticles showed controlled release and enhanced cytotoxic effects against breast cancer cells (MCF-7). These results indicated that EPS can effectively deliver chemotherapeutic agents, improve treatment efficacy while reducing side effects (Kalimuthu et al., 2023).

Alginate, a linear copolymer extracted from brown algae, stands out as a widely used EPS in drug delivery applications. Its biocompatibility, biodegradability, and remarkable ability to form hydrogels in the presence of divalent cations are key attributes contributing to its popularity (Anderson et al., 2018; Pathak, 2019). Alginate-based microspheres and microcapsules are commonly employed for the controlled release of a diverse range of drugs, providing a sustained therapeutic effect (Marković et al., 2016). Its use in wound healing applications has also been extensively investigated, highlighting its potential in regenerative medicine (Anderson et al., 2018). Furthermore, chemical derivatization of alginates offers opportunities to fine-tune its properties, enhancing its suitability for targeted drug delivery applications. The simplicity of alginate hydrogel formation, coupled with its versatility, makes it a robust and readily adaptable platform for drug delivery system development. Conversely, EPS nanoparticles can enhance the targeted delivery of drugs to specific tissues or organs, minimizing off-target effects and maximizing therapeutic efficacy (Pathak, 2019). This versatility, coupled with their inherent biocompatibility, makes EPS a highly promising platform for the development of next-generation drug delivery systems. Further research into the design and optimization of these systems is critical for translating their potential into clinical applications.

The potential of EPS as scaffolds for tissue engineering is considerable (Shih, 2010; Anderson et al., 2018; Ahuja et al., 2023). They provide a three-dimensional framework that supports cell growth and tissue regeneration, mimicking the natural extracellular matrix (Mano et al., 2007). Their inherent porosity, biocompatibility, and biodegradability are essential characteristics for this application (Shih, 2010; Ahuja et al., 2023). The specific structural features of different EPS can significantly influence their ability to support cell adhesion, proliferation, and differentiation, making the selection of the appropriate EPS crucial for the success of tissue engineering strategies (Shih, 2010; Zykwinska et al., 2019). The ability to tailor the properties of EPS through chemical modifications or blending with other biomaterials further enhances their versatility in tissue engineering applications. In wound healing applications, EPS-based materials have shown exceptional promise due to their moisture retention capabilities and antimicrobial properties. These materials create an optimal wound environment that promotes healing while preventing bacterial infection. The biodegradable nature of EPS ensures that they are gradually absorbed by the body as the wound heals, eliminating the need for removal (Hivechi et al., 2021; Muthukrishnan, 2021; Ashraf et al., 2022). Additionally, certain EPS possess inherent anti-inflammatory properties that can help reduce scarring and promote better healing outcomes. Many EPS exhibit exceptional biocompatibility and demonstrate a capacity to promote wound healing (Anderson et al., 2018; Gezginç et al., 2022). Their ability to effectively absorb wound exudate, maintain a moist wound environment conducive to healing, and stimulate cell proliferation makes them highly suitable for use in wound dressings (Anderson et al., 2018; Gezginç et al., 2022). EPS-based hydrogels and films can act as protective barriers while simultaneously promoting tissue regeneration, creating an ideal microenvironment for wound healing. The inherent biodegradability of many EPS also ensures that the dressing will eventually be resorbed as the wound heals.

In diagnostic applications, EPS have been utilized in developing biosensors and imaging agents. Their ability to function with various biological recognition elements enables the development of sensitive and specific diagnostic tools. Moreover, EPS-based materials can be used as contrast agents in imaging applications, providing better visualization of biological structures and processes. Li et al. isolated a naturally occurring spherical EPS nanoparticle (EPS-R503) from *Lactobacillus plantarum* R503, characterized by its heteropolysaccharide composition with functional groups and manipulatable morphological structure through solvent treatment. The EPS-R503 demonstrated practical application as a green reducing and stabilizing agent in the biosynthesis of manganese dioxide nanosheets, resulting in an EPS-MnO_2 nanocomposite. This nanocomposite, when combined with fluorescein, showed excellent potential for both in vitro and in vivo glutathione detection with high sensitivity (detection limit 0.16 μM) and a broad detection range (0.05–4 mM), offering a novel eco-friendly approach to biosensor development (Li et al., 2021).

1.5.2 Environmental and industrial applications

The remarkable properties of EPS, particularly their biodegradability, biocompatibility, and ability to interact with various molecules, have opened up exciting possibilities in environmental remediation. EPS are proving valuable in addressing pressing environmental challenges, such as heavy metal contamination and wastewater treatment.

Heavy metal contamination of water sources and soil poses a significant environmental and health hazard. EPS produced by various microorganisms offer a sustainable solution for removing these pollutants (Li et al., 2016; Concórdio-Reis et al., 2020; Ghosh et al., 2024). The negatively charged functional groups present in many EPS exhibit a strong affinity for positively charged heavy metal ions, allowing for efficient binding and removal through various mechanisms, including adsorption, complexation, and precipitation (Li et al., 2017; Banerjee et al., 2021; Balíková et al., 2022). *Pseudomonas* species, known for their diverse EPS production, have been extensively studied in this context (Li et al., 2017; Balíková et al., 2022). The EPS produced by these bacteria effectively binds heavy metals, facilitating their removal from contaminated water. Several studies have demonstrated the effectiveness of different EPS types in heavy metal removal, with high binding capacities reported for various metals (Li et al., 2017). The use of EPS-producing microorganisms, particularly extremophiles thriving in metal-rich environments, offers a sustainable and cost-effective approach to heavy metal remediation (Delbarre-Ladrat et al., 2014; Goswami et al., 2022). The ability of these extremophiles to produce EPS tailored to their specific environments enhances the efficiency of the remediation process.

The degradation of hydrocarbon pollutants, which are prevalent environmental contaminants, is another crucial application of EPSs. Extremophilic bacteria, particularly those inhabiting environments like hot springs and hydrothermal vents, produce EPS with remarkable properties that enhance the bioavailability of hydrocarbons for degradation (Goswami et al., 2022). These EPS often form protective biofilms, creating a microenvironment that facilitates the activity of hydrocarbon-degrading enzymes. The EPS matrix not only protects the bacteria from harsh environmental conditions but also helps to concentrate the hydrocarbons, making them more accessible to the degrading enzymes (Nguyen et al., 2020; Staninska-Pieta et al., 2020). The synergistic interaction between EPS and hydrocarbon-degrading microorganisms enhances the overall efficiency of bioremediation processes.

Beyond heavy metal removal and hydrocarbon degradation, EPS find applications in various other environmental contexts. In wastewater treatment, EPS can be used as bio-flocculants, facilitating the aggregation and separation of suspended solids (Subudhi et al., 2016; An et al., 2023; Chen et al., 2024; Yuan et al., 2024). Their ability to enhance the settling of suspended particles improves the efficiency of wastewater treatment processes. In soil improvement, EPS can enhance soil structure and water retention capacity, improve soil fertility, and reduce erosion (Talbi et al., 2023). Bioflocculation is a process where EPS molecules bind to suspended particles in wastewater, causing them to aggregate into larger flocs, which can then be easily removed through sedimentation or filtration. The effectiveness of bioflocculation depends on several factors, including EPS structure, concentration, and the characteristics of the pollutants. The use of EPS-producing microorganisms, such as *Klebsiella sp.*, has shown promising results in microalgae harvesting and could be further developed for broader wastewater treatment applications (Yuan et al., 2024). Moreover, the use of extremophiles for EPS production could enable the development of bioflocculants effective in harsh environmental conditions, expanding the applicability of this remediation technique to a wider range of wastewater types (Banerjee et al., 2021).

EPS are also finding increasing applications in the cosmetics and personal care industry (Mouro et al., 2024). Their ability to stabilize emulsions, improve viscosity, and enhance skin hydration makes them valuable ingredients in various cosmetic products. EPS can be used in lotions, creams, and other skincare formulations to provide a smooth texture, improve spreadability, and enhance the moisturizing effect. Furthermore, the antioxidant and antimicrobial properties of certain EPS can contribute to the preservation of cosmetic products and the protection of skin from environmental damage (Panchev et al., 2012; Carrión et al., 2015; Radchenkova et al., 2015; Zheng et al., 2024). The use of EPS in cosmetics aligns with the growing consumer demand for natural and sustainable products, further enhancing the appeal of these biopolymers in this market.

Their unique properties are being explored in a variety of industrial contexts. In the textile industry, EPS can be used as a sizing agent to improve the weaving properties of fabrics

(Ates, 2015). Their ability to form films has led to their exploration as coatings for packaging materials, contributing to the development of biodegradable and sustainable packaging solutions (Guerrero et al., 2024). In the petroleum industry, EPS are being investigated for their potential in enhanced oil recovery, where their rheological properties can improve the flow of oil through porous formations (Sutherland, 1990). Furthermore, the use of EPS in bioremediation and the green synthesis of nanoparticles opens up additional avenues for industrial applications, underscoring the versatility and sustainability of these biopolymers in a wide range of sectors (Li et al., 2017; Escárcega-González et al., 2018).

The ability of certain EPS to form films has led to their exploration as a sustainable alternative to synthetic polymers in food packaging (Mishra, 2020). EPS-based films offer several advantages, including biodegradability, edibility, and potential bioactivity. These films can provide a barrier against moisture, oxygen, and microorganisms, thereby extending the shelf life of packaged food products. The properties of the EPS film, such as its mechanical strength, barrier properties, and water solubility, depend on factors such as the type of EPS, the processing method, and the presence of any additives (Guerrero et al., 2024; Upadhyaya et al., 2024). The use of halophilic EPS offers advantages in terms of stability and potential cost-effectiveness. The incorporation of bioactive compounds into EPS films can further enhance their functionality and provide additional benefits (Upadhyaya et al., 2024).

1.5.3 Food and nutraceutical applications

EPSs, produced by a diverse range of microorganisms, including lactic acid bacteria, represent a class of high-molecular-weight polysaccharides with burgeoning applications across the food and nutraceutical sectors (Pourjafar et al., 2023; Tho and Thanh, 2023; Liang et al., 2024). Their remarkable versatility stems from their unique structural and functional characteristics. EPS possess inherent gelling, thickening, and emulsifying properties, making them invaluable as food additives for enhancing texture, stability, and overall sensory appeal (Basiri, 2021; Guerrero et al., 2024; Prabhu and Kannan, 2024). Beyond their technological contributions, a significant subset of EPS demonstrates notable bioactivity, thereby fueling their increasing prominence in nutraceutical formulations (Abdalla et al., 2021; Prete et al., 2021; Hussein et al., 2024). These bioactive properties encompass a broad spectrum of effects, including antioxidant, antimicrobial, prebiotic, and immunomodulatory activities (Abdalla et al., 2021; Prete et al., 2021; Pourjafar et al., 2023). The physicochemical properties of EPS are crucial in determining their applications. These properties are highly dependent on the monosaccharide composition, the type and frequency of glycosidic linkages, and the presence of any substituent groups, such as acetyl groups or sulfate groups (Banerjee et al., 2022; Ben Mansour et al., 2024; Prabhu and Kannan, 2024). The MW and degree of branching also significantly influence the rheological properties of EPS, such as viscosity, gelling ability, and emulsifying capacity (Banerjee et al., 2022; Pourjafar et al., 2023). These properties are exploited in food applications to modify texture, stability, and sensory characteristics (Basiri, 2021; Pourjafar et al., 2023). In addition to their technological properties, many EPS possess significant bioactivities (Nguyen et al., 2020; Prete et al., 2021; Amrutha and Beena, 2023). These bioactivities are often linked to the specific structural features of the EPS molecule and can include antioxidant, antimicrobial, prebiotic, and immunomodulatory effects (Abdalla et al., 2021; Prete et al., 2021; Pourjafar et al., 2023). The antioxidant activity of EPS is often attributed to their ability to scavenge free radicals and chelate metal ions (Adebayo-Tayo et al., 2024; Ben Mansour et al., 2024). The prebiotic effect of some EPS stems from their ability to selectively stimulate the growth of beneficial gut bacteria (Shih, 2010; Erdem et al., 2023; Tho and Thanh, 2023). The antimicrobial activity of certain EPS may be due to their interaction with bacterial cell membranes or their ability to inhibit enzyme activity (Abdalla et al., 2021; Fashogbon et al., 2022). The immunomodulatory effects of EPS can involve interactions with immune cells and modulation of cytokine production (Liu et al., 2011; Amrutha and Beena, 2023).

The application of EPS in the food industry is driven by their ability to modify the physicochemical properties of food products, thereby enhancing their texture, stability, and sensory attributes (Basiri, 2021; Erdem et al., 2023; Pourjafar et al., 2023). Their gelling, thickening, and emulsifying properties are widely exploited to improve the consistency and mouthfeel of various food products. In dairy products such as yogurt, EPS from lactic acid bacteria contribute to a smoother, creamier texture and improved stability. The EPS matrix helps to retain moisture, prevent syneresis (why separation), and enhance the overall sensory experience (Prete et al., 2021; Adebayo-Tayo et al., 2024). In bakery products, EPS can improve dough rheology, resulting in enhanced bread volume, texture, and shelf life (Potoroko et al., 2018; Fashogbon et al., 2022). The ability of EPS to form films also finds application in food packaging. Edible and biodegradable films made by EPS offer a sustainable alternative to synthetic polymers, providing a barrier against moisture and oxygen, thereby extending the shelf life of food products (Guerrero et al., 2024; Upadhyaya et al., 2024). The selection of an appropriate EPS for a specific food application often requires careful consideration of its physicochemical properties and the desired functional outcome (Erdem et al., 2023; Prete et al., 2024).

Banerjee et al. characterized a novel EPS produced by the thermophilic bacterium *Bacillus haynesii* CamB6. The researchers determined EPS's physicochemical properties and evaluated its potential applications in the food industry. The study showed that the EPS possessed desirable functional properties, such as antioxidant activity and emulsification capacity, suggesting its potential as a sustainable food additive (Banerjee et al., 2022). This exemplifies the ongoing discovery of new EPS with tailored properties for specific industrial applications. A review by Lynch et al. highlighted the significant role of LAB-produced EPS in improving the texture and shelf life of various food products, particularly fermented dairy products (Lynch et al., 2018). The extensive use of these EPS in the food industry demonstrates the economic and practical viability of utilizing EPS in large-scale food production.

The incorporation of EPS-producing lactic acid bacteria (LAB) into dairy fermentations significantly impacts the texture and stability of the final product. The EPS produced by these LAB creates a three-dimensional network within the dairy matrix, resulting in increased viscosity and improved mouthfeel. This effect is particularly noticeable in yogurt, where EPS contributes to a creamier texture and reduced syneresis, preventing the separation of whey and maintaining a desirable consistency (Prete et al., 2021; Adebayo-Tayo et al., 2024). The enhanced stability provided by EPS also contributes to an extended shelf life, reducing the risk of spoilage and maintaining product quality over time (Anderson et al., 2018). The type and concentration of EPS produced can be influenced by factors such as bacterial strain, growth conditions, and fermentation parameters. Optimizing these factors is crucial for achieving the desired textural and stability characteristics in dairy products (Erdem et al., 2023; Prete et al., 2024). Furthermore, the potential for EPS to contribute to the health benefits of dairy products adds to their value in functional food applications (Prete et al., 2021).

In bakery applications, EPS can positively influence dough rheology and improve the quality of the final product. The addition of EPS to dough can increase its viscosity and water-holding capacity, leading to improved dough handling and reduced stickiness. The resulting changes in dough structure can enhance bread volume and improve its texture, resulting in a softer, more desirable crumb. Furthermore, EPS can contribute to improved shelf life by retarding staling and maintaining bread freshness. The specific effects of EPS on dough rheology and bread quality depend on several factors, including the type and concentration of EPS, the type of flour, and the baking process (Potoroko et al., 2018; Fashogbon et al., 2022). The use of EPS-producing starter cultures in sourdough fermentation can offer a natural and sustainable approach to enhancing bread quality. The incorporation of EPS also opens avenues for creating functional breads enriched with prebiotic or other bioactive compounds (Fashogbon et al., 2022).

The antioxidant properties of EPS are a key driver of their nutraceutical value. Oxidative stress, caused by an imbalance between free radical production and antioxidant defense, is implicated in numerous chronic diseases. EPS can effectively scavenge free radicals, thereby reducing oxidative damage to cells and tissues. The antioxidant capacity of EPS varies depending on their chemical structure and composition. The presence of certain monosaccharides, such as galactose and uronic acids, has been linked to enhanced antioxidant activity. The mechanism of antioxidant action involves direct scavenging of free radicals, chelation of metal ions, and modulation of antioxidant enzymes. The incorporation of EPS into functional foods and supplements could offer a natural and effective strategy for reducing oxidative stress and improving overall health (Angelova et al., 2022; Adebayo-Tayo et al., 2024; Ben Mansour et al., 2024).

The prebiotic effects of EPS contribute to their growing importance in nutraceuticals. Prebiotics are nondigestible food ingredients that selectively stimulate the growth and/or activity of beneficial bacteria in the gut (Bikric et al., 2022). Certain EPS can act as prebiotics by serving as substrates for the fermentation by beneficial gut bacteria, such as *Bifidobacteria* and *Lactobacilli* (Tsuda and Miyamoto, 2010; Bikric et al., 2022; Lee et al., 2022; Kang et al., 2023). This fermentation process produces short-chain fatty acids (SCFAs), which have various beneficial effects on gut health, including improved intestinal barrier function, enhanced immune response, and reduced inflammation. The selective stimulation of beneficial bacteria by EPS can contribute to a balanced gut microbiota, which is crucial for overall health and well-being. The type and concentration of EPS required to elicit prebiotic effects can vary depending on the specific EPS and the gut microbiota composition (Abdalla et al., 2021; Prete et al., 2021; Pourjafar et al., 2023).

1.6 Summary and future outlook

An extensive review of natural EPSs, including their origins, structural classification, characteristics, and uses as nanobiomaterials, is given in this chapter. These microbial biopolymers are ideal for industrial, environmental, and biological applications due to their exceptional physicochemical characteristics, which include biodegradability, biocompatibility, and bioactivity. EPSs are categorized according to their structural conformations, microbial origins, and monomer makeup, which affects how well they work in a variety of contexts. Their capacity to form nanostructures has made it possible for them to be used as carriers, hydrogels, and bioactive coatings in regenerative medicine, wound healing, tissue scaffolding, and drug delivery.

Because of their inherent adsorption and binding capabilities, EPSs also aid in wastewater treatment, heavy metal removal, bio-filled, and bioremediation. They serve as texturizing agents, stabilizers, and prebiotics in the food and nutraceutical industries, enhancing product quality and providing health advantages. In order to determine their MW, content, and structure, the chapter also discusses extraction, purification, and characterization techniques. These include chromatography, ultrafiltration, precipitation, and spectroscopy.

In the future, EPS-based nanobiomaterials have enormous promise for advancement in the industrial, environmental, and biomedical domains. Smart EPS hydrogels, nanocarriers, and bio-scaffolds for targeted drug delivery, tissue engineering, and regenerative medicine will be made possible by developments in chemical functionalization. Large-scale EPS production and economic viability will be improved by improved fermentation techniques, especially when using extremophilic microorganisms. Green nanotechnology, pollution treatment, and biodegradable EPS-based films will all help with sustainability initiatives in environmental applications. Furthermore, molecular simulations and AI-driven modeling will speed up the creation of customized EPS materials with ideal qualities. EPS-based technologies have the potential to transform healthcare, industry, and environmental management by combining developments in biotechnology, material science, and computational methods. This will accelerate the shift toward high-performance, sustainable biomaterials.

Credit author statement

Varghese, R.T., Antony, T., and Raju, S.G.: Conducted literature search, wrote the initial and intermediate drafts. Chirayil, C. J.: Reviewed the final version and provided editorial comments. Kargarzadeh, H., Conceptualization, supervised writing, and performed final critical revision, coordination with the publisher.

Conflict of interest

The authors state that there are no known financial or personal conflicts of interest that could have influenced the work reported in this chapter.

Acknowledgments

The authors gratefully acknowledge the support provided by their respective institutions: the Department of Chemistry, Newman College; the Department of Chemistry, Pavanatma College; the School of Energy Materials, Mahatma Gandhi University; and the Centre of Molecular and Macromolecular Studies of the Polish Academy of Sciences.

References

Abdalla, A.K., Ayyash, M.M., Olaimat, A.N., Osaili, T.M., Al-Nabulsi, A.A., Shah, N.P., Holley, R., 2021. Exopolysaccharides as antimicrobial agents: mechanism and spectrum of activity. Front. Microbiol. 12. https://doi.org/10.3389/fmicb.2021.664395.

Adebayo-Tayo, B.C., Ogundele, B.R., Ajani, O.A., Olaniyi, O.A., 2024. Characterization of lactic acid bacterium exopolysaccharide, biological, and nutritional evaluation of probiotic formulated fermented coconut beverage. Int. J. Food Sci. 2024. https://doi.org/10.1155/2024/8923217.

Ahuja, V., Bhatt, A.K., Banu, J.R., Kumar, V., Kumar, G., Yang, Y.H., Bhatia, S.K., 2023. Microbial exopolysaccharide composites in biomedicine and healthcare: trends and advances. Polymers 15 (7). https://doi.org/10.3390/polym15071801.

Amrutha, T.A., Beena, A.K., 2023. Microbial exopolysaccharides: a promising health booster. J. Phytopharm. 12, 265–271. http://dx.doi.org/10.31254/phyto.2023.12409.

An, Q., Deng, S.M., Zhao, B., Huang, C., Yang, J.X., 2023. Bioflocculation characteristics of bound extracellular polymers substances from *Pseudomonas* sp. XD-3 and behavior of polysaccharides. Colloids Surf. B. 228, 113436. https://doi.org/10.1016/j.colsurfb.2023.113436.

Anderson, L.A., Islam, M.A., Prather, K.L.J., 2018. Synthetic biology strategies for improving microbial synthesis of "green" biopolymers. J. Biol. Chem. 293 (14), 5053–5061. https://doi.org/10.1074/jbc.TM117.000368.

Angelova, G., Brazkova, M., Mihaylova, D., Slavov, A., Petkova, N., Blazheva, D., Krastanov, A., 2022. Bioactivity of biomass and crude exopolysaccharides obtained by controlled submerged cultivation of medicinal mushroom *Trametes* versicolor. J. Fungi 8 (7), 738. https://doi.org/10.3390/jof8070738.

Ashraf, S.S., Parivar, K., Roodbari, N.H., Mashayekhan, S., Amini, N., 2022. Fabrication and characterization of biaxially electrospun collagen/alginate nanofibers, improved with *Rhodotorula mucilaginosa* sp. GUMS16 produced exopolysaccharides for wound healing applications. Int. J. Biol. Macromol. 196, 194–203. https://doi.org/10.1016/j.ijbiomac.2021.11.132.

Ates, O., 2015. Systems biology of microbial exopolysaccharides production. Front. Bioeng. Biotechnol. 3. https://doi.org/10.3389/fbioe.2015.00200.

Balíková, K., Vojtková, H., Duborská, E., Kim, H., Matús, P., Urík, M., 2022. Role of exopolysaccharides of *Pseudomonas* in heavy metal removal and other remediation strategies. Polymers 14 (20). https://doi.org/10.3390/polym14204253.

Banerjee, A., Breig, S.J.M., Gómez, A., Sánchez-Arévalo, I., Gónzalez-Faune, P., Sarkar, S., Bandopadhyay, R., Vuree, S., Cornejo, J., Tapia, J., Bravo, G., Cabrera-Barjas, G., 2022. Optimization and characterization of a novel exopolysaccharide from *Bacillus haynesii* CamB6 for food applications. Biomolecules 12 (6). https://doi.org/10.3390/biom12060834.

Banerjee, A., Sarkar, S., Govil, T., Gonzalez-Faune, P., Cabrera-Barjas, G., Bandopadhyay, R., Salem, D.R., Sani, R.K., 2021. Extremophilic exopolysaccharides: biotechnologies and wastewater remediation. Front. Microbiol. 12. https://doi.org/10.3389/fmicb.2021.721365.

Basiri, S., 2021. Applications of microbial exopolysaccharides in the food industry. Avicenna J. Med. Biochem. 9 (2), 107–120. https://doi.org/10.34172/ajmb.2021.16.

Ben Mansour, F., Guermazi, W., Chamkha, M., Bellassoued, K., Salah, H.B., Harrath, A.H., Aldahmash, W., Rahman, M.A., Ayadi, H., 2024. Bioactive potential of the sulfated exopolysaccharides from the brown microalga *Halamphora* sp.: antioxidant, antimicrobial, and antiapoptotic profiles. Anal. Sci. Adv. 5 (9-10). https://doi.org/10.1002/ansa.202400030.

Bikric, S., Aslim, B., Dincer, İ., Yuksekdag, Z., Ulusoy, S., Yavuz, S., 2022. Characterization of exopolysaccharides (EPSs) obtained from *Ligilactobacillus salivarius* strains and investigation at the prebiotic potential as an alternative to plant prebiotics at poultry. Probiotics Antimicrob. Proteins 1–11. https://doi.org/10.1007/s12602-021-09790-8.

Birch, J., Van Calsteren, M.R., Pérez, S., Svensson, B., 2019. The exopolysaccharide properties and structures database: EPS-DB. Application to bacterial exopolysaccharides. Carbohydr. Polym. 205, 565–570. https://doi.org/10.1016/j.carbpol.2018.10.063.

Carrión, O., Delgado, L., Mercade, E., 2015. New emulsifying and cryoprotective exopolysaccharide from Antarctic *Pseudomonas* sp. ID1. Carbohydr. Polym. 117, 1028–1034. https://doi.org/10.1016/j.carbpol.2014.08.060.

Carvalho, L.T., Vieira, T.A., Zhao, Y.J., Celli, A., Medeiros, S.F., Lacerda, T.M., 2021. Recent advances in the production of biomedical systems based on polyhydroxyalkanoates and exopolysaccharides. Int. J. Biol. Macromol. 183, 1514–1539. https://doi.org/10.1016/j.ijbiomac.2021.05.025.

Charoenwongpaiboon, T., Wangpaiboon, K., Pichyangkura, R., Nepogodiev, S.A., Wonganan, P., Mahalapbutr, P., Field, R.A., 2021. Characterization of a nanoparticulate exopolysaccharide from *Leuconostoc holzapfelii* KM01 and its potential application in drug encapsulation. Int. J. Biol. Macromol. 187, 690–698. https://doi.org/10.1016/j.ijbiomac.2021.07.174.

Chen, L., Zhao, B., An, Q., Guo, Z.Q., Huang, C., 2024. The characteristics and flocculation mechanisms of SMP and B-EPS from a bioflocculant-producing bacterium *Pseudomonas* sp. XD-3 and the application for sludge dewatering. Chem. Eng. J. 479. https://doi.org/10.1016/j.cej.2023.147584.

Concórdio-Reis, P., Reis, M.A.M., Freitas, F., 2020. Biosorption of heavy metals by the bacterial exopolysaccharide FucoPol. Appl. Sci.-Basel 10 (19). https://doi.org/10.3390/app10196708.

Dahiya, D., Nigam, P.S., 2023. Dextran of diverse molecular-configurations used as a blood-plasma substitute, drug-delivery vehicle and food additive biosynthesized by *Leuconostoc*, *Lactobacillus* and *Weissella*. Appl. Sci.-Basel 13 (22). https://doi.org/10.3390/app132212526.

Delbarre-Ladrat, C., Sinquin, C., Lebellenger, L., Zykwinska, A., Colliec-Jouault, S., 2014. Exopolysaccharides produced by marine bacteria and their applications as glycosaminoglycan-like molecules. Front. Chem. 2. https://doi.org/10.3389/fchem.2014.00085.

Erdem, T.K., Tatar, H.D., Ayman, S., Gezginç, Y., 2023. Exopolysaccharides from lactic acid bacteria: A review on functions, biosynthesis and applications in food industry. Turk. J. Agric.-Food Sci. Technol. 11 (2), 414–423. https://doi.org/10.24925/turjaf.v11i2.414-423.5213.

Escárcega-González, C.E., Garza-Cervantes, J.A., Vázquez-Rodríguez, A., Morones-Ramírez, J.R., 2018. Bacterial exopolysaccharides as reducing and/or stabilizing agents during synthesis of metal nanoparticles with biomedical applications. Int. J. Polym. Sci. 2018. https://doi.org/10.1155/2018/7045852.

Fashogbon, R.O., Sanusi, J., Ogunleye, G.E., Akintunde, M.O., Adebayo-Tayo, B.C., 2022. Potential application of exopolysaccharides from *Lactobacillus delbrueckii* FASHADFF1 (LDYG2) and *Weissella confusa* FASHADFF1 (WCFF1) in sourdough bread production. Adv. Microbiol. 12 (8), 481–499. https://doi.org/10.4236/aim.2022.128033.

Gezginç, Y., Karabekmez-erdem, T., Tatar, H.D., Ayman, S., Ganiyusufoğlu, E., Dayısoylu, K.S., 2022. Health promoting benefits of postbiotics produced by lactic acid bacteria: exopolysaccharide. Biotech Stud. 31 (2), 61–70. https://doi.org/10.38042/biotechstudies.1159166.

Ghosh, A., Sah, D., Chakraborty, M., Rai, J.P.N., 2024. Mechanism and application of bacterial exopolysaccharides: an advanced approach for sustainable heavy metal abolition from soil. Carbohydr. Res. 544. https://doi.org/10.1016/j.carres.2024.109247.

Goswami, R., Bandyopadhyay, B., Sadhukhan, S., 2022. Thermophilic bacterial exopolysaccharides: from biophysicochemical characterization to biotechnological applications. Physiology, Genomics, and Biotechnological Applications of Extremophiles. IGI Global, pp. 334–361. https://doi.org/10.4018/978-1-7998-9144-4.ch016.

Guerrero, B.G., Santos, K.D., Kamimura, E.S., de Oliveira, C.A.F., 2024. Application of microbial exopolysaccharides in packaging films for the food industry: a review. Int. J. Food Sci. Technol. 59 (1), 17–29. https://doi.org/10.1111/ijfs.16789.

Hamidi, M., Okoro, O.V., Milan, P.B., Khalili, M.R., Samadian, H., Nie, L., Shavandi, A., 2022. Fungal exopolysaccharides: properties, sources, modifications, and biomedical applications. Carbohydr. Polym. 284. https://doi.org/10.1016/j.carbpol.2022.119152.

Hivechi, A., Milan, P.B., Modabberi, K., Amoupour, M., Ebrahimzadeh, K., Gholipour, A.R., Sedighi, F., Amini, N., Bahrami, S.H., Rezapour, A., Hamidi, M., Delattre, C., 2021. Synthesis and characterization of exopolysaccharide encapsulated PCL/gelatin skin substitute for full-thickness wound regeneration. Polymers 13 (6). https://doi.org/10.3390/polym13060854.

Hussein, K.A., Niamah, A.K., Majeed, K.R., 2024. Strategies and trends for application exopolysaccharides of lactic acid bacteria in the food and biomedical. In: IOP Conference Series: Earth and Environmental Science, p. 1371.

Kalimuthu, A.K., Pandian, S.R.K., Pavadai, P., Panneerselvam, T., Kabilan, S.J., Sankaranarayanan, M., Ala, C., Kunjiappan, S., 2023. Drug delivery applications of exopolysaccharides from endophytic bacteria *Pseudomonas otitidis* from *Tribulus terrestris* L. J. Polym. Env. 31 (8), 3632–3649. https://doi.org/10.1007/s10924-023-02848-4.

Kang, M.J., Jeong, H., Kim, S., Shin, J., Song, Y., Lee, B.H., Park, Y.S., 2023. Structural analysis and prebiotic activity of exopolysaccharide produced by probiotic strain *Bifidobacterium bifidum* EPS DA-LAIM. Food Sci. Biotechnol. 32 (4), 517–529. https://doi.org/10.1007/s10068-022-01213-w.

Laubach, J.M., Sani, R.K., 2023. 5-Fluorouracil-encapsulated films using exopolysaccharides from a *Thermophilic Bacterium Geobacillus* sp. WSUCF1 for topical drug delivery. Micromachines 14 (5). https://doi.org/10.3390/mi14051092.

Lee, M.G., Joeng, H., Shin, J., Kim, S., Lee, C., Song, Y., Park, Y.S., 2022. Potential probiotic properties of exopolysaccharide-producing *Lacticaseibacillus paracasei* EPS DA-BACS and prebiotic activity of its exopolysaccharide. Microorganisms 10 (12), 2431. https://doi.org/10.3390/microorganisms10122431.

Li, C.C., Zhang, X.D., Guo, Y.X., Seidi, F., Shi, X.T., Xiao, H.N., 2021. Naturally occurring exopolysaccharide nanoparticles: formation process and their application in glutathione detection. ACS Appl. Mater. Interfaces 13 (17), 19756–19767. https://doi.org/10.1021/acsami.1c03489.

Li, C.C., Zhou, L., Yang, H., Lv, R.J., Tian, P.L., Li, X., Zhang, Y.Q., Chen, Z., Lin, F.M., 2017. Self-assembled exopolysaccharide nanoparticles for bioremediation and green synthesis of noble metal nanoparticles. ACS Appl. Mater. Interfaces 9 (27), 22808–22818. https://doi.org/10.1021/acsami.7b02908.

Li, H.M., Wei, M., Min, W.H., Gao, Y.W., Liu, X.Q., Liu, J.S., 2016. Removal of heavy metal Ions in aqueous solution by exopolysaccharides from *Athelia rolfsii*. Biocatal. Agric. Biotechnol. 6, 28–32. https://doi.org/10.1016/j.bcab.2016.01.013.

Liang, S.N., Wang, X.Y., Li, C., Liu, L.B., 2024. Biological activity of lactic acid bacteria exopolysaccharides and their applications in the food and pharmaceutical industries. Foods 13 (11). https://doi.org/10.3390/foods13111621.

Liu, C.F., Tseng, K.C., Chiang, S.S., Lee, B.H., Hsu, W.H., Pan, T.M., 2011. Immunomodulatory and antioxidant potential of *Lactobacillus* exopolysaccharides. J. Sci. Food Agric 91 (12), 2284–2291. https://doi.org/10.1002/jsfa.4456.

Lynch, K.M., Zannini, E., Coffey, A., Arendt, E.K., 2018. Lactic acid bacteria exopolysaccharides in foods and beverages: isolation, properties, characterization, and health benefits. Annu. Rev. Food Sci. Technol. 9 (1), 155–176. https://doi.org/10.1146/annurev-food-030117-012537.

Mano, J.F., Silva, G.A., Azevedo, H.S., Malafaya, P.B., Sousa, R.A., Silva, S.S., Boesel, L.F., Oliveira, J.M., Santos, T.C., Marques, A.P., Neves, N.M., Reis, R.L., 2007. Natural origin biodegradable systems in tissue engineering and regenerative medicine: present status and some moving trends. J. R. Soc. Interface 4 (17), 999–1030. https://doi.org/10.1098/rsif.2007.0220.

Marković, D., Zarubica, A., Stojković, N., Vasić, M., Cakić, M., Nikolić, G., 2016. Alginates and similar exopolysaccharides in biomedical application and pharmacy: controlled delivery of drugs. Adv. Technol. 5 (1), 39–52. https://doi.org/10.5937/savteh1601039M.

Mishra, P., 2020. Cyanobacterial exopolysaccharide as natural sources for food packaging applications. Innovations in Food Technology: Current Perspectives and Future Goals. Springer, Singapore, pp. 171–184. https://doi.org/10.1007/978-981-15-6121-4_12.

Mouro, C., Gomes, A.P., Gouveia, I.C., 2024. Microbial exopolysaccharides: structure, diversity, applications, and future frontiers in sustainable functional materials. Polysaccharides 5 (3), 241–287. https://doi.org/10.3390/polysaccharides5030018.

Muthukrishnan, L., 2021. Imminent antimicrobial bioink deploying cellulose, alginate, EPS and synthetic polymers for 3D bioprinting of tissue constructs. Carbohydr. Polym. 260. https://doi.org/10.1016/j.carbpol.2021.117774.

Netrusov, A.I., Liyaskina, E.V., Kurgaeva, I.V., Liyaskina, A.U., Yang, G., Revin, V.V., 2023. Exopolysaccharides producing bacteria: A review. Microorganisms 11 (6). https://doi.org/10.3390/microorganisms11061541.

Nguyen, P.T., Nguyen, T.T., Bui, D.C., Hong, P.T., Hoang, Q.K., Nguyen, H.T., 2020. Exopolysaccharide production by lactic acid bacteria: the manipulation of environmental stresses for industrial applications. AIMS Microbiol. 6 (4), 451–469. https://doi.org/10.3934/microbiol.2020027.

Osemwegie, O.O., Adetunji, C.O., Ayeni, E.A., Adejobi, O.I., Arise, R.O., Nwonuma, C.O., Oghenekaro, A.O., 2020. Exopolysaccharides from bacteria and fungi: current status and perspectives in Africa. Heliyon 6 (6). https://doi.org/10.1016/j.heliyon.2020.e04205.

Padmanabhan, A., Shah, N.P., 2020. Structural characterization of exopolysaccharide from *Streptococcus thermophilus* ASCC 1275. J. Dairy Sci. 103 (8), 6830–6842. https://doi.org/10.3168/jds.2019-17439.

Panchev, I.N., Dobreva, S., Karashanova, D., Pavlova, K., Kuncheva, M., Georgieva, K., 2012. Physical properties of cosmetic creams containing exopolysaccharides synthesized from Antarctic yeast. EPS, pp. 0–999.

Pathak, K., 2019. Marine bio adhesives: opportunities and challenges. Ther. Deliv. 10 (12), 749–751. https://doi.org/10.4155/tde-2019-0070.

Potoroko, I.Y., Paimulina, A.V., Uskova, D.G., Kalinina, I.V., 2018. Scientific and practical aspects of functional food technology bulletin of the South Ural State University. Ser. Food Biotechnol. 6 (1), 49–59.

Pourjafar, H., Ansari, F., Sadeghi, A., Samakkhah, S.A., Jafari, S.M., 2023. Functional and health-promoting properties of probiotics' exopolysaccharides; isolation, characterization, and applications in the food industry. Crit. Rev. Food Sci. Nutr. 63 (26), 8194–8225. https://doi.org/10.1080/10408398.2022.2047883.

Prabhu, K., Kannan, N., 2024. Review on microbial production of exopolysaccharides and its applications. Res. J. Biotechnol. 19 (6), 149–158.

Prete, R., Alam, M.K., Perpetuini, G., Perla, C., Pittia, P., Corsetti, A., 2021. Lactic acid bacteria exopolysaccharides producers: a sustainable tool for functional foods. Foods 10 (7). https://doi.org/10.3390/foods10071653.

Prete, R., Dell'Orco, F., Sabatini, G., Montagano, F., Battista, N., Corsetti, A., 2024. Improving the antioxidant and anti-inflammatory activity of fermented milks with exopolysaccharides-producing *Lactiplantibacillus plantarum* strains. Foods 13 (11), 1663. https://doi.org/10.3390/foods13111663.

Radchenkova, N., Panchev, I., Vassilev, S., Kuncheva, M., Dobreva, S., Kambourova, M., 2015. Continuous cultivation of a thermophilic bacterium *Aeribacillus pallidus* 418 for production of an exopolysaccharide applicable in cosmetic creams. J. Appl. Microbiol. 119 (5), 1301–1309. https://doi.org/10.1111/jam.12944.

Shih, I.L., 2010. Microbial exo-polysaccharides for biomedical applications. Mini-Rev. Med. Chem. 10 (14), 1345–1355. https://doi.org/10.2174/138955710793564115.

Staninska-Pieta, J., Czarny, J., Piotrowska-Cyplik, A., Juzwa, W., Wolko, L., Nowak, J., Cyplik, P., 2020. Heavy metals as a factor increasing the functional genetic potential of bacterial community for polycyclic aromatic hydrocarbon biodegradation. Molecules 25 (2). https://doi.org/10.3390/molecules25020319.

Subudhi, S., Bisht, V., Batta, N., Pathak, M., Devi, A., Lal, B., 2016. Purification and characterization of exopolysaccharide bioflocculant produced by heavy metal resistant *Achromobacter xylosoxidans*. Carbohydr. Polym. 137, 441–451. https://doi.org/10.1016/j.carbpol.2015.10.066.

Sutherland, I.W., 1990. Industrial uses of microbial polysaccharides. Biotechnology of Microbial Exopolysaccharides. Cambridge University Press, Cambridge, United Kingdom.

Talbi, C., Elmarrkechy, S., Youssfi, M., Bouzroud, S., Belfquih, M., Sifou, A., Bourais, I., 2023. Bacterial exopolysaccharides: from production to functional features. Prog. Microbes Mol. Biol. 6 (1). https://doi.org/10.36877/pmmb.a0000384.

Tho, N.P., Thanh, N.H., 2023. Exopolysaccharides of lactic acid bacteria: the manipulation of environmental stresses to modulate biosynthesis for industrial applications. TVUJS 13 (1). https://doi.org/10.3934/microbiol.2020027.

Tiwari, O.N., Bobby, M.N., Kondi, V., Halder, G., Kargarzadeh, H., Ikbal, A.A., Bhunia, B., Thomas, S., Efferth, T., Chattopadhyay, D., Palit, P., 2024. Comprehensive review on recent trends and perspectives of natural exopolysaccharides: Pioneering nano-biotechnological tools. Int. J. Biol. Macromol. 265. https://doi.org/10.1016/j.ijbiomac.2024.130747.

Tsuda, H., Miyamoto, T., 2010. Production of exopolysaccharide by *Lactobacillus plantarum* and the prebiotic activity of the exopolysaccharide. Food Sci. Technol. Res. 16 (1), 87–92. https://doi.org/10.3136/fstr.16.87.

Upadhyaya, C., Patel, H., Patel, I., Ahir, P., Upadhyaya, T., 2024. Development of biological coating from novel halophilic exopolysaccharide exerting shelf-life-prolonging and biocontrol actions for post-harvest applications. Molecules 29 (3). https://doi.org/10.3390/molecules29030695.

Wang, W., Ju, Y.H., Liu, N., Shi, S.B., Hao, L.J., 2023. Structural characteristics of microbial exopolysaccharides in association with their biological activities: a review. Chem. Biol. Technol. Agric. 10 (1). https://doi.org/10.1186/s40538-023-00515-3.

Yadav, M.K., Song, J.H., Vasquez, R., Lee, J.S., Kim, I.H., Kang, D.K., 2024. Methods for detection, extraction, purification, and characterization of exopolysaccharides of lactic acid bacteria-a systematic review. Foods 13 (22). https://doi.org/10.3390/foods13223687.

Yuan, Y.C., Lu, J.X., Wang, Q., 2024. Evaluating bioflocculation harvesting of freshwater and marine microalgae using exopolysaccharides (EPSs) from *Klebsiella* sp. Separations 11 (12). https://doi.org/10.3390/separations11120355.

Zheng, L., Yang, F.R., Liang, J., Zhao, Y.L., Niu, J.H., Ma, Y.H., Meng, Q.J., Liu, Q.A., Gong, W.L., Wang, B.L., 2024. Research on the application of *Thelephora ganbajun* exopolysaccharides in antioxidant, anti-inflammatory and spot-fading cosmetics. Int. J. Biol. Macromol. 257. https://doi.org/10.1016/j.ijbiomac.2023.128713.

Zykwinska, A., Marquis, M., Godin, M., Marchand, L., Sinquin, C., Garnier, C., Jonchère, C., Chédeville, C., Le Visage, C., Guicheux, J., Colliec-Jouault, S., Cuenot, S., 2019. Microcarriers based on glycosaminoglycan-like marine exopolysaccharide for TGF-β1 long-term protection. Mar. Drugs 17 (1). https://doi.org/10.3390/md17010065.

CHAPTER

2

Exopolysaccharides: Natural sources, structure, physicochemical properties, and biological activities

Tomy Muringayil Joseph[a], *Martin George Thomas*[b,c], *Ajmi Ayoob*[d], *Blessy Babukkutty*[e], *Debarshi Kar Mahapatra*[f], *Aparna Beena Unni*[g] *and Sabu Thomas*[c,d,h,i,j]

[a]Department of Polymer Technology, Faculty of Chemistry, Gdańsk University of Technology, Gdańsk, Poland [b]Institute Charles Gerhardt, ICGM UMR, Montpellierm, France [c]Department of Physics & Electronics, CHRIST (Deemed to be University), Bengaluru, Karnataka, India [d]International and Inter University Centre for Nanoscience and Nanotechnology, Mahatma Gandhi University, Kottayam, Kerala, India [e]Federation University, Gippsland Campus, Melbourne, VIC, Australia [f]Chitkara College of Pharmacy, Chitkara University, Rajpura, Punjab, India [g]Institute of Physics, University of Silesia, Chorzów, Poland [h]School of Chemical Sciences, Mahatma Gandhi University, Kottayam, Kerala, India [i]Department of Chemical Sciences, University of Johannesburg, Johannesburg, Gauteng, South Africa [j]Trivandrum Engineering Science & Technology Research Park, Sreekariyam, Trivandrum, Kerala, India

2.1 Introduction

Exopolysaccharide represent a cadre of microbial macromolecules secreted into extracellular matrices by bacteria, fungi, algae, yeasts, and actinomycetes (Nguyen et al., 2024). Their molecular mass can range broadly from 0.5×10^6 to 2.0×10^6 Da, which depends on the producer genus

DOI: https://doi.org/10.1016/B978-0-443-36619-2.00002-4

and strain (Shu and Lung, 2004). These polymeric assemblies consist of recurring monosaccharide units linked via glycosidic bonds of variable stereochemistry and branching patterns. Their structural heterogeneity derives from variations in monosaccharide composition—glucose, galactose, mannose, rhamnose, xylose, fucose—and modifications such as acetylation, phosphorylation, and methylation. This molecular diversity governs macromolecular conformations capable of adopting helical, coiled, and ordered tertiary structures (Osemwegie et al., 2020). Exopolysaccharide confer physicochemical functionalities including gelation, high viscosity, shear-thinning behavior, moisture retention, metal ion binding, emulsification, and film formation. Natural evolutionary drivers for exopolysaccharide development include biofilm formation, cell aggregation, protection against environmental stressors, osmotic balance, and facilitation of microbial symbioses in complex ecosystems. Understanding exopolysaccharide demands integration of carbohydrate chemistry, polymer physics, microbial physiology, and structure–activity relationships (Gupta et al., 2019).

Investigations into microbial polysaccharides commenced with the identification of dextran from *Leuconostoc mesenteroides* in the late nineteenth century. Subsequent research on alginate from *Azotobacter* and *Pseudomonas* and xanthan from *Xanthomonas* campestris catalyzed developments in food, pharmaceutical, and material sciences (Zhou et al., 2019). More recent studies have dissected molecular pathways underpinning exopolysaccharide biosynthesis—glycosyltransferase functions, nucleotide sugar pools, postpolymer assembly modifications—and underlying regulatory circuits such as two-component systems, quorum sensing, and global transcriptional regulators. Structural elucidation of bacterial exopolysaccharide, such as kefiran from *Lactobacillus kefiranofaciens*, revealed heteropolysaccharide backbones with repeating disaccharide units and branching side chains (Badel et al., 2011). Algal exopolysaccharide, like carrageenan and fucoidan, manifest sulfated galactan frameworks with unique bioactive potentials. Fungal exopolysaccharide, including pullulan and β-glucans, exhibit immunomodulation and prebiotic properties. Cutting-edge studies employ high-resolution NMR, MALDI-TOF MS, SEC-MALS, and microfluidic rheometry to resolve molecular weight distribution, branching frequency, conformational transitions under shear or ionic conditions, and structure–function correlations (Nwodo et al., 2012).

Modern exopolysaccharide research addresses limitations of synthetic polymers by promoting biodegradability, biocompatibility, renewability, and reduced ecological footprint. Commercial applications increasingly exploit exopolysaccharide in food for texture and stability, in cosmetics for moisturization, in biomedicine for controlled drug delivery, wound dressing matrices, and scaffolds for tissue engineering (Pallikkunnel et al., 2023). Environmental deployments include exopolysaccharide-facilitated bioflocculation, heavy metal chelation, phosphate removal, enhanced oil recovery, and carbon sequestration (Suresh Kumar et al., 2007). Global market growth reflects surging demand across these sectors. Yet production bottlenecks persist due to low microbial yields, complex downstream purification, high cost of monosaccharide substrates, and regulatory hurdles. Metabolic engineering strategies—overexpression of key glycosyltransferases, knockout of competing pathways, and synthetic chassis applications—are emerging to overcome yield constraints. Fermentation process intensification—fed-batch, continuous cultures, and co-subculture systems—is enhancing volumetric productivity. Detailed techno-economic analyses guide the evaluation of scalability and sustainability. Current research endeavors focus on the custom design of exopolysaccharide with tailored functionalities through combinatorial biology, synthetic biology, and rational bioprocess design (Mohd Nadzir et al., 2021).

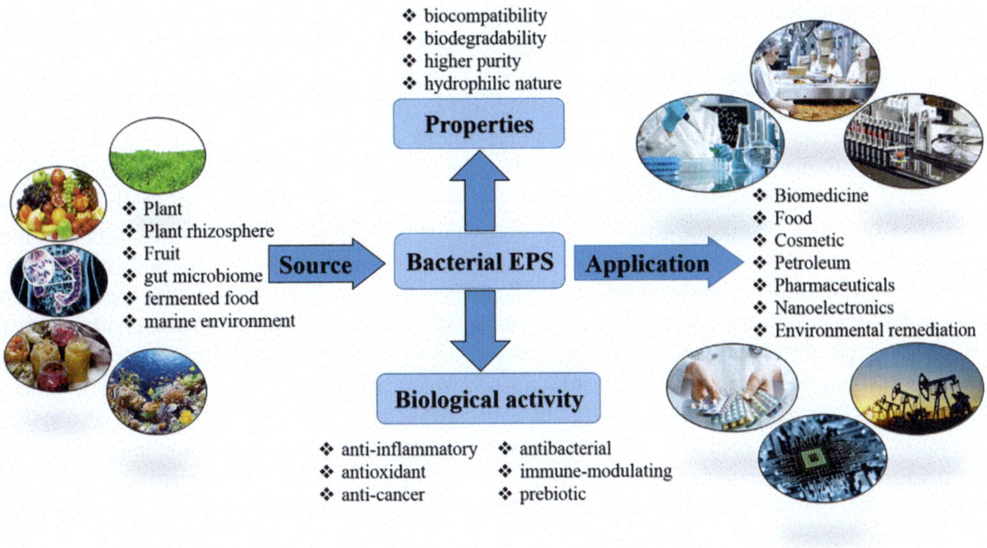

FIGURE 2.1 Bacterial exopolysaccharide: Properties, biological activity, application, and sources (Netrusov et al., 2023). *EPS*, Exopolysaccharide.

A recent case study on an artificial intelligence (AI)-driven multivariate optimization strategy, namely a hybrid support vector machine (h-SVM), was evaluated against the conventional response surface methodology (RSM) for optimizing medium composition to enhance exopolysaccharide production by a thermophilic strain of *Bacillus licheniformis*. Initially, a Plackett–Burman design was employed to screen critical factors influencing exopolysaccharide yield. Sucrose, KH_2PO_4, and NaCl were found to exert significant positive effects, whereas $NaNO_3$ had a significant negative impact. Subsequently, data from a central composite design (CCD) were analyzed using both RSM and the h-SVM approach. The h-SVM demonstrated superior predictive performance (R^2 = 0.9874; root mean square error = 0.693) compared to RSM. To further refine optimization, the SVM model was coupled with two metaheuristic algorithms: Teaching–learning-based optimization (TLBO) and particle swarm optimization (PSO). Both SVM–TLBO and SVM–PSO predicted an optimal exopolysaccharide concentration of 64 ± 1.1 g/L, which represented approximately a fivefold increase over the unoptimized medium (Rabiya and Sen, 2022) (Fig. 2.1).

2.2 Natural sources of exopolysaccharides

2.2.1 Bacterial sources

2.2.1.1 Major exopolysaccharides-producing bacteria

Bacterial species produce diverse classes of exopolysaccharides with distinct glycosidic linkages, monosaccharide configurations, molecular weights, and functional architectures. *Lactobacillus* species synthesize heteropolysaccharides consisting of repeating units of glucose, galactose, rhamnose, and mannose with β-1→3 and α-1→6 linkages (Ruas-Madiedo et al., 2002). These exopolysaccharide influence gastrointestinal adhesion and mucosal immunomodulation (Netrusov et al., 2023). *Bifidobacterium* strains excrete acidic heteropolysaccharides composed of glucose, galactose, and N-acetylglucosamine that facilitate anti-inflammatory pathways and inhibit pathogen colonization. *Pseudomonas aeruginosa* produces alginate Pel and Psl, which are critical for mucoid biofilm formation, antibiotic resistance, and immune evasion (Freitas et al., 2011). Alginate is a linear polymer of mannuronic and guluronic acids with extensive O-acetylation, conferring viscosity and ionic crosslinking potential. *Streptococcus thermophilus* and *Streptococcus mutans* generate dextran and mutan, respectively, through extracellular glucosyltransferases, which catalyze α-1→6 and α-1→3 glucosidic bonds essential for adhesion plaque formation and caries development (Xu et al., 2018). These bacterial exopolysaccharide exhibit diverse degrees of polymerization, branching frequencies, and structural persistence across environments and hosts. High-performance anion exchange chromatography pulsed amperometric detection, nuclear magnetic resonance (NMR), and mass spectrometry confirms compositional heterogeneity and strain specificity. The taxonomic diversity of exopolysaccharide-producing bacteria underpins their broad ecological roles and application potential in probiotics, therapeutics, biofilm engineering, and food matrices (Yildiz and Karatas, 2018). It is also important to isolate bacteria obtained from different strains. The serial dilution technique was employed to isolate bacteria from the collected samples. For this process, 1 g of soil and 1 ml of water sample was each transferred aseptically into separate containers holding 10 ml of sterile physiological saline. These mixtures were thoroughly homogenized to serve as stock solutions. Subsequently, 1 ml from each stock was added to 9 ml of fresh sterile saline, mixed well, and then serially diluted up to the sixth dilution. To estimate bacterial counts, 0.1 ml from each dilution was plated onto Bushnell-Hass agar and nutrient agar (supplemented with 2% sucrose as a carbon source). The plating was done in duplicate, and the plates were incubated at 37°C for 48 hours. Colonies that appeared thick and mucoid were selected and further purified using the streak plate method on newly prepared agar plates, ensuring isolation of distinct bacterial strains (Nwosu et al., 2019).

2.2.1.2 *Mechanisms of bacterial exopolysaccharides biosynthesis and regulation*

Exopolysaccharide biosynthesis in bacteria occurs through well-characterized pathways involving precursor activation, chain polymerization, and translocation. Glycosyltransferases utilize activated sugar donors such as UDP-glucose, GDP-mannose, and dTDP-rhamnose to assemble repeating oligosaccharide units (Sutherland, 2001). These units are transferred to undecaprenyl phosphate lipid carriers embedded in the cytoplasmic membrane. Polymer elongation proceeds through sequential glycosidic linkage formation, followed by flipping across the membrane via Wzx translocases or export through ABC transporter systems. Chain-length modulation is governed by polysaccharide co-polymerases, which control polymer size and termination. Translocation across the periplasm involves Wzy polymerases and accessory proteins. Regulation involves quorum-sensing systems, including acyl-homoserine lactone-mediated transcriptional regulation and two-component systems that sense nutrient availability, osmotic pressure and oxidative stress. Rcs phosphorelay and Csr global regulators modulate transcription of biosynthetic gene clusters in response to environmental stimuli. Second messengers such as cyclic-di-GMP activate transcriptional regulators and promote biofilm-associated exopolysaccharide expression. Sigma factors such as RpoS and RpoN participate in stress-induced exopolysaccharide production (Abdalla et al., 2021). Mutation of key biosynthetic genes or signal transduction elements alters exopolysaccharide output composition and functional

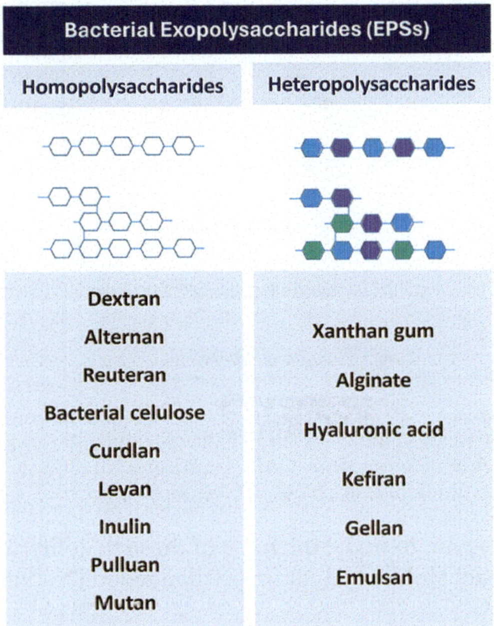

FIGURE 2.2 Illustration of the two types of bacterial exopolysaccharide: Examples of homopolysaccharides and heteropolysaccharides (Mouro et al., 2024).

performance. Omics-based profiling and gene knockout approaches confirm hierarchical control of exopolysaccharide biosynthesis, integrating transcriptional, metabolic, and environmental inputs into dynamic polysaccharide assembly systems (Welman and Maddox, 2003).

2.2.1.3 *Chemical diversity and structural variations in bacterial exopolysaccharides*

Bacterial exopolysaccharide exhibit extensive chemical diversity influenced by taxonomic lineage, environmental niche, and physiological state. Homopolysaccharides (HoPSs) such as levan from *Bacillus subtilis* or dextran from *Leuconostoc mesenteroides* comprise single monosaccharide types with specific linkages such as β-2→6 for fructose or α-1→6 for glucose, respectively (Fig. 2.2).

Bacteria are capable of making a wide variety of HoPSs, which are long chains built from single types of sugars like glucose, fructose, or galactose. The construction of these polymers is driven by enzymes called glycosyltransferases. How big the HoPS molecules get largely depends on what sugars are available in the bacterial food source, as well as which enzyme is doing the work. Other environmental factors—like how much nitrogen is present, the culture's pH, water content, temperature, and even how long the bacteria are grown—can all impact both the size and amount of HoPS produced. The genetic sequence of the polymer-building enzyme also matters, since it helps determine exactly how the sugar units are linked, affecting whether the chains branch out or stay straight. These structural details are important because they alter how the HoPSs behave, especially in terms of thickness and texture. Selecting the right enzyme and fine-tuning growth conditions provides a way to direct bacteria to make specific kinds of HoPSs, which opens up exciting possibilities for improving the texture of foods—whether the

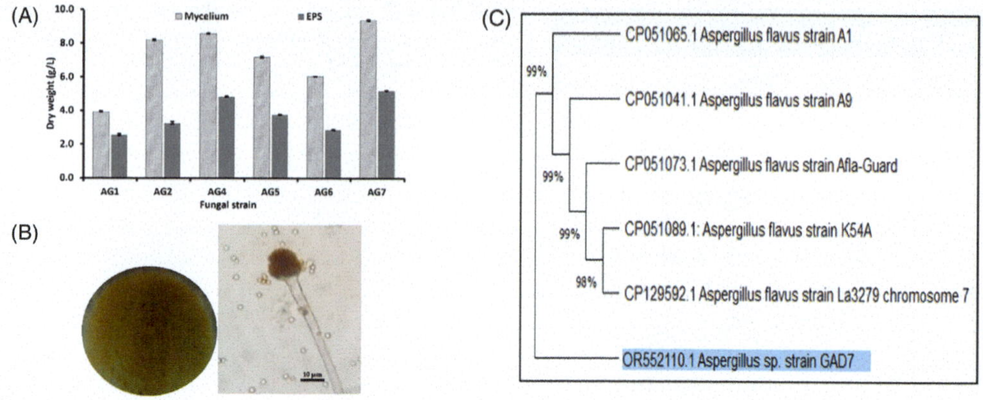

FIGURE 2.3 (A) *Mycelium* dry weights and exopolysaccharide yields of fungal isolates, (B) macrographs showing macroscopic features on SD agar (*left*) and microscopy details of *Aspergillus* sp. strain GAD7 (*right*), and (C) phylogenetic tree of *Aspergillus* sp. strain GAD7 (Strain designation/code) (Ibrahim et al., 2024). SD, Sabouraud Dextrose.

HoPSs are made right in the food or added later. However, to make the most of these benefits, a deeper understanding of how various conditions impact HoPS production is still needed (Nabot et al., 2022).

Recently, scientists have been taking a closer look at the heteropolysaccharides produced by both mesophilic and thermophilic lactic acid bacteria, especially due to their unique textures and behaviors. Detailed studies of their chemical structures and how they affect thickness have shown that these exopolysaccharide can differ widely—some are particularly good at thickening and show interesting shear-thinning properties, making them valuable for food applications like fermented dairy products. Many lactic acid bacteria that form slime create biopolymers with functional characteristics that could be put to use across various products, especially in naturally fermented milks such as yogurt. However, much is still unknown about how these complex polysaccharides are made, how they are organized at the molecular level, and what fermentation conditions help boost their production. Yields are often low, inconsistent, and the steps required to purify these biopolymers can be tricky. This review explores the different angles of heteropolysaccharide production by lactic acid bacteria—from their chemical makeup, genetics, and biological roles to the technology behind their manufacture and potential uses as food additives or agents to improve texture in products like yogurt. The discussion also touches on ways to improve bacterial strains, ramp up production, and modify processes using enzymes or advanced fermentation techniques that could help make commercial applications more practical and cost-effective for these promising natural biomolecules (De Vuyst and Degeest, 1999).

These polymers vary in chain length, branching degree, and molecular weight distribution. Heteropolysaccharides such as those from *Lactococcus lactis* include complex repeating units of three to eight sugars with multiple linkages, substitutions, and side chains. Structural variations arise from enzymatic modifications such as acetylation, methylation, phosphorylation, and sulfation, which affect charge density, solubility, and binding affinity. Strain-specific glycosyltransferase repertoires contribute to unique exopolysaccharide structures. Differences in glycosidic linkages such as α-1→4, α-1→6, β-1→2, β-1→3, and β-1→6 create diverse secondary conformations such as helices, coils, or extended chains. Branching introduces steric constraints and influences rheological behavior, including gel formation, shear-thinning, and viscosity modulation. Spectroscopic and chromatographic methods confirm that exopolysaccharide possess microheterogeneities which alter biological recognition by host immune cells or receptors. The exopolysaccharide chemical structure directly influences water-binding capacity, cation chelation, emulsification, and resistance to enzymatic degradation. These structural complexities confer adaptive and functional advantages in diverse habitats and host interactions.

2.2.1.4 Role in microbial symbiosis host interactions and biofilm formation

Exopolysaccharide play key roles in microbial adhesion, surface colonization, and host association across ecosystems. In plant-microbe symbioses, rhizobia such as *Sinorhizobium meliloti* produce succinoglycan and galactoglucans essential for root hair invasion, infection thread formation, and nodule organogenesis. Exopolysaccharide mutants exhibit impaired symbiosis and nitrogen fixation. In gastrointestinal ecosystems, exopolysaccharide from *Lactobacillus* and *Bifidobacterium* promote mucosal adhesion, epithelial barrier reinforcement, and dendritic cell modulation via carbohydrate-binding lectins. In pathogens such as *Klebsiella pneumoniae* and *Neisseria meningitidis*, exopolysaccharide capsules shield against complement-mediated lysis, neutrophil phagocytosis, and antimicrobial peptides. Biofilms incorporate exopolysaccharide as structural and functional matrix components. In *Pseudomonas aeruginosa* biofilms, Pel and Psl mediate cell-cell cohesion, hydration, and resistance to oxidative stress. Exopolysaccharide facilitates the retention of signaling molecules such as acyl-homoserine lactones, enabling quorum-sensing feedback loops and community coordination. They entrap extracellular DNA proteins and lipids, creating heterogeneous microenvironments with nutrient and oxygen gradients. Exopolysaccharide increases resistance to desiccation, antibiotics, and immune effectors. Matrix remodeling enzymes modulate exopolysaccharide composition during biofilm maturation, dispersion, and recolonization. Exopolysaccharide mediates interactions with fungi, protists, and viruses, contributing to polymicrobial biofilm complexity. Their structural modularity and chemical resilience enable long-term persistence in diverse host tissue surfaces and environmental interfaces (Hussain et al., 2017). Microbial biofilms are highly organized communities that allow disease-causing microbes to thrive while firmly attached to surfaces. In this state, the microorganisms gain several advantages—they are shielded from environmental stresses such as changes in temperature, oxygen, and pH, as well as from harmful chemicals and the host's immune defenses. This protection is mainly due to the extracellular matrix (ECM), a dense network of biopolymers produced by the microbes themselves. The ECM creates a supportive microenvironment that sustains cell survival, growth, and communication. Its composition differs depending on the species present, but it typically contains exopolysaccharide, proteins, lipids, extracellular DNA, and hydrolytic enzymes. Among these, exopolysaccharide serve as the structural framework, anchoring other biomolecules and maintaining the integrity of the biofilm. They may resemble or differ from the polysaccharides found on the microbial cell wall, and they play a crucial role in biofilm stability, pathogenic potential, drug resistance, and immune evasion. Because of their key role in disease processes, the pathways that synthesize EPSs are promising targets for new antimicrobial drugs. This chapter examines the functions, biosynthesis, and diversity of polysaccharides within microbial biofilms, the factors—both environmental and biological—that influence their formation, and the variations across different microbial species. It also outlines advanced microscopic and spectroscopic methods, such as field emission scanning electron microscopy (FESEM), confocal laser scanning microscopy (CLSM), Fourier transform infrared (FTIR) spectroscopy, and NMR, used to study and characterize these complex molecules (Gupta et al., 2019).

2.2.2 Fungal sources

2.2.2.1 *Exopolysaccharides production by fungi (Aspergillus, Saccharomyces, Penicillium)*

Fungal EPSs represent structurally complex high-molecular-weight biopolymers with significant biotechnological relevance and distinctive biosynthetic mechanisms (Fig. 2.3).

Filamentous fungi such as *Aspergillus niger, Aspergillus oryzae,* and *Aspergillus fumigatus* produce acidic polysaccharides such as galactomannan and heteroglycans enriched in galactose, glucose, and mannose with β-1→5 and α-1→6 linkages and frequent side-chain substitutions (Paul et al., 2024). These EPS exhibit high uronic acid content and negative surface charge, promoting water retention and ionic interaction. Yeasts such as *Saccharomyces cerevisiae* secrete mannoproteins and β-glucans during exponential growth and stress conditions. Their EPS contain α-mannan side chains and β-1→3-glucan backbones assembled in the Golgi and transported via secretory vesicles (Delbarre-Ladrat et al., 2014). *Penicillium* species synthesize acidic heteropolysaccharides incorporating rhamnose, xylose, galacturonic acid, and rare sugars. EPS synthesis in fungi involves cytoplasmic and vesicular glycosyltransferase activity, nucleotide sugar pools, and membrane-bound polysaccharide synthases. The composition of EPS is modulated by carbon source, nitrogen availability, metal ions, and pH. Batch and fed-batch cultivation of *Aspergillus* and *Penicillium* in aerated fermenters using glucose, sucrose, or molasses yield high EPS titers (Pereira et al., 2009). EPS recovery involves ethanol precipitation, ultrafiltration, and ion exchange chromatography. Analytical characterization by NMR methylation analysis and GPC confirms the presence of complex branched structures with molar masses exceeding 1000 kDa. Fungal EPS exhibit pseudoplastic behavior, thixotropy, and gelation properties influenced by branching degree and hydrophobicity. High molecular weight and sulfation patterns enhance bioactivity and physicochemical performance in industrial formulations.

2.2.2.2 *Comparative features of fungal versus bacterial exopolysaccharides*

Fungal EPS differ markedly from bacterial counterparts in monosaccharide composition, linkage type, molecular architecture, and biosynthetic pathway (Nicolaus et al., 2010). Fungal EPS are often rich in β-linked glucans and mannans with higher levels of acetylation, sulfation, and uronic acid substitution, leading to enhanced anionic character and water-holding capacity. Bacterial EPS, such as dextran and xanthan, are typically synthesized extracellularly by secreted enzymes, whereas fungal EPS are assembled intracellularly in Golgi-like compartments and secreted through exocytosis. Fungal EPS exhibit higher branching frequency and larger hydrodynamic volumes, contributing to superior viscoelasticity and gel strength. Bacterial EPS are often strain-specific with modular gene clusters encoding glycosyltransferases, whereas fungal EPS biosynthesis involves complex multienzyme systems with limited genetic modularity (Sutherland, 2001). Structural analyses show that fungal EPS have higher degrees of polymerization and wider molecular weight distribution than bacterial EPS. Functionally, fungal EPS have demonstrated superior antioxidant, immunostimulatory, and cholesterol-lowering effects attributed to β-1→3-glucan motifs and polyanionic domains. The fungus Rhizopus microsporus, which causes rice seedling blight, has a fascinating partnership with the bacterium Burkholderia rhizoxinica. This bacterium lives inside the fungal cells and produces a powerful toxin that damages crops. To better understand what makes this unusual symbiosis work, scientists closely examined the EPS produced by the bacterium—a key substance that helps cells interact with their environment. Using advanced two-dimensional NMR techniques, researchers were able to

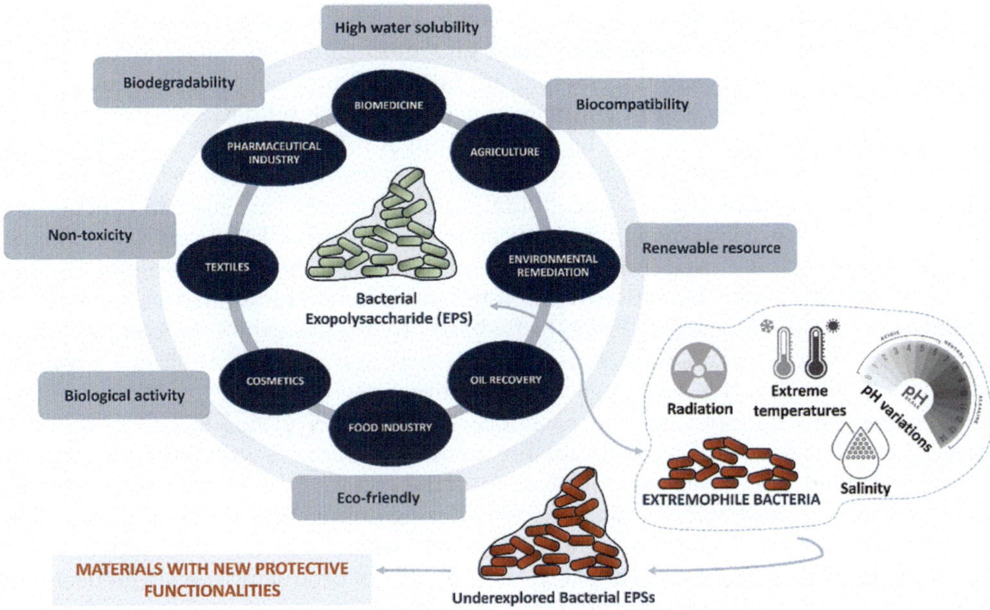

FIGURE 2.4 Microbial exopolysaccharides: Source, diversity, applications, and properties (Mouro et al., 2024).

uncover a unique EPS structure, made up of a repeating group of four sugars and featuring an acetyl group attached in a variable way to a mannose molecule. Further investigation involved mapping out the set of genes responsible for building this EPS and creating a mutant bacterium lacking these genes so that its behavior could be compared to the original. When viewed under a scanning electron microscope, the mutant showed a clear reduction in the ability to produce the sticky polymer matrix around clusters of cells. Genetic studies also revealed that these EPS-related genes have been conserved over time, highlighting their importance to the ongoing relationship between the fungus and its bacterial partner (Uzum et al., 2015).

Rheological studies indicate that fungal EPS form stronger gels and exhibit higher viscosity at lower concentrations due to extensive hydrogen bonding and network formation. Thermal analysis confirms higher thermal degradation onset temperatures for fungal EPS compared to bacterial analogs. Fungal EPS exhibit superior stability under acidic conditions and are less susceptible to enzymatic hydrolysis, enhancing their utility in harsh processing environments (Nguyen et al., 2024) (Fig. 2.4).

2.2.2.3 *Applications of fungal exopolysaccharides in medicine and the food industry*

Fungal EPSs have diverse applications in pharmaceutical, nutraceutical, and food sectors due to their immunomodulatory, antioxidant, hypocholesterolemic, and structural properties (Kaur and Dey, 2023). β-glucans from *Saccharomyces cerevisiae, Schizophyllum commune,* and *Lentinula edodes* stimulate innate immunity by binding to Dectin-1 and CR3 receptors on macrophages, neutrophils, and dendritic cells, leading to cytokine release, phagocytosis, and oxidative burst. Clinical studies confirm the efficacy of fungal β-glucans as adjuvants in cancer immunotherapy, infectious disease management, and metabolic regulation. EPS from *Aspergillus niger* reduces cholesterol absorption and modulates gut microbiota by acting as fermentable dietary fibers, promoting short-chain fatty acid production (Leroy and De Vuyst, 2016). *Penicillium* EPS shows anti-inflammatory activity by inhibiting COX-2 expression and suppressing NF-κB activation. In food technology, fungal EPS function as natural emulsifiers, stabilizers, gelling agents, and fat replacers. Mannoproteins from *S. cerevisiae* enhances wine and beer mouthfeel, foam stability, and colloidal clarity. β-glucans are used in functional food formulations for cholesterol reduction and glycemic index control. EPS from *Aspergillus oryzae* and *Rhizopus oligosporus* are used in fermented food matrices to improve rheology texture and moisture retention. Microencapsulation of probiotics using fungal EPS enhances viability during gastrointestinal transit and storage. Fungal EPS-based hydrogels exhibit biocompatibility and structural integrity for wound dressing, drug delivery, and tissue scaffolding applications. Industrial production involves submerged fermentation scale-up, downstream processing, and regulatory compliance for food and pharmaceutical use (Sørensen et al., 2022). Over the past two decades, fungal EPSs have emerged as valuable biomacromolecules with wide-ranging uses in food, medicine, pharmaceuticals, and various industrial sectors. Well-known examples include pullulan, scleroglucan, and botryosphaeran, each offering unique functional properties. Despite their importance, our understanding of how fungi synthesize EPSs remains limited, and there is still a strong need to identify new fungal species capable of producing novel and useful forms of these polymers. In many cases, the molecular weight and sugar composition of fungal EPSs are influenced by the ingredients in the growth medium and by the physical conditions during fermentation, such as temperature, pH, and aeration. This review provides a comprehensive overview of fungal EPS production, their structural features, and their diverse applications, with special attention given to cataloguing various fungal strains known for their EPS-producing abilities (Mahapatra and Banerjee, 2013).

2.2.3 Algal and marine sources

2.2.3.1 *Exopolysaccharides production in marine organisms (Chlorella, Spirulina, red, brown, and green algae)*

Marine organisms, including microalgae, cyanobacteria, and macroalgae, produce EPSs with unique structural features, high anionic density, and biological complexity (Satpute et al., 2010). *Chlorella vulgaris* secretes neutral and acidic heteropolysaccharides containing glucose, galactose, rhamnose, arabinose, and uronic acids. These EPS are produced during nitrogen starvation, light stress, and the stationary growth phase. *Spirulina platensis,* a filamentous cyanobacterium, synthesizes sulfated EPS with xylose, glucuronic acid, and galactose that exhibit high emulsifying capacity, antioxidant activity, and chelation potential (De Philippis et al., 2001). Marine macroalgae such as red algae (*Kappaphycus alvarezii, Gracilaria verrucosa*), brown algae (*Fucus vesiculosus, Laminaria japonica*), and green algae (*Ulva lactuca, Codium fragile*) yield commercially significant sulfated EPS, including carrageenan, agar, fucoidan, and ulvan. Carrageenans consist of alternating 3-linked β-D-galactose and 4-linked α-D-galactose or 3,6-anhydrogalactose with variable sulfate ester substitution (Hamidi et al., 2022). Fucoidans are L-fucose-rich polysaccharides with sulfate ester and uronic acid substitution and demonstrate potent bioactivity due to their high charge density and structural flexibility. Ulvans comprise rhamnose, glucuronic acid, iduronic acid, and xylose in branched repeating units with sulfation at C2 and C4 positions. Marine bacteria produce EPSs to support growth, adherence to surfaces, and survival in harsh environments, and these EPSs—often heteropolysaccharides with diverse, protective, and polyanionic properties—are found in bacteria from extreme marine niches like deep-sea vents, polar sea ice, and hypersaline ecosystems. These EPSs have significant commercial and biotechnological potential, with applications in pharmaceuticals, food processing, bioremediation, and more, prompting increased interest in their isolation and study, including from symbiotic bacteria in marine annelid worms from extreme habitats (Delbarre-Ladrat et al., 2014).

EPS biosynthesis in marine algae is driven by carbon overflow metabolism under nutrient-limiting phototrophic conditions. Biosynthetic pathways involve nucleotide sugar donors UDP-galactose, UDP-glucuronic acid, GDP-fucose, and sulfotransferases localized in the Golgi apparatus (Debnath et al., 2024). Environmental parameters, including salinity, light intensity, and micronutrient availability, influence EPS yield composition and degree of sulfation. In a study, a neutral, water-soluble polysaccharide (HPA) from the marine fungus *Hansfordia sinuosae* was found to be mainly composed of mannose, with small amounts of galactose and glucose, and had a molecular weight of about 22.5 kDa. Structural analysis revealed various α-d-mannopyranosyl linkages, including [→2)-, [→3)-, and [→2,6)-α-d-Manp(1→] types, with specific branching at the C-6 position. In vitro studies demonstrated that HPA significantly inhibited the growth of HeLa and MCF-7 cancer cells, promoted ROS generation, triggered apoptosis, reduced mitochondrial membrane potential, and activated caspase-3, indicating its promise as an antitumor agent (Li et al., 2018). Photobioreactor cultivation of marine strains under controlled trophic and light regimes enhances EPS productivity and structural consistency. Harvested EPS are recovered by filtration, ethanol precipitation, dialysis, and lyophilization, followed by characterization using FTIR, NMR, and HPSEC (Casillo et al., 2018).

2.2.3.2 *Unique properties and biotechnological potential of marine exopolysaccharides*

Marine EPS exhibit high molecular diversity, unique structural motifs, and multifunctional physicochemical properties not found in terrestrial EPS. High sulfate content contributes to a strong negative charge density that supports cation sequestration, antioxidant activity, and anticoagulant effects. Fucoidans with specific sulfation patterns inhibit thrombin formation and modulate heparin cofactor II interactions. Carrageenans form thermo-reversible gels with potassium ions and act as stabilizers in emulsions, suspensions, and pharmaceutical gels. Ulvans demonstrate immunomodulatory effects by activating macrophages through TLR signaling and enhancing nitric oxide production. Marine EPS have superior water retention, metal chelation, and bioadhesion properties due to their branched conformation and high hydrophilicity. Their amphiphilic nature enables emulsification and surfactant functions useful in cosmetics and marine antifouling coatings (Li et al., 2001). The marine environment, Earth's largest habitat, hosts diverse microorganisms with significant biotechnological potential. Polysaccharides from marine microbes are of particular interest due to their structural and functional diversity, shaped by extreme conditions such as temperature, pH, and salinity, which can lead to novel compounds. This review summarizes their unique features, potential applications in various industries, recent advances, and the challenges that limit their commercial use, aiming to guide future research in this field (Senni et al., 2013) (Fig. 2.5).

EPS from *Chlorella* and *Spirulina* act as antioxidants through radical scavenging, ferric-reducing power, and inhibition of lipid peroxidation. Sulfated EPS exhibits antiviral effects against enveloped viruses such as HIV, HSV, and coronaviruses by preventing viral adsorption

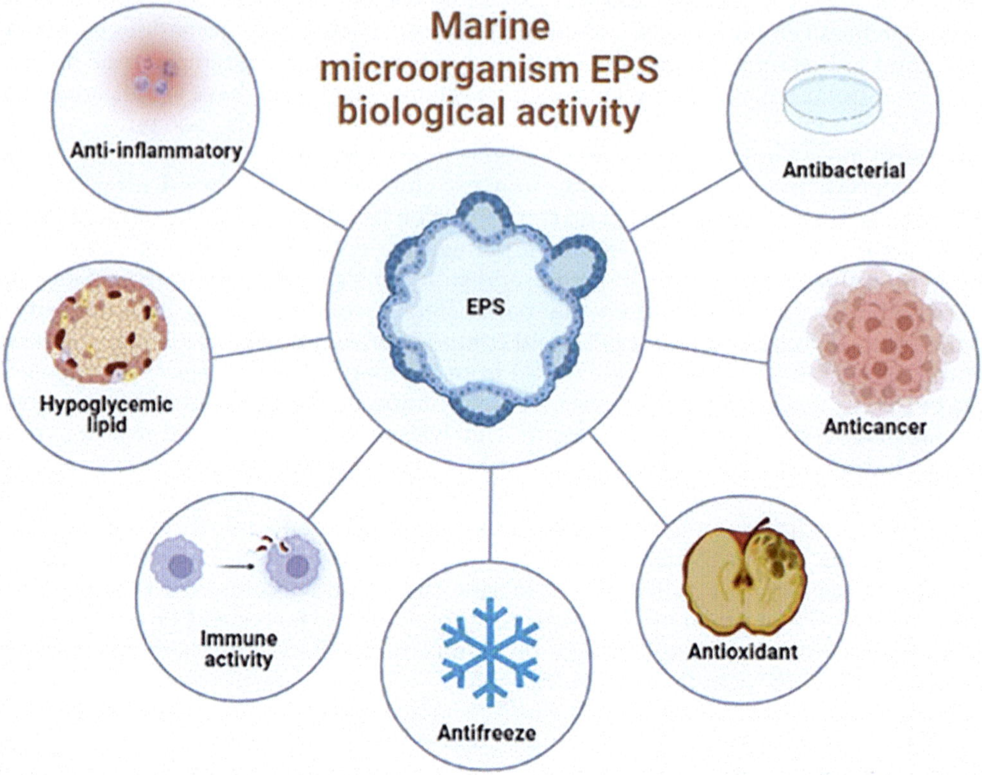

FIGURE 2.5 Main biological activities of EPS from marine microorganisms (Qi et al., 2022). *EPS*, Exopolysaccharides.

and fusion. Marine EPS serve as cryoprotectants due to their hydration shell formation, hydrogen bonding, and ice recrystallization inhibition (Daba et al., 2021). In biomedicine, they are used as drug carriers, bioadhesives, and scaffolds for tissue engineering. Bioprocessing approaches, including strain domestication, metabolic flux redirection, and high-light stress application, enhance marine EPS yield and functionality. Engineering of biosynthetic enzymes such as fucosyltransferases and sulfotransferases enables control over branching degree, sulfation position, and bioactivity. Marine EPS provide a sustainable source of functional biopolymers for pharmaceutical, food, and environmental industries (K et al., 2018).

2.2.3.3 *Role of marine exopolysaccharides in ocean ecosystems and carbon cycling*

Marine EPS play central roles in biogeochemical cycling, microbial ecology, and particle dynamics in ocean systems. The marine biosphere is a rich source of bioactive compounds, with biosurfactants (BS) and bioemulsifiers (BE) gaining attention for their diverse structures, functions, and wide industrial applications. Various marine microorganisms, including *Acinetobacter, Pseudomonas, Bacillus*, and others, have been explored for BS/BE and EPSs production, though much of the marine microbial diversity remains untapped. Discovering new BS/BE-producing strains could expand the use of biodegradable surfactants and reduce reliance on synthetic chemical alternatives (Satpute et al., 2010).

Microalgal EPS form transparent exopolymer particles (TEP) composed of acidic polysaccharides that aggregate with minerals and detritus to form marine snow. TEP contributes to vertical carbon flux by transporting organic carbon from euphotic zones to deep ocean layers, facilitating carbon sequestration and sedimentation (Mende et al., 2016). Diatoms and coccolithophores secrete EPS that promote flocculation and sinking of phytoplankton aggregates, enhancing biological carbon pump efficiency. EPS provide attachment surfaces and energy sources for heterotrophic bacteria, modulating microbial loop dynamics and nutrient turnover. EPS from cyanobacteria such as *Trichodesmium* and *Synechococcus* enhance nitrogen fixation through colony formation and nutrient retention (Gheorghita et al., 2023). EPS facilitates iron scavenging by binding ferric ions in siderophore-like structures, enhancing bioavailability and primary productivity. In oligotrophic gyres, EPS reduces grazing pressure by acting as a physical barrier and altering palatability. During algal bloom collapse, EPS stabilizes dissolved organic matter into colloidal forms, limiting remineralization and promoting export. EPS modulates oxygen and pH microgradients in the phycosphere, shaping microbial symbioses and interkingdom interactions (Tiwari et al., 2021). Biodegradation of marine EPS by glycoside hydrolases, sulfatases, and lyases regulates carbon turnover rates and community succession. Integration of EPS dynamics into Earth system models improves the prediction of carbon flux, nutrient cycling, and climate feedbacks. EPS turnover in ocean systems is regulated by bacterial enzymatic capacities, temperature, pH, and stoichiometric ratios of nitrogen and phosphorus to carbon. EPS-mediated aggregation and sedimentation influence deep-sea biogeochemistry, benthic carbon burial, and long-term carbon storage (Salimi and Farrokh, 2023).

2.2.4 Other microbial sources

2.2.4.1 *Exopolysaccharides synthesis in yeasts, actinomycetes, and other microbes*

Yeasts, actinomycetes, and extremophilic archaea represent nonconventional microbial sources of structurally distinct and functionally potent EPSs with high-value biotechnological potential (Kavitake et al., 2023). Yeast species such as *Cryptococcus neoformans* produce capsular polysaccharides like glucuronoxylomannans that are composed of mannose, xylose, and glucuronic acid with O-acetylation and branching at C2 and C6 positions. These EPS confer resistance to host immune responses through steric hindrance, complement inhibition, and downregulation of proinflammatory cytokines (Grechi et al., 2011). *Candida albicans* secretes β-glucans and mannoproteins that contribute to biofilm structure, antifungal resistance, and adhesion. Their EPS synthesis is controlled by Golgi-associated glycosyltransferases and regulated by carbon flux, oxidative stress, and cell wall integrity signaling (Morcillo and Manzanera, 2021). Yeast EPSs are a less-studied metabolite group, with industrial production limited by low yields and poorly understood biosynthetic pathways. They exhibit diverse properties, including antitumor, immunostimulatory, antioxidant, heavy metal-absorbing, and food additive functions. Their production requires strong oxygenation, low temperatures, sucrose as a carbon source, and a C:N ratio of 15:1, with maximum yield after 4–5 days, and their structures—often rich in mannose and glucose—vary with strain and culture conditions (Gientka et al., 2015).

Actinomycetes such as *Streptomyces, Micromonospora,* and *Nocardia* produce acidic heteropolysaccharides with high uronic acid content composed of glucose, galactose, mannose, arabinose, and rhamnose, along with noncarbohydrate decorations including pyruvyl groups, sulfate esters, and phosphates. These EPS have emulsifying, metal-chelating, and immunostimulatory properties. Thermophilic and halophilic archaea, including *Halobacterium salinarum, Sulfolobus acidocaldarius,* and *Methanothermobacter thermautotrophicus,* produce stable EPS with methylated mannose glucuronic acid and sulfated galactose (Selim et al., 2018). Their EPS withstand high temperatures, acidic pH, and extreme salinity due to a rigid backbone structure, dense branching, and electrostatic interactions. EPS synthesis in archaea is poorly understood but involves archaeal glycosylation pathways, unique glycosyltransferases, and UDP-sugar interconversion enzymes. Many of these microbes synthesize EPS constitutively or in response to nutrient limitation, oxidative stress, or surface attachment (Patel et al., 2012; Zhou et al., 2019; Kavitake et al., 2023). Their EPS production is often associated with extracellular enzyme secretion, antibiotic biosynthesis, and cell aggregation. The structural characterization of these polymers reveals high molecular weights reaching 2000 kDa, complex substitution patterns, and unique functional groups not commonly observed in bacterial or fungal EPS (Wang et al., 2019). High-performance anion-exchange chromatography, multiangle laser light scattering, and NMR spectroscopy is used to delineate compositional and structural features. These EPS exhibit high thermal stability, resistance to enzymatic degradation, and activity retention under extreme physicochemical conditions, making them ideal for harsh industrial processes, including oil recovery, acid mine drainage treatment, and thermophilic biocatalysis (Ates, 2015).

2.2.4.2 *Specific roles in environmental resilience and microbial communication*

EPS produced by nonconventional microbes play essential roles in enhancing resilience under abiotic stresses, facilitating microbial interactions, and structuring complex ecosystems. In soil and desert environments, EPS from cyanobacteria such as *Microcoleus vaginatus* and *Nostoc commune* stabilize soil particles, promote crust formation, and maintain hydration during desiccation (Angelin and Kavitha, 2020). These EPS form dense sheaths composed of xylose, glucose, galacturonic acid, and fucose that bind soil aggregates and retain micronutrients. In halophilic environments, EPS from archaea such as *Haloferax mediterranei* protect cells from osmotic lysis by forming hydrated matrices that trap water and buffer ionic stress. Thermoacidophilic archaea produce EPS with high uronic acid and sulfate content, which chelate heavy metals, reduce oxidative damage, and protect enzymes from denaturation. In bioleaching consortia, EPS from *Acidithiobacillus ferrooxidans* and *Leptospirillum ferrooxidans* facilitate colonization of mineral surfaces, enhance ferric ion generation, and mediate electron transfer during oxidation of sulfide ores. EPS also serve as scaffolds for retention of extracellular DNA proteins and signaling molecules, thereby modulating microbial community assembly and function (Poli et al., 2010). In polymicrobial communities, EPS enable horizontal gene transfer by stabilizing plasmid DNA and promoting competence development. Intermicrobial signaling involving autoinducer-2 and diffusible small molecules is facilitated by EPS matrices, which modulate diffusion rates, spatial localization, and threshold-dependent activation. EPS regulates stress-responsive pathways, including two-component signal transduction, quorum sensing, and oxidative stress response by providing temporal and spatial signal stability (Zannini et al., 2016). Rhizobia are bacteria that can live freely in soil or partner with legume plants to fix nitrogen inside root nodules. Their surface sugars, especially the EPS, are vital for successfully invading plant roots, mainly in plants that form indeterminate nodules. EPS protects rhizobia from stress and plant defenses, helps them attach to surfaces, and acts as a signal in plant-bacteria interactions. Producing EPS involves complex regulation by many genes and proteins, and its synthesis is influenced by

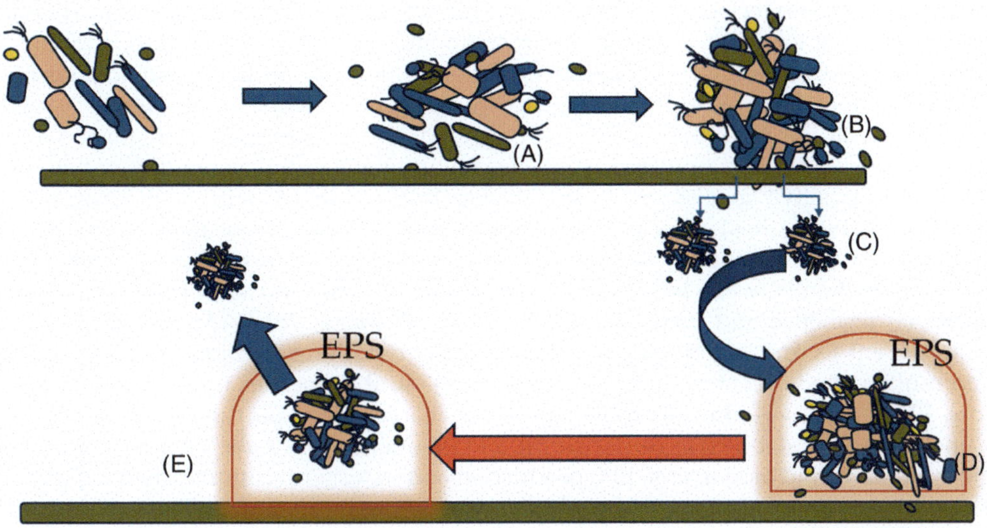

FIGURE 2.6 General mechanism of biofilm formation and maturation (Coppola et al., 2025).

environmental conditions like nutrient levels, stress, and plant molecules. Studies in species like *Sinorhizobium meliloti* and *Rhizobium leguminosarum* show that this process is tightly controlled and connected to bacterial communication and movement, helping rhizobia thrive and colonize plant roots (Janczarek, 2011).

EPS also acts as a carbon reservoir during starvation periods and enables long-term survival under nutrient-limited conditions. In rhizospheric environments, EPS from actinobacteria and yeasts modulate plant root architecture, stimulate beneficial microbial colonization, and contribute to induced systemic resistance. These functional attributes make EPS from rare and extremophilic microorganisms essential for microbial survival, adaptation and cross-kingdom communication in dynamic and harsh ecosystems (Barcelos et al., 2020).

2.2.5 Environmental and ecological significance

2.2.5.1 *Contribution of exopolysaccharides to microbial communities and biofilms*

EPSs constitute the major structural component of microbial biofilms and govern architecture, cohesion, stability, and metabolic functionality of microbial communities. EPS produced by biofilm-forming bacteria such as *Pseudomonas aeruginosa*, *Staphylococcus epidermidis*, and *Escherichia coli* form hydrated viscoelastic matrices that encapsulate microbial cells, trap nutrients, maintain redox gradients, and protect against mechanical shear (Fig. 2.6).

Biofilm EPS are composed of diverse neutral and acidic polysaccharides, including alginate, cellulose, Pel Psl, and poly-β-1→6-N-acetylglucosamine with molecular heterogeneity tailored to ecological function. In aquatic environments, EPS matrices promote colonization of abiotic surfaces, enable floatation of planktonic aggregates, and mediate formation of microcolonies with nutrient zonation and syntrophic interactions. Biofilm EPS binds metal ions, enzymes, and extracellular DNA, enabling horizontal gene transfer and resistance phenotype dissemination.

EPS matrices provide selective permeability, modulate the diffusion of antibiotics and antimicrobials, and confer tolerance through electrostatic interactions, sequestration, and chemical neutralization (Laws et al., 2001). Biofilm maturation involves dynamic remodeling of EPS composition regulated by quorum-sensing two-component systems, cyclic-di-GMP signaling, and environmental feedback. EPS from gram-negative and gram-positive organisms differ in chemical complexity, with gram-negative species favoring mixed linkage branched heteropolymers and gram-positive species producing thicker linear matrices with high charge density (Cugini et al., 2019). In host-associated microbiota such as oral, gut or skin ecosystems, EPS dictate spatial arrangement, resistance to host defenses, and colonization persistence. EPS mediates co-aggregation between bacterial species, fungal hyphae, and viral particles, enhancing polymicrobial resilience and virulence (Oleksy and Klewicka, 2018). Advanced microscopy using fluorescent lectins combined with Raman spectroscopy and atomic force imaging has revealed multilayered EPS structures with nanomechanical properties tuned to environmental stressors. EPS diversity within microbial consortia enables niche partitioning, metabolic coupling, and ecological stability in both natural and engineered systems, including wastewater treatment biofilters and biobarriers. EPS production correlates with community-level resistance to desiccation, nutrient limitation, and oxidative stress and is fundamental to the emergence and persistence of spatially structured microbial ecosystems.

Quorum sensing acts as a communication system in microbes, triggering collective behaviors such as biofilm formation—a major contributor to antibiotic resistance. Biofilms, especially those formed by the notorious ESKAPE pathogens, pose serious risks in hospital-acquired infections and are a key driver of antimicrobial resistance (AMR), which the WHO considers a global health threat. Targeting these microbial communities with antibiofilm agents offer a promising way to combat AMR. The choice of such agents depends on their chemical properties and the unique biofilm mechanisms of each pathogen, often requiring tailored or combined approaches with antibiotics to effectively disrupt their growth and improve treatment outcomes (Rajput et al., 2020).

2.2.5.2 *Exopolysaccharides in extreme environments (hydrothermal vents acidic pools)*

In extremophilic habitats such as hydrothermal vents, hypersaline lakes, acidic hot springs, and polar deserts, EPS provides essential functional protection to microbial inhabitants by stabilizing cells against physicochemical extremes (Jurášková et al., 2022). In deep-sea hydrothermal vents, thermophilic bacteria and archaea such as *Thermococcus, Aquifex,* and *Desulfurococcus* synthesize thermostable EPS with uronic acid and sulfate substitutions that resist hydrolysis and thermal denaturation. These polymers form cell-anchored or released matrices that facilitate metal ion binding, promote adherence to mineral substrates, and modulate pH and redox fluctuations. EPS from *Sulfolobus solfataricus* and *Metallosphaera sedula* mediate colonization of pyrite, chalcopyrite, and other metal sulfides and support sulfur oxidation and bioleaching under high-temperature acidic conditions (Galle and Arendt, 2014). In a case study, researchers investigated microbial life in two volcanic mud pools at Solfatara Pisciarelli, Naples, to discover heat-loving enzymes capable of breaking down plant materials. Using DNA sequencing, they found that both pools were dominated by Archaea, especially Crenarchaeota, and that many genetic sequences did not match known organisms, suggesting the presence of new species. Carbohydrate metabolism genes were the most common, and hundreds of carbohydrate-active enzymes (CAZymes), especially glycosyltransferases and glycoside hydrolases, were identified—some unique to this environment. Two novel enzymes from an unknown archaeon in the hotter, more acidic pool (GH5_Pool2 and GH109_Pool2) showed exceptional heat stability and unique catalytic abilities. Their properties, along with the overall diversity of enzymes found, highlight the potential of this ecosystem as a source of thermostable biocatalysts for future industrial applications, particularly in biofuel production and processing of plant biomass under extreme conditions (Strazzulli et al., 2020).

EPS from haloarchaea such as *Haloferax volcanii* and *Halobacterium salinarum* maintain osmotic balance through hydration shells and cation binding, preventing protein denaturation and membrane destabilization under hypersaline stress. In glacial ice and polar soils, psychrophilic microbes produce EPS with high fucose and glucuronic acid content, which function as antifreeze molecules to reduce ice crystal formation and preserve enzyme activity at subzero temperatures (Jolly et al., 2002). EPS enhances survival under UV irradiation, oxidative damage, and limited nutrient availability by creating microenvironments enriched in antioxidants and metal chelators. In acid mine drainage, EPS from *Acidithiobacillus ferrooxidans* and *Leptospirillum ferrooxidans* form biofilms that enable iron and sulfur oxidation while stabilizing cell-mineral interfaces and facilitating electron transport. These EPS exhibit remarkable structural resistance to low pH and high redox potential and act as electron shuttles in bioelectrochemical gradients. Functional stability of extremophilic EPS is driven by unique monosaccharide patterns, methylation of aromatic residues, and cross-linked domains that enhance rigidity and ionic tolerance (Andrew and Jayaraman, 2020). Proteomic and glycomic analyses confirm co-localization of EPS with stress proteins, hydrophobic peptides, and metal-responsive enzymes forming integrated protective complexes. EPS-mediated colonization of extreme environments contributes to biogeochemical cycling, metal transformation, and biomineralization under conditions incompatible with most life forms (Schmid et al., 2015).

2.2.5.3 *Ecological functions of exopolysaccharides in soil and water ecosystems*

EPS play a central role in microbial survival by creating a microenvironment that supports chemical reactions, stores nutrients, and shields cells from stresses such as high salinity, drought, and other harsh conditions. In soils, microbial EPS act like natural "bioglues," binding soil particles together, improving structure, and helping retain moisture and nutrients—benefits that extend to plant health and growth. Beyond their ecological role, EPS possess valuable properties, including biocompatibility and the ability to gel or thicken, which have attracted interest for industrial and agricultural uses. Yet, despite decades of study, only a limited range of these substances has been commercialized, particularly for agricultural applications. Understanding the ecological functions of EPS is therefore essential, not only for appreciating their role in natural soil systems but also for harnessing their potential to improve soil aggregation, fertility, and sustainability (Costa et al., 2018). EPS plays critical ecological roles in soil structure stability, nutrient cycling, and microbial–plant interactions in terrestrial and aquatic ecosystems (Fig. 2.7).

Cyanobacterial soil crusts are communities of microorganisms found on the soil surface in many habitats around the world. The cyanobacteria in these crusts produce exopolysaccharides (EPS) through photosynthesis, adding carbohydrates to the soil's carbon pool and helping bind soil particles together to reduce erosion. Although these crusts are important in dryland ecosystems, our knowledge of their roles and potential uses is limited. This review summarizes what is currently known about cyanobacterial EPS in dry soils and suggests directions for future research (Mager and Thomas, 2011).

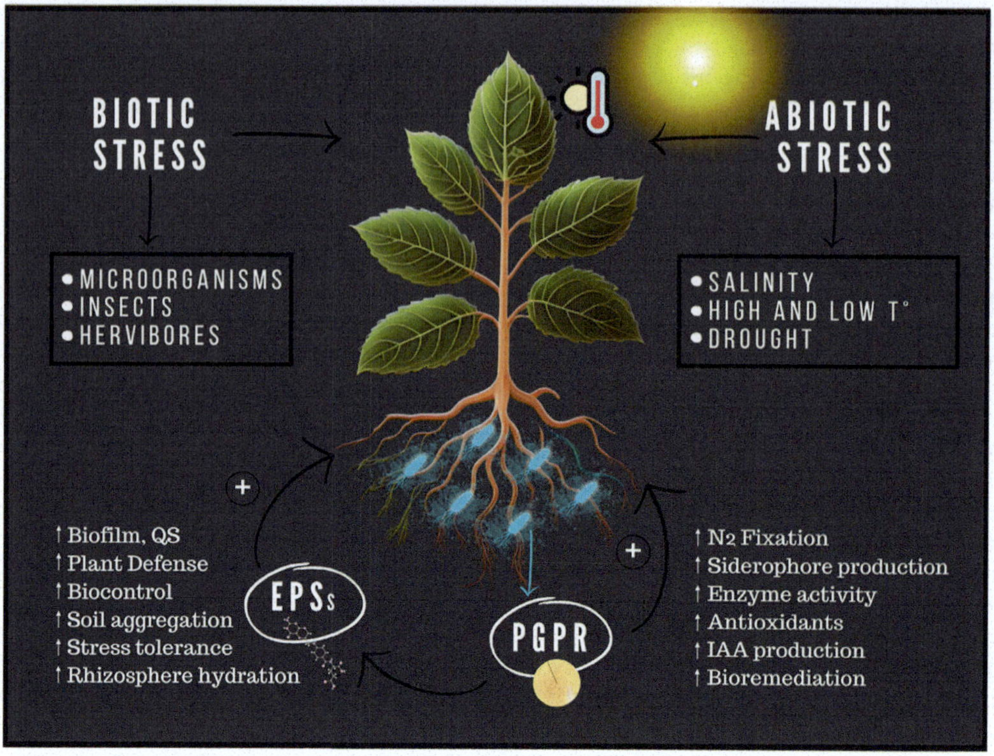

FIGURE 2.7 Diagram that summarizes the mechanisms of action of PGPR on plant growth, and the role of their EPSs in counteractingss the effects of biotic and abiotic stress. The *blue arrow* indicates the presence of PGPR bacteria in the rhizosphere. The *black arrows* indicate, on the *right side*, the positive effects produced by PGPR bacteria and, on the *left side*, the beneficial actions produced by the synthesis of EPSs by PGPR bacteria (Carezzano et al., 2023). *EPS*, Exopolysaccharides; PGPR, Plant Growth Promoting Rhizobacteria.

Soil-dwelling microbes, including *Bacillus subtilis, Azospirillum brasilense,* and cyanobacteria, secrete EPS that aggregate soil particles, enhance porosity, and prevent erosion by forming stable soil macroaggregates. EPS-mediated soil aggregation improves aeration, water retention, and microbial habitat complexity (Whitfield et al., 2015). EPS binds divalent cations such as Ca^{2+}, Mg^{2+}, and Fe^{3+} and enhances cation exchange capacity and micronutrient bioavailability. In plant rhizospheres, EPS from plant-growth-promoting rhizobacteria modulates root surface properties, enhances adhesion, and mediates colonization by beneficial microbes (Whitfield et al., 2015). EPS acts as a signaling interface that influences plant immune responses, stimulates root hair elongation, and induces systemic resistance. EPS forms protective sheaths around root-associated bacteria, buffering against desiccation, salinity, and pH stress and enabling persistence under drought or heavy metal contamination. In aquatic ecosystems, EPS regulates the formation of flocs and suspended solids that influence sedimentation, light penetration, and water clarity (Riaz Rajoka et al., 2020).

EPS from freshwater and marine cyanobacteria contributes to algal bloom stability formation of gelatinous matrices, and resistance to grazing. EPS binds dissolved organic matter and metals and regulates particulate organic carbon dynamics and microbial food web structure. EPS-mediated aggregation supports the removal of colloids and pollutants in wastewater treatment systems (Shukla et al., 2019). In wetlands and riparian systems, EPS contributes to biofilm development on plant roots and sediments, facilitating denitrification, phosphorus retention, and pollutant degradation. EPS participates in nitrogen and sulfur cycling through microbially mediated oxidation-reduction reactions facilitated by stable electron-transfer environments (Prete et al., 2021). Soil and aquatic EPS pools are influenced by land use seasonal changes and nutrient inputs, and function as reservoirs of slow-turnover carbon that contribute to long-term sequestration and soil organic matter formation. EPS interactions with clays, organic acids, and plant exudates determine biogeochemical gradients and microbial community assembly. EPS, therefore, represents a key molecular determinant of ecosystem structure, function, and resilience across terrestrial and aquatic interfaces (Zhang et al., 2024) (Fig. 2.8; Tables 2.1 and 2.2).

2.3 Chemical and physicochemical characteristics of exopolysaccharides

2.3.1 Molecular composition and variability

EPSs possess a broad spectrum of molecular architectures governed by monosaccharide composition, glycosidic linkages, degree of branching, and presence of noncarbohydrate substituents. EPS may be composed of HoPSs containing repeating units of a single monosaccharide, such as glucose in dextran or fructose in levan, or may exhibit heteropolysaccharide structures consisting of diverse monosaccharides, including galactose, rhamnose, mannose, fucose, xylose, arabinose, glucuronic acid, and galacturonic acid arranged in linear or branched configurations (Castro-Bravo et al., 2018). Structural analysis reveals the frequent occurrence of rare sugars such as 3,6-anhydrogalactose, 2-keto-3-deoxy-octulosonic acid, and 2-acetamido-2-deoxy sugars. Glycosidic linkage diversity spans α and β configurations at multiple positions, including 1→2, 1→3, 1→4, 1→6, creating structural motifs with distinct torsional angles, hydrogen bonding patterns, and conformational constraints (Liu et al., 2017). EPS are commonly substituted with acetyl phosphate sulfate or pyruvyl groups at C2, C3, or C6 positions, which alter net charge, hydrophilicity, and binding properties. Researchers discovered a marine fungus, *Aspergillus sp.* strain GAD7, from the Egyptian Mediterranean coast that produces a unique acidic, sulfated extracellular polysaccharide (EPS) with strong anticoagulant and antioxidant effects. Among six fungal strains screened, GAD7 stood out for its high EPS yield (~5.2 g/L). Detailed analysis showed the EPS is made mostly of carbohydrates and contains a significant amount of sulfate. It consists of multiple sugar units, including mannose, glucose, arabinose, galacturonic acid, galactose, and lyxose in specific ratios, and features both α- and β-glycosidic bonds. The structure of this polysaccharide forms a stable triple-helix shape, maintaining its integrity under various conditions. Functionally, EPS-AG7 was able to double blood clotting time at low concentrations and effectively neutralize free radicals, suggesting strong biological activity. These results point to potential medical uses for this compound, especially as an anticoagulant and antioxidant (Ibrahim et al., 2024).

Molecular weight of EPS ranges from 10 kDa to several million Da, depending on microbial source, cultivation conditions, and polymerization mechanisms. Some EPS are polydisperse, exhibiting broad molecular weight distribution, while others form narrowly defined oligomeric

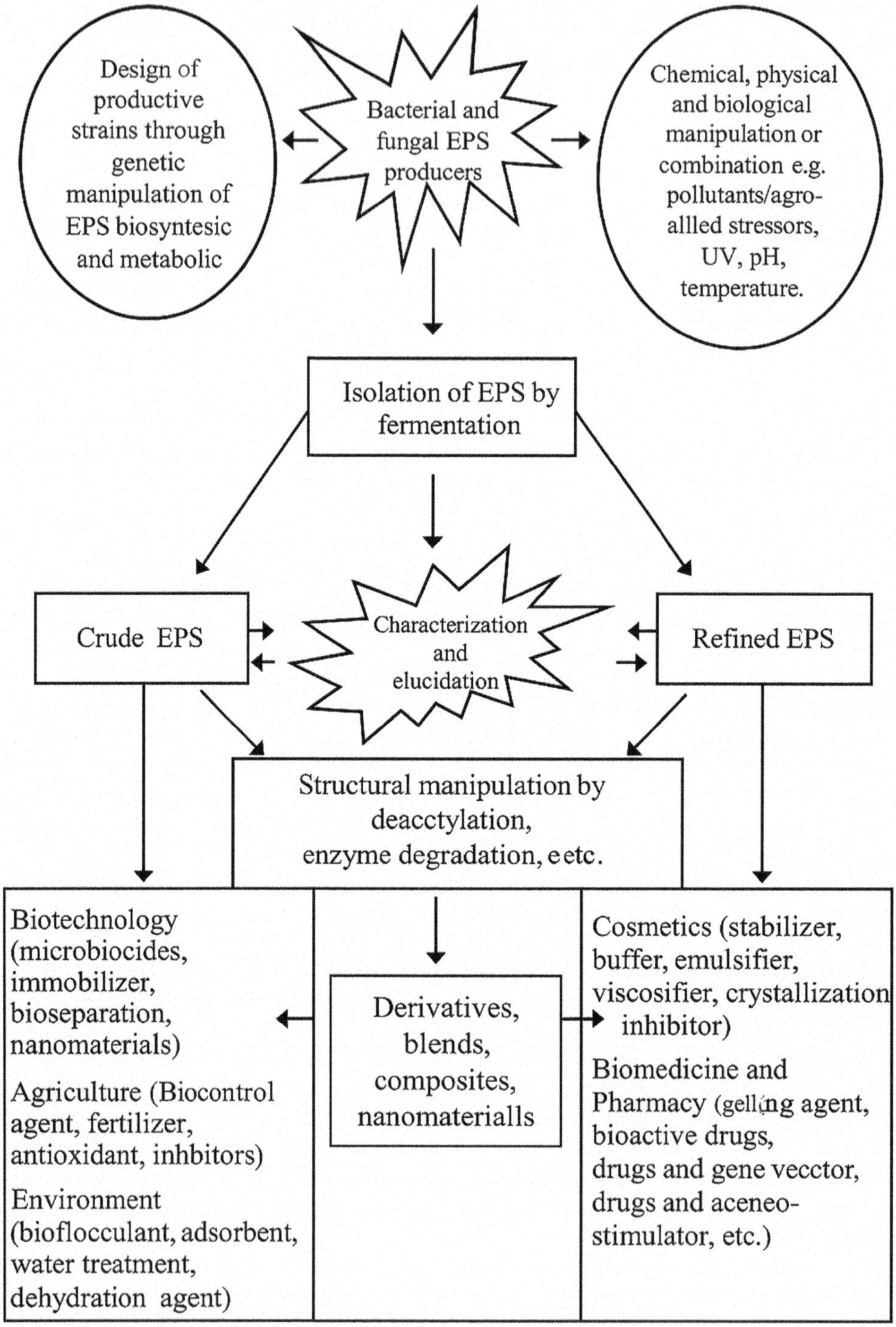

FIGURE 2.8 Overview of the method for the production of exopolysaccharides.

TABLE 2.1 Classification of exopolysaccharides based on source, type, and structural features.

Source	Representative microorganisms	Type of EPS	Major monosaccharides	Linkage and structural features	Unique modifications
Bacteria	*Lactobacillus*, *Pseudomonas*, *Streptococcus*	Homopolysaccharide/heteropolysaccharide	Glucose, galactose, rhamnose, mannose	α-1→6, β-1→3, α-1→4, branched, linear	Acetylation, phosphorylation
Fungi	*Aspergillus*, *Saccharomyces*, *Penicillium*	β-glucans, galactomannans	Glucose, mannose, galactose	β-1→3, β-1→6, with variable side chains	High uronic acid content
Algae and cyanobacteria	*Chlorella*, *Spirulina*, *Ulva*, *Fucus*	Sulfated EPS (e.g., carrageenan, fucoidan, ulvan)	Fucose, rhamnose, xylose, glucuronic acid	Alternating β- and α-linkages, highly branched	Sulfation, pyruvylation
Yeasts and actinomycetes	*Cryptococcus*, *Streptomyces*, *Nocardia*	Acidic heteropolysaccharides	Glucuronic acid, mannose, xylose	Branched, high MW, highly anionic	Methylation, sulfation, and rare sugar content
Archaea	*Haloferax*, *Sulfolobus*	Thermoacidophilic EPS	Mannose, glucuronic acid, galactose	Highly branched, thermostable	Sulfation, metal-binding motifs

EPS, exopolysaccharides.

TABLE 2.2 Structural features of EPS.

EPS	Monosaccharide units	Glycosidic linkages	Branching	Molecular weight (kDa)
Dextran	Glucose	α-(1→6) main chain	Yes (α-1→3)	10–50,000
Hyaluronic acid	Glucuronic acid, N-acetylglucosamine	β-(1→3), β-(1→4)	No	100–8000
Curdlan	Glucose	β-(1→3)	No	50–500
Pullulan	Maltotriose	α-(1→4), α-(1→6)	No	50–2000
Schizophyllan	Glucose	β-(1→3) main chain	Yes (β-1→6)	100–1000

EPS, exopolysaccharides.

or polymeric populations (Naseem et al., 2018). Degree of branching and molecular size influence solubility, viscosity, and diffusion properties. Anionic EPS exhibit strong interactions with divalent cations, leading to gel formation or crosslinked networks. High-performance size exclusion chromatography, matrix-assisted laser desorption ionization mass spectrometry, and gas chromatography following hydrolysis are employed for detailed compositional and structural elucidation. Isotope labeling with ^{13}C and ^{15}N, coupled with NMR, enables the determination of monosaccharide sequence substitution site and linkage configuration with atomic resolution.

2.3.2 Physicochemical properties

EPS exhibit a wide range of physicochemical behaviors dictated by molecular structure, concentration, ionic environment, and external stressors. Solubility of EPS in water varies with the degree of branching side-chain polarity and the presence of hydrophobic substituents (Jiang and Yang, 2018). Linear HoPSs tend to form clear viscous solutions while highly branched or substituted heteropolysaccharides may form colloidal suspensions or insoluble gels. Viscosity of EPS solutions increases nonlinearly with concentration due to entanglement and formation of transient hydrogen-bonded networks. EPS exhibits shear-thinning or pseudoplastic behavior characterized by decreased viscosity with increasing shear rate due to disruption of physical associations. This behavior is relevant to flowability and processability in industrial applications. Some EPS exhibit thixotropy, defined by time-dependent viscosity reduction and recovery under constant shear conditions. Gelation behavior of EPS depends on molecular weight, charge density, degree of sulfation, and presence of crosslinking ions such as calcium or potassium. Anionic EPS, such as alginate or carrageenan, form stable gels via ionic crosslinking with multivalent cations, creating egg-box or double helix structures (Joseph et al., 2024). EPS exhibits high water retention capacity due to hydrophilic hydroxyl, carboxyl, and sulfate groups capable of hydrogen bonding and osmotic stabilization. Emulsifying properties are attributed to amphiphilic regions within EPS molecules that stabilize oil-water interfaces through steric and electrostatic repulsion. EPS also demonstrates high metal ion binding capacity through complexation with carboxyl sulfate or phosphate groups, enabling heavy metal sequestration and detoxification. Surface-active properties of EPS enable bioadhesion, flocculation, and emulsion stabilization. Rheological characterization using rotational rheometry, dynamic oscillatory tests, and creep-recovery experiments quantifies viscoelastic moduli and flow curves under varying temperature, frequency, and shear conditions. Thermal stability of EPS assessed by thermogravimetric analysis and differential scanning calorimetry reveals degradation temperatures ranging from 180°C to 280°C, depending on glycosidic linkage stability and side-chain substitution. These physicochemical attributes define the functional utility of EPS in biotechnology, food, pharmaceuticals, and environmental engineering (Mollakhalili and Mohammadifar, 2015; Pachekrepapol et al., 2017; Pirog et al., 2004; Yildiz and Karatas, 2018).

2.3.3 Structural features and modifications

The EPS structure extends beyond the primary monosaccharide sequence to include secondary, tertiary, and quaternary conformational arrangements dictated by glycosidic linkage stereochemistry, hydrogen bonding, and intramolecular interactions. Primary structure is defined by the identity, order, and linkage of monosaccharide residues forming linear or branched chains. Secondary structure involves regular motifs such as helices, loops, or extended chains stabilized by hydrogen bonds and dipole interactions. Tertiary structure refers to the three-dimensional folding of polymer chains influenced by branching density, substitution pattern, and interaction with water and ions. Quaternary structure may involve association of multiple polymer chains via ionic bridges, hydrophobic interactions, or covalent crosslinks forming gel films or networks. Chemical modifications such as acetylation, methylation, sulfation, or phosphorylation occur naturally or may be introduced synthetically to alter EPS functionality. Acetylation of hydroxyl groups reduces hydrogen bonding, increases hydrophobicity, and lowers solubility. Sulfation introduces a negative charge, enhances water solubility, and imparts anticoagulant and antiviral properties. Phosphorylation enables coordination with metal ions and influences bioactivity. Pyruvylation imparts rigidity and enhances emulsifying properties. Controlled chemical derivatization using reagents such as acetic anhydride, chlorosulfonic acid, or orthophosphoric acid enables tuning of EPS properties for targeted applications. Enzymatic modification using specific hydrolases or transferases enables site-specific alterations without degrading the polymer backbone. Structural characterization of modified EPS requires advanced techniques such as Fourier-transform infrared spectroscopy for functional group analysis, solid-state NMR for chemical environment, and X-ray diffraction for crystallinity. Molecular modeling and simulation provide insight into conformational dynamics and interaction with ligands or receptors. Modified EPS demonstrate enhanced rheological behavior, bioactivity, thermal resistance, and compatibility with hydrophobic drug molecules, supporting their use in functional materials and biomedical systems (Tuinier et al., 2001; Nwodo et al., 2012; Salimi and Farrokh, 2023; Wang et al., 2023).

2.3.4 Environmental influences on exopolysaccharides properties

Environmental parameters exert a strong influence on the molecular structure, physicochemical properties, and biosynthetic yield of microbial EPSs (Sengupta et al., 2018; Dar et al., 2021). Temperature affects enzymatic activity, glycosyltransferase stability, and sugar nucleotide availability, thereby influencing polymer chain length, degree of branching, and monosaccharide ratio (Mestrom et al., 2019). High cultivation temperatures often reduce molecular weight and branching, leading to decreased viscosity and altered solubility. Low temperatures may enhance the production of cryoprotective EPS with high hydrophilicity and flexible linkages to resist cold-induced damage. pH modulates the ionization state of acidic groups in EPS, altering conformation charge distribution and metal binding affinity (Alavarse et al., 2022; Babu et al., 2024). Acidic pH may increase EPS yield in acidophilic strains by stabilizing glycosyltransferases and maintaining proton motive force. Alkaline pH can lead to deacetylation and hydrolysis of labile linkages, affecting polymer integrity. Salinity affects osmoregulatory EPS production, particularly in halophilic and marine organisms, which produce highly charged polymers with uronic acid and sulfate substitutions to maintain hydration and ionic balance. Increased NaCl concentration triggers upregulation of EPS biosynthetic genes and alters monosaccharide composition toward fucose, rhamnose, and glucuronic acid. Nutrient availability, especially carbon and nitrogen sources, determines EPS yield and composition (Hereher et al., 2018). Carbon-rich nitrogen-poor media favor EPS accumulation as a storage and protective mechanism. The type of carbon source, such as glucose, fructose, lactose, or glycerol, affects sugar nucleotide pools and influences polymer structure. Nitrogen limitation induces the production of EPS with higher branching and molecular weight due to prolonged stationary phase synthesis. Oxygen availability regulates aerobic respiration, energy supply, and redox balance for nucleotide sugar biosynthesis. Microaerobic or anaerobic conditions may alter the flux through glycolytic or pentose phosphate pathways, thereby affecting monosaccharide precursors and polymer branching. The presence of metal ions influences EPS structure through coordination and crosslinking. Divalent cations such as Ca^{2+}, Mg^{2+}, and Fe^{3+} bind carboxyl phosphate or sulfate groups, altering gelation properties and matrix stiffness. Environmental stressors such as oxidative stress, UV radiation, or antimicrobial exposure stimulate EPS production as a defense mechanism. Stress-induced EPS exhibit increased molecular weight, higher sulfation, and enhanced antioxidant capacity. These adaptive modifications confer resilience and support survival in fluctuating ecological niches. Detailed control of cultivation parameters allows tailoring of EPS for specific physicochemical attributes and industrial performance requirements (Fig. 2.9).

2.3.5 Analytical techniques for exopolysaccharides characterization

Comprehensive characterization of EPSs requires multidimensional analytical approaches targeting molecular composition, structural features, physical behavior, and functional performance. Monosaccharide composition is determined through acid hydrolysis, followed by derivatization and chromatographic separation using high-performance liquid chromatography, gas chromatography, or capillary electrophoresis. Linkage analysis is performed via methylation followed by mass spectrometry to determine glycosidic bond types and branching sites. NMR spectroscopy, both one-dimensional and two-dimensional, including ^{1}H NMR, ^{13}C NMR, COSY, TOCSY, and HSQC, is used for elucidating monosaccharide configuration sequence and substitution patterns. Fourier-transform infrared spectroscopy identifies functional groups such as hydroxyl, carboxyl, sulfate, phosphate, and acetyl groups based on characteristic vibrational frequencies. Elemental analysis provides quantification of carbon, hydrogen, nitrogen, sulfur, and phosphorus. Molecular weight and distribution are determined using high-performance size exclusion chromatography combined with multiangle light scattering, refractive index

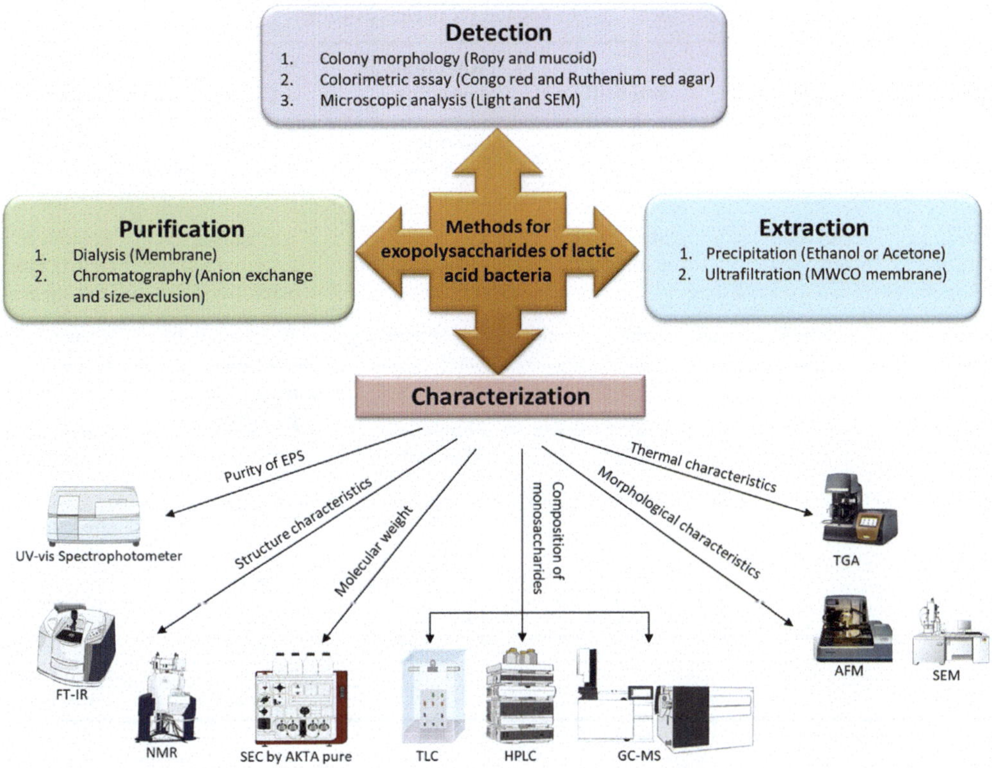

FIGURE 2.9 Methods for detection, extraction, purification, and characterization of exopolysaccharides of lactic acid bacteria (Yadav et al., 2024).

detection, or viscometry. Rheological properties such as viscosity, elasticity, thixotropy, and gel strength are assessed using rotational and oscillatory rheometry under controlled shear rate, temperature, and frequency conditions. Thermal stability is analyzed using thermogravimetric analysis to determine degradation onset temperature and differential scanning calorimetry to evaluate melting transition enthalpy and crystallinity. Surface activity and emulsification index are measured by drop shape analysis, interfacial tension, and emulsion stability tests. Metal binding capacity is quantified using atomic absorption spectroscopy or inductively coupled plasma mass spectrometry after incubation with metal solutions. Zeta potential and dynamic light scattering provide insight into colloidal behavior and surface charge. Microscopic imaging using scanning electron microscopy, transmission electron microscopy, and atomic force microscopy visualizes EPS morphology, surface texture, and nanostructure. CLSM combined with fluorescently labeled lectins or dyes enables in situ visualization of EPS in biofilms and microbial aggregates. Mass spectrometry techniques such as matrix-assisted laser desorption ionization, time-of-flight, and electrospray ionization tandem MS provide high-resolution molecular fingerprinting and oligosaccharide profiling. Advanced characterization enables structure-function correlation quality control and optimization of EPS for targeted applications

TABLE 2.3 Structural, physicochemical, and biological properties of EPS.

EPS type/source	Molecular weight (kDa)	Solubility	Rheological behavior	Biological activity	Functional attributes
Dextran (bacterial)	500–2000	Water-soluble	Pseudoplastic, high viscosity	Prebiotic, immunostimulatory	Stabilizer, viscosity modifier
β-glucans (Fungal)	200–1000	Moderately soluble	Gel-forming, shear-thinning	Anticancer, Immune cell activation	Bioadhesive, matrix in wound dressings
Fucoidan (brown algae)	100–800	Anionic, soluble	Ion-sensitive gel	Anticoagulant, antiviral, antioxidant	Drug carrier, tissue engineering
Pullulan (fungal)	100–600	Highly soluble	Nongelling, film-forming	Cholesterol-lowering, antioxidant	Edible films, encapsulation
EPS from extremophiles	>1000	Salt-stable	Highly stable under pH/temperature	Metal chelation, biofilm resistance	Bioremediation, oil recovery
Algal TEP (marine)	500–1500	Colloidal	Viscous, flocculating	Carbon sequestration, particle aggregation	Ocean biogeochemistry, marine snow formation

EPS, exopolysaccharides.

in food, pharmaceuticals environment and material science (Madhuri and Vidya Prabhakar, 2014; Rühmann et al., 2015; Shukla et al., 2019; Velamakanni et al., 2021) (Table 2.3).

2.4 Biological activities and therapeutic potential of exopolysaccharides

2.4.1 Immunomodulatory and anti-inflammatory effects

EPSs possess potent immunomodulatory activities governed by molecular structure, monosaccharide composition, degree of branching, and surface charge. EPS from *Lactobacillus plantarum, Bifidobacterium breve, Saccharomyces cerevisiae,* and *Ganoderma lucidum* activate innate and adaptive immune responses via interaction with pattern recognition receptors such as Toll-like receptors Dectin-1 and mannose-binding lectin. These interactions trigger signaling cascades involving nuclear factor kappa B, mitogen-activated protein kinases, and Janus kinase/signal transducers and activators of transcription, leading to transcriptional activation of cytokines, chemokines, and antimicrobial peptides. Sulfated EPS stimulates macrophage activation and enhances phagocytic activity, nitric oxide production, and reactive oxygen species generation. β-glucans induce maturation of dendritic cells and increase major histocompatibility complex class II expression and co-stimulatory molecules, enhancing antigen presentation and T cell priming. Acidic heteropolysaccharides with uronic acid and sulfate groups modulate Th1/Th2 balance and promote secretion of interleukin-10, transforming growth factor-beta, and interferon-gamma, contributing to anti-inflammatory and regulatory immune phenotypes. EPS from *Bacillus subtilis* and *Rhodotorula mucilaginosa* attenuate lipopolysaccharide-induced inflammation in macrophage and epithelial cell models by suppressing cyclooxygenase-2, inducible nitric oxide synthase, and pro-inflammatory cytokines. Molecular size and degree of sulfation directly influence receptor binding affinity and downstream signaling strength. EPS nanoformulations enhance mucosal delivery, immune cell targeting, and systemic bioavailability. Oral and parenteral administration of immunomodulatory EPS reduces hypersensitivity, improves vaccine efficacy, and restores immune homeostasis in models of allergy, autoimmunity, and infection. Functionalization of EPS with bioactive ligands or encapsulation into delivery vehicles augments immunomodulatory potency and selectivity. Immunological assays, including flow cytometry, cytokine bead arrays, enzyme-linked immunosorbent assays, and Western blotting, confirm the cellular and molecular targets of EPS immunomodulation.

2.4.2 Antimicrobial properties

EPS exhibits broad-spectrum antimicrobial activities through multiple mechanisms, including direct interaction with microbial membranes, nutrient sequestration, interference with quorum sensing, and inhibition of adhesion and biofilm formation. EPS from *Pseudomonas fluorescens, Streptomyces sp.*, and *Aspergillus niger* demonstrate bactericidal activity against gram-positive and gram-negative pathogens, including *Escherichia coli, Staphylococcus aureus, Listeria monocytogenes*, and *Salmonella enterica*. Anionic EPS with high uronic acid or sulfate content destabilize bacterial cell membranes, disrupt the electrostatic balance, and induce membrane depolarization leading to cell lysis. EPS acts as chelators of essential metal ions such as iron, magnesium, and zinc, depriving pathogens of cofactors critical for growth and enzyme activity. EPS interferes with autoinducer signaling molecules such as acyl-homoserine lactones and peptide pheromones, disrupting quorum sensing and virulence factor production. EPS reduces biofilm integrity by inhibiting initial adhesion, blocking receptor-ligand interactions, and promoting dispersion of established biofilms. EPS synergizes with conventional antibiotics by enhancing membrane permeability, inhibiting efflux pumps, or preventing biofilm-mediated resistance. EPS from marine organisms such as *Spirulina platensis* and *Ulva lactuca* show antiviral activity by blocking viral adsorption, fusion, and replication. Sulfated EPS from red algae inhibits the replication of enveloped viruses, including HIV, HSV, hepatitis C, and influenza, by mimicking heparan sulfate and binding to viral envelope glycoproteins. EPS exhibits antifungal effects by disrupting fungal membrane integrity, inhibiting hyphal growth, and inducing oxidative stress. Antimicrobial activity is evaluated using microdilution assays, minimum inhibitory concentration determination, scanning electron microscopy, and confocal imaging of treated pathogens. EPS-loaded hydrogel films and nanoparticles serve as antimicrobial coatings and delivery platforms for infection control and wound healing (Fig. 2.10).

2.4.3 Anticancer activity

EPS demonstrate anticancer activity through inhibition of tumor cell proliferation, induction of apoptosis, disruption of angiogenesis, and enhancement of antitumor immunity. EPS from *Ganoderma lucidum, Grifola frondosa, Schizophyllum commune*, and *Lactobacillus rhamnosus* reduce the viability of various cancer cell lines, including MCF-7, HeLa, HT-29, and A549, by activating intrinsic and extrinsic apoptotic pathways. EPS induces mitochondrial membrane depolarization, release of cytochrome c, activation of caspase-3, caspase-8, and caspase-9, and upregulation

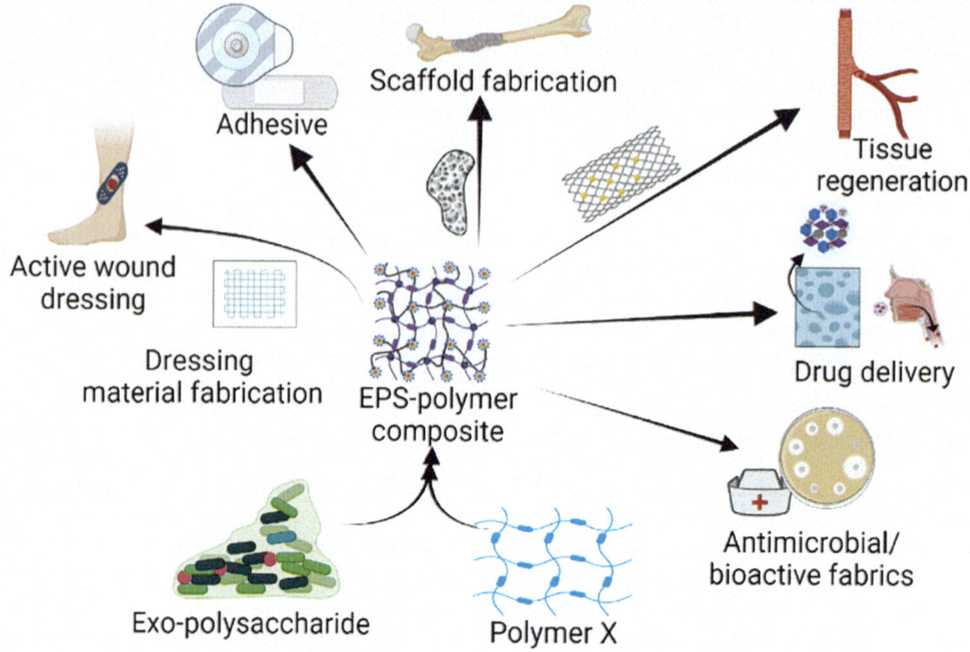

FIGURE 2.10 Exopolysaccharide composites with natural and synthetic polymers and their application in the health sector (Ahuja et al., 2023).

of pro-apoptotic genes such as BAX and p53. EPS inhibit cyclin-dependent kinases, downregulate cyclin D1, and arrest cell cycle progression at G0/G1 or G2/M phases. Sulfated EPS interferes with epidermal growth factor receptor signaling suppress phosphorylation of Akt, MAPK, and STAT3, and reduces cell migration, invasion, and metastasis. EPS modulate expression of angiogenic factors such as vascular endothelial growth factor, matrix metalloproteinases, and hypoxia-inducible factor-1α, reducing tumor vascularization and oxygen supply. In vivo studies using murine xenograft models confirm tumor growth inhibition, reduction in tumor weight, and increased survival upon EPS treatment. EPS enhances infiltration of cytotoxic T lymphocytes, natural killer cells, and macrophages into the tumor microenvironment and stimulates production of interferon-gamma and tumor necrosis factor-alpha. Combination therapy of EPS with chemotherapeutic drugs or immune checkpoint inhibitors shows additive or synergistic effects with reduced systemic toxicity. Mechanistic studies using transcriptomic, proteomic, and metabolomic analyses reveal global shifts in tumor cell signaling energy metabolism and immune recognition upon EPS exposure. EPS-based nanocarriers improve tumor-targeted drug delivery, enhance endosomal escape, and reduce off-target effects in solid tumor therapy.

2.4.4 Other biological functions

EPS possess additional biological activities, including antioxidant, antidiabetic, cholesterol-lowering, wound healing, and regenerative effects. EPS scavenges reactive oxygen species, including superoxide, hydroxyl radicals, and hydrogen peroxide, and enhances cellular antioxidant enzyme activities such as superoxide dismutase, catalase, and glutathione peroxidase (Wang et al., 2023). EPS inhibit lipid peroxidation and DNA damage and protect against oxidative stress-induced apoptosis. EPS modulate glucose metabolism by inhibiting α-glucosidase and α-amylase, reducing postprandial glucose spikes, and enhancing insulin sensitivity. EPS increases hepatic glycogen storage, upregulates GLUT4 expression, and regulates phosphoenolpyruvate carboxykinase and glucokinase activity in liver and muscle tissues. EPS lowers serum cholesterol by binding bile acids, reducing micelle formation, and promoting faecal excretion (Sutherland, 1982). EPS modulates gut microbiota composition, increases short-chain fatty acid production, and reduces systemic inflammation, contributing to metabolic homeostasis. In wound healing, EPS promote fibroblast proliferation, keratinocyte migration, angiogenesis, and ECM remodeling. EPS from *Aureobasidium pullulans* and *Leuconostoc mesenteroides* form protective films that maintain moisture and reduce bacterial colonization on wound surfaces. EPS stimulates the expression of growth factors such as vascular endothelial growth factor, fibroblast growth factor, and transforming growth factor-beta, enhancing granulation tissue formation and epithelialization. In tissue engineering, EPS-based scaffolds support cell adhesion, differentiation, and neovascularization (Ruijgrok et al., 2024). EPS hydrogels, composite matrices, and bioinks are used in 3D bioprinting of functional tissues, including skin, cartilage, and bone. EPS shows potential in neuroprotection, fertility enhancement, hepatoprotection, and detoxification based on emerging preclinical evidence. Biological activities are confirmed using in vitro assays, including DPPH, ABTS, FRAP, MTT, and scratch wound assay, and in vivo studies in animal models of diabetes, inflammation, and tissue injury (Qi et al., 2022) (Fig. 2.11).

2.4.5 Safety and toxicity of exopolysaccharides

Toxicological assessment of EPS is critical for their regulatory approval and safe application in pharmaceuticals, food, cosmetics, and biomedical devices. Acute subchronic and chronic toxicity studies in rodent models determine LD50, body weight changes, organ weight, hematological, biochemical, and histopathological parameters following oral dermal or parenteral EPS administration (Caggianiello et al., 2016). EPS from *Lactobacillus plantarum, Bifidobacterium longum, Saccharomyces cerevisiae,* and *Ganoderma lucidum* exhibit no genotoxicity, mutagenicity, or teratogenicity at therapeutic doses. Ames test, micronucleus assay, and comet assay confirm absence of DNA damage. In vitro cytotoxicity on normal human fibroblasts, keratinocytes, and epithelial cells using MTT, XTT, or live/dead staining confirms biocompatibility. EPS with high sulfate content or residual protein contaminants may cause immunogenicity, hypersensitivity, or inflammation requiring purification and structural refinement (Lynch et al., 2018). Regulatory agencies such as the FDA, EFSA, and JECFA require comprehensive safety dossiers including chemical composition, manufacturing process, residual solvents, endotoxin levels, and stability data. Acceptable daily intake, no-observed-adverse-effect levels, and maximum residue limits are established based on toxicokinetic and pharmacokinetic studies (Maunders and Welch, 2017). EPS with GRAS status is approved for food use, while biomedical applications require ISO 10993-based biocompatibility testing. Safety enhancement strategies include enzymatic deacetylation, dialysis, ultrafiltration, and endotoxin removal. Stability testing under various pH, temperature, and storage conditions ensures product integrity and reproducibility (Rühmann et al., 2015) (Table 2.4).

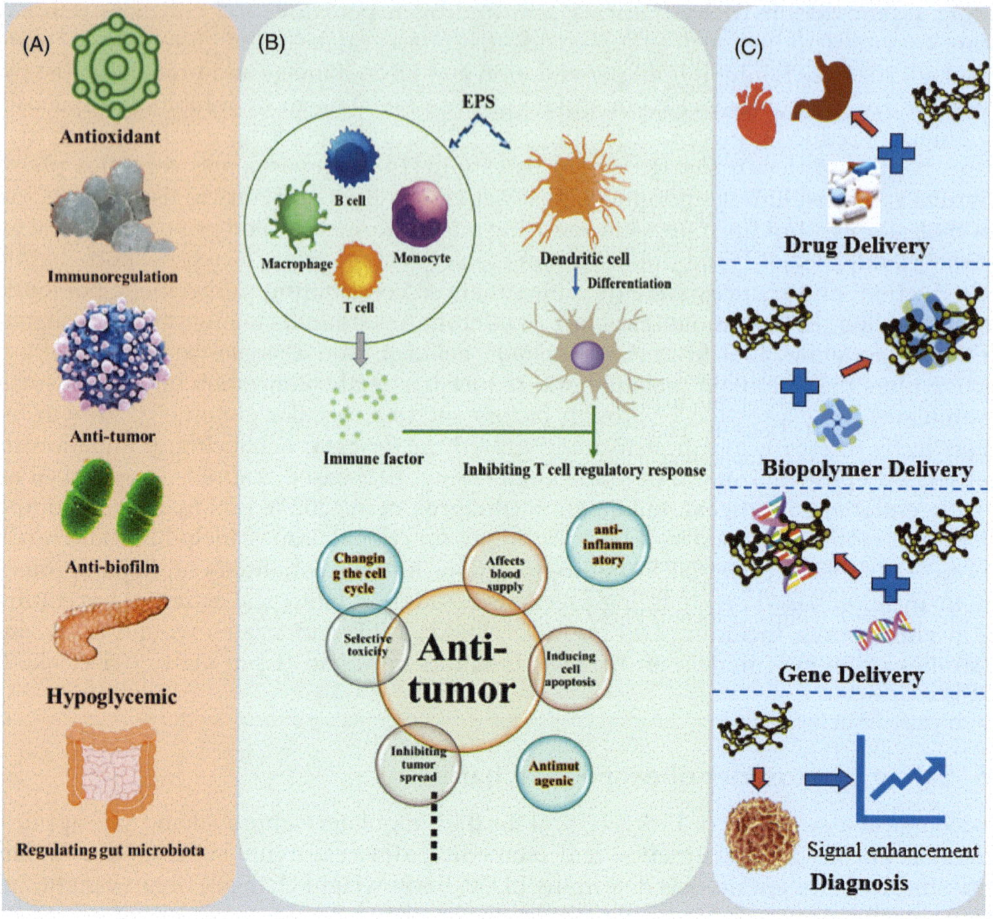

FIGURE 2.11 (A) The biological activity of EPS. (B) The mechanism of EPS regulating immune and antitumor activity. (C) The application of EPS produced by LAB in medicine (Liang et al., 2024). *EPS*, Exopolysaccharides; *LAB*, lactic acid bacteria.

TABLE 2.4 Biological activities of EPS.

EPS	Antioxidant	Antimicrobial	Immunomodulatory	Prebiotic	Antitumor
Kefiran	Yes	Yes (gram+)	Yes	Yes	No
Fucoidan	Strong	Moderate	Yes	No	Yes
Bacterial cellulose	No	No	Weak	No	No
Levan	Moderate	Yes (fungal)	Yes	Yes	Yes
Xanthan gum	No	No	No	No	No

EPS, exopolysaccharides.

2.5 Biotechnological approaches for exopolysaccharides production and optimization

2.5.1 Microbial cultivation techniques

Controlled microbial cultivation is essential for scalable EPS production and is determined by species-specific growth kinetics, carbon and nitrogen assimilation rates, oxygen transfer capacity, and environmental stress responses (Rühmann et al., 2015). Batch fermentation represents a standard approach where EPS yield is limited by substrate depletion and product inhibition. Fed-batch fermentation extends productivity by intermittent or continuous feeding of carbon sources such as glucose, sucrose, or lactose, and nitrogen sources such as ammonium sulfate or yeast extract to avoid catabolite repression and nitrogen starvation. Continuous fermentation using chemostat systems allows precise control over growth rate, dilution rate, and nutrient flux, leading to consistent EPS composition and molecular weight. Stirred tank reactors with automated pH, temperature, and dissolved oxygen control systems are employed for aerobic bacterial and fungal EPS production (Mouro et al., 2024). Oxygen transfer rate is critical for oxidative metabolism in species such as *Xanthomonas campestris* and *Aspergillus niger,* which require high aeration and agitation rates. Anaerobic or microaerophilic conditions are used for lactic acid bacteria and bifidobacteria. Shear-sensitive organisms are cultured using airlift reactors, bubble columns, or membrane bioreactors to reduce mechanical stress and foaming. Immobilized cell systems using alginate beads, loofa sponges, or biofilm reactors enhance cell stability, volumetric productivity, and continuous operation (Rahbar Saadat et al., 2019). Co-culture systems involving EPS-producing strains and metabolically synergistic microbes improve precursor availability, reduce inhibitory metabolite accumulation, and enhance total yield. Solid-state fermentation using agro-industrial residues supports fungal EPS synthesis under moisture-limited conditions and minimizes downstream processing costs. Optimization of media composition using RSM, factorial design, and artificial neural networks enables prediction of yield-maximizing conditions with statistical rigor. Critical parameters include carbon-to-nitrogen ratio, micronutrient concentration, trace element availability, and pH buffering capacity. Culture time and temperature profiles are adjusted to match EPS biosynthetic enzyme expression peaks for maximal productivity (Sharma et al., 2024).

2.5.2 Genetic and metabolic engineering

Genetic and metabolic engineering strategies enhance EPS yield composition and functionality by manipulating biosynthetic gene clusters, regulatory elements, and metabolic fluxes. Heterologous expression of glycosyltransferase genes in fast-growing chassis organisms such as *Escherichia coli, Corynebacterium glutamicum,* or *Bacillus subtilis* enables high-level production of target EPS (Gauri et al., 2012). Deletion of competing pathways for lactic acid, ethanol, or acetate production redirects carbon flux toward sugar nucleotide synthesis and polysaccharide assembly. Overexpression of key enzymes such as UDP-glucose pyrophosphorylase, GDP-mannose dehydratase, and phosphoglucomutase increases precursor pool availability. Promoter engineering using synthetic constitutive or inducible systems modulates transcription of EPS biosynthetic operons in response to environmental or chemical cues (Paul et al., 2023). CRISPR-Cas9 genome editing enables targeted knockout or activation of specific genes involved in EPS polymerization, transport, or degradation. Transcriptional regulators such as RcsA, RpoS, and CcpA are manipulated to enhance EPS biosynthesis under stress or stationary phase conditions.

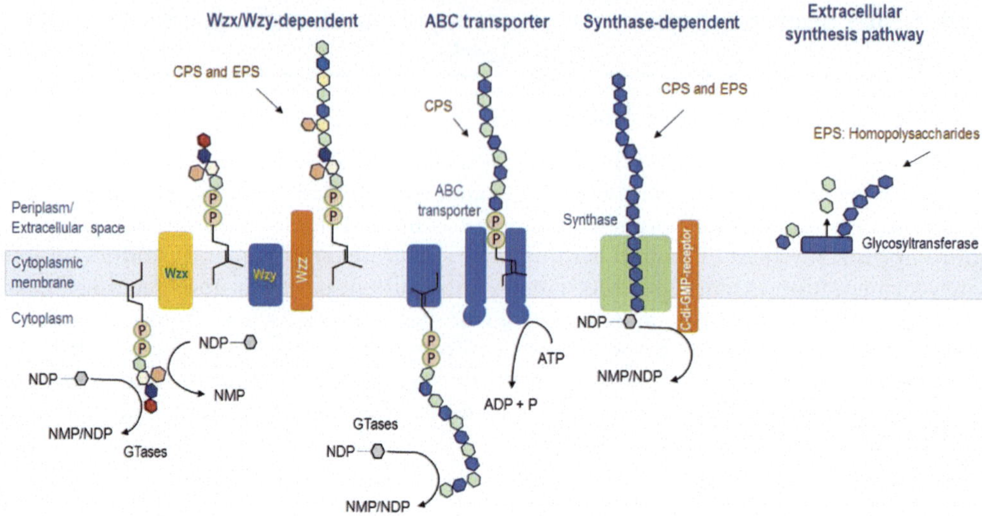

FIGURE 2.12 Biosynthesis of polysaccharides in microorganisms. Wzx/Wzy-dependent pathway: Responsible for synthesizing lipopolysaccharide O-antigen polysaccharides in gram-negative bacteria, as well as capsular polysaccharides (CPSs) and EPSs in both gram-negative and gram-positive bacteria. ATP-binding ABC transporter pathway: Facilitates the synthesis of CPS specifically in gram-negative bacteria. Synthase-dependent pathway: Involved in the synthesis of both CPSs and EPSs in gram-negative and gram-positive bacteria. Extracellular synthesis via a single glycosyltransferase: Responsible for the synthesis of EPSs that fall under the category of HoPS (Nguyen et al., 2024). *EPS*, Exopolysaccharides; *HoPS*, homopolysaccharides; ATP, Adenosine Triphosphate; ABC, ATP-Binding Cassette.

Synthetic biology platforms support the design of modular biosynthetic pathways with orthogonal regulation, codon optimization, and scaffold-based enzyme clustering to improve polymer yield and uniformity (Huang et al., 2024). Metabolic modeling using flux balance analysis and 13C-metabolic flux analysis predicts rate-limiting steps and metabolic bottlenecks for targeted interventions. Biosensors responsive to EPS intermediates or final products enable real-time monitoring and dynamic control of gene expression. Adaptive laboratory evolution combined with genome resequencing identifies beneficial mutations that improve EPS productivity under industrial conditions. Systems biology integration of transcriptomic, proteomic, and metabolomic data supports rational strain design and performance validation (Kavitake et al., 2020).

2.5.3 Isolation, purification, and characterization

Efficient downstream processing is critical for the recovery of high-purity EPS with intact molecular structure and biological functionality (Nguyen et al., 2024). EPS isolation begins with centrifugation or microfiltration to remove microbial biomass, followed by precipitation of soluble EPS using cold ethanol, isopropanol, or acetone under controlled ionic strength and pH. Precipitated EPS are redissolved in deionized water and subjected to deproteinization using the Sevag method, trichloroacetic acid, or protease digestion (Fig. 2.12).

Dialysis or ultrafiltration removes low-molecular-weight impurities such as salts, peptides, and residual media components (Yadav et al., 2024). Anion exchange chromatography using DEAE-Sepharose or Q-Sepharose separates EPS based on charge density and sulfation level.

Size exclusion chromatography provides molecular weight fractionation and polydispersity assessment. Precipitation with cationic detergents such as cetyltrimethylammonium bromide enables the selective isolation of acidic EPS. Final purification may include lyophilization, spray drying, or freeze drying, depending on thermal sensitivity and intended application. Structural integrity is confirmed using spectroscopic methods as outlined in previous sections (Baptista et al., 2022). Yield recovery, purity, and bioactivity are quantified using dry weight analysis, phenol-sulfuric acid assay, Bradford protein assay, and biological activity assays. Process parameters such as solvent-to-sample ratio, precipitation temperature, mixing time, and filtration membrane pore size are optimized to minimize structural degradation and maximize recovery. Scale-up requires validation of batch-to-batch consistency, removal of endotoxins, compliance with food or pharmaceutical-grade standards, and adaptation to continuous processing or inline purification systems (Beaumont et al., 2021).

2.5.4 Scale-up and commercial production

Commercial production of EPS requires translation from laboratory-scale protocols to pilot and industrial scales with emphasis on process robustness, cost-effectiveness, and regulatory compliance. Bioreactor design must ensure homogenous mixing, efficient aeration, temperature and pH control, and foam suppression. Airlift reactors and stirred tank bioreactors equipped with baffles, Rushton turbines, and spargers are employed for aerobic fermentation. Large-scale production utilizes fed-batch or continuous mode with real-time monitoring of biomass, sugar concentration, oxygen uptake rate, and viscosity. Use of inexpensive carbon sources such as molasses, whey, glycerol, or lignocellulosic hydrolysates reduces production cost but requires optimization of pretreatment and detoxification protocols. Media formulation is adapted to support high cell density and stable EPS production under nonsterile or semi-continuous conditions. Downstream processing at scale involves crossflow ultrafiltration, centrifugation, industrial ethanol precipitation, and spray drying or fluidized bed drying. Process integration with waste valorization enables the conversion of agricultural residues into EPS and reduces the environmental footprint. Compliance with current good manufacturing practices, hazard analysis, and critical control points, and ISO certification is essential for commercialization. Regulatory registration for EPS as a food additive, cosmetic ingredient, or pharmaceutical excipient involves detailed submission of safety, efficacy, and manufacturing data. Market segments include food thickening and stabilizing agents, biomedical hydrogels, wound dressings, drug delivery matrices, and environmental bioflocculants. Techno-economic analysis assesses capital investment, operational expenditure, product yield, and profit margins. Life cycle assessment evaluates environmental impact, including water use, energy consumption, carbon emissions, and waste generation. Commercial success of EPS bioprocesses depends on strain performance, process scalability, purification cost, regulatory acceptance, and alignment with market needs (Seviour et al., 2011; Schmid et al., 2015; Freitas et al., 2017; Kanimozhi et al., 2018; Barcelos et al., 2020) (Fig. 2.13).

2.6 Industrial applications of exopolysaccharides

2.6.1 Food industry

EPSs serve critical roles in the food industry as rheological modifiers, stabilizers, texturizers, emulsifiers and prebiotic dietary components. EPS from *Lactobacillus delbrueckii subsp.*

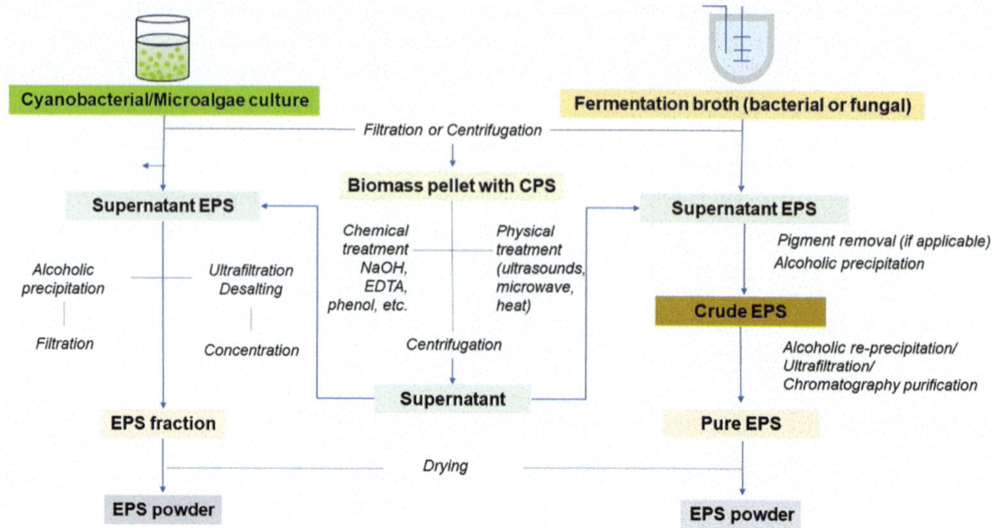

FIGURE 2.13 Workflow of EPS production (Nguyen et al., 2024). *EPS*, exopolysaccharides.

Bulgaricus, Streptococcus thermophilus, and *Leuconostoc mesenteroides* are extensively used in fermented dairy products for enhancing mouthfeel, water-holding capacity, and phase stability. HoPSs such as dextran and levan contribute to viscosity control and reduce syneresis in yoghurt, sour cream, and cheese formulations. Heteropolysaccharides from *Xanthomonas campestris* and *Weissella cibaria* exhibit pseudoplastic behavior and shear-thinning properties suitable for salad dressings, sauces, and low-fat spreads. EPS from *Lactobacillus rhamnosus* improves freeze-thaw stability and structural integrity in frozen desserts. EPS acts as a natural emulsifier stabilizing oil-water interfaces through steric and electrostatic repulsion (Tieking and Gänzle, 2005). They replace synthetic stabilizers in clean-label food systems. EPS with prebiotic functionality selectively stimulates the growth of beneficial gut microbiota, including *Bifidobacterium* and *Lactobacillus,* leading to increased production of short-chain fatty acids and modulation of host immune function. Structural attributes such as branching degree, charge density, and molecular weight influence viscosity, gelling capacity, and digestive resistance (Ruas-Madiedo and de los Reyes-Gavilán, 2005). EPS interacts with milk proteins, starch granules, and lipids, forming complex networks that improve sensory properties and nutrient retention. Regulatory approval of food-grade EPS includes GRAS status for xanthan dextran, gellan gum, and microbial β-glucans with well-defined purity specifications and acceptable daily intake levels. Process parameters such as pH, temperature, ionic strength, and storage conditions influence EPS functionality in food matrices (Donot et al., 2012). EPS are included in formulations for gluten-free bakery products, probiotic beverages, dairy analogues, plant-based emulsions, and edible coatings. Controlled production enables tailoring of EPS structure to desired rheological and nutritional properties. Analytical validation includes texture profile analysis, rheometry, organoleptic testing, and microbiological stability (Abarquero et al., 2022).

2.6.2 Pharmaceutical and biomedical applications

EPS are integral to pharmaceutical and biomedical applications due to their biocompatibility, nonimmunogenicity, biodegradability, and functional versatility. EPS from *Ganoderma lucidum, Saccharomyces cerevisiae, Lactobacillus plantarum,* and *Aureobasidium pullulans* are developed as immunomodulator vaccine adjuvants and bioactive nutraceuticals. EPS-based hydrogels form moisture-retaining matrices that accelerate wound healing by supporting fibroblast proliferation, angiogenesis, and collagen deposition. EPS from *Leuconostoc kimchii* and *Bacillus subtilis* form stable hydrocolloid films for wound dressings and surgical patches. EPS functions as a carrier in drug delivery systems through encapsulation, surface modification, and controlled release. Sulfated and carboxylated EPS enable ionic interactions with cationic drugs, enhancing loading efficiency and sustained release. EPS microspheres and nanoparticles deliver antibiotics, peptides, and chemotherapeutics across mucosal barriers and into target tissues. EPS from marine algae and cyanobacteria are explored for mucoadhesive and pH-sensitive oral delivery systems. EPS-based scaffolds support chondrocyte, osteoblast, and stem cell adhesion and differentiation in tissue engineering. EPS composites with collagen gelatin or synthetic polymers exhibit tailored porosity degradation and mechanical strength. EPS functional groups enable conjugation with targeting ligands, enzymes, and diagnostic probes for theranostic applications.

EPS modulate inflammation, oxidative stress, and immune activation in models of diabetes, cardiovascular disease, and neurodegeneration. EPS derivatives such as phosphorylated and sulfated analogues are tested for antiviral, anticoagulant, and anticancer activities. Regulatory classification of EPS includes excipient status in pharmacopoeias and ISO standards for biocompatible materials. Safety testing includes cytotoxicity, hemolysis, pyrogenicity, irritation, and sensitization. Biomedical EPS are sterilized via gamma irradiation, ethylene oxide, or filtration without compromising structural integrity or functionality. Manufacturing follows good laboratory practices and validation protocols for endotoxin removal, identity purity, and bioactivity (Shih, 2010; Vasiliu et al., 2016; Mohd Nadzir et al., 2021; Ahuja et al., 2023; Waoo et al., 2023).

2.6.3 Environmental and biotechnological applications

EPS have wide applications in environmental and biotechnological sectors due to their high metal binding capacity, flocculation performance, film-forming ability, and resilience under variable physicochemical conditions. EPS from *Bacillus megaterium, Paenibacillus polymyxa,* and *Pseudomonas putida* are effective bioflocculants in wastewater treatment for sedimentation of suspended solids, removal of turbidity, and decolorization of industrial effluents (Banerjee et al., 2021). EPS adsorb heavy metals such as Pb^{2+}, Cd^{2+}, Cu^{2+}, and Zn^{2+} through coordination with carboxyl sulfate and phosphate groups, enabling biosorption and detoxification of contaminated water. EPS from acidophilic bacteria mediate metal precipitation and stabilization in acid mine drainage systems. EPS-coated biofilms serve as active layers in biobarriers for nitrate, phosphorus, and pesticide retention. EPS facilitates microbial adhesion in biofilters, membrane bioreactors, and trickling filters, enhancing pollutant degradation rates (Qin et al., 2024). EPS stabilizes soil aggregates, reduces erosion, and improves water infiltration in sustainable agriculture. EPS-producing plant growth-promoting rhizobacteria enhance root colonization, nutrient acquisition, and stress tolerance. EPS-modified bioplastics serve as biodegradable alternatives to petroleum-derived polymers with applications in packaging, agriculture, and biomedical fields. EPS reduces surface tension and promotes emulsification for oil recovery, bioremediation, and surfactant replacement. EPS from marine bacteria protect surfaces from biofouling and are explored as eco-friendly antifouling coatings. EPS promotes aggregation of microalgae and flocs for biomass harvesting and biofuel production. Immobilization of enzyme cells or nanoparticles in EPS matrices supports biocatalysis, biosensing, and biotransformation processes. Biotechnological applications require optimization of yield purity and structural properties tailored to system-specific requirements (K et al., 2018)

2.6.4 Emerging and novel applications

Emerging applications of EPS span advanced materials, regenerative medicine, agricultural biostimulants, and cosmeceuticals. EPS-based smart hydrogels exhibit stimuli-responsive behavior, including thermo-reversibility, pH sensitivity, and ionic strength dependence, enabling use in sensors, actuators, and controlled delivery. EPS composites with graphene, carbon nanotubes, or metal oxides impart electroconductivity, piezoresistivity, and photocatalytic activity for electronic and environmental applications. EPS-derived nanoparticles stabilize emulsions and act as nanocarriers for targeted therapy and imaging. EPS from extremophiles and marine organisms is developed for high-performance coatings, adhesives, and textiles. In agriculture, EPS act as soil conditioners, drought mitigators, and biocontrol agents, enhancing root growth, nutrient retention, and pathogen resistance. EPS encapsulates biopesticides and biofertilizers, enabling sustained release and protection from environmental degradation. EPS functions as a carbon source and cryoprotectant in microbial inoculants, improving shelf life and efficacy. In cosmetics, EPS provides hydration, antiageing, antioxidant, and film-forming properties for skin care, hair care, and dermocosmetic formulations. EPS replaces synthetic polymers in biodegradable personal care products due to their natural origin and environmental compatibility. Functionalization with peptides, vitamins, or antioxidants enhances cosmeceutical value. EPS contribute to the formulation of edible films, biolubricants, eco-friendly inks, and flame-retardant materials. EPS-coated nanoparticles and scaffolds are applied in regenerative ophthalmology, dental materials, and cardiovascular implants. Innovation in EPS engineering, synthetic biology, and nanotechnology expands application domains and supports the development of multifunctional, sustainable, and high-value bioproducts (Sutherland, 1998; Moscovici, 2015; Barcelos et al., 2020; Mahmoud et al., 2021; Nguyen et al., 2024) (Table 2.5).

2.7 Technological innovations and advancements in exopolysaccharides research

2.7.1 Advances in fermentation technology

Fermentation technology for EPS production has evolved from traditional batch systems to highly controlled and optimized industrial-scale processes incorporating automation, real-time monitoring, and dynamic feedback control. Advanced bioreactor designs, including airlift reactors, stirred-tank reactors, membrane-integrated bioreactors, and fluidized bed reactors, enable improved mass transfer, oxygenation, mixing, and nutrient distribution, which are essential for high-yield EPS synthesis. Scale-up of EPS fermentation requires precise control of agitation, aeration, pH, temperature, and antifoam strategies to maintain culture stability and structural

TABLE 2.5 Applications and technological aspects of EPS production.

Application sector	Functional role of EPS	Key microbial EPS	Production strategy	Challenges
Food industry	Thickening, emulsifying, texture enhancement	Xanthan, dextran, levan, gellan	Fed-batch fermentation, use of GRAS strains	Batch variability, regulatory approval
Pharmaceutical and medical	Drug delivery, immunomodulation, wound healing	β-glucans, pullulan, sulfated EPS	Microencapsulation, scaffold synthesis, hydrogel fabrication	Biocompatibility, sterility, structure–function tuning
Environmental technology	Flocculation, heavy metal chelation, soil stability	EPS from *Pseudomonas*, *Bacillus*, Archaea	Solid-state fermentation, immobilized cells	Low yield, contaminant separation
Agriculture	Biofertilizer, drought resistance, root colonization	EPS from PGPR and cyanobacteria	Fermentation from waste substrates, consortium-based cultures	Field stability, cost-efficiency
Cosmeceuticals	Moisture retention, antiageing, film-forming	EPS from *Aureobasidium*, *Halomonas*, and Algae	Membrane bioreactors, downstream enzyme-assisted purification	Purity, sensory consistency
Advanced materials	Biofilm, nanocarriers, conductive films	Functionalized bacterial and algal EPS	Synthetic biology, CRISPR-engineered strains	Customization, IP and patent complexity

EPS, exopolysaccharides.

integrity of secreted polysaccharides. Inline sensors for dissolved oxygen, redox potential, sugar concentration, and pH facilitate closed-loop control systems using programmable logic controllers and supervisory control and data acquisition interfaces. Use of computational fluid dynamics assists in optimizing reactor geometry, flow regimes, and shear profiles to reduce energy consumption and shear-induced degradation of EPS. A study focused on isolating lactic acid bacteria (LAB) from Danish flowers to develop new starter cultures for plant-based fermentation. Researchers collected 46 flower samples from 34 genera and isolated 61 LAB strains belonging to 24 species, mainly *Leuconostoc mesenteroides*, *Fructobacillus fructosus*, and *Apilactobacillus* species. Many strains produced slimy EPSs made of glucose, fructose, or galactose, linked to their ability to create glucans, fructans, and galactans. When tested in almond, oat, and soy milk, all strains acidified the plant milks successfully. However, only soy milk fermented with certain LAB strains showed improved texture. These results suggest flower-derived LAB have promising potential for enhancing the texture of soy-based dairy alternatives. Innovations in substrate engineering include the use of low-cost agro-industrial residues such as molasses, whey, corn steep liquor, and lignocellulosic hydrolysates after detoxification and enzymatic saccharification (Khiabani et al., 2024).

Co-substrate feeding strategies balance carbon fluxes between biomass accumulation and EPS biosynthesis, minimizing accumulation of inhibitory byproducts. Oxygen vectors perfluorocarbons and nanobubbles enhance oxygen solubility and transfer in viscous fermentations with high cell densities. Adaptive control strategies using soft sensors and machine learning algorithms predict metabolic shifts and optimize feeding profiles. Use of immobilized and encapsulated cells support continuous and semi-continuous EPS production, reducing downtime and contamination risk. Bioreactor systems are now integrated with downstream processing modules such as membrane filtration, centrifugation, and spray drying to streamline product recovery and reduce cost.

2.7.2 Novel extraction and purification techniques

Extraction and purification of EPS have transitioned from solvent-intensive batch protocols to more sustainable, scalable, and structure-preserving techniques based on membrane separation, precipitation, and chromatographic fractionation. Ultrafiltration using tangential flow and cross-flow systems concentrates and purifies EPS based on molecular weight cut-off while retaining macromolecular integrity. Diafiltration and microfiltration remove salts, small molecules, and protein contaminants. Membrane materials, including polysulfone, polyether sulfone, and ceramic membranes, are selected based on EPS composition, viscosity, and fouling potential. Electro-ultrafiltration and dynamic filtration enhance separation efficiency and reduce fouling. Ethanol and isopropanol precipitation remains a primary step for polysaccharide recovery, but is now supplemented by phase separation techniques such as aqueous two-phase systems using polyethene glycol and salts for selective partitioning of EPS. Chromatographic purification using ion-exchange, size-exclusion, affinity, and hydrophobic interaction resins provide high-resolution separation of EPS fractions based on charge, hydrodynamic volume, and chemical affinity. Monolithic and membrane chromatography offer high flow rates and reduced pressure drop for large-scale operations. Integration of magnetic separation functionalized beads and aqueous biphasic extraction reduces the use of hazardous solvents and simplifies purification. Enzyme-assisted purification employs proteases, nucleases, and glycosidases to degrade impurities without altering the EPS backbone. Process analytical technology tools, including near-infrared spectroscopy, ultraviolet-visible spectrophotometry, and refractometry, monitor EPS concentration, purity, and structural features in real time. Green extraction technologies, including supercritical fluid extraction, pulsed electric field, and ultrasound-assisted extraction, are under development for EPS recovery from solid substrates and marine biomass. These methods preserve molecular weight distribution, branching pattern, and functional groups critical for bioactivity and performance (Rühmann et al., 2015; More et al., 2021; Srivastava et al., 2022) (Fig 2.14).

A case study: Among ten lactic acid bacterial strains tested for EPS production on MRS broth with 4% sucrose, *Lactobacillus* sp. Ca6 emerged as the top performer, yielding 2.4 g/L of EPS. Analytical studies using TLC and HPLC confirmed that the polymer was made entirely of glucose, while FTIR spectroscopy detected characteristic carboxyl, hydroxyl, and amide groups. Beyond its EPS output, strain Ca6 exhibited notable probiotic qualities, including antimicrobial activity, sensitivity to several antibiotics, and strong tolerance to acidic conditions (pH 2) and high bile levels (5% bovine bile). These combined properties make *Lactobacillus* sp. Ca6 is

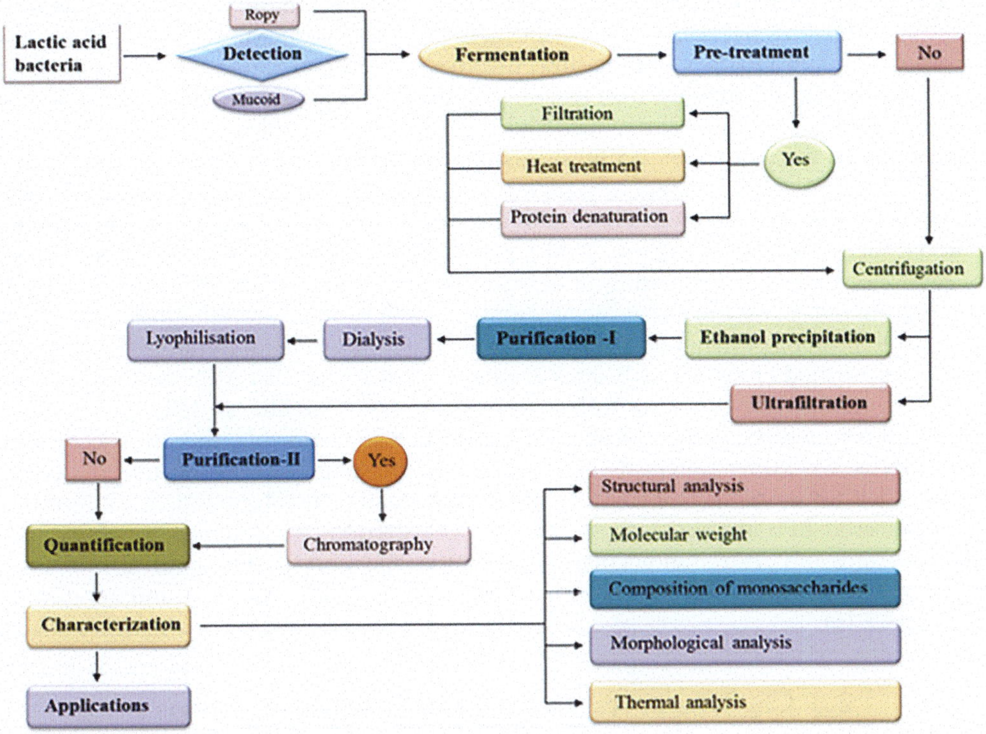

FIGURE 2.14 A schematic diagram of various steps involved in the detection, production, extraction, purification, and characterization of exopolysaccharides (EPSs) of lactic acid bacteria (LAB) (Yadav et al., 2024).

a promising candidate for use as a functional food additive, with potential applications in improving the texture, stability, and health benefits of fermented foods and beverages (Trabelsi et al., 2015).

2.7.3 Synthetic biology and exopolysaccharides engineering

EPSs from a wide variety of microorganisms are gaining rapid recognition as valuable biomaterials for industrial use. Thanks to their intricate chemical make-up and distinctive physicochemical and rheological properties, these biopolymers offer unique functionalities that make them attractive across many sectors—including food and feed production, packaging, chemicals, textiles, cosmetics, pharmaceuticals, agriculture, and medicine. Their natural biocompatibility, biodegradability, and safety for both the environment and humans have made them the subject of significant research over the past few decades, particularly for high-value applications (Ates, 2015).

Synthetic biology provides a platform for rational design and modular engineering of microbial hosts and biosynthetic pathways to produce tailor-made EPS with specific structures and functions. Modular cloning strategies, including Golden Gate Gibson assembly and CRISPR-enabled multiplex genome editing, enable precise construction of synthetic operons encoding glycosyltransferases, transporters, and regulatory proteins. Designer microbes such as *E. coli*, *Saccharomyces cerevisiae*, and *Corynebacterium glutamicum* are engineered to express heterologous EPS gene clusters from natural producers such as *Xanthomonas*, *Leuconostoc*, and marine bacteria with optimized codon usage, promoter strength, and ribosome binding sites. Pathway refactoring eliminates native feedback inhibition and enables redirection of metabolic flux toward sugar nucleotide biosynthesis using optimized flux balance analysis and ^{13}C-tracer experiments. Artificial metabolic pathways are constructed using synthetic enzymes and non-native cofactors for the production of non-natural monosaccharide residues, including fluorinated or azido-modified sugars. Orthogonal regulatory systems such as quorum-sensing circuits, synthetic riboswitches, and light-activated promoters provide spatiotemporal control over EPS expression. Scaffolded enzyme assemblies using protein-protein interaction domains or RNA scaffolds enhance channeling of intermediates and improve catalytic efficiency. Cell-free systems using purified enzymes or lysates are developed for in vitro EPS synthesis, enabling rapid prototyping and structural control. Chassis optimization includes deletion of proteases, nucleases, and competitive pathways as well as insertion of stress-tolerance and secretion systems. Biofoundry workflows integrate automated DNA assembly, high-throughput screening, and machine learning-driven strain selection for accelerated development of EPS-producing strains. Engineered EPS are customized for high charge density branched topology specific glycosidic linkages and desired molecular weight for applications in therapeutics, biomaterials, and food structuring (Ates, 2015; Freitas et al., 2017; Schmid, 2018; Broeker and Schmid, 2025) (Figs. 2.15 and 2.16).

2.8 Challenges, research gaps, and future directions

2.8.1 Commercialization and economic challenges

Commercial-scale EPS production faces significant technical, regulatory, and economic barriers that limit market penetration and application expansion. High production costs arise from low microbial yield, long fermentation times, high media input, viscosity-related oxygen transfer limitations, and complex downstream processing. EPS titers often range between 1 and 10 g/L, which necessitates large fermenter volumes and extended process durations to meet industrial demand. High viscosity of fermentation broth reduces mass transfer, increases energy consumption, and complicates scale-up. Purification procedures involving multiple steps, such as centrifugation, solvent precipitation, and dialysis, reduce product recovery and add cost. Use of food-grade and pharmaceutical-grade media increases cost and limits competitiveness with petrochemical-derived polymers. EPS exhibits batch-to-batch variability in molecular weight composition and functional performance, which challenges quality assurance standardization and regulatory approval. Market competitiveness is constrained by the availability of synthetic analogues with predictable behavior, low cost, and well-established supply chains. Regulatory hurdles include safety validation, toxicological testing, and compliance with international standards such as those set by the FDA, GRAS, EFSA's novel food regulation, REACH, and ISO biocompatibility standards. Lack of harmonized regulatory frameworks across regions hinders international commercialization. Market perception of microbial EPS is often limited by unfamiliarity and insufficient consumer awareness. The intellectual property landscape is complex due to overlapping patents on microbial strains, production processes, and applications. Economic feasibility analysis requires assessment of raw material availability, energy consumption, waste generation, equipment depreciation, and return on investment. Integration with circular

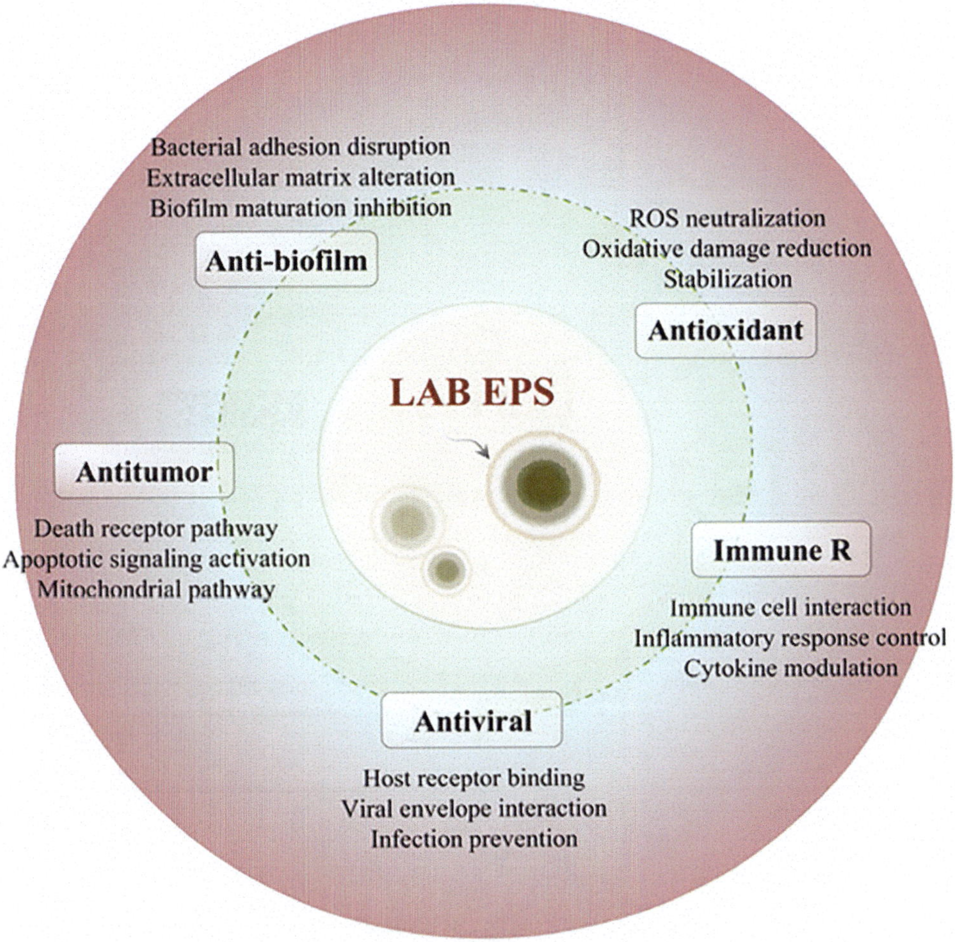

FIGURE 2.15 Biological functions of LAB EPSs, including antioxidant, immunomodulatory, and antibiofilm activities. ROS: reactive oxygen species, Immune R: immune response, Antibiofilm: Antibiofilm activity (Liu et al., 2025). *EPS*, Exopolysaccharides; *LAB*, lactic acid bacteria.

economy models, valorization of by-products, and coupling EPS production with high-value secondary metabolites such as enzymes, pigments, or bioplastics can improve profitability. Public-private partnerships targeted subsidies and regulatory incentives for bio-based products are essential to accelerate commercialization and industrial deployment (Andhare et al., 2014; Barcelos et al., 2020; Rana and Upadhyay, 2020).

2.8.2 Technological and methodological gaps

Scientific and technical limitations persist in understanding and controlling EPS biosynthesis structure-function relationships and application-specific performance. Most EPS biosynthetic pathways remain incompletely characterized with poorly defined roles of glycosyltransferases,

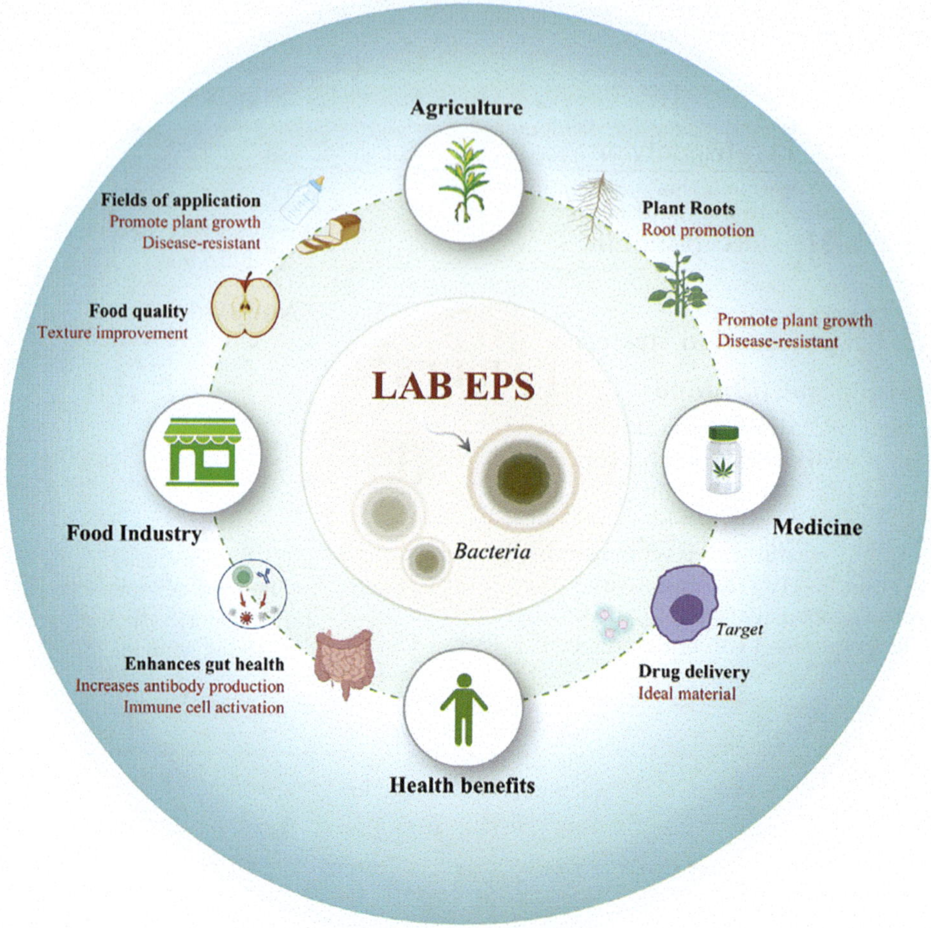

FIGURE 2.16 Applications of LAB EPSs in food, pharmaceuticals, and agriculture (Liu et al., 2025). *EPS*, Exopolysaccharides; *LAB*, lactic acid bacteria.

polymerases, transporters, and regulators. Lack of high-throughput functional annotation tools hinders gene discovery and pathway mapping. Structural analysis of complex branched and substituted EPS remains challenging due to spectral overlap, low abundance of rare sugars, and the need for multistep derivatization. Current analytical methods lack the throughput and reproducibility required for industrial screening. Existing models inadequately correlate EPS structure to physicochemical properties such as gel strength, emulsifying capacity, and viscosity. Predictive models for bioactivity based on primary and secondary structure are underdeveloped. Understanding of EPS interactions with proteins, lipids, and other biopolymers remains fragmented. Most studies rely on in vitro conditions that do not reflect real product matrices or biological environments. Lack of in situ analytical tools limits the spatial and temporal resolution of EPS production and function within biofilms or natural habitats. Standardization of EPS characterization protocols, including molecular weight determination, monosaccharide composition, and bioactivity assays, is lacking. Process monitoring tools for real-time EPS quantification and quality control are underdeveloped. Current bioprocess models often ignore the dynamic regulation of post-translational modification and metabolic cross-talk that influence EPS synthesis. The integration of omics data with metabolic and kinetic models remains limited by computational power, algorithmic constraints, and data heterogeneity. Scalable and eco-friendly extraction methods that preserve structural integrity are still under optimization. Absence of robust and validated databases for EPS sequence structures and functions limits data sharing and machine learning applications (Barcelos et al., 2020; Dey et al., 2023; Ahuja et al., 2023).

2.8.3 Future research opportunities

Research should focus on elucidating complete EPS biosynthetic networks using systems biology tools, including genome sequencing, transcriptomics, proteomics, and metabolomics integrated with pathway modeling and flux analysis. Identification of new EPS-producing strains from extreme environments such as hydrothermal vents, hypersaline lakes, polar regions, and acid mines can expand the structural and functional diversity of microbial EPS. Metagenomic and single-cell sequencing approaches are essential for accessing unculturable microbial diversity and discovering novel glycosyltransferases and transport systems. Directed evolution and adaptive laboratory evolution can improve microbial robustness, yield, and EPS structure. Functional genomics using CRISPR-Cas tools can dissect gene-function relationships and regulatory circuits. Advanced analytical platforms, including ion mobility mass spectrometry, NMR, cryo-electron microscopy, and atomic force microscopy, should be used for high-resolution EPS structure determination. Structure-activity relationship studies should incorporate biophysical modeling, artificial intelligence, and machine learning for predictive EPS design. Biosensor and microfluidic technologies should be developed for real-time monitoring of EPS synthesis in bioprocesses and environmental systems. Research into EPS interaction with host cells, microbial communities, and abiotic surfaces is needed to design functional biomaterials, probiotics, and environmental remediation systems. Interdisciplinary research combining synthetic biology, materials science, polymer physics, and microbiology will accelerate the development of tailor-made EPS for specific applications. Integration of EPS into smart materials, self-healing coatings, stimuli-responsive hydrogels, and bioelectronic devices represents an emerging frontier. Development of biodegradable EPS-based composite films and coatings that replace synthetic polymers is essential for the circular economy. Exploration of EPS-based prebiotics in personalized nutrition and microbiome engineering will drive innovation in health and wellness. Novel encapsulation systems using EPS for targeted drug delivery, vaccine carriers, and cell therapy support will advance precision medicine (Mouro et al., 2024; Nguyen et al., 2024).

2.9 Future outlook

2.9.1 Summary of key insights

EPSs constitute a structurally diverse class of microbial macromolecules with critical ecological, biological, physicochemical, and industrial significance. EPS are synthesized by a wide spectrum of microorganisms, including bacteria, fungi, algae, yeasts, actinomycetes, and archaea through complex biosynthetic pathways involving glycosyltransferases, sugar nucleotide precursors, polymerases, and transporters. Their structural variability encompasses homopolymers and heteropolymers composed of monosaccharides such as glucose, galactose, mannose, rhamnose, xylose, fucose, glucuronic acid, and galacturonic acid linked by glycosidic bonds in linear or branched configurations. Functional properties are influenced by molecular weight, charge density, degree of substitution, and environmental conditions during biosynthesis. EPS exhibit broad biological activities, including immunomodulation, antimicrobial activity, anticancer properties, antioxidant effects, antidiabetic action, cholesterol-lowering effects, wound healing, and regenerative capacity. Industrial applications span food, pharmaceuticals, biomedicine, environmental remediation, agriculture, cosmeceuticals, and advanced materials. Biotechnological strategies for enhanced production include optimization of fermentation conditions, genetic and metabolic engineering, isolation and purification protocols, and scale-up using advanced bioreactor systems. Analytical characterization using chromatography, spectroscopy, mass spectrometry, microscopy, and rheology enable structural and functional elucidation. Synthetic biology approaches enable customized EPS production through modular pathway engineering, strain design, and in vitro synthesis. Technological advancements support the development of EPS-based nanomaterials, scaffolds, drug carriers, bioinks, and functional coatings. Interdisciplinary integration of microbiology, materials science, systems biology, and bioprocess engineering is essential to unlock the full potential of EPS as sustainable biopolymers.

2.9.2 Sustainable potential of exopolysaccharides

EPS offers a sustainable and eco-friendly alternative to petrochemical-derived polymers due to their biodegradability, renewability, and nontoxic nature. Their biosynthesis utilizes renewable feedstocks, including agricultural residues, industrial by-products, and lignocellulosic waste. EPS production contributes to the circular bioeconomy by converting low-value substrates into high-value functional materials. EPS-based materials are biodegradable under natural and industrial composting conditions and exhibit minimal ecotoxicity. Replacement of synthetic stabilizers, emulsifiers, thickeners, and encapsulants with microbial EPS reduces environmental footprint and supports green chemistry principles. EPS-based composites and coatings offer biodegradable alternatives for packaging, agriculture, and personal care applications. EPS supports carbon sequestration in marine ecosystems, soil stabilization, and water purification, contributing to ecosystem services and climate resilience. Their role in sustainable agriculture includes the enhancement of soil structure, drought tolerance, nutrient retention, and plant-microbe interactions. Polysaccharides from microalgae, produced through photosynthesis, are safe, biodegradable, and versatile natural carbohydrates with complex structures that vary by species. They offer useful properties like emulsifying ability and biological benefits, including antioxidant, anti-inflammatory, antitumor, and antimicrobial effects. Some microalgae naturally have high carbohydrate levels, while others boost sugar production in response to environmental factors like light, nutrients, or stress. These organisms can even grow using industrial waste as nutrients. Microalgal polysaccharides serve roles in cell structure, energy storage, and protection. Thanks to their beneficial physical and biological traits, they are increasingly used in food, agriculture, and animal feed industries, making them a sustainable option (Moreira et al., 2022).

Integration of EPS into sustainable production systems requires process intensification, energy-efficient downstream processing, and waste minimization strategies. Development of life cycle assessment frameworks, carbon footprint models, and environmental impact metrics enables comparative analysis with synthetic counterparts. Policy frameworks, regulatory incentives, public awareness, and investment in bio-based technologies are essential for scaling EPS-based innovations. Industrial symbiosis models coupling EPS production with biorefineries, wastewater treatment, or enzyme manufacturing enhance resource efficiency and economic viability. The structural diversity and wide-ranging functions of EPSs make them valuable across many industries—including food, pharmaceuticals, biomedicine, cosmetics, textiles, oil recovery, and environmental clean-up. Well-studied examples such as HoPSs (like dextran, bacterial cellulose, curdlan, and levan) and heteropolysaccharides (such as xanthan gum, alginate, gellan, and kefiran) have already made significant contributions in these areas. Recently, scientists have begun to explore EPSs produced by extremophile bacteria, which thrive in harsh environments, as these unusual biopolymers may offer new protective and biological functions not seen in conventional EPSs. This review highlights the main functional qualities and industrial uses of widely recognized EPSs, underlining their importance to both industry and scientific research. It also calls attention to the growing interest in discovering EPSs with novel compositions and enhanced abilities—including special protective features. At the same time, the review addresses ongoing challenges that limit broader adoption and considers future opportunities for innovation. By focusing on the development of new, sustainable, and high-performance bio-based materials, the discussion points to promising directions for research and practical applications in the years ahead (Mouro et al., 2024).

2.9.3 Implications for future research and applications

Future research must address existing challenges in EPS commercialization by improving yield process economics, structural control, functional predictability, and regulatory compliance. Discovery of novel EPS from unexplored microbial niches using metagenomics, single-cell analysis, and synthetic ecology is a priority. Development of universal tools for EPS gene cluster annotation and functional prediction will accelerate pathway characterization. Structure-function relationship models using artificial intelligence and machine learning will enable rational EPS design for targeted applications. Microbial EPSs are diverse natural polymers produced mainly by lactic acid bacteria and probiotics, attracting growing interest for their wide range of chemical structures, functions, and applications. This review covers their composition, biosynthesis, production methods, and characterization. It also highlights their biological roles and health benefits, including prebiotic, anticancer, and immune-modulating effects. Beyond medicine and pharmaceuticals, microbial EPS are valuable in food industries as natural thickeners, stabilizers, and functional ingredients, and they also have applications in agriculture and other sectors, showing great promise for current and future uses (Nguyen et al., 2024).

Development of biofoundry platforms with integrated high-throughput screening robotic workflows and data analytics will reduce development timelines and cost. Real-time monitoring tools, biosensors, and smart bioreactors will improve process control and consistency. Development of EPS-based intelligent materials, including stimuli-responsive hydrogels, bioelectronic interfaces, and self-healing coatings, opens new frontiers in smart manufacturing, regenerative medicine, and wearable biosensors. Functionalization of EPS with ligands, peptides, and nanoparticles supports innovation in diagnostics, drug delivery, and precision health. EPS-based prebiotics, microbiome modulators, and immunonutrition agents will support personalized nutrition and metabolic therapy. The integration of EPS with digital manufacturing, additive manufacturing, and nanotechnology will revolutionize the design and deployment of functional biomaterials. Collaborative consortia between academia, industry, and regulatory bodies are essential for translating laboratory research into commercial products with societal impact. EPS will continue to play a pivotal role in advancing sustainable development, health innovation, and biotechnological transformation.

2.10 Conclusion

This chapter provides an in-depth overview of EPSs from natural microbial sources with emphasis on their origin, physicochemical attributes, structural diversity, and biological functions. It covers classification into HoPSs and heteropolysaccharides, molecular architecture including glycosidic linkages and branching patterns, and the role of noncarbohydrate substituents in defining solubility, charge density, and functional properties. The chapter details physicochemical behaviors such as gelation, thixotropy, pseudoplasticity, water retention, metal ion chelation, and emulsification, alongside the role of EPS in biofilm formation, host–microbe interactions, and environmental adaptation. Biological functions span immunomodulatory, antioxidant, anticancer, antiviral, and anticoagulant activities, with marine and extremophilic EPS highlighted for their stability under harsh conditions. Biosynthetic pathways, enzyme systems, and regulatory mechanisms, including quorum sensing and cyclic-di-GMP signaling, are outlined, together with the influence of environmental parameters on structure. The concluding remarks emphasize the multifunctionality, ecological importance, and industrial potential of EPS in food, pharmaceuticals, biomedicine, and environmental engineering. Challenges include controlling structural heterogeneity, achieving high-yield production with tailored properties, and scaling processes while preserving functional integrity. Future perspectives point toward precision metabolic engineering, advanced structural elucidation, and application-driven customization to develop high-performance EPS-based materials for energy, health, environmental, and marine technologies.

Competing interests

The authors declare that they have no competing interests.

References

Abarquero, D., Renes, E., Fresno, J.M., Tornadijo, M.E., 2022. Study of exopolysaccharides from lactic acid bacteria and their industrial applications: a review. Int. J. Food Sci. Technol. 57 (1), 16–26. https://doi.org/10.1111/ijfs.15227.

Abdalla, A.K., Ayyash, M.M., Olaimat, A.N., Osaili, T.M., Al-Nabulsi, A.A., Shah, N.P., Holley, R., 2021. Exopolysaccharides as antimicrobial agents: mechanism and spectrum of activity. Front. Microbiol. 12. https://doi.org/10.3389/fmicb.2021.664395.

Ahuja, V., Bhatt, A.K., Banu, J.R., Kumar, V., Kumar, G., Yang, Y.-H., Bhatia, S.K., 2023. Microbial exopolysaccharide composites in biomedicine and healthcare: Trends and advances. Polymers 15 (7), 1801. https://doi.org/10.3390/polym15071801.

Alavarse, A.C., Frachini, E.C.G., da Silva, R.L.C.G., Lima, V.H., Shavandi, A., Petri, D.F.S., 2022. Crosslinkers for polysaccharides and proteins: synthesis conditions, mechanisms, and crosslinking efficiency, a review. Int. J. Biol. Macromol. 202, 558–596. https://doi.org/10.1016/j.ijbiomac.2022.01.029.

Andhare, P., Chauhan, K., Dave, M., Pathak, H., 2014. Microbial exopolysaccharides: advances in applications and future prospects. Biotechnology, p. 25. https://doi.org/10.13140/RG.2.1.3518.4484.

Andrew, M., Jayaraman, G., 2020. Structural features of microbial exopolysaccharides in relation to their antioxidant activity. Carbohydr. Res. 487, 107881. https://doi.org/10.1016/j.carres.2019.107881.

Angelin, J., Kavitha, M., 2020. Exopolysaccharides from probiotic bacteria and their health potential. Int. J. Biol. Macromol. 162, 853–865. https://doi.org/10.1016/j.ijbiomac.2020.06.190.

Ates, O., 2015. Systems biology of microbial exopolysaccharides production. Front. Bioeng. Biotechnol. 3. https://doi.org/10.3389/fbioe.2015.00200.

Babu, A., Shams, R., Dash, K.K., Shaikh, A.M., Kovács, B., 2024. Protein-polysaccharide complexes and conjugates: structural modifications and interactions under diverse treatments. J. Agric. Food Res. 18, 101510. https://doi.org/10.1016/j.jafr.2024.101510.

Badel, S., Bernardi, T., Michaud, P., 2011. New perspectives for *Lactobacilli* exopolysaccharides. Biotechnol. Adv. 29 (1), 54–66. https://doi.org/10.1016/j.biotechadv.2010.08.011.

Banerjee, A., Sarkar, S., Govil, T., González-Faune, P., Cabrera-Barjas, G., Bandopadhyay, R., Salem, D.R., Sani, R.K., 2021. Extremophilic exopolysaccharides: biotechnologies and wastewater remediation. Front. Microbiol. 12. https://doi.org/10.3389/fmicb.2021.721365.

Baptista, S., Torres, C.A.V., Sevrin, C., Grandfils, C., Reis, M.A.M., Freitas, F., 2022. Extraction of the bacterial extracellular polysaccharide FucoPol by membrane-based methods: efficiency and impact on biopolymer properties. Polymers 14 (3), 390. https://doi.org/10.3390/polym14030390.

Barcelos, M.C.S., Vespermann, K.A.C., Pelissari, F.M., Molina, G., 2020. Current status of biotechnological production and applications of microbial exopolysaccharides. Crit. Rev. Food Sci. Nutr. 60 (9), 1475–1495. https://doi.org/10.1080/10408398.2019.1575791.

Beaumont, M., Tran, R., Vera, G., Niedrist, D., Rousset, A., Pierre, R., Shastri, V.P., Forget, A., 2021. Hydrogel-forming algae polysaccharides: from seaweed to biomedical applications. Biomacromolecules 22 (3), 1027–1052. https://doi.org/10.1021/acs.biomac.0c01406.

Broeker, J., Schmid, J., 2025. Engineering microbial exopolysaccharides for food applications. Curr. Opin. Biotechnol. 95, 103339. https://doi.org/10.1016/j.copbio.2025.103339.

Caggianiello, G., Kleerebezem, M., Spano, G., 2016. Exopolysaccharides produced by lactic acid bacteria: from health-promoting benefits to stress tolerance mechanisms. Appl. Microbiol. Biotechnol. 100 (9), 3877–3886. https://doi.org/10.1007/s00253-016-7471-2.

Carezzano, M.E., Alvarez Strazzi, F.B., Pérez, V., Bogino, P., Giordano, W., 2023. Exopolysaccharides synthesized by rhizospheric bacteria: a review focused on their roles in protecting plants against stress. Appl. Microbiol. 3 (4), 1249–1261. https://doi.org/10.3390/applmicrobiol3040086.

Casillo, A., Lanzetta, R., Parrilli, M., Corsaro, M.M., 2018. Exopolysaccharides from marine and marine extremophilic bacteria: structures, properties, ecological roles and applications. Mar. Drugs 16 (2), 69. https://doi.org/10.3390/md16020069.

Castro-Bravo, N., Wells, J.M., Margolles, A., Ruas-Madiedo, P., 2018. Interactions of surface exopolysaccharides from bifidobacterium and *Lactobacillus* within the intestinal environment. Front. Microbiol. 9. https://doi.org/10.3389/fmicb.2018.02426.

Coppola, F., Fratianni, F., Bianco, V., Wang, Z., Pellegrini, M., Coppola, R., Nazzaro, F., 2025. New methodologies as opportunities in the study of bacterial biofilms, including food-related applications. Microorganisms 13 (5), 1062. https://doi.org/10.3390/microorganisms13051062.

Costa, O.Y.A., Raaijmakers, J.M., Kuramae, E.E., 2018. Microbial exopolysaccharides: ecological function and impact on soil aggregation. Front. Microbiol. 9. https://doi.org/10.3389/fmicb.2018.01636.

Cugini, C., Shanmugam, M., Landge, N., Ramasubbu, N., 2019. The role of exopolysaccharides in oral biofilms. J. Dent. Res. 98 (7), 739–745. https://doi.org/10.1177/0022034519845001.

Daba, G.M., Elnahas, M.O., Elkhateeb, W.A., 2021. Contributions of exopolysaccharides from lactic acid bacteria as biotechnological tools in food, pharmaceutical, and medical applications. Int. J. Biol. Macromol. 173, 79–89. https://doi.org/10.1016/j.ijbiomac.2021.01.110.

Dar, A., Zahir, Z.A., Iqbal, M., Mehmood, A., Javed, A., Hussain, A., Bushra, Ahmad, M., 2021. Efficacy of rhizobacterial exopolysaccharides in improving plant growth, physiology, and soil properties. Env. Monit. Assess 193 (8), 515. https://doi.org/10.1007/s10661-021-09286-6.

Debnath, S., Muthuraj, M., Bandyopadhyay, T.K., Bobby, M.d.N., Vanitha, K., Tiwari, O.N., Bhunia, B., 2024. Engineering strategies and applications of cyanobacterial exopolysaccharides: a review on past achievements and recent perspectives. Carbohydr. Polym. 328, 121686. https://doi.org/10.1016/j.carbpol.2023.121686.

Delbarre-Ladrat, C., Sinquin, C., Lebellenger, L., Zykwinska, A., Colliec-Jouault, S., 2014. Exopolysaccharides produced by marine bacteria and their applications as glycosaminoglycan-like molecules. Front. Chem. 2. https://doi.org/10.3389/fchem.2014.00085.

De Philippis, R., Sili, C., Paperi, R., Vincenzini, M., 2001. Exopolysaccharide-producing cyanobacteria and their possible exploitation: a review. J. Appl. Phycol. 13 (4), 293–299. https://doi.org/10.1023/A:1017590425924.

De Vuyst, L., Degeest, B., 1999. Heteropolysaccharides from lactic acid bacteria. FEMS Microbiol. Rev. 23 (2), 153–177. https://doi.org/10.1111/j.1574-6976.1999.tb00395.x.

Dey, G., Patil, M.P., Banerjee, A., Sharma, R.K., Banerjee, P., Maity, J.P., Singha, S., Taharia, M.d., Shaw, A.K., Huang, H.-B., Kim, G.-D., Chen, C.-Y., 2023. The role of bacterial exopolysaccharides (EPS) in the synthesis of antimicrobial silver nanomaterials: a State-of-the-Art review. J. Microbiol. Methods 212, 106809. https://doi.org/10.1016/j.mimet.2023.106809.

Donot, F., Fontana, A., Baccou, J.C., Schorr-Galindo, S., 2012. Microbial exopolysaccharides: main examples of synthesis, excretion, genetics and extraction. Carbohydr. Polym. 87 (2), 951–962. https://doi.org/10.1016/j.carbpol.2011.08.083.

Freitas, F., Alves, V.D., Reis, M.A.M., 2011. Advances in bacterial exopolysaccharides: from production to biotechnological applications. Trends Biotechnol. 29 (8), 388–398. https://doi.org/10.1016/j.tibtech.2011.03.008.

Freitas, F., Torres, C.A.V., Reis, M.A.M., 2017. Engineering aspects of microbial exopolysaccharide production. Bioresour. Technol. 245, 1674–1683. https://doi.org/10.1016/j.biortech.2017.05.092.

Galle, S., Arendt, E.K., 2014. Exopolysaccharides from sourdough lactic acid bacteria. Crit. Rev. Food Sci. Nutr. 54 (7), 891–901. https://doi.org/10.1080/10408398.2011.617474.

Gauri, S.S., Mandal, S.M., Pati, B.R., 2012. Impact of Azotobacter exopolysaccharides on sustainable agriculture. Appl. Microbiol. Biotechnol. 95 (2), 331–338. https://doi.org/10.1007/s00253-012-4159-0.

Gheorghita, A.A., Wozniak, D.J., Parsek, M.R., Howell, P.L., 2023. Pseudomonas aeruginosa biofilm exopolysaccharides: assembly, function, and degradation. FEMS Microbiol. Rev. 47 (6), fuad060. https://doi.org/10.1093/femsre/fuad060.

Gientka, I., Błażejak, S., Stasiak-Różańska, L., Chlebowska-Śmigiel, A., 2015. Exopolysaccharides from yeast: insight into optimal conditions for biosynthesis, chemical composition and functional properties - review. Acta Sci. Pol. Technol. Aliment. 14 (4), 283–292. https://doi.org/10.17306/J.AFS.2015.4.29.

Grechi, J., Marinho-Carvalho, M., Zancan, P., Cinelli, L.P., Gomes, A.M.O., Rodrigues, M.L., Nimrichter, L., Sola-Penna, M., 2011. Glucuronoxylomannan from *Cryptococcus neoformans* down-regulates the enzyme 6-phosphofructo-1-kinase of macrophages. J. Biol. Chem. 286 (17), 14820–14829. https://doi.org/10.1074/jbc.M110.177030.

Gupta, P., Pruthi, P.A., Pruthi, V., 2019. Role of exopolysaccharides in biofilm formation. Introduction to Biofilm Engineering, 1323. American Chemical Society, pp. 17–57. https://doi.org/10.1021/bk-2019-1323.ch002.

Hamidi, M., Okoro, O.V., Milan, P.B., Khalili, M.R., Samadian, H., Nie, L., Shavandi, A., 2022. Fungal exopolysaccharides: properties, sources, modifications, and biomedical applications. Carbohydr. Polym. 284, 119152. https://doi.org/10.1016/j.carbpol.2022.119152.

Hereher, F., ElFallal, A., Abou-Dobara, M., Toson, E., Abdelaziz, M.M., 2018. Cultural optimization of a new exopolysaccharide producer "*Micrococcus roseus*". Beni-Suef. Univ. J. Basic Appl. Sci. 7 (4), 632–639. https://doi.org/10.1016/j.bjbas.2018.07.007.

Huang, X.-Y., Ye, X.-P., Hu, Y.-Y., Tang, Z.-X., Zhang, T., Zhou, H., Zhou, T., Bai, X.-L., Pi, E.-X., Xie, B.-H., Shi, L.-E., 2024. Exopolysaccharides of *Paenibacillus polymyxa*: a review. Int. J. Biol. Macromol. 261, 129663. https://doi.org/10.1016/j.ijbiomac.2024.129663.

Hussain, A., Zia, K.M., Tabasum, S., Noreen, A., Ali, M., Iqbal, R., Zuber, M., 2017. Blends and composites of exopolysaccharides; properties and applications: a review. Int. J. Biol. Macromol. 94, 10–27. https://doi.org/10.1016/j.ijbiomac.2016.09.104.

Ibrahim, M.I.A., Ibrahim, H.A.H., Haga, T., Ishida, A., Nehira, T., Matsuo, K., Gad, A.M., 2024. Potential bioactivities, chemical composition, and conformation studies of exopolysaccharide-derived *Aspergillus* sp. strain GAD7. J. Fungi 10 (9), 659. https://doi.org/10.3390/jof10090659.

Janczarek, M., 2011. Environmental signals and regulatory pathways that influence exopolysaccharide production in rhizobia. Int. J. Mol. Sci. 12 (11), 7898–7933. https://doi.org/10.3390/ijms12117898.

Jiang, Y., Yang, Z., 2018. A functional and genetic overview of exopolysaccharides produced by *Lactobacillus plantarum*. J. Funct. Foods 47, 229–240. https://doi.org/10.1016/j.jff.2018.05.060.

Jolly, L., Vincent, S.J.F., Duboc, P., Neeser, J.-R., 2002. Exploiting exopolysaccharides from lactic acid bacteria. Antonie Leeuwenhoek 82 (1), 367–374. https://doi.org/10.1023/A:1020668523541.

Joseph, T.M., Sathian, A., Joshy, K.S., Mahapatra, D.K., Haponiuk, J.T., Thomas, S., 2024. Chemical modifications of alginate-based biopolymers. In: Sadasivan, S.M., Ravindran, L., Goda, K., Thomas, S. (Eds.), Handbook Natural Polymers. Elsevier, pp. 97–122. https://doi.org/10.1016/B978-0-323-99856-7.00016-1.

Jurášková, D., Ribeiro, S.C., Silva, C.C.G., 2022. Exopolysaccharides produced by lactic acid bacteria: from biosynthesis to health-promoting properties. Foods 11 (2), 156. https://doi.org/10.3390/foods11020156.

K, K.R., Sardar, U.R., Bhargavi, E., Devi, I., Bhunia, B., Tiwari, O.N., 2018. Advances in exopolysaccharides based bioremediation of heavy metals in soil and water: a critical review. Carbohydr. Polym. 199, 353–364. https://doi.org/10.1016/j.carbpol.2018.07.037.

Kanimozhi, J., Sivasubramanian, V., Achary, A., Vasanthi, M., Vinson, S.P., Sivashankar, R., 2018. Bioprocessing of agrofood industrial wastes for the production of bacterial exopolysaccharide. Bioprocess Engineering for a Green Environment. CRC Press. https://doi.org/10.1201/b22021-4.

Kaur, N., Dey, P., 2023. Bacterial exopolysaccharides as emerging bioactive macromolecules: from fundamentals to applications. Res. Microbiol. 174 (4), 104024. https://doi.org/10.1016/j.resmic.2022.104024.

Kavitake, D., Devi, P.B., Delattre, C., Reddy, G.B., Shetty, P.H., 2023. Exopolysaccharides produced by *Enterococcus* genus—An overview. Int. J. Biol. Macromol. 226, 111–120. https://doi.org/10.1016/j.ijbiomac.2022.12.042.

Kavitake, D., Devi, P.B., Shetty, P.H., 2020. Overview of exopolysaccharides produced by *Weissella* genus—a review. Int. J. Biol. Macromol. 164, 2964–2973. https://doi.org/10.1016/j.ijbiomac.2020.08.185.

Khiabani, A., Xiao, H., Wätjen, A.P., Tovar, M., Poulsen, V.K., Hansen, E.B., Bang-Berthelsen, C.H., 2024. Exploring the diversity and potential use of flower-derived lactic acid bacteria in plant-based fermentation: insights into exo-cellular polysaccharide production. Foods 13 (18), 2907. https://doi.org/10.3390/foods13182907.

Laws, A., Gu, Y., Marshall, V., 2001. Biosynthesis, characterisation, and design of bacterial exopolysaccharides from lactic acid bacteria. Biotechnol. Adv. 19 (8), 597–625. https://doi.org/10.1016/S0734-9750(01)00084-2.

Leroy, F., De Vuyst, L., 2016. Advances in production and simplified methods for recovery and quantification of exopolysaccharides for applications in food and health. J. Dairy Sci. 99 (4), 3229–3238. https://doi.org/10.3168/jds.2015-9936.

Li, H., Cao, K., Cong, P., Liu, Y., Cui, H., Xue, C., 2018. Structure characterization and antitumor activity of the extracellular polysaccharide from the marine fungus *Hansfordia sinuosae*. Carbohydr. Polym. 190, 87–94. https://doi.org/10.1016/j.carbpol.2018.02.077.

Li, P., Harding, S.E., Liu, Z., 2001. Cyanobacterial exopolysaccharides: their nature and potential biotechnological applications. Biotechnol. Genet. Eng. Rev. 18 (1), 375–404. https://doi.org/10.1080/02648725.2001.10648020.

Liang, S., Wang, X., Li, C., Liu, L., 2024. Biological activity of lactic acid bacteria exopolysaccharides and their applications in the food and pharmaceutical industries. Foods 13 (11), 1621. https://doi.org/10.3390/foods13111621.

Liu, J., Wang, X., Pu, H., Liu, S., Kan, J., Jin, C., 2017. Recent advances in endophytic exopolysaccharides: production, structural characterization, physiological role and biological activity. Carbohydr. Polym. 157, 1113–1124. https://doi.org/10.1016/j.carbpol.2016.10.084.

Liu, W., Wei, Y., Xiang, R., Dong, B., Yang, X., 2025. Lactic acid bacteria exopolysaccharides unveiling multifaceted insights from structure to application in foods and health promotion. Foods 14 (5), 823. https://doi.org/10.3390/foods14050823.

Lynch, K.M., Zannini, E., Coffey, A., Arendt, E.K., 2018. Lactic acid bacteria exopolysaccharides in foods and beverages: isolation, properties, characterization, and health benefits. Annu. Rev. Food Sci. Technol. 9 (2018), 155–176. https://doi.org/10.1146/annurev-food-030117-012537.

Madhuri, K., Vidya Prabhakar, K., 2014. Recent trends in the characterization of microbial exopolysaccharides. Orient. J. Chem. 30, 895–904. https://doi.org/10.13005/ojc/300271.

Mager, D.M., Thomas, A.D., 2011. Exopolysaccharides (EPS) from cyanobacterial soil crusts: a review of their role in dryland soil processes. J. Arid Env. 75 (2), 91–97. https://doi.org/10.1016/j.jaridenv.2010.10.001.

Mahapatra, S., Banerjee, D., 2013. Fungal exopolysaccharide: production, composition and applications. Microbiol. Insights. 6, 1–16. https://doi.org/10.4137/MBI.S10957.

Mahmoud, Y.A.-G., El-Naggar, M.E., Abdel-Megeed, A., El-Newehy, M., 2021. Recent advancements in microbial polysaccharides: Synthesis and applications. Polymers 13 (23), 4136. https://doi.org/10.3390/polym13234136.

Maunders, E., Welch, M., 2017. Matrix exopolysaccharides; the sticky side of biofilm formation. FEMS Microbiol. Lett. 364 (13), fnx120. https://doi.org/10.1093/femsle/fnx120.

Mende, S., Rohm, H., Jaros, D., 2016. Influence of exopolysaccharides on the structure, texture, stability and sensory properties of yoghurt and related products. Int. Dairy J. 52, 57–71. https://doi.org/10.1016/j.idairyj.2015.08.002.

Mestrom, L., Przypis, M., Kowalczykiewicz, D., Pollender, A., Kumpf, A., Marsden, S.R., Bento, I., Jarzębski, A.B., Szymańska, K., Chruściel, A., Tischler, D., Schoevaart, R., Hanefeld, U., Hagedoorn, P.-L., 2019. Leloir glycosyltransferases in applied biocatalysis: a multidisciplinary approach. Int. J. Mol. Sci. 20 (21), 5263. https://doi.org/10.3390/ijms20215263.

Mohd Nadzir, M., Nurhayati, R.W., Idris, F.N., Nguyen, M.H., 2021. Biomedical applications of bacterial exopolysaccharides: a review. Polymers 13 (4), 4. https://doi.org/10.3390/polym13040530.

Mollakhalili, N., Mohammadifar, 2015. Microbial exopolysaccharides: a review of their function and application in food sciences. J. Food Qual. Hazards Control 2, 112–117.

Morcillo, R.J.L., Manzanera, M., 2021. The effects of plant-associated bacterial exopolysaccharides on plant abiotic stress tolerance. Metabolites 11 (6), 337. https://doi.org/10.3390/metabo11060337.

More, V.S., Ebinesar, A., Prakruthi, A., Praveen, P., Fasim, A., Rao, A., Zameer, F., Anantharaju, K.S., More, S., 2021. Isolation and purification of microbial exopolysaccharides and their industrial application. In: Vaishnav, A., Choudhary, D.K. (Eds.), Microbial Polymers: Applications and Ecological Perspectives. Springer, pp. 69–86. https://doi.org/10.1007/978-981-16-0045-6_3.

Moreira, J.B., Vaz, B.S., Cardias, B.B., Cruz, C.G., Almeida, A.C.A., Costa, J.A.V., Morais, M.G., 2022. Microalgae polysaccharides: an alternative source for food production and sustainable agriculture. Polysaccharides 3 (2), 441–457. https://doi.org/10.3390/polysaccharides3020027.

Moscovici, M., 2015. Present and future medical applications of microbial exopolysaccharides. Front. Microbiol. 6. https://doi.org/10.3389/fmicb.2015.01012.

Mouro, C., Gomes, A.P., Gouveia, I.C., 2024. Microbial exopolysaccharides: structure, diversity, applications, and future frontiers in sustainable functional materials. Polysaccharides 5 (3), 241–287. https://doi.org/10.3390/polysaccharides5030018.

Nabot, M., Guérin, M., Sivakumar, D., Remize, F., Garcia, C., 2022. Variability of bacterial homopolysaccharide production and properties during food processing. Biology 11 (2), 171. https://doi.org/10.3390/biology11020171.

Naseem, H., Ahsan, M., Shahid, M.A., Khan, N., 2018. Exopolysaccharides producing rhizobacteria and their role in plant growth and drought tolerance. J. Basic Microbiol. 58 (12), 1009–1022. https://doi.org/10.1002/jobm.201800309.

Netrusov, A.I., Liyaskina, E.V., Kurgaeva, I.V., Liyaskina, A.U., Yang, G., Revin, V.V., 2023. Exopolysaccharides producing bacteria: a review. Microorganisms 11 (6), 6. https://doi.org/10.3390/microorganisms11061541.

Nguyen, H.-T., Pham, T.-T., Nguyen, P.-T., Le-Buanec, H., Rabetafika, H.N., Razafindralambo, H.L., 2024. Advances in microbial exopolysaccharides: present and future applications. Biomolecules 14 (9), 9. https://doi.org/10.3390/biom14091162.

Nicolaus, B., Kambourova, M., Oner, E.T., 2010. Exopolysaccharides from extremophiles: from fundamentals to biotechnology. Env. Technol. 31 (10), 1145–1158. https://doi.org/10.1080/09593330903552094.

Nwodo, U.U., Green, E., Okoh, A.I., 2012. Bacterial exopolysaccharides: functionality and prospects. Int. J. Mol. Sci. 13 (11), 11. https://doi.org/10.3390/ijms131114002.

Nwosu, I.G., Abu, G.O., Agwa, K.O., 2019. Isolation, screening and characterization of exopolysaccharide producing bacteria. Microbiol. Res. J. Int. 1–9. https://doi.org/10.9734/mrji/2019/v29i530174.

Oleksy, M., Klewicka, E., 2018. exopolysaccharides produced by *Lactobacillus* sp.: biosynthesis and applications. Crit. Rev. Food Sci. Nutr. 58 (3), 450–462. https://doi.org/10.1080/10408398.2016.1187112.

Osemwegie, O.O., Adetunji, C.O., Ayeni, E.A., Adejobi, O.I., Arise, R.O., Nwonuma, C.O., Oghenekaro, A.O., 2020. Exopolysaccharides from bacteria and fungi: current status and perspectives in Africa. Heliyon 6 (6), e04205. https://doi.org/10.1016/j.heliyon.2020.e04205.

Pachekrepapol, U., Lucey, J.A., Gong, Y., Naran, R., Azadi, P., 2017. Characterization of the chemical structures and physical properties of exopolysaccharides produced by various *Streptococcus thermophilus* strains. J. Dairy Sci. 100 (5), 3424–3435. https://doi.org/10.3168/jds.2016-12125.

Pallikkunnel, M.L., Joseph, T.M., Haponiuk, J.T., Thomas, S., 2023. Alginate-based wound-healing dressings. Foundation and Growth of Macromolecular Science. Apple Academic Press. https://doi.org/10.1201/9781003370505-16.

Patel, S., Majumder, A., Goyal, A., 2012. Potentials of exopolysaccharides from lactic acid bacteria. Indian J. Microbiol. 52 (1), 3–12. https://doi.org/10.1007/s12088-011-0148-8.

Paul, P., Nair, R., Mahajan, S., Gupta, U., Aalhate, M., Maji, I., Singh, P.K., 2023. Traversing the diverse avenues of exopolysaccharides-based nanocarriers in the management of cancer. Carbohydr. Polym. 312, 120821. https://doi.org/10.1016/j.carbpol.2023.120821.

Paul, S., Parvez, S.S., Goswami, A., Banik, A., 2024. Exopolysaccharides from agriculturally important microorganisms: conferring soil nutrient status and plant health. Int. J. Biol. Macromol. 262, 129954. https://doi.org/10.1016/j.ijbiomac.2024.129954.

Pereira, S., Zille, A., Micheletti, E., Moradas-Ferreira, P., De Philippis, R., Tamagnini, P., 2009. Complexity of cyanobacterial exopolysaccharides: composition, structures, inducing factors and putative genes involved in their biosynthesis and assembly. FEMS Microbiol. Rev. 33 (5), 917–941. https://doi.org/10.1111/j.1574-6976.2009.00183.x.

Pirog, T.P., Kovalenko, M.A., Kuzminskaya, Y.V., Votselko, S.K., 2004. Physicochemical properties of the microbial exopolysaccharide ethapolan synthesized on a mixture of growth substrates. Microbiology 73 (1), 14–18. https://doi.org/10.1023/B:MICI.0000016361.71744.6f.

Poli, A., Anzelmo, G., Nicolaus, B., 2010. Bacterial exopolysaccharides from extreme marine habitats: production, characterization and biological activities. Mar. Drugs 8 (6), 1779–1802. https://doi.org/10.3390/md8061779.

Prete, R., Alam, M.K., Perpetuini, G., Perla, C., Pittia, P., Corsetti, A., 2021. Lactic acid bacteria exopolysaccharides producers: a sustainable tool for functional foods. Foods 10 (7), 1653. https://doi.org/10.3390/foods10071653.

Qi, M., Zheng, C., Wu, W., Yu, G., Wang, P., 2022. Exopolysaccharides from marine microbes: source, structure and application. Mar. Drugs 20 (8), 512. https://doi.org/10.3390/md20080512.

Qin, H., Wang, Z., Sha, W., Song, S., Qin, F., Zhang, W., 2024. Role of plant-growth-promoting rhizobacteria in plant machinery for soil heavy metal detoxification. Microorganisms 12 (4), 700. https://doi.org/10.3390/microorganisms12040700.

Rabiya, R., Sen, R., 2022. Artificial intelligence driven advanced optimization strategy vis-à-vis response surface optimization of production medium: bacterial exopolysaccharide production as a case-study. Biochem. Eng. J. 178, 108271. https://doi.org/10.1016/j.bej.2021.108271.

Rahbar Saadat, Y., Yari Khosroushahi, A., Pourghassem Gargari, B., 2019. A comprehensive review of anticancer, immunomodulatory and health beneficial effects of the lactic acid bacteria exopolysaccharides. Carbohydr. Polym. 217, 79–89. https://doi.org/10.1016/j.carbpol.2019.04.025.

Rajput, A., Bhamare, K.T., Mukhopadhyay, A., Rastogi, A., Sakshi, Kumar, M., 2020. Efficacy of anti-biofilm agents in targeting ESKAPE Pathogens with a focus on antibiotic drug resistance. Quorum Sensing: Microbial Rules of Life, 1374. American Chemical Society, pp. 177–199. https://doi.org/10.1021/bk-2020-1374.ch010.

Rana, S., Upadhyay, L.S.B., 2020. Microbial exopolysaccharides: synthesis pathways, types and their commercial applications. Int. J. Biol. Macromol. 157, 577–583. https://doi.org/10.1016/j.ijbiomac.2020.04.084.

Riaz Rajoka, M.S., Wu, Y., Mehwish, H.M., Bansal, M., Zhao, L., 2020. *Lactobacillus* exopolysaccharides: new perspectives on engineering strategies, physiochemical functions, and immunomodulatory effects on host health. Trends Food Sci. Technol. 103, 36–48. https://doi.org/10.1016/j.tifs.2020.06.003.

Ruas-Madiedo, P., de los Reyes-Gavilán, C.G., 2005. Invited review: methods for the screening, isolation, and characterization of exopolysaccharides produced by lactic acid bacteria. J. Dairy Sci. 88 (3), 843–856. https://doi.org/10.3168/jds.S0022-0302(05)72750-8.

Ruas-Madiedo, P., Hugenholtz, J., Zoon, P., 2002. An overview of the functionality of exopolysaccharides produced by lactic acid bacteria. Int. Dairy J. 12 (2), 163–171. https://doi.org/10.1016/S0958-6946(01)00160-1.

Rühmann, B., Schmid, J., Sieber, V., 2015. Methods to identify the unexplored diversity of microbial exopolysaccharides. Front. Microbiol. 6. https://doi.org/10.3389/fmicb.2015.00565.

Ruijgrok, G., Wu, D.-Y., Overkleeft, H.S., Codée, J.D.C., 2024. Synthesis and application of bacterial exopolysaccharides. Curr. Opin. Chem. Biol. 78, 102418. https://doi.org/10.1016/j.cbpa.2023.102418.

Salimi, F., Farrokh, P., 2023. Recent advances in the biological activities of microbial exopolysaccharides. World J. Microbiol. Biotechnol. 39 (8), 213. https://doi.org/10.1007/s11274-023-03660-x.

Satpute, S.K., Banat, I.M., Dhakephalkar, P.K., Banpurkar, A.G., Chopade, B.A., 2010. Biosurfactants, bioemulsifiers and exopolysaccharides from marine microorganisms. Biotechnol. Adv. 28 (4), 436–450. https://doi.org/10.1016/j.biotechadv.2010.02.006.

Schmid, J., 2018. Recent insights in microbial exopolysaccharide biosynthesis and engineering strategies. Curr. Opin. Biotechnol. 53, 130–136. https://doi.org/10.1016/j.copbio.2018.01.005.

Schmid, J., Sieber, V., Rehm, B., 2015. Bacterial exopolysaccharides: biosynthesis pathways and engineering strategies. Front. Microbiol. 6, 496. https://doi.org/10.3389/fmicb.2015.00496.

Selim, M.S., Amer, S.K., Mohamed, S.S., Mounier, M.M., Rifaat, H.M., 2018. Production and characterisation of exopolysaccharide from *Streptomyces carpaticus* isolated from marine sediments in Egypt and its effect on breast and colon cell lines. J. Genet. Eng. Biotechnol. 16 (1), 23–28. https://doi.org/10.1016/j.jgeb.2017.10.014.

Sengupta, D., Datta, S., Biswas, D., 2018. Towards a better production of bacterial exopolysaccharides by controlling genetic as well as physico-chemical parameters. Appl. Microbiol. Biotechnol. 102 (4), 1587–1598. https://doi.org/10.1007/s00253-018-8745-7.

Senni, K., Gueniche, F., Changotade, S., Septier, D., Sinquin, C., Ratiskol, J., Lutomski, D., Godeau, G., Guezennec, J., Colliec-Jouault, S., 2013. Unusual glycosaminoglycans from a deep sea hydrothermal bacterium improve fibrillar collagen structuring and fibroblast activities in engineered connective tissues. Mar. Drugs 11 (4), 1351–1369. https://doi.org/10.3390/md11041351.

Seviour, R.J., McNeil, B., Fazenda, M.L., Harvey, L.M., 2011. Operating bioreactors for microbial exopolysaccharide production. Crit. Rev. Biotechnol. 31 (2), 170–185. https://doi.org/10.3109/07388551.2010.505909.

Sharma, P., Sharma, A., Lee, H.-J., 2024. Antioxidant potential of exopolysaccharides from lactic acid bacteria: a comprehensive review. Int. J. Biol. Macromol. 281 (Pt 2), 135536. https://doi.org/10.1016/j.ijbiomac.2024.135536.

Shih, I.-L., 2010. Microbial exo-polysaccharides for biomedical applications. Mini Rev. Med. Chem. 10 (14), 1345–1355. https://doi.org/10.2174/138955710793564115.

Shu, C.-H., Lung, M.-Y., 2004. Effect of pH on the production and molecular weight distribution of exopolysaccharide by *Antrodia camphorata* in batch cultures. Process Biochem. 39 (8), 931–937. https://doi.org/10.1016/S0032-9592(03)00220-6.

Shukla, A., Mehta, K., Parmar, J., Pandya, J., Saraf, M., 2019. Depicting the exemplary knowledge of microbial exopolysaccharides in a nutshell. Eur. Polym. J. 119, 298–310. https://doi.org/10.1016/j.eurpolymj.2019.07.044.

Sørensen, H.M., Rochfort, K.D., Maye, S., MacLeod, G., Brabazon, D., Loscher, C., Freeland, B., 2022. Exopolysaccharides of lactic acid bacteria: production, purification and health benefits towards functional food. Nutrients 14 (14), 2938. https://doi.org/10.3390/nu14142938.

Srivastava, N., Kumari, S., Kurmi, S., Pinnaka, A.K., Choudhury, A.R., 2022. Isolation, purification, and characterization of a novel exopolysaccharide isolated from marine bacteria *Brevibacillus borstelensis* M42. Arch. Microbiol. 204 (7), 399. https://doi.org/10.1007/s00203-022-02993-9.

Strazzulli, A., Cobucci-Ponzano, B., Iacono, R., Giglio, R., Maurelli, L., Curci, N., Schiano-di-Cola, C., Santangelo, A., Contursi, P., Lombard, V., Henrissat, B., Lauro, F.M., Fontes, C.M.G.A., Moracci, M., 2020. Discovery of hyperstable carbohydrate-active enzymes through metagenomics of extreme environments. FEBS J. 287 (6), 1116–1137. https://doi.org/10.1111/febs.15080.

Suresh Kumar, A., Mody, K., Jha, B., 2007. Bacterial exopolysaccharides – a perception. J. Basic Microbiol. 47 (2), 103–117. https://doi.org/10.1002/jobm.200610203.

Sutherland, I., 2001. Biofilm exopolysaccharides: a strong and sticky framework. Microbiol. (Read. Engl.) 147 (Pt 1), 3–9. https://doi.org/10.1099/00221287-147-1-3.

Sutherland, I.W., 1982. Microbial exopolysaccharides—their role in microbial adhesion in aqueous systems. CRC Crit. Rev. Microbiol. 10 (2), 173–201. https://doi.org/10.3109/10408418209113562.

Sutherland, I.W., 1998. Novel and established applications of microbial polysaccharides. Trends Biotechnol. 16 (1), 41–46. https://doi.org/10.1016/S0167-7799(97)01139-6.

Tieking, M., Gänzle, M.G., 2005. Exopolysaccharides from cereal-associated *Lactobacilli*. Trends Food Sci. Technol. 16 (1), 79–84. https://doi.org/10.1016/j.tifs.2004.02.015.

Tiwari, S., Kavitake, D., Devi, P.B., Halady Shetty, P., 2021. Bacterial exopolysaccharides for improvement of technological, functional and rheological properties of yoghurt. Int. J. Biol. Macromol. 183, 1585–1595. https://doi.org/10.1016/j.ijbiomac.2021.05.140.

Trabelsi, I., Slima, S.B., Chaabane, H., Riadh, B.S., 2015. Purification and characterization of a novel exopolysaccharides produced by *Lactobacillus* sp. Ca_6. Int. J. Biol. Macromol. 74, 541–546. https://doi.org/10.1016/j.ijbiomac.2014.12.045.

Tuinier, R., van Casteren, W.H.M., Looijesteijn, P.J., Schols, H.A., Voragen, A.G.J., Zoon, P., 2001. Effects of structural modifications on some physical characteristics of exopolysaccharides from *Lactococcus lactis*. Biopolymers 59 (3), 160–166. https://doi.org/10.1002/1097-0282(200109)59:3⟨160::AID-BIP1015⟩3.0.CO;2-V.

Uzum, Z., Silipo, A., Lackner, G., De Felice, A., Molinaro, A., Hertweck, C., 2015. Structure, genetics and function of an exopolysaccharide produced by a bacterium living within fungal hyphae. ChemBioChem. 16 (3), 387–392. https://doi.org/10.1002/cbic.201402488.

Vasiliu, S., Racovita, S., Lungan, M.A., Desbrieres, J., Popa, M., 2016. In: Pillay, V., Choonara, Y.E., Kumar, P. (Eds.), Frontiers in biomaterials: Unfolding the biopolymer landscape, (Vol. 2,. Bentham Science Publishers, pp. 180–238. https://doi.org/10.2174/97816810819531160200 11.

Velamakanni, R.P., Vuppugalla, P., Merugu, R., 2021. Techniques used for characterization of microbial exopolysaccharides. In: Nadda, A.K., V, S.K., Sharma, S. (Eds.), Microbial Exopolysaccharides as Novel and Significant Biomaterials. Springer International Publishing, pp. 19–43.

Wang, J., Salem, D.R., Sani, R.K., 2019. Extremophilic exopolysaccharides: a review and new perspectives on engineering strategies and applications. Carbohydr. Polym. 205, 8–26. https://doi.org/10.1016/j.carbpol.2018.10.011.

Wang, W., Ju, Y., Liu, N., Shi, S., Hao, L., 2023. Structural characteristics of microbial exopolysaccharides in association with their biological activities: a review. Chem. Biol. Technol. Agric. 10 (1), 137. https://doi.org/10.1186/s40538-023-00515-3.

Waoo, A.A., Singh, S., Pandey, A., Kant, G., Choure, K., Amesho, K.T.T., Srivastava, S., 2023. Microbial exopolysaccharides in the biomedical and pharmaceutical industries. Heliyon (8) 9. https://doi.org/10.1016/j.heliyon.2023.e18613.

Welman, A.D., Maddox, I.S., 2003. Exopolysaccharides from lactic acid bacteria: perspectives and challenges. Trends Biotechnol. 21 (6), 269–274. https://doi.org/10.1016/S0167-7799(03)00107-0.

Whitfield, G.B., Marmont, L.S., Howell, P.L., 2015. Enzymatic modifications of exopolysaccharides enhance bacterial persistence. Front, Microbiol. 6. https://doi.org/10.3389/fmicb.2015.00471.

Xu, R.-R., Yang, W.-D., Niu, K.-X., Wang, B., Wang, W.-M., 2018. An update on the evolution of glucosyltransferase (Gtf) genes in streptococcus. Front. Microbiol. 9. https://doi.org/10.3389/fmicb.2018.02979.

Yadav, M.K., Song, J.H., Vasquez, R., Lee, J.S., Kim, I.H., Kang, D.-K., 2024. Methods for detection, extraction, purification, and characterization of exopolysaccharides of lactic acid bacteria—a systematic review. Foods 13 (22), 3687. https://doi.org/10.3390/foods13223687.

Yildiz, H., Karatas, N., 2018. Microbial exopolysaccharides: resources and bioactive properties. Process Biochem. 72, 41–46. https://doi.org/10.1016/j.procbio.2018.06.009.

Zannini, E., Waters, D.M., Coffey, A., Arendt, E.K., 2016. Production, properties, and industrial food application of lactic acid bacteria-derived exopolysaccharides. Appl. Microbiol. Biotechnol. 100 (3), 1121–1135. https://doi.org/10.1007/s00253-015-7172-2.

Zhang, K., Liu, S., Liang, S., Xiang, F., Wang, X., Lian, H., Li, B., Liu, F., 2024. Exopolysaccharides of lactic acid bacteria: structure, biological activity, structure-activity relationship, and application in the food industry: a review. Int. J. Biol. Macromol. 257, 128733. https://doi.org/10.1016/j.ijbiomac.2023.128733.

Zhou, Y., Cui, Y., Qu, X., 2019. Exopolysaccharides of lactic acid bacteria: structure, bioactivity and associations: a review. Carbohydr. Polym. 207, 317–332. https://doi.org/10.1016/j.carbpol.2018.11.093.

CHAPTER

3

Latest advancements in green technologies for the eco-friendly extraction and processing of biomaterials from natural extracellular polysaccharides (EPS): Benefits and advantages in sustainability and renewability

Meena Rathore, Pushpendra Sharma, Kirti Suman and Rajeev Kaushik

Division of Microbiology, ICAR-Indian Agricultural Research Institute, New Delhi, India

3.1 Introduction

3.1.1 Microbial exopolysaccharides: Nature's versatile biopolymers

A wide range of microorganisms produce exopolysaccharides (EPS), highly hydrated polymers that are mainly composed of polysaccharides, proteins, and DNA. EPS are fundamental for microbial life and provide an ideal environment for chemical reactions, nutrient entrapment, and protection against environmental stresses. Microbial EPS can enhance the aggregation of soil particles and benefit plants by maintaining the moisture of the environment and trapping nutrients (Costa et al., 2018). Polysaccharides are a major fraction of the EPS matrix (Wingender et al., 2001). Most are long molecules, linear or branched, with a molecular mass of 0.5×10^6 daltons to 2×10^6 daltons (Flemming and Wingender, 2010). Microbial exopolysaccharides

DOI: https://doi.org/10.1016/B978-0-443-36619-2.00003-6

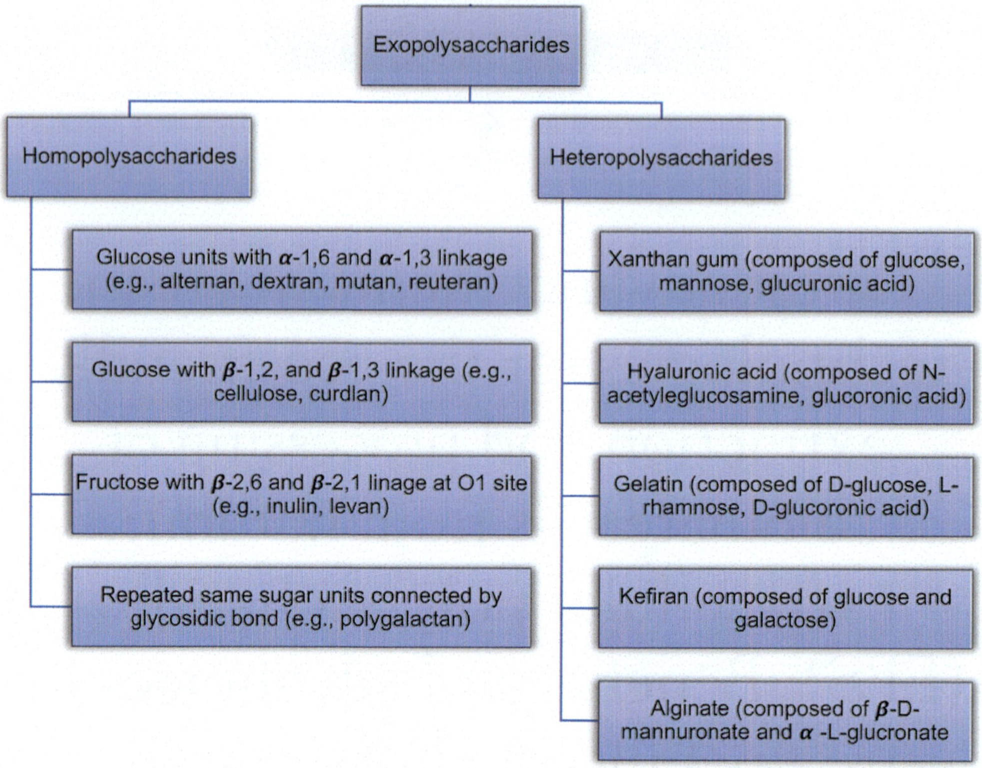

FIGURE 3.1 Classification of microbial polymers. Bacterial exopolysaccharides can be broadly classified based on their location within or outside the cells. Structurally, these exopolysaccharides could be composed of repeating units of the same monosaccharides (homopolysaccharide) or two or more units of different monosaccharides (heteropolysaccharide).

(EPS) are a diverse group of high-molecular-weight carbohydrate polymers secreted by microorganisms, including bacteria, cyanobacteria, microalgae (Boonchai et al., 2014), yeasts (Pavlova and Grigorova, 1999), fungi (Elisashvili et al., 2009), and protists (Jain et al., 2005), into their surrounding environment. These biopolymers are not essential for primary metabolism but play crucial roles in survival, facilitating cell aggregation, adhesion to surfaces, formation of protective biofilms, and resistance against environmental stresses such as desiccation, pH fluctuations, predation, and antimicrobial agents (Flemming and Wingender, 2010). Chemically, EPS can be classified as homopolysaccharides, composed of a single type of monosaccharide unit (e.g., dextran from *Leuconostoc mesenteroides*, cellulose from *Komagataeibacter xylinus*, curdlan from *Alcaligenes faecalis*) (Yildiz and Karatas, 2018), or, more commonly, as heteropolysaccharides, containing two or more different monosaccharide units in a repeating sequence (e.g., xanthan from *Xanthomonas campestris*, gellan from *Sphingomonas elodea*, alginate from *Azotobacter vinelandii*) (Fig. 3.1) (Sutherland, 1998; Salimi and Farrokh, 2023; Waoo et al., 2023). The remarkable structural diversity, arising from variations in monomer composition, glycosidic linkage types (α or β, 1→2, 1→3, 1→4, 1→6), and branching patterns, endows EPS with a wide spectrum of functional properties, including thickening, gelling, emulsifying, and stabilizing capabilities (Kaur and Dey, 2023). This versatility has led to their widespread application across various industrial sectors, from food technology (as additives in sauces, dairy, and baked goods), and cosmetics (as moisturizing agents), to agriculture (as soil conditioners) and biomedicine (in drug delivery, wound healing, and tissue engineering) (Ahuja et al., 2023; Netrusov et al., 2023; Yaşar Yıldız and Radchenkova, 2023; Guerrero et al., 2024).

3.1.2 The environmental imperative: Moving beyond conventional methods

Historically, the recovery of EPS from microbial fermentation broths has relied on conventional methods that are often harsh, inefficient, and environmentally taxing. The most common approach involves precipitating the EPS by adding 2–4 volumes of an organic solvent like ethanol, isopropanol, or acetone to the cell-free supernatant (Murugu and Narayanan, 2024; Nguyen et al., 2024; Yadav et al., 2024). While effective for separating the polymer, this method entails significant drawbacks. It requires vast quantities of flammable and volatile organic compounds (VOCs), which pose safety risks and necessitate costly, energy-intensive solvent recovery and disposal procedures. Furthermore, the use of harsh chemicals, such as strong acids or alkalis (e.g., trichloroacetic acid, sodium hydroxide) for cell lysis or viscosity reduction, can lead to the partial degradation of the polysaccharide chains via hydrolysis. This alters their molecular weight, conformation, and, consequently, their functional properties and bioactivity (Yadav et al., 2024). These processes are not only energy-intensive but also generate substantial waste streams, contributing to a significant environmental footprint that is increasingly at odds with global sustainability goals and stricter environmental regulations.

3.1.3 The dawn of green technologies in biomaterial science

In response to these challenges, the fields of biomaterial science and biochemical engineering are undergoing a profound transformation, guided by the principles of "Green Chemistry" and "Green Engineering" (Anastas and Warner, 1998). These principles, first articulated by Anastas and Warner, promoter for the design of chemical products and processes that reduce or eliminate the use and generation of hazardous substances. In the context of EPS processing, this deciphers to developing extraction and purification methods that are more energy-efficient, utilize safer solvents (ideally water), minimize waste, preserve the native structure of the biopolymer, and are derived from renewable resources (Chemat et al., 2019; Yadav et al., 2024). This shift is not merely an environmental consideration but also a strategic one. Green technologies can enhance product quality by preserving delicate structures, improve process efficiency through reduced steps and time, and unlock new economic opportunities by creating higher-value products with fewer environmental compliance costs.

3.1.4 Chapter objectives and scope

This chapter aims to provide a detailed and critical overview of the latest advancements in eco-friendly technologies for the extraction, purification, and modification of microbial EPS. We will begin by exploring a suite of innovative green extraction techniques, analyzing their underlying scientific mechanisms, key operational parameters, advantages, and limitations with

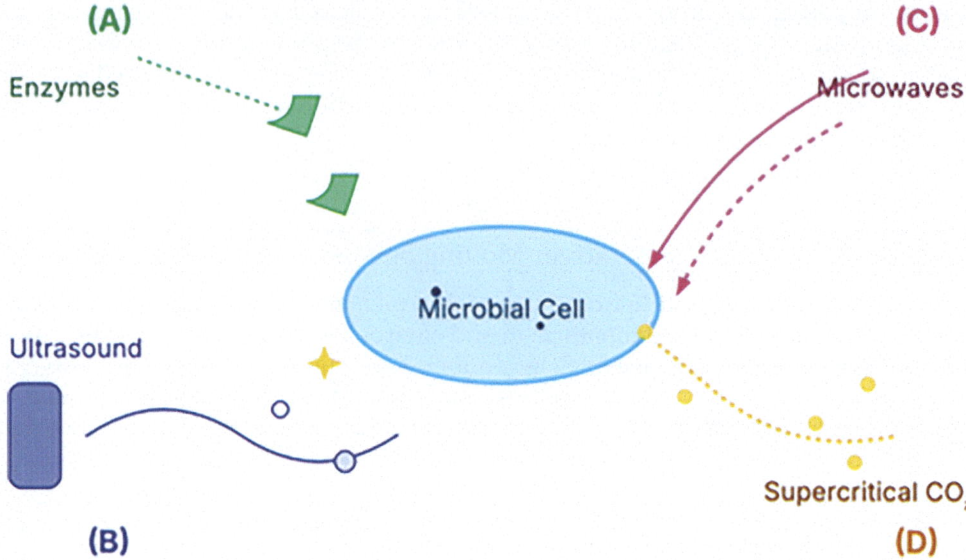

FIGURE 3.2 Schematic representation of various green extraction technologies for microbial EPS. (A) Enzymes specifically breaking down the cell wall, (B) ultrasound waves creating cavitation bubbles that implode on the cell surface, (C) microwaves causing internal heating, and (D) supercritical CO_2 penetrating the cell to extract EPS. *EPS*, Exopolysaccharides.

supporting case studies. Subsequently, we will discuss sustainable downstream processing strategies that align with green principles, focusing on purification and functional modification. The chapter will then quantify the benefits of this green transition in terms of sustainability, renewability, and economic feasibility, framing it within the context of the circular bioeconomy. Finally, we will address the remaining challenges to industrial-scale adoption and highlight future research directions that promise to further solidify the role of microbial EPS as a cornerstone of the future bio-based economy.

3.2 Advanced green extraction technologies for microbial exopolysaccharides

The development of nonconventional extraction methods has been pivotal in advancing the green processing of EPS. These technologies leverage physical forces, enzymatic action, or green solvents to achieve efficient extraction while minimizing environmental impact (Fig. 3.2).

3.2.1 Enzyme-assisted extraction

Principle: EAE utilizes the high specificity of enzymes to selectively degrade microbial cell walls, facilitating the release of intracellular or cell-associated EPS into the fermentation broth without damaging the target polymer (Lakmal et al., 2015; Marathe et al., 2017). The choice of enzyme is critical; for instance, lysozyme is effective against the peptidoglycan layer of gram-positive bacteria, while proteases like papain or trypsin can disrupt protein-polysaccharide complexes on the cell surface (Nadar et al., 2018a). The process is conducted under mild conditions of pH (typically 5–8) and temperature (30°C–60°C), which fall within the optimal range for most enzymes and are gentle enough to preserve the delicate tertiary structure and bioactivity of the EPS (Puri et al., 2012).

Advantages: EAE offers exceptional specificity, leading to higher purity extracts and improved yields by ensuring complete cell lysis. The gentle conditions ensure the preservation of high molecular weight and complex structural features, which are often crucial for desired functionalities like viscosity or immunomodulatory activity (Nadar et al., 2018a). It significantly reduces the need for harsh chemical treatments and organic solvents.

Recent advances and case studies: Recent studies have demonstrated the efficacy of EAE for various EPS. For instance, the use of papain was shown to increase the extraction yield of EPS from *Lactobacillus plantarum* by 1.5-fold compared to conventional methods, by disrupting the protein-polysaccharide complexes on the cell surface (Li et al., 2014). Similarly, an effective method involving alkaline and acidic treatments followed by enzymatic hydrolysis with α-amylase (Termamyl) has been successfully applied to release chitosan and chitin–glucan complexes from the mycelia of *Gongronella butleri*, resulting in more than a 40% increase in yield compared to chemical extraction alone (New et al., 2008).

3.2.2 Ultrasound-assisted extraction

Principle: Ultrasound-assisted extraction (UAE) employs high-frequency sound waves (typically >20 kHz) that propagate through a liquid medium, generating a phenomenon known as acoustic cavitation. This involves the formation, growth, and violent implosion of microscopic bubbles. The collapse of these bubbles creates localized hot spots with intense pressure (up to 1000 atm), high temperatures (up to 5000 K), and powerful shear forces. This combination of physical effects effectively disrupts microbial cell walls, creates micro-fractures, and enhances the mass transfer of EPS from the cell or biofilm matrix into the solvent (usually water) (Kentish and Ashokkumar, 2011).

Advantages: The primary benefits of the UAE are a dramatic reduction in extraction time (from hours to minutes) and lower operating temperatures, which help prevent thermal degradation of the EPS. It is a highly efficient method that often leads to increased yields and reduces solvent consumption. The physical nature of the process can also reduce the viscosity of highly viscous fermentation broths, simplifying subsequent processing steps (Chemat et al., 2017).

Process Optimization: The efficiency of the UAE is dependent on several parameters. Higher ultrasonic power and longer sonication times generally improve yields, but excessive application can lead to polymer scission and a reduction in molecular weight. Therefore, optimizing frequency, power density, temperature, and processing time is critical for each specific EPS system. This was demonstrated for the extraction of EPS from *Ganoderma lucidum*, where optimized UAE conditions (80 W, 20 minute) resulted in a 34% higher yield than conventional hot water extraction (4 hour) (Ma et al., 2013).

3.2.3 Microwave-assisted extraction

Principle: Microwave-assisted extraction (MAE) utilizes microwave energy (typically at 2.45 GHz) to heat the solvent and microbial biomass. Unlike conventional heating, which relies on slow conduction and convection, microwaves cause efficient and rapid internal heating through two main mechanisms: Ionic conduction and dipole rotation. The interaction of the electromagnetic field with polar molecules (like water) in the sample causes them to rapidly oscillate, generating heat. This rapid increase in internal pressure within the microbial cells leads to the rupture of the cell membrane and the explosive release of EPS (Woo et al., 2000).

Advantages: MAE is characterized by its speed (often completed in minutes), high efficiency, and reduced solvent requirements. The targeted and uniform heating minimizes the thermal gradient and the risk of overheating the bulk solution, thereby preserving the structural integrity of the EPS better than conventional heating methods (Martins et al., 2019).

Case studies and comparison: MAE has been successfully used to extract EPS from various cyanobacteria and microalgae. In a study on *Nostoc* species, MAE yielded 2.5 times more EPS in 15 minutes than conventional heating did in 5 hours, with the resulting polysaccharide showing higher antioxidant activity (Quan et al., 2015). It can also be used synergistically with EAE, where a short microwave pulse can enhance cell permeability before enzymatic treatment, reducing the required enzyme dosage.

3.2.4 Supercritical fluid extraction

Principle: Supercritical fluid extraction (SFE) uses a supercritical fluid—a substance at a temperature and pressure above its critical point, where it exhibits properties of both a liquid (high density, solvating power) and a gas (low viscosity, high diffusivity)—as the extraction solvent. Carbon dioxide (CO_2) is the most common supercritical fluid used due to its mild critical conditions (31.1°C, 73.8 bar), nontoxicity, nonflammability, and low cost. While supercritical CO_2 itself is nonpolar and not ideal for extracting polar polysaccharides, its polarity can be tuned by adding a small amount of a polar co-solvent (modifier), such as ethanol or water, which significantly enhances its ability to solubilize EPS (Gallego et al., 2019; Dias et al., 2021).

Advantages: The key advantage of SFE is the ability to obtain exceptionally pure products. After extraction, the CO_2 can be easily and completely removed from the extract by simple depressurization, leaving no solvent residue. This makes it highly attractive for producing high-value EPS for biomedical and pharmaceutical applications where purity is a strict requirement (Gallego et al., 2019).

Challenges and niche applications: The primary drawback of SFE is the high capital and operating costs associated with the high-pressure equipment. Therefore, its application is currently limited to high-value EPS where purity is paramount, such as those with specific therapeutic activities, or for analytical sample preparation (Khaw et al., 2017).

3.2.5 Pressurized liquid/subcritical water extraction

Principle: This technique uses water as the extraction solvent at elevated temperatures (100°C–374°C) and pressures high enough to maintain it in a liquid state. Under these subcritical conditions, the hydrogen bonds in water weaken, and its dielectric constant decreases significantly, making it behave like a less polar solvent (similar to ethanol). This altered state enhances its ability to solubilize a wide range of organic molecules, including polysaccharides, while also promoting cell lysis (Dias et al., 2021).

Advantages: Pressurized liquid (PLE) is a powerful green technology as it uses water, the most environmentally benign solvent. It is fast and efficient, and by precisely tuning the temperature, the extraction process can be made selective for different types of compounds from the same biomass (Dias et al., 2021).

3.2.6 Aqueous two-phase system

Principle: Aqueous two-phase system (ATPS) involves creating two immiscible aqueous phases, typically by mixing a polymer (e.g., polyethylene glycol, [PEG]) with another polymer or a high concentration of a salt (e.g., phosphate, citrate) in water above their critical concentrations. When a crude fermentation broth is added to this system, the various components will partition differently between the two phases based on their surface properties, size, and charge. The EPS will preferentially partition into one phase, while cells, proteins, and other impurities partition into the other phase or accumulate at the interface (Albertsson, 1995).

Advantages: ATPS is an extremely gentle method that operates in a high-water-content (80%–90%) environment, which is ideal for preserving the bioactivity of sensitive biopolymers. It offers the unique potential to integrate extraction, concentration, and purification into a single, continuous, and highly scalable step, making it very attractive for industrial applications (Albertsson et al., 2020).

3.2.7 Comparative analysis and synergistic approaches

No single green technology is universally superior; the optimal choice depends on the specific microbial source, the nature of the EPS, the desired purity, and economic constraints. Hybrid methods, which combine the advantages of two or more techniques (e.g., enzyme-ultrasound or microwave-ATPS), are a promising frontier, often resulting in synergistic effects that lead to higher yields and purity in shorter times (Table 3.1).

3.3 Eco-friendly downstream processing and modification

Green principles must extend beyond extraction to the entire processing chain to ensure a truly sustainable product.

3.3.1 Green purification strategies

3.3.1.1 Membrane filtration

Techniques like ultrafiltration (UF) and diafiltration (DF) are increasingly replacing solvent precipitation for purification and concentration. UF uses a semi-permeable membrane with a specific molecular weight cut-off (MWCO) to separate the high-molecular-weight EPS from low-molecular-weight impurities like salts, monosaccharides, and residual media components. DF is a subsequent washing step where fresh solvent (water) is continuously added while filtration continues, effectively washing away remaining impurities. These methods avoid organic solvents entirely, are easily scalable, can be operated continuously, and are far less energy-intensive than distillation-based solvent recovery (Mould and Synge, 1954; Cheryan, 1998).

TABLE 3.1 Comparative analysis of green EPS extraction technologies.

Technology	Principle	Key advantages	Key disadvantages	Best suited for
EAE	Enzymatic cell lysis	High specificity, mild conditions, preserves bioactivity (Nadar et al., 2018b)	Enzyme cost, process optimization required (Nadar et al., 2018a)	High-value, bioactive EPS; structurally sensitive polymers
UAE	Acoustic cavitation	Very fast, low temperature, high efficiency (Chemat et al., 2017)	Potential for polymer degradation, scalability challenges (Ma et al., 2013)	Rapid screening, bulk EPS production, viscous broths
MAE	Dielectric heating	Extremely fast, reduced solvent, uniform heating (Woo et al., 2000; Martins et al., 2019)	Requires polar solvent, potential for localized overheating	Lab-scale extraction, heat-stable EPS
SFE	Supercritical CO_2	Ultra-high purity, no solvent residue (Gallego et al., 2019; Dias et al., 2021)	High capital and operating cost, requires co-solvent (Khaw et al., 2017)	High-purity pharmaceutical/medical grade EPS
PLE/SWE	Hot, pressurized water	Uses water as solvent, fast, tunable selectivity (Dias et al., 2021)	High-pressure equipment, potential for thermal degradation	Thermally stable polysaccharides, biorefinery concepts
ATPS	Phase partitioning	Gentle, integrates purification, highly scalable (Albertsson et al., 2020)	Cost of phase-forming chemicals, complex optimization (Albertsson, 1995)	Large-scale production, recovery from complex broths

ATPS, aqueous two-phase systems; *EAE*, enzyme-assisted extraction; *EPS*, exopolysaccharides; *MAE*, microwave-assisted extraction; *PLE/SWE*, pressurized liquid/subcritical water extraction; *SFE*, supercritical fluid extraction; *UAE*, ultrasound-assisted extraction.

3.3.1.2 *Advanced chromatography*

While chromatography is a standard purification tool, its "greenness" can be improved by using eco-friendly mobile phases (e.g., water/ethanol gradients instead of acetonitrile or methanol) and developing durable, reusable column materials like monoliths, which offer lower backpressure and higher throughput (Plotka et al., 2013).

3.3.2 Sustainable modification and functionalization of exopolysaccharides

Often, the native properties of EPS need to be tailored for specific applications. Green modification techniques aim to achieve this with minimal environmental impact (Ahmad et al., 2015).

3.3.2.1 *Physical modification*

High-pressure homogenization (HPH) and ultrasonication can be used to modify the physical properties (e.g., viscosity, solubility, emulsifying capacity) of EPS solutions by altering their conformation, particle size, and chain entanglement through intense shear forces, without the use of any chemical reagents (Chen et al., 2024).

3.3.2.2 *Enzymatic modification*

Enzymes can be used as highly specific biocatalysts to precisely tailor the structure of EPS. For example, glycoside hydrolases can be used under controlled conditions to partially depolymerize an EPS to a desired molecular weight range, while transferases can be used to add new functional groups (e.g., acyl groups) to create novel biomaterials with enhanced properties (Hara and Hutchinson, 1992; Golkar et al., 2018; Arreola-Barroso et al., 2021).

3.3.2.3 *Green chemical pathways*

When chemical modification is necessary, green chemistry principles guide the choice of nontoxic reagents, the use of water as a solvent, and the development of solvent-free reaction conditions (e.g., solid-state reactions) to minimize environmental impact. An example is the esterification of EPS using citric acid (a natural, nontoxic acid) instead of harsher chemical agents (Jain et al., 2005; Halpern et al., 2014; Wu et al., 2020).

3.4 Quantifying the benefits: Sustainability, renewability, and economic advantages

The adoption of green technologies for EPS processing offers profound, interconnected benefits that span environmental, social, and economic domains (Fig. 3.3).

3.4.1 Environmental sustainability gains

3.4.1.1 *Life cycle assessment*

LCA is a powerful tool for holistically quantifying the environmental benefits of green technologies. Studies comparing conventional solvent precipitation with membrane filtration for EPS purification consistently show that the green alternative leads to significant reductions in energy consumption (>50%), global warming potential (>60%), and ecotoxicity (>80%), primarily due to the elimination of the solvent production and recovery stages (Shen and Patel, 2008; Hottle et al., 2017; Kogie et al., 2025).

3.4.1.2 *Metrics*

The key sustainability metrics improved by green processing include a reduced carbon footprint (due to lower energy and solvent use), lower water usage (especially with recycling in membrane systems), elimination of VOC emissions, and the generation of biodegradable waste streams that can often be repurposed as biofertilizers or animal feed (Jain et al., 2005; Beena and Muringayil, 2024).

3.4.2 Enhancing renewability and the circular bioeconomy

Microbial EPS are inherently renewable, as they are produced by microorganisms that can be cultivated on a variety of feedstocks. The green paradigm takes this a step further by promoting the use of low-cost, nonfood agro-industrial waste streams (e.g., molasses from sugar

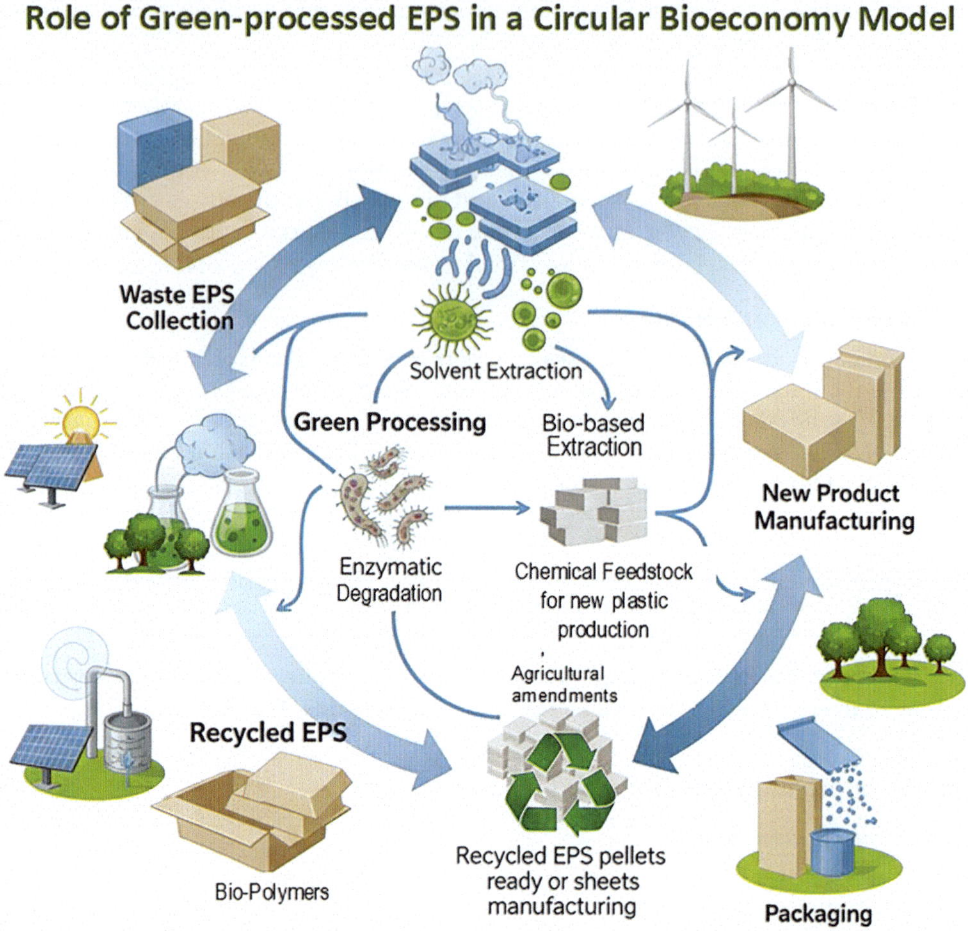

FIGURE 3.3 The role of green-processed EPS in a circular bioeconomy model: The process begins with agricultural and industrial waste (such as molasses or whey) being used as feedstock for microbial fermentation to produce EPS, which is then extracted and processed using green methods like ultrasound-assisted extraction (UAE) or membrane filtration; the resulting EPS is applied across biomedical, food, and agricultural sectors, and at the end of their life, these products biodegrade, returning nutrients back into the system. *EPS*, Exopolysaccharides.

refining, whey from the dairy industry, lignocellulosic hydrolysates from forestry or agricultural residues) as fermentation substrates (Nikodinovic-Runic et al., 2013; Salari et al., 2019; Aziz et al., 2022). This approach not only reduces the cost of EPS production by up to 40%–60% but also creates value from waste, mitigates disposal problems, and perfectly aligns with the principles of a circular bioeconomy (Abol-Fotouh et al., 2020).

3.4.3 Economic viability and industrial feasibility

While some green technologies require a higher initial investment (capital expenditure), they often lead to significant long-term economic advantages through lower operating expenditures.

The savings come from drastically reduced expenditure on organic solvents (both purchasing and disposal), lower energy consumption, and decreased water treatment costs. Furthermore, the ability of green technologies to produce higher quality, purer, and novel functionalized EPS can open up new, high-value markets (e.g., in biomedicine), enhancing the overall economic viability and competitiveness of the production process (Donot et al., 2012; Nguyen et al., 2024).

3.5 Challenges, regulatory aspects, and future outlook

Despite the clear advantages, several hurdles remain for the widespread industrial adoption of these technologies.

3.5.1 Hurdles to industrial adoption

a. Scalability: Translating the success of lab-scale extraction methods (especially UAE and MAE) to large, industrial-scale continuous processes remains a significant engineering challenge. Issues include ensuring uniform energy distribution in large vessels and the durability of equipment under continuous operation (Leonelli and Mason, 2010; Chemat et al., 2012).
b. Initial investment: The high capital cost of specialized equipment, particularly for SFE, PLE, and HPH, can be a barrier for many companies, especially small and medium-sized enterprises (SMEs) (Fahad et al., 2022; Wicaksono, 2025).
c. Process standardization: There is a need for robust, standardized protocols and quality control measures to ensure batch-to-batch consistency in EPS yield, molecular weight distribution, and functionality (Shen and Patel, 2008; Kogie et al., 2025). This is critical for meeting the strict specifications of commercial applications, particularly in the food and pharmaceutical sectors (Wang et al., 2015; Zhao et al., 2021; Banerjee et al., 2022).

3.5.2 Regulatory considerations

For EPS intended for food, cosmetic, or biomedical applications, stringent regulatory approval is required from bodies like the U.S. Food and Drug Administration (FDA) or the European Food Safety Authority (EFSA). Many well-established EPS, like xanthan gum, have "Generally Recognized as Safe" (GRAS) status (Stoica et al., 2022). The use of green technologies can be advantageous in the regulatory process for new EPS, as the elimination of toxic solvents simplifies the demonstration of product safety and purity, potentially accelerating the path to market (Valvanidis, 2016; Panda and Gorantla, 2025).

3.5.3 Future perspectives and research directions

The future of EPS processing is bright and points toward even greater integration, intelligence, and biological innovation.

a. Integrated biorefineries: The ultimate goal is to develop integrated biorefineries where microbial biomass is used to produce not only EPS but also other valuable co-products (e.g., proteins for food, lipids for biofuels, pigments for dyes), ensuring that no part of the feedstock is wasted and maximizing economic value from a single process (Olguin et al., 2022; Singh et al., 2025).

b. Process intensification: Research into continuous processing, process automation, and the development of more efficient hybrid technologies will be key to improving efficiency and reducing the footprint and cost of production plants (Muller et al., 2022).

c. Artificial intelligence (AI): The use of AI and machine learning to model complex extraction processes can rapidly predict optimal parameters (e.g., temperature, time, power) without extensive experimentation. AI-driven sensors and control systems can monitor processes in real-time and make adjustments to maintain optimal conditions, ensuring consistent quality and maximizing yield (Alloun and Calvio, 2024; Mehta et al., 2024).

d. Synthetic biology and metabolic engineering: Advances in genetic engineering tools like CRISPR-Cas9 will allow for the precise design of microorganisms that overproduce specific EPS with tailored structures (e.g., specific molecular weights or branching patterns) and functionalities. It is also possible to engineer strains that are easier to lyse, or that secrete 100% of their EPS, dramatically simplifying the downstream extraction process itself (Rutering et al., 2017; Wang et al., 2019; Park et al., 2024).

3.6 Conclusion

The transition from conventional, solvent-heavy processes to clean, efficient, and sustainable green technologies represents a critical and necessary evolution in the field of microbial exopolysaccharide production. Advanced methods such as UAE, EAE, membrane filtration, and ATPS are not merely incremental improvements; they are transformative approaches that significantly reduce the environmental footprint of EPS manufacturing while often enhancing product yield, purity, and functionality. By preserving the native structure and bioactivity of these versatile biopolymers, green processing unlocks their full potential for high-value applications in medicine, food, and advanced materials. While challenges in scalability and cost remain, the compelling benefits in sustainability, renewability, and long-term economic viability are powerful drivers for innovation. The continued development and integration of these green technologies, coupled with advances in biotechnology and artificial intelligence, will undoubtedly cement the role of microbial EPS as essential building blocks in a sustainable, circular, and bio-based global economy.

Credit author statement

All authors contributed to the conceptualization, literature review, and writing of this chapter. Meena Rathore led the initial draft; Pushpendra Sharma and Kirti Suman contributed to critical revisions and data visualization; and Rajeev Kaushik conceptualized, supervised and finalized the manuscript.

Conflict of interest

The authors declare that there is **no conflict of interest** related to the content, authorship, or publication of this chapter. All authors have contributed to the work in accordance with their respective roles and have approved the final version of the manuscript for publication.

Ethical statement

This chapter is a review and does not involve any studies with human participants or animals performed by any of the authors. All information, data, and literature cited have been duly acknowledged and referenced following standard academic and ethical publication practices.

Acknowledgment

The authors gratefully acknowledge Division of Microbiology, ICAR- Indian Agricultural Research Institute, New Delhi for research infrastructure and encouragement in advancing studies on sustainable bioprocessing and microbial biotechnology.

References

Abol-Fotouh, D., Hassan, M.A., Shokry, H., Roig, A., Azab, M.S., Kashyout, A.E.-H.B., 2020. Bacterial nanocellulose from agro-industrial wastes: low-cost and enhanced production by *Komagataeibacter saccharivorans* MD1. Sci. Rep. 10, 3491.

Ahmad, N.H., Mustafa, S., Man, C., Y., B., 2015. Microbial polysaccharides and their modification approaches: a review. Int. J. Food Prop. 18 (2), 332–347.

Ahuja, V., Bhatt, A.K., Banu, J.R., Kumar, V., Kumar, G., Yang, Y.H., Bhatia, S.K., 2023. Microbial exopolysaccharide composites in biomedicine and healthcare: trends and advances. Polymers 15, 1801.

Albertsson, P.A., 1995. Aqueous biphasic systems. Properties and applications in bioseparation. Springer, Boston, MA, USA, pp. 21–30.

Albertsson, P.Å., Johansson, G., Tjerneld, F., 2020. Aqueous Two-Phase Separations. Separation processes in Biotechnology, pp. 287–328.

Alloun, W., Calvio, C., 2024. Bio-driven sustainable extraction and AI-optimized recovery of functional compounds from plant waste: a comprehensive review. Fermentation 10 (3), 126.

Anastas, P.T., Warner, J.C., 1998. Green Chemistry: Theory and Practice. Oxford University Press, Oxford, UK.

Arreola-Barroso, R.A., Llopiz, A., Olvera, L., Saab-Rincón, G., 2021. Modulating glycoside hydrolase activity between hydrolysis and transfer reactions using an evolutionary approach. Molecules 26 (21), 6586.

Aziz, T., Farid, A., Haq, F., Kiran, M., Ullah, A., Zhang, K., Li, C., Ghazanfar, S., Sun, H., Ullah, R., Ali, A., 2022. A review on the modification of cellulose and its applications. Polymers 14 (15), 3206.

Banerjee, A., Mohammed Breig, S.J., Gómez, A., Sánchez-Arévalo, I., González-Faune, P., Sarkar, S., Cabrera-Barjas, G., 2022. Optimization and characterization of a novel exopolysaccharide from *Bacillus haynesii* CamB6 for food applications. Biomolecules 12 (6), 834.

Beena Unni, A., Muringayil Joseph, T., 2024. Enhancing polymer sustainability: eco-conscious strategies. Polymers 16 (13), 1769.

Boonchai, R., Kaewsuk, J., Seo, G., 2014. Effect of nutrient starvation on nutrient uptake and extracellular polymeric substance for microalgae cultivation and separation. Desalin. Water Treat 55, 360–367.

Chemat, F., Abert-Vian, M., Fabiano-Tixier, A.S., Strube, J., Uhlenbrock, L., Gunjevic, V., Cravotto, G., 2019. Green extraction of natural products. Origins, current status, and future challenges. TrAC Trends Anal. Chem. 118, 248–263.

Chemat, F., Rombaut, N., Sicaire, A.G., Meullemiestre, A., Fabiano-Tixier, A.S., Abert-Vian, M., 2017. Ultrasound assisted extraction of food and natural products. Mechanisms, techniques, combinations, protocols and applications. A review. Ultrason. Sonochem. 34, 540–560.

Chemat, F., Vian, M.A., Cravotto, G., 2012. Green extraction of natural products: concept and principles. Int. J. Mol. Sci. 13 (7), 8615–8627.

Chen, Y., Zhang, N., Chen, X., 2024. Structurally modified polysaccharides: physicochemical properties, biological activities, structure–activity relationship, and applications. J. Agric. Food Chem. 72 (7), 3259–3276.

Cheryan, M., 1998. Ultrafiltration and Microfiltration Handbook. CRC press, Boca Raton, Florida, USA.

Costa, O.Y., Raaijmakers, J.M., Kuramae, E.E., 2018. Microbial extracellular polymeric substances: ecological function and impact on soil aggregation. Front. Microbiol. 9, 1636.

Dias, A.L.B., de Aguiar, A.C., Rostagno, M.A., 2021. Extraction of natural products using supercritical fluids and pressurized liquids assisted by ultrasound: current status and trends. Ultrason. Sonochem. 74, 105584.

Donot, F., Fontana, A., Baccou, J.C., Schorr-Galindo, S., 2012. Microbial exopolysaccharides: main examples of synthesis, excretion, genetics and extraction. Carbohydr. Polym. 87 (2), 951–962.

Elisashvili, V.I., Kachlishvili, E.T., Wasser, S.P., 2009. Carbon and nitrogen source effects on basidiomycetes exopolysaccharide production. Appl. Biochem. Microbiol., 45, 531–535.

Fahad, S., Alnori, F., Su, F., Deng, J., 2022. Adoption of green innovation practices in SMEs sector: evidence from an emerging economy. Econ. res.-Ekon. istraž. 35 (1), 5486–5501.

Flemming, H.C., Wingender, J., 2010. The biofilm matrix. Nat. Rev. Microbiol. 8 (9), 623–633.

Gallego, R., Bueno, M., Herrero, M., 2019. Sub-and supercritical fluid extraction of bioactive compounds from plants, food-by-products, seaweeds and microalgae—an update. TrAC Trends Anal. Chem. 116, 198–213.

Golkar, T., Zieliński, M., Berghuis, A.M., 2018. Look and outlook on enzyme-mediated macrolide resistance. Front. Microbiol. 9, 1942.

Guerrero, B.G., Santos, K.d.L., Kamimura, E.S., de Oliveira, C.A.F., 2024. Application of microbial exopolysaccharides in packaging films for the food industry: a review. Int. J. Food Sci. Technol. 59, 17–29.

Halpern, J.M., Urbanski, R., Weinstock, A.K., Iwig, D.F., Mathers, R.T., Von Recum, H.A., 2014. A biodegradable thermoset polymer made by esterification of citric acid and glycerol. J. Biomed. Mater. Res. A 102 (5), 1467–1477.

Hara, O., Hutchinson, C.R., 1992. A macrolide 3-O-acyltransferase gene from the midecamycin-producing species *Streptomyces mycarofaciens*. J. Bacteriol. 174, 5141–5144.

Hottle, T.A., Bilec, M.M., Landis, A.E., 2017. Biopolymer production and end of life comparisons using life cycle assessment. Resour. Conserv. Recycl. 122, 295–306.

Jain, R., Raghukumar, S., Tharanathan, R., Bhosle, N.B., 2005. Extracellular polysaccharide production by Thraustochytrid protists. Mar. Biotechnol. 7, 184–192.

Kaur, N., Dey, P., 2023. Bacterial exopolysaccharides as emerging bioactive macromolecules: from fundamentals to applications. Res. Microbiol. 174 (4), 104024.

Kentish, S., Ashokkumar, M., 2011. The physical and chemical effects of ultrasound. In: Feng, H., Barbosa-Cánovas, G.V., Weiss, J. (Eds.), Ultrasound technologies for food and bioprocessing. Springer, pp. 1–12. https://doi.org/10.1007/978-1-4419-7472-3_1.

Khaw, K.Y., Parat, M.O., Shaw, P.N., Falconer, J.R., 2017. Solvent supercritical fluid technologies to extract bioactive compounds from natural sources: a review. Molecules 22 (7), 1186.

Kogje, M., Satdive, A., Mestry, S., Mhaske, S.T., 2025. Biopolymers: a comprehensive review of sustainability, environmental impact, and lifecycle analysis. Iran. Polym. J. 34, 1481–1524.

Lakmal, H.H., Lee, J.H., Jeon, Y.J., 2015. Enzyme-assisted extraction of a marine algal polysaccharide, fucoidan and bioactivities. In: Ramawat, K.G., Mérillon, J.-M. (Eds.), Polysaccharides: Bioactivity and Biotechnology. Springer, Cham, Switzerland, pp. 1065–1077.

Leonelli, C., Mason, T.J., 2010. Microwave and ultrasonic processing: now a realistic option for industry. Chem. Eng. Process 49 (9), 885–900.

Li, W., Ji, J., Rui, X., Yu, J., Tang, W., Chen, X., 2014. Production of exopolysaccharides by *Lactobacillus plantarum* R315 and its immune-enhancing activity in RAW 264.7 macrophages. Food Funct. 5 (11), 2874–2882.

Ma, C.W., Feng, M., Zhai, X., Hu, M., You, L., Luo, W., Zhao, M., 2013. Optimization for the extraction of polysaccharides from Ganoderma lucidum and their antioxidant and antiproliferative activities. J. Taiwan Inst. Chem. Eng. 44 (6), 886–894.

Marathe, S.J., Jadhav, S.B., Bankar, S.B., Singhal, R.S., 2017. Enzyme-assisted extraction of bioactives. In: Egbuna, C., Ng, K.S.H., Sawhney, S.K. (Eds.), Bioactives: Extraction and Biotechnology Applications. Springer International Publishing, Cham, Switzerland, pp. 171–201.

Martins, C.P., Cavalcanti, R.N., Couto, S.M., Moraes, J., Esmerino, E.A., Silva, M.C., Raices, R.S., Gut, J.A., Ramaswamy, H.S., Tadini, C.C., Cruz, A.G., 2019. Microwave processing: current background and effects on the physicochemical and microbiological aspects of dairy products. Compr. Rev. Food Sci. Food Saf. 18 (1), 67–83.

Mehta, A., Niaz, M., Adetoro, A., Nwagwu, U., 2024. Advancements in manufacturing technology for the biotechnology industry: the role of artificial intelligence and emerging trends. Int. J. Chem. Math. Phys. 8 (2), 12–18.

Mould, D.L., Synge, R.L.M., 1954. Separations of polysaccharides related to starch by electrokinetic ultrafiltration in collodion membranes. Biochem. J. 58, 571–585.

Müller, D., Klein, L., Lemke, J., Schulze, M., Kruse, T., Saballus, M., Matuszczyk, J., Kampmann, M., Zijlstra, G., 2022. Process intensification in the biopharma industry: Improving efficiency of protein manufacturing processes from development to production scale using synergistic approaches. Chem. Eng. Process.-Process Intensif. 171, 108727.

Murugu, J., Narayanan, R., 2024. Production, purification, and characterization of a novel exopolysaccharide from probiotic *Lactobacillus amylovorus*: MTCC 8129. Indian J. Microbiol. 64, 1355–1365.

Nadar, S.S., Rao, P., Rathod, V.K., 2018. Enzyme assisted extraction of biomolecules as an approach to novel extraction technology: a review. Food Res. Int. 108, 309–330.

Nadar, S.S., Rao, P., Rathod, V.K., 2018. Enzyme assisted extraction of biomolecules as a greener approach: a review. J. Food Process Eng. 41 (4), e12698.

Netrusov, A.I., Liyaskina, E.V., Kurgaeva, I.V., Liyaskina, A.U., Yang, G., Revin, V.V., 2023. Exopolysaccharides producing bacteria: a Review. Microorganisms 11, 1541.

Nguyen, H.T., Pham, T.T., Nguyen, P.T., Le-Buanec, H., Rabetafika, H.N., Razafindralambo, H.L., 2024. Advances in microbial exopolysaccharides: present and future applications. Biomolecules 14 (9), 1162.

Nikodinovic-Runic, J., Guzik, M., Kenny, S.T., Babu, R., Werker, A., Connor, K.E., 2013. Carbon-rich wastes as feedstocks for biodegradable polymer (polyhydroxyalkanoate) production using bacteria. Adv. Appl. Microbiol. 84, 139–200.

Nwe, N., Furuike, T., Tamura, H., 2008. An effective method for the extraction of chitosan and chitin-glucan complexes from the mycelia of *Gongronella butleri*. Carbohydr. Polym. 73 (2), 244–251.

Olguín, E.J., Sánchez-Galván, G., Arias-Olguín, I.I., Melo, F.J., González-Portela, R.E., Cruz, L., De Philippis, R., Adessi, A., 2022. Microalgae-based biorefineries: challenges and future trends to produce carbohydrate enriched biomass, high-added value products and bioactive compounds. Biology 11 (8), 1146.

Panda, S., Gorantla, S., 2025. Green analytical approaches and eco-friendly solvents: Advancing industrial applications and environmental sustainability: a comprehensive review. Orient. J. Chem. 41 (2), 616–627.

Park, H., Zhang, L., He, H., Yan, X., Scrutton, N.S., Chen, G.Q., 2024. Engineering EPS in halomonas bluephagenesis leads to self-flocculation and filamentation for convenient downstream processing. Chem. Eng. J. 495, 153374.

Pavlova, K., Grigorova, D., 1999. Production and properties of exopolysaccharide by Rhodotorula acheniorum MC. Food Res. Int., 32, 473–477.

Płotka, J., Tobiszewski, M., Sulej, A.M., Kupska, M., Górecki, T., Namieśnik, J., 2013. Green chromatography. J. Chromatogr. A 1307, 1–20.

Puri, M., Sharma, D., Barrow, C.J., 2012. Enzyme-assisted extraction of bioactives from plants. Trends Biotechnol. 30 (1), 37–44.

Quan, Y., Yang, S., Wan, J., Su, T., Zhang, J., Wang, Z., 2015. Optimization for the extraction of polysaccharides from Nostoc commune and its antioxidant and antibacterial activities. J. Taiwan Inst. Chem. Eng. 52, 14–21.

Rütering, M., Cress, B.F., Schilling, M., Rühmann, B., Koffas, M.A., Sieber, V., Schmid, J., 2017. Tailor-made exopolysaccharides—CRISPR-Cas9 mediated genome editing in Paenibacillus polymyxa. Synth. Biol. 2 (1), ysx007.

Salari, M., Sowti Khiabani, M., Ghanbarzadeh, B., Samadi Kafil, H., 2019. Preparation and characterization of cellulose nanocrystals from bacterial cellulose produced in sugar beet molasses and cheese whey media. Int. J. Biol. Macromol. 122, 280–288.

Salimi, F., Farrokh, P., 2023. Recent advances in the biological activities of microbial exopolysaccharides. World J. Microbiol. Biotechnol. 39, 213.

Shen, L., Patel, M.K., 2008. Life cycle assessment of polysaccharide materials: a review. J. Polym. Env. 16 (2), 154–167.

Singh, A., Javeria, Singh, K., Kumar, U., 2025. Microbial bio-refineries: engineering microbes for the production of versatile short-chain organic acids. Indian J. Microbiol. 65, 645–667.

Stoica, M., Stoica, D., Ivan, A.S., Dragomir, C.M.B., 2022. Biopolymers: regulatory and legislative issues. In: Nadda, A.K., Sharma, S., Bhat, R. (Eds.), Biopolymers: Recent Updates, Challenges and Opportunities. Springer International Publishing, Cham, Switzerland, pp. 55–71.

Sutherland, I.,W., 1998. Novel and established applications of microbial polysaccharides. Trends Biotechnol. 16 (1), 41–46.

Valavanidis, A., 2016. Green chemistry and new technological developments. New Ave. Green Econ. Sustain. Future Sci. Technol. 1, 1–33.

Wang, K., Li, W., Rui, X., Li, T., Chen, X., Jiang, M., Dong, M., 2015. Chemical modification, characterization and bioactivity of a released exopolysaccharide (r-EPS1) from Lactobacillus plantarum 70810. Glycoconj. J. 32 (1), 17–27.

Wang, Y., Ling, C., Chen, Y., Jiang, X., Chen, G.Q., 2019. Microbial engineering for easy downstream processing. Biotechnol. Adv. 37 (6), 107365.

Waoo, A.A., Singh, S., Pandey, A., Kant, G., Choure, K., Amesho, K.T.T., Srivastava, S., 2023. Microbial exopolysaccharides in the biomedical and pharmaceutical industries. Heliyon 9, e18613.

Wicaksono, T., 2025. Green entrepreneurship: transforming challenges into profitable solutions. J. Pengabdi. dan Kewirausahaan 9 (1).

Wingender., J., Strathmann, M., Rode, A., Leis, A., Flemming, H.-C., 2001. Isolation and biochemical characterization of extracellular polymeric substances from *Pseudomonas aeruginosa*. Methods Enzym. 336, 302–314.

Woo, I.S., Rhee, I.K., Park, H.D., 2000. Differential damage in bacterial cells by microwave radiation on the basis of cell wall structure. Appl. Env. Microbiol. 66 (5), 2243–2247.

Wu, X., Li, X., Yang, Q., Xu, Q., Tao, Z., Huang, X., Wang, D., 2020. Effect of citric acid on extracellular polymeric substances disruption and cell lysis in the waste activated sludge by pH regulation. Bioresour. Technol. 302, 122859.

Yadav, M.K., Song, J.H., Vasquez, R., Lee, J.S., Kim, I.H., Kang, D.K., 2024. Methods for detection, extraction, purification, and characterization of exopolysaccharides of lactic acid bacteria—a systematic review. Foods 13 (22), 3687.

Yaşar Yıldız, S., Radchenkova, N., 2023. Exploring extremophiles from Bulgaria: biodiversity, biopolymer synthesis, functional properties, applications. Polymers 16, 69.

Yildiz, H., Karatas, N., 2018. Microbial exopolysaccharides: resources and bioactive properties. Process Biochem. 72, 41–46.

Zhao, D., Jiang, J., Liu, L., Wang, S., Ping, W., Ge, J., 2021. Characterization of exopolysaccharides produced by Weissella confusa XG-3 and their potential biotechnological applications. Int. J. Biol. Macromol. 178, 306–315.

CHAPTER

4

Green extraction and production of high-yielded exopolysaccharides: Optimization of regulatory process parameters for industrial scale-up

Jothieswari Dhamotharan[a], *Radha Rayi*[b], *Shaik Naveen Taj*[c], *Mopuri Deepa Reddy*[d] *and Sibbala Subramanyam*[e]

[a]Department of Pharmaceutical Analysis, Sri Venkateswara College of Pharmacy (Autonomous), Chittoor, Andhra Pradesh, India [b]Department of Pharmaceutical Chemistry, Mohan Babu University, Tirupati, Andhra Pradesh, India [c]Department of Pharmaceutics, Sri Padmavathi Mahila Visvavidyalayam, Tirupati, Andhra Pradesh, India [d]Department of Pharmaceutical Chemistry, Annamacharya College of Pharmacy (Autonomous), Kadapa, Andhra Pradesh, India [e]Department of Pharmaceutical Chemistry, Vignan Foundation for Science Technology and Research, Guntur, Andhra Pradesh, India

4.1 Introduction

Exopolysaccharides (EPS) are biopolymers produced by microorganisms that are one of the most valuable of them for multiple sectors such as food, pharmacy, cosmetics, and biotechnology; their demand is enlarged by the increasing search for sustainability and environmentally friendly extraction ways that generated the necessity for the development of green technologies. This chapter focuses on green extraction and high-yield EPS production regarding the optimization of regulating process parameters for large-scale production.

DOI: https://doi.org/10.1016/B978-0-443-36619-2.00004-8

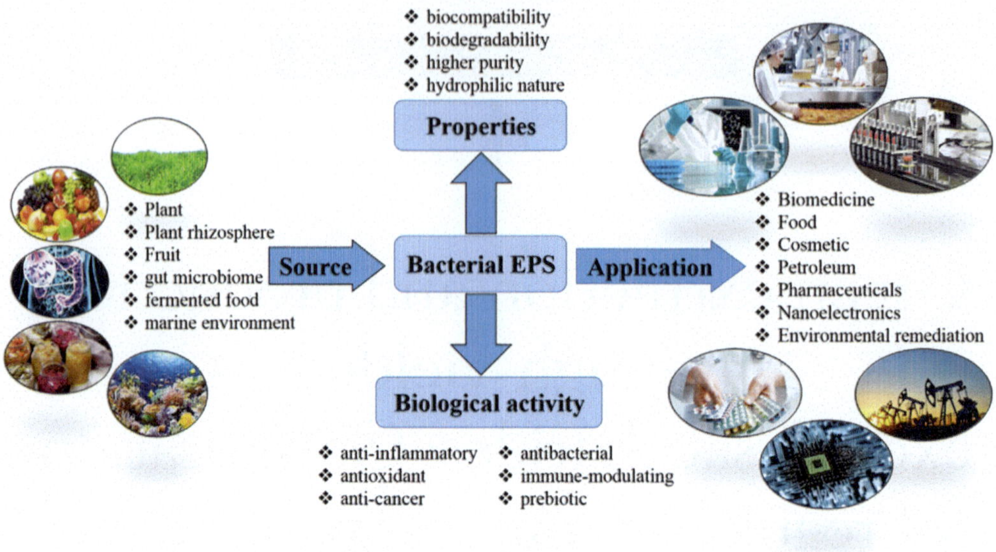

FIGURE 4.1 Sources, properties, biological activities, and application of bacterial EPS. *EPS*, Exopolysaccharides.

In general, polysaccharides are those long-chain polymers acquired by microorganisms, plants, and mammals. They display several biological roles and are expelled outside the cell wall, in which they are known as EPS, or EPSs (Tang et al., 2017; Yue et al., 2023). They are generally considered as high-molecular-weight bio-polymers secreted by numerous microbes, fungi, and bacterial cells into the surrounding environment in the course of their metabolism (Sutherland et al., 1972; Asianezhad et al., 2023). The first use of the term "exopolysaccharide" was used in 1972 by Sutherland to encompass the immense variety of polysaccharides associated with microbes beyond the confines of cell walls and, based on interactions with the cell's structure and function, either loosely or firmly bound within the surrounding environment (Tiwari et al., 2020).

4.2 Exopolysaccharides: An overview

4.2.1 Definition and properties

Extracellular polysaccharides, as shown in Fig. 4.1, are biopolymers that are secreted by microorganisms, specifically bacteria, fungi, and algae. These biopolymers vary widely in structural and functional properties, for example, being viscous, gelatinous, and biologically active.

4.2.2 Industrial applications

EPS finds applications in multiple sectors:

- Food industry: As thickeners, emulsifiers, and stabilizers.

- Pharmaceuticals: For drug delivery and wound healing.
- Biotechnology: As biofilm-forming agents and bioremediation aids.

4.3 Green extraction methods for exopolysaccharides

Some of the best performing green extraction techniques that provide maximum EPS are: As its name implies, green extraction is following all 12 principles of green chemistry—it is reducing the environment by maximizing the efficiency (Anastas and Warner, 1998; Singh and Patel, 2020; Roy, 2021). Recent developments that are taking place in green extraction have also attracted interest among researchers in the food industry, who are envisioning improving sustainability and reducing environmental impact. The extraction techniques include high-voltage electrical discharge (HVED), pressurized liquid extraction (PLE), pulse electric field (PEF), enzyme-assisted extraction (EAE), microwave-assisted extraction (MAE), ultrasound-assisted extraction (UAE), and supercritical fluid extraction (SFE). Green extraction is characterized by its increased selectivity in extracting important bioactive components, reduced energy consumption, and decreased dependence on the use of harmful solvents. Extracting chemicals that the regulatory authorities have deemed "Generally Recognized as Safe" (GRAS) substances—such as essential oils, polyphenols, vitamins, and functional ingredients—is one of the areas of focus. These GRAS compounds are well-known and used in food industry applications, like flavorings, antioxidants, and natural preservatives, because they contain potential beneficial values.

4.3.1 Enzymatic extraction

Without requiring costly chemical treatment, EPS is released using enzymatic extraction, which uses specific enzymes to break down cell walls (Whitfield et al., 2015). The integrity and usefulness of polysaccharides are preserved by this incredibly effective and eco-friendly extraction method.

For example, streptomyces griseus-derived chitin from shrimp shells demonstrated significant solubility and a high deproteination value.

General procedure for enzymatic extraction of EPS

Enzymatic extraction is a quick, environmentally benign, and gentle process that uses enzymes to liberate and extract EPS from microbes or plant materials. Following is a step-by-step procedure in detail:

4.3.1.1 Microbial cultivation or sample preparation

1. Selection of source
   a. Microbial sources: Cultivate microorganisms known for EPS production (e.g., *Lactobacillus, Pseudomonas, Xanthomonas*).
   b. Plant materials: Preparation of plant samples (e.g., seaweed or other polysaccharide-rich sources) by washing and chopping.
2. Culture media preparation (for microbial sources)
   a. Preparation of nutrient-rich media to promote EPS secretion.
   b. Optimization of parameters like pH, temperature, and carbon source for high yield.

3. Fermentation
   a. Fermentation of the culture under optimal conditions to achieve maximum EPS production.
   b. Harvesting the culture at the peak EPS secretion phase.

4.3.1.2 *Enzymatic treatment*

1. Preparation of crude extract
   a. Separation of the biomass or plant material from the culture medium or harvested material by centrifugation or filtration.
2. Selection of enzymes
   a. Choose appropriate enzymes to release EPS from the source:
      i. Cellulases: Break down cell walls to release intracellular EPS.
      ii. Amylases: Degrade starch impurities.
      iii. Proteases: Remove protein contaminants.
3. Enzymatic reaction setup
   a. Dissolve the separated material in a suitable buffer (e.g., phosphate buffer, pH 6.0–8.0).
   b. Add enzymes at optimized concentrations (e.g., 1%–5% w/v).
   c. Incubate the mixture under specific conditions:
      i. Temperature: 30°C–50°C (optimal for enzyme activity).
      ii. Time: 2–6 hours, depending on the material and enzyme.
4. Control of conditions
   a. Maintain agitation (e.g., 100–150 rpm) for uniform enzyme-substrate interaction.
   b. Monitor pH and temperature throughout the process.

4.3.1.3 *Separation of enzyme-treated exopolysaccharides*

1. Filtration or centrifugation
   a. Filter or centrifuge (10,000 × g for 20 minutes) the enzymatically treated mixture to remove cell debris or undigested materials.
   b. Collect the supernatant containing the released EPS.
2. Inactivation of enzymes
   a. Heat the supernatant (e.g., 80°C–90°C for 5–10 minutes) to deactivate residual enzymes.

4.3.1.4 *Purification of exopolysaccharides*

1. Precipitation of EPS
   a. Precipitate EPS using cold ethanol (2–3 volumes of ethanol per 1 volume of supernatant).
   b. Incubate the mixture at 4°C for 12–24 hours for complete precipitation.
2. Recovery of precipitate
   a. Centrifuge the precipitated EPS and wash with cold ethanol or acetone to remove residual impurities.
3. Dialysis
   a. To eliminate contaminants with low molecular weight, dissolve the EPS pellet in distilled water and dilute it with deionized water.

4. Activated carbon treatment (optional)
   a. Add activated carbon to remove pigments or other colored impurities, followed by filtration.

4.3.1.5 *Final drying and storage*

1. Lyophilization (freeze-drying)
   a. Freeze the purified EPS solution and lyophilize it to obtain a dry powder.
2. Vacuum drying
   a. Alternatively, use vacuum drying at low temperatures to avoid thermal degradation.
3. Storage
   a. Store the dried EPS powder in an airtight container at a cool, dry place to prevent degradation.

4.3.2 Microwave-assisted extraction and production of high-yielding exopolysaccharides

By using microwave energy to speed up extraction procedures, MAE produces EPS much faster than with conventional techniques (Deamici et al., 2021). Good yields of EPS from various biomass sources have been reported when microwave conditions are used with ethanol as a solvent. Utilizing microwave energy in conjunction with hydrothermal treatment improved the extraction efficiency of EPS. Studies have shown that optimizing parameters like pressure and extraction time led to significantly higher yields.

The most widely used technique is MAE, which uses microwave energy as a medium to accelerate extraction operations. This method can yield EPS in a lesser time when compared to conventional methods by heating and agitation. Reports on the application of microwave heating for ethanol as a solvent gave promising results for EPS extraction from various biomass sources. A combination of microwave energy with hydrothermal handling has proved to be a good technique for extraction improvement. Studies have also revealed how optimization of parameters such as pressure and period of extraction could increase the yield.

4.3.2.1 *Principle of microwave-assisted extraction*

- Microwave energy: Uses electromagnetic waves (0.3–300 GHz) to generate localized heating, disrupting cell walls and enhancing the solubility and diffusion of EPS.
- Green technology: Reduces solvent use, energy consumption, and extraction time, making it an eco-friendly option.

4.3.2.2 *General procedure for microwave-assisted extraction*

Preparation of source material

- Microbial cultures or plant materials
  - Use microbial cultures (e.g., *Lactobacillus, Pseudomonas, Xanthomonas*) or polysaccharide-rich plant materials (e.g., seaweed).
  - Harvest the biomass after cultivation or collect plant materials.

- Pretreatment
  - Wash and dry the biomass or plant material to remove impurities.
  - Grind the material into a fine powder (for plant sources).
  - Suspend the material in a suitable solvent (e.g., water or aqueous buffer).

Microwave treatment

1. Microwave apparatus setup
   a. Use a microwave reactor designed for laboratory-scale or industrial-scale extractions.
   b. Select appropriate vessel size and material to withstand microwave heating.
2. Optimization of microwave parameters
   a. Power level: 300–800 W for laboratory-scale systems.
   b. Time: 5–20 minutes, depending on the material.
   c. Temperature: 50°C–120°C, optimized for EPS stability.
   d. Solvent ratio: Use 10–20 mL of solvent per gram of material.
3. Microwave extraction process
   a. Add the prepared material and solvent to the microwave reactor.
   b. Set the desired parameters and initiate microwave treatment.
   c. Stirring or intermittent agitation can improve extraction efficiency.
4. Separation of extract
5. Filtration or centrifugation
   a. After microwave treatment, filter or centrifuge the mixture to separate solid residues.
   b. Collect the supernatant containing dissolved EPS.

4.3.2.3 *Precipitation and purification of exopolysaccharides*

- Ethanol precipitation
  - Add cold ethanol (2–3 times the volume of supernatant) to the extract.
  - Stand the mixture at 4°C for 12–24 hours to precipitate.
- Recovery of EPS
  - Centrifuge the precipitated EPS at 10,000 × g for 20 minutes.
  - Wash the pellet with ethanol or acetone to remove impurities.
- Dialysis
  - To get rid of low-molecular-weight contaminants, dissolve the EPS pellet in distilled water and dialyze it against deionized water.
- Lyophilization
  - Freeze purified EPS solution and lyophilize it to obtain a dry powder.

4.3.2.4 *Advantages of microwave-assisted extraction in exopolysaccharides production*

- Efficiency: Reduces extraction time significantly compared to conventional methods.
- Eco-friendly: Requires less solvent and energy.
- Preservation of bioactivity: Operates under controlled conditions to maintain the functional properties of EPS.
- Scalability: Suitable for industrial-scale production.

4.3.2.5 *Applications of microwave-assisted extraction-derived exopolysaccharides*

Pharmaceuticals: Bioactive agents for drug delivery, immunomodulation, and wound healing.

- Food industry: Thickeners, emulsifiers, and prebiotics.
- Cosmetics: Hydrating agents and antiaging products.
- Agriculture: Biostimulants and soil conditioners.

4.3.3 *Ultrasound-assisted extraction and production of high-yielded exopolysaccharides*

The application of the UAE uses ultrasonic waves for making cavitation bubbles in the solvent so that it can destruct the cell structures and displace EPS out from the cell (Ying et al., 2011). The method is generally well known for efficient extraction with short time consumption, meeting requirements for large-scale application: The UAE has been confirmed to be capable of extracting EPS effectively from microalgae and other sources.

- Employ ultrasonic waves to disintegrate cell walls and release EPS.
- Benefits: Reduced solvent consumption and short extraction time.
- UAE is another nature-friendly approach with less time and better yield over the conventional methods. The technique has been successfully utilized for the extraction of polysaccharides from brown seaweeds.

4.3.3.1 *Principle of ultrasound-assisted extraction*

- Ultrasonic waves: In a very similar manner, sonic waves (20–100 kHz) produce cavitation bubbles in a liquid medium that mechanically disrupt cell walls and improve mass transfer.
- Green technology: Minimizing the consumption of energy and chemicals in pursuit of sustainability.

4.3.3.2 *Preparation of source material*

The following organismic cultures or plant materials can be used to obtain exopolymer substances: EPS-producing microorganisms (for example, *Lactobacillus*, *Xanthomonas*, and *Pseudomonas*) or polysaccharide-rich plants (e.g., seaweed and various medicinal plants). The biomass of fermented microorganisms should be collected. Clean, dry, and grind the plant source into fine powder for application in further tests.

Suspend in a solvent: Hold the materials in a suitable solvent, such as distilled water or an aqueous buffer, in the most optimized ratios, for example, 1:10 w/v.

4.3.3.3 *Ultrasound treatment*

- Ultrasound set-up
  - Using a sonicator, either with a probe or using a bath system, select a suitable frequency (20–40 kHz for nearly all EPS applications).
- Setting optimization parameters
  - Power density: 100–500 W, depending on the equipment.
  - Temperature: 20°C–50°C, controlled so as not to produce thermal degradation from the EPS.

  - Time: Depending on the type of material, from 10 to 30 minutes.
  - Solvent ratio: Maintain enough solvent volume to promote cavitation.
- The extraction process
  - Place the sample suspension within the ultrasonic bath or treat by probe.
  - Control and monitor the temperature to prevent overheating.
  - Stirring during sonication will enhance extraction efficiency.
- Separating and filtering
- Centrifugation or filtration
  - Filter or centrifuge the sonicated mixture (10,000 × g for 15–20 minutes) to separate the solid residue.
  - Collect the supernatant with the extracted EPS.
- Clarification (optional)
  - Activated charcoal or a very fine filter can be used to remove pigments or other impurities.

4.3.3.4 *Purification and concentration of exopolysaccharides*

- Ethanol precipitation:
  - Add 2–3 volumes of cold ethanol to the supernatant containing EPS.
  - Stand the sample for precipitation at 4°C for 12–24 hours.
- Pellet recovery:
  - Centrifuge the precipitated EPS at 10,000 × g for 15 minutes and wash the pellet with ethanol or acetone.
- Dialysis:
  - Dialyze the EPS against deionized water after dissolving it in distilled water to eliminate low-molecular-weight contaminants.
- Lyophilization:
  - Freeze-purified EPS solution and lyophilize to obtain a dry powder.

4.3.4 Supercritical fluid extraction for extraction of exopolysaccharides at high yields

Bioactive substances, such as EPS, can be extracted using SFE, an environmentally safe and sophisticated technology, from microbial cultures, plants, or other natural sources, using supercritical fluids, in most cases CO_2. The detailed procedure for SFE application in the production of high-yield EPS is outlined (Sagar et al., 2018; Herzyk et al., 2024).

Supercritical Fluid Extraction

This advanced extraction method, especially based on supercritical carbon dioxide (SC-CO_2) as solvent, is capable of directly extracting certain nonpolar as well as some polar compounds from the materials. When applying cosolvents, such extraction also increases the solubility of EPS and, thereby, increases its yield. The extraction is also scalable and can be applied in the food and pharmaceutical industries (Tang et al., 2017).

- Suitable for extraction under supercritical conditions using CO_2 as solvent.
- Advantages: Nontoxic, recyclable, and highly selective.

4.3.4.1 *Principles of supercritical fluid extraction*

- Supercritical condition: A fluid is in a supercritical condition when it exceeds both its critical temperature and pressure, and is characterized by superior solvent power, high diffusivity, and low viscosity, among other special qualities.
- Common fluid: The supercritical fluid most widely used is carbon dioxide (CO_2), because of its nontoxic nature, inertness, and mild critical conditions (e.g., 31.1°C and 73.8 bar).
- Modifiers: Co-solvents such as ethanol or water can be used to improve the solubility of polar types of EPS.

4.3.4.2 *Generalized procedure of supercritical fluid extraction*

Preparation of source material

- Microbial cultures or plant materials
  - Microbial sources: Harvest the biomass after fermentation from EPS-producing microorganisms- *Lactobacillus*, *Pseudomonas*, or *Xanthomonas*.
  - Plant sources: Clean and dry the polysaccharide-rich plant materials, like seaweed or fungi.
- Pretreatment
  - For microbial biomass, it is cell separation from the culture broth through centrifugation.
  - For plants, grinding into fine powders increase surface area for extraction.
- SFE apparatus set-up
  - Adoption of a SFE apparatus for operation with a CO_2 pump, an extraction vessel, and a collection chamber is also among the activities.
  - The system shall be clean and calibrated to ensure precise pressure and temperature controls.
  - Above set conditions, use for loading material.
  - Pack the biomass or plant powder into the extraction chamber for supercritical fluid flow to simply through it, but lightly compact it.
- Operating conditions
  - Pressure: 100–300 bar (optimized for materials from source and EPS solubility).
  - Temperature: 35°C–50°C will ensure keeping the supercritical condition and integrity of EPS.
  - Flow rate: steady flow of CO_2 will be maintained (e.g., 10–20 mL/min for laboratory-scale systems).
- Use of modifiers:
  - Add polar co-solvents like ethanol or water (5%–20%), which enhance the removal of EPS polar substances from the matrix.
- Extraction process
  - Run the system for about 30–90 minute depending on the material and efficiency of extraction.

Recovery of EPS

- Decompression:
  - Slowly decrease the pressure so as not to thermally degrade EPS and completely recover it.

- Collection
  - The EPS extract can be collected in the collecting chamber or dissolved for further purification in water or buffer.

4.3.4.3 *Purification of exopolysaccharides*

1. Ethanol precipitation
   Ethanol decreases the solubility of polysaccharides, leading to their precipitation from solution. It is a common method of EPS extraction because it is efficient and easy.
2. Centrifugation and washing
   Centrifugation separates the EPS from the supernatant, while ethanol washing eliminates remaining impurities, increasing the purity of the EPS and PMC.
3. Dialysis
   Dialysis eliminates low-molecular-weight contaminants, including salts and small molecules, to yield a purer sample of EPS.
4. Lyophilization
   Lyophilization eliminates water at low pressure without distorting the structural integrity and bioactivity of the EPS. The dry powder obtained is amenable to storage and subsequent analysis.

4.3.4.4 *Pros of using supercritical fluid extraction in the production of exopolysaccharides*

- Eco-friendliness: CO_2 serves as a solvent that is completely nontoxic, recyclable, and environmentally safe indeed.
- Selective extraction: Adjusting the temperature and pressure leads to specificity for the target compound.
- Bioactivity retention: Extraction occurs under mild conditions, so the structural integrity of EPS is preserved.
- Growth potential: Easily adapted for industrial-scale production.

4.3.5 Pressurized liquid extraction for highly-yielding exopolysaccharides production

PLE or accelerated solvent extraction (ASE) is an efficient method for the extraction of bioactive compounds such as EPS from microbial cultures, plants, or any other natural source (Chipaca-Domingos et al., 2023). This method operates under high pressure and moderate temperature, which favors the solubility and mass transfer of the target compound. Wet extraction for production: Which involves the inclusion of bioactive compound obtain yield extrapolation extraction methods PLE.

4.3.5.1 *The principle of pressurized liquid extraction*

- Pressure and temperature: PLE involves the application of high pressure (10–20 MPa) at moderate temperatures (50°C–200°C) to keep the solvent above its boiling point in a liquid state.

- Improved solubility: Under such conditions, the matrix permeability to the solvent improves with molecule dissolution occurring by increasing the polarizing effect, even though reducing viscosity, affecting the efficiency of extraction.
- Environmentally-friendly solvent: Uses mostly water, ethanol, or aqueous mixtures.

4.3.5.2 *General procedure for pressurized liquid extraction*

Preparation of source material

- Microbial cultures or plant materials
  - Use EPS-producing microorganisms (for example, *Lactobacillus, Pseudomonas,* or *Xanthomonas*) or polysaccharide-rich plant materials like seaweeds and fungi.
  - Harvest the microbial biomass after fermentation or prepare the plant material by drying and grinding it before disintegration into a fine powder.
- The preparation of the extraction cell
  - Weigh prepared biomass or powder into the PLE extraction cell and load it.
  - Place inert fillers (e.g., glass beads) if necessary to ensure that solvent flows uniformly.

PLE process

- System setup
  - Use a PLE apparatus that has a solvent reservoir, pump, extraction cell, and collection system. Ensure the cleanliness of all components and calibration so that they can operate precisely.
- Solvent selection and preparation
  - Choose a suitable solvent (water, ethanol, or an aqueous mixture). Preheat the solvent to the desired extraction temperature.
- Operating conditions
  - Pressure: 10–20 MPa. THE PLE process is optimized according to the solvent and material. Temperature: 50°C–120°C. EPS should be safe, but it has to dissolve easily.
  - Solvent flow rate: Must be a continuous flow (for example, 1–5 mL/minute in lab-scale systems).
  - Extraction time: 15–60 minutes, adjusted according to material type and EPS yield.
- Extraction cycle
  - Run several extraction cycles if necessary to maximize EPS recovery.
  - Collect the extract in the collection vial or chamber.

Processing after extraction

- Filtration and separation
  - Filter or centrifuge the extracted solution to remove solid residues.
- Concentration (optional)
  - Concentrate the filtrate using a rotary evaporator if necessary.

4.3.5.3 *Purification of exopolysaccharides*

- Ethanol precipitation
  - By adding cold ethanol (preferably about 2–3 times the volume of extract) to the extract, the EPS is precipitated.
  - Let it be incubated for 12–24 hours at 4°C for further precipitation.

- Pellet recovery
  - Collect the EPS pellet by centrifuging at 10,000 × g for 15 minutes.
  - Wash the pellet with ethanol or acetone for the removal of impurities.
- Dialysis
  - Proceed by dissolving the pellet in distilled water, followed by dialysis against deionized water for the removal of low-molecular-weight impurities.
- Lyophilization
  - Subject the purified solution to freezing, followed by lyophilization, to yield dry EPS powder.

4.3.5.4 *Advantages of pressurized liquid extraction for exopolysaccharides production*

- Efficiency:
  - Under high-pressure conditions, solvents gain better access, and the EPS is released more efficiently.
  - Extraction time is shorter than that of the traditional methods of extraction.
- Eco-friendly:
  - Uses solvents like ethanol and water that are safe for the environment.
  - Produces low waste and runs in controlled conditions.
- Scalability:
- Factory customizable for mass production.
- Preservation of bioactivity:
  - Operates the EPS function while maintaining a controlled temperature.

4.4 Novel green extraction techniques

Constantly reinvented, the progressive landscape of extraction methodologies is under continuous development of new methods for resolving the problems of traditional methods (Alara et al., 2021; Peng et al., 2021; Pinto et al., 2021). This evolutionary effort is situated in the enhancement of efficiency, strength, and the selectivity underlying drawing-out of precious items from diversified sources.

4.4.1 Supercritical water extraction

SWE, its complement, subcritical water extraction (SCWE) are exemplary manifestations of the shown trend (Lampakis et al., 2021). Both types of waters, above or below the critical point, can be used as the condition for the simultaneous energy and solvent savings, and for the extraction from compounds liable for thermal degradation

The instant controlled pressure drop method, characterized by the fast pressure shock, induces cracking of the plant cell walls and thus releases intracellular components. In fact, this yields elimination of thermal degradation and solvent waste while it also generates rapid and efficient extraction. Especially important with regard to heat-sensitive bioactive materials, DIC is going so far as to revolutionize processes.

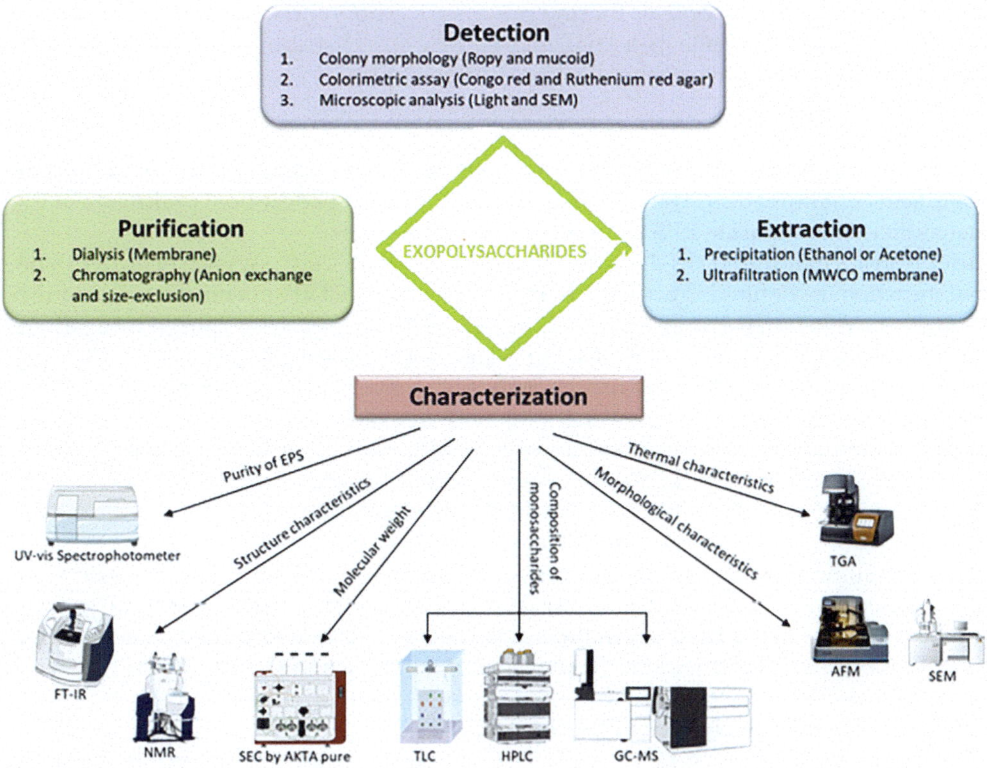

FIGURE 4.2 Characterization, extraction, purification, and detection of exopolysaccharides.

4.4.2 Pressurized hot water extraction

As an environmentally friendly substitute for conventional solvent-based extraction methods, pressurized hot water extraction (PHWE) combines the extraction of bioactive chemicals with pressurized hot water.

4.5 Methods for characterization of exopolysaccharides

There is currently a lack of clarity regarding the relationship between the structure and functions of EPS, as shown in Fig. 4.2. Thus, before examining functional characteristics, the chemical and structural characterization of LAB-EPS, that is, its functional groups, monosaccharide composition, types of sugar subunits, ring conformation, morphology, molecular weight, degree of widely diverse branching, and types of glycosidic linkages, becomes a primary goal (Jurášková et al., 2022; Kavitake et al., 2023; Zhang et al., 2023).

Many methodologies are employed for the analysis and identification of EPS molecular weight, structure, and composition, but no one method alone can measure all of these factors. Therefore, methods must usually be combined. The different categories of spectroscopic, chromatographic, microscopical, thermogravimetric, differential scanning calorimetric, polarimetric measurement, zeta potential, and particle size distributions were employed in the characterization of the physicochemical properties of EPSs.

- UV-V is spectroscopy:
  Define the physical features of EPSs, and confirm that the sample is devoid of protein and nucleic acid contamination. The UV-Vis approach can detect additional pollutants in EPS, but it is insufficient to separate EPS from other polysaccharides.
- Fourier transform infrared (FT-IR) spectroscopy (Van et al., 2008; Ayyash et al., 2020) because EPSs' function is defined by their chemical structure, FT-IR spectroscopy is utilized to distribute the functional groups in pure EPS. EPS FT-IR spectra are acquired using attenuated total reflectance (ATR) in the 4000–400 cm^{-1} area, with a resolution of 4 cm^{-1}. The hydroxyl groups contained in the EPS structure contribute to scavenging activity due to their ability to donate electrons or hydrogen when engaging with free radicals. The glycosidic linkage has an almost direct impact on the bioactivities of EPS, although other physicochemical properties play a limited contribution. EPS's scavenging activity is enhanced by α-(1 → 6) glycosidic linkages, which provide flexible areas for increased accessibility.
- X-ray diffraction (XRD) spectroscopy (Wei et al., 2023).
  The crystalline and amorphous qualities of EPS are assessed using an X-ray diffractometer, XRD spectroscopy, which provides information on the physical properties of EPSs associated with food systems. XRD from crystalline materials is based on two fundamental principles: X-ray photon wave properties and the presence of atom planes in crystals. XRD is a technique that employs X-rays to determine the geometry or shape of a substance.
- Analysis of molecular weight of EPS (Ali et al., 2023)
  The molecular weights of EPS have a large influence on their functions; lower molecular weights correlate with the bioactivity (in contrast to antioxidant activity) of compounds; higher molecular weights result in increased viscosities in aqueous solutions, demonstrating their strong anticancer properties. As a result, the lower the molecular weight of an EPS molecule, the better its antioxidant activity, because the smaller EPS complex might infiltrate the cell and protect it from free radicals.
- Analysis of monosaccharide composition of EPSs
  - Either thin layer chromatography (TLC) or high-performance thin layer chromatography (HPTLC)
  - High-pressure liquid chromatography
  - High-performance anion-exchange chromatography with pulsed amperometric detection
  - Gas chromatography
  - Gas chromatography with mass spectrometry
  - Polarimetry

Because of the reasons of various factors such as source material, purity desired, and scalability requirements, the types of high-exopolysaccharide extraction depend on these factors. There are also effective alternative green methods for increased yield with lesser environmental impact by using techniques like enzymatic extraction, MAE, UAE, SFE, PLE, ethanol precipitation, and dialysis. Continued research in optimizing these methods, however, will enhance their applicability on a large scale in the industrial circuit.

4.6 Optimization strategies

4.6.1 Process analytical technology

Green extraction and production of high-yield EPS examines the employing a new process for modern performance analysis to isolate and produce EPS in a more sustainable and efficient way, analyzing these applications of advanced analytical tools. A framework introduced by the FDA to ensure product quality through real-time monitoring. The systems fitted with the above technology shall guarantee the product acceptance by process monitoring in identified parameters like DO, pH, substrate, and EPS concentration. Furthermore, uses feedback loops for automated control. Integration with process analytical technology (PAT) and green extraction actually keeps real-time monitoring working under the condition that optimal interviewing is regarding the end output. PAT tools can guide adjustments to improve yield, purity, and reproducibility while ensuring less waste and adherence to sustainability principles.

4.6.2 Mathematical modeling and simulation

Mathematical modeling in EPS production

It plays a pivotal role in mathematical modeling in optimum extraction as well as green production of EPS. It refines the extraction and fermentation processes in such a way as to maximize yield and minimize waste. Predictive models simulate the behavior of systems under various conditions; hence, it can largely reduce the time and costs that would be incurred in experimentation. These models also favor the seamless transition from lab-scale to industrial production without compromising the efficiency and sustainability of manufacturing.

Components of mathematical modeling in green extraction

- Reaction kinetics: Models of actions that are either known to be enzymatic or due to microbial intervention, that ultimately lead to the formation of EPS. Major parameters account for the substrate consumption rate, the rate of establishment of a particular product, and the yield coefficients.
- Mass transfer: Movement of solute during extraction processes such as ultrasound or MAE through models evaluating the effect of that parameter on extraction efficiency, for example, particle size, type of solvent, and temperature.
- Heat transfer: Models energy transfer across a system using unconventional methods such as SFE and provides the accuracy desired in the conduct and yield of experiments.

Simulation techniques

- Computational fluid dynamics (CFD): It simulates the flow patterns of the bioreactor or extraction system in order to optimize the operative conditions.
- Kinetic and thermodynamic modeling: Modeling equilibrium conditions for solvation will improve the process efficiency and solvent selection.
- Artificial neural networks (ANN): Predict complicated nonlinear relationships of the response variables, which include temperature, pH, and yield of extraction.

Green extraction processes into modeling

- MAE: Mathematical models will predict the energy efficiencies and extraction rates while optimizing the variable.

4.6.3 Bioreactor design

Advanced bioreactor systems are crucial for ensuring optimal green extraction and maximally productive yield of EPS. These largely include stirred-tank, air-lift, membrane, packed-bed, and photo-bioreactors; such systems bring in the State-of-the-Art designs and sustainable technology with highly accurate control. Each bioreactor type provides its advantages: Energy-efficient agitators, reduced infection risks, high cell density, or uses renewable resources such as solar energy and CO_2 sequestration.

In addition, recyclable membranes and substrate wastage have generated environmentally friendly care, contributing to sustainability in operations and maintaining high levels of productivity from different scales and across various microbes. Design robust cell retention systems in chemostats (e.g., membrane filtration) to support the maximum retention time required for cell division within the bioreactor and to prevent cell losses.

4.7 Recent advances in optimization strategies

1. The selection of bacterial strains will improve productivity through genetic engineering and the identification of bacterial high-yield strains. For example, co-cultivation methods have come forward as promising approaches to further improvements in yield.
2. Waste utilization: Agricultural and industrial wastes waste as substrates for EPS production reducing the cost and also promoting sustainability in the bioeconomy.
3. Technological innovations: Improvements in bioreactor design and fermentation technology lead to more efficient large-scale production systems, with better control of environmental conditions affecting EPS biosynthesis.

4.8 Optimization of process parameters

Optimization of process parameters for EPS production

This roadmap will now provide clear paths for improving the regulatory processes so that they support the sustainable and highly productive industrial manufacture of EPS, with emphasis on green extraction and quality-oriented process development.

4.8.1 Key factors for exopolysaccharides production

Environmental condition (Mahapatra et al., 2013; Netrusov et al., 2023; Revin et al., 2023):

Critical variables: Temperature, pH, O_2 levels, and agitation rate optimization for optimum yield production.

Optimization techniques: Systematic optimization could be carried out using techniques such as response surface methodology (RSM) and Plackett-Burman designs.

4.8.1.1 Source material selection

Renewable sources: Sources should be from microbial, algal, or plant-based raw materials.

Strain optimization: Select or design high-output producing strains that are robust and high in productivity.

4.8.1.2 Green extraction principles

Solvent selection: Use green solvents (i.e., for example, water, ethanol) or solvent-free techniques such as the supercritical CO_2 method for extraction.

Energy efficiency: Energy-saving techniques from ultrasonication or MAE.

Waste minimization: Develop systems for the recycling of the solvents and utilization of by-products.

4.8.1.3 Critical process parameters

Fermentation conditions, temperature, pH, aeration, as well as agitation speeds. This includes an optimum ratio between carbon and nitrogen sources to obtain maximum yield.

Extraction variables: Time, temperature, and solvent concentration as fine-tuning parameters. Recovery methods like alcohol or salt precipitation led to a good extraction of EPS.

Purification parameters: Optimization of filtration, centrifugation, and dialysis processes for the production of highly pure EPS.

4.8.1.4 Process scale up

Design of bioreactor: A stirred-tank or air-lift bioreactor is chosen, which has the advantage of perfect mixing at uniform control of temperature. Mass transfer optimization: In process design, costs and specific capital investments are defined carefully. The anaerobic digester processes typically transfer organic strains into methane using the anaerobic fermentation process under optimized and defined conditions.

4.8.1.5 Quality by design approach

Critical quality attributes: Such attributes include molecular weight, composition, purity, viscosity, and bioactivity.

Risk assessment: CPPs affecting CQAs shall be identified through any of the following: Failure mode and effects analysis (FMEA).

4.8.1.6 Regulations

Guidelines: Adhere to both the ICH Q7, Q8, and Q9 guidelines as well as ensure compliance with FDA, EMA, and local regulations.

Documentation: Prepare dossiers (e.g., eCTD) with proper records on source materials, processes, and product data.

GRAS status: Establish EPS sources as GRAS for food or nutraceutical applications.

4.8.1.7 *Analytical method development*

EPS characterization: Using GPC or HPLC for molecular weight determination, FTIR, NMR, or mass spectroscopy for structural analysis.

Process monitoring: Use real-time tools like Raman spectroscopy or online HPLC to monitor fermentation and extraction processes.

4.8.1.8 *Sustainability and cost optimization*

Energy use: Minimize energy by fermentation and extraction protocols (Anastas and Warner, 1998; Singh and Patel, 2020; Roy, 2021; Okoro et al., 2021).

Waste management: Recycle process water and upcycle by-products.

Life cycle assessment: Environmental impact assessment to ensure green compliance.

4.8.1.9 *Training and documentation*

Employee development: Training of employees on environmental stewardship, process control, and regulatory compliance.

Regulatory backing: Complete audit trails and product registration.

4.8.1.10 *Tools for optimization*

Simulation software: Aspen Plus, MATLAB, or SIMCA for process simulations.

Data analysis: Utilization of Minitab or JMP for DOE and CPP optimization.

Energy monitoring: Use energy monitoring tools to keep a check on the usage during scale-up.

4.9 Conclusion

Green extraction methods for EPS offer environment-friendly efficient extraction techniques to recover bioactive polysaccharides from microbial and plant-based cultures. Such Enzymatic extraction has its great merits of integrity and activity of the extracted EPS. It has also excluded the defiance of strong chemicals to secure the integrity of bioactivity. MAE is fast because microwave energy disrupts the cell walls and promotes diffusion, facilitating greater scalability and shorter processing time. UAE is efficacious yet sustainable when combined with the use of ultrasonic waves. This method is very suitable for large-scale applications.

These advanced techniques all point toward the sustainable production of EPS for several applications in food and pharmaceuticals, cosmetics, and agriculture. With green extraction incorporated into the mainstream industry, greater yields will be achieved, and bioactivity will be retained and promoted for the sake of sustainability in the environment behind the green approach to creating innovative health-promoting products. Green extraction and green production of highly yielding EPS are apostate objectives for any industrial application to be sustainable. By optimizing process parameters supplemented with advanced green technologies, industries will have a promising future in efficient and eco-friendly production, complying with global sustainability goals.

Conflict of interest

All the authors declared no Conflict of interest

Ethical statement

This chapter does not involve studies with humans or animals. Hence, no ethical approval was required.

Acknowledgments

The authors gratefully acknowledge the support of Sri Venkateswara College of Pharmacy, Chittoor, and collaborating institutions for providing the necessary infrastructure and academic support in preparing this chapter.

Credit author statement

D. Jothieswari: Supervision; Writing—Original draft (abstract, introduction, discussion and conclusion); oversight of exopolysaccharides: An overview and green extraction methods for eps; project administration.
R. Radha: Writing—Review and editing (all sections); preparation of novel green extraction techniques and optimization strategies; assistance with methods for characterization of EPS.
S. Naveen Taj: Drafting methods for characterization of EPS, recent advances in optimization, and optimization of process parameters, writing—Review and editing; support in references, ethical statement, acknowledgements.
M. Deepa: Contribution to optimization strategies and recent advances in optimization; technical support in green extraction methods for EPS; writing—review and editing.
Sibbala Subramanyam: Critical revision of the manuscript; writing—review and editing (abstract, keywords, discussion); oversight of references, conflict of interest, and ethical statement sections.

References

Alara, O.R., Abdurahman, N.H., Ukaegbu, C.I., 2021. Extraction of phenolic compounds: a review. Curr. Res. Food Sci. 4, 200–214. https://doi.org/10.1016/j.crfs.2021.03.011.

Ali, A.H., Bamigbade, G., Tarique, M., Esposito, G., Obaid, R., Abu-Jdayil, B., Ayyash, M., 2023. Physicochemical, rheological, and bioactive properties of exopolysaccharide produced by *Enterococcus faecalis* 84B. Int. J. Biol. Macromol. 240, 124425.

Anastas, P.T., Warner, J.C., 1998. Green Chemistry: Theory and Practice. Oxford University Press, Walton Street in Oxford, United Kingdom.

Asianezhad, A., Bari, M.R., Amiri, S., 2023. Bio-producing and characterizing biochemical and physicochemical properties of a novel antioxidant exopolysaccharide by *Bacillus coagulans* IBRC-M 10807. J. Polym. Env. 31, 4338–4352.

Ayyash, M., Abu-Jdayil, B., Olaimat, A., Esposito, G., Itsaranuwat, P., Osaili, T., Obaid, R., Kizhakkayil, J., Liu, S.Q., 2020. Physicochemical, bioactive and rheological properties of an exopolysaccharide produced by *Pediococcus pentosaceus* M41. Carbohydr. Polym. 229, 115462.

Chipaca-Domingos, H.S., Ferreres, F., Fornari, T., Gil-Izquierdo, A., Pessela, B.C., Villanueva-Bermejo, D., 2023. Pressurized liquid extraction for the production of extracts with antioxidant activity from borututu (*Cochlospermum angolense* Welw.). Foods 12, 1186. https://doi.org/10.3390/foods12061186.

Deamici, K., Morais, M., Santos, L., Muylaert, K., Gardarin, C., Costa, J.A., Laroche, C., 2021. Static magnetic fields effects on polysaccharide production by different microalgae strains. Appl. Sci. 11, 5299. https://doi.org/10.3390/app11115299.

Herzyk, F., Piłakowska-Pietras, D., Korzeniowska, M., 2024. Supercritical extraction techniques for obtaining biologically active substances from plant byproducts. Foods 13, 1713. https://doi.org/10.3390/foods13111713.

Jurášková, D., Ribeiro, S.C., Silva, C.C.G., 2022. Exopolysaccharides produced by lactic acid bacteria: from biosynthesis to health-promoting properties. Foods 11, 156.

Kavitake, D., Devi, P.B., Delattre, C., Reddy, G.B., Shetty, P.H., 2023. Exopolysaccharides produced by *Enterococcus* genus: an overview. Int. J. Biol. Macromol. 226, 111–120.

Lampakis, D., Skenderidis, P., Leontopoulos, S., 2021. Technologies and extraction methods of polyphenolic compounds from pomegranate peels: a mini review. Processes 9, 236. https://doi.org/10.3390/pr9020236.

Mahapatra, S., Banerjee, D., 2013. Optimization of a bioactive exopolysaccharide production from endophytic *Fusarium solani* SD5. Carbohydr. Polym. 97, 627–634. https://doi.org/10.1016/j.carbpol.2013.05.039.

Netrusov, A.I., Liyaskina, E.V., Kurgaeva, I.V., Liyaskina, A.U., Yang, G., Revin, V.V., 2023. Exopolysaccharide-producing bacteria: a review. Microorganisms 11 (6), 1541. https://doi.org/10.3390/microorganisms11061541.

Okoro, O., Gholipour, A., Sedighi, F., Shavandi, A., Hamidi, M., 2021. Optimization of exopolysaccharide (EPS) production by *Rhodotorula mucilaginosa* sp. GUMS16. ChemEngineering 5 (3), 39. https://doi.org/10.3390/chemengineering5030039.

Peng, J., Bi, J., Yi, J., Wu, X., Zhou, M., Zhao, Y., Liu, J., 2019. Characteristics of cell wall pectic polysaccharides affect textural properties of instant controlled pressure drop dried carrot chips from different tissue zones. Food Chem. 293, 358–367. https://doi.org/10.1016/j.foodchem.2019.05.008.

Pinto, D., Vieira, E.F., Peixoto, A.F., Freire, C., Freitas, V., Costa, P., Delerue-Matos, C., Rodrigues, F., 2021. Optimizing the extraction of phenolic antioxidants from chestnut shells by subcritical water extraction using response surface methodology. Food Chem. 334, 127521. https://doi.org/10.1016/j.foodchem.2020.127521.

Revin, V.V., Liyaskina, E.V., Parchaykina, M.V., Kurgaeva, I.V., Efremova, K.V., Novokuptsev, N.V., 2023. Production of bacterial exopolysaccharides: Xanthan and bacterial cellulose. Int. J. Mol. Sci. 24 (19), 14608. https://doi.org/10.3390/ijms241914608.

Roy, et al., 2021. Green Extraction of Bioactive Compounds: Principles and Bioprocess Advancements. J. Sust. Biotechnol. 14 (3), 112–125.

Sagar, N.A., Pareek, S., Sharma, S., Yahia, E.M., Lobo, M.G., 2018. Fruit and vegetable waste: bioactive compounds, their extraction and possible utilization. Compr. Rev. Food Sci. Food Saf. 17, 512–531. https://doi.org/10.1111/1541-4337.12330.

Singh, S., Patel, A.K., 2020. Optimization of exopolysaccharide production: A review. Biopr. Eng. Rev. 8 (1), 45–58.

Sutherland, I.W., 1972. Bacterial exopolysaccharides. Adv. Microb. Physiol. 8, 143–213.

Tang, W., Dong, M., Wang, W., Han, S., Rui, X., Chen, X., Jiang, M., Zhang, Q., Wu, J., Li, W., 2017. Structural characterization and antioxidant property of released exopolysaccharides from *Lactobacillus delbrueckii* ssp. *bulgaricus* SRFM-1. Carbohydr. Polym. 173, 654–664.

Tiwari, O.N., Sasmal, S., Kataria, A.K., Devi, I., 2020. Application of microbial extracellular carbohydrate polymeric substances in food and allied industries. 3 Biotech 10, 221.

Van, Q.N., Issaq, H.J., Jiang, Q., Li, Q., Muschik, G.M., Waybright, T.J., Lou, H., Dean, M., Uitto, J., Veenstra, T.D., 2008. Comparison of 1D and 2D NMR spectroscopy for metabolic profiling. J. Proteome Res. 7, 630–639.

Wei, G., Dai, X., Zhao, B., Li, Z., Tao, J., Wang, T., Huang, A., 2023. Structure–activity relationship of exopolysaccharides produced by *Limosilactobacillus fermentum* A51 and the mechanism contributing to the textural properties of yogurt. Food Hydrocoll. 144, 108993.

Whitfield, G.B., Marmont, L.S., Howell, P.L., 2015. Enzymatic modifications of exopolysaccharides enhance bacterial persistence. Front. Microbiol. 6, 471. https://doi.org/10.3389/fmicb.2015.00471.

Ying, Z., Han, X., Li, J., 2011. Ultrasound-assisted extraction of polysaccharides from mulberry leaves. Food Chem. 127, 1273–1279. https://doi.org/10.1016/j.foodchem.2011.01.083.

Yue, Y., Wang, Y., Han, Y., Zhang, Y., Cao, T., Huo, G., Li, B., 2023. Genome analysis of *Bifidobacterium bifidum* E3, structural characteristics, and antioxidant properties of exopolysaccharides. Foods 12, 2988.

Zhang, J., Xiao, Y., Wang, H., Zhang, H., Chen, W., Lu, W., 2023. Lactic acid bacteria-derived exopolysaccharide: formation, immunomodulatory ability, health effects, and structure–function relationship. Microbiol. Res. 274, 127432.

CHAPTER

5

Role of exopolysaccharides as prebiotics in health-promoting nutraceutical formulation development

Sahanaz Khatun[a], *Sohini Chatterjee*[a], *Mousumi Tudu*[a], *Suparna Garai*[a], *Pijush Kanti Shit*[a], *Abhishek Mohanta*[a], *Amalesh Samanta*[a] *and Arnab De*[b]

[a]Division of Microbiology and Pharmaceutical Biotechnology, Department of Pharmaceutical Technology, Jadavpur University, Kolkata, West Bengal, India

[b]Department of Pharmaceutics, Bharat Technology, Uluberia, Howrah, West Bengal, India

5.1 Introduction

Exopolysaccharides (EPS) are biopolymers produced by various microorganisms such as bacteria, fungi, and yeasts, and released into their adjoining surroundings. EPS is sugar-based, and its units are cross-linked with glycosidic bonds (Osemwegie, 2020; Santra and Banerjee, 2021). Recently, EPS has gained huge importance in the food, pharmaceutical(Hazra, 2024; Chatterjee, 2025), and nutraceutical sectors as potential agents for prebiotics (Tudu, 2025), functional additives, and health-beneficial compounds. This is what has triggered immense interest in their applications in the formulation of nutraceuticals to promote health. A prebiotic is defined as any compound that specifically promotes the development or activity of beneficial gut microbes and, thereby, improves overall well-being in a host organism (Sanders, 2019). Incorporation of EPS as prebiotics in nutraceutical formulations can enhance the performance of the immune system, intestinal health, and general well-being (Kareb and Aïder, 2019). EPS are high-molecular-weight compounds, and they display a highly diverse range of structures and compositions. Though EPS are generally composed of repeating sugar units but they also may include additional functional groups, such as acyl, sulfate, or phosphate groups. Other EPS

DOI: https://doi.org/10.1016/B978-0-443-36619-2.00005-X

include complex branching structures that improve their functional properties (Suresh Kumar et al., 2007).

EPS is first synthesized from sugar monomers by the microbe, and further polymerization takes place and is finally secreted into the extracellular space. The release of EPS can be in response to environmental stresses, availability of nutrients, or mechanisms associated with biofilm development (Landini, 2009). Microorganisms like *Streptococcus, Enterococcus, Bifidobacterium, Lactobacillus, and Lactococcus* are known to generate different types of EPS (Harutoshi, 2013). Certain EPS bind to water strongly and allow the formation of viscoelastic fluids or gels (Flemming, 2016). Thus, it has played a beneficial role in the food industry, specifically in dairy products, sauces, and beverages. Some EPS exhibit emulsion-stabilizing properties, thus proving useful for products needing a combination of the oil and water phases (de Melo Teixeira, 2024). Some EPS also modulate the immune response and thereby affect the function of immune cells such as lymphocytes and macrophages (Pourjafar, 2023). Certain EPS have antioxidant properties that reduce oxidative stress and hinder the damage caused due to free radicals in the body (Andrew and Jayaraman, 2020). EPS has also played an essential role as prebiotics and can selectively stimulate the growth and activity of beneficial gut microbiota, such as *Lactobacilli* and *Bifidobacterium*.

Beneficial gut microorganisms frequently ferment EPS and use the polysaccharides as fuel. Short-chain fatty acids (SCFAs), one of the fermentation products, are good for gut health because they decrease the pH of the colon, which stops harmful microbes from growing and provides a healthy microbial ecology (Markowiak-Kopeć and Śliżewska, 2020). By encouraging the activity of immune cells such as T-lymphocytes, dendritic cells, and macrophages, EPS generated by probiotics may have immunomodulatory effects and improve the immunological response(Angelin and Kavitha, 2020). This may result in better defense against intestinal irritation and infections. EPS might defend against pathogenic microbes and enhance the general stability of the microbial community by encouraging the formation of biofilms by beneficial bacteria in the gut (Hooshdar, 2020).

5.2 Type of exopolysaccharides (EPS)

EPS are classified according to their origin, composition, structure, functionality, etc.

a. Based on microbial origin
Algal EPS: Algal EPS obtained from various types of microalgae like *Phaeodactylum tricornutum, Nannochloropsis oculate, Picochlorum sp.*, etc. Examples of some algal EPS are *Spirulina, Porphyridium*, etc.(de Jesús Paniagua-Michel et al., 2014).

Bacterial EPS: Bacterial EPS are obtained from various types of bacterial species like *Lactobacillus, Pseudomonas, Xanthomonas*, etc. As an example, xanthan gum is obtained from *Xanthomonas campestris*.

Fungal EPS: Fungal EPS are also synthesized from various types of fungi species like *Cryptococcus, Aspergillus*, etc. Example: Pullulan obtained from *Aureobasidium pullulans* (Hazra, 2023).

b. Based on composition
Homopolysaccharides (HoPS): Contain only one type of monosaccharide, like dextran and curdlan (made of glucose), levan and inulin (made of fructose) (Fig. 5.1).

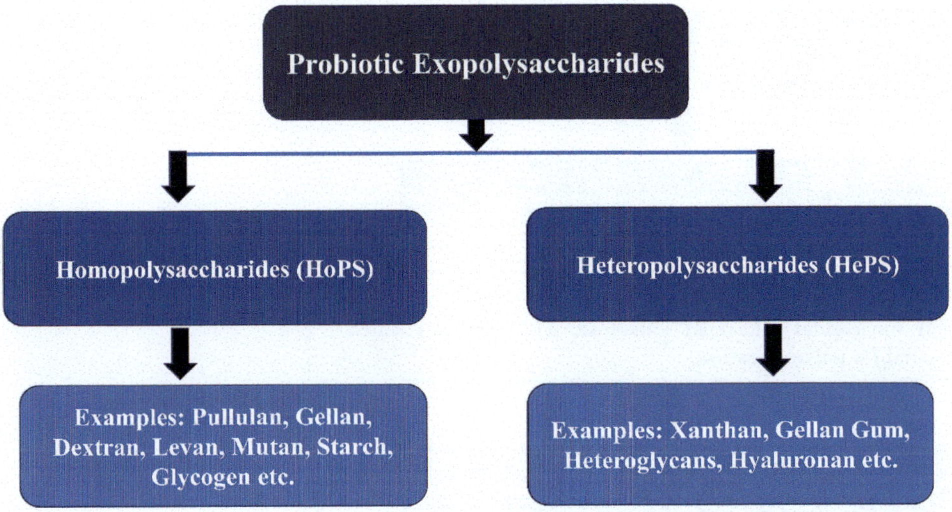

FIGURE 5.1 Schematic representation about the classification of probiotic exopolysaccharides (Pourjafar, 2023).

Heteropolysaccharides (HePS): HePS are composed of two or more than two types of monosaccharides, like gellan gum, xanthan (contains mannose, glucose, glucuronic acid, etc.) (Fig. 5.1).

c. Based on charge
Neutral EPS: Neutral EPS contain only neutral sugars and no charged groups. Example: Dextran.

Cationic EPS: Cationic EPS are positively charged EPS that contain cationic functional groups (imino, amino groups) in their structure. Example: Chitosan, diutan gum, etc (You, 2009).

Anionic EPS: The presence of negatively charged groups is observed in anionic EPS, such as the sulfate group, etc. Examples of anionic EPS are alginate, xanthan gum, etc.

d. Based on biological uses
Adhesive EPS: Adhesive EPS contribute to microorganisms' adhesion in various platform including industrial, medical, and natural bionetworks. Example: Cellulose (obtained from *Salmonella spp.)*, Alginate (obtained from *Pseudomonas aeruginosa)*, Dextran (obtained from *Leuconostoc mesenteroides)*, etc.

Storage EPS: They are obtained from microorganisms and serve as an energy reserve. Storage EPS are used in the production of bioplastics, medical applications, industrial purposes, etc. Example: Polyhydroxyalkanoates (PHA), Fructans like inulin, Glycogen, etc., (Nambiar, 2024) (Fig. 5.1).

Biosynthesis

Biosynthesis of EPS obtained from probiotic microorganisms occurs in extracellular as well as intracellular. HoPS are normally formed by an extracellular pathway, but in the case of HePS are produced from both intracellular and extracellular pathways. Biosynthesis of EPS can be easily demonstrated by a simple schematic flowchart (Figs. 5.2 and 5.3).

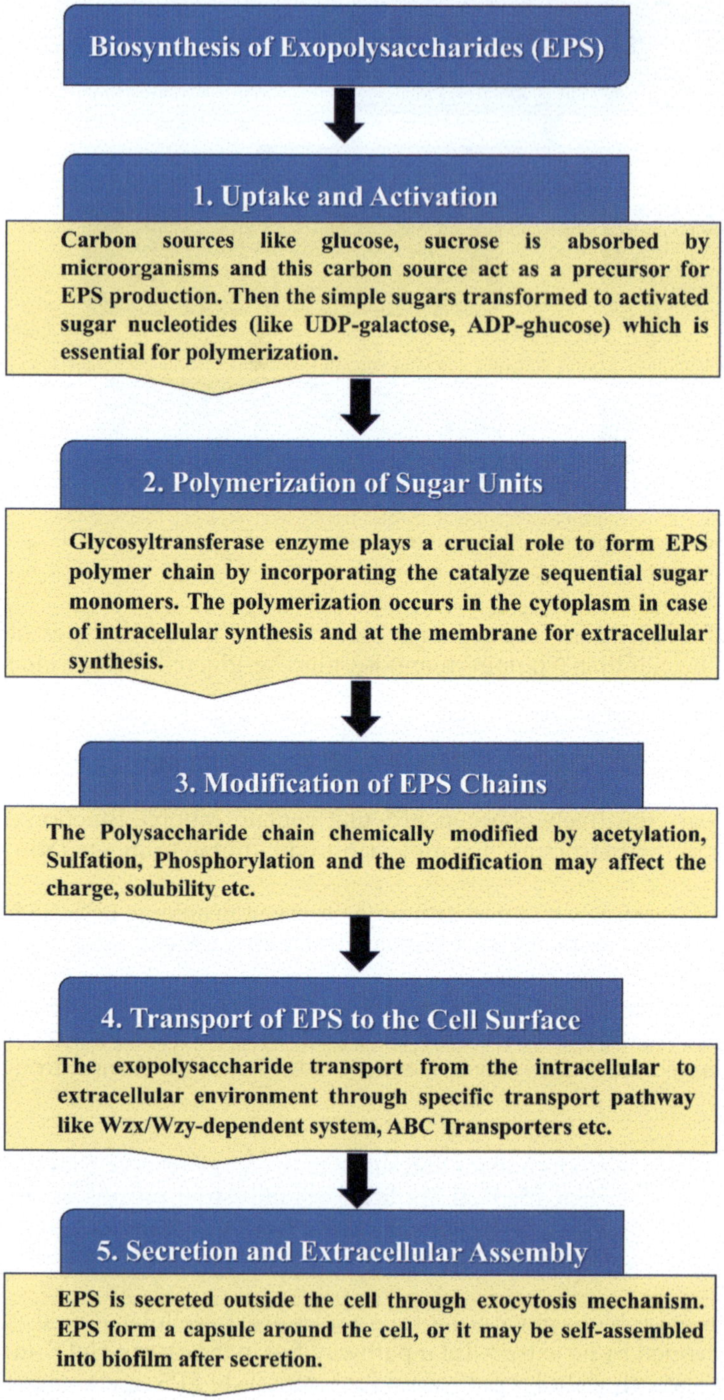

FIGURE 5.2 Step-by-step schematic representation of biosynthesis of exopolysaccharides (Becker., 1998; Whitfield and Roberts, 1999).

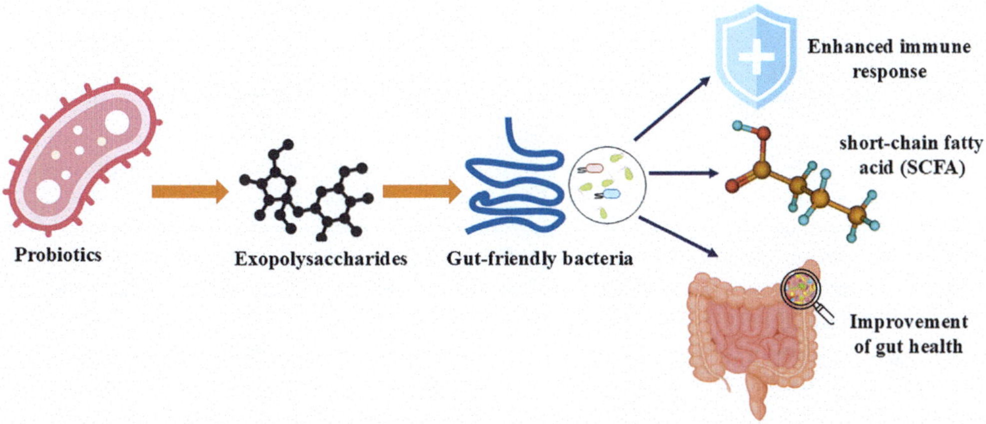

FIGURE 5.3 Microbial EPS produced by probiotics promote the growth of beneficial gut bacteria, contributing to improved digestion, immune response, and overall health. *EPS*, Exopolysaccharides.

5.3 The health potential of exopolysaccharides originated from probiotics

EPS extracted from probiotics serves a pivotal role in enhancing gut health by acting as prebiotics (Tudu, 2023), which selectively enhance the growth of beneficial microbiota in the gut. Research shows that EPS produced by *Bifidobacterium* and *Lactobacillus* species increases the proliferation of commensal bacteria like *Akkermansia muciniphila* and *Bifidobacterium spp.* thereby improves gut homeostasis(Delgado, 2020). EPS from *Bifidobacterium breve* and *Lactobacillus delbrueckii* have proven to enhance macrophage activity, increase production of cytokine, and activate the toll-like receptors (TLRs), specifically TLR2, which plays a crucial role in immune response(Hidalgo-Cantabrana, 2012). Many EPS generated from probiotics have exhibited antioxidant properties, which are crucial in ameliorating oxidative stress, which is a key factor in long-standing diseases. EPS produced by *Pediococcus pentosaceus* and *Lactobacillus plantarum* exhibit strong free radicals and scavenging activity that reduces cell oxidative damage (Li, 2012). EPS derived from probiotic have shown to decrease the levels of serum cholesterol by sequestering bile acids and enhancing their excretion, resulting in enhanced bile acid production from reserves of cholesterol. EPS from *Bifidobacterium bifidum* and *Lactobacillus acidophilus* have exhibited hypocholesterolemic effects by regulating lipid metabolism (Habib, 2022).

5.4 Biotechnological approaches for exopolysaccharides production

5.4.1 Isolation and purification of exopolysaccharides from probiotics

Isolation and purification of EPS from probiotic microorganisms include several steps to obtain pure polysaccharides with fewer contaminations. To generate EPS, the probiotic strains able to produce such polysaccharides should be isolated initially. Such methods generally involve screening probiotic strains on selective media with specific carbohydrates, which reports the production of EPS (Abid. 2018; Angelin and Kavitha, 2020). The media used for growth culture, both in solid and liquid forms, containing the basic nutrients, sources of nitrogen, and vitamins, aid in selecting high-yield EPS-producing strains. If colonies possess a mucoid, slimy consistency, then it indicates the formation of HoPS or HePS, which are the two principal classes of microbial EPS (Lynch, 2018).

The first step in EPS production analysis involves identifying probiotic strains that will produce EPS, but optimizing the growth conditions of the identified probiotic strains is important for the higher yield of EPS. This involves choosing a suitable culture medium that facilitates probiotics' metabolism without unwanted by-products (Oleksy-Sobczak and Klewicka, 2020). For EPS production, growth media are supplemented with carbohydrates like sucrose, glucose, lactose, etc., such as MRS broth (Lynch, 2018). The measurement and characteristics of EPS may be hindered by the inclusion of non-EPS carbohydrates such as glucomannans in yeast extract or peptone (Ale et al., 2019).

The EPS purification process consists of several steps to separate polysaccharides from microbial cells and undesirable substances. The process involves several stages, which include the removal of cells, the precipitation of proteins, and the last step, the precipitation of EPS. The culture medium with EPS is centrifuged to eliminate water, microbial cells, and precipitated proteins from extracellular polysaccharides. Trichloroacetic acid (TCA) or proteolytic enzymes like proteases are added to the supernatant to precipitate the proteins (Shao, 2014). This step removes contaminating proteins and ensures a purer extract of EPS. The EPS is then precipitated from the culture supernatants using organic solvents like acetone or ethanol (malang, 2015; tudu, 2025). the main methods of purification include dialysis and chromatographic techniques. For the complete removal of undesirable low-molecular-weight materials, the precipitated EPS is dissolved in distilled water and dialyzed against water using membranes with a molecular weight cutoff of 12–14 kDa (Wang, 2009). Purified EPS is preserved in the final stage by lyophilization, or freeze-drying, which guarantees its stability and suitability for additional uses (Fontana, 2015).

5.4.2 Microbial fermentation techniques for producing exopolysaccharides

The EPS production capability of probiotics is influenced by the process parameters, composition of the medium, and fermentation techniques (Bibi, 2021). Batch fermentation is a common technique used for the production of EPS in which microbial cultures are cultivated under closed conditions with a predetermined volume of the nutrient medium (Devi, 2024). The fermentation process is monitored carefully and terminated when the desired yield of EPS is achieved. The EPS is then isolated from the culture supernatant and purified to obtain the best product. With this approach, small-scale production can be done and is ideal for the preliminary screening and evaluation of highly productive EPS-producing strains. In contrast to batch fermentation, fed-batch fermentation is a form of processing that takes place in batches with fresh nutrient supplies to prolong microbial growth and support the production of EPS. This method increases the yield of EPS through a continuous supply of substrates and reduces substrate inhibition effects. These attributes mark fed-batch fermentation as ideal for the production of industrial EPS (De Baets, 2002).

Optimization of fermentation parameters is essential to maximize the yield and quality of EPS, including carbon and nitrogen sources like sucrose, glucose, lactose, whey, yeast extract, and peptone (Oleksy-Sobczak and Klewicka, 2020). Maintaining optimal pH (5.5–7.0) and temperature (30°C–37°C), ensuring proper aeration and agitation, and adjusting inoculum size are crucial for enhancing microbial metabolism and EPS synthesis.

5.4.3 Genetic engineering of exopolysaccharides-producing microorganisms

EPS are high-molecular-weight microbial polymers with diverse applications in industries such as food, pharmaceuticals, cosmetics, and bioremediation. The yield and functional properties of EPS are largely influenced by the genetic and metabolic pathways of the producing microorganisms. Recent advances in genetic and metabolic engineering have enabled the targeted manipulation of these pathways to enhance EPS production and tailor its properties for specific applications. Genetic engineering techniques, such as CRISPR-dCas9 (Kong et al., 2022), TALENs (Yadav, 2018), and homologous recombination (Kong et al., 2022), have been employed to modify the genomes of EPS-producing microorganisms. The major key strategies include overexpression of EPS biosynthesis genes and regulatory gene manipulation (Dewanjee, 2014). Enhancing the expression of genes involved in EPS biosynthesis, such as glycosyltransferases and polysaccharide synthases, can significantly increase EPS yield.

5.4.4 Exopolysaccharides production by novel probiotics

Over the past few years, the increasing importance of probiotics has driven extensive research toward the discovery and isolation of novel strains with special capabilities, including resistance to high and low temperatures, pH fluctuations, and increased metabolite production. Accordingly, researchers are constantly on the lookout for novel probiotic strains that not only have strong survival under stress conditions but also show effective production of EPS, especially when cultivated in cost-effective nutrient sources. Among different groups of probiotics, lactic acid bacteria (LAB) have gained significant interest, with several studies reporting the isolation of EPS-producing LAB from a variety of fermented food products (Mitra, 2016; Samanta, 2019; Almalki, 2020). One such research was conducted by Wang (2015) was on the isolation and characterization of an EPS-producing strain, *Lactiplantibacillus plantarum* SKT109, derived from Tibetan Kefir (Wang, 2015). The purified EPS of this strain consisted mainly of glucose and fructose in a molar ratio of about 3:1 and had an estimated molecular weight of about 2.1×10^6 Da. Similarly, Abid (2018) also explored the potential of EPS production by newly isolated LAB strains from Tunisian fermented foods. The EPS isolated from this strain had a molecular weight of around 9.85×10^4 Da and consisted of a HePS with D-mannose and D-glucose repeating units (You, 2020). Further advancement in EPS research was achieved by Ale (2020), who investigated the statistical optimization of EPS production in *Limosilactobacillus fermentum Lf2*. The strain demonstrated the ability to synthesize high amounts of EPS, and thus it is a good candidate for application as a functional ingredient in several dairy products (Ale et al., 2019).

5.5 Exopolysaccharides in health-promoting nutraceutical formulation development

Several EPS have been discovered lately and documented, of which a few have found profound interest in the field of food and pharmaceuticals. The major limitation include their production cost, which is being addressed lately by using cheaper substrates or genetically modifying the microbial strain with higher yields along with an optimized fermentation process. Again, worth mentioning that a few particular strains are responsible for the production of high-quality EPS, where no compromise can be made with the production cost due to the extremely

TABLE 5.1 Properties of some bacterial EPS in the food industry.

EPS	Polysaccharide component	Properties	Microbial strain
Glucuronan	Glucuronic acid	Gelling agent	*Gluconacetobacter hansenii*
Cellulose	Glucose	Good mechanical properties	*Acetobacter sp.*
Curdlan	Glucose	Edible and gel former.	*Rhizobium meliloti, Agrobacterium radiobacter*
Xanthan	Glucose, mannose, and glucuronic acid	Highly viscous, water-soluble, stable, resistant to degradation at high temperatures.	*Xanthomonas sp.*
Hyaluronan	Glucuronic acid and N-acetylglucosamine	Biocompatible, high water retention capacity, viscous behavior.	*Streptococcus sp.*
Levan	Sucrose and fructans	Neutral charge, low viscosity, high water solubility, nonswelling, along with antiviral properties, tumor, antioxidant, and anti-inflammatory properties.	*Bacillus, Zymomonas, Pseudomonas, Aerobacter,* and *Microbacterium*
Pullulan	Maltotriose	Water soluble, film forming and adhesion properties.	*Aureobasidium pullulans*

EPS, exopolysaccharides.

useful byproduct it produces, having immense application in respective fields (Costerton et al., 1999) (Table 5.1).

5.6 Safety and toxicological considerations

EPS (Mitra, 2016) are a potential and adaptable element for nutraceutical formulations that promote health, especially when used as prebiotics (Korcz et al., 2018). To make sure they are devoid of dangerous substances and negative effects, however, a comprehensive analysis of their safety and toxicological profiles is necessary. EPS can be safely added to nutraceutical goods by depending on the microbiological source, structural features, dosage, and thorough testing for impurities (Bibi, 2021). The European Food Safety Authority (EFSA) and the United States Food and Drug Administration (FDA) have created standards to assess the safety of novel prebiotic compounds (Deehan, 2024). Evaluation of EPS requires both acute toxicity studies (single-dose treatment) and sub-chronic toxicity studies (repeated dosing). Chronic toxicity and carcinogenicity testing are examples of long-term research that help ascertain whether long-term EPS intake carries dangers such as organ toxicity or the development of cancer (Pinto, 2016). However, hypersensitivity reactions may result from excessive immunological activation in certain people. Thus, it is essential to evaluate EPS's immunogenicity, especially in groups with autoimmune disorders, allergies, or weakened immune systems (Wong, 2013).

The possible presence of endotoxins is a significant toxicological concern for EPS generation, particularly when the EPS is produced by gram-negative bacteria. In large concentrations, these endotoxins can cause fever, systemic inflammation, and other harmful symptoms. Therefore, to guarantee that EPS products are free of these contaminants, appropriate purification techniques and endotoxin testing are required (Sotoudegan, 2019).

The safety profile of EPS is dependent on the source, chemical structure, dose, etc. Most of the microbes that produce EPS have been identified from naturally occurring and have been safe in the past (Rana and Upadhyay, 2020). A few species of *Bifidobacterium* and *Lactobacillus* that produce EPS have also been generally recognized as safe and the qualified presumption of safety. Additionally, their capacity to build biofilms causes hygienic issues in the food industry (Martínez-Viedma., 2008). Previously, it was reported that bacterial EPS are being safely used as food additives, for tissue regeneration in the biomedical field as well as in the pharmaceutical field, like dextran, levan, curdlan, succinoglycan, glucuronan, colonic acid, *Spirulina*, etc (Nwodo et al., 2012). When using EPS as prebiotics (Mitra, 2023) in nutraceutical formulations that promote health, regulatory guidelines must be carefully investigated. The regulatory agencies that oversee the use of EPS as prebiotics (Mitra, 2017) in nutraceutical formulations that promote health differ by geographical location. These groups make certain that EPS-based goods are developed, marketed, and distributed with the utmost safety, effectiveness, and quality (Rao, 2024). The aim of ongoing research is to improve the EPS preparation with enhanced quality. For large-scale production, it is very important to standardize the production and purification processes to achieve the desired quality EPS and defeat obstacles. One of the main concerns in the production of EPS is cost-effectiveness. To achieve cost effectiveness, cheap raw materials (industrial and agricultural waste) can be used, which lowers the cost along with minimizing environmental concerns.

5.7 Conclusion

EPS have arisen as promising prebiotic agents with an extensive potential in developing formulations that are health-promoting nutraceuticals. Their bioactive properties, structural complexity, and ability to regulate the gut microbiota place them as valuable ingredients in dietary supplements and functional foods. EPS has several operational advantages in nutraceutical formulations, which include their capacity to increase improve texture, product stability, and act as natural emulsifiers and thickeners. These specifications enable EPS to serve as a dual role, both as bioactive agents and stabilizers of formulation. Moreover, EPS that are derived from *Leuconostoc mesenteroides*, *Lactiplantibacillus plantarum*, and *Bacillus subtilis* showed antioxidant, immunomodulatory properties, and anti-inflammatory, making them invaluable components in nutraceuticals which are designed for the management and prevention of chronic disease. With advancing research, integrating EPS into nutraceuticals and functional foods can significantly contribute to global health and preventive health strategies.

Credit author statement

Sahanaz Khatun: Conceptualized and outlined the chapter, performed the literature review and manuscript writing, software. Sohini Chatterjee: Manuscript writing and drafting. Mousumi Tudu: Editing and software, Suparna Garai: Manuscript writing and drafting, Pijush Kanti Shit:

Editing and software, Abhishek Mohanta: Manuscript writing and drafting, Amalesh Samanta: Guidance and conceptualization, Arnab De: Guidance and conceptualization.

Acknowledgment

The authors are grateful to the Department of Pharmaceutical Technology, Jadavpur University, for providing essential research facilities.

References

Abid, Y., Casillo, A., Gharsallah, H., Joulak, I., Lanzetta, R., Corsaro, M.M., Attia, H., Azabou, S., 2018. Production and structural characterization of exopolysaccharides from newly isolated probiotic lactic acid bacteria. Int. J. Biol. Macromol. 108, 719–728.

Ale, E.C., Reinheimer, J.A., Binetti, A.G., 2019. Exopolysaccharides from *Lactobacillus*: Biosynthesis, characterization, functional aspects and applications in food industry. In: LeBlanc, J.G., deMorenodeLeBlanc, A. (Eds.), The Many Benefits of Lactic Acid Bacteria. Santa Fe, Argentina, pp. 253–292.

Almalki, M.A., 2020. Exopolysaccharide production by a new *Lactobacillus lactis* isolated from the fermented milk and its antioxidant properties. J. King Saud Univ.-Sci. 32 (2), 1272–1277.

Andrew, M., Jayaraman, G., 2020. Structural features of microbial exopolysaccharides in relation to their antioxidant activity. Carbohydr. Res. 487, 107881.

Angelin, J., Kavitha, M., 2020. Exopolysaccharides from probiotic bacteria and their health potential. Int. J. Biol. Macromol. 162, 853–865.

Becker, A., Katzen, F., Pühler, A., Ielpi, L.J.A.M., 1998. Xanthan gum biosynthesis and application: a biochemical/genetic perspective. Appl. Microbiol. Biotechnol. 50 (2), 145–152.

Bibi, A., Xiong, Y., Rajoka, M.S.R., Mehwish, H.M., Radicetti, E., Umair, M., Shoukat, M., Khan, K., Aadil, R.M., 2021. Recent advances in the production of exopolysaccharide (EPS) from *Lactobacillus spp.* and its application in the food industry: A review. Sustainability 13 (22), 12429.

Chatterjee, S., Mohanta, A., De, A., Mukherjee, A., Hazra, A., Niloy, P.P., Tudu, M., Chattopadhyay, K., Samanta, A., 2025. Evaluation of gum odina/carbopol composite mucoadhesive hydrogel on pharmaceutical performance: Focusing on potential periodontal treatment. Int. J. Biol. Macromol. 288, 138708.

De Baets, S., Du Laing, S., Francois, C., Vandamme, E.J., 2002. Optimization of exopolysaccharide production by *Tremella mesenterica* NRRL Y-6158 through implementation of fed-batch fermentation. J. Ind. Microbiol. Biotechnol. 29 (4), 181–184.

de Jesús Paniagua-Michel, J., Olmos-Soto, J., Morales-Guerrero, E.R., 2014. Algal and microbial exopolysaccharides: New insights as biosurfactants and bioemulsifiers. Adv. Food. Nutr. Res. 73, 221–257.

de Melo Teixeira, L., da Silva Santos, É., Dos Santos, R.S., Ramos, A.V.G., Baldoqui, D.C., Bruschi, M.L., Gonçalves, J.E., Gonçalves, R.A.C., de Oliveira, A.J.B., 2024. Production of exopolysaccharide from *Klebsiella oxytoca*: Rheological, emulsifying, biotechnological properties, and bioremediation applications. Int. J. Biol. Macromol. 278, 134400.

Deehan, E.C., Witwer, R.S., Guerra, P., John, T., Monheit, L., 2024. Revisiting the concepts of prebiotic and prebiotic effect in light of scientific and regulatory progress—a consensus paper from the Global Prebiotic Association. Adv. Nutr. 15 (12), 100329.

Delgado, S., Sánchez, B., Margolles, A., Ruas-Madiedo, P., Ruiz, L., 2020. Molecules produced by probiotics and intestinal microorganisms with immunomodulatory activity. Nutrients 12 (2), 391.

Devi, M., Thakur, K., Kumari, D., Behl, M., Rathour, R.K., Bhatt, A.K., Bhatia, R.K., 2024. Advances in fermentation technology for EPS production. Microbial Exopolysaccharides. CRC Press, England & Wales, Howick Place, London, pp. 37–76.

Dewanjee, S., Gangopadhyay, M., Das, U., Sahu, R., Samanta, A., Banerjee, P., 2014. Signal transducer and oxidative stress mediated modulation of phenylpropanoid pathway to enhance rosmarinic acid biosynthesis in fungi elicited whole plant culture of *Solenostemon scutellarioides*. Enzyme Microb. Technol. 66, 1–9.

Flemming, H.C., 2016. EPS—then and now. Microorganisms 4 (4), 41.

Fontana, C., Li, S., Yang, Z., Widmalm, G., 2015. Structural studies of the exopolysaccharide from *Lactobacillus plantarum* C88 using NMR spectroscopy and the program CASPER. Carbohydr. Res. 402, 87–94.

Habib, B., Vaid, S., Bangotra, R., Sharma, S., Bajaj, B.K., 2022. Bioprospecting of probiotic lactic acid bacteria for cholesterol lowering and exopolysaccharide producing potential. Biologia 77 (7), 1931–1951.

Harutoshi, T., 2013. Exopolysaccharides of lactic acid bacteria for food and colon health applications. Lactic Acid Bacteria-R&D for Food, Health and Livestock Purposes. IntechOpen.

Hazra, A., Chatterjee, S., Mohanta, A., Paul, P., Samanta, A., 2023. A review on fungal pullulan as a natural polymer focusing on targeted therapy for colon cancer and other pharmaceutical applications. Polym. Adv. Technol. 34 (10), 3183–3192.

Hazra, A., Tudu, M., Mohanta, A., Samanta, A., 2024. Gum odina prebiotic induced gut modulation for the treatment of colon cancer on Swiss albino mice: By using capecitabine loaded biopolymeric microsphere. Int. J. Biol. Macromol. 267, 131410.

Hidalgo-Cantabrana, C., López, P., Gueimonde, M., de Los Reyes-Gavilán, C.G., Suárez, A., Margolles, A., Ruas-Madiedo, P., 2012. Immune modulation capability of exopolysaccharides synthesised by lactic acid bacteria and bifidobacteria. Probiotics Antimicrob. Proteins 4 (4), 227–237.

Hooshdar, P., Kermanshahi, R.K., Ghadam, P., Khosravi-Darani, K., 2020. A review on production of exopolysaccharide and biofilm in probiotics like *Lactobacilli* and methods of analysis. Biointerface Res. Appl. Chem. 10, 6058–6075.

Jw, C., 1999. Bacterial biofilms: a common cause of persistent infections. Science 284, 1318–1322.

Kareb, O., Aïder, M., 2019. Whey and its derivatives for probiotics, prebiotics, synbiotics, and functional foods: A critical review. Probiotics Antimicrob. Proteins 11 (2), 348–369.

Kong, L., Song, X., Xia, Y., Ai, L., Xiong, Z., 2022. Construction of a CRISPR/nCas9-assisted genome editing system for exopolysaccharide biosynthesis in *Streptococcus thermophilus*. Food Res. Int. 158, 111550.

Kong, L., Xiong, Z., Song, X., Xia, Y., Ai, L., 2022. CRISPR/dCas9-based metabolic pathway engineering for the systematic optimization of exopolysaccharide biosynthesis in *Streptococcus thermophilus*. J. Dairy Sci. 105 (8), 6499–6512.

Korcz, E., Kerényi, Z., Varga, L., 2018. Dietary fibers, prebiotics, and exopolysaccharides produced by lactic acid bacteria: potential health benefits with special regard to cholesterol-lowering effects. Food Funct. 9 (6), 3057–3068.

Landini, P., 2009. Cross-talk mechanisms in biofilm formation and responses to environmental and physiological stress in *Escherichia coli*. Res. Microbiol 160 (4), 259–266.

Li, S., Zhao, Y., Zhang, L., Zhang, X., Huang, L., Li, D., Niu, C., Yang, Z., Wang, Q., 2012. Antioxidant activity of *Lactobacillus plantarum* strains isolated from traditional Chinese fermented foods. Food Chem. 135 (3), 1914–1919.

Lynch, K.M., Zannini, E., Coffey, A., Arendt, E.K., 2018. Lactic acid bacteria exopolysaccharides in foods and beverages: isolation, properties, characterization, and health benefits. Annu. Rev. Food Sci. Technol. 9 (1), 155–176.

Malang, S.K., 2015. Characterization of exopolysaccharide and ropy capsular polysaccharide formation by Weissella. Food Microbiol. 46, 418–427.

Markowiak-Kopeć, P., Śliżewska, K., 2020. The effect of probiotics on the production of short-chain fatty acids by human intestinal microbiome. Nutrients 12 (4), 1107.

Martínez-Viedma, P., Abriouel, H., Omar, N.B., Valdivia, E., López, R.L., Gálvez, A., 2008. Inactivation of exopolysaccharide and 3-hydroxypropionaldehyde-producing lactic acid bacteria in apple juice and apple cider by enterocin AS-48. Food Chem. Toxicol. 46 (3), 1143–1151.

Mitra, D., Basu, A., Das, B., Jena, A.K., De, A., Das, M., Bhattacharya, S., Samanta, A., 2017. Gum odina: an emerging gut modulating approach in colorectal cancer prevention. RSC Adv. 7 (46), 29129–29142.

Mitra, D., Jena, A.K., De, A., Das, M., Das, B., Samanta, A., 2016. Prebiotic potential of gum odina and its impact on gut ecology: In vitro and in vivo assessments. Food Funct. 7 (7), 3064–3072.

Mitra, D., Sikdar, S., Chakraborty, M., Das, O., Samanta, A., Dutta, S., 2023. Gum odina prebiotic prevents experimental colitis in C57BL/6 mice model and its role in shaping gut microbial diversity. Food Biosci. 53, 102509.

Nambiar, K., Devaraj, D., Sevanan, M., 2024. Development of biopolymers from microbes and their environmental applications. Phys. Sci. Rev. 9 (4), 1903–1929.

Nwodo, U.U., Green, E., Okoh, A.I., 2012. Bacterial exopolysaccharides: Functionality and prospects. Int. J. Mol. Sci. 13 (11), 14002–14015.

Oleksy-Sobczak, M., Klewicka, E., 2020. Optimization of media composition to maximize the yield of exopolysaccharides production by *Lactobacillus rhamnosus* strains. Probiotics Antimicrob. Proteins 12 (2), 774–783.

Osemwegie, O.O., Adetunji, C.O., Ayeni, E.A., Adejobi, O.I., Arise, R.O., Nwonuma, C.O., Oghenekaro, A.O., 2020. Exopolysaccharides from bacteria and fungi: Current status and perspectives in Africa. Heliyon 6 (6), 6e04205.

Pinto, F.C.M., De-Oliveira, A.C.A., De-Carvalho, R.R., Gomes-Carneiro, M.R., Coelho, D.R., Lima, S.V.C., Paumgarten, F.J.R., Aguiar, J.L.A., 2016. Acute toxicity, cytotoxicity, genotoxicity and antigenotoxic effects of a cellulosic exopolysaccharide obtained from sugarcane molasses. Carbohydr. Polym. 137, 556–560.

Pourjafar, H., Ansari, F., Sadeghi, A., Samakkhah, S.A., Jafari, S.M., 2023. Functional and health-promoting properties of probiotics' exopolysaccharides; isolation, characterization, and applications in the food industry. Crit. Rev. Food Sci. Nutr. 63 (26), 8194–8225.

Rana, S., Upadhyay, L.S.B., 2020. Microbial exopolysaccharides: Synthesis pathways, types and their commercial applications. Int. J. Biol. Macromol. 157, 577–583.

Rao, T.J.M., Kesharwani, R.K., Keservani, R.K., Sharma, A.K., 2024. Formulations, Regulations, and Challenges of Nutraceuticals. CRC Press, Florida, USA.

Samanta, A., De, A., Hasnain, M.S., Bera, H., Nayak, A.K., 2019. Gum odina as pharmaceutical excipient. Natural Polysaccharides in Drug Delivery and Biomedical Applications. Academic Press, Cambridge, Massachusetts, USA, pp. 327–337.

Sanders, M.E., Merenstein, D.J., Reid, G., Gibson, G.R., Rastall, R.A., 2019. Probiotics and prebiotics in intestinal health and disease: from biology to the clinic. Nat. Rev. Gastroenterol. Hepatol. 16 (10), 605–616.

Santra, H.K., Banerjee, D., 2021. Microbial exopolysaccharides: structure and therapeutic properties. Microbial Polymers: Applications and Ecological Perspectives. Singapore: Springer, Singapore, pp. 375–420.

Shao, L.I., Wu, Z., Zhang, H., Chen, W., Ai, L., Guo, B., 2014. Partial characterization and immunostimulatory activity of exopolysaccharides from *Lactobacillus rhamnosus* KF5. Carbohydr. Polym. 107, 51–56.

Sotoudegan, F., Daniali, M., Hassani, S., Nikfar, S., Abdollahi, M., 2019. Reappraisal of probiotics' safety in human. Food Chem. Toxicol. 129, 22–29.

Suresh Kumar, A., Mody, K., Jha, B., 2007. Bacterial exopolysaccharides–a perception. J. Basic Microbiol. 47 (2), 103–117.

Tudu, M., Chatterjee, S., Mohanta, A., Priya, K., Samanta, A., 2025. Employing biotechnology for the treatment of organic refuse. AI Technologies for Enhancing Recycling Processes. IGI Global Scientific Publishing, Chocolate Avenue, Hershey, PA, USA, pp. 113–134.

Tudu, M., Hazra, A., Paul, P., Mohanta, A., Chatterjee, S., Samanta, A., 2025. Effect of different ratios of polyvinyl alcohol-gum odina for the preparation and characterization of biodegradable composite films. Biopolymers 116 (1), e23630.

Tudu, M., Samanta, A., 2023. Natural polysaccharides: chemical properties and application in pharmaceutical formulations. Eur. Polym. J. 184, 111801.

Wang, H.J., Di, L., Ren, Q.S., Wang, J.Y., 2009. Applications and degradation of proteins used as tissue engineering materials. Materials 2 (2), 613–635.

Wang, J., Fang, X., Wu, T., Min, W., Yang, Z., 2018. Exopolysaccharide producing *Lactobacillus plantarum* SKT109 as adjunct culture in cheddar cheese production. Lwt 97, 419–426.

Whitfield, C., Roberts, I.S., 1999. Structure, assembly and regulation of expression of capsules in *Escherichia coli*. Mol. Microbiol. 31 (5), 1307–1319.

Wong, V.W.S., Wong, G.L.H., Chim, A.M.L., Chu, W.C.W., Yeung, D.K.W., Li, K.C.T., Chan, H.L.Y., 2013. Treatment of nonalcoholic steatohepatitis with probiotics. A proof-of-concept study. Ann. Hepatol 12 (2), 256–262.

Yadav, R., Kumar, V., Baweja, M., Shukla, P., 2018. Gene editing and genetic engineering approaches for advanced probiotics: a review. Crit. Rev. Food Sci. Nutr. 58 (10), 1735–1746.

You, L., Lu, F., Li, D., Qiao, Z., Yin, Y., 2009. Preparation and flocculation properties of cationic starch/chitosan crosslinking-copolymer. J. Hazard. Mater. 172 (1), 38–45.

You, X., Li, Z., Ma, K., Zhang, C., Chen, X., Wang, G., Yang, L., Dong, M., Rui, X., Zhang, Q., Li, W., 2020. Structural characterization and immunomodulatory activity of an exopolysaccharide produced by *Lactobacillus helveticus* LZ-R-5. Carbohydr. Polym. 235, 115977.

CHAPTER

6

Probiotics potential of novel natural exopolysaccharides: Prospects and scope

Kaushiki Priya, Suparna Garai, Mousumi Tudu, Sohini Chatterjee, Abhishek Mohanta, Md. Nahid Zaman and Amalesh Samanta

Division of Microbiology and Pharmaceutical Biotechnology, Department of Pharmaceutical Technology, Jadavpur University, Kolkata, West Bengal, India

6.1 Introduction

Probiotics consist of live microorganism which, when administered in an appropriate amount, propose various health benefits to the host organism (Zhou, 2019). The primary recognition is based on their ability to balance gut microbiota and prevent various gastrointestinal disorders. Exopolysaccharides (EPS) recently gained attention as an efficacious part of probiotics. EPS are high molecular weight polysaccharides secreted by lactic acid bacteria (LAB) in the surrounding environment. These biopolymers have a variety of functions, including facilitating microbial adhesion, modulating immune responses, and protective barrier (Tudu et al., 2025a) (Fig. 6.1).

The exploration of novel natural sources for the production of EPS and probiotics is important for advancement in the field of therapeutics and functional food. Fermented dairy products were considered a traditional source of probiotics. In recent researches bacteria, fungi, and marine species were highlighted as potential sources of probiotics with unique structural and functional properties (Cai, 2020). These sources help in the development of sustainable and biocompatible probiotic formulations. Apart from probiotics' structural diversity, it has ability to maintain gut microbiota and intestinal barrier function. This activity helps in treating various chronic diseases like inflammatory bowel syndrome and metabolic syndrome (Zhou, 2019). EPS also has antimicrobial and antioxidant properties, making it an important factor in promoting health. This chapter will explore the latest advancement in the isolation and characterization of EPS from novel sources.

DOI: https://doi.org/10.1016/B978-0-443-36619-2.00006-1

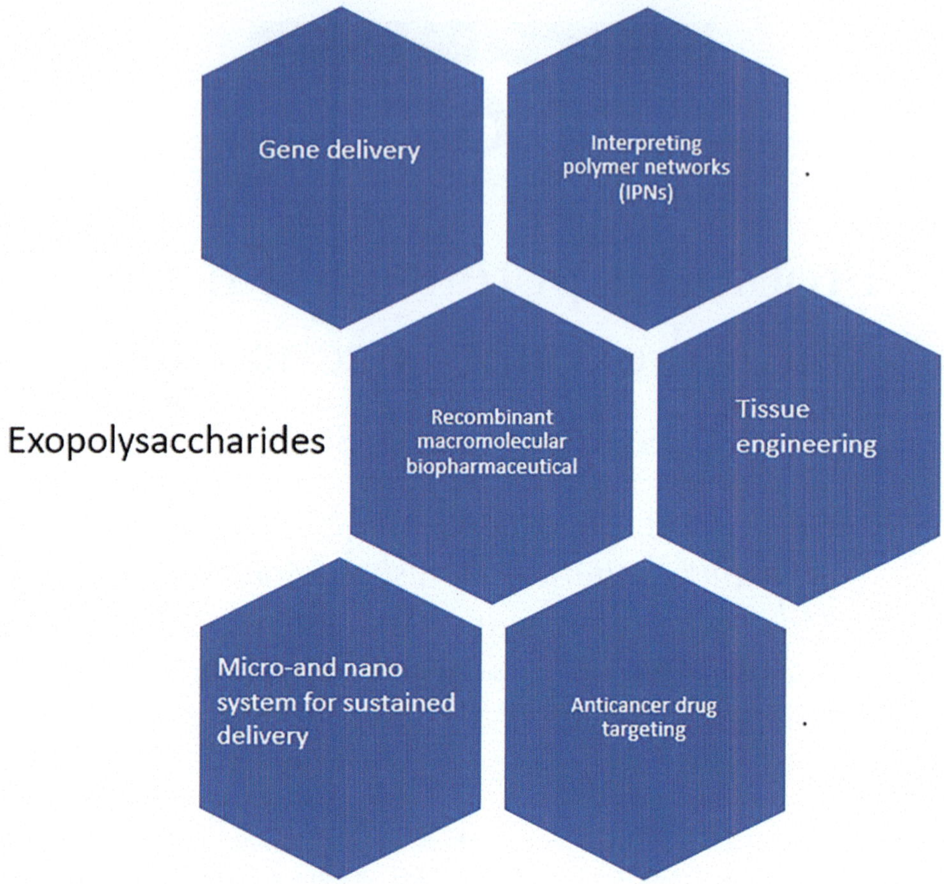

FIGURE 6.1 Diverse application of exopolysaccharide.

6.2 Sources of exopolysaccharides

6.2.1 Bacteria, fungi, and marine organisms as exopolysaccharides sources

Marine and terrestrial microorganisms are key producers of EPS, with diverse applications ranging from environmental remediation to biomedical uses (Tudu and Samanta, 2023) (Table 6.1).

6.3 Role of lactic acid bacteria in exopolysaccharides production

Marine LAB such as *Pediococcus pentosaceus* M41 have been reported as EPS producers. These polymers have a diverse structure due to the presence of active biochemical units such as glucose, galactose, and mannose, along with significant bioactivities including: Antioxidant activity (e.g., radical scavenging) and biofilm inhibition (e.g., EPS273 from *Pseudomonas stutzeri* inhibiting *P. aeruginosa* biofilms) (Zhou, 2019) (Table 6.2).

TABLE 6.1 Sources of exopolysaccharides.

Category	Examples	Key organisms/strains
Bacteria	Terrestrial and general marine species.	*Pseudomonas, Bacillus, Corynebacteria.*
Marine Bacteria	Extreme environments (hydrothermal vents, deep-sea).	*Pseudoalteromonas, Vibrio, Halomonas.*
Fungi	Terrestrial and marine-associated species.	*Aspergillus versicolor, Fusarium solani, Penicillium commune.*
Marine fungi	Deep-sea and coral-associated species.	*Aspergillus terreus, Fusarium oxysporum, Botryosphaeria rhodina.*

TABLE 6.2 EPS-based probiotic formulations have diverse applications.

Application	Mechanism	Examples
Food Industry	Works as stabilizers, emulsifiers, and texturizers.	Improves yogurt viscosity and cheese texture.
Pharmaceuticals	Effective as anti-inflammatory, antioxidant, and immunomodulatory.	*L. reuteri* EPS reduces TNF-α secretion by 63% in macrophages.
Synbiotics	Act as prebiotics to enhance probiotic viability.	*L. plantarum* EPS boosts *Bifidobacterium adolescentis* growth by 300%.

EPS, exopolysaccharides.

Outside the marine scope, *Lactobacillus* and *Streptococcus* are recognized EPS producers in fermented products for their beneficial texture. Nevertheless, due to their severe condition's structural modifications, marine LAB-derived EPS are now recognized as new candidate materials for biomedical and environmental fields (Cai, 2020).

Uses of LAB-EPS

1. The food industry—as thickeners or stabilizers.
2. Biomedicine—for oxidative and inflammatory diseases.
3. Environmental engineering—n controlling biofilms in water bodies.

In regard to marine EPS, *Flavobacteria* (phylum bacteroidetes) have an important role in the degradation of bacterial EPS and therefore play a role in the carbon cycle in oceans (Zhao et al., 2018).

6.4 Functional benefits of exopolysaccharides

High-molecular-weight polysaccharides called EPS are released by bacteria, fungi, and algae. They offer a range of practical advantages in a number of sectors, such as biotechnology (Tudu et al., 2025a). The following are the main advantages of EPS:

6.4.1 Enhancement of gastrointestinal health

Modifying carbohydrate metabolism causes the breakdown of EPS for nutrition uptake by intestinal commensals. Polysaccharide-specific breakdown of carbohydrates was triggered when gut commensal *Bacteroides thetaiotaomicron* broke down *L. reuteri* strain 121 fructan EPS. Since prebiotic activity results indicate how well a substrate stimulates the proliferation of probiotic organisms alone (Mitra et al., 2023), emerging information about the microbial biology and diversity in the gut has increased interest in examining the growth-promoting histories of other genera, such as *Eubacterium, Faecalibacterium*, and *Roseburia*. By selectively stimulating the growth of probiotic microbes, EPS over *Leuconostoc pseudomesenteroides* XG5 altered the organization of the gastrointestinal microbiota (Oerlemans et al., 2021) in C57BL/6J mice by reducing the phylum-level proportion of bacterium to Firmicutes genera . Since their functions in postbiotic products and subsequent effects on host health have been discovered, knowledge of the EPS fermenting community has become quite significant (Zhang et al., 2016).

6.4.2 Modulation of the immune system

The chemical makeup of monosaccharides also affects the immunomodulatory impact of EPSs. In an investigation using several EPS fractions isolated from *L. reuteri* Mh-001 EPS, the portions with the highest galactose concentration had the strongest anti-inflammatory effects on macrophages, followed by the portions that were rich in glucose and rhamnose (Chen, 2019). The *Lactobacillus kiferi* WXD029 glucose-rich extracellular polymer (EPS) demonstrated increased T-cell proliferation. The immunoadjuvant capability of EPS against *S. aureus* infections was demonstrated by the much-enhanced preventive activity of EPS adjunct with *S. aureus* antigen compared to *S. aureus* antigen alone . Two portions of *Lactobacillus rhamnosus* KF5 EPS were separated by purification (Shao et al., 2014); the composition of portions 1 (1.36×10^4 Da) and 2 (1.23×10^6 Da) differed. By promoting the growth of splenic lymphocytes, both portions demonstrated immunostimulatory activity, with portion 1 displaying more mitogenic activity (Surayot, 2014). *Lactobacillus confusus* TISTR1498 EPS, which contains D-glucose connected by a-(1–6) glycosidic bond, demonstrated that RAW264.7 cells produced proinflammatory compounds such as nitric oxide (NO) and cytokine. These reactions were consistent with the findings of *L. rhamnosus* KL37 EPS, which demonstrated its (Th1-type) immunoregulatory pro-inflammatory capacity in peritoneal macrophages in vitro by producing more TNF-a and IL-12 than IL-10. It is commonly assumed that low molecular weight anionic EPS has better immunostimulatory action than neutralized polysaccharides with high molecular weight (Lopez, 2012). Although the immune system modulation capacity of EPS and EPS metabolites in the gut is well recognized, the extent of immune-modulation by upward or downward mRNA regulation and intermediary component interventions in the relevant pathways differed greatly depending on the content, molecular size, and type of EPS.

6.4.3 Antioxidant and antimicrobial properties

EPS are a complex mixture of proteins, lipids, nucleic acids, and polylactides. EPS have caught significant attention due to their diverse biological activities like antioxidant, antimicrobial, and antibacterial. These diverse activities make EPS as a great candidate for its application in food, medicine, cosmetics, and environmental remediation (Vishwakarma, 2018).

EPS consists of antioxidant activity through various mechanism through several mechanisms, such as scavenging free radicals, preventing the formation of reactive oxygen species, and chelating metal ions. Studies have suggested that EPS have many sources, such as marine actinobacteria (*Micromonospora sp.*) and fungi (*Fusarium culmorum*) (Ruiz-Ruiz et al., 2020). The EPS extracted from *Micromonospora sp* species has shown higher antioxidant properties than ascorbic acid. EFP obtained from the PGPF strain of *F. culmorum* exhibits high phenolic content, contributing toward its high antioxidant property. The EPS obtained extends the shelf life of food products and also protects against oxidative stress-related diseases.

The antimicrobial activity of EPSs is proven by properties like the ability to permeabilize microbial cell membranes, disrupt biofilm formation, and alter immune response. (Xu et al., 2022) cited other authors who state that EPSs derived from microalgae showed antiviral and anti-inflammatory effects. Additionally, EPSs can be beneficial in restricting the growth of bacteria and fungi, which is essential for the control of infections and the putrefaction of foods.

These EPSs' properties enable them to be applied in numerous fields. For example, in medicine, EPSs can be utilized for producing new antimicrobial formulations to fight multidrug-resistant infections. In the food industry, EPSs may serve as bio-preservatives to enhance the longevity of food items and prevent the decay of microorganisms. Moreover, EPSs can be applied in other disciplines such as the control of hazardous microbial growth in bioremediation or cleansing of polluted areas.

6.4.4 Maintenance of gut microbiota composition

Natural EPS with probiotic efficiency play an important role in the management of gut microbiota dysbiosis. This refers to the imbalance in the microbial communities normally associated with digestive systems, leading to digestive problems and health issues. Natural EPSs help in maintaining microbial composition in the gastrointestinal tract (Pourjafar, 2023). Probiotic EPS modulates gut health in various ways, such as immunity enhancement, competing with pathogenic bacteria, synthesizing small-chain fatty acids (SCFAs), supporting good bacteria, etc. Previous research shows that several EPS from different microbial strains can enhance immune response, e.g., *Lactobacillus reuteri, Lactobacillus rhamnosus, Lactobacillus kiferi*, etc. The prevention of pathogenic bacterial adhesion to the intestinal epithelial cells is dependent on the competitive attachment of EPS and pathogenic surface molecules. Probiotic EPS competes with pathogenic surface molecules and gets attached to the binding site of intestinal epithelial cells. Thus, hinders the pathogenic adhesion. SCFAs are important for maintaining gut health. It was reported that *Lactobacillus sp.* and *Bifidobacterium sp.* can interact with probiotic EPS and break into smaller components like sugars. These smaller components, upon fermentation by gut bacteria, produce SCFAs. These smaller components, including SCFAs, serve as nutrients for other bacteria in the gut, promoting their growth (Mitra et al., 2017).

Dextran, a type of EPS produced by *Leuconostoc mesenteroides* have immense prebiotic potential, helping in improving gut health.

Kefiran and EPS produced by *Lactobacillus kefiri* have anti-inflammatory properties along with prebiotic potential improving gut health and reducing symptoms of irritable bowel syndrome (IBS) (Lee et al., 2017).

6.4.5 Support for intestinal cell barrier function

EPS provides support to the intestinal cell barrier by maintaining the integrity of the epithelium cell lining and modulating tight junction (TJ) proteins. EPS are capable of developing a physical barrier that plays an important role in protecting the intestinal mucosal layer and maintaining its integrity. EPS encounters intestinal mucosa, it forms a biofilm-like layer, which inhibits the adhesion of toxic substances, pathogenic substrates, etc., e.g., *Lactobacillus paracasei subsp. Paracasei* BGSJ2 can prevent adhesion of *E. coli* (Živković, 2016), TJ in the intestinal cell lining refers to the cell–cell junction between epithelial cells for selective passage of substances through paracellular pathways to provide nutrition. This TJ consists of various transmembrane proteins like claudin, occludin, junctional adhesion molecules (JAMs), tricellulin (Chiba, 2008). These proteins are important for maintaining membrane integrity by sealing the gap between cells, allowing selective permeation. Among these claudins are those responsible for maintaining the permeation. Claudin 1, 4, 5, 7, 8, and 14 reduce the permeability, and claudin 2 and 16 increase the permeability. EPS can enhance the expression of claudins responsible for reducing permeability, which seals the gap and restricts the passage of harmful substances, thus maintaining the integrity. Moreover, EPS can repair the damaged cells and restore the function. EPS from *Streptococcus thermophilus* MN-BM-A01 can restore membrane integrity, and EPS from *Bifidobacterium infantis* can increase claudin 4 expression and decrease claudin 2 expressionAnderson et al., 2018 (JB et al., 2008).

6.4.6 Management of chronic diseases

Anti-inflammatory activity: EPS also has anti-inflammatory properties, assisting in diseases like arthritis, allergies, etc. EPS from *Serratia marcescens* reduces inflammation and tissue damage as seen in the colitis mouse model.

Antidiabetic activity: EPS has the ability to improve insulin sensitivity and glucose metabolism, allowing better management of type 2 diabetes. *Streptococcus thermophilus* releases EPS that can assist in tissue regeneration, collagen synthesis, and even angiogenesis, which can help to heal diabetic wounds (Wang, 2018).

Cardiovascular health: Reduction in cholesterol levels, improved lipid profiles, and reduction in the chances of atherosclerosis are all activities of EPS in maintaining cardiovascular health. *Bifidobacterium bifidum* releases such EPS responsible for the above functions (Chen et al., 2019; Ewaschuk et al., 2008).

6.5 Structural and functional diversity of exopolysaccharides

6.5.1 Adaptability to functional foods, nutraceuticals, and pharmaceuticals

In recent years, EPS gained substantial attention because of their potential applications in functional food production, the nutraceutical and pharma sector.

Functional foods: EPS can be used in energy drinks, juice, beverages industry to provide nutrition and good immunity. It can be used as a prebiotic component (Mitra et al., 2016), which is used in probiotic food for better health. In the case of dairy product EPS plays a key role in the production of cheese and yogurt to improve their taste and nutritional value.

Nutraceuticals: EPS is also used in dietary supplement and health drinks production, which can provide good antioxidant and immune-boosting benefits (Wang, 2013).
Pharmaceuticals: Formulations like Hydrogel, Spongy scaffolds, hydrogel films containing EPS provide good wound healing activity, they can reduce inflammation in the wound site and provides good soothing sensation along with tissue regeneration property. EPS can be used as an adjuvant in the preparation of vaccines, nanoparticles, nanoemulsions, in situ gel, etc. (Chatterjee, 2025)
Comparison with traditional probiotics: EPS are more stable and have longer shelf-life than traditional probiotics. EPS are more flexible in various types of formulation development, and they can be administered in numerous forms like capsules, liquids, powders, etc. EPS can be engineered more precisely for specific health benefits like antioxidant and anti-inflammatory activity, and they are safer and less toxic compared to traditional probiotics.

EPS are the latest research area and need more research and studies to understand their efficacy, side effects, and their compositions.

6.6 Advancements in exopolysaccharides research

6.6.1 Isolation from the distinctive sources

- Marine bacteria and algae: Bacteria from marine sources, such as *Pseudoalteromonas spp and Alteromonas spp.* have demonstrated sturdy EPS production with distinctive bioactivities.
- Extreme environments: Halophilic and Thermophilic bacteria, like *Halomonas spp.* and *Thermococcus spp.*, have been recognized as efficient EPS producers, offering idiosyncratic rheological properties and improved thermal stability.
- Plant and soil rhizosphere: Bacteria originating from nutrient-rich rhizospheres, specifically *Bacillus spp.* and *Pseudomonas spp.*, have produced EPS with antimicrobial and strong biofilm-forming properties(Singh et al., 2016) .
- Fermentation by novel strains: Adaptive evolution techniques and genetic engineering have been applied to improve EPS yield in recognized strains like *Streptococcus spp.* and *Lactobacillus spp.*

6.7 Production and formulations of exopolysaccharides

Techniques for EPS production. There are numerous uses for these biopolymers in commercial biotechnology, food, and pharmaceuticals. The production of EPS involves various techniques:

6.7.1 Production of exopolysaccharides using biotechnology

Compared to chemical and plant-based manufacturing, biotechnological EPS production has the following benefits: Energy efficiency (when generated by algae), rapid production in a matter of days regardless of location or season, and the ability to employ waste from agriculture and manufacturing as a substrate. But in addition to the necessary requirements, several obstacles, such as production costs, may delay the process (Rutering, 2016). Recent studies concentrate on nontoxic EPS and "more sustainable" synthesis, where a single-step in-cell functioning and polymerization are preferred at an optimal level, avoiding the usage of catalysts and ocean collection for alginate. Thus, the microbial generation of certain amino polysaccharides, such as chitin and chitosan, is investigated to prevent their acquisition by hydrolysis of acids from shellfish, and hyaluronic acid is investigated to prevent their acquisition through the purification of animal tissues. Recent benign EPS discoveries, such as microbactan, have the ability to bioabsorb cadmium, offering an alternative for heavy metal remediation, particularly in industrial waterways. Additional examples include the EPS generated by *Rhodotorula mucilaginosa* UANL-001L co-cultured with *E. coli*, which has several applications in both medical and industrial settings and inhibits bacterial colonization. Nowadays, biotechnology is used to manufacture a large amount of EPS. Every EPS has a unique biosynthesis pathway, and as a result, there are variables that could affect these pathways. The production and accumulation of EPS typically take place following the microorganism's development phase, and it may be broken down into four distinct phases: (1) Absorption of the carbon supply, (2) the production of oligosaccharide-repeating units or direct synthesis through the continuous or sequential action of glycosyltransferases, (3) assembling the repeating units to form the polysaccharide, and (4) the transportation of EPS into the extracellular environment. Four general methods can also be used to describe how bacteria produce EPS: (1) The so-known Wzx/Wzy dependent mechanism, in which several glycosyltransferases combine separate recurrent units to form heteropolymers; (2) capsular polysaccharides is produced via the ATP-binding cassettes (ABC) transporter-dependent process; (3) the process that requires synthase to produce homopolymers; and (4) solitary sucrase protein-mediated extracellular synthesis (Schmid, 2015). Enzyme control, genetic alterations, and an increase in biosynthesis rate are required to get greater EPS production values.

6.7.2 Production of exopolysaccharides using fermentation techniques

The main technique for producing EPS is microbial fermentation, which involves cultivating bacteria, fungi, or algae under particular circumstances. There are two types of fermentation processes: Solid-state fermentation (SSF) and submerged fermentation (SmF). The most often used method is SmF, in which microorganisms are grown in a liquid medium that is rich in carbon and nitrogen sources, hence encouraging the production of high-yield EPS. SSF, which is frequently employed for fungi and actinomycetes, entails the culture of microorganisms on solid substrates with little free water. The 250 ml shake flask was used to hold the batch fermentation for 24 hours at 150 rpm and 35°C. By adding 1N NaOH/1N HCl, the pH of the fermentation medium was brought down to 6.5 ± 0.3. Flasks were examined for the formation of EPS at the conclusion of fermentation (Desai, 2006).

6.7.3 Production of exopolysaccharides using the genetic engineering method

One possible alternative is the biotechnological production of microbial polysaccharides by genetic engineering; a continuous process aimed at increasing commercial-scale production and the potential applications in a variety of industries. Stingele et al. (1999) extracted and transported the gene cluster of Streptococcus thermophilus Sfi6, a species that may make an EPS consisting of glucose, galactose, and N-acetylgalactosamine, to Lactococcus lactis MG1363, a nonproducer of EPS. The authors' ability to create a novel EPS with a high molecular weight and distinct structure suggests that polymerase can polymerize the repeating unit in many ways and that glycosyltransferases can have numerous specificities for the donor and acceptor (Barcelos, 2020). In a different study, the expression of the gene cluster pNZ4000 in *Lactococcus lactis* NIZO B40, a potential strain for polysaccharide production, was elevated, increasing the production of EPS. Rather than the quantity of EPS precursors, this discovery implies that the activity level of expression of the pNZ4000 gene cluster limits the production of EPS (Boles et al., 2003).

6.7.4 Development of probiotics formulations using exopolysaccharides

EPS from probiotic bacteria are increasingly utilized in functional food and pharmaceutical formulations due to their health-promoting properties and techno-functional benefits. The development of these formulations involves careful strain selection based on EPS characteristics. High-yield EPS producers like *Lactobacillus reuteri* Mh-001 and *Lactiplantibacillus plantarum* HDC-01 are prioritized for their bioactive EPS. The monosaccharide composition of EPS, such as galactose-rich EPS, correlates with enhanced anti-inflammatory effects, while structural stability ensures viability during food processing (Vazquez-Rodriguez et al., 2018; Zhou et al., 2019).

These formulations offer several health benefits, including anti-inflammatory effects, antimicrobial properties, and improved gut health. However, challenges such as low production yields and scalability issues need to be addressed through strain optimization and emerging techniques like CRISPR editing. Future directions include targeted delivery systems, clinical validation, and multifunctional blends combining EPS with other bioactive compounds for synergistic effects.

6.8 Future perspectives

6.8.1 Potential of exploring newer exopolysaccharides sources

Extracellular polymeric substances (EPSs) produced by microorganisms are gaining attention for their variety of applications, especially in the field of probiotic bacteria. EPS produced from *Lactobacillus* strains possesses qualities of being antioxidant, antidiabetic, and antitumor as well. These EPS have the ability to act upon the immune response and also reduce cholesterol levels. Traditionally, dairy sources act as a great reservoir for EPS, but nondairy sources like cereals, vegetable and fruits are now also offering as a potential probiotics source. Nonvegetarian sources offer benefits for individuals having milk allergies and lactose intolerance (Fig. 6.2).

EPS plays a crucial role in safeguarding microbial communities within biofilms by offering compositional support and shielding them from harsh environmental conditions. As biofilms mature, EPS enhances physical stability, resistance to mechanical removal, antimicrobials, and host immunity. The EPS and environmental DNA contribute to the viscoelasticity of mature biofilms, making detachment challenging, even under fluid shear stress or high mechanical pressure.

Recent studies have explored new sources of EPS-producing bacteria, like infant feces (*Enterococcus faecium* WEFA23), the gastrointestinal tract of indigenous poultry (*Lactobacillus* sp. Ca6), and the marine environment (*Pediococcus pentosaceus* M41). These sources exhibit various biological activities like antioxidant, antimicrobial, antitumor, and immunomodulatory

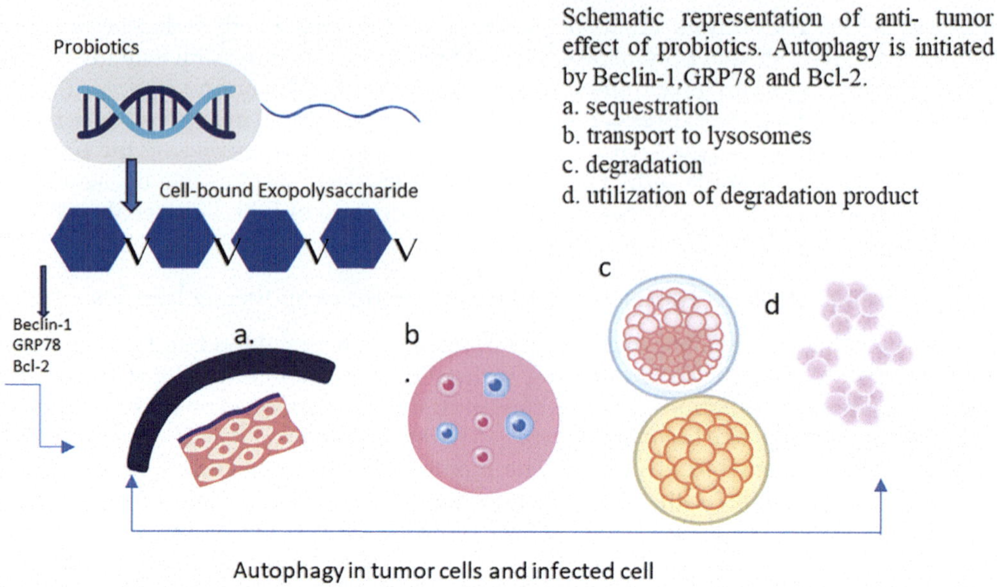

FIGURE 6.2 Schematic representation of antitumor effect of probiotics.

(Dewanjee, 2014). Exploring novel uses of EPS is important in expanding their industrial and therapeutic applications, hence replacing polysaccharides of animal and plant origin. Exploring in-vivo studies of EPS will help in exploring its full potential. EPS is used in cosmetic, therapeutic industries, bioremediations and food.

EPS has also demonstrated its ability as an anti-inflammatory and reducing cholesterol level ability as well. Due to its feature to reduce the cholesterol level, EPS can be effectively used as a functional food. The EPS from various probiotic organisms can be incorporated as spores for the creation of functional food (De Melo, 2021; Hazra, 2023). The EPS feature to act as an anti-inflammatory substance, and it has gained recognition in being used as a cosmetic or as a pharmaceutical substance.

To meet the increasing demand for sustainable sources, investment in capacity projects are necessary step. The need of development of waste bio feedstock, like cooking oil and agricultural waste, is required to reduces heavy cost of funding toward feedstocks. This necessity makes it important to explore nondairy sources of probiotic EPS production, leading to the promotion of sustainability and reducing the reliability on traditional sources. Future research should focus on developing sustainable probiotics incorporated with EPS from novel sources to enhance health-promotion and rheological properties.

6.9 Conclusion

Recently, EPS from probiotic bacteria have received considerable attention in food, nutraceuticals, and pharmacology as a biopolymer of broad application. The notable health benefits of high molecular weight polysaccharides produced by LAB, such as antioxidant, antimicrobial, and immunomodulatory effects, are well-known. EPS are also important for the preservation of microbiota balance, intestinal barrier integrity, and management of chronic conditions like metabolic syndrome and inflammatory bowel disease. Their immunomodulatory as well as prebiotic activities make them the most attractive candidate for probiotics. The diverse structure of EPS allows them to be multifunctional, from stabilizers in the food industry to therapeutic agents in biomedicine. With new sources of EPS such as marine bacteria, fungi, and other microorganisms, new possibilities open up for developing truly sustainable and biocompatible formulations. Improving EPS isolation, characterization, and production has addressed the low yield and scalability problems while expanding the usability scope.

Further studies ought to examine the new microbial sources of EPS, refine production processes through the use of CRISPR, and carry out the clinical trials needed to confirm the claim. Next generation probiotics could be developed with the help of EPS designed to cope with particular health diseases, paving new paths toward integration of EPS, solving areas of innovation and personalized nutrition as well as therapy.

Credit author statement

All authors contributed substantially to the conception, design, drafting, and critical revision of this book chapter. Specific contributions are as follows: Kaushiki Priya: Conceptualized and outlined the chapter, performed the literature review, and manuscript writing. Suparna Garai: Literature review, manuscript writing, and drafting. Mousumi Tudu: Literature review, manuscript writing, and drafting. Sohini Chatterjee: Literature review, manuscript writing, and drafting. Abhishek Mohanta: Literature review, manuscript writing, and editing. Md. Nahid Zaman: Literature review and manuscript writing. Amalesh Samanta: Guidance and conceptualization.

Conflict of interest

The authors declare they have no known competing interests or personal relationships that could have appeared to influence the work reported in this paper.

Ethical statement

Ethics approval and consent to participate not applicable.

Acknowledgments

Authors are Grateful toward the Department of Pharmaceutical Technology, Jadavpur University for providing essential research facilities.

References

Anderson, L.A., Islam, M.A., Prather, K.L., 2018. Synthetic biology strategies for improving microbial synthesis of "green" biopolymers. J. Biol. Chem. 293 (14), 5053–5061.

Barcelos, M.C., Vespermann, K.A., Pelissari, F.M., Molina, G., 2020. Current status of biotechnological production and applications of microbial exopolysaccharides. Crit. Rev. Food Sci. Nutr. 60 (9), 1475–1495.

Boels, I.C., Van Kranenburg, R., Kanning, M.W., Chong, B.F., De Vos, W.M., Kleerebezem, M., 2003. Increased exopolysaccharide production in *Lactococcus* lactis due to increased levels of expression of the NIZO B40 EPS gene cluster. Appl. Env. Microbiol. 69 (8), 5029–5031.

Cai, Z., Wang, P., Guo, Y., Zeng, X., Wu, Z., Pan, D., 2020. S-layer protein modulates the stimulatory effects of *Lactobacillus acidophilus* CICC 6074 by triggering PKC signaling cascade in RAW 264.7 cells. J. Funct. Foods 67, 103841.

Chatterjee, S., Mohanta, A., De, A., Mukherjee, A., Hazra, A., Niloy, P.P., 2025. Evaluation of gum odina/carbopol composite mucoadhesive hydrogel on pharmaceutical performance: Focusing on potential periodontal treatment. Int. J. Biol. Macromol. 288, 138708.

Chen, Y.C., Wu, Y.J., Hu, C.Y., 2019. Monosaccharide composition influence and immunomodulatory effects of probiotic exopolysaccharides. Int. J. Biol. Macromol. 133, 575–582.

Chiba, H., Osanai, M., Murata, M., Kojima, T., Sawada, N., 2008. Transmembrane proteins of tight junctions. Biochim. Biophys. Acta. (BBA)-Biomembr. 1778 (3), 588–600.

De Melo Pereira, G.V., Karp, S.G., Letti, L.A.J., Soccol, V.T., Soccol, C.R., 2021. Current status and perspectives of probiotic cheese. Trends Food Sci. Technol. 110, 417–430.

Desai, K.M., Akolkar, S.K., Badhe, Y.P., Tambe, S.S., Lele, S.S., 2006. Optimization of fermentation media for exopolysaccharide production from Lactobacillus plantarum using artificial intelligence-based techniques. Process Biochem. 41 (8), 1842–1848.

Dewanjee, S., Gangopadhyay, M., Das, U., Sahu, R., Samanta, A., Banerjee, P., 2014. Signal transducer and oxidative stress mediated modulation of phenylpropanoid pathway to enhance rosmarinic acid biosynthesis in fungi elicited whole plant culture of *Solenostemon scutellarioides*. Enzyme Microb. Technol. 66, 1–9.

Ewaschuk, J.B., Diaz, H., Meddings, L., Diederichs, B., Dmytrash, A., Backer, J., 2008. Secreted bioactive factors from *Bifidobacterium infantis* enhance epithelial cell barrier function. Am. J. Physiol.-Gastrointest. Liver Physiol. 295 (5), G1025–G1034.

Hazra, A., Chatterjee, S., Mohanta, A., Paul, P., Samanta, A., 2023. A review on fungal pullulan as a natural polymer focusing on targeted therapy for colon cancer and other pharmaceutical applications. Polym. Adv. Technol. 34 (10), 3183–3192.

Lopez, P., Monteserin, D.C., Gueimonde, M., de los Reyes-Gavilan, C.G., Margolles, A., Suarez, A., 2012. Exopolysaccharide-producing *Bifidobacterium* strains elicit different in vitro responses upon interaction with human cells. Food Res. Int. 46 (1), 99–107.

Mitra, D., Basu, A., Das, B., Jena, A.K., De, A., Das, M., 2017. Gum odina: an emerging gut modulating approach in colorectal cancer prevention. RSC Adv. 7 (46), 29129–29142.

Mitra, D., Jena, A.K., De, A., Das, M., Das, B., Samanta, A., 2016. Prebiotic potential of gum odina and its impact on gut ecology: in vitro and in vivo assessments. Food Funct. 7 (7), 3064–3072.

Mitra, D., Sikdar, S., Chakraborty, M., Das, O., Samanta, A., Dutta, S., 2023. Gum odina prebiotic prevents experimental colitis in C57BL/6 mice model and its role in shaping gut microbial diversity. Food Biosci. 53, 102509.

Oerlemans, M.M., Akkerman, R., Ferrari, M., Walvoort, M.T., de Vos, P., 2021. Benefits of bacteria-derived exopolysaccharides on gastrointestinal microbiota, immunity and health. J. Funct. Foods 76, 104289.

Pourjafar, H., Ansari, F., Sadeghi, A., Samakkhah, S.A., Jafari, S.M., 2023. Functional and health-promoting properties of probiotics' exopolysaccharides; isolation, characterization, and applications in the food industry. Crit. Rev. Food Sci. Nutr. 63 (26), 8194–8225.

Ruiz-Ruiz, C., de Souza Nascimento, A., Weiss, R.G., Goycoolea, F.M., 2020. Differences in production, composition, and antioxidant activity of exopolymeric substances produced by *Fusarium culmorum* grown on optimized culture media. J. Fungi 6 (1), 21.

Rütering, M., Schmid, J., Rühmann, B., Schilling, M., Sieber, V., 2016. Controlled production of polysaccharides–exploiting nutrient supply for levan and heteropolysaccharide formation in *Paenibacillus sp.* Carbohydr. Polym. 148, 326–334.

Schmid, J., Sieber, V., Rehm, B., 2015. Bacterial exopolysaccharides: Biosynthesis pathways and engineering strategies. Front. Microbiol. 6, 496.

Shao, L.I., Wu, Z., Zhang, H., Chen, W., Ai, L., Guo, B., 2014. Partial characterization and immunostimulatory activity of exopolysaccharides from *Lactobacillus rhamnosus* KF5. Carbohydr. Polym. 107, 51–56.

Singh, R.P, et al., 2016. Characterization of EPS by SEC and their role in biofilm formation. Colloids Surf B Biointerfaces 146, 133–140.

Stingele, F., Vincent, S.J., Faber, E.J., Newell, J.W., Kamerling, J.P., Neeser, J.R., 1999. Introduction of the exopolysaccharide gene cluster from Streptococcus thermophilus Sfi6 into Lactococcus lactis MG1363: production and characterization of an altered polysaccharide. Molecular microbiology 32 (6), 1287–1295.

Surayot, U., Wang, J., Seesuriyachan, P., Kuntiya, A., Tabarsa, M., Lee, Y., 2014. Exopolysaccharides from lactic acid bacteria: Structural analysis, molecular weight effect on immunomodulation. Int. J. Biol. Macromol. 68, 233–240.

Tudu, M., Chatterjee, S., Mohanta, A., Priya, K., Samanta, A., 2025a. Employing biotechnology for the treatment of organic refuse. In AI Technologies for Enhancing Recycling Processes. IGI Global Scientific Publishing, pp. 113–134.

Tudu, M., Hazra, A., Paul, P., Mohanta, A., Chatterjee, S., Samanta, A., 2025b. Effect of different ratios of polyvinyl alcohol-gum odina for the preparation and characterization of biodegradable composite films. Biopolymers 116 (1), e23630.

Tudu, M., Samanta, A., 2023. Natural polysaccharides: chemical properties and application in pharmaceutical formulations. Eur. Polym. J. 184, 111801.

Vazquez-Rodriguez, A., Vasto-Anzaldo, X.G., Barboza Perez, D., Vázquez-Garza, E., Chapoy-Villanueva, H., García-Rivas, G., et al., 2018. Microbial competition of *Rhodotorula mucilaginosa* UANL-001L and *E. coli* increase biosynthesis of non-toxic exopolysaccharide with applications as a wide-spectrum antimicrobial. Sci. Rep. 8 (1), 798.

Vishwakarma, V., Prasanna, R., Joshi, R., Nain, L., Kumar, A., 2018. Microbial extracellular polymeric substances: ecological functions and biotechnological potential. Front. Microbiol. 9, 1636.

Wang, K., Li, W., Rui, X., Chen, X., Jiang, M., 2018. Activation of immune receptors by exopolysaccharides from *Streptococcus thermophilus*. J. Appl. Microbiol. 125 (4), 931–941.

Wang, Y., Zhang, L., 2013. Exopolysaccharides from lactic acid bacteria: structure, production, and applications. J. Food Sci. 88 (5), S1448–S1456.

Xu, J., Li, M., Liao, K., Song, Y., Ma, Y., Sun, H., 2022. Extracellular polymeric substances produced by the cyanobacterium *Geitlerinema* sp. strain CCC726: Characteristics and potential application. Polymers 14 (7), 1462.

Zhang, Z., Chen, W., Li, Q., Lin, Y., 2016. Production of short-chain fatty acids from exopolysaccharides by gut bacteria. J. Agric. Food Chem. 64 (2), 533 541.

Zhao, Y., Peng, Y., Chen, L., Ding, Y., 2018. HPLC analysis of monosaccharide composition in bacterial EPS. J. Agric. Food Chem. 66 (22), 5788–5796.

Zhou, Y., Cui, Y., Qu, X., 2019. Exopolysaccharides of lactic acid bacteria: structure, bioactivity and associations: a review. Carbohydr. Polym 207, 317–332.

Živković, M., Miljković, M.S., Ruas-Madiedo, P., Markelić, M.B., Veljović, K., Tolinački, M., 2016. EPS-SJ exopolisaccharide produced by the strain *Lactobacillus paracasei* subsp. *paracasei* BGSJ2-8 is involved in adhesion to epithelial intestinal cells and decrease on *E. coli* association to Caco-2 cells. Front, Microbiol. 7, 286.

CHAPTER

7

Exopolysaccharide-producing probiotics and their applications in translational medicine

Aiswarya S.V.[a], *Roopam Kumawat*[a], *Merugu Shashank Goud*[a], *Kesavan Bhaskar Reddy*[b], *Avijeet Singh Oinam*[c] and *Onkar Nath Tiwari*[a]

[a]Centre for Conservation and Utilization of Blue Green Algae (CCUBGA), Division of Microbiology, ICAR-Indian Agricultural Research Institute (PUSA), New Delhi, India [b]Apollo Institute of Pharmaceutical Sciences, The Apollo University, Chittoor, Andhra Pradesh, India [c]Department of Botany, Regional College, Imphal, Manipur, India

7.1 Introduction

Exopolysaccharides (EPS) are complex polysaccharides synthesized and secreted by a variety of microorganisms, including bacteria, yeasts, algae, and fungi, into their external environment. These extracellular polymers can be either linear or branched, and based on their composition, they are classified as homopolysaccharides, in which polysaccharide is made of a single type of sugar, or heteropolysaccharides, in which polysaccharide is made of multiple types of sugars (Angelin and Kavitha, 2020). The sugar components commonly found in EPS include glucose, galactose, fructose, rhamnose, mannose, xylose, and arabinose that are present in varying proportions. Additionally, sugar derivatives are frequently part of the structure (Saadat et al., 2019). Beyond carbohydrates, EPS molecules may also be chemically modified by the attachment of phospholipids, nucleic acids, or nonsugar substituents such as sulfate, carboxylate, phosphate, and succinate groups, further enhancing their structural diversity and functional complexity. These polysaccharides exhibit a wide range of industrial and health-related applications. These biopolymers, with significant structural and functional characteristics, are secreted by diverse prokaryotic and eukaryotic microorganisms such as fungi and algae in response to various biotic and abiotic stress conditions, helping them adapt and survive in challenging environments. Due to their unique material properties, along with being nontoxic, completely biodegradable, and

DOI: https://doi.org/10.1016/B978-0-443-36619-2.00007-3

biocompatible, microbial EPSs have gained considerable attention across multiple industrial sectors in recent years.

According to the definitions provided by the WHO and the FAO, probiotics are known as microorganisms that are live when taken in quantity that are sufficient to provide various benefits in health to the host (Morelli, 2013). Among various microbial polysaccharides, EPS synthesized by probiotic LAB are particularly notable for their diverse applications, because of their generally recognized as safe (GRAS) status and proven biological activity under both under lab conditions and in the body (Badel et al., 2011). Microbial polysaccharides exhibit unique and valuable properties that have led to their exploitation in the food, cosmetic, and pharmaceutical sectors, where they are used as bio-flocculants, bio-adsorbents, and in drug delivery systems (Silva et al., 2019). As these natural polysaccharides offer significant advantages over artificial ones, they are also used in the green production of silver nanoparticles. Efficient EPS producing lactic acid bacteria (LAB) include *Lactobacillus, Enterococcus, Lactococcus, Bifidobacterium, Leuconostoc, Weissella,* and *Pediococcus* species (Sanalibaba and Çakmak, 2016). Among the diverse microorganisms ingested by humans, probiotic bacteria are distinguished by their ability to withstand harsh gastrointestinal conditions, including low pH, bile salts, and digestive enzymes, and to adhere to and inhabit the epithelial layer of the GI tract.

Antibiotic usage most of the time reduces the protective functions of the native gut microbiota against pathogenic organisms. On the other hand, probiotics possess the ability to counteract some of the adverse effects of antibiotics, which, if consumed regularly, may suppress the immune system. Probiotics enhance host immune responses, thus aiding in the defense against diseases triggered by pathogens (Surayot et al., 2014). Due to these reasons, probiotic-derived EPS are being investigated and used in the treatment of various human health conditions, such as inflammatory bowel disease (IBD), autoimmune disorders, cancer in the colon, gastric ulcers, heart diseases, and obesity (London et al., 2014; El-Deeb et al., 2018). EPS are also gaining attention in pharmacological research for their roles as anticoagulants, antithrombotics, immunomodulators, lowering cholesterol, and nutraceuticals. Additionally, EPS exhibit remarkable water-binding capacity, moisture retention, swelling, and gelation properties. EPS types such as dextran, xanthan, cellulose, and alginate contribute to the formation of biofilm on the surface of the bacteria, functioning as a protective layer (Kodali et al., 2009).

EPS produced by strains such as *Lactobacillus* sp. isolated from distinct sources have been reported to show notable antioxidant and antitumor properties (Sungur et al., 2017). Recent investigations into EPS synthesized by *Lactobacillus plantarum* strains C70 and RJF4, derived from diverse origins, demonstrated antidiabetic potential through the inhibition of α-amylase and α- glucosidase enzymes under laboratory conditions (Ayyash et al., 2020). Additionally, EPS isolated from *Lactobacillus* sp Ca6 has shown effective dermal wound healing in vivo when tested on male Wistar rats (Trabelsi et al., 2017). EPS produced by *Lactobacillus* strains has also been found to enhance the native immune system. Specifically, EPS from *Lactobacillus plantarum* JLK0142, *Lactobacillus johnsonii* 142, *Leuconostoc citreum*, and certain *Bifidobacterium* species have demonstrated notable immunoregulatory activities (Domingos-Lopes et al., 2017).

This chapter focuses on the EPS secreted by probiotic bacteria, highlighting their origin, nature categories, mechanism, effect on gut microbiota, and their potential applications in translational medicine. A wide spectrum of health-promoting properties of EPS is thoroughly discussed, integrating the most current studies in the field. Additionally, the correlation between the physical features of EPS and their biological functionality is explored in detail, offering new insights into how structural features influence therapeutic outcomes, and also the challenges along with future directions in which research under this aspect can be done will be discussed.

7.2 Exopolysaccharide production in probiotics

7.2.1 Source of probiotic exopolysaccharide

Generally, probiotic strains are isolated from a variety of dairy-based and nondairy-based sources. Milk and milk obtained products from dairy origin, which are widely studied for their probiotic ability and health benefits. However, such sources are not suitable for individuals who cannot tolerate lactose, milk protein allergies, or hypercholesterolemia (Shori, 2016). On the other hand, nondairy source including cereals, fruits, and vegetables, offer promising alternatives. These are good in vitamins, antioxidants, minerals, dietary fibers, and other phytochemicals, which are useful for the growth and survival of various probiotic bacteria and other beneficial microbes (Homayouni et al., 2012). As a matter of fact, any food matrix that contains fermentable sugars can potentially support probiotic growth. Recently, fermented foods have gained significant attention for the isolation of novel probiotic LAB, due to their extended shelf life, improved flavor, enhanced texture, and functional value. These foods serve not only as reservoirs of diverse microbial communities but also offer natural environments for EPS producing bacteria, which are frequently used as starter cultures in the industries. Furthermore, animal-derived sources such as meat and meat-based fermented products also serve as habitats for probiotic strains. Probiotic bacteria isolated from these foods often display robust exopolysaccharide production, contributing to both their functional characteristics and health-promoting properties (Table 7.1) (Patel et al., 2012).

7.2.2 Mechanisms underlying exopolysaccharide production in probiotics

EPS are made of repeating units of polymerized oligosaccharides. Probiotic microbes, particularly LAB, which include various genera such as *Lactobacillus, Lactococcus, Pediococcus, Leuconostoc,* and *Enterococcus,* can naturally synthesize different types of polysaccharides or glycans that function as carbohydrate-based biopolymers. Several enzymes and proteins are key players in EPS biosynthesis, and their activity is often governed by the regulation of specific genes. A central enzyme involved is glycosyltransferase (GTF), which uses sugar substrates such as galactose, glucose, fructose, and sugar derivatives to form the EPS structures. The strain type, environmental conditions, and medium ingredients significantly influence the properties of the EPS generated. Structural diversity in EPS is primarily shaped by the types of monosaccharides involved, the nature of their linkages, the presence of side chains, charge variations, and chemical modifications (Saadat et al., 2019). The biosynthetic process of EPS in probiotics can be extracellular or intracellular. Generally, homopolysaccharides are synthesized extracellularly, whereas heteropolysaccharides can be synthesized via both pathways. The extracellular synthesis of homopolysaccharides involves the activity of fructosyltransferases and GTFs during polymerization. In this mechanism, sugar residues are first hydrolyzed into monomers outside the cell and then polymerized by GTFs into larger polysaccharides (Lynch et al., 2018; Angelin and Kavitha, 2020). The primary substrate used is sucrose, which is broken down to glucose and fructose, and they undergo polymerization, which leads to the formation of glucans and fructans by glucan and fructan sucrase, respectively. These are further classified into various types based on sources. These homopolysaccharides are secreted directly into the external space (Zarour et al., 2017; Angelin and Kavitha, 2020).

On the other hand, the intracellular EPS synthesis begins when sugars are transported into the cytoplasm, converted into various monomeric sugar units, and then polymerized. These

TABLE 7.1 EPS producing probiotic strains with their key roles.

Probiotic strain	Type of EPS produced	Key impacts	References
Lactobacillus plantarum	Heteropolysaccharide, high molecular weight	Antioxidant, antimicrobial, anti-inflammatory, promotes wound healing and gut barrier integrity.	Wang et al. (2022)
Lactobacillus delbrueckii subsp. *bulgaricus*	Heteropolysaccharide	Improves texture in dairy, immunomodulation, and cholesterol-lowering.	Xu et al. (2015)
Lacticaseibacillus rhamnosus GG	Galactose-rich heteropolysaccharide	Enhances gut mucosal barrier, protects against pathogens, and immunomodulation.	Lebeer et al. (2011)
Weissella cibaria	Branched heteropolysaccharide	Antioxidant, biofilm inhibition, oral health promotion, tissue repair.	Kang and Park (2021)
Bifidobacterium longum	Neutral heteropolysaccharide	Anti-inflammatory, supports gut microbiota, enhances immune regulation.	Inturri et al. (2017)
Streptococcus thermophilus	Homopolysaccharide and heteropolysaccharide	Improves dairy product texture, antioxidant, and anti-inflammatory.	Ruas-Madiedo et al. (2014)
Levilactobacillus brevis	Heteropolysaccharide	Antioxidant, antimicrobial, stress tolerance, gut microbiota modulation.	Ahmad et al. (2024)
Pediococcus pentosaceus	Neutral heteropolysaccharide	Antioxidant, antimicrobial, probiotic formulation stability.	Tingirikari et al. (2014)

EPS, exopolysaccharide

polymers are anchored to membrane-bound isoprenoid lipid carriers. Following necessary structural modifications, the polysaccharides are exported outside the cell and assembled into final EPS structures (Angelin and Kavitha, 2020). The enzymes involved in this process fall into four major categories: Hexokinase, UDP-glucose pyrophosphorylase, GTFs, and membrane proteins, which perform each step in a proper manner.

The heteropolysaccharide synthesis relies primarily on the Wzx/Wzy-based pathway, where all stages of everything happen in the cytoplasm except polymerization. The key steps include sugar import, synthesis of sugar nucleotides, same unit formation, and polymerization. Sugar uptake depends on the sugar type and can occur via passive or active transport mechanisms, including the phosphoenolpyruvate-phosphotransferase system. Once inside, glucose is phosphorylated by hexokinase to glucose-6-phosphate and then converted to glucose-1-phosphate. UDP-glucose pyrophosphorylase then converts this intermediate to UDP-glucose (Werning et al., 2012; Angelin and Kavitha, 2020).

Precursors for EPS formation are derived from core carbon metabolism and include nucleoside diphosphate sugars like NDP-glucose, GDP-mannuronic acid, and sugar derivatives such as UDP-GlcNAc, dTDP-rhamnose, and UDP-galactose. In the Wzx/Wzy system, these sugar nucleotides are connected to an undecaprenyl diphosphate molecule in the membrane by GTFs, and subsequently transported along the membrane by means of membrane proteins like flippases (Rehm, 2010; Islam and Lam, 2014). The lipid carrier, an isoprenoid alcohol, binds the sugar unit via a pyrophosphate linkage. Postpolymerization, EPS may undergo enzymatic modifications such as methylation, acetylation, or sulfation before being released as either a capsular layer or a slime/EPS layer through transporters like ABC-type transporters, permeases, or flippases (Meng et al., 2021).

7.2.3 Factors affecting exopolysaccharide production

The overall EPS generated by probiotics are highly impacted by multiple factors that involve species, medium of growth, internal and external growth conditions. Key environmental parameters include pH, activity of water, oxidation-reduction potential, and medium composition. Presence of antimicrobials, temperature, relative humidity, gas composition, and microbial competition (Lynch et al., 2018; Scheuerl et al., 2020). Though probiotic strains have inherent EPS-producing capabilities, these conditions can greatly alter the yield. For instance, adjusting temperature and pH has been reported to significantly improve EPS output. Low temperatures are generally favorable for homopolysaccharides production, while heteropolysaccharides synthesis is closely tied to carbon source availability and microbial growth. Maintaining stable pH levels during fermentation helps preserve the enzymatic activity involved in EPS formation, ultimately boosting the yield (Liu et al., 2017).

Several reports show that EPS production is a key adaptive strategy used by probiotics to cope with environmental stress (Ozturk and Aslim, 2010; Nguyen et al., 2020). The transition of bacteria from a plankton stage to the formation of biofilm enhances their ability to survive in harsh conditions. Biofilm formation has been shown to increase the resistance of bacteria against organic acids and antibiotics. Specific stress materials common to the gastrointestinal tract, like bile salts and high osmolarity, have been found to stimulate biofilm development in probiotic *Lactobacillus* species during in vitro studies (Nguyen et al., 2020). The LuxS gene, which performs a vital role in quorum sensing, is connected to biofilm formation and cell adhesion, and has also been identified as important for acid and bile tolerance as well as for enhancing the survival of *Lactobacilli* during GI transit (Xu et al., 2006). EPS, a fundamental component of the biofilm matrix, contributes to increased resilience of probiotics under challenging situations like different pH GI environments, immune responses, and the presence of harmful substances like ethanol, lysozyme, and oxidative agents. This protective effect of EPS is attributed to its molecular composition and location on the cell surface, where it forms a shielding barrier that can trap toxins (Nguyen et al., 2020). However, it is not the same in all cases. Some research suggests that certain strains do not enhance their resistance to gastrointestinal stress through EPS production, and in some cases, exposure to bile has been shown to decrease EPS synthesis (Ruiz et al., 2013; Liu et al., 2017).

7.3 Health benefits of exopolysaccharide from probiotics

The biological activity of EPS is closely linked to their chemical structure, which in turn is shaped by the fermentation conditions used during the growth of probiotic bacteria. EPS derived from various sources can vary significantly based on factors such as monosaccharide composition, glycosidic bond types, and chemical modifications. These elements influence key structural features like molecular weight, chain conformation, and branching patterns. Such structural diversity plays a crucial role in determining the health-enhancing properties of EPS. EPS synthesized by these strains has garnered significant attention due to their therapeutic ability, antimicrobial potential, immunomodulatory, anti-inflammatory, antioxidant, antiulcer, antiviral, antidiabetic, and cholesterol-lowering effects.

7.3.1 Immunomodulatory activity

EPS produced by probiotic bacteria have been shown to act a notable role in changing both native and adaptive immunity through various mechanisms. These include stimulating the expansion of lymphocytes, immune cells activation and enhancing the body's protection against foreign bodies, tumors, and inflammation conditions such as IBD (Angelin and Kavitha, 2020). EPS are reported to boost the tumoricidal activity of natural killer cells, increase the phagocytic activity of mononuclear cells, which increases the production of cytokines and antibodies by B cells.

The immunomodulatory ability of EPS is strongly impacted by their physical and chemical properties, including molecular weight, solubility in H_2O, electrical charge, and stereochemistry. Research indicates that low molecular weight EPS or those with negative charges, such as heteropolysaccharides containing phosphate groups, are more effective in stimulating host immunity. In contrast, neutral EPS with higher molecular weights tend to exhibit either weak immune-stimulating effects or even immunosuppressive activity (Laiño et al., 2016). Nevertheless, those EPS that have a higher molecular weight have demonstrated immunogenic properties (Poli et al., 2010). Numerous studies have documented the immunomodulatory properties of EPS derived from LAB. For instance, EPS from *Lactobacillus paracasei* induced the production of pro-inflammatory cytokines IL-6, IL-1β, and TNF-α in murine RAW macrophages. EPS obtained from *Lactobacillus plantarum* NTU102 activated the NF-κB pathway, resulting in enhanced secretion of IL-6, IL-1β, and TNF-α. The phosphopolysaccharide produced by *Lactobacillus bulgaricus* OLL1073-R1 has been found to improve macrophage activity both in vitro and in vivo (Laiño et al., 2016). *Bifidobacterium* species also produce strain-specific EPS that elicit distinct immune responses. For example, EPS from *Bifidobacterium longum* BCRC 14634 promotes the generation against inflammation cytokine IL-10. In contrast, EPS from *Bifidobacterium breve* UCC2003 stimulates the release of pro-inflammatory cytokines such as IL-12, INF-γ, and TNF-α. In another study exploring EPS derived from thermophilic bacilli isolated from hydrothermal vents, peripheral blood mononuclear cells produced elevated levels of Th1-type cytokines, including IFN-γ, IFN-α, TNF-α, IL-12, and IL-18, but Th2-type cytokines were not induced (Gugliandolo et al., 2015). Moreover, EPS from *Weissella confusa* has demonstrated the ability to increase the levels of IgG, IgM, and IgA antibodies in mice (Adebayo-Tayo et al., 2018), further confirming the immunostimulatory potential of EPS-producing probiotics.

7.3.2 Antioxidant potential

Maintaining a stable, reduced cellular environment is vital for the survival of living organisms. However, the generation of ROS, such as peroxides and free radicals, can disrupt this equilibrium, leading to damage of cellular components like proteins, lipids, and DNA. These oxidative stresses may result in serious outcomes, including cell injury, apoptosis, or necrosis. Typically, ROS are produced in minimal amounts, and their harmful effects are mitigated by the body's defense mechanisms, which contain enzymes that act against oxygen, such as catalase, superoxide dismutase, and lactoperoxidase. Oxidative stress caused by free radicals has been involved in the evolution of various diseases, including neurological disorders, genetic disorders (such as sickle cell anemia), metabolic conditions (e.g., diabetes), heart ailments, chronic fatigue syndrome, and cardiovascular diseases (Ye et al., 2015). Recently, there has been a new interest is found in using natural antioxidants due to their perceived safety, higher consumer acceptance, and easier regulatory approval. These natural compounds are fully metabolized by the body, minimizing the risk of accumulation in fat tissues. However, there is still a need for more data regarding their potential carcinogenic or mutagenic effects (Sarkar and Ghosh, 2016). EPS have emerged as promising natural antioxidant agents due to their distinctive biological properties, molecular structure, and surface characteristics. Their polyelectrolytic nature and high binding capacity make them particularly suitable for medical and food applications (Andrew and Jayaraman, 2020). Notably, certain EPS can sustain their antioxidant activity at higher temperatures, a crucial characteristic needed for the food industry, where products often undergo heat processing. For example, EPS from *Enterococcus faecium* K1, which contains various sugars, retains stability up to 237°C (Bhat and Bajaj, 2018). EPS exerts antioxidant effects through mechanisms similar to those of other polysaccharides. One key pathway is their ability to donate hydrogen atoms. Their structural features include functional groups capable of electron donation or withdrawal, as well as monosaccharides rich in ketone and aldehyde groups that contribute to antioxidant activity (Andrew and Jayaraman, 2020). Furthermore, structural modifications can increase their antioxidant properties. For instance, sulfonation of β-D-glucan, an EPS from the fungus *Lasiodiplodia theobromae*, resulted in significant changes in morphology and improved hydroxyl radical scavenging capacity (Calegari et al., 2017).

EPS also exhibit metal-binding capabilities, making them effective in counteracting cell toxicity caused by heavy metals and in the bioremediation of contaminated environments (Andrew and Jayaraman, 2020). While many in vitro studies support the antioxidant potential of EPS, in vivo evidence remains limited. Some studies have investigated their biological effects using model organisms. For example, EPS from *Lactobacillus fermentum* S1 increased total antioxidant capacity, enhanced SOD activity, and reduced malondialdehyde levels in *Caenorhabditis elegans*. Similarly, EPS YW11 derived from *Lactobacillus plantarum* and *Bifidobacterium animalis* reduced oxidative stress induced by D-galactose in aging mice by increasing antioxidant enzyme levels (Andrew and Jayaraman, 2020). To date, there are no known clinical studies directly evaluating the antioxidant properties of EPS in humans, which pressurize the need for future study to fully understand their therapeutic potential.

7.3.3 Anti-inflammatory effects

To explore the anti-inflammatory potential of EPS, it's important to first understand the basics of inflammation. This is defined as a complex biological reaction that the body initiates in response to deleterious agents such as pathogens or allergens. It is broadly categorized into acute and chronic inflammation, depending on its duration and the type of immune response involved.

Acute inflammation typically unfolds in three main phases: Increased blood flow to the affected area, vasodilation, and an increase in permeability of vascular tissues, followed by the infiltration of plasma and leukocyte transfer to the site of injury or infection. The body uses specific receptors and signaling pathways depending on the nature of the trigger. For example, bacterial infections activate toll-like receptors (TLRs), a part of the innate immune system, resulting in the upregulation of genes associated with inflammation. This leads to the release of cytokines and mediators, including nitric oxide, TNF-α, IL-1β, and chemokines like CCL2 and CXCL8. At the same time, the production of anti-inflammatory cytokines like IL-10 is often downregulated.

These mediators stimulate tissue responses such as antibody production, C-reactive protein synthesis, and complement system activation. When the stimulus persists, chronic inflammation may ensue. This prolonged immune activation can contribute to various chronic diseases, including diabetic complications, rheumatoid arthritis, certain cancers, cardiovascular conditions, and endocrine or metabolic disorders (Arulselvan et al., 2016).

Recent research suggests that EPS produced by probiotics can modulate inflammatory responses. For instance, EPS derived from *Lactobacillus* species isolated from the healthy human vagina was found to reduce TNF-α production while boosting IL-10 levels in HeLa cells. Similarly, in *Bifidobacterium longum* BCRC 14634 demonstrated immunomodulatory properties in J77A. one macrophage, again decreasing TNF-α and enhancing IL-10 production. Other strains, such as *Bacillus licheniformis* BioE-BL11 and *Leuconostoc mesenteroides* BioE-LMD18, suppressed IL-6 secretion in LPS-induced RAW264. Mouse macrophages, while increasing IL-10 levels in a dose-dependent manner (Angelin and Kavitha, 2020). These findings suggest a strong anti-inflammatory effect of EPS via cytokine modulation. Additionally, EPS has been shown to promote the production of fatty acids with short chains like butyrate and acetate that have properties like anti-inflammation. This is noted in the case of *Lactobacillus fermentum* Lf2, when administered with yogurt to BALB/c mice, the level of these fatty acids increased. Butyrate, in particular, helps reduce inflammation and supports healthy colon motility (Angelin and Kavitha, 2020). Interestingly, EPS can also trigger a balanced immune response. For instance, EPS from *Lactobacillus rhamnosus* provoked the secretion of pro-inflammatory cytokines and anti-inflammatory cytokine IL-10 in mouse peritoneal macrophages (Angelin and Kavitha, 2020). This dual role implies that EPS may help fine-tune immune responses rather than simply suppressing inflammation. Furthermore, EPS may exert indirect anti-inflammatory effects by activating TLRs, enhancing the body's ability to detect and fight pathogens. A study demonstrated that EPS from *Bifidobacterium animalis* subsp. lactis stimulated TLR-4 in a mouse colitis model, suggesting its potential in controlling intestinal inflammation (Castro-Bravo et al., 2018).

7.3.4 Anticancer potential

Cancer is characterized by the multiplication of cells in an uncontrolled manner, which often results in irreversible genetic damage to proto-oncogenes and tumor suppressor genes. These changes not only lead to damage at the site of the tumor but also affect surrounding tissues. Furthermore, cancerous cells can metastasize, spreading to distant organs via the bloodstream or lymphatic system, causing widespread harm throughout the body. Conventional cancer treatments, such as chemotherapy, while effective in killing tumor cells, also tend to harm normal healthy cells, leading to significant side effects. This has driven interest in developing alternative cancer therapies that are highly accessible, less cost and have reduced adverse effects (Saadat et al., 2019; Wu et al., 2021).

Probiotic-derived EPS have shown promise in cancer prevention and therapy through a range of mechanisms. These include the induction of programmed cell death (apoptosis), suppression of cell proliferation, selective toxicity toward cancer cells, interference with the cell cycle, inhibition of mutations, antioxidative effects, antiangiogenesis, modulation of inflammatory responses, and regulation of key signaling pathways involved in cancer progression. EPS produced by *Lactobacillus plantarum, L. acidophilus,* and *L. helveticus* have demonstrated particularly strong anticancer properties among LAB. However, the efficacy of EPS in cancer inhibition can vary significantly between different strains. Properties such as the EPS's chemical structure, water solubility, adherence to tumor cells, and degree of purification all influence its anticancer potential (Wu et al., 2021).

Numerous in vitro studies have documented the anticancer activities of EPS. For instance, EPS G10 from *L. gasseri* enhanced cell death in Henrietta lacks (HeLa) cells by upregulating the pro-apoptotic BAX gene and increasing protein levels. Similarly, EPS derived from the marine probiotic *Pediococcus pentosaceus* M41 reduced the growth of cancer cell lines in humans, such as breast cancer and colon cancer. Another example is EPS from *L. plantarum* NCU116, which activated certain genes like *FasL, Fas,* and *c-Jun* by means of TLR that triggers apoptosis in mouse having cancer cells.

In addition to lab-based studies, those which conducted inside the body for checking anticancer properties of EPS have been studied. EPS from both wild-type and mutant strains of *L. delbrueckii* enhanced immunoglobulin levels (IgG, IgM) in mice with experimentally induced tumors, ultimately extending their survival time (Adebayo-Tayo and Fashogbon, 2020). Similarly, EPS from *Lactobacillus plantarum* NRRL B-4496 exhibited antitumor effects against Ehrlich Ascites Carcinoma in animal models, resulting in prolonged survival and improved overall health outcomes (Haroun et al., 2013).

7.4 Applications in translational medicine

Probiotic EPS are those polymers that are secreted by various probiotic bacteria, especially LAB such as *Lactiplantibacillus plantarum, Lacticaseibacillus rhamnosus, Levilactobacillus brevis, Bifidobacterium* spp., and *Weissella* spp. These polymers exhibit diverse monosaccharide compositions, branching patterns, and molecular weights that influence their bioactivity. In translational medicine, probiotic EPS bridges laboratory discoveries and real-world healthcare applications, acting through antimicrobial, immunomodulatory, antioxidant, and barrier-enhancing mechanisms. (Fig. 7.1) shows the application of probiotic eps in translational medicine.

7.4.1 Use in treating digestive disorders

Probiotic EPS helps in digestive disorders by following ways: (1) Modulating mucosal and systemic immune responses, (2) strengthening the intestinal barrier and tight junctions, (3) improving mucoadhesion and colonization, (4) shaping gut microbial communities, and (5) increasing beneficial metabolites (like short-chain fatty acids [SCFA]). These all evidences are in the preclinical (cell culture, mouse colitis models) stage; in the case of human clinical data,

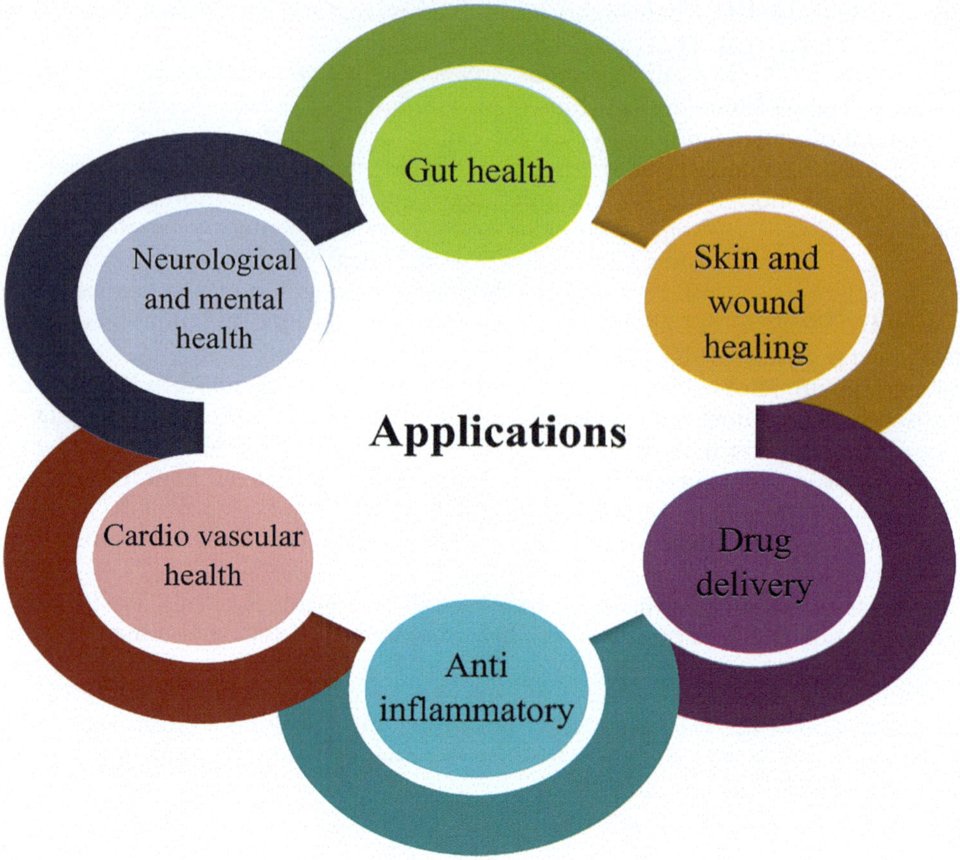

FIGURE 7.1 Applications of probiotic EPS in translational medicine. *EPS*, Exopolysaccharide.

specifically on isolated EPS are limited so far. Certain researches are conducted in this direction. A study by (Zamora-Pineda et al., 2022) shows that EPS can downregulate cytokines that are pro-inflammatory in nature and upregulate anti-inflammatory markers, often via modulation of macrophages, dendritic cells, and T cell responses. This reduces mucosal inflammation in colitis models. EPS-producing strains increase expression of tight junction proteins (occludin and claudins) and mucin production, reducing epithelial permeability, which is also known as leaky gut, that is commonly seen in irritable bowel disorder and other gut disorders. Multiple probiotic formulations and EPS from isolates show this in vitro and in animal models (Gou et al., 2022). EPS also modifies cell surface properties to improve mucin binding and transient colonization, which helps probiotics persist longer in the gut and interact with the host mucosa. This physical interaction fosters barrier and immunoregulatory effects (di Vito et al., 2022). EPS can modify gut microbiota as they can be fermented by resident microbiota, promote growth of beneficial taxa, alter community composition, and increase SCFA (such as acetate, propionate, and butyrate) that feed colonocytes and have anti-inflammatory effects. Animal colitis studies showed that microbiota shifts and SCFA increase after EPS treatment (Liu et al., 2021). Certain types of EPS reduce pathogen adherence or interfere with pathogen-driven inflammation. e.g., EPS from *Lactobacillus rhamnosus* GG reduced *Salmonella*-induced inflammation in model systems (Li et al., 2023).

7.4.2 Role in metabolic disorders

Metabolic diseases such as obesity, diabetes, type II and dyslipidemia share common pathological features like gut dysbiosis, increased intestinal permeability, chronic low-grade inflammation, and dysregulated lipid/glucose metabolism. EPS can act at several of these nodes by modulating gut microbiota by means of increasing SCFA, which in turn reduces inflammation; improved barrier by less endotoxemia; and direct signaling by means of improved hepatic/insulin signaling, so they are mechanistically plausible adjuncts to therapy or prevention. Many LAB and bifidobacterial EPS are fermentable by resident gut microbes, promoting SCFA. SCFAs modulate host energy metabolism, improve gut barrier function, and influence appetite. In vitro fermentation and animal studies show EPS can be metabolized by fecal microbiota and increase SCFA production (Miyamoto et al., 2023). EPS producing strains or isolated EPS can selectively enrich beneficial taxa and reduce pathobiont changes associated with improved metabolic outputs in animal models, such as less fat gain and better glucose tolerance (Zhao et al., 2025). Some EPS generating microbes are tend to reduce cholesterol in serum and change hepatic lipid metabolism in animals through bile-acid binding/entrapment, interference with micelle formation, or indirect effects via microbiota. In vitro cholesterol removal assays and in vivo animal reductions of serum lipids have been reported (London et al., 2014).

7.4.3 Cardiovascular benefits

Probiotic EPS from *Lactobacillus*, *Bifidobacterium*, and related lactic-acid bacteria show cholesterol-lowering, anti-inflammatory, antioxidant, antihypertensive, and antiatherosclerotic activities in vitro and in animal models; human evidence for EPS specifically is still limited and often indirect. EPS producing LAB can reduce serum total cholesterol and low-density lipoprotein in animal studies and some in vitro assays. Proposed mechanisms include: Direct cholesterol assimilation by cells, binding/precipitation of cholesterol or bile salts by EPS, modulation of host bile-acid metabolism, and induction of SCFA that affect hepatic lipid metabolism (London et al., 2014; Tsai et al., 2014). Some probiotic interventions reduce vascular inflammation, improve nitric oxide availability, and slow atherosclerotic lesion development in animal studies (Zhai et al., 2022). Certain bacterial metabolites and EPS fractions show angiotensin converting enzyme inhibitory activity in vitro; this could translate to modest blood-pressure effects, although direct human EPS evidence is limited (Kwon et al., 2020).

7.4.4 Use in skin and wound healing

Bacterial cellulose (BC) is synthesized through bacterial fermentation, allowing precise control over its three-dimensional architecture via strain selection or optimization of parameters for growth (Lin et al., 2013). That leads to the formation of a nanofibrillar, water-insoluble network that closely resembles collagen in structure and exhibits excellent biocompatibility (Torres et al., 2012). BC has gained attention for wound healing and skin repair applications, serving as an absorbent for wound exudates while remaining permeable to gases. Compared to plant-derived cellulose, BC offers higher crystallinity, superior water retention, and greater resistance to degradation of enzymes and microbes. Its performance as dressing for wounds or engineering of tissue scaffolds can be further tailored through chemical modification or by incorporating synthetic or mineral materials, other biopolymers (such as hyaluronan, alginate, chitosan, collagen), or growth factors (Torres et al., 2012; Fu et al., 2013).

When compared to plant-cellulose dressings, clinical trials utilizing BC-based dressings for both acute and chronic wounds, including foot ulcers in diabetic patients, showed better healing results (Fu et al., 2013). With their demonstrated moisturizing properties, BC-glycerin membranes may be especially helpful in treating dryness brought on by dermatological disorders, including psoriasis and atopic dermatitis (Almeida et al., 2014). Furthermore, cell adhesion and incorporation were improved by the more symmetric nanoporous structure that electrospun acetylated BC generated compared to casting films (Costa et al., 2011). Acrylic acid hydrogel dressings produced by electron beam irradiation accelerated the healing of burn wounds in rats in animal models (Mohamad et al., 2014).

In tissue engineering, BC and other polysaccharides have been investigated for regenerative applications. For cardiac tissue engineering, the objective is to create structures that support, repair, or replace damaged myocardium, particularly postinfarction. Products such as pullulan, hyaluronan, and dextran have been extensively used for these purposes (Silva et al., 2015). For example, photopolymerizable hyaluronic acid hydrogels have been designed to mimic the extracellular matrix of valves in heart constructs (Aravamudhan et al., 2014). Alginate origin hybrid copolymers crosslinked with acrylates have demonstrated improved mechanical strength, with some formulations enhancing cardio myoblast proliferation, suggesting their potential in cardiac regeneration (Thankam and Muthu, 2014). Other polysaccharide-based materials also show promise: Oxidized dextran combined with poly-L-lysine forms bioadhesives for surgical use (Matsumura et al., 2014), while pullulan developed carboxymethyl conjugates have been used for tissue engineering sectors (Mishra et al., 2011). Nanofibers with electrospun gelatin crosslinked to oxidized dextran reported favorable scaffold properties for L-929 fibroblast culture (Jalaja et al., 2014). This has proven to be effective for chondrocyte and stem cell cultures, making them promising candidates for cartilage regeneration in osteoarthritis treatment and postarthroplasty recovery. When combined with chitosan, calcium, or antibiotics, alginate scaffolds may also enhance antibacterial properties and heal the wound (Ivanova et al., 2014). Furthermore, these systems have been applied in tissue engineering for generating artificial pancreas, bone, vascular and liver cell constructs. Notably, certain EPS from bacteria isolated from extremophilic environments have shown potential for bone tissue engineering (Nichols et al., 2005). In regenerative medicine, there is increasing interest in integrating stem cells with bioactive, nanopatterned scaffolds made from diverse polymers that include EPS, which bind with specific ligands to direct stem cell behavior and differentiation (Evans et al., 2006).

7.5 Impact on gut microbiota

The human gut harbors a diverse and dynamic microbial ecosystem known as the gut microbiota, which plays a crucial part in balancing host health. This complex microbial community assists in digestion, modulates the immune system, and contributes in synthesis of major metabolites like vitamins and fatty acids with short chains. In recent years, the focus on dietary and microbial interventions to modulate gut microbiota has intensified, particularly with probiotics and prebiotics. Among the various microbial-derived compounds, EPS have gained increasing attention for their multifaceted roles in modulating gut health. These polymers are secreted by microorganisms, those belonging to the LAB category. These compounds can directly affect the composition of microbes in the gut by acting as a selective source for beneficial microbes, influencing microbial competition, and shaping the gut environment to favor symbiosis over dysbiosis.

7.5.1 How exopolysaccharide helps beneficial microbes grow

EPS function not only as structural components or protective layers for the producing bacteria but also serve as important nutrient sources for other commensal microbes residing in the gut. Many beneficial bacteria, such as *Bifidobacterium, Lactobacillus*, and *Streptococcus*, can enzymatically break down EPS into simpler oligosaccharides or monosaccharides that they utilize for energy and growth. This selective utilization creates a nutritional niche that supports the proliferation of health-promoting microbes over harmful ones. For instance, EPS produced by strains like *Lactobacillus plantarum, Lactobacillus delbrueckii*, and *Fructilactobacillus sanfranciscensis* have been found to significantly enhance *Bifidobacterium infantis* and *Lactobacillus acidophilus* growth under lab conditions (Prete et al., 2021). Moreover, EPS can withstand harsh gastric conditions and reaches colon, where they undergo fermentation by the microbes present in the gut, which leads to the production of SCFA like butyrate and acetate metabolites that are known for their anti-inflammatory, immunomodulatory, and gut barrier-supporting properties. Some EPS have also demonstrated antimicrobial activity by inhibiting pathogenic species such as *E. coli* and other members of Enterobacteriaceae. In essence, EPS act as growth enhancers and protective agents that help beneficial microbes establish and maintain dominance within the gut ecosystem (Zhou et al., 2019).

7.5.2 Prebiotic effects of exopolysaccharide

The concept of prebiotics has evolved over time. Initially defined as "non-digestible food ingredients that selectively stimulate the growth or activity of beneficial bacteria in the colon," the modern definition, according to the International Scientific Association for Probiotics and Prebiotics (ISAPP), now includes "any substrate that is selectively utilized by host microorganisms, conferring a health benefit" (Gibson et al., 2017). This broader definition recognizes that prebiotics may act beyond the gut, influencing microbial communities at multiple body sites. For a compound to be considered as prebiotic, it should not undergo digestion in the upper gastrointestinal tract, no absorption in the small intestine, be fermentable by colonic microbes, and positively impact host health by modulating the microbial community or its metabolic activity (Prete et al., 2021). EPS secreted by LAB has shown promising prebiotic potential similar to that of plant-derived polysaccharides. Studies have confirmed that EPS are highly strain-specific in their resistance to digestive processes and in their ability to modulate gut microbiota composition (Ioannou et al., 2025). Research states that EPS from LAB, such as *L. delbrueckii* ssp. *bulgaricus* SRFM-1 and *L. plantarum* not only withstand gastric acidity but also selectively promote the growth of beneficial microbes. These studies additionally report increased SCFA production and inhibition of undesirable microbes, including *E. coli*. A notable example is the

EPS from *L. plantarum* DM5, which significantly enhanced the growth of beneficial strains in controlled experiments (Xiong et al., 2023). Similarly, EPS from *L. sanfranciscensis* showed bifidogenic properties and supported healthy gut flora development (Pourjafar et al., 2023). However, not all results have been consistent. For example, although *Pediococcus parvulus* produces β-D-glucan that increases the adherence and growth of probiotics in vitro, it failed to demonstrate any significant physiological benefit in vivo using mouse models (Pérez-Ramos et al., 2017). This highlights the challenge of translating in vitro findings to complex living systems. Most prebiotic studies involving EPS rely on simulated gastrointestinal systems that lack the full complexity of host-microbe interactions. To bridge this gap, advanced gut fermentation models have been proposed that allow real-time tracking of microbial changes in response to EPS fermentation (Dhiman et al., 2025).

Furthermore, EPS degradation often results in oligosaccharides and monosaccharides that are utilized by other gut microbes in cross-feeding interactions, leading to a cascade of beneficial metabolic outcomes such as increased SCFA and lactic acid production. These metabolic by-products are critical in maintaining intestinal pH, regulating immunity, and supporting epithelial health (Pan et al., 2021). Despite promising data, a lack of human clinical trials remains a limitation, and more in vivo research is essential to validate the prebiotic effects of EPS conclusively.

7.5.3 Balancing the gut microbiome

One of the most crucial roles of EPS is their ability to maintain or restore balance within the gut known as eubiosis. Disrupting it leads to dysbiosis, which causes a lot health related issues, including IBD, obesity, metabolic syndrome, and type 2 diabetes. EPS can promote a balanced microbial environment by inhibiting pathogens, encouraging the growth of SCFAs-producing microbes, and protecting gut epithelial cells. Certain LAB strains, such as *Weissella, Pediococcus, L. plantarum,* and *L. reuteri*, produce EPS that not only improve the adhesion of beneficial microbes to gut epithelium but also enhance epithelial regeneration and immune modulation (Moscovici, 2015; Pan et al., 2021). EPS-mediated adhesion is critical for biofilm formation, which helps beneficial microbes colonize the gut more effectively and prevents the attachment of harmful bacteria. These biofilms act as physical barriers, shielding intestinal cells from toxins, carcinogens, and opportunistic pathogens (Zhou et al., 2021). Additionally, EPS modulate response of the host by influencing cytokine secretion and boosting the natural killer cells activity. They may also lower cholesterol levels by enhancing bile acid excretion. Some EPS have demonstrated antioxidant, anticancer, antiulcer, and antibiofilm properties, further enhancing their functional value in both gut health and therapeutic applications (Moscovici, 2015; Zhou et al., 2021).

Modern omics technologies have begun to unravel the complex mechanisms by which EPS influence microbial and host physiology. These tools reveal that EPS-producing strains can target specific gut taxa, especially SCFA producers, making them promising candidates for managing metabolic diseases linked to high-fat diets such as NAFLD and metabolic syndromes (Yang et al., 2024). Moreover, their application as drug carriers in medical therapies, especially for delivering molecules to the brain or targeting inflammatory conditions, is currently being explored due to their high biocompatibility and ability to form hydrogels (Moscovici, 2015).

Overall, EPS play a multifaceted role in gut microbiota modulation by promoting beneficial microbial growth, exerting prebiotic effects, and maintaining microbial balance. While more human studies are required, current findings support their inclusion in functional foods and therapeutic interventions that aim at improving gut and overall health.

7.6 Product development and industry use

7.6.1 Exopolysaccharide in probiotic foods and supplements

EPS are secreted glycan polymers, either attached loosely to microbial cell surfaces. These biopolymers are synthesized by various bacteria such as LAB, *Bifidobacterium* and *Propionibacterium* during their growth phases (Sørensen et al., 2022). EPS play multiple protective roles for bacterial cells by shielding them from environmental stresses, including desiccation, extreme pH, osmotic changes, UV light, toxic metals, bacteriocins, phagocytic cells, protozoa, and harmful compounds (Tiwari et al., 2020). Although EPS can serve as a carbon source for some organisms, the majority of EPS-producing microbes lack the enzymatic tools to degrade and utilize them (Sørensen et al., 2022).

When applied as functional food ingredients, EPS are categorized as postbiotics that are nonviable microbial products or metabolic byproducts, including vitamins, peptides, acids, proteins, and EPS, that contribute to health benefits upon ingestion. Unlike probiotics, postbiotics do not require viable cells, offering advantages in terms of processing, storage, and shelf-life stability (Cuevas-González et al., 2020). Meanwhile, prebiotics refer to "a nondigestible food ingredient that beneficially affects the host by selectively stimulating the growth and/or activity of one or a limited number of bacteria in the colon, and thus improves host health" (Bevilacqua et al., 2024). These include compounds like dietary fibers, oligosaccharides, polyphenols, and SCFA.

In food applications, the incorporation of EPS influences key characteristics such as texture, viscosity, and rheology. The extent of this impact depends on structural attributes like molecular weight, charge, and branching of the EPS. Various applications of EPS producing probiotics are depicted in Fig. 7.2. Their application can enhance creaminess and thickness, reduce fat content, and substitute traditional additives like stabilizers and emulsifiers (Korcz and Varga, 2021). For example, the addition of EPS from LAB increased the zeta potential of sodium caseinate, indicating improved colloidal stability in dairy matrices (Abid et al., 2018).

7.6.2 Use in fermented dairy, beverages, capsules, and plant-based products

7.6.2.1 *Yoghurt*

In dairy products like yoghurt, EPS-producing LAB are crucial for improving texture and reducing syneresis, especially during low pH conditions caused by fermentation. In situ EPS production is preferred over externally added EPS for enhanced rheological performance (Korcz and Varga, 2021). Under normal condition yoghurt has *Lactobacillus delbrueckii* subsp. *bulgaricus* and *Streptococcus thermophilus* are notable EPS producers, typically yielding 60–150 mg L^{-1} and 30–890 mg L^{-1}, respectively (Daba et al., 2021). Han et al. (2016) screened 19 *Streptococcus thermophilus* strains from traditional Chinese fermented milk and identified strain SH-1 for its superior EPS production, improved texture, and reduced syneresis. Similarly, *Limosilactobacillus mucosae* DPC 6426 has shown improvements in low-fat yoghurt's functional qualities (London et al., 2015). Numerous LAB species that produce EPS are incorporated into various formulations

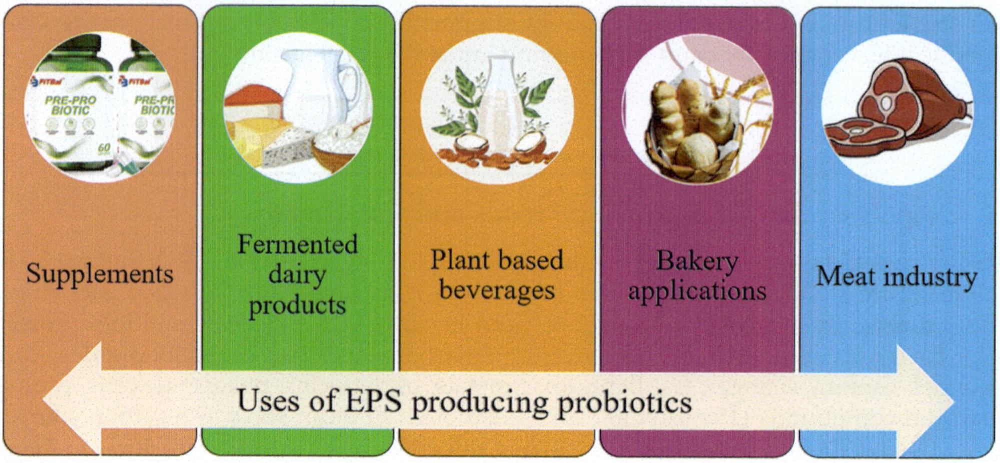

FIGURE 7.2 Use of EPS producing probiotics in various products and industries. *EPS*, Exopolysaccharide.

in dairy products, and multiple reports have examined how these strains influence the texture and stability of fermented dairy products (Table 7.2).

7.6.2.2 Cheese

EPS-producing LAB are also increasingly used in cheese production, particularly for low-fat varieties. These strains help simulate the texture and moisture content of full-fat cheeses (Daba et al., 2021). For instance, *Lactobacillus* strains have increased the melting and moisture profile of fatless mozzarella (Jurášková et al., 2022), while *Lactococcus lactis* improved yield in semi-fat cheddar cheese without compromising flavor (Kondrotiene et al., 2023). *Lactobacillus plantarum* increased moisture, proteolysis, and sensory appeal in cheddar cheese (Wang et al., 2019). EPS-producing cultures also enhanced the structure of low-fat Kasar cheese (Jurášková et al., 2022), while different EPS types influenced texture differently—ropy EPS decreased creaminess, and capsular EPS reduced syneresis (Surber et al., 2021). Full-fat cheeses such as Prato have also benefited from EPS, leading to increased yield and moisture (Chourasia et al., 2021). For example, *L. kefiranofaciens* improved chewiness in mozzarella (Rehman et al., 2018), and EPS enhanced sour whey cheese-requesón's elasticity and adhesiveness (Carrero-Puentes et al., 2022).

7.6.2.3 Kefir

Kefir, a traditional fermented beverage, relies on kefir grains containing symbiotic consortia of LAB, yeasts, and acetic acid bacteria. During fermentation, these microbes produce kefiran, a glucogalactan responsible for the drink's viscosity, primarily synthesized by *Lactobacillus kefiranofaciens* (Werning et al., 2022). Other EPS-producing LAB from kefir include *Lactiplantibacillus plantarum, Lacticaseibacillus paracasei, Lactobacillus helveticus*, and *Leuconostoc mesenteroides* (You et al., 2020). EPS production has been employed to enhance the rheological and sensory properties of dairy and nondairy kefir products (Hamet et al., 2015). However, variability in microbial composition poses challenges in standardizing industrial kefir production (Egea et al., 2022).

TABLE 7.2 EPS-producing microorganisms in fermented dairy products and their impact on techno-functional characteristics.

EPS-producing strains	Food product	Observed functional effects	Reference
Streptococcus thermophilus	Yogurt drink (Ayran)	Enhanced apparent viscosity.	Yilmaz et al. (2015)
Streptococcus thermophilus	Low-fat Italian Cacciotta cheese	Better flavor, taste, and chewiness.	Di Cagno et al. (2014)
Streptococcus thermophilus ST-5581, ST-4239, ST-PH; *Lactococcus lactis* subsp. *cremoris* JFR1	Fermented milk	Reduced syneresis, improved viscosity, and decreased risk factors linked to colon tumors via lowered cyclooxygenase-2 enzyme activity.	Cruz et al. (2020a)
Streptococcus thermophilus ASCC 1275	Set and stirred yogurt	Reduced firmness and syneresis.	Zeidan et al. (2017)
Streptococcus thermophilus zlw TM11	Yogurt	Enhanced texture and minimized whey separation.	Han et al. (2016)
Lactobacillus delbrueckii subsp. *bulgaricus* 147, 2214	Fermented milk	Increased viscosity, reduced syneresis, and improved water retention.	Bancalari et al. (2020)
Lactobacillus delbrueckii subsp. *bulgaricus* CNRZ 1187, 416	Fermented milk	Higher viscosity	Khanal (2017)
Lactobacillus plantarum KX881772, KX881779	Low-fat Akawi cheese	Better water retention, more elastic texture, reduced hardness/adhesiveness, and improved visual appeal and overall quality.	Ayyash et al. (2018)
Limosilactobacillus fermentum Lf2	Nonfat yogurt	Creamier consistency, greater hardness, better water retention, and reduced syneresis.	Khan et al. (2024)
Limosilactobacillus mucosae DPC 6426	Low-fat yogurt	Increased viscosity and decreased syneresis.	Celligoi et al. (2021)

EPS, exopolysaccharide

7.6.2.4 *Plant-based beverages*

With the rise in plant-based diets, EPS-producing LAB offer solutions to improve the sensory properties of plant-based milk alternatives (Korcz and Varga, 2021). Studies show that fermentation with EPS-producing strains improves viscosity, mouthfeel, and syneresis resistance in soy milk, lupin-based yoghurt, and quinoa-based beverages (Pua et al., 2022). For example, *L. plantarum* maintained EPS levels and viscosity in soy milk over 21 days (Werning et al., 2022), while lupin-based yoghurt fermented with *L. plantarum* TMW 1.460 achieved high sensory scores (Hickisch et al., 2016).

7.6.2.5 Bakery applications

EPS-producing LAB are gaining traction as functional additives in sourdough and gluten-free baked goods. In situ EPS production enhances dough stability and final bread texture (Tinzl-Malang et al., 2020). It also offsets quality loss in reduced-sugar or fat-free products and preserves the structure of frozen doughs (Zhang et al., 2021). In gluten-free bread, EPS can retain water, form networks with dough components, and extend shelf life (Akobeng et al., 2020). *L. reuteri* strains producing fructans and glucans significantly improved gluten-free bread quality (Raj et al., 2022).

7.6.2.6 Meat industry

Though less explored than dairy applications, the use of EPS-producing LAB in meat products shows promise, especially for low-fat variants. LAB can enhance texture, color, and food safety through acidification (Grujović et al., 2022). Consumer concerns over food additives have made in situ EPS production a desirable alternative. Studies using EPS-forming *L. plantarum* and *L. curvatus* in sausages have demonstrated improved fat reduction and texture (Hilbig et al., 2019).

7.6.3 Challenges in production and shelf life

Despite their wide-ranging applications, the large-scale use of EPS in the food and supplement sectors faces several challenges. The yield and quality of EPS production are influenced by multiple factors, including microbial strain selection, fermentation parameters, and nutrient availability (Jyoti et al., 2024). Industrial production often requires optimization of growth conditions such as pH, temperature, and C/N sources to achieve consistent EPS quality and functionality (Cruz et al., 2020b). However, scaling up from laboratory to industrial fermenters can lead to variability in EPS composition and molecular weight, affecting its technological and bioactive properties (Mouro et al., 2025).

Another critical challenge lies in shelf-life stability. Although EPS can protect probiotics during storage, environmental factors such as temperature, moisture, and oxygen exposure may still reduce viability over time (Gurram et al., 2021). For products with live cultures, maintaining both EPS integrity and microbial viability requires advanced preservation methods, such as freeze-drying or spray-drying with protective carriers (Wang et al., 2024). Cost is also a limiting factor; the fermentation process for EPS production can be expensive due to the need for controlled conditions and downstream purification (Devi et al., 2024). Future innovations may focus on metabolic engineering of probiotic strains for higher EPS yields, as well as integrating EPS production into standard food fermentation processes to reduce additional costs.

7.7 Challenges and future directions

7.7.1 Microbial strain diversity in exopolysaccharide synthesis

The amount, structure, and functionality of EPS produced by probiotics can vary greatly between species and even among strains of the same species. This variability arises from genetic differences, regulatory mechanisms, and environmental influences such as nutrient availability, pH, and temperature during fermentation (Salazar et al., 2016). Such heterogeneity poses a challenge for standardizing EPS yield and functionality in translational medicine, where reproducibility is critical for therapeutic outcomes.

7.7.2 Safety and regulatory issues

Although many EPS-producing probiotics belong to GRAS or qualified presumption of safety (QPS) categories, certain strains may harbor transferable antibiotic resistance genes or produce immunogenic polysaccharides (Lynch et al., 2014). Regulatory agencies require comprehensive safety assessments, including genomic screening and in vivo testing, before approving probiotics for therapeutic use. Additionally, EPS intended for pharmaceutical applications may face stricter manufacturing and purity standards than those used in foods, increasing development costs and timelines.

7.7.3 Research gaps

Despite promising results in vitro and in preclinical models, clinical evidence supporting EPS-mediated therapeutic effects in humans remains limited. Key gaps include insufficient understanding of EPS metabolism in the human gut, mechanisms underlying immunomodulatory effects, and the interaction of EPS with host cell receptors. Moreover, the influence of host genetics and microbiome composition on EPS efficacy is underexplored, limiting personalized applications.

7.7.4 Future research in exopolysaccharide-based therapies

Future work should prioritize well-designed, placebo-controlled clinical trials to validate the health benefits of EPS-producing probiotics in targeted conditions such as IBD, metabolic syndrome, and cancer adjuvant therapy (Caggianiello et al., 2016). Advances in omics-based tools, such as metagenomics and glycomics, can help identify biomarkers for patient stratification and monitor EPS bioactivity in vivo. Integration of EPS studies with systems biology approaches may also reveal synergistic effects when used in combination with prebiotics or other functional biomolecules.

7.7.5 Genetic engineering for enhanced exopolysaccharide production

Synthetic biology and metabolic engineering offer opportunities to increase EPS yield, alter structural features, and enhance bioactivity. Genetic modification of probiotic strains can optimize precursor availability, activate dormant biosynthetic pathways, or introduce novel GTFs to produce designer EPS with tailored therapeutic properties. However, genetically engineered probiotics face additional biosafety, ethical, and regulatory hurdles, particularly in clinical settings, necessitating robust containment and risk assessment strategies.

7.8 Conclusions

EPS produced by probiotic bacteria are emerging as multifunctional biopolymers with a broad range of applications in human health, translational medicine and industrial product development. These carbohydrate polymers, secreted mainly by LAB and bifidobacteria, exhibit remarkable structural diversity in terms of monosaccharide composition, branching, molecular weight, and chemical modifications, which directly influences their bioactivity. As a result, probiotic EPS have demonstrated immunomodulatory, antioxidant, anti-inflammatory, anticancer, antidiabetic, antiulcer, cholesterol-lowering, and antimicrobial properties, as well as beneficial effects on gut barrier integrity and microbiota composition.

From a therapeutic view, EPS act through multiple mechanisms, such as modulating host immune responses, downregulating pro-inflammatory cytokines, enhancing antioxidant enzyme activity, inducing apoptosis in cancer cells, and promoting the growth of beneficial gut microbes through selective fermentation and SCFA production. These bioactivities highlight their relevance in managing digestive disorders, metabolic diseases, cardiovascular conditions, skin repair, and wound healing. In regenerative medicine, EPS-based biomaterials, such as BC scaffolds and polysaccharide hydrogels, have shown promise in tissue engineering, drug delivery, and as bioactive dressings for chronic wounds. The current bottleneck lies in the translation of promising in vitro and animal data into robust human clinical outcomes. Research gaps include incomplete understanding of EPS metabolism in the human gut, strain-specific functional differences, host-microbiome interactions, and long-term safety profiles. Addressing these gaps requires standardized EPS characterization protocols, advanced omics-based analysis, metabolic engineering for enhanced yields, and carefully designed placebo-controlled clinical trials to validate therapeutic claims.

In the future, integrating EPS-based strategies into personalized nutrition, functional foods, pharmaceuticals, and biomaterial applications could transform them into reliable tools for disease prevention, adjunct therapy, and overall health promotion. With continued scientific innovation and regulatory alignment, probiotic-derived EPS have the potential to evolve from promising laboratory findings into mainstream, evidence-based solutions that bridge microbial biotechnology and human health care.

Credit statement

Aiswarya S V: Writing—original draft, investigation, and conceptualization. Roopam Kumawat: Validation, and investigation. Merugu Shashank Goud: Methodology. K. Bhaskar Reddy: Investigation and formal analysis. Avijeet Singh Oinam: Editing. Onkar Nath Tiwari: Writing—original draft, visualization, investigation, funding acquisition, formal analysis, conceptualization, supervision.

Conflict of interest

The authors state that they have no known personal relationships or financial interests that would have seemed to affect the work this paper reports.

Ethical statement

This article contains no studies with human participants or animals performed by any authors.

Acknowledgments

All authors express their gratitude to the Head, Division of Microbiology and The Graduate School, ICAR-Indian Agricultural Research Institute, New Delhi, India, for the encouragement and moral support.

References

Abid, Y., Joulak, I., Amara, C.B., Casillo, A., Attia, H., Gharsallaoui, A., Azabou, S., 2018. Study of interactions between anionic exopolysaccharides produced by newly isolated probiotic bacteria and sodium caseinate. Colloids Surf. B. 167, 516–523.

Adebayo-Tayo, B., Fashogbon, R., 2020. In vitro antioxidant, antibacterial, in vivo immunomodulatory, antitumor and hematological potential of exopolysaccharide produced by wild type and mutant *Lactobacillus delbureckii* subsp. *bulgaricus*. Heliyon 6 (2), e03268.

Adebayo-Tayo, B., Ishola, R., Oyewunmi, T., 2018. Characterization, antioxidant and immunomodulatory potential on exopolysaccharide produced by wild type and mutant *Weissella confusa* strains. Biotechnol. Rep. 19, e00271.

Ahmad, S., Ahmed, T., Samad, A., Rehman, S.U., Ghalib, M., 2024. Innovative strategies in fermented food production: harnessing genetic and metabolic engineering of lactic acid bacteria. Dialogue Soc. Sci. Rev. (DSSR) 2 (4), 165–198.

Akobeng, A.K., Singh, P., Kumar, M., Al Khodor, S., 2020. Role of the gut microbiota in the pathogenesis of coeliac disease and potential therapeutic implications. Eur. J. Nutr. 59 (8), 3369–3390.

Almeida, I.F., Pereira, T., Silva, N.H.C.S., Gomes, F.P., Silvestre, A.J.D., Freire, C.S.R., Lobo, J.S., Costa, P.C., 2014. Bacterial cellulose membranes as drug delivery systems: an in vivo skin compatibility study. Eur. J. Pharm. Biopharm. 86 (3), 332–336.

Andrew, M., Jayaraman, G., 2020. Structural features of microbial exopolysaccharides in relation to their antioxidant activity. Carbohydr. Res. 487, 107881.

Angelin, J., Kavitha, M., 2020. Exopolysaccharides from probiotic bacteria and their health potential. Int. J. Biol. Macromol. 162, 853–865.

Aravamudhan, A., Ramos, D.M., Nada, A.A., Kumbar, S.G., 2014. Natural polymers: Polysaccharides and their derivatives for biomedical applications. Natural and Synthetic Biomedical Polymers. Elsevier, Amsterdam, pp. 67–89.

Arulselvan, P., Fard, M.T., Tan, W.S., Gothai, S., Fakurazi, S., Norhaizan, M.E., Kumar, S.S., 2016. Role of antioxidants and natural products in inflammation. Oxid. Med. Cell. Long. 2016 (1), 5276130.

Ayyash, M., Abu-Jdayil, B., Hamed, F., Shaker, R., 2018. Rheological, textural, microstructural and sensory impact of exopolysaccharide-producing *Lactobacillus plantarum* isolated from camel milk on low-fat akawi cheese. Lwt 87, 423–431.

Ayyash, M., Abu-Jdayil, B., Itsaranuwat, P., Galiwango, E., Tamiello-Rosa, C., Abdullah, H., Esposito, G., Hunashal, Y., Obaid, R.S., Hamed, F., 2020. Characterization, bioactivities, and rheological properties of exopolysaccharide produced by novel probiotic *Lactobacillus plantarum* C70 isolated from camel milk. Int. J. Biol. Macromol. 144, 938–946.

Badel, S., Bernardi, T., Michaud, P.J.B.A., 2011. New perspectives for *Lactobacilli* exopolysaccharides. Biotechnol. Adv. 29 (1), 54–66.

Bancalari, E., Alinovi, M., Bottari, B., Caligiani, A., Mucchetti, G., Gatti, M., 2020. Ability of a wild Weissella strain to modify viscosity of fermented milk. Front. Microbiol. 10, 3086.

Bevilacqua, A., Campaniello, D., Speranza, B., Racioppo, A., Sinigaglia, M., Corbo, M.R., 2024. An update on prebiotics and on their health effects. Foods 13 (3), 446.

Bhat, B., Bajaj, B.K., 2018. Hypocholesterolemic and bioactive potential of exopolysaccharide from a probiotic *Enterococcus faecium* K1 isolated from kalarei. Bioresour. Technol. 254, 264–267.

Caggianiello, G., Kleerebezem, M., Spano, G., 2016. Exopolysaccharides produced by lactic acid bacteria: from health-promoting benefits to stress tolerance mechanisms. Appl. Microbiol. Biotechnol. 100 (9), 3877–3886.

Calegari, G.C., Santos, V.A., Teixeira, S.D., Barbosa-Dekker, A.M., Dekker, R.F., Cunha, M.A.A., 2017. Sulfonation of (1→ 6)-β-D-glucan (lasiodiplodan) and its antioxidant and antimicrobial potential. J. Pharm. Pharmacol. 5 (12), 850–863.

Carrero-Puentes, S., Fuenmayor, C., Jiménez-Pérez, C., Guzmán-Rodríguez, F., Gómez-Ruiz, L., Rodríguez-Serrano, G., Alatorre-Santamaría, S., García-Garibay, M., Cruz-Guerrero, A., 2022. Development and characterization of an exopolysaccharide-functionalized acid whey cheese (requesón) using *Lactobacillus delbrueckii* ssp. *bulgaricus*. J. Food Process. Preserv. 46 (6), e16095.

Castro-Bravo, N., Wells, J.M., Margolles, A., Ruas-Madiedo, P., 2018. Interactions of surface exopolysaccharides from *Bifidobacterium* and *Lactobacillus* within the intestinal environment. Front. Microbiol. 9, 2426.

Celligoi, M.A.P.C., Bersanetti, G.T., Baldo, C., da Silva, R.T., Silveira, V.A.I., 2021. Exopolysaccharides in fermented foods and their potential health benefits. Bioactive Compounds in Fermented Foods. CRC Press Boca Raton, FL, USA, pp. 159–184.

Chourasia, R., Abedin, M.M., Chiring Phukon, L., Sahoo, D., Singh, S.P., Rai, A.K., 2021. Biotechnological approaches for the production of designer cheese with improved functionality. Compr. Rev. Food Sci. Food Saf. 20 (1), 960–979.
Costa, L.M.M., de Olyveira, G.M., Basmaji, P., Xavier Filho, L., 2011. Nanopores structure in electrospun bacterial cellulose. J. Biomater. Nanobiotechnol. 3 (1), 92–96.
Cruz, B.C., Sarandy, M.M., Messias, A.C., Gonçalves, R.V., Ferreira, C.L., Peluzio, M.C., 2020a. Preclinical and clinical relevance of probiotics and synbiotics in colorectal carcinogenesis: a systematic review. Nutr. Rev. 78 (8), 667–687.
Cruz, D., Vasconcelos, V., Pierre, G., Michaud, P., Delattre, C., 2020b. Exopolysaccharides from cyanobacteria: strategies for bioprocess development. Appl. Sci. 10 (11), 3763.
Cuevas-González, P.F., Liceaga, A.M., Aguilar-Toalá, J.E., 2020. Postbiotics and paraprobiotics: From concepts to applications. Food Res. Int. 136, 109502.
Daba, G.M., Elnahas, M.O., Elkhateeb, W.A., 2021. Contributions of exopolysaccharides from lactic acid bacteria as biotechnological tools in food, pharmaceutical, and medical applications. Int. J. Biol. Macromol. 173, 79–89.
Devi, M., Thakur, K., Kumari, D., Behl, M., Rathour, R.K., Bhatt, A.K., Bhatia, R.K., 2024. Advances in fermentation technology for EPS production. Microbial Exopolysaccharides. CRC Press Boca Raton, FL, USA, pp. 37–76.
Dhiman, S., Kaur, S., Thakur, B., Singh, P., Tripathi, M., 2025. Nutritional enhancement of plant-based fermented foods: microbial innovations for a sustainable future. Fermentation 11 (6), 346.
Di Cagno, R., De Pasquale, I., De Angelis, M., Buchin, S., Rizzello, C.G., Gobbetti, M., 2014. Use of microparticulated whey protein concentrate, exopolysaccharide-producing *Streptococcus thermophilus*, and adjunct cultures for making low-fat Italian Caciotta-type cheese. J. Dairy Sci. 97 (1), 72–84.
di Vito, R., Conte, C., Traina, G., 2022. A multi-strain probiotic formulation improves intestinal barrier function by the modulation of tight and adherent junction proteins. Cells 11 (16), 2617.
Domingos-Lopes, M.F.P., Nagy, A., Stanton, C., Ross, P.R., Gelencsér, E., Silva, C.C.G., 2017. Immunomodulatory activity of exopolysaccharide producing *Leuconostoc citreum* strain isolated from Pico cheese. J. Funct. Foods 33, 235–243.
Egea, M.B., Santos, D.C.D., Oliveira Filho, J.G.D., Ores, J.D.C., Takeuchi, K.P., Lemes, A.C., 2022. A review of nondairy kefir products: their characteristics and potential human health benefits. Crit. Rev. Food Sci. Nutr. 62 (6), 1536–1552.
El-Deeb, N.M., Yassin, A.M., Al-Madboly, L.A., El-Hawiet, A., 2018. A novel purified *Lactobacillus acidophilus* 20079 exopolysaccharide, LA-EPS-20079, molecularly regulates both apoptotic and NF-κB inflammatory pathways in human colon cancer. Microb. Cell Fact. 17 (1), 29.
Evans, N.D., Gentleman, E., Polak, J.M., 2006. Scaffolds for stem cells. Mater. Today 9 (12), 26–33.
Fu, L., Zhang, J., Yang, G., 2013. Present status and applications of bacterial cellulose-based materials for skin tissue repair. Carbohydr. Polym. 92 (2), 1432–1442.
Gibson, G.R., Hutkins, R., Sanders, M.E., Prescott, S.L., Reimer, R.A., Salminen, S.J., Scott, K., Stanton, C., Swanson, K.S., Cani, P.D., Verbeke, K., 2017. Expert consensus document: the international scientific association for probiotics and prebiotics (ISAPP) consensus statement on the definition and scope of prebiotics. Nat. Rev. Gastroenterol. Hepatol 14 (8), 491–502.
Gou, H.Z., Zhang, Y.L., Ren, L.F., Li, Z.J., Zhang, L., 2022. How do intestinal probiotics restore the intestinal barrier? Front. Microbiol. 13, 929346.
Grujović, M.Ž., Mladenović, K.G., Semedo-Lemsaddek, T., Laranjo, M., Stefanović, O.D., Kocić-Tanackov, S.D., 2022. Advantages and disadvantages of non-starter lactic acid bacteria from traditional fermented foods: potential use as starters or probiotics. Compr. Rev. Food Sci. Food Saf. 21 (2), 1537–1567.
Gugliandolo, C., Spanò, A., Maugeri, T.L., Poli, A., Arena, A., Nicolaus, B., 2015. Role of bacterial exopolysaccharides as agents in counteracting immune disorders induced by herpes virus. Microorganisms 3 (3), 464–483.
Gurram, S., Jha, D.K., Shah, D.S., Kshirsagar, M.M., Amin, P.D., 2021. Insights on the critical parameters affecting the probiotic viability during stabilization process and formulation development. AAPS PharmSciTech. 22 (5), 156.
Hamet, M.F., Piermaria, J.A., Abraham, A.G., 2015. Selection of EPS-producing Lactobacillus strains isolated from kefir grains and rheological characterization of the fermented milks. LWT-Food Sci. Technol. 63 (1), 129–135.
Han, X., Yang, Z., Jing, X., Yu, P., Zhang, Y., Yi, H., Zhang, L., 2016. Improvement of the texture of yogurt by use of exopolysaccharide producing lactic acid bacteria. BioMed. Res. Int. 2016 (1), 7945675.
Haroun, B.M., Refaat, B.M., El-Menoufy, H.A., Amin, H.A., El-Waseif, A.A., 2013. Structure analysis and antitumor activity of the exopolysaccharide from probiotic *Lactobacillus plantarum* NRRL B-4496 in vitro and in vivo. J. Appl. Sci. Res. 9 (1), 425–434.

Hickisch, A., Beer, R., Vogel, R.F., Toelstede, S., 2016. Influence of lupin-based milk alternative heat treatment and exopolysaccharide-producing lactic acid bacteria on the physical characteristics of lupin-based yogurt alternatives. Food Res. Int. 84, 180–188.

Hilbig, J., Loeffler, M., Herrmann, K., Weiss, J., 2019. The influence of exopolysaccharide-producing lactic acid bacteria on reconstructed ham. Int. J. Food Sci. Technol. 54 (9), 2763–2769.

Homayouni, A., Payahoo, L., Azizi, A., 2012. Effects of probiotics on lipid profile: a review. Am. J. Food Technol. 7 (5), 251–265.

Inturri, R., Molinaro, A., Di Lorenzo, F., Blandino, G., Tomasello, B., Hidalgo-Cantabrana, C., De Castro, C., Ruas-Madiedo, P., 2017. Chemical and biological properties of the novel exopolysaccharide produced by a probiotic of *Bifidobacterium longum*. Carbohydr. Polym. 174, 1172–1180.

Ioannou, A., Berkhout, M.D., Geerlings, S.Y., Belzer, C., 2025. *Akkermansia muciniphila*: Biology, microbial ecology, host interactions and therapeutic potential. Nat. Rev. Microbiol. 23 (3), 162–177.

Islam, S.T., Lam, J.S., 2014. Synthesis of bacterial polysaccharides via the Wzx/Wzy-dependent pathway. Can. J. Microbiol. 60 (11), 697–716.

Ivanova, E.P., Bazaka, K., Crawford, R.J., 2014. Natural polymer biomaterials: advanced applications. New Funct. Biomater. Med. Healthcare 1, 32–70.

Jalaja, K., Kumar, P.A., Dey, T., Kundu, S.C., James, N.R., 2014. Modified dextran cross-linked electrospun gelatin nanofibres for biomedical applications. Carbohydr. Polym. 114, 467–475.

Jurášková, D., Ribeiro, S.C., Silva, C.C., 2022. Exopolysaccharides produced by lactic acid bacteria: from biosynthesis to health-promoting properties. Foods 11 (2), 156.

Jyoti, K., Soni, K., Chandra, R., 2024. Optimization of the production of exopolysaccharide (EPS) from biofilm-forming bacterial consortium using different parameters. Microbe 4, 100117.

Kang, M.S., Park, G.Y., 2021. In vitro evaluation of the effect of oral probiotic *Weissella cibaria* on the formation of multi-species oral biofilms on dental implant surfaces. Microorganisms 9 (12), 2482.

Khan, M.R., Khan, S., Islam, J., Mahmud, S., Biki, S.P., Ahmed, M., 2024. Influence of microencapsulated *Streptococcus thermophilus* ST-HSTU-FPP and *Limosilactobacillus fermentum* LS-HSTU-FPP strains on biochemical features of yogurt following in-vitro simulated gastrointestinal conditions. J. Food Meas. Charact. 18 (2), 1229–1236.

Khanal, S.N., 2017. Exopolysaccharides Produced by *Streptococcus thermophilus* During Fermentation of Milk, and Their Impact on the Properties of Fermented Milk Gels. The University of Wisconsin-MadisonWisconsin, WI, USA.

Kodali, V.P., Das, S., Sen, R., 2009. An exopolysaccharide from a probiotic: biosynthesis dynamics, composition and emulsifying activity. Food Res. Int. 42 (5-6), 695–699.

Kondrotiene, K., Zavistanaviciute, P., Aksomaitiene, J., Novoslavskij, A., Malakauskas, M., 2023. *Lactococcus lactis* in dairy fermentation—Health-promoting and probiotic properties. Fermentation 10 (1), 16.

Korcz, E., Varga, L., 2021. Exopolysaccharides from lactic acid bacteria: techno-functional application in the food industry. Trends Food Sci. Technol. 110, 375–384.

Kwon, M., Lee, J., Park, S., Kwon, O.H., Seo, J., Roh, S., 2020. Exopolysaccharide isolated from *Lactobacillus plantarum* L-14 has anti-inflammatory effects via the toll-like receptor 4 pathway in LPS-induced RAW 264.7 cells. Int. J. Mol. Sci. 21 (23), 9283.

Laiño, J., Villena, J., Kanmani, P., Kitazawa, H., 2016. Immunoregulatory effects triggered by lactic acid bacteria exopolysaccharides: New insights into molecular interactions with host cells. Microorganisms 4 (3), 27.

Lebeer, S., Claes, I., Tytgat, H.L., Verhoeven, T.L., Marien, E., von Ossowski, I., Reunanen, J., Palva, A., de Vos, W.M., De Keersmaecker, S.C., Vanderleyden, J., 2012. Functional analysis of *Lactobacillus rhamnosus* GG pili in relation to adhesion and immunomodulatory interactions with intestinal epithelial cells. Appl. Env. Microbiol. 78 (1), 185–193.

Li, J., Li, Q., Wu, Q., Gao, N., Wang, Z., Yang, Y., Shan, A., 2023. Exopolysaccharides of *Lactobacillus rhamnosus* GG ameliorate *Salmonella typhimurium*-induced intestinal inflammation via the TLR4/NF-κB/MAPK pathway. J. Anim. Sci. Biotechnol. 14 (1), 23.

Lin, S.P., Loira Calvar, I., Catchmark, J.M., Liu, J.R., Demirci, A., Cheng, K.C., 2013. Biosynthesis, production and applications of bacterial cellulose. Cellulose 20 (5), 2191–2219.

Liu, Y., Zheng, S., Cui, J., Guo, T., Zhang, J., Li, B., 2021. Alleviative effects of exopolysaccharide produced by *Lactobacillus helveticus* KLDS1. 8701 on dextran sulfate sodium-induced colitis in mice. Microorganisms 9 (10), 2086.

Liu, Z., Zhang, Z., Qiu, L., Zhang, F., Xu, X., Wei, H., Tao, X., 2017. Characterization and bioactivities of the exopolysaccharide from a probiotic strain of *Lactobacillus plantarum* WLPL04. J. Dairy Sci. 100 (9), 6895–6905.

London, L.E., Chaurin, V., Auty, M.A., Fenelon, M.A., Fitzgerald, G.F., Ross, R.P., Stanton, C., 2015. Use of *Lactobacillus mucosae* DPC 6426, an exopolysaccharide-producing strain, positively influences the techno-functional properties of yoghurt. Int. Dairy J. 40, 33–38.

London, L.E., Kumar, A.H., Wall, R., Casey, P.G., O'Sullivan, O., Shanahan, F., Hill, C., Cotter, P.D., Fitzgerald, G.F., Ross, R.P., Caplice, N.M., 2014. Exopolysaccharide-producing probiotic *Lactobacilli* reduce serum cholesterol and modify enteric microbiota in ApoE-deficient mice. J. Nutr. 144 (12), 1956–1962.

Lynch, K.M., McSweeney, P.L., Arendt, E.K., Uniacke-Lowe, T., Galle, S., Coffey, A., 2014. Isolation and characterisation of exopolysaccharide-producing *Weissella* and *Lactobacillus* and their application as adjunct cultures in cheddar cheese. Int. Dairy J. 34 (1), 125–134.

Lynch, K.M., Zannini, E., Coffey, A., Arendt, E.K., 2018. Lactic acid bacteria exopolysaccharides in foods and beverages: Isolation, properties, characterization, and health benefits. Annu. Rev. Food Sci. Technol. 9 (1), 155–176.

Matsumura, K., Nakajima, N., Sugai, H., Hyon, S.H., 2014. Self-degradation of tissue adhesive based on oxidized dextran and poly-L-lysine. Carbohydr. Polym. 113, 32–38.

Meng, Q., Lu, C., Gao, H., Chen, G., Wu, L., Wu, J., Li, S., He, B.F., 2021. Efficient biosynthesis of exopolysaccharide from Jerusalem artichoke using a novel strain of *Bacillus velezensis* LT-2. Bioresour. Technol. 320, 124346.

Mishra, B., Vuppu, S., Rath, K., 2011. The role of microbial pullulan, a biopolymer in pharmaceutical approaches: a review. J. Appl. Pharm. Sci. 45–50.

Miyamoto, J., Shimizu, H., Hisa, K., Matsuzaki, C., Inuki, S., Ando, Y., Nishida, A., Izumi, A., Yamano, M., Ushiroda, C., Irie, J., 2023. Host metabolic benefits of prebiotic exopolysaccharides produced by *Leuconostoc mesenteroides*. Gut Microbes 15 (1), 2161271.

Mohamad, N., Amin, M.C.I.M., Pandey, M., Ahmad, N., Rajab, N.F., 2014. Bacterial cellulose/acrylic acid hydrogel synthesized via electron beam irradiation: Accelerated burn wound healing in an animal model. Carbohydr. Polym. 114, 312–320.

Morelli, L., 2013. Probiotics: Definition and taxonomy 10 years after the FAO/WHO guidelines. Probiotic Bact. Their Eff. Hum. Health Well-Being 107, 1–8.

Moscovici, M., 2015. Present and future medical applications of microbial exopolysaccharides. Front. Microbiol. 6, 1012.

Mouro, C., Gomes, A.P., Gouveia, I.C., 2024. Microbial exopolysaccharides: structure, diversity, applications, and future frontiers in sustainable functional materials. Polysaccharides 5 (3), 241–287.

Nguyen, P.T., Nguyen, T.T., Bui, D.C., Hong, P.T., Hoang, Q.K., Nguyen, H.T., 2020. Exopolysaccharide production by lactic acid bacteria: the manipulation of environmental stresses for industrial applications. AIMS Microbiol. 6 (4), 451.

Nichols, C.M., Guezennec, J., Bowman, J.P., 2005. Bacterial exopolysaccharides from extreme marine environments with special consideration of the Southern Ocean, sea ice, and deep-sea hydrothermal vents: a review. Mar. Biotechnol. 7 (4), 253–271.

Ozturk, S., Aslim, B., 2010. Modification of exopolysaccharide composition and production by three cyanobacterial isolates under salt stress. Env. Sci. Pollut. Res. 17 (3), 595–602.

Pan, L., Zhou, Z., Han, Y., 2021. Exopolysaccharide from *Leuconostoc pseudomesenteroides* XG5 delay the onset of autoimmune diabetes by modulating gut microbiota and its metabolites SCFAs in NOD mice. J. Funct. Foods 79, 104427.

Patel, A., Lindström, C., Patel, A., Prajapati, J., Holst, O., 2012. Probiotic properties of exopolysaccharide producing lactic acid bacteria isolated from vegetables and traditional Indian fermented foods. Int. J. Fermented Foods 1 (1), 87–101.

Pérez-Ramos, A., Mohedano, M.L., López, P., Spano, G., Fiocco, D., Russo, P., Capozzi, V., 2017. In situ β-glucan fortification of cereal-based matrices by *Pediococcus parvulus* 2.6: technological aspects and prebiotic potential. Int. J. Mol. Sci. 18 (7), 1588.

Poli, A., Anzelmo, G., Nicolaus, B., 2010. Bacterial exopolysaccharides from extreme marine habitats: production, characterization and biological activities. Mar. Drugs 8 (6), 1779–1802.

Pourjafar, H., Ansari, F., Sadeghi, A., Samakkhah, S.A., Jafari, S.M., 2023. Functional and health-promoting properties of probiotics' exopolysaccharides; isolation, characterization, and applications in the food industry. Crit. Rev. Food Sci. Nutr. 63 (26), 8194–8225.

Prete, R., Alam, M.K., Perpetuini, G., Perla, C., Pittia, P., Corsetti, A., 2021. Lactic acid bacteria exopolysaccharides producers: a sustainable tool for functional foods. Foods 10 (7), 1653.

Pua, A., Tang, V.C.Y., Goh, R.M.V., Sun, J., Lassabliere, B., Liu, S.Q., 2022. Ingredients, processing, and fermentation: addressing the organoleptic boundaries of plant-based dairy analogues. Foods 11 (6), 875.

Raj, T., Chandrasekhar, K., Kumar, A.N., Kim, S.H., 2022. Recent biotechnological trends in lactic acid bacterial fermentation for food processing industries. Syst. Microbiol. Biomanufacturing 2 (1), 14–40.

Rehm, B.H., 2010. Bacterial polymers: biosynthesis, modifications and applications. Nat. Rev. Microbiol. 8 (8), 578–592.

Rehman, R.U., Wang, Y., Wang, J., Geng, W., 2018. Physicochemical analysis of mozzarella cheese produced and developed by the novel EPS-producing strain *Lactobacillus kefiranofaciens* ZW3. Int. J. Dairy Technol. 71, 90–98.

Ruas-Madiedo, P., 2014. Biosynthesis and bioactivity of exopolysaccharides produced by probiotic bacteria. Food Oligosaccharides: Production, Analysis and BioactivityChichester. John Wiley & Sons, Ltd, West Sussex, pp. 118–133.

Ruiz, L., Margolles, A., Sánchez, B., 2013. Bile resistance mechanisms in *Lactobacillus* and *Bifidobacterium*. Front. Microbiol. 4, 396.

Saadat, Y.R., Khosroushahi, A.Y., Gargari, B.P., 2019. A comprehensive review of anticancer, immunomodulatory and health beneficial effects of the lactic acid bacteria exopolysaccharides. Carbohydr. Polym. 217, 79–89.

Salazar, N., Gueimonde, M., De, L., Reyes-Gavilán, C.G., Ruas-Madiedo, P., 2016. Exopolysaccharides produced by lactic acid bacteria and bifidobacteria as fermentable substrates by the intestinal microbiota. Crit. Rev. Food Sci. Nutr. 56 (9), 1440–1453.

Sanalibaba, P., Çakmak, G.A., 2016. Exopolysaccharides production by lactic acid bacteria. Appl. Microbiol. 2 (2), 10–4172.

Sarkar, A., Ghosh, U., 2016. Natural antioxidants-the key to safe and sustainable life. Int. J. Latest Trends Eng. Technol. 6 (3), 460–466.

Scheuerl, T., Hopkins, M., Nowell, R.W., Rivett, D.W., Barracloug, T.G., Bell, T., 2020. Bacterial adaptation is constrained in complex communities. Nat. Commun. 11 (1), 754.

Shori, A.B., 2016. Influence of food matrix on the viability of probiotic bacteria: a review based on dairy and non-dairy beverages. Food Biosci. 13, 1–8.

Silva, A.K., Juenet, M., Meddahi-Pellé, A., Letourneur, D., 2015. Polysaccharide-based strategies for heart tissue engineering. Carbohydr. Polym. 116, 267–277.

Silva, L.A., Lopes Neto, J.H.P., Cardarelli, H.R., 2019. Exopolysaccharides produced by *Lactobacillus plantarum*: technological properties, biological activity, and potential application in the food industry. Ann. Microbiol. 69 (4), 321–328.

Sørensen, H.M., Rochfort, K.D., Maye, S., MacLeod, G., Brabazon, D., Loscher, C., Freeland, B., 2022. Exopolysaccharides of lactic acid bacteria: production, purification and health benefits towards functional food. Nutrients 14 (14), 2938.

Sungur, T., Aslim, B., Karaaslan, C., Aktas, B., 2017. Impact of exopolysaccharides (EPSs) of *Lactobacillus gasseri* strains isolated from human vagina on cervical tumor cells (HeLa). Anaerobe 47, 137–144.

Surayot, U., Wang, J., Seesuriyachan, P., Kuntiya, A., Tabarsa, M., Lee, Y., Kim, J.K., Park, W., You, S., 2014. Exopolysaccharides from lactic acid bacteria: structural analysis, molecular weight effect on immunomodulation. Int. J. Biol. Macromol. 68, 233–240.

Surber, G., Schäper, C., Wefers, D., Rohm, H., Jaros, D., 2021. Exopolysaccharides from *Lactococcus lactis* affect manufacture, texture and sensory properties of concentrated acid milk gel suspensions (fresh cheese). Int. Dairy J. 112, 104854.

Thankam, F.G., Muthu, J., 2014. Alginate based hybrid copolymer hydrogels—Influence of pore morphology on cell–material interaction. Carbohydr. Polym. 112, 235–244.

Tingirikari, J.M.R., Kothari, D., Goyal, A., 2014. Superior prebiotic and physicochemical properties of novel dextran from *Weissella cibaria* JAG8 for potential food applications. Food Funct. 5 (9), 2324–2330.

Tinzl-Malang, S.K., Grattepanche, F., Rast, P., Fischer, P., Sych, J., Lacroix, C., 2020. Purified exopolysaccharides from *Weissella confusa* 11GU-1 and *Propionibacterium freudenreichii* JS15 act synergistically on bread structure to prevent staling. LWT 127, 109375.

Tiwari, O.N., Sasmal, S., Kataria, A.K., Devi, I., 2020. Application of microbial extracellular carbohydrate polymeric substances in food and allied industries. 3 Biotech 10 (5), 221.

Torres, F.G., Commeaux, S., Troncoso, O.P., 2012. Biocompatibility of bacterial cellulose based biomaterials. J. Funct. Biomater. 3 (4), 864–878.

Trabelsi, I., Ktari, N., Slima, S.B., Triki, M., Bardaa, S., Mnif, H., Salah, R.B., 2017. Evaluation of dermal wound healing activity and in vitro antibacterial and antioxidant activities of a new exopolysaccharide produced by *Lactobacillus* sp Ca6. Int. J. Biol. Macromol. 103, 194–201.

Tsai, C.C., Lin, P.P., Hsieh, Y.M., Zhang, Z.Y., Wu, H.C., Huang, C.C., 2014. Cholesterol-lowering potentials of lactic acid bacteria based on bile-salt hydrolase activity and effect of potent strains on cholesterol metabolism in vitro and in vivo. Sci. World J. 2014 (1), 690752.

Wang, A., Zhong, Q., 2024. Drying of probiotics to enhance the viability during preparation, storage, food application, and digestion: a review. Compr. Rev. Food Sci. Food Saf. 23 (1), e13287.

Wang, B., Song, Q., Zhao, F., Xiao, H., Zhou, Z., Han, Y., 2019. Purification and characterization of dextran produced by *Leuconostoc pseudomesenteroides* PC as a potential exopolysaccharide suitable for food applications. Process Biochem. 87, 187–195.

Wang, J., Li, M., Gao, Y., Li, H., Fang, L., Liu, C., Liu, X., Min, W., 2022. Effects of exopolysaccharides from *Lactiplantibacillus plantarum* JLAU103 on intestinal immune response, oxidative stress, and microbial communities in cyclophosphamide-induced immunosuppressed mice. J. Agric. Food Chem. 70 (7), 2197–2210.

Werning, M.L., Hernández-Alcántara, A.M., Ruiz, M.J., Soto, L.P., Dueñas, M.T., López, P., Frizzo, L.S., 2022. Biological functions of exopolysaccharides from lactic acid bacteria and their potential benefits for humans and farmed animals. Foods 11 (9), 1284.

Werning, M.L., Notararigo, S., Nácher, M., Fernández de Palencia, P., Aznar, R., López, P., 2012. Biosynthesis, purification and biotechnological use of exopolysaccharides produced by lactic acid bacteria. Food Addit. 83–114.

Wu, J., Zhang, Y., Ye, L., Wang, C., 2021. The anti-cancer effects and mechanisms of lactic acid bacteria exopolysaccharides in vitro: a review. Carbohydr. Polym. 253, 117308.

Xiong, J., Liu, D.M., Huang, Y.Y., 2023. Exopolysaccharides from *Lactiplantibacillus plantarum*: isolation, purification, structure–function relationship, and application. Eur. Food Res. Technol. 249 (6), 1431–1448.

Xu, L., Li, H., Vuong, C., Vadyvaloo, V., Wang, J., Yao, Y., Otto, M., Gao, Q., 2006. Role of the luxS quorum-sensing system in biofilm formation and virulence of *Staphylococcus epidermidis*. Infect. Immun. 74 (1), 488–496.

Xu, Z., Li, S., Gong, G., Liu, Z., Wu, Z., Ma, C., 2015. Influence of different acidifying strains of *Lactobacillus delbrueckii* subsp. *bulgaricus* on the quality of yoghurt. Food Sci. Technol. Res. 21 (2), 263–269.

Yang, X., Hong, J., Wang, L., Cai, C., Mo, H., Wang, J., Fang, X., Liao, Z., 2024. Effect of lactic acid bacteria fermentation on plant-based products. Fermentation 10 (1), 48.

Ye, Z.W., Zhang, J., Townsend, D.M., Tew, K.D., 2015. Oxidative stress, redox regulation and diseases of cellular differentiation. Biochim. Biophys. Acta (BBA)-Gen. Subj. 1850 (8), 1607–1621.

Yilmaz, M.T., Dertli, E., Toker, O.S., Tatlisu, N.B., Sagdic, O., Arici, M., 2015. Effect of in situ exopolysaccharide production on physicochemical, rheological, sensory, and microstructural properties of the yogurt drink ayran: an optimization study based on fermentation kinetics. J. Dairy Sci. 98 (3), 1604–1624.

You, X., Yang, L., Zhao, X., Ma, K., Chen, X., Zhang, C., Wang, G., Dong, M., Rui, X., Zhang, Q., Li, W., 2020. Isolation, purification, characterization and immunostimulatory activity of an exopolysaccharide produced by *Lactobacillus pentosus* LZ-R-17 isolated from Tibetan kefir. Int. J. Biol. Macromol. 158, 408–419.

Zamora-Pineda, J., Kalinina, O., Osborne, B.A., Knight, K.L., 2022. Probiotic molecules that inhibit inflammatory diseases. Appl. Sci. 12 (3), 1147.

Zarour, K., Llamas, M.G., Prieto, A., Ruas-Madiedo, P., Dueñas, M.T., de Palencia, P.F., Aznar, R., Kihal, M., Lopez, P., 2017. Rheology and bioactivity of high molecular weight dextrans synthesised by lactic acid bacteria. Carbohydr. Polym. 174, 646–657.

Zeidan, A.A., Poulsen, V.K., Janzen, T., Buldo, P., Derkx, P.M., Øregaard, G., Neves, A.R., 2017. Polysaccharide production by lactic acid bacteria: from genes to industrial applications. FEMS Microbiol. Rev. 41 (Supp_1), S168–S200.

Zhai, T., Wang, P., Hu, X., Zheng, L., 2022. Probiotics bring new hope for atherosclerosis prevention and treatment. Oxid. Med. Cell. Long. 2022 (1), 3900835.

Zhang, B., Omedi, J.O., Zheng, J., Huang, W., Jia, C., Zhou, L., Zou, Q., Li, N., Gao, T., 2021. Exopolysaccharides in sourdough fermented by *Weissella confusa* QS813 protected protein matrix and quality of frozen gluten-red bean dough during freeze-thaw cycles. Food Biosci. 43, 101180.

Zhao, C., Xie, L., Shen, J., He, H., Zhang, T., Hao, L., Sun, C., Zhang, X., Chen, M., Liu, F., Li, Z., 2025. *Lactobacillus acidophilus* YL01 and its exopolysaccharides ameliorate obesity and insulin resistance in obese mice via modulating intestinal specific bacterial groups and AMPK/ACC signaling pathway. Int. J. Biol. Macromol. 300, 140287.

Zhou, X., Zhang, D., Qi, W., Hong, T., Xiong, T., Wu, T., Geng, F., Xie, M., Nie, S., 2021. Exopolysaccharides from *Lactobacillus plantarum* NCU116 facilitate intestinal homeostasis by modulating intestinal epithelial regeneration and microbiota. J. Agric. Food Chem. 69 (28), 7863–7873.

Zhou, Y., Cui, Y., Qu, X., 2019. Exopolysaccharides of lactic acid bacteria: Structure, bioactivity and associations: a review. Carbohydr. Polym. 207, 317–332.

CHAPTER

8

Microbial exopolysaccharides: A potential source for bio-fertilizers in eco-friendly green farming for crop improvement and productivity

Kuppam Chandrasekhar[a], *Md. Nazneen Bobby*[a], *Davuluri Syam Babu*[a], *Alugunulla Venkata Narayana*[a], *Ambati Ranga Rao*[a] *and Prasun Kumar*[b]

[a]Department of Biotechnology, Vignan's Foundation for Science, Technology and Research, Vadlamudi, Guntur, Andhra Pradesh, India [b]Department of Biotechnology, School of Engineering and Technology, Sharda University, Greater Noida, Uttar Pradesh, India

8.1 Introduction

Microbial exopolysaccharides (EPS) are high-molecular-weight, high-viscosity polysaccharides produced and released by microorganisms into the local microenvironment. These biopolymers are central determinants of the microbial community's living and survival. Both homopolysaccharides and heteropolysaccharides will be considered EPS (Wang et al., 2023; Chandrasekhar et al., 2025). This structural heterogeneity enables functional diversity from a high throughput. They have significant functional roles in many ecological processes such as biofilm formation, cell adhesion, and environmental stress tolerance (Mouro et al., 2024). Besides, EPS enhances microbe interactions, especially positive correlations, and restores contributions to the health and productivity of the composite ecosystem. Analyzing the structure-activity relationships of EPS is complicated since its structure is both complex and multifaceted.

DOI: https://doi.org/10.1016/B978-0-443-36619-2.00008-5

Very few distinctive features impacting biological activities are molecular weight, monosaccharide composition, type of glycosidic bond, branching, and spatial configuration (Fu and Yan, 2023). EPS can be present as the linear polysaccharide pullulan or branched polysaccharides such as EPS-W1 from *Lactobacillus plantarum W1*. It has been observed that certain acidic heteropolysaccharides are very potent immune inducers, like the high-molecular-weight sulfated heteropolysaccharides from *Lactobacillus paracasei* VL8 that have potent immunomodulating activities (Fu and Yan, 2023).

EPS is associated with its structural features and delivers health-related advantages such as immunomodulation and anticancer properties. They enable higher nutrient acquisition by plants, emulsion stabilization, and stability to microbial communities. Used increasingly as thickening substances in the production of pharmaceuticals, food products, and cosmetics, EPS further supports eco-friendly agriculture through superior soil and plant growth promotion as a result of enhanced microbial activities (Cania et al., 2019; Singh et al., 2025). EPS is being increasingly used in food, pharmaceuticals, and cosmetics as stabilizers and thickening agents, besides supporting soil health and plant growth in sustainable agriculture by promoting enhanced microbial activity (Cania et al., 2019; Fu and Yan, 2023; Tiwari et al., 2024). Green farming practices are essential for sustainable agriculture; they not only improve environmental health but also enhance soil health (Fig. 8.1). Soil structure and fertility can be improved, from the soil-product chains to composting, through methods such as reduced tillage, organic farming, and microbial EPS and lipopolysaccharides (LPS) (Mouro et al., 2024). Having up to one-third of the earth's surface damaged by bad farming practices like repeated tillage use, "Compassionate agriculture" with reduced tillage, and doing mixed farming, at a sustainable economic rate, will help prevent this. However, reduced tillage is a less common practice in organic farming, partly due to pull factors such as weed pressure and topsoil compaction, which can affect crop yields (Fu and Yan, 2023).

Soil bacteria produce EPS and LPS, which play an important role in soil physical properties, such as improving soil structure, aggregate stability, reducing erodibility, and buffering against environmental disturbances. Tillage practices break down soil aggregates, which change physical and chemical components, such as bulk density and the availability of water (Fu and Yan, 2023). Advocating green farming practices that encourage microbial communities will help to rejuvenate soil health and agricultural productivity (Srivastava et al., 2022; Fu and Yan, 2023). For example, the beneficial bacteria *Paraburkholderia phytofirmans* PsJN promotes crop growth under drought conditions by inducing the plant's production of extracellular polysaccharides, resulting in enhanced water retention and drought resistance (Prihatna and Yan, 2024). Senser's focus on regenerative agriculture exemplifies how sustainable practices can maximize crop production and environmental resilience (Yang et al., 2020).

Microbial EPS shows a significant impact in promoting sustainable agricultural performance and increasing soil fertility (Tiwari et al., 2024). These EPS are biopolymers produced by several soil microorganisms, which in one way or another contribute to enhancing the farming ecosystem, predominantly under green farming practices (Etesami, 2024). EPS contributes to the stabilization of soil aggregates, which is crucial for preventing soil erosion and enhancing crop growth (Mouro et al., 2024). EPS binds soil particles together by acting as a natural "glue," improving soil structure and porosity, thus facilitating better water retention and nutrient availability (Cania et al., 2019). This is especially important in contexts where conventional tillage disrupts the integrity of soil aggregates, as reduced tillage combined with organic farming practices can foster a healthier microbial community capable of producing EPS. Certain EPS

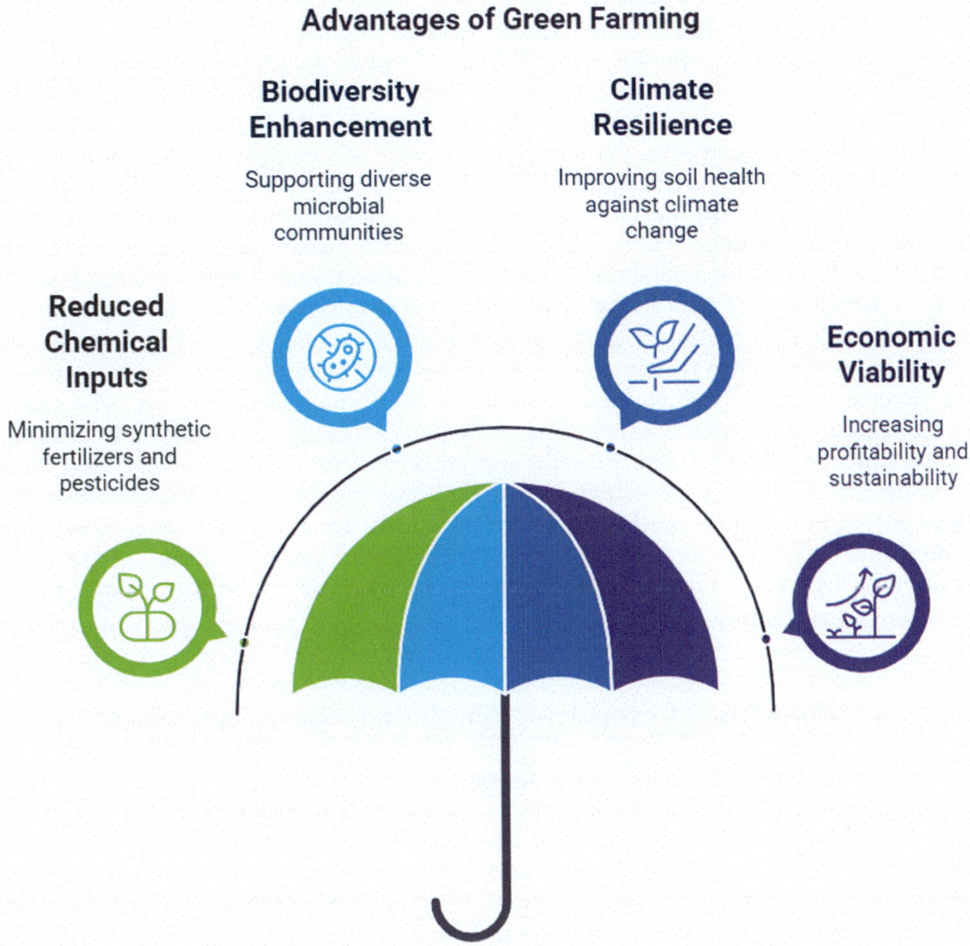

FIGURE 8.1 Potential advantages of green forming.

have been shown to enhance plant growth by acting as signaling molecules that promote nutrient uptake and stimulate beneficial microbial interactions (Cania et al., 2019; Mouro et al., 2024). For example, EPS can improve the survival of beneficial prokaryotes in harsh environments by forming protective biofilms around microbial cells, thus aiding in their resilience against environmental stresses like desiccation and nutrient scarcity. Moreover, some EPS-producing bacteria have been identified as plant growth-promoting rhizobacteria (PGPR), capable of improving crop productivity under stressful conditions such as salinity and drought (Injamum-Ul-Hoque et al., 2024). EPS fosters symbiotic relationships among microbial species, encouraging collaboration for resource sharing and nutrient uptake. This interaction enhances the robustness of microbial communities toward environmental stressors, ultimately aiding microbes and their symbiotic plants (Injamum-Ul-Hoque et al., 2024).

Research indicates that bacterial communities in agricultural soils are stable over time, showing little variation in response to different tillage regimes. This stability suggests that regulatory mechanisms may operate at the RNA level rather than through observable changes in community structure. EPS production plays a vital role in plant colonization and survival against various biotic and abiotic stresses such as desiccation, UV damage, and salinity (Mouro et al., 2024). Low organic matter content and adverse soil conditions limit microbes' performance, thus limiting EPS synthesis. Advancing microbial resilience and function requires the integration of holistic approaches that recognize the complexity of soil ecosystems under conditions or drivers hallmarked by reliance on microbial communities (Injamum-Ul-Hoque et al., 2024). Even though the crucial role that EPSs play in sustainable agriculture is recognized, the precise contribution of EPS to the myriads of plant-microbe interactions remain poorly understood. Genetic information and biochemical pathways related to EPS production should be explored to improve crop resilience under different environments. Several researchers are working extremely hard on EPS and their potential role as biofertilizers to develop eco-friendly forming strategies. This chapter explores the potential of EPS as bio-fertilizers, emphasizing their role in promoting soil health, plant growth, and stress tolerance. EPS improves soil structure by enhancing water retention, nutrient availability, and microbial activity. Additionally, the synergistic interaction of EPS-producing microbes with plant roots facilitates better nutrient uptake and offers resilience to abiotic stresses. This chapter provides insights into how EPS influences crop physiology, outlines practical applications in green farming, and evaluates its environmental and economic impact.

8.2 Properties of exopolysaccharides relevant to agriculture

Soils have an essential role in agricultural practices, and one of the biomolecules produced by microorganisms that substantially influences their health is the EPS (Behl et al., 2024). Soil organic matter (SOM) is comprised of natural polymers that improve critical soil properties, such as water-holding capacity, soil aggregation, and nutrient-binding capacity, making it essential for sustainable crop productivity (Guo et al., 2018). The EPS's excellent performance in soil dynamics has attracted attention in agricultural research and practice to deal with several difficulties presented in modern farming, especially in terms of soil moisture management and nutrient availability (Guo et al., 2018; Zhang et al., 2024). EPS possess unique properties, for example, enhancing soil structure, improving water retention, and supporting nutrient availability that directly contribute to the sustainable and resilient agriculture practices. Further, these benefits help strengthen rural livelihoods and reduce poverty with high crop yield, especially in resource-limited areas. Promoting EPS-based practices also supports quality agricultural education and skill development and encourages inclusive participation of women in eco-friendly farming technologies. By providing affordable, low-input alternatives to chemical fertilizers, EPS can reduce agricultural inequalities while fostering transparent, science-driven, and environmentally responsible agricultural decision-making. Further, by binding soil particles together, EPS also promotes soil aggregation, forming stable aggregates that increase aeration, nutrient availability, and overall soil health (Costa et al., 2018) (Fig. 8.2).

In addition, the nutrient-attaching nature of EPS improved nutrient turnover and retention, which could decrease the reliance on the use of chemical fertilizers and contribute to sustaining agriculture (Awasthi, 2017; Wang et al., 2023). While EPS benefits range from their

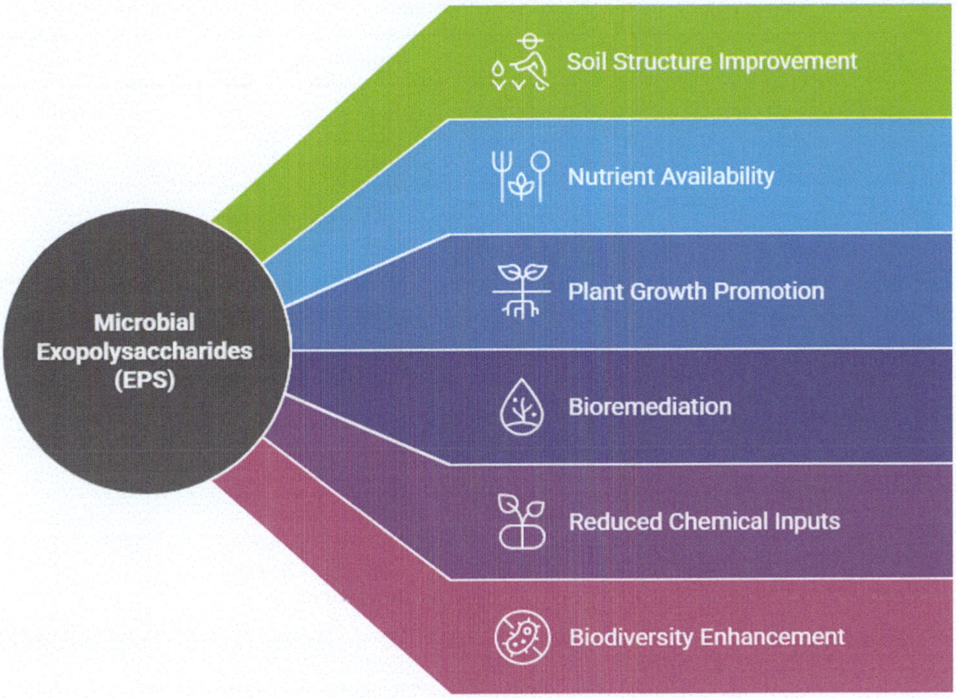

FIGURE 8.2 Multifaceted applications of exopolysaccharides (EPS) relevant to agriculture.

roles as biofertilizers, biostimulants, and growth-promoting molecules, EPS application in agriculture does face challenges such as soil contaminants, variable EPS efficacy in different soils, and complexities in certifying it as an organic farming practice (Wang et al., 2023). Such variations of EPS provide us with great opportunities to apply EPS in many agricultural fields; however, a lack of research on EPS diversity and specific functional mechanisms presents challenges that need more investigation. Tackling these issues is critical to ensuring the efficacy of EPS as a tool to improve soil health and ultimately increase agricultural productivity to meet global food security needs in the context of climate change and increasing food demand (Singh et al., 2025). Further, different types of EPS and their potential applications are listed in the (Table 8.1).

8.3 Mechanisms of exopolysaccharides-plant interaction

EPS-Plant interaction mechanisms are critical for promoting plant development, optimizing nutrient acquisition, and augmenting stress tolerance (Behl et al., 2024). In the rhizosphere, EPS are both structural and functional components that play a crucial role in the interaction amid microbes and plant roots, thus greatly affecting plant health and productivity in a range of ecosystems (Fu and Yan, 2023). One of the major roles of EPS in plant interactions is the enhancement of nutrient acquisition by plants. EPS promotes the attachment of various beneficial

TABLE 8.1 Different types of EPS-producing microbial species and their potential applications.

S. no	EPS	Source	Application	References
1	Curdlan	*Paenibacillus polymyxa*	It involves bioremediating wastewater by removing cadmium, cobalt, copper, and lead.	Vardanyan et al. (2019)
2	EPS biofilm	*Bacillus mojavensis*	Remediation of wastewater effluents.	Mnif et al. (2023)
3	Glucose galacturonic acid	*Bacillus mojavensis*	Remediation of wastewater.	Mnif et al. (2023)
4	Dead cell-bound EPS	*Bacillus cereus, B. pumilus, Pantoea agglomerans*	As a chromium remover from activated sludge.	Yadav et al. (2019)
5	Colloidal EPS	*Acidithiobacillus ferrooxidans*	Helps organisms to bind with minerals and extract metals from sulfide ores.	Vardanyan et al. (2019)
6	Fungal EPS	*Flavodon flavus*	Catalyst for the breakdown of hazardous chemical substances.	Negi and Das (2023)
7	EPS biofilm	*Bacillus mojavensis*	Remediation of wastewater effluents.	Mnif et al. (2023)
8	Fungal EPS	*Colletotrichum Sp.*	Elimination of Cd and Pb metals in the water treatment process.	da Silva et al. (2014)
9	Xanthan	*Xanthomonas Sp.*	Emulsifier and gelling agent.	Abu Elella et al. (2021)
10	Glucomannan	*Rhizobium leguminosarum, R. meliloti*	Aerogels, dietary fiber, food additives, emulsifiers, and thickeners.	Shi et al. (2020)
11	Gellan	*Sphingomonas Sp., S. paucimobilis*	Solidifying and gelling agent.	Díaz-Cornejo et al. (2023)
12	α-D-glucans	*Leuconostoc pseudomesenteroides*	Mutanase inducers.	Li et al. (2020)
13	Fructans	*Streptococcus salivarius, Leuconostoc Sp.*	Improve lipid metabolism, laxatives.	Meng et al. (2020)

(*continued on next page*)

TABLE 8.1 Different types of EPS-producing microbial species and their potential applications—cont'd

S. no	EPS	Source	Application	References
14	Cellulose	*Salmonella, E. coli, Gram-positive bacteria Komagataeibacter xylinus*	Key ingredients of artificial skin products, natural nondigestible fibers, hollow fibers or membranes, and acoustic membrane material in audiovisual equipment.	Díaz-Cornejo et al. (2023)
15	Succinoglycan	*Alcaligenes faecalis var. myxogenes Sinorhizobium fredii*	Gelling agent and immobilizing agent.	Jeong et al. (2022)
16	Emulsan	*Acinetobacter calcoaceticus*	Emulsifying agent and immobilizing agent.	Liu et al. (2021)
17	Clavan	*Clavibacter Sp.*	Therapeutic Agents and key ingredients of cosmeceuticals.	Roca et al. (2015)

EPS, exopolysaccharides.

microbes to the root surfaces, which then promotes root colonization and biofilm formation, further facilitating the uptake of vital nutrients and water from the soil (Zhang et al., 2024). For instance, the nitrogen-fixing bacterium, *Paraburkholderia phymatum*, uses EPS to enhance root attachment and nodulation in common beans, while *Gluconacetobacter diazotrophicus* provides similar advantages to rice when the plants are exposed to stress (Fu and Yan, 2023; Zhang et al., 2024). EPS also helps endophytic bacteria and their plant hosts afford tolerance to stress.

EPS generation is important for the survival of endophytes such as the environmental isolate, *Paraburkholderia phytofirmans* PsJN, during drought stress, as EPS-producing strains of PsJN survived stress in greater numbers than their EPS-deficient mutants (Fu and Yan, 2023). EPS protects microbial communities in the rhizosphere from environmental upsets (such as drought and salinity) and promotes plant overall growth and health. Addressing the growing pressures of climate change and environmental degradation, EPS–plant interactions are key to creating sustainable agricultural practices and improving crop resilience (Fu and Yan, 2023) (Fig. 8.3). This new investigation into the dynamics of EPS stress through local visits is another strong example of ecological interactions and the ways through which plants can be more productive when challenged.

8.4 Applications of exopolysaccharides in green farming

EPS has various applications in green agriculture as bio-fertilizers, organic pest control, and soil remediation. EPS usage in green farming is sustainable, as it promotes eco-friendly farming practices. As traditional farming practices begin to escalate into a more concerning manner, the

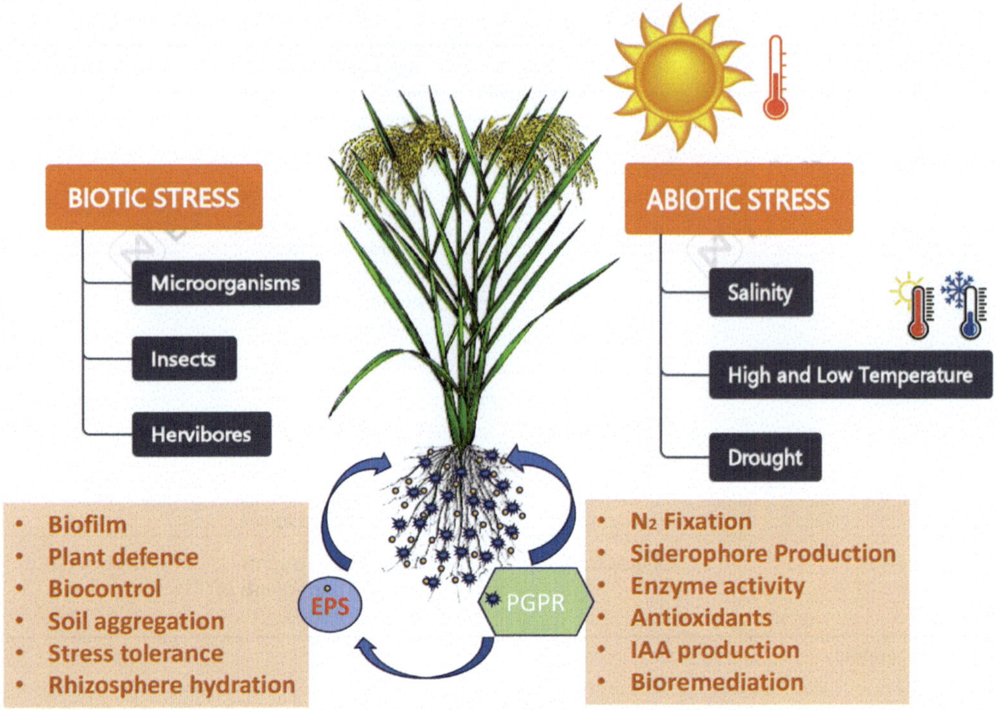

FIGURE 8.3 Diagram summarizing how plant growth-promoting rhizobacteria (PGPR) promote plant growth and how their exopolysaccharides (EPS) counteract biotic and abiotic stress.

use of EPS within agriculture is an excellent proposal for soil and productivity improvement, alongside minimizing the ecological footprint (Waoo et al., 2023; Zhang et al., 2024). In biofertilizers, EPS plays a vital part in the provision of microbial EPS by encapsulating a wide range of beneficial and dominant microbes, which enhances their survival and colonization in the soil. Coupled with this is the fact that it helps improve soil fertility through the sustained release of nutrients, making it extremely economical since agricultural waste is utilized in its production. Furthermore, a spectacular improvement in soil and a thriving microbial ecosystem needed for regenerative agriculture is shown to be fostered by EPS (Markam et al., 2024; Zhang et al., 2024). EPS also has astounding benefits within soil remediation due to its ability to adsorb heavy metals and detoxify contaminated soils.

In addition to the support of EPS in soil environmental remediation, the material itself can adsorb metal ions either through physical adsorption processes or complexation, rendering soil capable of retaining nutrients again and facilitating healthier plant as well as microbial life through that (Markam et al., 2024; Zhang et al., 2024). This permeates the competence of EPS, a promising technique for the remediation of soil pollution and food crop sustainability. In addition to being a mulch, EPS provides organic pest control via a physical barrier that helps weed suppression and lessens the use of chemical herbicides that are part of its functioning (Jia et al., 2025). Moisture retention ability enhances the resilience of crops against pests and diseases, which leads to eco-friendly agriculture methods. The cited uses of EPS underpin the critical role it plays in bridging the gap for green farming solutions to present reality-based challenges (Zhang et al., 2024; Jia et al., 2025). EPS in bio-fertilizer and organic pest control applications have opened various uses for green farming with environmentally friendly practices (EPS).

8.4.1 Soil remediation

The application of EPS in soil remediation focused on adsorptive retention of heavy metals and soil remediation. Functional groups in EPS secreted from microorganisms can prevent metal ions by trapping and binding them; consequently, it facilitates the degradation of soil containing heavy metals. The biosorption process with adjustment of these factors results in the enhancement of soil to retain nutrients and also microbial life, utilizing several physical mechanisms together with complexation-physical adsorption (Markam et al., 2024; Zhang et al., 2024). Hence, not only EPS is assistive in the clean-up of polluted sites, but it also rapidly develops healthy microbial populations crucial to restoring soil (Markam et al., 2024).

8.4.2 Exopolysaccharides as a source of organic pest control

Research on EPS is also being conducted on organic pest management strategies. EPS mulch, when spread over crops, will suppress weed growth by acting as a physical barrier and therefore can reduce chemical herbicides. This is a way to support more sustainable forms of agriculture that opt for environmentally lean solutions. In addition, EPS mulch keeps the ground moist, allowing for more favorable growing environments for crops that, in turn, are more resistant to pests and diseases (Zhang et al., 2024).

8.4.3 In bio-fertilizers

The bio-EPS enhances the efficiency of biofertilizers. Biofilms formed by these endophytic bacteria encapsulate microbial strains within EPS matrices, thereby achieving high colonization rates and survival of potential soil-beneficial microorganisms (Tiwari et al., 2024). This encapsulation improves the soil structure and facilitates the gradual release of nutrients, thus enhancing soil fertility. In addition, EPS can be synthesized from agricultural waste materials, which is a cost-effective process for bio-fertilizer production and solves the problem of waste treatment (Zhang et al., 2024). Since EPS technologies reduce reliance on costly chemical inputs, they help minimize agricultural inequalities. Furthermore, adopting EPS-driven, science-based practices supports transparent, accountable, and environmentally responsible agricultural systems.

8.5 Case studies and success stories

EPS is used extensively as biostimulants and biocontrol agents, showing emerging agroecological features such as sustainable replacements to synthetic fertilizers with soil fertility enhancement, improved tolerance to environmental stresses, and nutrient uptake. Many case studies have shown their effectiveness and are related to the promotion of plant development and disease suppression in different crops, indicating the importance of such compounds in contemporary agriculture (Wang et al., 2023; Chen et al., 2024). Interestingly, the production of EPSs by bacteria such as *Lactobacillus plantarum* and *Burkholderia ambifaria* confer significant benefits on plant performance. In what follows, *L. plantarum* EPS was co-cultured with beneficial

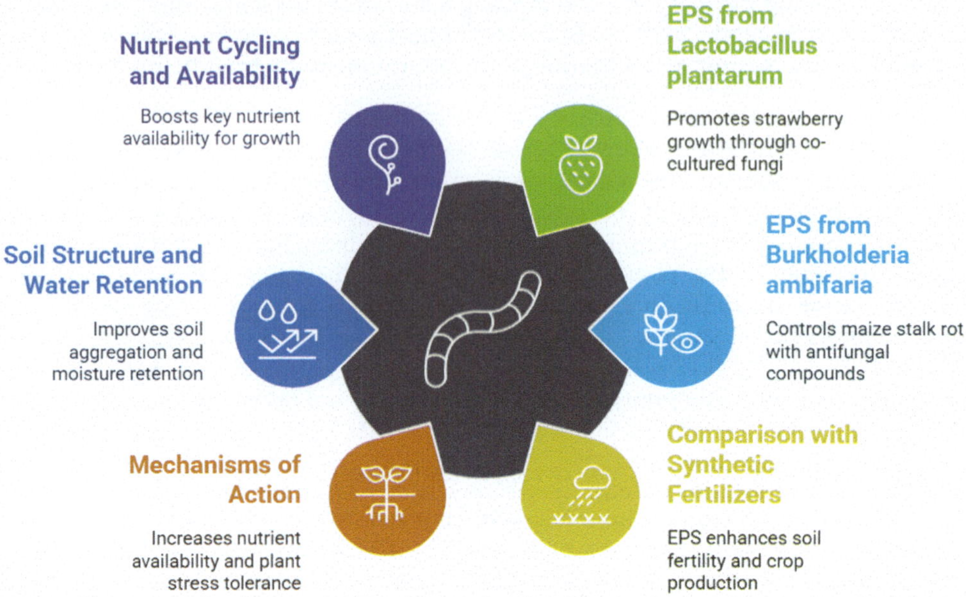

FIGURE 8.4 Role of exopolysaccharides in sustainable agriculture.

fungi to foster strawberry growth, and *B. ambifaria* EPS showed significant activity in controlling maize decortication, utilizing rot due to biocontrol antifungal compounds. These case studies illustrate the dual function of EPS at the interface of plant health and soil quality with imperative implications for sustainable farming (Chaudhary et al., 2024) (Fig. 8.4).

EPS versus traditional synthetic fertilizers have been listed clearly in a comparative analysis, which highlights their merits. Studies have shown that EPS uptake increased yields and benefited soil health-related concepts such as biofumigation, beneficial microbial communities, and decreased nutrient leaching, unlike traditional synthetic fertilizers (Chamizo et al., 2020; Carezzano et al., 2023). Moving away from chemical solutions into microorganisms lessens the influence of environmental factors and provides a much greater dose of resilience to climate change for agricultural systems; EPS demonstrates a clear future focus for intensified agricultural research and development. However, EPS's commercial and market penetration across agriculture faces challenges tied to their insights (economic viability) and industrial relevance (Schmid et al., 2015; Chamizo et al., 2020). The high production costs and regulatory barriers to the incorporation of EPS into existing agricultural systems have limited potential. Nonetheless, as sustainable practices become more widely adopted and the demand for environmentally acceptable alternatives grows increasingly urgent, EPS could be pivotal in reshaping the future of agricultural sustainability (Schmid et al., 2015).

8.5.1 Case studies of microbial exopolysaccharides

Microbial EPS is emerging as a suitable material for agricultural purposes, especially as biostimulants and biocontrol agents. Different case studies underline the potential of EPSs not only in plant health but also in the benefits related to their application in comparison with synthetic fertilizers.

8.5.1.1 *Application in crop production*

A typical case study is EPS from *Lactobacillus plantarum,* which promotes strawberries' growth. As EPS co-cultured with fungi, these demonstrated additive effects; i.e., biocontrol and growth promotion, consequently healthier plants resulting in higher yields (Wang et al., 2023). This underscores the polyfunctional nature of microbial EPS in agricultural systems. Another study explored the application of EPS from the bacterium *Burkholderia ambifaria* H8 to manage maize stalk rot. The EPS molecules were found to be able to generate volatile molecules that are beneficial for the suppression of fungal pathogens, resulting in a double benefit of plant defense and better tolerance/resilience (Chen et al., 2024).

8.5.1.2 *Comparative analysis with synthetic fertilizers*

Microbial EPS as a promising renewable substitute for inorganic fertilizer applications has succeeded in agriculture, increasing soil fertility and crop production by enhancing nutrient holding, water holding, and soil structure. New studies show that EPS is very effective compared to conventional fertilizers and helps develop the root system, stress tolerance (e.g., drought, salinity), and microbiological activity. Furthermore, EPS can also contribute to minimizing environmental issues such as chemical runoff: An eco-friendly, long-term soil health and plant productivity solution for a wide range of climatic and soil conditions.

8.5.1.3 *Mechanisms of action*

Plant growth and soil aggregation of microbial EPS, porosity increase/moisture retention—Enhances plant health through microbial-engineered EPS. The biopolymers increase nutrient availability through chelation and mineral solubilization as well as enhance plant stress tolerance using biofilm formation, osmoregulation, and induction of defense genes (Nastro et al., 2025). Moreover, EPS supports the establishment of microbes in the rhizosphere that are beneficial to participants. Together, they suppress pathology sustainably and reduce dependence on synthetic inputs while enabling agriculture systems to cope better with environmental stressors like drought or salinity.

8.5.1.4 *Soil structure and water retention*

Soil aggregation and aggregation stability are enhanced by EPS produced by microbial inoculants such as cyanobacteria. The EPS allows soil particles to aggregate, creating aggregates that are sticky and stay that way, leading to improved soil structure and soil erosion prevention (Chamizo et al., 2020). This binding improves the water-holding capacity of sandy soils, so water is not lost and becomes accessible for plant roots, especially in arid regions (Chamizo et al., 2020). Deng et al. (2020) research shows that cyanobacteria inoculation of sandy soils causes improved water-holding capacity and enhances moisture conservation, ensuring plant survival during dry spells.

8.5.1.5 *Nutrient cycling and availability*

Microbial EPS enables the build-up of nutrient-cycling-friendly multimicrobial communities, which in turn boost the availability of key nutrients for growth. For instance, in organic farming systems, more diverse microbial communities support faster cycling of nutrients (nitrogen and phosphorus being especially important to plants), which is one of the many functions beyond yield organic (Lehman et al., 2015). Organic soybean plants have a larger root system and nitrogen-fixing nodules than conventional crops, suggesting that the EPS-stimulated microbial activity is essential in these processes, as studies demonstrated (Hartmann et al., 2015).

8.5.1.6 *Biochemical interactions*

Apart from the beneficial plant health, as a result of EPS interactions with microorganisms, they are also involved in biochemical interactions between microbes and plants. EPS production can modify root architecture and microbial diversity, therefore impacting plant health and yield (Schmid et al., 2015). Specific noncoding small RNAs activated by two-component systems can control the expression of metabolic pathways in several members of EPS synthesis, highlighting more complex regulatory networks mediating these interactions (Schmid et al., 2015).

8.6 Challenges and future directions

8.6.1 Challenges and limitations

Several challenges limit the widespread adoption of microbial EPSs in agricultural practices, including the production and application of microbial EPSs (Fig. 8.5).

8.6.1.1 *Economic viability*

The chief problem is the economic comparability of microbial EPS versus plant and algal polysaccharides. One of the limiting factors at present is the high production yield of microbial EPS like xanthan, levan, and pullulan, all produced under optimized conditions, whereas even then, they are expensive to extract industrially (Chamizo et al., 2020; Singh et al., 2022b; Yadavalli et al., 2022, 2024; Raj et al., 2023; Chandrasekhar et al., 2025). Thus far, only a handful of bacterial EPSs are commercially viable, considering how expensive they are to produce and, too often, not worth the potential product value in that particular market (Ates, 2015).

8.6.1.2 *Production efficiency*

Although microbial EPSs have potential advantages, their production efficiency is also affected by factors such as metabolic pathways and regulatory mechanisms that control biosynthesis. The regulation of EPS biosynthesis is an intricate and multifaceted example would be; curdlan, the pH, and the specific gene knockout can lead to a dozen-fold difference in yield (Schmid et al., 2015; Singh et al., 2022a).

8.6.1.3 *Regulatory and technical hurdles*

Microbial EPS commercialization could also hit roadblocks through the regulatory frameworks. Major regulatory roadblocks are new agricultural products; it is not an easy or quick process, and you really do have to prove your stuff. This may lack the potential to be innovative microbial applications of EPS in farming system introduction (Chamizo et al., 2020). Further, the technical ability to actually do this with microbial EPS in established agricultural systems is not as prevalent yet. Sometimes, the farmers and agronomists may not be trained/aware of how productive those materials can be when utilized in practice (Chamizo et al., 2020).

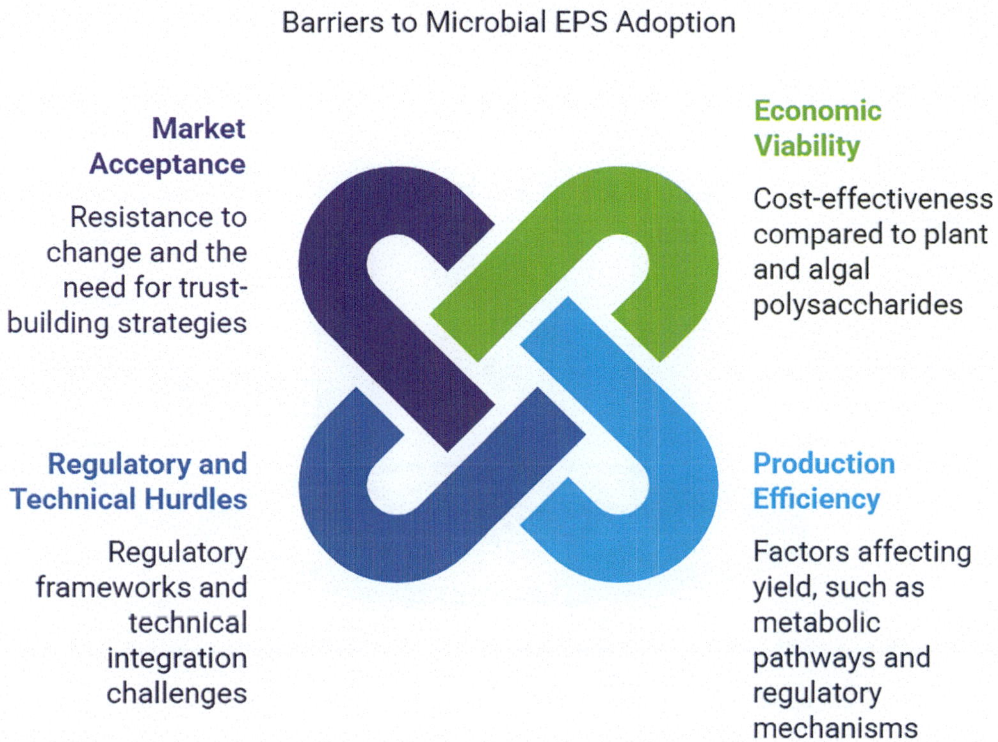

FIGURE 8.5 Schematic illustration of potential barriers to microbial exopolysaccharides (EPS) adaptation to green forming.

8.6.1.4 Market acceptance

The level of market acceptance for microbial EPSs as a real alternative that is able to synthesize or replace synthetic fertilizers and traditional agricultural chemicals is still problematic. The agricultural sector can be quite resistant to change—especially when time-honored practices have been hard-wired for years. A special strategy about how to make farmers trust and get the clear benefits of EPS from microbes through case studies and successful examples will be required for its future adoption.

8.6.2 Future prospects

Microbial EPS in agriculture has a vast future potential effect on sustainable agriculture by promoting soil fertility and crop yields, especially for degraded lands exposed to harsh climatic conditions. Findings from additional studies demonstrating the role of EPS in different agricultural systems further illustrate why they can be viewed as an alternative to synthetic fertilizers (Fig. 8.6).

8.6.2.1 Advancements in research

A field study on a semi-arid agriculture ecosystem in India found that EPS not only controlled soil erosion by 60%–88%, but it also increased yields of soybean by 57.7% and maize by 44.8%

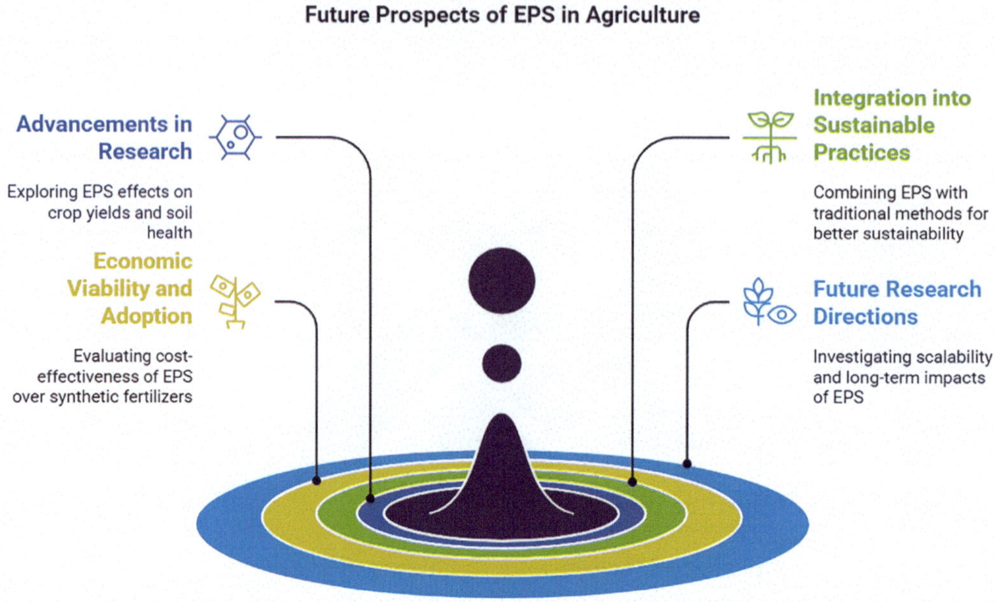

FIGURE 8.6 Schematic illustration of the future potential effect of microbial exopolysaccharides (EPS) in agriculture.

over the controls, showing potential for enhancing soil health and alleviating some abiotic stress (Chaudhary et al., 2024). Scientists will resolve engineered structure and activity relationships of EPS, and thus, expectations are that this could yield further information concerning mechanisms as well as applications specific to different host environments where customized formulations could be optimized for a given crop and/or set of conditions (Chamizo et al., 2020).

8.6.2.2 *Integration into sustainable practices*

EPSs are becoming essential to sustainable agriculture in general and organic farming systems in particular, which increase microbial diversity and soil health. EPS promoted an increase in beneficial microbes, allowing for nutrient recycling and aggregation, thereby helping to sustain soil fertility in the long term and improving soil functionality under threatened environments (Chandrasekhar et al., 2022; Dattatraya Saratale et al., 2022). Beneficial synergism between EPS and conventional practices like crop rotation and cover cropping may provide additional system-level benefits that improve total productivity and sustainability (Wang et al., 2023).

8.6.2.3 *Economic viability and adoption*

The importance of the EPS application is to do economic analyses to facilitate farmers' adoption of this technology. With the rising price of synthetic fertilizers and increased recognition of the environmental footprint of their production, EPS may well represent a cost-effective and eco-friendly solution. To show their economic relevance, research on the cost-benefit dynamics of various EPS applications will continue to be important in overcoming the challenges of smallholder farmers in developing regions (Chaudhary et al., 2024).

8.7 Conclusions

The study of microbial EPS shows their potential to transform agriculture through sustainable, eco-friendly green farming systems. EPS is highly capable of improving soil and plant health and stress tolerance, and it effectively solves some of the vital problems of agriculture today. The benefits of EPS from better nutrient absorption, more effective water retention, and greater microbial activity greatly increase soil fertility and agricultural productivity. The EPS functionalities can be extended to various areas of agriculture, including bio-fertilizers, organic pest control, and soil remediation, which display their overall environmental benefits. In the course of the use of EPS-based strategies that were all but confirmed to be in the right, nobody doubted the results that came about, like better yield, a more rigid soil texture, and a stable condition of the plants against stress made via the use of this biopolymer; nevertheless, the economic practicability and the large production horizon have been posing major problems, and therefore, the topic of further research in that respect has become very actual. The central theme of this paragraph is showcasing the use of EPS in agricultural practices as one of the most promising ways of ensuring sustainable food production and minimizing adverse environmental impact. As we approach a time of scientific advancement and changed implementations, microbial EPS is likely to become a driving force in edible food security worldwide without harming nature. This unique solution is perfectly matched with the increasing need for sustainable agricultural solutions that can cater to both productivity and environmental conservation needs.

Acknowledgments

We sincerely thank the Advanced Research Centre, Department of Biotechnology, Vignan's Foundation for Science and Technology (Deemed to be University), Vadlamudi, Guntur, Andhra Pradesh, for providing the research facilities.

References

Abu Elella, M.H., et al., 2021. Xanthan gum-derived materials for applications in environment and eco-friendly materials: a review. J. Env. Chem. Eng. 9 (1), 104702. https://doi.org/10.1016/J.JECE.2020.104702.

Ates, O., 2015. Systems biology of microbial exopolysaccharides production. Front. Bioeng. Biotechnol. 3 (1), 162636. https://doi.org/10.3389/fbioe.2015.00200.

Awasthi, S., 2017. Application of EPS in agriculture: an important natural resource for crop improvement. Agric. Res. Technol.: Open Access J. 8 (2). https://doi.org/10.19080/artoaj.2017.08.555731.

Behl, K., Jaiswal, P., Pabbi, S., 2024. Recent advances in microbial and nano-formulations for effective delivery and agriculture sustainability. Biocatal. Agric. Biotechnol. 58, 103180. https://doi.org/10.1016/j.bcab.2024.103180.

Cania, B., et al., 2019. A long-term field experiment demonstrates the influence of tillage on the bacterial potential to produce soil structure-stabilizing agents such as exopolysaccharides and lipopolysaccharides. Env. Microbiomes 14 (1), 1–14. https://doi.org/10.1186/s40793-019-0341-7.

Carezzano, M.E., et al., 2023. Exopolysaccharides synthesized by rhizospheric bacteria: a review focused on their roles in protecting plants against stress. Appl. Microbiol. 3 (4), 1249–1261. https://doi.org/10.3390/applmicrobiol3040086.

Chamizo, S., et al., 2020. Exopolysaccharide features influence growth success in biocrust-forming cyanobacteria, moving from liquid culture to sand microcosms. Front. Microbiol. 11, 568224. https://doi.org/10.3389/fmicb.2020.568224.

Chandrasekhar, K., et al., 2022. Regulation and augmentation of anaerobic digestion processes via the use of bioelectrochemical systems. Bioresour. Technol. 346, 126628. https://doi.org/10.1016/J.BIORTECH.2021.126628.

Chandrasekhar, K., Nastro, R.A., Kumar, P., et al., 2025. Exploring the potential of microbial exopolysaccharides as innovative and eco-friendly cleansing agents: A review. J. Biochem. Technol. 16 (3), 51–62. https://doi.org/10.51847/RQGqjwM4nG.

Chandrasekhar, K., Nastro, R.A., Narayana, A.V., et al., 2025. Microbial cell factory engineering for scalable production of bio-commodities: emphasis on robustness. J. Biochem. Technol. 16 (3), 19–29. https://doi.org/10.51847/6z0mtOiyyc.

Chaudhary, A., et al., 2024. Assessment of microbial exopolysaccharides for improving soil fertility and crop yield in semi-arid India. Emergent Life Sci. Res. 10 (01), 140–153. https://doi.org/10.31783/elsr.2024.101140153.

Chen, Q., et al., 2024. Mechanisms and impact of rhizosphere microbial metabolites on crop health, traits, functional components: a comprehensive review. Molecules 29 (24), 5922. https://doi.org/10.3390/molecules29245922.

Costa, O.Y.A., Raaijmakers, J.M., Kuramae, E.E., 2018. Microbial extracellular polymeric substances: ecological function and impact on soil aggregation. Front. Microbiol. 9 (Jul.), 337094. https://doi.org/10.3389/fmicb.2018.01636.

da Silva, L.J., et al., 2014. Biosorption of cadmium (II) and lead (II) from aqueous solution using exopolysaccharide and biomass produced by *Colletotrichum sp*. Desalin. Water Treat. 52 (40–42), 7878–7886. https://doi.org/10.1080/19443994.2013.833871.

Dattatraya Saratale, G., et al., 2022. Bioelectrochemical systems in aid of sustainable biorefineries for the production of value-added products and resource recovery from wastewater: a critical review and future perspectives. Bioresour. Technol. 359, 127435. https://doi.org/10.1016/j.biortech.2022.127435.

Deng, S., et al., 2020. Biological soil crust succession in deserts through a 59-year-long case study in China: How induced biological soil crust strategy accelerates desertification reversal from decades to years. Soil Biol. Biochem. 141, 107665. https://doi.org/10.1016/j.soilbio.2019.107665.

Díaz-Cornejo, S., et al., 2023. Biological properties of exopolysaccharides produced by *Bacillus spp*. Microbiol. Res. 268, 127276. https://doi.org/10.1016/J.MICRES.2022.127276.

Etesami, H., 2024. Enhancing crop disease management through integrating biocontrol bacteria and silicon fertilizers: challenges and opportunities. J. Env. Manage. 371, 123102. https://doi.org/10.1016/j.jenvman.2024.123102.

Fu, B., Yan, Q., 2023. Exopolysaccharide is required for motility, stress tolerance, and plant colonization by the endophytic bacterium *Paraburkholderia phytofirmans* PsJN. Front. Microbiol. 14, 1218653. https://doi.org/10.3389/fmicb.2023.1218653.

Guo, Y.S., et al., 2018. Bacterial extracellular polymeric substances amplify water content variability at the pore scale. Front. Environ. Sci. 6 (SEP), 379664. https://doi.org/10.3389/fenvs.2018.00093.

Hartmann, M., et al., 2015. Distinct soil microbial diversity under long-term organic and conventional farming. ISME J. 9 (5), 1177–1194. https://doi.org/10.1038/ismej.2014.210.

Injamum-Ul-Hoque, M., et al., 2024. Isolation and identification of multi-traits PGPR for sustainable crop productivity under salinity stress. Sustainability 16 (21), 9263. https://doi.org/10.3390/su16219263.

Jeong, J.P., et al., 2022. Bacterial succinoglycans: structure, physical properties, and applications. Polymers 14, 276. https://doi.org/10.3390/POLYM14020276.

Jia, X., et al., 2025. Research hotspots and application of algal extracellular polymeric substances (EPS) in wastewater treatment and resources recovery. Desalination 600, 118510. https://doi.org/10.1016/j.desal.2024.118510.

Lehman, R.M., et al., 2015. Understanding and enhancing soil biological health: the solution for reversing soil degradation. Sustainability 7 (1), 988–1027. https://doi.org/10.3390/su7010988.

Li, X., et al., 2020. Structures, physico-chemical properties, production and (potential) applications of sucrose-derived α-D-glucans synthesized by glucansucrases. Carbohydr. Polym. 249. https://doi.org/10.1016/J.CARBPOL.2020.116818.

Liu, J., et al., 2021. Pectin modified with phenolic acids: evaluation of their emulsification properties, antioxidation activities, and antibacterial activities. Int. J. Biol. Macromol. 174, 485–493. https://doi.org/10.1016/J.IJBIOMAC.2021.01.190.

Markam, S.S., et al., 2024. Microbial biosurfactants: green alternatives and sustainable solution for augmenting pesticide remediation and management of organic waste. Curr. Res. Microb. Sci. 7, 100266. https://doi.org/10.1016/j.crmicr.2024.100266.

Meng, Y., et al., 2020. Extraction, characterization and anti-inflammatory activities of an inulin-type fructan from *Codonopsis pilosula*. Int. J. Biol. Macromol. 163, 1677–1686. https://doi.org/10.1016/J.IJBIOMAC.2020.09.117.

Mnif, I., et al., 2023. Identification of different lipopeptides isoforms produced by *Bacillus mojavensis* BI2 and evaluation of their surface activities for potential environmental application. J. Polym. Env. 31 (6), 2668–2685. https://doi.org/10.1007/S10924-022-02752-3/FIGURES/5.

Mouro, C., Gomes, A.P., Gouveia, I.C., 2024. Microbial exopolysaccharides: structure, diversity, applications, and future frontiers in sustainable functional materials. Polysaccharides 5 (3), 241–287. https://doi.org/10.3390/polysaccharides5030018.

Nastro, R.A., et al., 2025. Bio-electrosynthesis of polyhydroxybutyrate and surfactants in microbial fuel cells: a preliminary study. Front. Microbiol. 16, 1372302. https://doi.org/10.3389/fmicb.2025.1372302.

Negi, B.B., Das, C., 2023. Mycoremediation of wastewater, challenges, and current status: a review. Bioresour. Technol. Rep. 22, 101409. https://doi.org/10.1016/J.BITEB.2023.101409.

Prihatna, C., Yan, Q., 2024. Exopolysaccharide is required by *Paraburkholderia phytofirmans* PsJN to confer drought-stress tolerance in pea. Front. Microbiol. 15, 1442001. https://doi.org/10.3389/fmicb.2024.1442001.

Raj, T. et al., (2023)., Marine microalgae in food and health applications. in A. Pandey, A. K. Patel, R. R. Singhania, C.-D. Dong (Eds.) Sustainable Marine Food and Feed Production Technologies. Boca Raton: CRC Press, 45–60. https://doi.org/10.1201/9781003326946-4.

Roca, C., et al., 2015. Exopolysaccharides enriched in rare sugars: bacterial sources, production, and applications'. Front. Microbiol. 6 (APR), 137116. https://doi.org/10.3389/FMICB.2015.00288/BIBTEX.

Schmid, J., Sieber, V., Rehm, B., 2015. Bacterial exopolysaccharides: biosynthesis pathways and engineering strategies. Front. Microbiol. 6 (MAY), 147519. https://doi.org/10.3389/fmicb.2015.00496.

Shi, X.D., et al., 2020. Plant-derived glucomannans: sources, preparation methods, structural features, and biological properties. Trends Food Sci. Technol. 99, 101–116. https://doi.org/10.1016/J.TIFS.2020.02.016.

Singh, M., et al., 2025. An overview of the role of algae-fortified foods in nutraceutical industries: synthesis pathway of value-added bioproducts and co-products. Food Biosci. 63, 105568. https://doi.org/10.1016/j.fbio.2024.105568.

Singh, M., Unadkat, K., Kapoor, S., et al., (2022a). Potential applications of biopolymers in fisheries industry. 199–221. https://doi.org/10.1007/978-3-030-98392-5_10.

Singh, M., Unadkat, K., Parikh, P., 2022b. Algae mediated pesticides bioremediation: mechanisms, approaches, limitations, and prospects for future research. Pestic. Bioremediation 353–380. https://doi.org/10.1007/978-3-030-97000-0_13.

Srivastava, K., et al., 2022. Algal biofilms: potential wastewater treatment applications and biotechnological significance. Appl. Biofilms Appl. Microbiol. 203–233. https://doi.org/10.1016/B978-0-323-90513-8.00014-5.

Tiwari, O.N., et al., 2024. Comprehensive review on recent trends and perspectives of natural exo-polysaccharides: pioneering nano-biotechnological tools. Int. J. Biol. Macromol. 265, 130747. https://doi.org/10.1016/j.ijbiomac.2024.130747.

Vardanyan, A., et al., 2019. Adhesion to mineral surfaces by cells of *Leptospirillum*, *Acidithiobacillus* and *Sulfobacillus* from Armenian Sulfide Ores. Minerals 9, 69. https://doi.org/10.3390/MIN9020069.

Wang, W., et al., 2023. Structural characteristics of microbial exopolysaccharides in association with their biological activities: a review. Chem. Biol. Technol. Agric. 10 (1), 1–16. https://doi.org/10.1186/s40538-023-00515-3.

Waoo, A.A., et al., 2023. Microbial exopolysaccharides in the biomedical and pharmaceutical industries. Heliyon 9 (8), e18613. https://doi.org/10.1016/j.heliyon.2023.e18613.

Yadav, A.N., et al., 2019. Metabolic engineering to synthetic biology of secondary metabolites production. New Future Dev. Microb. Biotechnol. Bioeng.: Microb. Second. Metab. Biochem. Appl. 279–320. https://doi.org/10.1016/B978-0-444-63504-4.00020-7.

Yadavalli, R., et al., 2022. Simultaneous production of astaxanthin and lipids from *Chlorella sorokiniana* in the presence of reactive oxygen species: a biorefinery approach. Biomass Convers. Biorefinery 12 (3), 881–889. https://doi.org/10.1007/s13399-021-01276-5.

Yadavalli, R., et al., 2024. Potential applications of microalgae and its derivatives in various industries. Algal Biotechnol. 223–237. https://doi.org/10.1201/9781003219194-18.

Yang, A., et al., 2020. Burkholderia phytofirmans PsJN stimulate growth and yield of quinoa under salinity stress. Plants 9 (6), 672. https://doi.org/10.3390/plants9060672.

Zhang, M., et al., 2024. Microbial extracellular polymeric substances (EPS) in soil : from interfacial behaviour to ecological multifunctionality. Geo-Bio Interfaces 1, 1–11. https://doi.org/10.1180/gbi.2024.4.

CHAPTER

9

Exopolysaccharides as raw nanobiomaterials for 3D printing technology

Blessy Maruthukunnel Mani[a], *Sherin Shaji*[a], *Blessy Babukutty*[b], *Martin George Thomas*[c] and *Sabu Thomas*[a,b,d]

[a] International and Inter University Centre for Nanoscience and Nanotechnology, Mahatma Gandhi University, Kottayam, Kerala, India [b] School of Chemical Sciences, Mahatma Gandhi University, Kottayam, Kerala, India [c] Department of Physics & Electronics, CHRIST (Deemed to be University), Bengaluru, Karnataka, India [d] School of Energy Materials, Mahatma Gandhi University, Kottayam, Kerala, India

9.1 Introduction

3D bioprinting utilizes biomaterials as bioinks that function as bioactive matrices that promote cell proliferation, structural integrity, and printability. The fabrication of these bioinks is intended to facilitate the spatial deposition of cells and biomaterials using computer-aided design, enabling the formation of tissue constructs that closely mimic the native tissue hierarchy and architecture (Khoeini et al., 2021). Biopolymers are the main ingredient in bioink formulations because they successfully replicate the mechanical and biological characteristics of the native extracellular matrix (ECM) and offer crucial structural support for a variety of cell types (Li et al., 2021). Microorganisms like bacteria, yeasts, and molds produce and release microbial exopolysaccharides (EPS), a distinctive and flexible class of biopolymers, into the surrounding environment. These naturally occurring polymers support microbial defense and biofilm structure, and their innate biocompatibility, adjustable viscosity, and hydrogel-forming properties make them very appealing for use in 3D bioprinting applications (Keshavarz et al., 2025).

DOI: https://doi.org/10.1016/B978-0-443-36619-2.00009-7

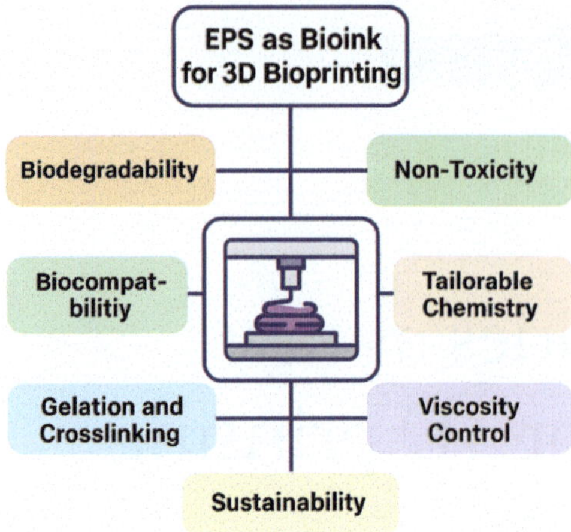

FIGURE 9.1 Advantages of microbial EPS as bioinks for 3D bioprinting. *EPS*, Exopolysaccharides.

Microbial EPS are primarily carbohydrate-based polymers made up of monosaccharides like rhamnose, galactose, and glucose, while EPS's structural complexity and functional diversity are enhanced by the presence of noncarbohydrate components such as proteins, enzymes, lipids, and nucleic acids in addition to these saccharide units. The microbial species and environmental growth circumstances can have a substantial impact on the exact composition and physicochemical characteristics of EPS. Particularly, they are ideal for biomedical applications, especially in the preparation of bioinks for 3D bioprinting (Fig. 9.1), due to their nontoxic nature, innate biocompatibility, and capacity to create hydrated gel-like networks (Ahuja et al., 2023). Promising rheological properties like shear-thinning behavior, specified yield stress, and quick structural recovery following extrusion are provided by microbially generated hydrogels, such as EPS based hydrogels.

These characteristics are essential for preserving cell viability and shape integrity during the printing process. The viscoelastic and self-healing properties necessary for the creation of soft scaffolds can be tailored into these materials by ionic cross-linking or supramolecular interactions. Because of their natural origin, adaptability, and tunability, EPS-based materials are excellent prospects for the creation of sustainable bioinks of the future (Fragal et al., 2025). This chapter focuses on the promising properties of microbial EPS as a bioink by examining the different microbial sources of EPS and their functional traits that make them feasible choices for 3D bioprinting while addressing the sustainability concerns, as well as potential future uses for EPS-based bioinks in tissue engineering and regenerative medicine.

9.2 Microbial exopolysaccharides: Sources and functional properties

Microbial EPS are high-molecular-weight polymers produced by a broad array of microorganisms into the extracellular environment with desirable properties for 3D bioprinting applications (Table 9.1). These compounds play protective and structural functions in microbial communities by facilitating biofilm formation, hydration, and desiccation or hostile condition resistance. In contrast with plant or animal polysaccharides, microbial EPSs are

TABLE 9.1 Summary of EPS-producing microorganisms and their functional applications in 3D printing.

Microorganism	EPS produced	Key properties	Application in 3D printing
Pseudomonas aeruginosa	Alginate	Biocompatible, ion-induced gelation, nonimmunogenic.	Tissue scaffolds, cell encapsulation.
Xanthomonas campestris	Xanthan gum	Shear-thinning, high viscosity, stable under pH and temp. variation.	Extrusion-based bioinks, structural support matrix.
Aureobasidium pullulans	Pullulan	Film-forming, transparent oxygen-barrier, water-soluble.	Biodegradable films, mucoadhesive drug delivery layers.
Gymnogongrus tenuis	Carrageenan	Sulfated, thermoreversible gelation, biocompatible.	Hydrogels for drug delivery and tissue mimics.
Gelidium spp.	Agar	Strong gelling agent, thermoplastic, nontoxic.	Gel-based 3D print structures, controlled drug release.
Nostoc commune	Cyanobacterial EPS	Antioxidant, metal-binding, sticky matrix.	Bioink additives for antimicrobial/antioxidant 3D constructs.
Lactobacillus plantarum	Heteropolysaccharide	Biocompatible, immunomodulatory.	Probiotic-loaded bioprintable scaffolds.
Leuconostoc mesenteroides	Dextran	Water-soluble, adhesive, nontoxic, viscosity enhancer.	Scaffold support, wound dressing ink components.
Rhizobium leguminosarum	Rhizobial EPS	Biofilm forming, nitrogen-fixing compatible, flexible.	Soil-simulation biomodels, environmental sensor printing.
Sphingomonas elodea	Gellan gum	Transparent, anionic, ion-responsive gelation.	Cell-laden hydrogels, soft tissue structure printing.
Saccharomyces cerevisiae	β-glucan	Immunostimulatory, viscous, fiber-forming.	Bioactive print layers for dermal and oral applications.
Bacillus subtilis	Levan	Fructan-based, flexible, antioxidant.	Encapsulation films, customizable viscoelastic inks.

EPS, exopolysaccharides.

structurally and functionally highly diversified and could be produced under controlled fermentation processes, representing renewable and scalable alternatives. Based on the microbial source and environmental conditions, EPS are either homopolysaccharides or heteropolysaccharides of varying branching, solubility, and functional groups—hence accommodating a plethora of industrial, environmental, and biomedical applications (Rehm, 2010; Jose et al., 2022)

Pseudomonas aeruginosa is one of the best-studied EPS-producing bacteria as it produces alginate, a linear anionic polysaccharide consisting of β-D-mannuronic acid (M) and α-L-guluronic acid (G) units. Alginate is a key contributor to *Pseudomonas* biofilm structural stability and resistance, especially in the clinical scenario of cystic fibrosis infections. Biosynthesis of alginate in *P. aeruginosa* is a complex process, regulated by numerous genes like algD, alg8, and alg44 involved in the synthesis and exportation of the polymer (Thi et al.,2020). Alginate's gelling property, biocompatibility, and ability to form hydrogels under mild conditions have rendered it an ageless option for wound dressings, drug delivery, and tissue engineering scaffolds (Lee and Mooney, 2012). Genetic engineering developments have also enabled higher alginate yields and customized monomer ratios to achieve varying biological and mechanical aspects for specific applications (Hay et al., 2013).

The alginate from *Pseudomonas* has become industrially significant not just for its usage in medicine but also in food, cosmetics, and agriculture. In comparison to alginate obtained from seaweed, the microbial production of alginate presents increased purity, reproducibility, and lower contamination risk. It serves as a gelling and thickening agent in foods, an emulsion stabilizer, and a material for matrices for cell or active ingredient encapsulation. Recent advances also investigate its application in 3D bioprinting, whereby ionotropic gelation by calcium ions allows layer-by-layer printing of biocompatible scaffolds. The sustainability and fast microbial alginate biosynthesis also make it consistent with green chemistry and circular bioeconomy principles (Tomić et al., 2023; Farshidfar et al ., 2023)

Xanthomonas campestris, a gram-negative phytopathogenic bacterium, is used as the major industrial source of xanthan gum. Xanthan gum is an EPS consisting of a high-molecular-weight heteropolysaccharide having a cellulose-like backbone (β-D-glucose) with trisaccharide side chains made up of mannose and glucuronic acid. Xanthan gum has excellent viscosity at low concentrations and is stable under broad ranges of pH, temperature, and salt levels (Chaturvedi et al., 2021). These are due to its semi-rigid helix structure and the capability to build networks of entangled structures in an aqueous environment. In microbial cultures, xanthan production is closely regulated and depends on carbon sources, pH, and aeration, and such genes as gumB–gumK govern the biosynthetic process (Becker et al., 1998).

Due to its rheological characteristics, xanthan gum finds wide usage in food industries as a stabilizer, thickener, and emulsifier, particularly for gluten-free food. Xanthan acts as a matrix for controlled delivery of drugs, mucoadhesive gels, and wound dressings in pharmaceutical and biomedical applications. Its pseudoplastic (shear-thinning) property renders it well-suited for extrusion-based 3D bioprinting, under which conditions it can be employed singularly or combined with other polymers to enhance the printability and mechanical stability of bioinks (Piola et al., 2022). Modified xanthan derivatives have also been demonstrated to have potential in environmental applications like heavy metal adsorption and bio-remediation (Kashaudhan et al., 2024).

These are widely used in food, microbiology, and systems for controlled drug delivery (Dayala at al., 2008). Cyanobacteria such as *Nostoc, Anabaena*, and *Spirulina* release EPS containing uronic acids and sulfated sugars with antioxidant, antimicrobial, and metal-binding properties. These bioactive polymers are being widely explored for biofertilizers, biosorbents, and biomedical coatings (Gheorghita et al., 2023). The heterogeneity in microbial EPS synthesis gives not only the sources of renewable materials but also opportunities for biotechnological innovation in different industries. Except for bacteria, other fungi and algae synthesize structurally and functionally heterogeneous EPS. *Aureobasidium pullulans*, a yeast-like fungus, synthesizes pullulan, a water-soluble, nonionic polysaccharide consisting of maltotriose units, for edible films, oral delivery of drugs, and biodegradable packages. Marine red algae, such as gelidium and gracilaria, yield agar and carrageenan, which are high-gelling and stabilizing sulfated galactans.

9.3 Functional properties of exopolysaccharides relevant to 3D printing

Because of their excellent biocompatibility, microbial EPS make good choices for biomedical 3D printing of tissue scaffolds, wound dressings, and drug delivery systems. Nonimmunogenic in nature, EPS, like gellan gum, pullulan, and alginate, can also encourage cellular adhesion, migration, and proliferation when properly modified or combined with bioactive molecules (Chen et al., 2020). Their biodegradability reduces the need for surgical resection by ensuring that printed constructs will break down naturally over time. Alginate scaffolds, for instance, are candidates for temporary tissue engineering scaffolds because they break down through ion exchange in physiological settings (Liu et al., 2018). Additionally, by controlling the crosslinking density, EPS composition, and mixing with other biopolymers, degradation rates can be precisely regulated (Lim et al., 2020).

Extrusion printing is still the recommended method for EPS-based hydrogels because it can create larger, structurally sound structures and work with high-viscosity bioinks. For example, calcium chloride may ionically crosslink alginate during or soon after deposition, causing scaffold development to proceed quickly. When combined with cellulose or gelatin, xanthan gum and gellan gum improve shape fidelity and lessen printing collapse (Wu et al., 2018). Cell-laden printing, which incorporates viable cells into the EPS matrix and maintains cell viability above 85% in most studies, is made easier by the soft extrusion process (Sahana and Rekha, 2020). Extrusion also makes it easier to integrate with coaxial or multinozzle equipment, producing the multimaterial or multicellular constructs needed for complex tissue regeneration.

Maintaining shape and accuracy in EPS based printing becomes challenging once laid down. Important factors, including nozzle size, ink distribution, gel setting speed, and thickness, all contribute to maintaining form properly. Certain EPS bioinks that set fast, such as gellan gum and alginate, aid maintain form by preventing it from breaking apart (Sathishkumar et al., 2021). Making the shape more evident depends on changing the print path, heat, and printing speed. Furthermore, employed to make EPS more robust and prevent it from spreading or shrinking are materials that affect the ink flow, such as PEG, silk fibroin, or cellulose (Asgher et al., 2021). EPS forms that respond to light can be made under light printing installations, obtaining extremely delicate shape details with the suitable light and starter combination.

It is crucial to alter the biochemical and mechanical signals of EPS- based bioinks for constructions that are multicellular and multilayered. Bioinks must aid in cell adhesion, nutrient diffusion, and spatial arrangement. Because native extracellular matrices (EPS) typically do not contain cell adhesion motifs (such as RGD), it is common to functionalize them or combine them with proteins such as gelatin, fibrin, or collagen (Charoenwongpaiboon et al., 2021). The inclusion of growth factors, microparticles, or vasculature-mimicking channels aids in the support of various tissues, such as liver, skin, and cartilage. Multicellular printing methods such as core-shell printing, modular assembly, and sequential deposition are increasingly being used to replicate tissue heterogeneity and complexity (McCarthy et al., 2019).

Although EPS based printing has shown enormous promise in biomedicine, scale-up, repeatability, and standardization still face problems. Clinical translation calls for solutions for elements including EPS purity, microbial variation, and batch-to-batch consistency. Dynamic tissue models show potential with emerging technologies like 4D printing, in which EPS hydrogels react to stimuli (pH, enzymes, temperature). Furthermore, prospective directions to improve performance and functionality are biofabrication automation, machine learning for print optimization, and biosynthetic engineering of customized EPS (McCarthy et al., 2019). Ultimately, EPS based bioinks offer a sustainable, customizable platform for next-generation bioprinting.

9.4 Exopolysaccharides bioink formulation for 3D bioprinting

Biomedical research relies on bioactive polymers as bioink for bioprinting, as they support diverse cell types and mimic the ECM to trigger a systemic response (Ramiah et al., 2020). The key attributes that an optimal bioink should have for bioprinting applications are printability and biocompatibility.

Choosing materials is contingent upon their physicochemical characteristics, which play a crucial role in developing an appropriate bioink (Butler et al., 2021). An optimal bioink ought to exhibit necessary physical and chemical attributes, including appropriate rheological, mechanical, chemical, and biological aspects (Lee et al., 2015). These characteristics is required to facilitate: (1) The development of tissue constructs that possess adequate mechanical integrity and durability, while also sustaining the properties that conform to the natural tissue, optimally in a controllable way; (2) customizable gelation and stabilization to support the bioprinting of structures with enhance structure accuracy; (3) compatibility with biological systems and, where required, biodegradation that replicates the natural tissue microenvironment; (4) the ability for chemical refinement to address tissue-specific requirements; and (5) feasibility for large-scale production with minimal variations between batches (Loo et al., 2015). As determining the ideal cell-laden bioink formulation is crucial for successful bioprinting, a variety of natural and synthetic biomaterials with specific attributes have been explored as bioinks (Lee et al., 2015).

Natural hydrogels are widely utilized in tissue engineering due to their biological compatibility and structural strength that can be adjusted for extrusion-driven 3D bioprinting. Various polysaccharides, including xanthan gum, gellan gum, hyaluronic acid, and alginate are incorporated into bioink formulations for 3D bioprinting due to their shear thinning characteristics (Lee et al., 2015; Chen et al., 2020; Garcia-Cruz et al., 2021). Additionally, materials such as nanocellulose, gelatin, chitin, silk, fibrinogen, agarose and k-carrageenan have been investigated for bioprinting purposes due to their favorable gelation behavior, dual crosslinking potential and compatibility with the living systems (Liu et al., 2018; Wu et al., 2018; Lim et al., 2020). Among biopolymers, EPS secreted by microorganisms have acquired interest in both the scientific and industrial domains as a sustainable resource due to their diverse biological properties, compatibility with biological systems, and ability to decompose naturally (Asgher et al., 2020; Sahana and Rekha, 2020; Keshavarz et al., 2025; Sathishkumar et al., 2021). In addition, EPS resembles the morphology and constituents of the ECM, which facilitates cell attachment, growth, and proliferation while also demonstrating the ability to self-repair (McCarthy et al., 2019; Charoenwongpaiboon et al., 2021). Bacterial polysaccharides are recognized as a key element in the inks utilized in various bioprinting methods, affecting important characteristics like mechanical and thermal properties, printability, biocompatibility, and biodegradability. Nonetheless, introducing any foreign structure into the body raises the likelihood of bacterial infections and immune responses. In recent times, this concern has been heightened by the increase in multidrug-resistant bacterial infections acquired in healthcare settings. To address this issue, bioprinting inks are being enhanced with antimicrobial agents (McCarthy et al., 2019).

Considering the essential properties and potential limitations of bioinks, numerous studies have explored the formulation of bioinks using various EPS to enhance printability and functional performance. Markstedt et al. (2015) developed a bacterial cellulose and alginate-based bioink for regenerative engineering of soft tissues. The nanofibrillated cellulose and alginate constructs embedded with human nasal chondrocytes (hNC) showed remarkable stability in shape and size, along with enhanced cell viability and proliferation during in vitro cultivation. Additionally, the bioink facilitated the redifferentiation of hNCs and the production of cartilage-specific ECM components. This indicates that the bioink not only aided in the redifferentiation of hNCs but also provided suitable printability within a biologically relevant aqueous 3D environment, positioning it as a prospective resource for auricular cartilage tissue engineering and various other biomedical uses (Martínez Ávila et al., 2016). Lozano et al. developed a brain-like structure utilizing a bioink made from a novel peptide-modified biopolymer gellan gum produced by *Sphingomonas elodea*, paired with primary cortical neurons. The bioink was refined for a tailored reactive printing technique and designed for application in conventional cell culture facilities without the necessity for an advanced bioprinting machine. Additionally, the peptide alteration of the gellan gum hydrogel significantly enhanced both primary cell multiplication and network development. The viability of neural cells, along with the facilitation of neural network formation, illustrated the supportive qualities of the hydrogel matrix for cells (Lozano et al., 2015). Nagaraj et al. developed a bioink combining alginate and succinoglycan-like exopolysaccharide (EPS-RL7), are produced and secreted by a bacterial strain, Rhizobium sp. PRIML7. An extrusion bioprinting experiment, simulating bioplotting, demonstrated the sustained viability of cells within the EPS-RL7-alginate polymer matrix. These findings suggest that EPS-RL7 can be effectively utilized along with alginate for bioinks intended for the application of bioprinting, providing both beneficial physical and bioactive properties (Nagaraj and Rekha, 2023). Succinoglycans synthesized by various species of rhizobium are heteropolysaccharides primarily made up of galactose and glucose, which are interconnected through beta-1,3, beta-1,4, and beta-1,6 glycosidic bonds. Succinyl, acetyl, and pyruvyl groups are further used to transform these molecules (Halderet al., 2017). The glycan component of polysaccharides serves crucial functions in biological systems and is involved in major processes such as intercellular signaling, immune regulation, detection of antigens, and interactions between cells and the ECM (Varki, 2017).

Another study by Deidda et al. (2024) focused on developing an extrudable GelMA-based bioink combining xanthan gum (XG). The GelMA- xanthum gum composite hydrogel revealed an elastic modulus of around 9 kPa, which is similar to that of cardiac tissue, and facilitated 3D printing with impressive levels of accuracy (83%) and printability (0.98). Experiments conducted using hiPSCs revealed the hydrogel's potential to enhance hiPSC proliferation in both two-dimensional and three-dimensional cell culture environments. The experiments then showed that hiPSCs situated within the hemispherical structures of the hydrogel were capable of differentiating into cardiomyocytes that exhibited unstimulated contractions (Deidda et al., 2024). Lim et al. (2020) designed a bioink constituting alginate, gelatin, dextran, and aldehyde (AGDA) to achieve an optimal balance between structural stability and cellular function. Fibroblasts that were encapsulated in the bioink, whether as individual cells, spheroids, or a mix of both, demonstrated high levels of viability and the capability of proliferation. Interestingly, the combination approach led to the highest density of cells and the most significant morphological changes specific to fibroblasts, exceeding those observed with the commercially available GelXA bioink. These results underscore AGDA based bionk[3]s promise as a flexible bioink for creating complex and scalable tissue constructs (Lim et al., 2020).

While various studies have demonstrated the potential of EPS-based bioinks in 3D bioprinting applications, the success of these formulations heavily relies on the use of suitable crosslinking strategies and the bioink's ability to support cellular processes such as attachment, multiplication, and viability. Bioinks generally consist of cell-laden hydrogels formulated with specific characteristics for biofabrication, primarily composed of biopolymers and solvents.

Biopolymers provide a secure and supportive matrix for cells, maintaining optimal conditions throughout formulation and processing, while solvents—most commonly water—facilitate appropriate rheological properties and create a favorable environment during printing. The viscosity behavior of hydrogels is heavily influenced by the concentration and molecular weight of the biopolymers, and higher concentrations typically lead to increased viscosity. Even though this factor can be advantageous for extrusion accuracy, it can adversely affect cell growth, migration, and differentiation. Consequently, many formulations rely on lower concentrations of high molecular weight biopolymers to balance printability and biological compatibility. The quality of the printed structure depends on the ratio and combination of ingredients in the bioink formulation, complemented by the influence of printing parameters such as pressure and temperature. Bioinks are subjected to internal forces like shear and compression during the process of printing, and that can impose stress on the cells and impact the viscosity and flow behavior of the ink. Therefore, bioinks must be engineered to flow eficiently while maintaining their structural integrity to minimize cell damage and maintain the precision of the printed construct. A critical phase in bioprinting is the stabilization of the bioink, during which it undergoes a solgel transition, where a solvated liquid exhibiting pseudoplastic behavior transforms to a viscoelastic gel, forming a stable or pseudo-stable three-dimensional network through the interaction of macromolecular chains. This gelation process can occur via physical or chemical crosslinking.

Physical crosslinking mainly depends on reversible interactions such as hydrogen bonding, hydrophilic interactions, or electrostatic forces and is often preferred due to its rapid gelation speed and ability to maintain shape accuracy. In contrast, chemical crosslinking involves the formation of permanent covalent bonds through the addition of natural or synthetic crosslinking agents. The mechanical properties imparted by different crosslinking techniques can directly affect cell survival and functionality. Moreover, the bioink composition can be modified for different mechanical and rheological properties, which can enhance bioactivity such as cell adhesion, proliferation, and tissue-specific architecture. Ideally, the bioink components should not only support cell viability within the three-dimensional structure but should also enable functionalization of the biopolymers for the incorporation and controlled release of biomolecules or biochemical signals (Lee et al., 2012; Liu et al., 2018; Chen et al., 2020; Lim et al., 2020).

9.5 Processing and modification techniques for exopolysaccharides

EPS are a class of large molecular weight biopolymers that are produced and secreted into the environment by numerous microorganisms, including bacteria and fungi, as part of their metabolic processes (Asianezhad et al., 2023). Because of their rapid rate of multiplication and weak adhesion of EPS to the mucoid surface, they can be easily separated from cells using different isolation techniques. Various types of bacteria are most frequently used for the production of EPS (Angelin and Kavitha, 2020). In addition, each bacterial strain produces a different type of EPS that serve a variety of biological purposes. These EPSs can be either used alone or in combination with other substances for a wide range of pharmacological and biomedical applications (Kavitake et al., 2023). The generating strain, the content of the media, and the culture conditions are some of the variables that can contribute to the amount, chemical composition, and biological activity of EPSs (Jurášková, Ribeiro and Silva, 2022; Zaghloul and Ibrahim, 2022). In order to produce novel EPSs with significant biological activity, strains that produce EPS must be carefully chosen, and production conditions must be optimized. After the manufacturing process, the EPS must be separated and purified from impurities such as proteins, biomass of microbes, nucleic acids like DNA and RNA, and medium pollutants (Sørensen et al., 2022). The extraction of EPS involves a number of steps that can change based on the microorganisms' source, the EPS types, the complexity of the production media, and the necessary purity (Jurášková et al., 2022; Kavitake et al., 2023).

EPS is extracted using two distinct techniques: Precipitation and ultrafiltration. Precipitation, protein degradation, and biomass removal are some of the phases involved in the precipitation method of exopolysaccharide extraction (Sørensen et al., 2022). The initial step in this process is centrifugation, which is done to extract the biomass of the cell and separate the cell-free supernatant (CFS) that contains EPS (Yu et al., 2023). A membrane filter (0.45 μm) is then used to filter the CFS (Zaghloul et al., 2023). According to some research, the bacterial cells are killed, and the enzymes are inactivated by heating the growing culture for 10–15 minutes at 90°C–100°C. The cells are then extracted by centrifugation (Yu et al., 2023). The second step in the process of isolation of pure EPS involves protein precipitation and degradation (Sørensen et al., 2022). Trichloroacetic acid (TCA) is the most widely used agent for protein precipitation and denaturation, typically utilized in concentrations ranging from 2% to 80% (w/v) at 4°C for a duration of 2 hours to overnight (Angelov et al., 2023). TCA promotes the precipitation of proteins by dehydrating the hydration shells surrounding them, while some researchers suggest that its acidic properties disrupt the protein structure, leading to precipitation; however, the underlying mechanism behind this process remains unclear. Additionally, other chemicals, such as Sevag reagent—a blend of chloroform and n- butanol (4:1)—and protease enzyme hydrolysis (at 50 μg/mL), are also employed to remove protein contamination from EPSs (Rahnama Vosough et al., 2021; Zhan et al., 2023). The third essential step is the process of precipitation, which is employed to isolate EPS by using one, two, three, or four volumes of ice-cooled ethanol at 95% or 100% (v/v), with an overnight incubation at 4°C (Kavitake et al., 2023). Ethanol interferes with the hydration shell formed by water molecules that interact with the hydrophilic components of EPS, as it is less polar than water, which decreases the solubility of EPS (Avinash and Jha, 2013). Occasionally, alternative precipitating agents such as propanol, acetone, and isopropanol are utilized to precipitate EPSs (Murugu and Narayanan, 2024; Nachtigall et al., 2021). The precipitated EPS is then gathered through centrifugation and re-dissolved in double-distilled water or Milli-Q water (Pourjafar et al., 2023). The precipitation method is commonly utilized because of its straightforwardness, effectiveness, low chemical contamination, and capacity to process large quantities, which makes it appropriate for extracting EPS. However, some drawbacks of this particular approach include incomplete precipitation, the co-precipitation of contaminants (such as proteins, nucleic acids, and other biopolymers), and the significant amounts of ethanol needed (Yadav et al., 2024).

Ultrafiltration is frequently suggested as a quicker and more accurate method for isolating EPS, serving as an alternative to TCA and ethanol treatments (Sørensen et al., 2022). The principle behind ultrafiltration relies on size exclusion and separation driven by pressure (Shoshaa et al., 2023). This technique effectively distinguishes EPS from small contaminants, such as salts and small organic compounds, by size (Zhang et al., 2016).

Molecules that exceed the pore size are captured by the membrane, while smaller molecules and solvents are allowed to pass through the membrane (Zhang et al., 2016). In this process, the centrifuged liquid medium yields coagulated proteins and cell components, while the EPS is isolated through ultrafiltration (Sørensen et al., 2022). Initially, the cell free supernatant is filtered using 0.20 μm filters, followed by ultrafiltration utilizing a membrane with a molecular weight selectivity of L0–500 kDa, which is generally appropriate depending on the size of the EPS molecules, and TCA is used for protein degradation (Delattre et al., 2016). The concentrated EPS can then be made to undergo processes like drying, freeze-drying, or re-dissolution in a preferred buffer or water for future analysis or applications. Ultrafiltration is a membrane-based separation technique that uses semipermeable membranes to concentrate and decontaminate EPS, providing an effective, scalable, and chemical-free approach for these processes. Nevertheless, challenges like membrane fouling, high upfront expenses, and possible restrictions in molecular weight-based selectivity must be addressed for productive application (Yadav et al., 2024).

The process of extraction is followed by purification, which is essential for the characterization of EPS. The required EPS is extracted and separated from a microbial culture using a multistep procedure called purification. Depending on the specific characteristics and level of purity needed for EPS applications, different purification techniques may be used (Talbi et al., 2023). The first stage of purification is dialysis, which is a widely used technique for EPS purification utilizing a molecular weight cut-off membrane (Kavitake et al., 2023). The second stage of purification that is required to verify the EPS's exceptional purity, involves anion-exchange chromatography and size-exclusion chromatography (Pourjafar et al., 2023; Feng et al., 2023)

One of the key areas of polysaccharide chemistry is thought to be molecular modification. The bioactivities of naturally occurring polysaccharides with lower biological activity can be improved or increased by these chemical changes. The general methods of alteration include chemical, biological, and physical changes. These alterations to their structure, molecular weight, and functional groups change the bioactivity of polysaccharides. Chemical changes can change polysaccharide substituents, while biological and physical changes only change the molecular weight of polysaccharides (Nachtigall et al.,2021). Physical modification refers to the use of external energy to break down glycosidic connections to reduce the molecular weight of polysaccharides (Chaisuwan et al., 2020). Common methods include thermal, alkaline, radiation-induced reaction, microwave exposure, and ultrasonic disruption (Viñarta et al., 2013). In biological modification, enzymes speed up the breakdown of polysaccharides to their molecular weight forms. The eficiency and selectivity of these biological changes are very high. However, a disadvantage of the enzyme was that it could only catalyze specific types of polysaccharides (Chaisuwan et al., 2020). Transferases and hydrolases can add or remove functional groups from EPS, such as succinyls, acetyls, glyceryls, pyruvyls, and dactyls, which can alter the polymer's surface electrostatics and solubility.

Chemical modification is a common modification technique that entails adding substituent groups to polysaccharide structures in order to increase their bioactivities (Zhang et al., 2016). Although a polysaccharide's structure is linked to its properties and functions, chemical modifications can alter the microstructures of polysaccharides to increase biological activity. Sulfation, acetylation, carboxymethylation, phosphorylation, benzoylation, and selenium are examples of chemical alterations (Chaisuwan et al., 2020). In sulfonation, sulfated EPS were produced by substituting sulphate groups for the hydroxyl, carboxyl, and amino terminals of polysaccharides (Andrew and Jayaraman, 2020). The sulfur trioxide-pyridine method (N, N-dimethylformamide (DMF), SO_3•Py) (Sun et al., 2015), the chlorosulfonic acid-pyridine method (Chlorosulfonic acid (CAS), anhydrous pyridine) (Xu et al., 2016), and the sulfuric acid method (Sulfuric acid, butanol complex, ammonium sulphate) (Jiang et al., 2014) are frequently employed techniques for the synthesis of sulfonated derivatives. According to a study, *Streptococcus thermophilus* EPS showed improved solubility, a decrease in degradation temperature, and a little change in microstructure after being exposed to the sulfonation process (Li and Shah, 2024). Additionally, Zhang et al. (2016) reported antibacterial, antiviral, antitumor, anticoagulation, and antioxidant qualities of sulphated EPSs (Zhang et al., 2016). In acetylation, through the esterification reaction, the process changes the hydroxyl (-OH) groups on the EPS sugar unit into acetyl groups (-COCH3). Proper acetylation can guarantee hydrogen bond weakening and anomeric carbon activation (Wang et al., 2015). This change, which is more noticeable in plant polysaccharides, modifies the structure and function of cell walls in particular, shielding them from harm brought on by microbial hydrolytic enzymes. Acetylation may alter the bioavailability and bioactivity of polysaccharides, reveal more polar groups, alter chain conformation, and improve water solubility.

In carboxymethylation, the main reaction is Williamson synthesis, where the polymer reacts with mono-chloroacetic acid to convert the hydroxyl (-OH) groups into carboxylic groups (Theis et al., 2019). Research on carboxymethylation of polysaccharides is quite widespread because the polymer's combination with chloroacetic acid results in a straightforward chemical change. In short, at room temperature, EPS can be quickly spun while suspended in isopropanol. Chloroacetic acid and NaOH solution are gradually added, and the chemical process is carried out for three hours at 60°C. Neutralization, which achieved a neutral pH at normal temperature, marks the process's conclusion. It is possible to concentrate, dialyze, precipitate, and lyophilize the resulting derivative to powder. Acid-base titration can be used to determine the degrees of substitution (Zhang et al.,2016). Many polymers are insoluble in water, which makes it dificult for them to develop bioactivities. One way to change this is by carboxymethylation, which improves the polymers' solubility, bioactivities, and ability to stop the growth of cancer cells (Qi et al., 2022).

In phosphorylation, hydroxyl groups in the polymer chain structure are replaced by phosphate groups (Lai et al., 2024; Zhang et al., 2024). In nature, phosphorylated polysaccharides are rare. Phosphorylation can activate monosaccharides such as fructose and glucose, which lack bioactivity (Zhang et al., 2024). Phosphorylation can be accomplished by means of phosphate, phosphorus pentoxide, phosphoric acid, and their anhydride (Ahuja, 2023). According to Andrew and Jayaraman (2020), this process may improve the polysaccharide's solubility in water, assign electron donation, give it nucleophile characteristics, change its molecular weight, and improve its chain conformation (Andrew, n. d.)The putative antioxidant, antiviral, anticancer, immunomodulatory, and hepatoprotective qualities of these modifications have drawn attention (Xiong et al., 2019).

Benzoylation modification studies have shown that the addition of benzyl groups to EPS made of levan enhanced its antitumor and antioxidant properties. This increase in bioactivity can be attributed to its selectivity for immune cell receptors and its capacity to donate electrons. The capacity to decrease substances, scavenge free radicals, and build connections with transition metal ion catalysts are just a few of the ways that compounds' antioxidant qualities have been connected. Following these integrated categories, these characteristics might serve as prospective markers for unknown antioxidant capability. The greater reducing power of the levan benzoylated derivative compared to the native form indicated that chemical changes had greatly boosted the antioxidant activity (Liu et al.,2012)

Selenization-based recent studies suggest that selenium can improve polysaccharide's biological benefits. The trace element selenium is a component of the enzyme systems glutathione peroxidase and superoxide dismutase in humans. With the use of selenium, glutathione peroxidase can catalyze the inhibition and elimination of free radicals (Li et al., 2016). Under suitable conditions, selenite or selenium acid can form selenite ester bonds with the hydroxyl groups on polysaccharides to produce selenium polysaccharides (Li et al., 2016). After selenium alteration, Enterobacter cloacae Z0206's EPS, composed of glucose, galactose, and mannose monomers, increased bioactivity and boosted innate, cellular, and humoral immunity in mice who had experienced immunosuppression from cyclophosphamide. In addition, it has been shown to have immunomodulatory effects in broilers, mice, and crayfish (Xu et al., 2009). Various studies proved that selenium-EPS exhibited enhanced immunomodulatory activity than EPS, as evidenced by increased multiplication of mouse spleen lymphocytes and noticeably higher concentration of IL-2, IL-6, and TNF in spleen cells. According to Pan et al. (2015), these findings imply that EPS and Selenium-EPS can boost the immune response by stimulating spleen lymphocytes and increasing macrophages (Ahuja et al., 2023). In conclusion, modification of EPS with functional chemical groups can alter its structural and biological properties, making it more suitable for incorporation into bioinks for various biomedical applications.

9.6 3D printing techniques for exopolysaccharides-based materials

Due to their biocompatibility, gelling properties, and rheological adjustability, microbial EPS have emerged as appealing biomaterials for 3D printing. Extrusion-based printing works best with EPS-based materials among the frequently used 3D printing techniques. Due to their shear thinning and viscoelastic behavior, EPSs such as alginate, gellan gum, xanthan gum, and pullulan may be easily extruded, ensuring a smooth flow through the nozzle and a quick recovery following deposition (Martínez Ávila et al., 2016). It is possible to utilize the inkjet printing with EPS, though less frequently, if the materials are diluted and modified to meet the criteria for viscosity and droplet production. However, the employment of stereolithography (SLA) in EPS is still restricted by the technology's inherent absence of photo reactivity, even if recent advances in photo-crosslinkable modifications (such as methacrylated alginate) show promise (Lozano et al., 2015; Nagaraj and Rekha, 2023).

Extrusion printing is still the best method for hydrogels based on EPS since it can handle bioinks with high viscosity and produce larger, more structurally sound structures. For example, alginate can be ionically crosslinked with calcium chloride either during or immediately following deposition, resulting in quick scaffold formation. The addition of gelatin or cellulose to xanthan and gellan gums improves shape accuracy and lowers print failure (Nagaraj and Rekha, 2023). The gentle extrusion process aids in cell-laden printing, which encapsulates live cells within the EPS matrix and maintains cell viability in most studies above 85% (Varki, 2017). Additionally, extrusion enables integration with coaxial or multinozzle systems, which further supports the multimaterial or multicellular assembly required for complex tissue repair.

One of the challenges of EPS-based printing is maintaining accuracy and structural stability following deposition. Surface tension, nozzle diameter, gelation kinetics, and viscosity are some of the main variables affecting fidelity. By preventing structural collapse (Deidda et al., 2024), alginate and gellan gum, which are faster in situ crosslinking EPS bioinks, produce higher fidelity. One can improve structural resolution by changing the printing route, extrusion speed, and temperature. In addition, rheology modifiers such as nanocellulose, silk fibroin, or PEG are used to strengthen EPS matrices and reduce shrinkage or spreading (Lim et al., 2020). Methacrylated EPS derivatives can achieve micron-scale accuracy in stereolithographic applications if the photoinitiator and exposure parameters are carefully calibrated.

For multicellular organisms and multilayered structures, precise control of their mechanical and biochemical signals is essential for creating bioinks based on EPS. These bioinks must support spatial organization, nutrient delivery, and cell attachment. Because native EPS lacks cell adhesion motifs (e.g., RGD), it is often functionalized or combined with proteins such as collagen, fibrin, or gelatin (Asianezhad et al., 2023). Incorporating growth factors, microparticles, or channels that mimic vasculature provides support for specific tissues like skin, liver, and cartilage. The complexity and diversity of tissues are increasingly simulated through multicellular printing methods such as sequential deposition, core- shell printing, and modular assembly (Angelin and Kavitha, 2020).

Despite the potential of EPS-based printing in biomedicine, it remains challenging to achieve standardization, reproducibility, and scaling. Issues like EPS purity, microbial variability, and batch-to-batch consistency must be addressed in order for clinical translation to occur. Innovative methods like 4D printing, in which EPS hydrogels respond to stimuli such as pH, enzymes, and temperature, are promising for dynamic tissue models. Additionally, future roads to enhance utility and performance include biofabrication automation, machine learning for print optimization, and biosynthetic design of tailored EPS (Kavitake et al., 2023). Bioinks based on EPS offer an adaptable, environmentally friendly foundation for the bioprinting of the future.

9.7 Future challenges and consideration

Microbial EPS must overcome several obstacles before they can be widely used in tissue engineering and 3D bioprinting. The challenges that must be methodically resolved to realize their full potential in clinical and industrial contexts include material properties, processing methods, scalability, and regulatory issues. Although there are many intriguing potentials use for microbial EPS as bioink materials, there are drawbacks to their practical use in 3D bioprinting. Development of bioinks that balance printability, mechanical strength, and biological compatibility remain a major challenge for researchers. Developing a material that facilitates tissue growth and cell survival while also holding its shape after deposition and flowing easily through a printer nozzle is a demanding task (Qi et al., 2022). In applications where cell interaction is essential, such as skin or cartilage regeneration, some EPS may not be as effective at promoting cell attachment. In order to overcome this, it is frequently needed to combine EPS with other biopolymers, such as collagen, gelatin, or even other polysaccharides, or reinforce them with nanomaterials like chitin nanowhiskers, cellulose nanofibers, or starch nanocrystals. These mixtures can promote cell support, strength, and viscosity without adding toxic substances (Lee et al., 2015).

Another area of growing attention is the chemical modification of EPS to facilitate crosslinking, which helps solid structures develop after printing. UV or light-based crosslinking is made feasible by including groups like methacrylate or acrylate, which produce scaffolds that are stronger and more resilient. But there are drawbacks to this strategy. The chemical processes commonly involve toxic chemicals, hazardous solvents, and labor-intensive procedures, all of which run counter to the sustainability principles that initially attracted people to microbial EPS. More importantly, printing can harm cells by exposing them to UV radiation or reactive chemicals, while certain formulations have been successful in maintaining high cell viability (Lai et al., 2024).

Another challenge is maintaining the consistency of EPS produced by each batch of production (Zhang et al., 2024). The composition and properties of microbial EPS can vary depending on the microorganism, the fermentation process, and even slight changes in environmental conditions. This batch-to-batch variability can make it dificult to predict how a particular EPS will behave in a bioink or how it will affect cells. For bioinks to be used clinically, reproducibility is key—and that's an area where microbial EPS still needs work. Lastly, scaling up production for commercial or clinical use brings its own set of challenges. Producing large amounts of sterile, high-purity EPS under good manufacturing practice (GMP) conditions is costly and technically demanding. And even when the materials are ready, regulatory approvals can be complex, especially when the EPS are chemically modified or produced using genetically engineered microbes. Despite these hurdles, researchers have made impressive progress in this direction (Chen et al., 2019; Zhang et al., 2024). By developing smarter formulations, improving fermentation processes, and exploring greener crosslinking methods, were getting closer to realizing the full potential of microbial EPS in 3D bioprinting.

9.8 Addressing environmental impact and sustainability in exopolysaccharides production and application

Microbial EPS are a sustainable substitute for conventional, animal-derived bioink materials as biomedical engineering shifts toward greener and more responsible innovation (Ahuja, 2023). In contrast to collagen, gelatin, or fibrin, which are frequently derived from animal tissues, EPS can be made through microbial fermentation using renewable feedstocks, reducing ethical and environmental impacts, hence stands as a desirable source of bioink for 3D bioprinting (Table 9.2), where we can customize the EPS source based on the type of printing technology (Table 9.3).

Because of this, EPS is not only appropriate for use in biomedicine but also supports global objectives to lessen environmental damage and advance circular bioeconomy models. Numerous microorganisms used in the production of EPS, such as *Komagataeibacter* species, *Pseudomonas*, and *Xanthomonas*, can be grown on inexpensive agricultural waste or food industry waste, converting possible waste streams into useful biomaterials (Andrew, 2020). Additionally, compared to the production of synthetic polymers, microbial fermentation uses less energy because it is usually conducted in mild conditions of ambient temperature and pressure (Xiong and Huang, 2019).

However, the downstream processing of EPS, such as extraction, purification, drying, and sterilization, must also become more sustainable and "green" if they are to fully realize their potential as environmentally friendly materials. Even though many methods are used for the synthesis of EPS from various sources of microbes (Rehm et al, 2010; Jose et al.,2022). Present-day techniques frequently entail energy-intensive drying procedures like spray drying,

TABLE 9.2 Comparative study of microbial EPS with animal-derived bioink.

EPS material	Extrusion-based printing	Inkjet Printing	Stereolithography (SLA)	Notable features
Alginate	Highly suitable	Limited by viscosity	Requires photo-functionalization	Fast gelation with Ca^{2+}, good for cell encapsulation
Xanthan gum	Suitable (with blending)	Not ideal due to viscosity	Not compatible	Shear-thinning improves viscosity in blends
Gellan gum	Highly suitable	Limited	Rarely used	Ionotropic gelation, transparent hydrogel
Pullulan	Suitable (needs modifiers)	Good (low viscosity)	Methacrylation needed	Film-forming, printable with additives
Levan	Soft hydrogels	Possible	Not explored	Antioxidant, flexible EPS
Curdlan	Thermo-gelling	Not compatible	Not suitable	Heat-induced gelation is ideal for wound scaffolds
Dextran	Low viscosity; needs blending	Inkjet-compatible	Needs a photoreactive group	Mucoadhesive, flexible
Succinoglycan	Viscous and stable	Droplet control needed	Not yet demonstrated	High MW, shear-resistant

EPS, exopolysaccharides.

lyophilization, or solvent-intensive steps. To make the entire life cycle of EPS-based bioinks more environmentally friendly, greener alternatives like enzyme-assisted extraction, solvent-free purification, and solar drying technologies are to be employed more. Furthermore, caution must be taken to prevent the use of toxic reagents and to make sure that any leftovers from chemically altering EPS (for example, for photo-crosslinking or structural tuning) do not enter the environment or compromise biocompatibility.

EPS-based materials are safer for the environment and human health because they are biodegradable and usually decompose into nontoxic byproducts (Li et al., 2016). Their application in 3D bioprinting, specifically in tissue scaffolds, implantable devices, and in vitro models, provides a viable path forward for regenerative medicine without adding to long-term environmental waste.

Integrating environmentally friendly fermentation, low-impact processing, and responsible disposal procedures will be essential as research and industrial practices develop to guarantee that EPS-based bioinks continue to be not only cutting-edge but also genuinely sustainable.

9.9 Future directions and potential applications

Despite being biocompatible and gel-forming, EPS, like xanthan gum or alginate, often lack natural cell adhesion domains like RGD motifs. In biomedical engineering, the integration of EPS with hybrid biomaterials, which combine natural and synthetic polymers to enhance biofunctionality, mechanical endurance, and cellular response, is the way of the future (Xu et al., 2009).

TABLE 9.3 Sustainability comparison of microbial EPS-based bioinks versus traditional animal-derived bioinks.

Parameter	Microbial EPS-based bioinks	Animal-derived bioinks (e.g., collagen, gelatin)
Source	Microbial fermentation (e.g., *Xanthomonas*, *Komagataeibacter*).	Animal tissues (skin, bones, cartilage, tendons).
Ethical concerns	Minimal; no animal use.	High; involves animal harvesting and ethical limitations.
renewability of raw material	High (uses renewable feedstocks, even agro-waste).	Limited, dependent on the livestock industry.
Production scalability	Scalable via bioreactors and controlled fermentation.	Limited by animal tissue supply and processing.
Energy consumption	Generally low to moderate (ambient fermentation conditions).	High (thermal extraction, long purification processes).
Toxic byproduct risk	Low, especially with green processing methods.	Moderate; depends on preservation and extraction chemicals.
Biodegradability	Excellent; breaks down into nontoxic sugars or acids.	Good, but may vary with crosslinkers or blends.
Purification requirements	Moderate; can be enzyme-assisted and solvent-free.	High; often requires harsh treatments to remove pathogens.
Batch consistency	Moderate; can be improved via strain engineering.	Variable; depends on tissue source and animal health.
Environmental impact	Low (if optimized).	High (land use, water consumption, emissions from livestock).
Green chemistry compatibility	Highly amenable to enzyme-based or solvent-free processes.	Lower; harsh chemicals are often used in preparation.

EPS, exopolysaccharides.

Combining gelatin with silk fibroin, collagen, or synthetic hydrogels (such as polyethylene glycol, Pluronic FL27) allows researchers to produce composite bioinks that are better able to promote cell attachment, proliferation, and tissue-specific activity. Particularly promising are these hybrids in replicating soft tissues like the liver and cartilage, which need biologically active environments and variable elasticity. Additionally, nanocomposites composed of EPS and nanocellulose, graphene oxide, or hydroxyapatite are being studied for improved mechanical integrity and performance in hard tissue applications like bone scaffolds (Ahuja et al., 2023).

The ongoing development of bioink formulas based on EPS is making it possible to produce novel materials for a wide range of tissue engineering applications, such as skin, nerve, cartilage, and vascular tissue. By customizing the EPS concentration, crosslinking methods, and rheological properties, one may achieve a high level of printability, shape accuracy, and degradation rate. For example, dual-crosslinking approaches that combine ionic and covalent bonds have created materials that are both biologically biodegradable and structurally robust. Pregelation methods, cell spheroid integration, and stimulus-responsive EPS-based systems that dynamically respond to cellular environments are some of the most recent advances. In skin and wound healing applications, EPS such as levan and dextran offer promise for delivering bioactive substances, antibiotics, or stem cells while promoting tissue regeneration and antimicrobial protection.

Using EPS-based bioinks for multicellular and organoid-based bioprinting aims to replicate the complexity of functional tissue, thus highlighting a significant future path (Clevers and Tuveson, 2024). Bioinks that encourage spatial organization, diverse cellular development, and intercellular communication are necessary for multicellular structures. EPS hydrogels provide a favorable microenvironment for such applications due to their controllable porosity and water retention. When utilized with microfluidic systems or coaxial printing processes, EPS-based inks can generate vascularized structures, which are necessary for high-volume tissue production. Organoid culture systems that combine EPS matrices are currently being developed to aid with organ-level activities for disease modeling, drug screening, and personalized medicine, providing a link between in vitro models and clinical translation.

In addition to tissue engineering, microbial EPS are gaining attention as renewable, environmentally beneficial biomaterials. Because of their microbial origin, renewable feedstocks can be used to manufacture low-energy, low-emission products, which aligns with the goals of the circular bioeconomy[L08]. Pullulan, gellan gum, and curdlan are examples of EPS that are biodegradable, nontoxic, and compatible with the principles of green chemistry, making them ideal for medical applications that are environmentally friendly. Materials like this are used in regenerative medicine to make biodegradable implants, wound-healing membranes, and medication delivery systems that have minimal environmental effect, in addition to reducing environmental footprint. Additionally, EPS may be genetically modified for particular uses, resulting in designer biopolymers with superior biological and mechanical characteristics.

EPS have a lot of promise, but there are many challenges to overcome in order to turn them into clinical and commercial products. These include scalability, sterilization stability, regulatory concerns, and batch-to-batch fluctuation. The establishment of uniform processing criteria and GMP-compliant EPS manufacturing is necessary for widespread acceptance. Interdisciplinary collaborations between microbiologists, materials scientists, and clinicians will be necessary to bring EPS-based advances from the lab to the clinic. The development of next-generation intelligent bioinks will be accelerated even more by the convergence of 4D printing, AI-driven material design, and synthetic engineering. In addition to potentially revolutionizing regenerative medicine in the not-too-distant future, microbial EPS may also have significant applications in wearable medications, environmental biosensors, flexible electronics, and soft robotics.

9.10 Conclusion

The development of bioinks based on microbial EPS holds great promise for the future of 3D bioprinting (Lee et al., 2012). Microbial EPS will be crucial in forming next-generation biofabrication technologies as the need for sustainable, cell-friendly, and adaptable materials increases. Global priorities in biomedical innovation and sustainability are in line with their biocompatibility, tunable chemistry, and potential for large-scale, environmentally friendly production. We predict a boom in research in the upcoming years focused on improving EPS's functional characteristics via chemical modification, molecular engineering, and hybrid formulations. For example, bioinks with enhanced mechanical strength, bioactivity, and targeted cellular responses may be produced by combining EPS with nanomaterials, peptides, growth factors, or even extracellular vesicles (Lee et al., 2012). Dynamic scaffolds that can adjust to the changing requirements of regenerating tissues could result from the creation of intelligent EPS-based bioinks that react to physiological stimuli like pH, temperature, or enzymatic activity.

The combination of AI-driven material design and omics technologies to screen and customize EPS compositions for particular tissue types is another fascinating avenue (Kim et al., 2024). This would speed up the process of finding new EPS variations that are appropriate for precision bioprinting and have distinctive biological or rheological characteristics. The development of microbial strains that generate EPS with regulated molecular weight, branching patterns, or functional groups may be made possible concurrently by developments in synthetic biology and metabolic engineering, providing more control over the ultimate performance of the bioink (Kiran and Raveendran, 2023).

Another important factor will be scalability and regulatory acceptance. Standardized procedures for EPS extraction, purification, and characterization will be required as bioink formulations approach clinical translation in order to guarantee safety and reproducibility. A scalable and legal method of creating clinical-grade materials may be provided by the bioreactor-based production of microbial EPS using food-grade or generally recognized as safe (GRAS) organisms. In addition, EPS-based bioinks may find use in organ-on-chip platforms, drug screening, and disease modeling besides tissue engineering. They are perfect for creating physiologically accurate 3D in vitro models because of their capacity to replicate the ECM, maintain cell viability, and facilitate high-resolution bioprinting.

Microbial EPS has revolutionary potential for 3D bioprinting in the future. They serve as a link between the development of environmentally friendly materials and State-of-the-Art regenerative technologies by fusing sustainability with biomedical functionality. To reach their full potential and advance EPS-based bio-fabrication toward clinical and industrial applications, more interdisciplinary work in the fields of microbiology, materials science, and bioengineering will be needed. Given their intrinsic biocompatibility, biodegradability, and adjustable physical characteristics, microbial EPS, including bacterial cellulose, hyaluronic acid, alginate, xanthan gum, dextran, gellan gum, and xanthan gum, have become important ingredients in the creation of bioinks for 3D bioprinting. Because of their biological significance in mammalian systems and their capacity to form hydrogels in mild environments, hyaluronic acid and alginate are the most widely used of them. These polymers are ideal for use in tissue engineering and regenerative medicine because they can mimic the ECM, support cellular viability and proliferation, and encapsulate cells without causing cytotoxicity.

In addition to its basic gelation properties, a desirable bioink must fulfill a complex set of requirements. Maintaining cell viability both during and after printing, giving the printed structure enough mechanical support, and enabling the exchange of waste products, nutrients, and oxygen through suitable porosity and diffusivity are all crucial components. Additionally, the bioink's rheological behavior—which includes its viscosity, yield stress, and shear-thinning characteristics—is essential in determining its printability and structural integrity after printing. It is possible to optimize EPS-based bioinks to display these essential rheological properties through appropriate formulation and processing.

Significantly, a large number of EPS also exhibit inherent bioactive qualities. Hyaluronic acid, for instance, is involved in angiogenesis, wound healing, and cell signaling. Despite being biologically inert, alginate can be mixed with other bioactive substances or chemically altered to promote cell adhesion. Although their mechanical and viscosity-modifying qualities make xanthan gum and gellan gum valuable, they can also be used as secondary agents to enhance printability and structural integrity. Another microbial EPS, bacterial cellulose, has a nanofibrous structure that is advantageous for tissue repair and cellular integration, as well as exceptional mechanical strength. Because of these biofunctional qualities and their capacity to modify their physical behavior, EPS is a promising option for both active and passive scaffolding materials in the regenerative process.

Another noteworthy advantage of microbial EPS is its sustainability. Microbial polysaccharides, as opposed to biopolymers derived from animals, can be made through controlled fermentation processes with low-cost substrates, such as waste from the agro-industrial domain. As a result, EPS sourcing is more morally, practically, and sustainably done. It is feasible to improve EPS yields and fine-tune their composition for particular biomedical applications by creating ideal fermentation conditions and genetically modifying microbial strains. Additionally, using microbial EPS provides a low-carbon footprint substitute for traditional bioink materials and is consistent with the circular bioeconomy and green chemistry principles. A smart and progressive method for producing high-performance biomedical materials and ecological sustainability is the incorporation of microbial EPS into 3D bioprinting techniques. By improving mechanical stability, cell responsiveness, and long-term structural integrity, EPS modification innovations like chemical functionalization, blending with nanoparticles, or nanostructuring techniques can increase their applicability even more. With further research, EPS-based bioinks could be developed to actively direct cellular behavior through microenvironmental cues and biochemical signaling, in addition to supporting cells.

In conclusion, microbial EPS are formidable competitors in the continuing development of 3D bioprinting materials due to their distinct blend of biological, mechanical, and sustainability-related benefits. Their adaptability and versatility offer enormous potential, even though there are still dificulties in maximizing their qualities for particular applications, especially when it comes to striking a balance between printability and biological functionality. EPS will probably play a bigger part in bio-fabrication in the future, particularly as efforts to create sustainable, cell-responsive, and adaptable bioinks continue to spur advancements in tissue engineering and regenerative medicine.

Credit author statement

Blessy M. Mani: Conceptualization, writing—Original draft, writing—review and editing. Sherin Shaji: Writing, editing. Blessy Babukutty: Writing—original draft. Martin George Thomas: Writing—original draft. Prof. Sabu Thomas: Supervision, writing—review and editing.

Ethical statement

This chapter is a literature review and does not involve human or animal subjects; therefore, ethical approval was not required.

Acknowledgment

The authors thank Dr. Hanna J Maria for coordinating and supporting the writing process.

Conflict of interest statement

The authors declare no conflict of interest.

References

Ahuja, V., Bhatt, A.K., Banu, J.R., Kumar, V., Kumar, G., Yang, Y.H., Bhatia, S.K., 2023. Microbial exopolysaccharide composites in biomedicine and healthcare: trends and advances. Polymers 15 (7), 1801. https://doi.org/10.3390/polym15071801.

Andrew, M., Jayaraman, G., 2020. Structural features of microbial exopolysaccharides in relation to their antioxidant activity. Carbohydr. Res. 487, 107881. https://doi.org/10.1016/j.carres.2019.107881.

Angelin, J., Kavitha, M., 2020. Exopolysaccharides from probiotic bacteria and their health potential. Int. J. Biol. Macromol. 162, 853–865. https://doi.org/10.1016/j.ijbiomac.2020.06.190.

Angelov, A., Georgieva, A., Petkova, M., Bartkiene, E., Rocha, J.M., Ognyanov, M., Gotcheva, V., 2023. On the molecular selection of exopolysaccharide-producing lactic acid bacteria from indigenous fermented plant-based foods and further fine chemical characterization. Foods 12 (18), 3346. https://doi.org/10.3390/foods12183346.

Asgher, M., Qamar, S.A., Iqbal, H.M.N., 2020. Microbial exopolysaccharide-based nanocarriers with multifunctionalities for biomedical sectors. Biologia 76 (2), 673–685. https://doi.org/10.2478/s11756-020-00588-7.

Asgher, M., Rani, A., Khalid, N., Qamar, S.A., Bilal, M., 2021. Bioconversion of sugarcane molasses waste to high-value exopolysaccharides by engineered Bacillus licheniformis. Case Stud. Chem. Environ. Eng. 3, 100084. https://doi.org/10.1016/j.cscee.2021.100084.

Asianezhad, A., Bari, M.R., Amiri, S., 2023. Bio-producing and characterizing biochemical and physicochemical properties of a novel antioxidant exopolysaccharide by *Bacillus coagulans* IBRC-M 10807. J. Polym. Env. 31, 4338–4352. https://doi.org/10.1007/s10924-023-02892-0.

Avinash, M., and Jha B. 2013. Microbial Exopolysacchrides. In: Rosenberg E, DeLong EF, Thompson F, Lory S, Stackebrandt E (Eds.), The Prokaryotes: Applied Bacteriology and Biotechnology, 4th ed. Springer Berlin Heidelberg, pp. 179-192. https://doi.org/10.1007/978-3-642-31331-8_25. ISBN 978-3-642-31330-1[Print];978-3-642-31331-8.

Becker, A., Katzen, F., Pühler, A., Ielpi, L., 1998. Xanthan gum biosynthesis and application: a biochemical and genetic perspective. Appl. Microbiol. Biotechnol. 50, 145–152. https://doi.org/10.1007/s002530051269.

Butler, H.M., Naseri, E., MacDonald, D.S., Tasker, R.A., Ahmadi, A., 2021. Rheology, printability, and biocompatibility of N,O-carboxymethyl chitosan and agarose bioinks for neuronal 3D bioprinting. Materialia 18, 101169. https://doi.org/10.1016/j.mtla.2021.101169.

Chaisuwan, W., Jantanasakulwong, K., Wangtueai, S., Phimolsiripol, Y., Chaiyaso, T., Techapun, C., Phongthai, S., You, S.G., Regenstein, J.M., Seesuriyachan, P., 2020. Microbial exopolysaccharides for immune enhancement: fermentation, modifications and bioactivities. Food Biosci. 35, 100564. https://doi.org/10.1016/j.fbio.2020.100564.

Charoenwongpaiboon, T., Wangpaiboon, K., Pichyangkura, R., Nepogodiev, S.A., Wonganan, P., Mahalapbutr, P., Field, R.A., 2021. Nanoparticulate exopolysaccharide from *Leuconostoc holzapfelii* KM01 for drug encapsulation. Int. J. Biol. Macromol. 187, 690–698. https://doi.org/10.1016/j.ijbiomac.2021.07.174.

Chaturvedi, S., Kulshrestha, S., Bhardwaj, K., Jangir, R., 2021. Properties and applications of xanthan gum. Microbial Polymers: Applications and Ecological Perspectives. Springer, pp. 87–107. https://doi.org/10.1007/978-981-16-0045-6_4.

Chen, F., Huang, G., Yang, Z., Hou, Y., 2019. Antioxidant activity of *Momordica charantia* polysaccharide and its derivatives. Int. J. Biol. Macromol. 138, 673–680. https://doi.org/10.1016/j.ijbiomac.2019.07.129.

Chen, Y., Xiong, X., Liu, X., Cui, R., Wang, C., Zhao, G., Zhi, W., Lu, M., Duan, K., Weng, J., Qu, S., Ge, J., 2020. Shear-thinning hybrid gellan/alginate bioinks for 3D bioprinting. J. Mater. Chem. B. 8, 5500–5514. https://doi.org/10.1039/D0TB00060D.

Clevers, S., Tuveson, H., 2024. Organoid models for cancer research. Annu. Rev. Cancer Biol. 8, 213–234. https://doi.org/10.1146/annurev-cancerbio-052023-101456.

D'Ayala, G.G., Malinconico, M., Laurienzo, P., 2008. Marine-derived polysaccharides for biomedical applications: chemical modification approaches. Molecules 13 (9), 2069–2106. https://doi.org/10.3390/molecules13092069.

Deidda, V., Ventisette, I., Langione, M., Giammarino, L., Pioner, J.M., Credi, C., Carpi, F., 2024. 3D-printable gelatin methacrylate–xanthan gum hydrogel bioink enabling human induced pluripotent stem cell differentiation into cardiomyocytes. J. Funct. Biomater. 15 (10), 297. https://doi.org/10.3390/jfb15100297.

Delattre, C., Pierre, G., Laroche, C., Michaud, P., 2016. Production, extraction and characterization of microalgal and cyanobacterial exopolysaccharides. Biotechnol. Adv. 34, 1159–1179. https://doi.org/10.1016/j.biotechadv.2016.08.001.

Farshidfar, N., Iravani, S., Varma, R.S., 2023. Alginate-based biomaterials in tissue engineering and regenerative medicine. Mar. Drugs 21 (3), 189. https://doi.org/10.3390/md21030189.

Feng, L., Qian, T., Yang, G., Mu, J., 2023. Characteristics of exopolysaccharides produced by isolates from natural bioflocculant of Ruditapes philippinarum conglutination mud. Front. Microbiol. 13, 1068922. https://doi.org/10.3389/fmicb.2022.1068922.

Fragal, E.H., Poirier, A., Bleses, D., Silva, Y.F.G., Baccile, N., Rharbi, Y., 2025. Microbial biosurfactant hydrogels with tunable rheology for precision 3D printing. Soft Matter 21, 4476–4487. https://doi.org/10.1039/D5SM00248F.

Garcia-Cruz, M.R., Postma, A., Frith, J.E., Meagher, L., 2021. Methacrylated xanthan–gelatin composite bioink. Biofabrication 13, 045008. https://doi.org/10.1088/1758-5090/abec2d.

Gheorghita, A.A., Wozniak, D.J., Parsek, M.R., Howell, P.L., 2023. Pseudomonas aeruginosa biofilm exopolysaccharides. FEMS Microbiol., Rev. 47, fuad060. https://doi.org/10.1093/femsre/fuad060.

Halder, U., Banerjee, A., Bandopadhyay, R., 2017. Structural and functional properties, biosynthesis, and patenting trends of bacterial succinoglycan: A review. Indian J. Microbiol. 57, 278–284. https://doi.org/10.1007/s12088-017-0655-3.

Hay, I.D., Rehman, Z.U., Moradali, M.F., Wang, Y., Rehm, B.H.A., 2013. Microbial alginate production and applications. Microb., Biotechnol. 6 (6), 637–650. https://doi.org/10.1111/1751-7915.12076.

Jiang, C., Yang, M., Liu, J., Ju, X., Ding, R., 2014. Sulfated modification of longan polysaccharide and its immunomodulatory and antitumor activity in vitro. Int. J. Biol. Macromol. 67, 323–329. https://doi.org/10.1016/j.ijbiomac.2014.03.030.

Jose, A.A., Hazeena, S.H., Lakshmi, N.M., Madhavan, A., Sirohi, R., Tarafdar, A., Sindhu, R., Awasthi, M.K., Pandey, A., Binod, P., 2022. Bacterial biopolymers: From production to biomedicine. Sustain. Chem. Pharm. 25, 100582. https://doi.org/10.1016/j.scp.2021.100582.

Jurášková, D., Ribeiro, S.C., Silva, C.C.G., 2022. Exopolysaccharides produced by lactic acid bacteria: From biosynthesis to health-promoting properties. Foods 11 (2), 156. https://doi.org/10.3390/foods11020156.

Kashaudhan, K., Pande, P.P., Sharma, J., Shankar, R., Nath, A., 2024. Modified xanthan gum for wastewater treatment: a review. Env. Prog. Sustain. Energy 43, e14419. https://doi.org/10.1002/ep.14419.

Kavitake, D., Devi, P.B., Delattre, C., Reddy, G.B., Shetty, P.H., 2023. Exopolysaccharides produced by enterococcus genus—an overview. Int. J. Biol. Macromol. 226, 111–120. https://doi.org/10.1016/j.ijbiomac.2022.12.042.

Keshavarz, S., Azizian, R., Malakootikhah, J., Fathizadeh, H., Hamidi, M., 2025. Microbial exopolysaccharides in additive manufacturing. Encyclopedia of Green Materials. Springer. https://doi.org/10.1007/978-981-16-4921-9_225-3.

Khoeini, R., Nosrati, H., Akbarzadeh, A., Eftekhari, A., Kavetskyy, T., Khalilov, R., Ahmadian, E., Nasibova, A., Datta, P., Roshangar, L., Cucchiarini, M., Ozbolat, I.T., 2021. Natural and synthetic bioinks for 3D bioprinting. Adv. Nanobiomed Res. 1, 2000097. https://doi.org/10.1002/anbr.202000097.

Kim, J., Kim, J., Lee, M., 2024. Artificial intelligence in biomaterials design and tissue engineering. Adv. Funct. Mater. 34, 2310456. https://doi.org/10.1002/adfm.202310456.

Kiran, M., Raveendran, N., 2023. Engineering microbial cell factories for tailor-made exopolysaccharides. Biotechnol. Adv. 62, 108093. https://doi.org/10.1016/j.biotechadv.2022.108093.

Lai, Y., Yu, X., Yang, S., Zhuang, Y., Zhou, Z., Zhou, H., 2024. Photocrosslinkable biomaterials for 3D bioprinting. Bioact. Mater., 33, 487–508. https://doi.org/10.1016/j.bioactmat.2023.10.030.

Lee, H.J., Kim, Y.B., Ahn, S.H., Lee, J.S., Jang, C.H., Yoon, H., Chun, W., Kim, G.H., 2015. Collagen/ECM-based bioinks. Adv. Healthc. Mater. 4, 1359–1368. https://doi.org/10.1002/adhm.201500193.

Lee, K.Y., Mooney, D.J., 2012. Alginate: properties and biomedical applications. Prog. Polym. Sci. 37, 106–126. https://doi.org/10.1016/j.progpolymsci.2011.06.003.

Li, N., Qiao, D., Zhao, S., Lin, Q., Zhang, B., Xie, F., 2021. 3D printing to innovate biopolymer materials for demanding applications: a review. Mater. Today Chem. 20, 100459. https://doi.org/10.1016/J.MTCHEM.2021.100459.

Li, S., Shah, N.P., 2017. Sulphonated modification of polysaccharides from Pleurotus eryngii and Streptococcus thermophilus ASCC 1275 and antioxidant activities investigation using CCD and Caco-2 cell line models. Food Chem. 225, 246–257. https://doi.org/10.1016/j.foodchem.2017.01.037.

Lim, W., Kim, G.J., Kim, H.W., Lee, J., Zhang, X., Kang, M.G., Seo, J.W., Cha, J.M., Park, H.J., Lee, M.Y., Shin, S.R., Shin, S.Y., Bae, H., 2020. Kappa-carrageenan-based dual crosslinkable bioink for extrusion type bioprinting. Polymers 12 (10), 2377. https://doi.org/10.3390/polym12102377.

Liu, F., Chen, Q., Liu, C., Ao, Q., Tian, X., Fan, J., Tong, H., Wang, X., 2018. Natural polymers for organ 3D bioprinting. Polymers 10, 1278. https://doi.org/10.3390/polym10111278.

Liu, J., Luo, J., Ye, H., Zeng, X., 2012. Preparation, antioxidant and antitumor activities in vitro of different derivatives of levan from endophytic bacterium Paenibacillus polymyxa EJS-3. Food Chem. Toxicol. 50 (3–4), 767–772. https://doi.org/10.1016/j.fct.2011.11.016.

Loo, Y., Lakshmanan, A., Ni, M., Toh, L.L., Wang, S., Hauser, C.A.E., 2015. Peptide bioink self-assembling nanofibrous scaffolds. Nano Lett. 15, 6919–6925. https://doi.org/10.1021/acs.nanolett.5b02859.

Lozano, R., Stevens, L., Thompson, B.C., Gilmore, K.J., Gorkin, R., Stewart, E.M., in het Panhuis, M., Romero-Ortega, M., Wallace, G.G., 2015. 3D printing of layered brain-like structures using peptide modified gellan gum substrates. Biomaterials 67, 264–273. https://doi.org/10.1016/j.biomaterials.2015.07.022.

Markstedt, K., Mantas, A., Tournier, I., Martínez Ávila, H., Hägg, D., Gatenholm, P., Biomacromolecules 2015, 16(5) 1489-1496. https://doi.org/10.1021/acs.biomac.5b00188.

Martínez Ávila, H., Schwarz, S., Rotter, N., Gatenholm, P., 2016. 3D bioprinting of human chondrocyte-laden nanocellulose hydrogels for patient-specific auricular cartilage regeneration. Bioprinting (1–2) 22–35. https://doi.org/10.1016/j.bprint.2016.08.003.

McCarthy, R.R., Ullah, M.W., Booth, P., Pei, E., Yang, G., 2019. Bacterial polysaccharides in bioprinting. Biotechnol. Adv. 37, 107448. https://doi.org/10.1016/j.biotechadv.2019.107448.

Murugu, J., Narayanan, R., 2024. Production, purification, and characterization of a novel exopolysaccharide from probiotic Lactobacillus amylovorus: MTCC 8129. Indian J. Microbiol. 64 (3), 1355–1365. https://doi.org/10.1007/s12088-024-01346-y.

Nachtigall, C., Surber, G., Bulla, J., Rohm, H., Jaros, D., 2021. Pilot scale isolation of exopolysaccharides from Streptococcus thermophilus DGCC7710: Impact of methodical details on macromolecular properties and technofunctionality. Eng. Life Sci. 21 (3–4), 220–230. https://doi.org/10.1002/elsc.202000073.

Nagaraj, A., Rekha, P.D., 2023. Development of a bioink using exopolysaccharide from *Rhizobium* sp. PRIM17. Int. J. Biol. Macromol. 234, 123608. https://doi.org/10.1016/j.ijbiomac.2023.123608.

Pan, D., Liu, J., Zeng, X., Liu, L., Li, H., Guo, Y., 2015. Immunomodulatory activity of selenium exopolysaccharide produced by Lactococcus lactis subsp. lactis. Food Agric. Immunol. 26 (2), 248–259. https://doi.org/10.1080/09540105.2014.894000.

Piola, B., Sabbatini, M., Gino, S., Invernizzi, M., Renò, F., 2022. Gelatin–xanthan gum bioinks for skin cells. Int. J. Mol. Sci. 23, 539. https://doi.org/10.3390/ijms23010539.

Pourjafar, H., Ansari, F., Sadeghi, A., Samakkhah, S.A., Jafari, S.M., 2023. Functional and health-promoting properties of probiotics' exopolysaccharides; isolation, characterization, and applications in the food industry. Crit. Rev. Food Sci. Nutr. 63 (26), 8194–8225. https://doi.org/10.1080/10408398.2022.2047883.

Qi, Y., Zhang, J., Wu, Y., Dong, C., Zhou, C., Nie, J., He, X., 2022. A cell adhesion-promoting multi-network 3D printing bio-ink. Front. Bioeng. Biotechnol. 10, 1070566. https://doi.org/10.3389/fbioe.2022.1070566.

Rahnama Vosough, P., Habibi Najafi, M.B., Edalatian Dovom, M.R., Javadmanesh, A., Mayo, B., 2021. Evaluation of antioxidant, antibacterial and cytotoxicity activities of exopolysaccharide from Enterococcus strains isolated from traditional Iranian Kishk. J. Food Meas. Charact. 15, 5221–5230. https://doi.org/10.1007/s11694-021-01092-5.

Ramiah, P., du Toit, L.C., Choonara, Y.E., Kondiah, P.P.D., Pillay, V., 2020. Hydrogel-based bioinks for tissue regeneration. Front. Mater. 7, 506968. https://doi.org/10.3389/fmats.2020.506968.

Rehm, B.H.A., 2010. Bacterial polymers. Nat. Rev. Microbiol. 8, 578–592. https://doi.org/10.1038/nrmicro2354.

Sahana, T.G., Rekha, P.D., 2020. Marine bacterial EPS and wound healing. Carbohydr. Polym. 238, 116191. https://doi.org/10.1016/j.carbpol.2020.116191.

Sathishkumar, R., Kannan, R., Jinendiran, S., Sivakumar, N., Selvakumar, G., Shyamkumar, R., 2021. *Bacillus* subtilis EPS characterization. Int. J. Biol. Macromol. 166, 1471–1479. https://doi.org/10.1016/j.ijbiomac.2020.11.026.

Shoshaa, R., Ashfaq, M.Y., Al-Ghouti, M.A., 2023. Recent developments in ultrafiltration membrane technology for the removal of potentially toxic elements, and enhanced antifouling performance: a review. Environ. Technol. Innov. 31, 103162. https://doi.org/10.1016/j.eti.2023.103162.

Sørensen, H.M., Rochfort, K.D., Maye, S., MacLeod, G., Brabazon, D., Loscher, C., Freeland, B., 2022. Exopolysaccharides of lactic acid bacteria: production, purification and health benefits towards functional food. Nutrients 14 (14), 2938. https://doi.org/10.3390/nu14142938.

Sun, T., Yao, D., Shen, S.-R., He, G.-S., 2015. Sulfated modification of the water-soluble polysaccharides from Polyporus albicans mycelia and its potential biological activities. Int. J. Biol. Macromol. 44 (1), 412–418. https://doi.org/10.1016/j.ijbiomac.2015.01.211.

Talbi, C., Elmarrkechy, S., Youssfi, M., Bouzroud, S., Belfquih, M., Sifou, A., Bouhaddou, N., Badaoui, B., Balahbib, A., Bouyahya, A., Bourais, I., 2023. Bacterial exopolysaccharides: From production to functional features. Prog. Microbes Mol. Biol., a0000384. https://doi.org/10.36877/pmmb.a0000384.

Theis, T.V., Santos, V.A.Q., Appelt, P., Barbosa-Dekker, A.M., Vetvicka, V., Dekker, R.F.H., Cunha, M.A., 2019. Fungal exocellular (1→6)-β-D-glucan: Carboxymethylation, characterization, and antioxidant activity. Int. J. Mol. Sci. 20 (9), 2337. https://doi.org/10.3390/ijms20092337.

Thi, M.T.T., Wibowo, D., Rehm, B.H.A., 2020. *Pseudomonas aeruginosa* biofilms. Int. J. Mol. Sci. 21, 8671. https://doi.org/10.3390/ijms21228671.

Tomić, S.L., Babić Radić, M.M., Vuković, J.S., Filipović, V.V., Nikodinovic-Runic, J., Vukomanović, M., 2023. Alginate hydrogels and scaffolds. Mar. Drugs 21, 177. https://doi.org/10.3390/md21030177.

Varki, A., 2017. Biological roles of glycans. Glycobiology 27, 3–49. https://doi.org/10.1093/glycob/cww086.

Viñarta, S.C., Yossen, M.M., Vega, J.R., Figueroa, L.I.C., Fariña, J.I., 2013. Scleroglucan compatibility with thickeners, alcohols and polyalcohols and downstream processing implications. Carbohydr. Polym. 92 (2), 1107–1115. https://doi.org/10.1016/j.carbpol.2012.10.065.

Wang, J.-H., Zhang, Z., Wang, Y.-H., Zhao, H.-B., Shen, Y.-M., 2015. Chemical modification of an exopolysaccharide isolated from Lactobacillus plantarum KX041 and its biological activities. Int. J. Biol. Macromol. 78, 87–94. https://doi.org/10.1016/j.ijbiomac.2015.03.031.

Wu, D., Yu, Y., Tan, J., Huang, L., Luo, B., Lu, L., Zhou, C., 2018. Gellan gum-based hydrogels for bioprinting. Mater. Des. 160, 486–495. https://doi.org/10.1016/j.matdes.2018.09.040.

Xiong, X., Huang, G., Huang, H., 2019. Phosphorylated ginseng polysaccharide antioxidant activity. Int. J. Biol, Macromol. 129, 735–742. https://doi.org/10.1016/j.ijbiomac.2019.02.074.

Xu, C.L., Wang, Y.Z., Jin, M.L., Yang, X.Q., 2009. Preparation, characterization and immunomodulatory activity of selenium-enriched exopolysaccharide produced by bacterium Enterobacter cloacae Z0206. Bioresour. Technol. 100 (6), 2095–2097. https://doi.org/10.1016/j.biortech.2008.10.037.

Xu, Y., Song, S., Wei, Y., Wang, F., Zhao, M., Guo, J., Zhang, J., 2016. Sulfated modification of the polysaccharide from Sphallerocarpus gracilis and its antioxidant activities. Int. J. Biol. Macromol. 87, 180–190. https://doi.org/10.1016/j.ijbiomac.2016.02.037.

Yadav, M.K., Song, J.H., Vasquez, R., Lee, J.S., Kim, I.H., Kang, D.-K., 2024. Methods for detection, extraction, purification, and characterization of exopolysaccharides of lactic acid bacteria—a systematic review. Foods 13 (22), 3687. https://doi.org/10.3390/foods13223687.

Yu, L., Ye, G., Qi, X., Yang, Y., Zhou, B., Zhang, Y., Du, R., Ge, J., Ping, W., 2023. EPS from *Lactiplantibacillus plantarum*. Front. Microbiol. 14, 1210302. https://doi.org/10.3389/fmicb.2023.1210302.

Zaghloul, E.H., Ibrahim, M.I.A., 2022. Marine probiotic EPS with wound healing activity. Front. Microbiol. 13, 903363. https://doi.org/10.3389/fmicb.2022.903363.

Zaghloul, E.H., Ibrahim, M.I.A., Zaghloul, H.A.H., 2023. Antibacterial EPS from bee gut enterococcus. BMC Microbiol. 23, 277. https://doi.org/10.1186/s12866-023-02977-9.

Zhan, K., Ji, X., Luo, L., 2023. *Momordica charantia* polysaccharides. Chem. Biol. Technol. Agric. 10, 33. https://doi.org/10.1186/s40538-023-00433-4.

Zhang, J., Cao, Y., Wang, J., Guo, X., Zheng, Y., Zhao, W., Mei, X., Guo, T., Yang, Z., 2016. Sulphated EPS from *Streptococcus thermophilus*. Carbohydr. Polym. 146, 368–375. https://doi.org/10.1016/j.carbpol.2016.03.063.

Zhang, X.-X., Zhang, W.-W., Ni, Z.-J., Thakur, K., Zhang, J.-G., Khan, M.R., Xu, W.-D., Wei, Z.-J., 2024. Effects of different chemical modifications on physicochemical and antioxidation properties of Lycium barbarum seed dreg polysaccharides. Food Chem.: X 22, 101271. https://doi.org/10.1016/j.fochx.2024.101271Q33.

CHAPTER

10

Potential application of EPS in nano-biosensor development toward biomedical diagnosis

Bedanta Bhattacharjee[a], *Nayan Ranjan Ghose Bishwas*[b], *Hrishav Das Purkayastha*[c], *Ananya Bhattacharjee*[d], *Partha Palit*[c] *and Abhishek Bhattacharjee*[c]

[a] School of Pharmaceutical Sciences, Girijananda Chowdhury University, Tezpur Campus, Guwahati, Assam, India [b] Department of Pharmaceutical Sciences, Dibrugarh University, Dibrugarh, Assam, India [c] Department of Pharmaceutical Sciences, Assam University (A Central University), Silchar, Assam, India [d] Symbiosis Institute of Technology, Symbiosis International (Deemed University), Lavale, Pune, Maharashtra, India

10.1 Introduction

Exopolysaccharides (EPS) are high-molecular-weight extracellular polymers generated by a wide range of microbes (fungi, bacteria, or microalgae) that exhibit remarkable chemical diversity in monosaccharide composition, glycosidic linkage patterns, and branching (Nguyen et al., 2024). This structural versatility underlies their varied physicochemical properties such as hydrophilicity, viscoelasticity, multiple reactive functional groups, and the ability to form hydrogels and porous architectures, making EPS attractive materials for biomedical applications (Mohd Nadzir et al., 2021). Beyond their material properties, EPS often possess intrinsic bioactivities (e.g., immunomodulatory, antioxidant, antimicrobial) and are biocompatible and biodegradable, which reduces the risk of adverse responses when used in diagnostic platforms

DOI: https://doi.org/10.1016/B978-0-443-36619-2.00010-3

or as immobilization matrices for biomolecules. These biological and safety attributes position EPS as promising candidates for interfacing biological recognition elements (antibodies, enzymes, aptamers) with transducer surfaces (Picheth et al., 2017; Wang et al., 2023). Nano-biosensors combine specific biological recognition with nanostructured transducers (metal nanoparticles, graphene, carbon nanotubes (CNTs), metal oxides, quantum dots) to achieve high sensitivity, low detection limits, and rapid readout, essential for modern point-of-care (POC) and clinical diagnostics. Integrating EPS with nanomaterials can produce hybrid nanocomposites that (1) provide a benign immobilization scaffold preserving biomolecule activity, (2) present modifiable chemical groups for conjugation, (3) create porous/high-surface-area matrices that increase loading of recognition elements, and (4) enable synergistic electrical or plasmonic coupling with conductive nanomaterials to amplify signal transduction. These combined effects can markedly improve sensor stability, selectivity, and analytical performance (Lakshmi et al., 2021; Yadav et al., 2022). Recent literature highlights two converging trends that motivate a focused review on EPS-based nano-biosensors: (1) A surge in research on microbial EPS production, characterization, and tailorable functionalization, and (2) rapid advances in nanomaterial-based electrochemical and optical biosensing platforms for biomedical analytes. Together, these advances create realistic translational pathways for EPS-derived hydrogels, films, and nanocomposites as core materials in next-generation diagnostic devices from wearable sensors to POC electrochemical strips and microfluidic cartridges (Rahbar Saadat et al., 2019; Kim et al., 2025). This book chapter therefore examines (1) EPS structural and functional features relevant to biosensor design, (2) strategies for EPS-nanomaterial hybridization and surface functionalization, (3) recent EPS-integrated nano-biosensor examples for clinically relevant targets, and (4) challenges and future prospects for clinical translation.

10.2 Classification of exopolysaccharides

EPS can be broadly classified based on their chemical composition, source, and functional characteristics. They are primarily composed of polysaccharides but may also contain proteins, nucleic acids, lipids, and other organic substances (Angelin and Kavitha, 2020; Mehta et al., 2021). The specific composition largely depends on the microbial species and environmental conditions. For instance, cyanobacterial EPS, composed mainly of heteropolysaccharides, have gained attention due to their potential in biotechnological and biomedical fields. Genetic engineering and chemical modifications are strategies to enhance the functionalities of these EPS (Pereira et al., 2019). Understanding their classification, particularly into homo-polysaccharides and hetero-polysaccharides, is essential for exploring their ecological functions and industrial applications (Madsen et al., 2023). Table 10.1 highlights the classification of EPS based on origin and structural characteristics, with representative applications in bio-nanosensor development for biomedical diagnostics and food safety assessment.

10.2.1 Homo-polysaccharides

These are composed of repeating units of a single type of monosaccharide. While specific recent examples of microbial homo-polysaccharides are not explicitly detailed in the existing literature discussed here, EPS in general is noted for its varied polysaccharide content, which

TABLE 10.1 Classification and applications of EPS in nano-biosensor development.

Type of EPS	Sources	Key properties	Example nano-biosensor application	References
Alginate	Brown algae, bacterial (*Azotobacter*, *Pseudomonas*)	Biocompatible, biodegradable, hydrogel-forming, ionically cross-linkable.	Alginate hydrogel-based wearable biosensor for sweat lactate or glucose monitoring.	Gunatilake et al. (2021)
Carboxymethyl-dextran (CMD)	Chemical modification of dextran from *Leuconostoc* spp.	Hydrophilic, functionalizable, SPR-compatible surface chemistry.	CMD-coated SPR chips for protein–ligand interaction studies; FO-LSPR immunoassays.	Garcia-Rey et al. (2021)
Dextran	*Leuconostoc mesenteroides*, other lactic acid bacteria	Hydrophilic, nonimmunogenic, modifiable with functional groups.	Dextran-based hydrogel microneedle electrochemical sensors for continuous monitoring.	Chiang et al. (2023), Darmau et al. (2025)
Hyaluronic acid (HA)	Animal connective tissues, microbial fermentation (*Streptococcus*)	High water retention, biocompatible, viscoelastic.	HA-based electrochemical immunoplatforms; microneedle-based wearable biosensors.	Mobed et al. (2023)
Xanthan gum	*Xanthomonas campestris*	Shear-thinning, stable over wide pH/temp range, forms viscous solutions.	Xanthan-gum/polyacrylamide conductive hydrogels for wearable strain sensors.	Xu et al. (2025)

(*continued on next page*)

TABLE 10.1 Classification and applications of EPS in nano-biosensor development—cont'd

Type of EPS	Sources	Key properties	Example nano-biosensor application	References
Gellan gum	*Sphingomonas elodea*	Thermogelling, transparent, mechanically strong.	Gellan-gel integrated electrochemical immunosensors; optical biosensor matrices.	Gatti et al. (2025)
Pullulan	*Aureobasidium pullulans*	Film-forming, edible, oxygen-impermeable.	pH-sensitive pullulan/PVA nanofiber colorimetric biosensors.	Quazi et al. (2023), Shi et al. (2025)
Levan	*Zymomonas mobilis, Bacillus subtilis*	Highly branched fructan, bioactive, biocompatible.	Levan-modified electrochemical DNA biosensors.	Congur et al. (2021)
Spirulina-derived EPS	*Arthrospira platensis*	Sulfated polysaccharide, antioxidant, metal-binding.	Spirulina-mediated green synthesis of Au/AgNPs for biosensor transducers.	Harutyunyan et al. (2024)

EPS, exopolysaccharides.

can include homo-polysaccharides in certain microbial contexts (Salimi and Farrokh, 2023). Common examples include:

10.2.1.1 Dextran

Composed of glucose, dextran is produced by dental bacteria and has significant applications in medical and pharmaceutical fields due to its ability to act as a volume expander in blood and its use in chromatographic methods (Mishra et al., 2022).

10.2.1.2 Levan

Formed from fructose, levan is produced by specific strains of *Bacillus*. It is noted for its use as a food additive due to its beneficial effects on gut health and potential prebiotic properties.

10.2.1.3 Cellulose

Plant cell walls and certain bacteria, such as *Gluconacetobacter xylinus*, contain cellulose, a crucial structural element made of glucose. It is an essential component of paper and textile production due to its durability and biodegradability, and it may also be used to produce biofuels (Lee et al., 2022; Mishra et al., 2022).

10.2.2 Hetero-polysaccharides

Hetero-polysaccharides are made of repeating units of multiple monosaccharides. They are often larger than homo-polysaccharides, and may fall to many degrees of substituent, such as acetyl, pyruvyl, or succinyl groups. These features aid hetero-polysaccharides demonstrated functional capacity. Recent publications stress a variety of functions that hetero-polysaccharides provide for various ecosystems and industrial processes (Zhang et al., 2024). For example, they are responsible for regulating soil structure through aiding in aggregation, which is essential for ecosystem health and soil fertility (Tian et al., 2025). In aquatic environments, heteropolysaccharides in EPS will drive unique and complex reactions that facilitate the trapping of nutrients and degrading contaminants, further serving their ecological multifunctionality (Tian et al., 2025). Specific examples will include:

10.2.2.1 Alginate

Pseudomonas and *Azotobacter* are examples of species that synthesize alginate, which is derived from two acids: Mannuronic and guluronic acids. Alginate in the presence of calcium ions has the ability to form gels, making it important in food as a thickener or medicine to dress wounds.

10.2.2.2 Xanthan gum

Xanthan gum is produced from *Xanthomonas* campestris and is composed of glucose, mannose, and glucuronic acid. Xanthan gum has been a very important thickener and stabilizing agent in numerous industrial and food acting due to its unique rheological properties.

10.2.2.3 Gellan gum

Gellan gum is obtained from *Sphingomonas elodea* and contains rhamnose, glucuronic acid, and glucose. Gellan gum possesses significant gelling properties and serves as a gel-forming agent and stabilizer in the food and cosmetics industries (Abdl Aali and Al-Sahlany, 2024). Table 10.1: EPS categorization and use in the development of nano-biosensors.

10.3 Main application of exopolysaccharides in biomedical diagnosis and nano-biosensors

In the area of biomedical applications, EPS has received much attention, particularly for its role in diagnostic uses that strengthen its utility for developing efficacious diagnostic instruments (Tabernero and Cardea, 2020). The manufacture of nanostructured elements for targeted delivery and subsequent release is a major utility of EPS in diagnostic applications. These systems improve the sensitivity and specificity of diagnostic assays, particularly in relation to disease-related biomarkers. EPS-based nano-carriers are an excellent alternative to traditional diagnostic implementations for profile drug delivery and bioimaging applications. The capacity to modify EPS structures via different bio-conjugation modes also facilitates the production of multipurpose diagnostic devices that can accurately target and accumulate therapeutic agents (Falsini et al, 2023). In addition, EPS has also been incorporated with sophisticated technologies such as lab-on-a-chip (LOC) where they are affecting the miniaturization and automation of diagnostic processes. LOC integrates multiple laboratory functions on a chip and incorporates an array of testing formats with the advantages offered by EPS as structural components, donor/receptor elements, improved sensitivity, and reduction of assay times for POC diagnostics (Zolti et al., 2023).

Moreover, the potential integration of EPS in electronic-nose (e-nose) technology has been explored in the field of diagnostics. E-nose technology has been developed to take advantage of the sensory nature of EPS to detect volatile compounds and biochemical markers associated with different health states to improve diagnostic efficacy and offer rapid, noninvasive testing. Applications of e-nose technology have been shown to be useful in diagnostics and metabolic monitoring of disease and show the ability of EPS systems to expand the process of diagnostic methodologies (Li et al., 2023). Similarly, microfluidic systems have also adopted the use of EPS to take advantage of the different structures and functions of EPS to manipulate and detect biological analytes. The potential of integrating EPS into microfluidic platforms as an analytical solution is to improve diagnostics to high-throughput screening capabilities with accurate control of a biological experiment for personalized and precision medicine (Broks et al., 2023). There are many prospects for diagnostic technology breakthroughs and growth when EPS is used in biomedical testing. They are essential elements in the development of healthcare technology because of their versatility and multipurpose qualities, which enable improved specificity, sensitivity, and efficiency in a range of diagnostic applications.

10.4 Modality of fluorescence and luminescence property attributed to exopolysaccharides

EPS can present fluorescence and luminescence characteristics that are typically influenced by certain functional moieties (chromophore) present within their structural framework (Shen et al., 2023; Zang et al., 2025). Fluorescence and luminescence characteristics come about when light interacts with some functional group in EPS to exhibit light. In relation to chemistry, fluorescence is a form of photoluminescence in which a chemical substance absorbs light, then re-emits light at a longer wavelength, and usually will continue to illuminate for a brief time after the radiant light source is withdrawn (Zang et al., 2025). The fluorescence features found in EPS can be attributed to certain functional groups; for example, groups with amide (–CONH–), amino (–NH2), carboxylic acid (–COOH), and hydroxyl (–OH) groups are able to form fluorescence when light interacts with these functional groups. The functional groups may be part of glycone (sugar) residues that form the core of EPS, or connected to the EPS backbone as a noncarbohydrate constituent (Xu et al., 2024). For example, EPS related carbon dots (CDs) have recently been shown to have fluorescent properties because of functional groups, allowing them to be employed in viability studies on bacteria (Xu et al., 2024).

Research conducted by Lin et al. synthesized inexpensive CDs using EPS (CDs-EPS605), which have shown a simple means to quantify microbial survivability with no cytotoxicity. The fluorescent N-doped CDs-EPS605 were easily synthesized using bacterial amino EPS with a one-step hydrothermal carbonization process that was cost-effective and environmentally sustainable. The CDs-EPS605 were negatively charged and comprised of S, P, N, O, H, and C and contained functional moieties: –NH2, –CONH–, –OH, and –COOH. CDs-EPS605 preferentially stains dead microbes and effectively stains dead microbes to discriminate live from dead microbes. CDs-EPS605 also exhibited better photo-stability and had significantly lower cytotoxicity than the conventional version. In conclusion, CDs-EPS605 is an inexpensive but strong staining agent for microbial survivability assays and provides a new application of microbial EPS (Lin et al., 2018).

Shen et al. prepared fluorescent based CQDs using xanthan gum and oxidized xanthan gum with the help of a hydrothermal approach. In this study, the microstructure, optical characteristics, accuracy, and selectivity in Cr^{6+} ion sensing and fluorescent anticounterfeiting activities have been explored. They showed sustained fluorescence at high salt concentrations (4 mol/L NaCl) and a broad pH range, as well as selectivity for the Cr^{6+} ion, successful data encoding, and anticounterfeiting properties. Oxidized xanthan gum-polyetherimide hydrogel was developed by linking aldehyde groups of oxidized xanthan gum-CQDs with NH_2 groups of polyetherimide. Oxidized xanthan gum-polyetherimide hydrogel exhibited the identical photoluminescence potential as oxidized xanthan gum-CQDs and may have been paired with a smartphone to develop an RGB-based detecting system having a lower threshold of identification (3.67 mg/L) for Cr^{6+} ion monitoring. Depending on its luminescence abilities, oxidized xanthan gum-polyetherimide hydrogel might absorb over 99.0% of UV light (200–400 nm) for UV filtering purposes. It functions as a light-converting substance barrier and may be layered on light-emitting diode (LED) circuits to produce a clear green LED system. This study sheds a novel perspective on the creation of versatile xanthan gum-based CQDs components, which have exciting applications in fluorescent anticounterfeiting, LED building, UV protection, and environmental monitoring (Shen et al., 2023).

In a more general way, luminescence refers to fluorescence and other types of luminous projection. EPS may also demonstrate luminescence by other methods, which include the inclusion of metallic ions or the development of particular complexes that may react with light to generate a glow. For instance, a few EPS have been revealed to contain europium ions, which can generate light under specific circumstances (Gao et al., 2023).

10.5 Potential scheme and principle for nano-biosensor device development

Nano-biosensors combine a biological recognition element (antibody, enzyme, aptamer, molecularly imprinted polymer, etc.) with a nanoscale transducer to convert a specific biochemical interaction into a measurable signal (electrical, optical, mass, or luminescent). Integrating EPS into these platforms adds a biocompatible, chemically versatile matrix that can (1) preserve bimolecular activity, (2) provide abundant functional groups for conjugation, and (3) stabilize or template nanomaterials, collectively improving sensitivity, selectivity, and operational stability in complex biological media (Antonopoulos et al., 2023).

10.5.1 Key functional advantages of exopolysaccharides (load-bearing claims)

Biocompatible immobilization scaffold: EPS hydrogels and films maintain a hydrated microenvironment that preserves the native conformation and activity of immobilized biomolecules, reducing loss of function commonly seen on bare electrodes or rigid surfaces (Egorikhina et al., 2023).

Chemical handle for functionalization: EPS contains hydroxyl, carboxyl, sulfate, and other substituents that permit covalent coupling (EDC/NHS, click chemistry), electrostatic immobilization, or chelation to metal nanostructures, thus enabling robust attachment of recognition elements and nanomaterials (Shin et al., 2023).

Nanomaterial templation and stabilization: EPS can act as reducing/capping agents for green synthesis of metallic nanoparticles (e.g., AuNPs, AgNPs), producing stable EPS–nanoparticle hybrids with enhanced plasmonic/electrocatalytic properties useful for signal amplification (Scala et al., 2019).

Porous, high surface area matrices: EPS gels/films create porous architectures facilitating analyte diffusion and high loading of receptors, which directly improves analytical sensitivity and lowers limits of detection (LOD) (Gongi et al., 2022).

10.5.1.1 Detection modalities and exopolysaccharides-enabled innovations

Electrochemical sensing (amperometric/voltammetric/impedimetric): Electrochemical biosensing platforms, employing amperometric, voltammetric, and impedimetric techniques, have been significantly advanced through the integration of EPS–nanomaterial composites, such as EPS combined with graphene, gold nanoparticles (AuNPs), and CNTs. These hybrid systems have been demonstrated to enhance charge transfer kinetics and modulate electrode surface chemistry, thereby enabling low-noise, high-gain electrochemical readouts for a range of applications, including enzyme activity assays, antigen–antibody binding detection, and nucleic acid hybridization. Additionally, the intrinsic antifouling properties of EPS help to minimize biofouling effects in complex biological matrices such as human serum, thereby improving sensor reproducibility and operational longevity (Li et al., 2021).

Optical/plasmonic sensing (surface plasmon resonance [SPR], localized surface plasmon resonance [LSPR], colorimetric, fluorescence): Optical and plasmonic biosensing modalities have seen notable improvement with EPS-stabilized plasmonic nanoparticles comprised of gold (Au) and silver (Ag), including SPR, LSPR, colorimetric assays, and fluorescent-based detection modalities. The EPS coating stabilizes the nanoparticle and allows for a reproducible LSPR response, while simultaneously providing a biocompatible surface for immobilizing ligands, which are antibodies, aptamers, or peptides. These properties have facilitated the development of ultrasensitive biosensors capable of detecting protein biomarkers, nucleic acids, and pathogenic microorganisms in clinical, environmental, and food safety applications (Zhang et al., 2023).

Electrochemiluminescence (ECL) and dual-mode sensing: ECL biosensing, in which electrochemical excitation of a reaction emits light, can yield a very high signal-to-noise (S/N) ratio, a wide dynamic range of detection, and low detection limits. An EPS-based matrix has been utilized in combination with ECL luminophores and conductive nanostructures to develop label-free or minimally labeled ECL biosensors. Importantly, these designs have exhibited better photochemical stability, reduced levels of background interference, and improved signal reproducibility. Moreover, EPS-enabled ECL systems have also been applied to economically viable platforms such as screen-printed electrodes (SPEs) and microfluidic chips that have great relevance in the context of POC diagnostics (Sun et al, 2021).

Piezoelectric/mass-based detection (quartz crystal microbalance [QCM], microcantilever): The traditional biosensing technologies that are sensitive to mass, such as QCM or microcantilevers, have gained improvements from the use of EPS coatings. The EPS coatings increase the efficiency of capturing biorecognition molecules and reduce the impact of nonspecific adhesion. The target binding events cause detectable shifts in the oscillation frequency (QCM) or the bending deflection (microcantilever), where the objective is to quantify collection or sensing events in real-time (>3Hz), or to do so label-free with good accuracy. This demonstration has been very successful for pathogenic detections in viscous or protein-rich biological fluids (Skládal, 2024).

2-D-material synergism: The incorporation of two-dimensional (2D) nanomaterials, such as graphene oxide, black phosphorus, and transition metal carbides/nitrides (MXenes) into EPS matrices has been shown to provide synergistic benefits. These 2D-EPS hybrids exhibit exceptional electrical conductivity, large surface area, and improved mechanical stability. Such characteristics significantly enhance both electrochemical and optical transduction efficiency, while enabling the fabrication of flexible, stretchable, and wearable biosensors for continuous physiological monitoring (Mostafavi and Iravani, 2022).

10.5.1.2 Generalized exopolysaccharides-centric fabrication workflow (scheme)

A reproducible development pathway for an EPS-based bionanosensor is typically structured into sequential stages and is described as follows:

Selection and characterization of EPS: An appropriate EPS, either a homopolysaccharide or a heteropolysaccharide, is first selected based on its functional group availability, mechanical properties, and biocompatibility. Comprehensive characterization is then performed, including assessment of molecular weight (MW), monosaccharide composition, degree of substitution (e.g., acetylation, sulfonation), rheological profile, and surface chemistry properties to ensure suitability for sensor fabrication (Wang et al., 2023).

Functionalization of EPS: Bioreceptors and nanomaterials are stably conjugated to biopolymers by introducing carboxyl (–COOH), amino ($-NH_2$), thiol (–SH), or azide ($-N_3$) groups. Bonds by bioreceptors and nanomaterials, EPS can be constructed using EDC/NHS coupling, silanization of hydroxyls, or click reactions. EDC/NHS is known to activate –COOH groups for coupling to primary amines, forming amide bonds with a primary water-soluble byproduct, making the chemistry a standard for biomolecule immobilization (Shah et al., 2024).

Integration of nanomaterials: Conductive or plasmonic nanomaterials such as AuNPs, graphene, CNTs, two-dimensional (2D) materials, and metal oxides are combined with EPS to construct composite matrices. This can be accomplished by either in-situ reduction, where EPS acts as both reducing and capping agent, or by ex situ mixing, resulting in EPS-nanomaterial hybrids with precisely optimized properties of morphology, electrical conductivity, and compatibility (Ahuja et al., 2013).

Immobilization of bioreceptors: Biorecognition components like antibodies, aptamers, enzymes, nucleic acid probes, etc., are anchored to the EPS–nanomaterial composite by covalent or noncovalent associations. Suitable immobilization conditions, such as pH, ionic strength, temperature, and linker lengths, are kept in order to retain the biological activity and increase the sensing performance of the sensor (Yang et al., 2024).

Device integration and transducer coupling: The composite material is applied to an appropriate transducing platform (e.g., potentiometric or amperometric electrode; optical platform for SPR sensing; or QCM to capture mass-based detection). The device's interface is then connected to the necessary readout electronics (e.g., potentiostats for electrochemical signals; photodetectors for optical responses; and microfluidic modules for sample handling) (Butt, 2025). SPR is a label-free optical technique that records surface refractive index changes in real-time associated with analyte binding, and QCM is a piezoelectric mass-sensing platform capable of measuring nanogram-level changes by sensing frequency shift (Butt, 2025).

Calibration and analytical validation: The device is calibrated to identify critical analytical parameters such as the limit of detection (LOD), selectivity, sensitivity, response time, reproducibility, storage stability, and linear dynamic range. Validation is accomplished in both standard buffers as well as clinical matrices (e.g., serum, saliva, whole blood) while testing for interfering substances and shelf-life (Azeem et al., 2023).

Prototype optimization and translation: The biosensor is refined for practical application by miniaturizing through SPEs or paper-based formats, followed by validation using larger clinical cohorts. Attention is given to scalable manufacturing, regulatory compliance, and biocompatibility, particularly for POC and portable deployment (Hemdan et al., 2024).

10.6 Role of exopolysaccharides in nano-biosensor development

EPS, including microbial polysaccharides such as alginate, dextran, xanthan gum, and gellan gum, have emerged as critical functional materials in nano-biosensing platforms due to their inherent biocompatibility, tunable chemical functionalities, and structural versatility. EPS fulfil multiple roles ranging from antifouling surface engineering to nanomaterial stabilization, enabling enhanced sensitivity, reproducibility, and long-term operational stability in biomedical diagnostics (Kiran et al., 2024).

10.6.1 Biocompatible and antifouling interface

EPS coatings, owing to their hydrophilic and charge-neutral properties, form nonfouling hydrogel matrices or thin films that prevent nonspecific adsorption of proteins and other biomolecules, an indispensable feature for sensors operating in complex biological fluids such as serum, saliva, or wound exudate. For example, oxidized bacterial cellulose (a functionalized EPS) hydrogel demonstrated exceptional antifouling and antimicrobial activity when applied to an electrochemical biosensor for involucrin detection in wound exudate, achieving a detection limit as low as $0.45\ pg\bullet mL^{-1}$ (Lv et al., 2024). Reviews on antifouling biomaterials for biosensing have also identified EPS-based coatings as particularly effective in preserving signal integrity in challenging biological environments (Ahuja et al., 2024).

10.6.2 Structural scaffold and nanomaterial stabilizer

EPS provides a hydrated, porous network enriched with hydroxyl, carboxyl, and other reactive groups, enabling efficient immobilization of biomolecules and nanomaterials. This structure not only facilitates stable attachment of sensing elements but also supports the green synthesis and stabilization of metallic nanoparticles, improving their colloidal stability and sensor performance. Advances in microbial EPS composites highlight their ability to integrate with conductive materials and maintain cellular viability, offering a multifunctional scaffold for biosensing platforms (Ahuja et al., 2024).

10.6.3 Tunable mechanics for flexible sensing platforms

EPS, such as gellan gum, are mechanically tunable, thermally stable, and biocompatible, making them an excellent choice for flexible or wearable or implantable biosensing devices. Gellan gum gelation behavior can be adjusted via ionic crosslinking, allowing for precise adjustment of the mechanical compliance of the sensing interface, which is important for conformal skin or tissue contact without resultant performance loss (Ara et al., 2024).

10.6.4 Signal amplification and stability

Stabilizing plasmonic nanomaterials such as Au or AgNPs with EPS provides enhancement of optical signal strength and stability, as seen in interferometric sensing and colorimetric or LSPR sensing systems. EPS, through their regulation of dispersion and prevention of agglomeration (not to mention the influence of liquid tension of EPS), aid in better sensitivity and reproducibility. EPS also possess functional moieties providing opportunities for selective conjugation with recognition ligands that provide signal transduction mechanisms in electrochemical and optical biosensors (Sathiyanarayanan et al., 2017).

10.7 Role of exopolysaccharides based carbon quantum dots in the nano-biosensor device developments

EPS-derived nano-biosensors have demonstrated enormous promise in many biological fields because of their exceptional biocompatibility, photoluminescence, water solubility, intrinsic fluorescence, and transformation properties (Li et al., 2012). At present, clinical outcomes are being investigated in cancer research, gene therapy, photodynamic therapy, drug delivery, photo-thermal therapy, biosensing, and bio-imaging tools (Fig. 10.1). Such nano-biosensors, generated from EPS, include pectin, pullulan, dextran, carrageenan, starch, chitosan, hyaluronic acid, agarose, and cellulose, which have remarkable physicochemical capabilities that increase their efficacy in biomedical fields while maintaining sustainable development and low toxicity (Li et al., 2012).

10.7.1 Biosensing

In view of their outstanding biocompatibility, fluorescence strength, and luminescent characteristics, CDs provide a potential viewpoint as fluorescent carbon nanostructures for molecular

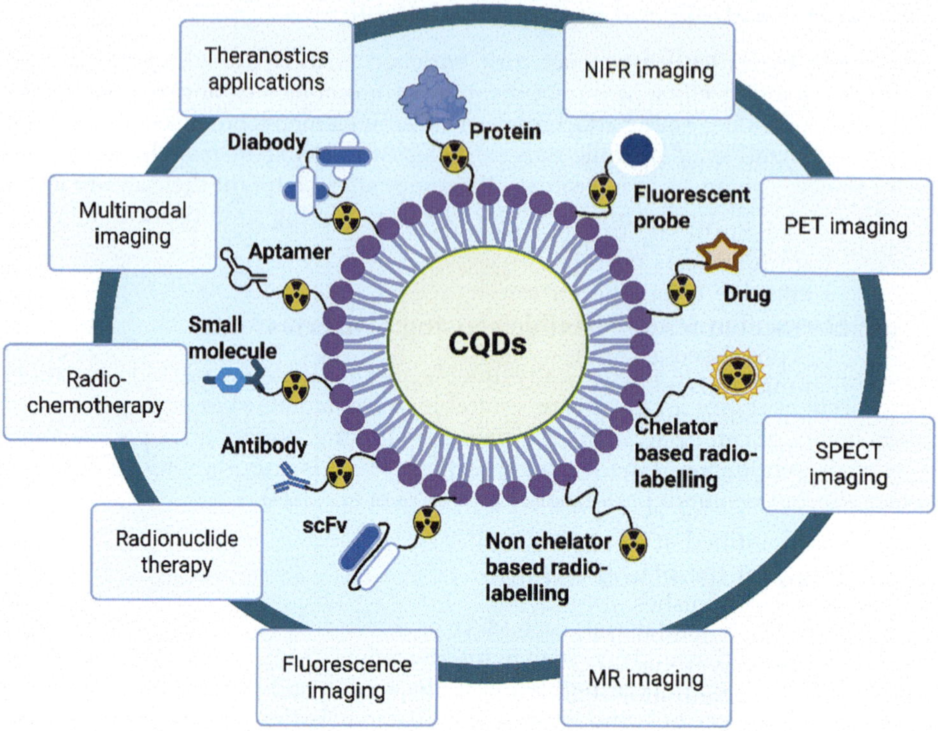

FIGURE 10.1 A diagrammatic illustration of CQDs in theranostic settings. *CQDs*, Carbon quantum dots.

biomedical sensing (Belkin, 2025). Based on the investigation, an elevated content of iron ion concentration and/or iron shortage in the circulatory system may exacerbate human medical conditions. For such an explanation, it is imperative to design reliable and simple techniques for monitoring ionic iron. Liu et al. used CDs fabricated from acrylamide and chitosan to develop a carbon sensor designed for quantitative and/or qualitative measurement of metal ions (Fe^{2+}); to pursue this, they measured the levels of fluorescence for the CDs that exhibited chelating metal characteristics at 300 μM concentration for a variety of metal ions inclusive of Fe^{2+}, Fe^{3+}, Hg^{2+}, Cu^{2+}, Cr^{3+}, and Pb^{2+}. The fluorescence level of CDs were concerned with Fe^{2+} was inhibited by 55% (Liu et al., 2019). Ascorbic acids and Fe^{3+} ions were detected concurrently, as demonstrated by Xu et al. using CDs fabricated with κ-carrageenan and chitosan treated with sulfur and nitrogen (Xu et al., 2021). Mu et al. developed a fluorescent sensor with high stability based on the ratiometric fluorescence method from CDs doped with chitosan, with ethylenediamine and chitosan as substrates and the metal ion biomarker, i.e., calcein, for the detection of alkaline phosphatase (AP). A number of testings was carried out to identify AP based on ratiometric fluorescence, using fluorescence emission spectra. Initially, they proved the calcein quenching action on CDs by exhibiting two fluorescence emission peaks. CDs had a peak of lower than 1000 a.u. at approximately emission wavelength of 412 nm, whereas calcein had a peak of 4000 a.u. at around the emission wavelength of 512 nm. The fluorescence peak of calcein was diminished when Eu^{3+} was introduced to the combination. Moreover, introducing the AP enzyme at various concentrations (from 0.09–2 mU/mL) to the combination of Eu^{3+}, calcein, and CDs, they detected an improvement in the fluorescence strength of calcein and a reduction in the fluorescence strength of CDs. The ratiometric fluorescence measurement of AP function was then validated using human serum specimens. They reported that they could detect 0.295 to 0.297 mU/mL of AP in 100-times diluted human serum (Mu et al., 2021). Das et al. created a fluorescence sensor employing sulfur and nitrogen-co-doped CDs to detect the acetone content in biological samples like urine and blood. The sulfur and nitrogen-co-doped CDs were produced by combining κ-carrageenan with urea. The sulfur and nitrogen-co-doped CDs showed excellent specificity for acetone compounds in biological specimens, with a detection threshold at 7.2×10^{-7} M. Acetone may quench sulfur and nitrogen-co-doped CDs in urine and blood specimens at concentrations ranging from 0–0.05 M to 0–0.01 M, depending on the fluorescence attributes (Das et al., 2018).

10.7.2 Bio-imaging Systems

The distinctly fluorescent properties of CDs have made it a center of attraction among various researchers who would like to design in-vitro and in-vivo surveillance and identification methods for cancerous cells diagnosis and pathological disorders (Maheshwari et al., 2025). For instance, the construction of sensors for determining abnormal concentrations of AP, an early indicator for the diagnosis of inflammatory diseases and cancer cells in human blood, is critical. Mu et al. developed a fluorescent sensor using chitosan-doped CDs as an initiator with calcein for identifying AP. For the detection of AP in cells, the investigators assessed the survival of CDs in HepG2 cells. The findings revealed that higher than 80% survivability at doses larger than 0.05 mg/mL of CDs. They investigated the internalized bio-imaging of AP in HepG2 cells due to their specificity and biocompatibility for AP recognition. HepG2 cells exposed to the CDs and calcein combination exhibited mild blue fluorescence from CDs and high green fluorescence from calcein. Furthermore, cells cultured with a combination of Eu^{3+}, calcein, and CDs revealed a reduction in calcein-induced green fluorescence. When pNPP was introduced to the culture media, endogenous alkaline phosphatase degraded it, producing PO_4^{3-}. This PO_4^{3-} molecule reacted with Eu^{3+}, releasing free calcein. As an outcome, the green fluorescent strength of the calcein enhanced in HepG2 cells, but the blue fluorescent strength of the CDs decreased nearly completely (Mu et al., 2021).

Sun et al. studied the bioimaging of *Bacillus* subtilis and Escherichia coli bacterium with a fluorescent sensor constructed from nitrogen-doped CDs. To explore the cytotoxicity of CDs produced from chitosan at various concentrations (20–400 μg/mL) on the two-bacterium employed in the study. The outcomes suggested over 93% of bacterial cell survivability, exhibiting that the nitrogen-doped CDs are minimal in cytotoxicity. Bacteria cultured with CDs exhibited red, green, and blue fluorescence at various excitation wavelengths: 560 nm, 470 nm, and 405 nm. The CDs may be able to cross cell membranes and activate bacteria cell, according to the fluorescence images (Sun et al., 2021a).

10.7.3 Drug delivery systems

The application of fluorescence carbon quantum dots (CQDs) combining with urea and sodium alginate were utilized to encapsulate the anticancer drug DOX (doxorubicin). The aim of combining urea and sodium alginate was to introduced various functional moieties (–$CONH_2$, –COOH, and –OH) connected to the interface of the urea-alginate CQDs which enhance couplings with drug components. As a consequence, the urea and sodium alginate-based CQDs at doses of 0.186–0.3 mg/mL did not prove potentially dangerous to the MCF-7 cell lines. At pH 7.4, the DOX encapsulation efficiency over the urea and sodium alginate-based CQDs interface was 70%. The in-vitro release kinetics of DOX at various pH values was examined during 80 hours. It was observed that approximately 20% of the release is complete within 72 hours. This slow release is related with superior management of DOX delivery, hence reducing the initial-burst of drug release when no CQDs transporter is utilized (95% after 20 minutes) (Ganguly et al., 2020).

Another investigation employed CQDs containing phenyl boronic acid and κ-carrageenan to encapsulate metformin, a biguanide class of oral antidiabetic drug. The encapsulation efficiency was 71.6%, whereas the drug release was around 60% over the initial 10 minutes. In contrast with the pristine drug's release kinetic study, which displayed burst-release effects (92% after 5 minute), the κ-carrageenan-mediated CQDs shown promise as a metformin delivery carrier (Das et al., 2019).

10.7.4 Sensing tools

It is imperative for evaluating water quality based on ion and metal levels, as well as ecologically undesirable pigments. Zhan et al. utilized CDs for detecting metallic ions. The fluorescence-quenching experiments with various metallic ions revealed that CQDs had a greater detection sensitivity for ferric ion at 1 mM. They also investigated the CDs quenching response at various ferric levels. As an outcome, they observed that as concentration expanded, the level of fluorescence dropped. The presence of a metalloid i.e., ferric ion decreased the fluorescence of the CDs due to ion-phenolic hydroxyl group chelation (Zhan et al., 2019). Chauhan et al. outlined using CQDs from agarose gel waste to develop a fluorometric sensor for detecting $CO_3^{2-}$and Zn^{2+} ions. From the results, the authors reported the detection limits for this agarose-waste CQDs sensor to be 0.17 nM for CO_3^{2-} and 0.26 nM for Zn^{2+} ions. Such low levels of detection offer a direct route for environmentally friendly, safe and economical means of water treatment (Chauhan et al., 2020). Wang et al. explored the use of κ-carrageenan-derived CQDs for metal ions detection in the form of a fluorescence-based sensor. The CQDs developed had a strong fluorescence and stability and which had a quantum yield of 20.6%. The fluorescence intensity of the CQDs were altered with the addition of oxytetracycline and ferric ion, showing the outstanding sensitivity and responsiveness for detection of metal ions. The detection limit for ions such Hg^{2+}, Na^{2+}, Pb^{2+}, Ca^{2+}, and Al^{3+} was 0.21 μM. After oxytetracycline was added, these CQDs demonstrated a 0.05 μM detection limit when exposed to ions including SO_4^-, Zn^{2+}, Ca^{2+}, Mg^{2+}, Na^+ along with other materials like oxalic acid, ascorbic acid, ethanol, urea, glycine, glucose, and sucrose (Wang et al., 2021).

10.8 Scientific instances of exopolysaccharides-based nano-biosensors, including patents and technology transfer

EPS-based nano-biosensors have garner various patent applications as they are generally biocompatible, fluorescent, and produced with environmentally-friendly methods (Kuppusamy et al., 2025). Many patents refer to the issue of low quantum efficiency dealing with dopants/sulfur/nitrogen, to improve fluorescence strength and photo-stability, which allows for use in bio-imaging and diagnosis (Yan et al., 2015). Surface passivation methods such as amino acid modification and PEGylation will increase durability and minimize photobleaching. There are patents on CQD-based drug delivery systems include coatings with selective ligands such as peptides, antibodies, or folic acid for delivery to the target sites and enhance bioavailability with reduced side effects (Niu et al., 2015). Theranostic uses of applying imaging and therapeutics had patents granted, particularly for treatment of cancer and management of neurological disease. CQD-based nano-biosensors include metal ion sensor for environmental monitoring, pH mediated CQDs for cancer diagnosis, and mapping glucose for treatment of diabetes. Patents emphasize CQD-based water filtering devices that use photocatalytic characteristics to degrade contaminants, as well as antimicrobial CQDs personalized with nanoparticles (NPs) or silver to promote wound recovery and antibacterial films (Niu et al., 2015). Such patent filings demonstrate the growing economic prospects of EPS-derived CQDs, opening the path for their incorporation into biological, environmental, and industrial purposes (Table 10.2).

10.9 Future outlook of exopolysaccharides based nano-biosensors

EPS are one of the prevalent and intricate organic substances in the ocean (Mardhekar et al., 2025). Combining a variety of polysaccharides as either substrates or doping elements can yield a broad range of CQDs. The main roadblocks preventing the complete establishment of a unique carbon dot are the manufacturing processes, surface properties, uneven size distributions, and photoluminescence properties of CQDs. To increase their effectiveness in biosensing and bio-imaging, the quantum efficiency must be enhanced by selective doping and surface modification (Hallaji et al., 2021). Furthermore, a better knowledge of the in-vivo responses of CQDs, comprising its clearance routes, cellular absorption, and bio-distribution, is essential for anticipating and reducing possible harmful consequences. Resolving challenges associated with fast renal clearance and organ build-up is critical for increasing the circulatory time while limiting toxicity. Innovative methods must permit the regulation of optical characteristics to create customizable emissions. The majority of research on CQDs has revealed that it is reasonably simple to produce green and blue fluorescent carbon NPs with good fluorescence quantum efficiencies (Hallaji et al., 2021). However, CQDs often display dominant emissions in the blue region and significant absorption in the ultraviolet (UV) band (Zhu et al., 2019). Although, for biological purposes, red-luminescent CQDs are required since green and blue emissions do not permeate tissue sufficiently and are probably to produce auto-fluorescence in biological samples (La Ferla and Vercelli, 2023). Thus, the fabrication of CQDs with significant red/near-infrared (NIR) excitation/emission is currently considered an essential characteristic in broadening their applicability in the biomedical area (Masha and Samuel Oluwafemi, 2022; Zhong et al., 2023).

Zhu et al. investigated the method of manufacturing red CQDs from different substrates. They showed that some of the substrates used in the synthesis of red CQDs possess heteroatoms that form new energy band gaps, leading to a transition to red-shifted bands (Zhu et al., 2019). Marine-derived EPS, like chitosan and carrageenan, contain nitrogen and sulfur heteroatoms that might be utilized for the synthesis of red CQDs with smaller electronic band gaps. Prospective studies must concentrate on finding methods of processing that utilize the heteroatoms prevalent in marine EPS to generate red-emitting CQDs.

10.10 Discussion and conclusions

EPS have emerged as multifunctional, sustainable biopolymers capable of transforming nano-biosensor design for biomedical diagnostics. Their structural diversity, biocompatibility,

TABLE 10.2 Relevant patents in bio-remediation, bio-imaging, and pharmaceutical delivery, demonstrate the extensive application of EPS-based nano-biosensors in a variety of sectors.

Title of the patent	Patent number	Country	References
Fluorescent CDs for optical and chemical sensing	EP3437196A1	European Patent Office	(Hou et al., 2016)
Fluorescent CQDs for biosensing applications	WO2020145862A1	International (WIPO)	(Yuan et al., 2018)
Folic acid functionalized CDs for targeted drug delivery	CN110283492A	China	(He et al., 2018)
Photodynamic therapy using CQDs	US20220098425A1	United States	(Boonsin et al., 2018)
Antimicrobial CDs for biomedical applications	US10342746B2	United States	(Xu et al., 2018)
Highly luminescent CQDs with enhanced quantum yield	US11094832B2	United States	(Zheng et al., 2020)
Green synthesis of CDs from plant-based polysaccharides	KR20210078834A	South Korea	(Wu et al., 2013)
CQDs for water purification and pollutant degradation	CN109355324A	China	(Shao et al., 2017)
Theranostic CQDs for cancer imaging and therapy	US10799492B2	United States	(Zheng et al., 2017)

CDs, carbon dots; *CQDs*, carbon quantum dots; *EPS*, exopolysaccharides.

and capacity for nanomaterial stabilization enable seamless integration into electrochemical, optical, piezoelectric, and hybrid sensing platforms, delivering high sensitivity and specificity even in complex biological environments. Continued progress will depend on elucidating EPS nanomaterial interaction mechanisms, optimizing large-scale production, and achieving clinically validated performance advances that could position EPS-based nano-biosensors as next-generation diagnostic technologies at the interface of microbiology, nanotechnology, and translational medicine.

Credit statement

BB, NRGB, and HDP: Writing-original draft, visualization, software, methodology, formal analysis, data curation, and conceptualization. AB: Writing-review and editing, investigation, formal analysis, and data curation. PP and AB: Writing-review and editing, supervision, resources, project administration, methodology, investigation, data curation, and conceptualization.

Acknowledgment

Bio render software is acknowledged for the artwork in this chapter.

References

Abdl Aali, R., Al-Sahlany, S., 2024. Gellan gum as a unique microbial polysaccharide: Its characteristics, synthesis, and current application trends. Gels 10 (3), 183. https://doi.org/10.3390/gels10030183.

Ahuja, M., Singh, S., Kumar, A., 2013. Evaluation of carboxymethyl gellan gum as a mucoadhesive polymer. Int. J. Biol. Macromol. 53, 114–121. https://doi.org/10.1016/j.ijbiomac.2012.10.033.

Ahuja, V., Chauhan, S., Dasgupta, D., Wadhwa, P., Raj, T., Yang, Y.-H., Bhatia, S.K., 2024. Microbial exopolysaccharide composites with inorganic materials and their biomedical applications: a review. Carbohydr. Polym. Technol. Appl. 7, 100482. https://doi.org/10.1016/j.carpta.2024.100482.

Angelin, J., Kavitha, M., 2020. Exopolysaccharides from probiotic bacteria and their health potential. Int. J. Biol. Macromol. 162, 853–865. https://doi.org/10.1016/j.ijbiomac.2020.06.190.

Antonopoulos, A.S., Papastamos, C., Cokkinos, D.V., Tsioufis, K., Tousoulis, D., 2023. Epicardial adipose tissue in myocardial disease: from physiology to heart failure phenotypes. Curr. Probl. Cardiol. 48 (10), 101841. https://doi.org/10.1016/j.cpcardiol.2023.101841.

Ara, L., Sher, M., Khan, M., Rehman, T.U., Shah, L.A., Yoo, H.-M., 2024. Dually-crosslinked ionic conductive hydrogels reinforced through biopolymer gellan gum for flexible sensors to monitor human activities. Int. J. Biol. Macromol. 276, 133789. https://doi.org/10.1016/j.ijbiomac.2024.133789.

Azeem, M.M., Shafa, M., Aamir, M., Zubair, M., Souayeh, B., Alam, M.W., 2023. Nucleotide detection mechanism and comparison based on low-dimensional materials: a review. Front. Bioeng. Biotechnol. 11, 1117871. https://doi.org/10.3389/fbioe.2023.1117871.

Belkin, S., 2025. Bioluminescent microbial bioreporters: a personal perspective. Biosensors 15 (2), 111. https://doi.org/10.3390/bios15020111.

Boonsin, R., Barros, A., Donat, F., Boyer, D., Chadeyron, G., Schneider, R., Boutinaud, P., Mahiou, R., 2018. Optical properties and reliability studies of gradient alloyed green emitting $(CdSe)_x(ZnS)_{1-x}$ and red emitting $(CuInS_2)_x(ZnS)_{1-x}$ quantum dots for white light-emitting diodes. ACS Photonics. 5 (2), 462–470. https://doi.org/10.1021/acsphotonics.7b00980.

Broks, V.M.A., Stegers-Jager, K.M., Fikrat-Wevers, S., Van Den Broek, W.W., Woltman, A.M., 2023. The association between how medical students were selected and their perceived stress levels in year-1 of medical school. BMC Med. Educ. 23 (1), 443. https://doi.org/10.1186/s12909-023-04411-0.

Butt, M.A., 2025. Surface plasmon resonance-based biodetection systems: principles, progress and applications—a comprehensive review. Biosensors 15 (1), 35. https://doi.org/10.3390/bios15010035.

Chauhan, P., Saini, J., Chaudhary, S., 2020. Agarose waste derived toxicologically screened carbon dots as dual sensor: a mechanistic insight into luminescence and solvatochromic behaviour. Nano-Struct. Nano-Objects 24, 100585. https://doi.org/10.1016/j.nanoso.2020.100585.

Chiang, C.-Y., Chen, C.-H., Wu, C.-W., 2023. Fiber optic localized surface plasmon resonance sensor based on carboxymethylated dextran modified gold nanoparticles surface for high mobility group box 1 (HMGB1) analysis. Biosensors 13 (5), 522. https://doi.org/10.3390/bios13050522.

Congur, G., Eksin, E., Erdem, A., 2021. Levan modified DNA biosensor for voltammetric detection of daunorubicin-DNA interaction. Sens. Actuators B. 326, 128818. https://doi.org/10.1016/j.snb.2020.128818.

Darmau, B., Sacchi, M., Texier, I., Gross, A.J., 2025. Self-extracting dextran-based hydrogel microneedle arrays with an interpenetrating bioelectroenzymatic sensor for transdermal monitoring with matrix protection. Adv. Healthc. Mater. 14 (2), 2403209. https://doi.org/10.1002/adhm.202403209.

Das, P., Ganguly, S., Mondal, S., Bose, M., Das, A.K., Banerjee, S., Das, N.C., 2018. Heteroatom doped photoluminescent carbon dots for sensitive detection of acetone in human fluids. Sens. Actuators B. 266, 583–593. https://doi.org/10.1016/j.snb.2018.03.183.

Das, P., Maity, P.P., Ganguly, S., Ghosh, S., Baral, J., Bose, M., Choudhary, S., Gangopadhyay, S., Dhara, S., Das, A.K., Banerjee, S., Das, N.C., 2019. Biocompatible carbon dots derived from κ-carrageenan and phenyl boronic acid for dual modality sensing platform of sugar and its anti-diabetic drug release behavior. Int. J. Biol. Macromol. 132, 316–329. https://doi.org/10.1016/j.ijbiomac.2019.03.224.

Egorikhina, M.N., Timofeeva, L.B., Linkova, D.D., Rubtsova, Y.P., Bugrova, M.L., Charykova, I.N., Ryabkov, M.G., Kobyakova, I.I., Farafontova, E.A., Aleynik, D.Y., 2023. Biocompatibility study of hydrogel biopolymer scaffold with encapsulated mesenchymal stem cells. Polymers 15 (6), 1337. https://doi.org/10.3390/polym15061337.

Falsini, S., Rosi, M.C., Ravegnini, E., Schiff, S., Gonnelli, C., Papini, A., Adessi, A., Urciuoli, S., Ristori, S., 2023. Nanoformulations with exopolysaccharides from cyanobacteria: enhancing the efficacy of bioactive molecules in the Mediterranean fruit fly control. Env. Sci. Pollut. Res. 30 (35), 83760–83770. https://doi.org/10.1007/s11356-023-28180-x.

Ganguly, S., Das, P., Itzhaki, E., Hadad, E., Gedanken, A., Margel, S., 2020. Microwave-synthesized polysaccharide-derived carbon dots as therapeutic cargoes and toughening agents for elastomeric gels. ACS Appl. Mater. Interfaces. 12 (46), 51940–51951. https://doi.org/10.1021/acsami.0c14527.

Gao, L., Li, L., Li, Y., He, C., Zhou, L., Qu, X., Fang, S., 2023. Effects of europium complex on thermal and photoluminescence properties of polyurethane-europium materials. Polymers 15 (5), 1064. https://doi.org/10.3390/polym15051064.

Garcia-Rey, S., Ojeda, E., Gunatilake, U.B., Basabe-Desmonts, L., Benito-Lopez, F., 2021. Alginate bead biosystem for the determination of lactate in sweat using image analysis. Biosensors 11 (10), 379. https://doi.org/10.3390/bios11100379.

Gatti, L., Sciutto, G., Cancelliere, R., Severini, L., Lisarelli, C., Mazzuca, C., Prati, S., Mazzeo, R., Micheli, L., 2025. Advanced label-free electrochemical immunosensor for a minimally invasive detection of proteins in paintings. Talanta 283, 127167. https://doi.org/10.1016/j.talanta.2024.127167.

Gongi, W., Touzi, H., Sadly, I., Ben Ouada, H., Tamarin, O., Ben Ouada, H., 2022. A novel impedimetric sensor based on cyanobacterial extracellular polymeric substances for microplastics detection. J. Polym. Env. 30 (11), 4738–4748. https://doi.org/10.1007/s10924-022-02555-6.

Gunatilake, U.B., Garcia-Rey, S., Ojeda, E., Basabe-Desmonts, L., Benito-Lopez, F., 2021. TiO2 nanotubes alginate hydrogel scaffold for rapid sensing of sweat biomarkers: lactate and glucose. ACS Appl. Mater. Interfaces 13 (31), 37734–37745. https://doi.org/10.1021/acsami.1c11446.

Hallaji, Z., Bagheri, Z., Kalji, S.-O., Ermis, E., Ranjbar, B., 2021. Recent advances in the rational synthesis of red-emissive carbon dots for nanomedicine applications: a review. FlatChem. 29, 100271. https://doi.org/10.1016/j.flatc.2021.100271.

Harutyunyan, A., Gabrielyan, L., Aghajanyan, A., Gevorgyan, S., Schubert, R., Betzel, C., Kujawski, W., Gabrielyan, L., 2024. Comparative study of physicochemical properties and antibacterial potential of cyanobacteria *Spirulina platensis* -derived and chemically synthesized silver nanoparticles. ACS Omega 9 (27), 29410–29421. https://doi.org/10.1021/acsomega.4c01604.

He, Y., He, J., Yu, Z., Zhang, H., Liu, Y., Hu, G., Zheng, M., Dong, H., Zhuang, J., Lei, B., 2018. Double carbon dot assembled mesoporous aluminas: solid-state dual-emission photoluminescence and multifunctional applications. J. Mater. Chem. C 6 (10), 2495–2501. https://doi.org/10.1039/C8TC00182K.

Hemdan, M., Ali, M.A., Doghish, A.S., Mageed, S.S.A., Elazab, I.M., Khalil, M.M., Mabrouk, M., Das, D.B., Amin, A.S., 2024. Innovations in biosensor technologies for healthcare diagnostics and therapeutic drug monitoring: applications, recent progress, and future research challenges. Sensors 24 (16), 5143. https://doi.org/10.3390/s24165143.

Hou, J., Wang, W., Zhou, T., Wang, B., Li, H., Ding, L., 2016. Synthesis and formation mechanistic investigation of nitrogen-doped carbon dots with high quantum yields and yellowish-green fluorescence. Nanoscale 8 (21), 11185–11193. https://doi.org/10.1039/C6NR02701F.

Kim, S.-U., Kim, Y.J., Lee, T.H., 2025. Polysaccharide hydrogel-assisted biosensing platforms for point-of-care use. Biosensors 15 (1), 13. https://doi.org/10.3390/bios15010013.

Kiran, N.S., Yashaswini, C., Singh, S., Prajapati, B.G., 2024. Revisiting microbial exopolysaccharides: a biocompatible and sustainable polymeric material for multifaceted biomedical applications. 3 Biotech 14 (4), 95. https://doi.org/10.1007/s13205-024-03946-3.

Kuppusamy, S., Bhattacharjee, B., Ghose, S., Tamilanban, T., Barman, D., Ahmed, A.B., Sahu, R.K., 2025. Current status and future prospect of bioremediation using green synthesis of nanoparticle/nanomaterials. In: Dave, V., Kuila, A. (Eds.), Nanomaterials as a Catalyst for Biofuel Production. Springer Nature, Singapore, pp. 295–327. https://doi.org/10.1007/978-981-96-1706-7_12.

La Ferla, B., Vercelli, B., 2023. Red-emitting carbon quantum dots for biomedical applications: synthesis and purification issues of the hydrothermal approach. Nanomaterials 13 (10), 1635. https://doi.org/10.3390/nano13101635.

Lakshmi, G.B.V.S., Yadav, A.K., Mehlawat, N., Jalandra, R., Solanki, P.R., Kumar, A., 2021. Gut microbiota derived trimethylamine N-oxide (TMAO) detection through molecularly imprinted polymer based sensor. Sci. Rep. 11 (1), 1338. https://doi.org/10.1038/s41598-020-80122-6.

Lee, J., Lee, K.H., Kim, S., Son, H., Chun, Y., Park, C., Yoo, H.Y., 2022. Microbial production of bacterial cellulose using chestnut shell hydrolysates by *Gluconacetobacter xylinus* ATCC 53524. J. Microbiol. Biotechnol. 32 (11), 1479–1484. https://doi.org/10.4014/jmb.2208.08022.

Li, H., Kang, Z., Liu, Y., Lee, S.-T., 2012. Carbon nanodots: synthesis, properties and applications. J. Mater. Chem. 22 (46), 24230. https://doi.org/10.1039/c2jm34690g.

Li, Y., Han, R., Chen, M., Zhang, L., Wang, G., Luo, X., 2021. Bovine serum albumin-cross-linked polyaniline nanowires for ultralow fouling and highly sensitive electrochemical protein quantification in human serum samples. Anal. Chem. 93 (9), 4326–4333. https://doi.org/10.1021/acs.analchem.1c00089.

Li, Y., Wei, X., Zhou, Y., Wang, J., You, R., 2023. Research progress of electronic nose technology in exhaled breath disease analysis. Microsyst. Nanoeng. 9 (1), 129. https://doi.org/10.1038/s41378-023-00594-0.

Lin, F., Li, C., Chen, Z., 2018. Exopolysaccharide-derived carbon dots for microbial viability assessment. Front. Microbiol. 9, 2697. https://doi.org/10.3389/fmicb.2018.02697.

Liu, G., Li, B., Liu, Y., Feng, Y., Jia, D., Zhou, Y., 2019. Rapid and high yield synthesis of carbon dots with chelating ability derived from acrylamide/chitosan for selective detection of ferrous ions. Appl. Surf. Sci. 487, 1167–1175. https://doi.org/10.1016/j.apsusc.2019.05.069.

Lv, M., Li, Y., Qiao, X., Zeng, X., Luo, X., 2024. An antifouling electrochemical biosensor based on oxidized bacterial cellulose and quaternized chitosan for reliable detection of involucrin in wound exudate. Anal. Chim. Acta. 1316, 342821. https://doi.org/10.1016/j.aca.2024.342821.

Madsen, M.A., Semerdzhiev, S., Twigg, J.D., Moss, C., Bavington, C.D., Amtmann, A., 2023. Environmental modulation of exopolysaccharide production in the cyanobacterium *Synechocystis* 6803. Appl. Microbiol. Biotechnol. 107 (19), 6121–6134. https://doi.org/10.1007/s00253-023-12697-9.

Maheshwari, S., Singh, A., Verma, A., 2025. Polysaccharide-based carbon quantum dots: synthesis and theranostic applications—a review. Biomed. Mater. Devices. https://doi.org/10.1007/s44174-025-00355-9.

Mardhekar, S., Luong, P., Seeberger, P.H., 2025. Exploring marine glycans: structure, function, and the frontier of chemical synthesis. RSC Chem. Biol. 6 (8), 1195–1213. https://doi.org/10.1039/D5CB00090D.

Masha, S., Samuel Oluwafemi, O., 2022. Cost-effective synthesis of red-emitting carbon-based quantum dots and its photothermal profiling. Mater. Lett. 323, 132590. https://doi.org/10.1016/j.matlet.2022.132590.

Mehta, K., Shukla, A., Saraf, M., 2021. Articulating the exuberant intricacies of bacterial exopolysaccharides to purge environmental pollutants. Heliyon 7 (11), e08446. https://doi.org/10.1016/j.heliyon.2021.e08446.

Mishra, S., Singh, P.K., Pattnaik, R., Kumar, S., Ojha, S.K., Srichandan, H., Parhi, P.K., Jyothi, R.K., Sarangi, P.K., 2022. Biochemistry, synthesis, and applications of bacterial cellulose: a review. Front. Bioeng. Biotechnol. 10, 780409. https://doi.org/10.3389/fbioe.2022.780409.

Mobed, A., Kohansal, F., Dolati, S., Hasanzadeh, M., Shakouri, S.K., 2023. An innovative electrochemical immunoplatform for monitoring chronic conditions using the biosensing of hyaluronic acid in human plasma samples. Chemosensors 11 (7), 367. https://doi.org/10.3390/chemosensors11070367.

Mohd Nadzir, M., Nurhayati, R.W., Idris, F.N., Nguyen, M.H., 2021. Biomedical applications of bacterial exopolysaccharides: a review. Polymers 13 (4), 530. https://doi.org/10.3390/polym13040530.

Mostafavi, E., Iravani, S., 2022. MXene-graphene composites: a perspective on biomedical potentials. Nano-Micro Lett. 14 (1), 130. https://doi.org/10.1007/s40820-022-00880-y.

Mu, X., Jiang, X., Zhang, Y., Liu, X., Zhang, S., Wang, W., Huang, Y., Ma, P., Song, D., 2021. Sensitive ratiometric fluorescence probe based on chitosan carbon dots and calcein for alkaline phosphatase detection and bioimaging in cancer cells. Anal. Chim. Acta. 1188, 339163. https://doi.org/10.1016/j.aca.2021.339163.

Nguyen, H.-T., Pham, T.-T., Nguyen, P.-T., Le-Buanec, H., Rabetafika, H.N., Razafindralambo, H.L., 2024. Advances in microbial exopolysaccharides: present and future applications. Biomolecules 14 (9), 1162. https://doi.org/10.3390/biom14091162.

Niu, W.-J., Li, Y., Zhu, R.-H., Shan, D., Fan, Y.-R., Zhang, X.-J., 2015. Ethylenediamine-assisted hydrothermal synthesis of nitrogen-doped carbon quantum dots as fluorescent probes for sensitive biosensing and bioimaging. Sens. Actuators B. 218, 229–236. https://doi.org/10.1016/j.snb.2015.05.006.

Pereira, S.B., Sousa, A., Santos, M., Araújo, M., Serôdio, F., Granja, P., Tamagnini, P., 2019. Strategies to obtain designer polymers based on cyanobacterial extracellular polymeric substances (EPS). Int. J. Mol. Sci. 20 (22), 5693. https://doi.org/10.3390/ijms20225693.

Picheth, G.F., Pirich, C.L., Sierakowski, M.R., Woehl, M.A., Sakakibara, C.N., De Souza, C.F., Martin, A.A., Da Silva, R., De Freitas, R.A., 2017. Bacterial cellulose in biomedical applications: a review. Int. J. Biol. Macromol. 104, 97–106. https://doi.org/10.1016/j.ijbiomac.2017.05.171.

Quazi, M.Z., Hwang, J., Song, Y., Park, N., 2023. Hydrogel-based biosensors for effective therapeutics. Gels 9 (7), 545. https://doi.org/10.3390/gels9070545.

Rahbar Saadat, Y., Yari Khosroushahi, A., Pourghassem Gargari, B., 2019. A comprehensive review of anticancer, immunomodulatory and health beneficial effects of the lactic acid bacteria exopolysaccharides. Carbohydr. Polym. 217, 79–89. https://doi.org/10.1016/j.carbpol.2019.04.025.

Salimi, F., Farrokh, P., 2023. Recent advances in the biological activities of microbial exopolysaccharides. World J. Microbiol. Biotechnol. 39 (8), 213. https://doi.org/10.1007/s11274-023-03660-x.

Sathiyanarayanan, G., Dineshkumar, K., Yang, Y.-H., 2017. Microbial exopolysaccharide-mediated synthesis and stabilization of metal nanoparticles. Crit. Rev. Microbiol. 43 (6), 731–752. https://doi.org/10.1080/1040841X.2017.1306689.

Scala, A., Piperno, A., Hada, A., Astilean, S., Vulpoi, A., Ginestra, G., Marino, A., Nostro, A., Zammuto, V., Gugliandolo, C., 2019. Marine bacterial exopolymers-mediated green synthesis of noble metal nanoparticles with antimicrobial properties. Polymers 11 (7), 1157. https://doi.org/10.3390/polym11071157.

Shah, I.A., Kavitake, D., Tiwari, S., Devi, P.B., Reddy, G.B., Jaiswal, K.K., Jaiswal, A.K., Shetty, P.H., 2024. Chemical modification of bacterial exopolysaccharides: antioxidant properties and health potentials. Curr. Res. Food Sci. 9, 100824. https://doi.org/10.1016/j.crfs.2024.100824.

Shao, J., Zhu, S., Liu, H., Song, Y., Tao, S., Yang, B., 2017. Full-color emission polymer carbon dots with quench-resistant solid-state fluorescence. Adv. Sci. 4 (12), 1700395. https://doi.org/10.1002/advs.201700395.

Shen, J., He, Z., Zhang, J., Guo, H., Lin, W., Gu, H., 2023. Green and sustainable preparation of functionalized carbon quantum dots and fluorescent hydrogel from exopolysaccharides for multiple fluorescence applications. Polymer 288, 126471. https://doi.org/10.1016/j.polymer.2023.126471.

Shi, J., Tang, J., Zhang, M., Zou, Y., Pang, J., Wu, C., 2025. Recent advances in polysaccharide-based electrospun nanofibers for food safety detection. Sensors 25 (7), 2220. https://doi.org/10.3390/s25072220.

Shin, M., Lim, J., An, J., Yoon, J., Choi, J.-W., 2023. Nanomaterial-based biohybrid hydrogel in bioelectronics. Nano Converg. 10 (1), 8. https://doi.org/10.1186/s40580-023-00357-7.

Skládal, P., 2024. Piezoelectric biosensors: shedding light on principles and applications. Microchim. Acta 191 (4), 184. https://doi.org/10.1007/s00604-024-06257-9.

Sun, L., Zhang, H., Wang, Y., Xiong, Z., Zhao, X., Xia, Y., 2021a. Chitosan-derived N-doped carbon dots for fluorescent determination of nitrite and bacteria imaging. Spectrochim. Acta A 251, 119468. https://doi.org/10.1016/j.saa.2021.119468.

Sun, Y., Huang, C., Sun, X., Wang, Q., Zhao, P., Ge, S., Yu, J., 2021b. Electrochemiluminescence biosensor based on molybdenum disulfide-graphene quantum dots nanocomposites and DNA walker signal amplification for DNA detection. Microchim. Acta 188 (10), 353. https://doi.org/10.1007/s00604-021-04962-3.

Tabernero, A., Cardea, S., 2020. Microbial exopolysaccharides as drug carriers. Polymers 12 (9), 2142. https://doi.org/10.3390/polym12092142.

Tian, J., Fan, Y., Li, X., Cui, K., Zhang, J., 2025. Optimising the production of exopolysaccharides from *Pediococcus pentosaceus* SSC-12 and characterizing its properties. LWT 226, 117967. https://doi.org/10.1016/j.lwt.2025.117967.

Wang, W., Ju, Y., Liu, N., Shi, S., Hao, L., 2023. Structural characteristics of microbial exopolysaccharides in association with their biological activities: a review. Chem. Biol. Technol. Agric. 10 (1), 137. https://doi.org/10.1186/s40538-023-00515-3.

Wang, Y., Liu, Y., Zhao, L., Sun, L., Zhao, X., Xia, Y., 2021. κ-Carrageenan-derived carbon dots for highly selective and sensitive detection of Fe3+ and oxytetracycline. J. Mater. Sci. 56 (2), 1272–1285. https://doi.org/10.1007/s10853-020-05301-2.

Wu, Z.L., Zhang, P., Gao, M.X., Liu, C.F., Wang, W., Leng, F., Huang, C.Z., 2013. One-pot hydrothermal synthesis of highly luminescent nitrogen-doped amphoteric carbon dots for bioimaging from Bombyx mori silk – natural proteins. J. Mater. Chem. B 1 (22), 2868. https://doi.org/10.1039/c3tb20418a.

Xu, J., Wang, Y., Sun, L., Qi, Q., Zhao, X., 2021. Chitosan and κ-carrageenan-derived nitrogen and sulfur co-doped carbon dots "on-off-on" fluorescent probe for sequential detection of Fe3+ and ascorbic acid. Int. J. Biol. Macromol. 191, 1221–1227. https://doi.org/10.1016/j.ijbiomac.2021.09.165.

Xu, M., Ge, R., Wang, W., Peng, Y., Li, Z., Liu, W., Wang, L., 2025. Xanthan gum-toughened nanocomposite conductive hydrogel for wearable sensors and smart livestock monitoring. Int. J. Biol. Macromol. 318, 144983. https://doi.org/10.1016/j.ijbiomac.2025.144983.

Xu, Q., Zhang, M., Liu, Y., Cai, W., Yang, W., He, Z., Sun, X., Luo, Y., Liu, F., 2018. Synthesis of multi-functional green fluorescence carbon dots and their applications as a fluorescent probe for Hg^{2+} detection and zebrafish imaging. New J. Chem. 42 (12), 10400–10405. https://doi.org/10.1039/C8NJ01639A.

Xu, X., Ding, Z., Pu, C., Kong, C., Chen, S., Lu, W., Zhang, J., 2024. The structural characterization and UV-protective properties of an exopolysaccharide from a *Paenibacillus* isolate. Front. Pharmacol. 15, 1434136. https://doi.org/10.3389/fphar.2024.1434136.

Yadav, A.K., Verma, D., Sajwan, R.K., Poddar, M., Yadav, S.K., Verma, A.K., Solanki, P.R., 2022. Nanomaterial-based electrochemical nanodiagnostics for human and gut metabolites diagnostics: recent advances and challenges. Biosensors 12 (9), 733. https://doi.org/10.3390/bios12090733.

Yan, Z., Yu, Y., Chen, J., 2015. Glycine-functionalized carbon quantum dots as chemiluminescence sensitization for detection of m-phenylenediamine. Anal. Methods 7 (3), 1133–1139. https://doi.org/10.1039/C4AY02124J.

Yang, R., Liu, L., Gao, D., Zhao, D., 2024. Purification, structural characterization, and bioactive properties of exopolysaccharides from *Saccharomyces cerevisiae* HD-01. Front. Bioeng. Biotechnol. 12, 1455708. https://doi.org/10.3389/fbioe.2024.1455708.

Yuan, B., Xie, Z., Chen, P., Zhou, S., 2018. Highly efficient carbon dots and their nanohybrids for trichromatic white LEDs. J. Mater. Chem. C 6 (22), 5957–5963. https://doi.org/10.1039/C8TC01659C.

Zang, J., Lin, T., Shi, Y., Lin, Y., Xu, C., Ma, K., Zhang, C., Rui, X., Gan, D., Li, W., 2025. Recent advances in exopolysaccharide–protein interactions in fermented dairy- and plant-based yogurts: mechanisms, influencing factors, health benefits, analytical techniques, and future directions. Compr. Rev. Food Sci. Food Saf. 24 (4), e70219. https://doi.org/10.1111/1541-4337.70219.

Zhan, J., Peng, R., Wei, S., Chen, J., Peng, X., Xiao, B., 2019. Ethanol-precipitation-assisted highly efficient synthesis of nitrogen-doped carbon quantum dots from chitosan. ACS Omega 4 (27), 22574–22580. https://doi.org/10.1021/acsomega.9b03318.

Zhang, H., Zhou, X., Li, X., Gong, P., Zhang, Y., Zhao, Y., 2023. Recent advancements of LSPR fiber-optic biosensing: combination methods, structure, and prospects. Biosensors 13 (3), 405. https://doi.org/10.3390/bios13030405.

Zhang, M., Wu, Y., Qu, C., Huang, Q., Cai, P., 2024. Microbial extracellular polymeric substances (EPS) in soil: from interfacial behaviour to ecological multifunctionality. Geo-Bio Interfaces 1, e4. https://doi.org/10.1180/gbi.2024.4.

Zheng, J., Wang, Y., Zhang, F., Yang, Y., Liu, X., Guo, K., Wang, H., Xu, B., 2017. Microwave-assisted hydrothermal synthesis of solid-state carbon dots with intensive emission for white light-emitting devices. J. Mater. Chem. C 5 (32), 8105–8111. https://doi.org/10.1039/C7TC01701D.

Zheng, J., Xie, Y., Wei, Y., Yang, Y., Liu, X., Chen, Y., Xu, B., 2020. An efficient synthesis and photoelectric properties of green carbon quantum dots with high fluorescent quantum yield. Nanomaterials 10 (1), 82. https://doi.org/10.3390/nano10010082.

Zhong, Y., Chen, A., Yin, X., Li, R., Deng, Q., Yang, R., 2023. Red emission carbon dots for mitoxantrone detection. Sens. Actuators B 382, 133535. https://doi.org/10.1016/j.snb.2023.133535.

Zhu, Z., Zhai, Y., Li, Z., Zhu, P., Mao, S., Zhu, C., Du, D., Belfiore, L.A., Tang, J., Lin, Y., 2019. Red carbon dots: optical property regulations and applications. Mater. Today 30, 52–79. https://doi.org/10.1016/j.mattod.2019.05.003.

Zolti, O., Suganthan, B., Ramasamy, R.P., 2023. Lab-on-a-chip electrochemical biosensors for foodborne pathogen detection: a review of common standards and recent progress. Biosensors 13 (2), 215. https://doi.org/10.3390/bios13020215.

CHAPTER

11

Potential application of EPS as a cleansing agent

Amresh Kumar[a], *Gaurav Gupta*[b], *Partha Palit*[c] *and Manabendra Dutta Choudhury*[d]

[a]Department of Zoology, M.L. Arya College, Purnia, Bihar, India [b]Department of Immunology, University of Manitoba, Winnipeg, MB, Canada [c]Department of Pharmaceutical Sciences, Assam University (A Central University), Silchar, Assam, India [d]Department of Life Sciences and Bioinformatics, Assam University & Ravindranath Tagore University, Hojai, Assam, India

11.1 Introduction

Exopolysaccharides (EPS) are characterized as natural polymers generated and excreted by microorganisms to maintain the structural and operational reliability of biofilms. Consequently, they are crucial for assessing physicochemical characteristics of biofilms that include 50%–90% of organic matter (Donlan, 2002; Staudt et al., 2004). Microbial polysaccharides are primarily categorized into diverse factions according to (1) cellular localization (cell wall polysaccharides, EPS, and endopolysaccharides), (2) structural configuration (linear and branched), (3) sugar composition, and (4) types of connection between monomers {β-(1→3), β-(1→6), and α-(1→3)} (Mahapatra and Banerjee, 2013). EPSs mostly provide defensive activities, with their specific tasks depending upon the natural setting of the microbes. They can safeguard microbial populations against severe temperatures, salinity, and nutrient scarcity by establishing a separation across the microbes and their environment (Mohamed et al., 2018). Furthermore, they are crucial for aggregation formation, surface adherence, biofilm development, biofouling, and nutritional absorption, among other functions (Nichols et al., 2005). Given their benign nature and biodegradability, microbial EPSs have garnered attention for their prospective application in the context of the environment, particularly as bioflocculants (Moghannem et al., 2018). Furthermore, these compounds may serve as effective cleaning agents within the petroleum and wastewater treatment conducted by industries (Schmid et al., 2015).

EPSs are distinguished by diversity of functional groups, including hydroxyl, amine, phosphate, and urinate. These groups augment the higher electrostatic charge of EPS, facilitating their

DOI: https://doi.org/10.1016/B978-0-443-36619-2.00011-5

interaction with metallic ions and enhancing the potentiality of microbial-based precipitation. They are also integral to their flocculation activities and ion exchange characteristics of the organism itself (Bhaskar and Bhosle, 2006). Additionally, the unstable characteristics of EPSs sourced from bacteria, along with their capacity to complex with heavy metals, result in the conveyance of these metals through the seafood chain, which helps clean the metal pollutants in organisms at elevated trophic levels (Abdel-Aziz, 2012).

The primary objective of this chapter is to emphasize the significance of the metabolite EPS as a pivotal element in attaining future environmental sustainability. This chapter will analyze the diverse dynamic and active roles of EPS to comprehend the bioremediation capabilities of different bacteria about environmental contaminants.

This discussion will encompass a range of understanding of a detailed examination of its architecture, the various compositions, and the genetic regulation of the biosynthesis of EPS. The innovative implementation of EPS formulations produced from *Pseudomonas* for enhancing plant growth and disease control has also been incorporated (Fig. 11.1). This chapter will also particularly assist individuals investigating the intersections of EPS with genera *Pseudomonas* sp. to create diverse bioformulations aimed at achieving environmental sustainability.

11.2 Genetic regulation of the synthesis of exopolysaccharides

The fundamental necessity for EPS synthesis involves monosaccharide units derived from catabolized pentose sugars, which include sugars such as uridine diphosphate (UDP)-N-acetylglucosamine and GDP-mannuronic acid (Balíková et al., 2022). Plasmid or chromosomal DNA are the locations that the bacteria use to target EPS genes. The organization of these genes takes the shape of gene clusters. Observations reveal the involvement of several gene clusters in bacteria that participate in the manufacture of EPS (Knoshaug et al., 2007). A thorough evaluation of the processes behind EPS biosynthesis and export has elucidated three main routes. The first route is essential for enabling both polymerization and exporting (Weigel and DeAngelis, 2007). In the second pathway, the ATP-binding cassette transporter is crucial for the export of the repeating biosynthesized unit. The involvement of Wzx–Wzy proteins is noted in the third pathway (Whitfield and Mainprize, 2010). The Wzy protein facilitates polymerization and aids in the elongation of the growing EPS chain, while the final translocation is carried out by Wza (Reid and Whitfield, 2005; Weigel and DeAngelis, 2007).

The Bacterial enzyme, such as tyrosine kinases are associated with EPS biosynthesis (Elsholz et al., 2014). Genes that code for bacterial tyrosine kinases frequently correlate with genes that play a role in EPS production. The kinases are recognized for their role in regulating EPS production through phosphorylation, which subsequently activates a biosynthetic enzyme involved in the EPS production pathway (Mijakovic et al., 2016).

11.3 Properties of exopolysaccharides suitable for cleaning purposes

The features of EPSs are enhanced by their chemical, mechanical, and physiological properties qualities. Assessing the characteristics of EPS facilitates the prediction of prospective usages and performance in diverse situations. The chemical analysis of EPS involves the detection of monosaccharide content and architecture, weight, pattern of glycosidic bonds, branching, and

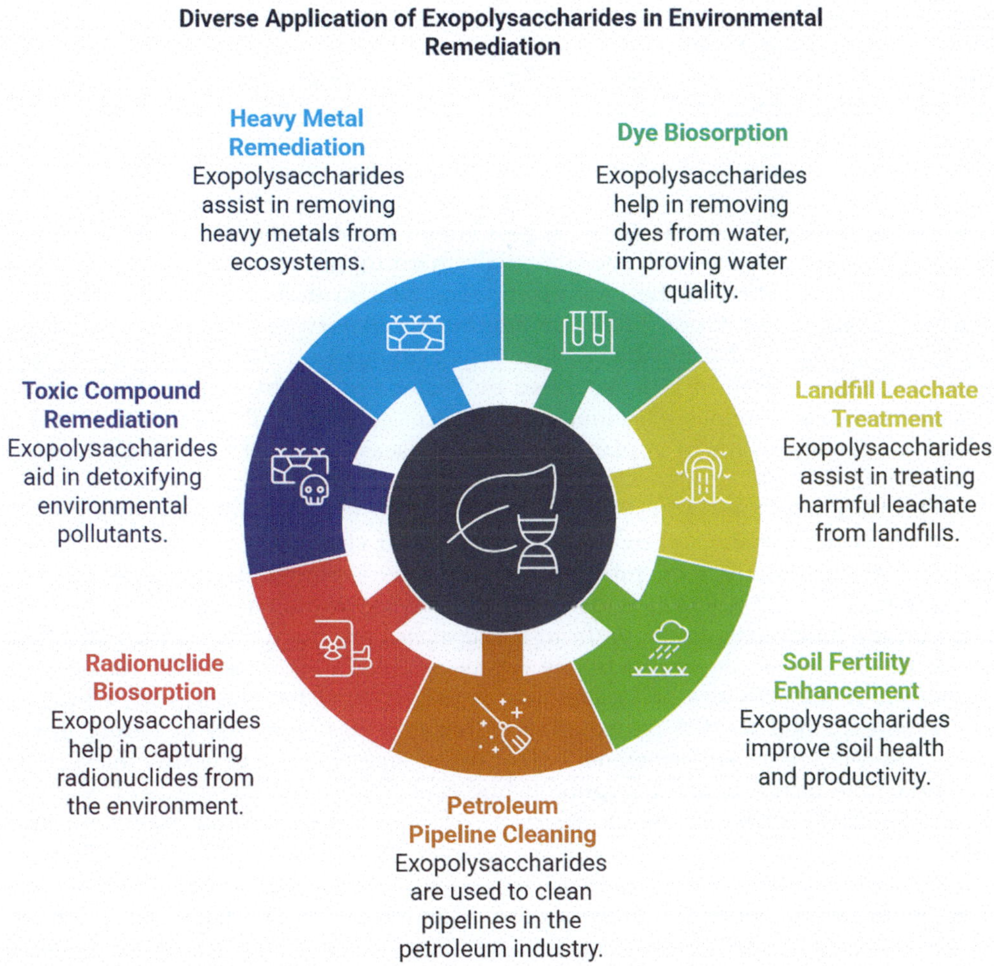

FIGURE 11.1 This diagram shows various application of EPS in the leaching of heavy metal, dye biosorption, landfill leachate, soil fertility enhancement, petroleum pipeline cleaning, radionuclide biosorption and toxic compound remediation indicating their diverse roles in environment clean-up. *EPS*, Exopolysaccharides.

chain arrangements components (Poli et al., 2011). The intricate structure of EPS complicates their characterization using a singular approach.

11.3.1 Bioactivity of exopolysaccharides

EPS are bioproducts that are naturally occurring, nontoxic, and demonstrate a diverse biological activity. To achieve environmental sustainability, these natural biopolymers are very transformational. Overall, the Alg operon encodes thirteen proteins in *P. aeruginosa* that are directly related to the alginate manufacturing pathway (excluding for AlgC) (Franklin et al., 1994). There are three key enzymes that are involved in the entire procedure of producing alginate precursors. These enzymes include phosphomannomutase (Alg C), guanosine 5′-diphospho-D-mannose pyrophosphorylase (Alg A), and GDP-mannose dehydrogenase (Alg D). These enzymes facilitate the catalysis of fructose-6-phosphate into GDP-mannuronic acid through four stages: Primarily, alginate is a synthetic polymer that is pure, environmentally sustainable, and eco-friendly, making it a chemical-free resource in order to the preparation of bioformulations, including biopesticides, biofertilizers, biostabilizers, bioadhesives, biosorbents, and bioplastics. Recent research conducted by (Tewari and Sharma, 2020) has focused on the function of exopolysaccharide (EPS) generated from microbes as a biostimulant. This biostimulant serves as an alternative to potentially harmful agrochemicals, hence preserving the survival and production of the vital pulse crop arhar in India's neglected salty soils. It was proven by (Fatima and Arora, 2021a), that the biostimulatory performance of *Pseudomonas*-derived EPS is similar, and their effectiveness was evaluated on the oilseed crop known as sunflower. The introduction of EPS generated by bacteria, either by itself or in a mixture of *Pseudomonas* cells, is a sustainable technique for the ecological rehabilitation of salinized land, which leads to the improvement in plant output and disease control (Tewari and Arora, 2014; Fatima and Arora, 2021b).

Diverse bacteria have been professionally utilized for the extraction of different EPS, including gellan, xanthan, alginate, and dextran, owing to their sticky and bioadhesive properties. These microbial compounds may present themselves as a powerful alternative to harmful chemical adhesives in the future (Freitas et al., 2011). Because they are recyclable and do not cause cancer, the bioemulsifiers made from *Pseudomonas* are more environmentally friendly than synthetic emulsifiers like heptadecane (Caruso et al., 2018). A notable characteristic of EPS formed from the activities of *Pseudomonas* is its potential for biosorption, which might potentially be exploited for the remediation of harmful heavy metals in nature. Microbe-mediated biosorption is an environmentally friendly strategy helpful to ensure the sustainable future compared to using chemical sorption (Chug et al., 2021). Several species of *Pseudomonas* have been identified as bacteria that are capable of sequestering harmful pollutants such as zinc, lead, mercury, and chromium from wastewater (Upadhyay et al., 2017; Chug et al., 2021). The multifunctional qualities of *Pseudomonas's* EPS render it a formidable option for the advancement of next-generation bioformulations. Biological products based on EPS have the potential to completely transform and create a new standard for environmentally responsible environmental sustainability.

11.3.2 Microbial adhesion activity

There is a possibility that one may research the significance of bacterial cellulose in order to gain an understanding of the relevance of cellulose in connection to the adhesion quality of EPS. In nature, bacterial cellulose serves as a sort of molecular glue to maintain contacts (Augimeri et al., 2015). For example, during symbiosis, bacterial cellulose facilitates the adherence of bacteria from the species rhizobium to leguminous plants, whereas the infection of agrobacterium and salmonella aids in plant colonization and offers defense against rivals (Fratty et al., 2022). The microbial adhesion to hydrocarbon (MATH) test is a criterion employed to assess the surface adhesion capability of environmental microbes. A test microbe is considered hydrophobic if its result is greater than 70%, hydrophilic if its result is less than 30%, and hydrophobic if its result is between 30% and 70% (Kwaszewska et al., 2006; Abasolo-Pacheco et al., 2016). Microorganisms are thought to have hydrophilic MATH values, which are considered to contribute to an enhancement in surface adhesion. Hydrophilic and hydrophobic traits may be found on surfaces, and microbes that possess these qualities can also be found on surfaces. It is possible for the surface chemical characteristics to change from one strain to the next. Conversely, the MATH values acquired for the examined microorganisms are predominantly in agreement with the findings documented in the writings for the corresponding species (Li and McLandsborough, 1999; Hamadi and Latrache, 2008). Due to the unavailability of MATH values for *L. garvieae* and *B. toyonensis* in the literature, their MATH values were compared with those of *L. lactis and B. subtilis*. Furthermore, in the published scientific report, the MATH value for *L. lactis* was reported as 34.0% (Marín et al., 1997), whereas for *B. subtilis* it was documented as 28.0% (Abasolo-Pacheco et al., 2016). Microorganisms' capacity to stick on the surfaces is inversely related to absolute zeta potential (Li and Logan, 2004). *P. aeruginosa and E. coli* had the absolute zeta potential values that were the lowest among the isolates, whereas *L. garvieae (M171.1.1. and M1713.1.2)* and *S. epidermidis* had the greatest values. The absolute zeta potential values (surface) of *L. garvieae* and *S. epidermidis* were the greatest, while those of *P. aeruginosa*, along with *E. coli*, were the lowest.

11.4 Utilization of exopolysaccharides as a cleaning agent in environmental contexts

11.4.1 Treatment of the dumpsite leachate

Over the last several decades, there has been a significant growth in population, urbanization, and industry, which has led to the buildup of municipal solid waste (MSW), which poses a significant hazard. Solid waste from municipal sites is disposed of at trash dumps due to the continuous input of refuse, which raises the demand for landfill space and, consequently, the costs of landfill treatment. The quantity of trash that is created can be reduced by the reuse of biological debris that is collected from municipal solid rubbish. Consequently, the most effective, substantial, and proficient approach is composting (Gutiérrez-Miceli et al., 2017). Nevertheless, the removal of water from the garbage produces a substantial quantity of the breakdown leachate during the stage of composting. Decomposing discharge comprises several hazardous compounds, thereby rendering it a considerable pollutant that poses risks to human well-being (Naveen et al., 2017). Various approaches exist for the proper management of digesting leachates, each with distinct constraints. The flocculation-coagulation procedure is a straightforward, economical, and efficient technique for eliminating organic substances along with leachate containing heavy metals. The chemical coagulants that are employed in the coagulation and flocculation methodology present significant risks to their own health as well as to the health of the environment. EPS concern with bacteria is being researched as a possible substitute in results (Guo et al., 2017). In addition, this procedure is utilized for the remediation of leachates from dumpsites that have been stabilized and aged (Renou et al., 2008). Early dumping leachate, which is abundant in organic materials, serves as a substrate for better growth of bacteria that produce EPS (More et al., 2014). A reduction of 45% in the COD characteristics of dumpsite leachate is brought about by EPS belonging to *Rhizomonas sp.* At a dosage of 20 mg/L and a pH =7–7.5, EPS achieves about 85% more efficient in eliminating humic acid so far. Furthermore, EPS demonstrates greater efficacy for the elimination of humic acid than that of synthetic coagulants with enhanced capacity. It is possible to achieve an elimination efficiency of 20% between the optimal and minimal concentrations. In COD elimination, EPS at 50 mg/L yields results comparable to alum at 500 mg/L. Furthermore, EPS as bioflocculant facilitates optimal results at the pH levels of leachates, thereby negating the necessity for any pH adjustments, which is shown to be more advantageous (Zouboulis et al., 2004). The principal issues associated with trash dumps are the intrusion of leachates as well as the contamination of groundwater. The EPS manufacturer can resolve this issue, as EPS has critical features that obstruct fluid supply, reduce the diameter of pores to restrict fluid movement, and diminish hydraulic conductive properties. EPS is an ideal choice for a prophylactic strategy against groundwater pollution as it possesses chemical features that are completely unique (More et al., 2014). The environmental utilization of EPS is illustrated in Fig. 11.1, encompassing the elimination of heavy metals, dye biosorption, landfill leachate, soil fertility enhancement, petroleum pipeline cleaning, radionuclide biosorption, and toxic compound remediation, thereby highlighting their multifaceted role in environmental cleanup.

11.4.2 Biosorption of dye of various effluents

Artificial colorants are widely employed for coloring products in cosmetics, textiles, cardboard and pulp, printing, pharmaceuticals, leather processing, plastics, along with processing of food. The discharges by such enterprises are significantly enriched with hazardous substances, heavy metals, and constituents of artificial dyes. When this effluent is discharged into water bodies without being treated, it would have a negative influence not only on people but also on aquatic species and the surroundings (Patel, 2016). Biosorption has emerged as a promising methodology in order to remove dyes and neutralize sewage discharges (Solís et al., 2012). Extracellular polymeric substances associated with microbes possess properties that promote adsorption and may remove both inorganic and organic substances from effluent (Zhang et al., 2009). It is because of the existence of a number of chemical groups, including phenolic, carboxyl, hydroxyl, phosphoric, and sulfhydryl in EPS, which have the potential of having high biosorption. These chemical groups are liable to favorable interactions that occur between the colorants and EPS (Zhang et al., 2009). When it comes to the decolorization process, the efficacy concern with EPS is affected by a number of different aspects. The efficacy of decolorization is affected by what kind of dye, heat, structural composition, pH, ionic strength, quantity of EPS, as well as duration of contact (Buthelezi et al., 2012). The EPS generated by *Ochrobactrum sp.* demonstrate significant potential, attaining a clearance effectiveness of 89.4% toward Remazol blue color after 72 hours of incubation kept at 30°C (Kiliç and Dönmez, 2012). Additionally, the EPS that are formed by *Acinetobacter baumannii* are effective in the neutralization of the activity of dyes. Following a 48-hour decolorization experiment, a decolorization effectiveness of 98.62% (Congo red dye) was found using the EPS of *Acinetobacter baumannii* (Li et al., 2015b). The bioflocculant synthesized by *Alteromonas sp.* exhibits dye removal efficiencies of 98.5%, 97.9%, and 72.3% for Congo red, straight black, and methylene blue, respectively (Chen et al., 2017). The EPS manufactured by *Paenibacillus elgii* exhibits a moderate potential for decolorizing cationic dyes, achieving the rate of decolorization by 65% for methylene blue and 72% for Red X-GRL (Li et al., 2013). Under ideal circumstances, the EPS that is formed by *Klebsiella sp.* has a clearance efficacy of 67.82% for sulfamethoxazole in the solution. The elimination efficiency in household wastewater decreases to 53.27% (Xing et al., 2013). Research indicates that EPS generated by bacteria, including *Bacillus, Exiguobacterium, Klebsiella, Pseudomonas,* and *Staphylococcus*, are highly effective in the elimination of colors from wastewater. EPS successfully eradicates hues such as fawn dye, whale dye, medi-blue, and mixed dyes, demonstrating bleaching efficiencies between 20% and 99.9% (Zhang et al., 2009). The efficiency with which the bacteria work together to remove dye is far higher than that of the bacteria working for themselves. The EPS formed due to the collective effort of *Staphylococcus sp.* and *Pseudomonas sp.* has potential uses in the neutralization and elimination of dyeing wastewater along with indigotin printing, achieving a decolorization efficacy of 86.5% (Zhang et al., 2007). *Bacillus cereus* B-11, grown on wastewater from molasses, generates EPS, a potent flocculating agent to eliminate reactive colors from aqueous medium. The efficacy of decolorization potential demonstrated by 25 mL of EPS for light-yellow K-4G and turquoise blue KN-G is 97.5% and 94.7%, respectively, in a 500 mL solution containing 100 mg/L dye (Mao et al., 2011). Consequently, EPS reveals its capability for the cleanup of wastewater and dye-laden fluids.

11.4.3 Remediation of heavy metal in aquatic bodies

The burning of fossil fuels, mining, and metallurgical operations are all examples of industrial activities that contribute to the introduction of metals into terrestrial ecosystems. The heavy metals, which are both poisonous and carcinogenic, pose a significant risk to the health of humans and aquatic species that inhabit it (Pal and Paul, 2008). Consequently, there is a significant amount of research being conducted on eliminating or cleaning up heavy metals. The predominant strategy employed for the retention of metal in the metal remediation process is the utilization of EPS derived from bacteria. The usage of biopolymers is advantageous, practicable, accessible, and safe, while their nonliving characteristics alleviate worries about pathogenicity (Diels et al., 2002). Since EPS contains a number of diverse structural aids, it provides locations for attachment to metals during the biosorption process (More et al., 2014). Furthermore, EPS demonstrates exceptional metal-binding capabilities with varying specificity. Various experimental parameters affect metal uptake, including pH, temperature, metal concentration, EPS concentration, stirring rate, and charge density (Salehizadeh and Shojaosadati, 2003). EPS derived from *Bacillus subtilis,* along with *Azotobacter beijreinckii* demonstrated chromium removal rates of 26% and 48%, respectively, from a water-based solution with a starting quantity of 10 ppm (Chug et al., 2016). Recent investigations indicate that EPS derived from bacteria *Bacillus mucilaginosus* exhibited significant removal of iron ions from aqueous solutions under optimal circumstances. The ideal biosorption conditions include a pH of 5, an ion concentration of 200–800 mg/L, and an EPS with 60 mg/L, resulting in an absorption rate of 90% (Yang et al., 2017). The recent study demonstrates that EPS generated by *Bacillus licheniformis* strain KX657843, extracted from the intestines of earthworms, serves as an excellent agent for the elimination of heavy metals. pH values are necessary for the metal cleanup process mediated by EPS. The EPS displayed a metal reduction performance of 54% and 42% at pH 5, which significantly rose to 86% and 81% at pH 8 for the efficient elimination of Cu (II) and Zn (II), respectively. This was accomplished throughout the process of removing metals from the solution. The metal elimination capacity of 94.8% for the specified metals is noted at an EPS concentration of 100 mg/L (Biswas et al., 2020). Electrostatic force of interactions often confers metal-binding capabilities to the carbohydrate and protein segment located in EPS, enabling their association with cations. Chromium negatively impacts human health; hence, pretreatment is essential prior to dispose in to environmental body. When it comes to the elimination of Cr (VI), the EPS that was derived from *Pseudochrobactrum saccharolyticum* LY10 displays an increased level of effectiveness. An examination of both the intracellular and extracellular fractions is carried out in order to ascertain the degree of decrease in Cr (VI). During the 16-hour experiment, the maximum lowering rate of Cr (VI) seen in EPS was 81.5%, whereas the minimum reduction rate recognized in the cell was 30.1%. The hierarchy of reduction capacity among various cellular

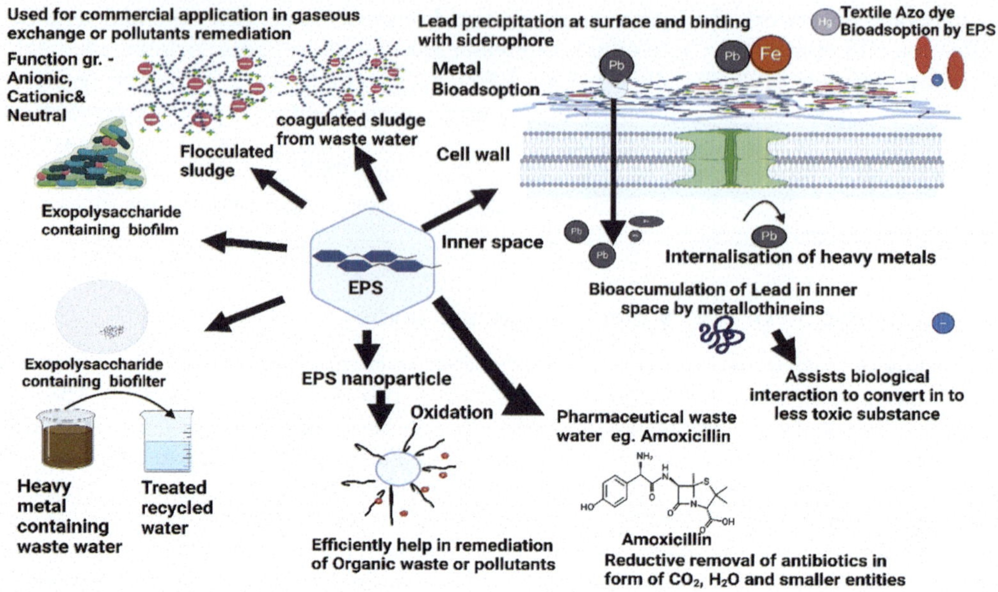

FIGURE 11.2 This flow chart describes the mechanism of action of exopolysaccharide in bioaccumulation, bioadsoption, bioleaching of heavy metals including Pb, Cr, Fe. Azo dye from textile to cleaning of waste water, along with dumpsite. This diagram also focusses on the enhanced action of remediation process by EPS based nanoparticle. *EPS*, Exopolysaccharides.

fractions is as follows: EPS (81.5%) > soluble cell fraction (69.2%) > cell membrane (49.3%) > cell wall (30.1%) (Long et al., 2019). Moreover, EPS obtained by bacterial consortia demonstrate significant efficacy in heavy metal cleanup than that of pure. The EPS manufactured utilizing bacterial populations from hydrocarbon-enriched water exhibit absorption rates of 87.12% for cadmium (Cd), 19.82% for zinc (Zn), and 37.64% for copper (Cu) inside the matrix (Martins et al., 2008). The biopolymer-based bioflocculant synthesized by the combination of *R. radiobacter* F2 and *B. sphaeicus F6* bacteria demonstrates substantial cadmium absorption in liquid medium within 60 minutes at its optimal pH of 7 (Wang et al., 2013). A number of metals have been successfully removed from aqueous solutions using EPS from gram negative bacterial consortia, with elimination performance of 77.15% (Zn), 78.18% (Pb), 74.48% (Cr), 66.63% (Ni), 71% (Cu), 72.71% (Cd), and 76.12% (Co) (Fig. 11.2) (Wang et al., 2018). The bio-elimination capacity for Ni, Cu, and Cr by EPS derived from a blend culture of bacterial with starting heavy metal level of 50 mg/L. A significant disparity in metal reduction effectiveness was recognized to be between the molasses-containing solution and the control devoid of molasses. The clearance rates are 32%, 75.7%, and 51.1%, respectively, in media with molasses, which are considerably greater than those in control media (Kiliç et al., 2015).

11.4.4 Remediation of toxic compounds

Polycyclic aromatic hydrocarbons remain in soil throughout prolonged durations owing to their hydrophobic properties, thermodynamic robustness, and a propensity to assemble.

The proper execution of EPS facilitates the natural breakdown of PAH. The isolation of PAH from polluted soil using EPS synthesized by bacteria is thoroughly documented (Zhang et al., 2015). EPS substances released by bacteria are pivotal in facilitating robust cell adhesion and biofilm formation. The enzymatic degradation of PHE and PYR was accomplished utilizing the strains *Micrococcus sp. PHE9,* along with *Mycobacterium sp. NJS-P,* both with or without EPS. Environmental bacterial possessing EPS form potential biofilms to escape stress condition. Degradation of phenanthrene and pyrene is significantly correlated with the continued existence of EPS associated with biofilm, which demonstrates a substantial connection (Zhang et al., 2015). Nonpolar groups found in EPS, such as those found in proteins, aromatic lipids, and polysaccharides, generate a binding pose for PHE and PYR, leading to the accumulation of these molecules inside the EPS. Additionally, EPS performs as a good contribution of a sorbent, which increment PHE and PYR that is transferred from water to the EPS. EPS in biofilms serves to promote the biological breakdown of PHE and PYR by enhancing the bioavailability of these chemicals. This is accomplished by increasing the interfacial space that exists between the cells and the substrate. Biofilm biodegrades phenanthrene at degree of 35.18 mg L^{-1} per day along with pyrene at 5.08 mg L^{-1} per day. The reaction that occurs between phenanthrene and EPS is one that occurs spontaneously and generates heat. Phenanthrene interacts with the constituents of EPS contingent on hydrophobic interactions (Pan et al., 2010). Phenanthrene's mass transfer is improved by the polysaccharide-induced increase in its solubility in a large quantity of water. *Pseudomonas aeruginosa's* EPS demonstrates substantial binding of 25.1 g mL^{-1} phenanthrene and 14.9 g mL^{-1} pyrene (Mangwani et al., 2017). Notwithstanding this, each and every application of EPS for the elimination of PAH to be documented has only been carried out to perform on an experimental scale.

11.4.5 Biosorption of radionuclides produced during anthropogenic activity

Human activities can result in the elimination of radioactive substances in the surroundings. These chemicals can be harmful to the ecosystem. The accidental discharge of radionuclides in nature is mostly caused by the operations of metallurgical mining, nuclear reactor testing, and the emissions from nuclear reactors. Each of these activities is responsible for a significant portion of the problem. Furthermore, the poisonous effect of radioactive materials has promised to have serious harmful effects on individuals as well as other organisms that are living in nature. The most effective technique for removing radionuclides from aquatic environments is biosorption. Biosorption is an economical, accurate, and effective method than traditional techniques. EPS has possessed a significant level of efficacy and may have a higher biosorption capacity than other materials when it comes to the disposal of harmful substances. Considering their high levels of radioactivity and toxicity, uranium are constituted among the most dangerous and intensively researched hazardous metals due to substantial radioactivity. The pH, heat, the amounts of EPS, the beginning amount of uranium, and incubation duration significantly influence metal absorption. EPS obtained from *Pseudomonas* bacteria exhibit uranium buildup, quantified at 96 μg/mg. The quantity of uranium to be absorbed by using EPS is contingent upon the pH, while remaining unaffected by variations in temperature. Uranium aggregation diminishes at pH levels below 5 (Marqués et al., 1991). EPS formed of anaerobic induced sludge demonstrates substantial uranium absorption from wastewater. The efficacy of uranium removal mediated by EPS is 98.3% at pH 6, with a uranium quantity of 20 mg/L and an EPS concentration of 56.1 g/L; nevertheless, the highest adsorption efficiency of uranium is at 0.891 mg/g (Li et al., 2015b). At an ideal pH of six and under other favorable conditions, EPS that is generated from aerobic sewage sludge has a removal of uranium efficacy of 93%. The maximum capacity for adsorption of uranium using EPS is 333.3 mg/g as per the Langmuir fitting standard. The immobilization of uranium (VI), followed by sorption-reduction processes, is demonstrated by biofilms that were generated from EPS generated by *Shewanella sp*. HRCR-1 that were acquired (Cao et al., 2011). The EPS of *P. aeruginosa* have the potential to be function as a biosorbent toward the leaching of uranium (Table 11.1). It has been discovered that the highest level of uranium acquisition by EPS is 985 mg U g^{-1} as per the investigation (Kazy et al., 2008). Iron and thorium create a considerable barrier that prevents EPS from absorbing uranium. EPS extracted and stored from the bacterium *Pseudomonas sp.* are integral to the redox mechanism to recover plutonium, leading to an influence in its environmental mobility. Cells possessing intact EPS exhibit superior sorption potential of plutonium (V) efficiency compared to those lacking EPS (Boggs et al., 2016). *Ochrobactrum intermedium* exhibits significant thorium tolerance. *O. intermedium* is a prospective biofilm producer, and the persistence of EPS does not impede thorium availability. The FT-IR spectroscopic study was performed on the EPS that was recovered from the EPS medium that was enriched with thorium. The FT-IR spectrum exhibits increased absorbance at 3000–3200, 1600–1700, and 1000–1200 cm, specifically signifying the existence of hydroxyl, amino, and carboxyl groups in EPS. The contribution of EPS generated by utilizing *O. intermedium* in thorium biosorption is therefore validated as a result of this phenomenon (Shukla et al., 2020). Furthermore, research has shown that metallic minerals such as uranium and arsenate, along with biological contaminants like dibenzothiophene present in effluent, could be treated effectively using electroactive bacteria such as *Shewanella oneidensis*. This approach leads to a reduction in the content of U(VI), particularly when combined with *Pseudomonas putida*. To improve electron transport for power production from waste materials, the redox properties of electro bacterial extracellular polymeric compounds that contain heme-mediated proteins are advantageous (Li et al., 2016. This recognition positions EPS as a promising biosorbent for addressing the challenges associated with radioactive waste treatment.

11.4.6 Cleaning pipelines in the petroleum sector

The rheological characteristics of EPS released by *Haloferax mediterranei,* belonging to a halophilic archaebacterium, exhibited pseudoplastic behavior and high apparent viscosity at minimal concentrations, with this viscosity demonstrating significant resistance to severe pH, temperature, and salinity conditions. The natural bacteria *E. cloacae* are utilized to generate benign by-products such as gases (CO_2, CH_4, and H_2), Acetic and propionic acids, along with acetone, ethanol, and butanone as solvents, which enhance oil flow by diminishing its viscosity. In the results, a bigger quantity of oil is recovered from the location (Chien et al., 2017). These characteristics make this EPS appropriate for applications requiring a highly durable thickening agent, such as increased oil recovery (Moppert et al., 2009). With their environmentally friendly nature, microbial extracellular polysaccharides (EPSs) are receiving much attention because they operate as biosurfactants and bioemulsifiers (Sun et al., 2011). Consequently, the possibility of a strategy for modifying the functional activity and structural makeup of microbial habitats in petroleum reserves is known as microbial enhanced oil recovery (MEOR), and details are mentioned in Table 11.1. and Fig. 11.1. The subfield of biotechnology makes use of microbes to enhance the quantity of oil extracted from wells that are already in operation, hence boosting the petroleum output from an oil reservoir (Sun et al., 2011).

TABLE 11.1 This table shows summary of the list of the action of EPS from various microorganism which enable the cleaning up of environmental pollutants.

Application	Process	Biological action of EPS	Bacteria involve
Water treatment	Bioflocculant, biosorbent	EPS can facilitate the elimination of suspended particles and natural organic materials. EPS may serve as flocculant when combined with multivalent cations (Ca2+, Mg2+, Al3+, Fe3+). Concerns regarding safety arise from its extensive utilization.	*Pseudochrobactrum saccharolyticum, B. sphaeicus F6* (Wang et al., 2013, 2018).
Wastewater flocculation and settling	Bioflocculant, biosorbent	Usage in the elimination of suspended particles, removal of organic substance. EPS can decrease reliance on chemical polymers by integrating bioflocculants with traditional chemical polymers (PAC, PAM). The utilization of EPS in wastewater cleanup is highly appealing.	*Alteromonas sp., Acinetobacter baumannii, Paenibacillus elgii* (Li et al., 2013, 2015a).
Dye removal from natural sources	Bioflocculant, biosorbent	Eliminate dye from wastewater, Dye buildup can be destroyed by microorganisms. EPS in dye elimination is highly encouraging.	*Pseudomonas spp., Citrobacter sp., Aeromonas hydrophila, Streptomyces bacillaris*
Sludge dewatering	Thickener, bioflocculant	The utilization of EPS for sludge thickening and dehydrating necessitates the incorporation of multivalent cations (Ca2+, Mg2+, Al3+, Fe3+) in conjunction with EPS. The utilization of EPS on elevated amounts of sludge solids has not been documented.	*Pseudochrobactrum saccharolyticum, B. sphaeicus F6* (Wang et al., 2013, 2018).
Metal removal from waste effluents	Bioabsorbent, bioleaching, bioflocculant	Elimination or extraction of heavy and hazardous metals from waste. Used for the removal or recovery of metals from waste effluents is highly appealing and promising.	*Pseudochrobactrum saccharolyticum* (Wang et al., 2018).
Removal of hazardous compounds	Bioabsorbent, biodegradation, Bioflocculant	EPS for the elimination of hazardous organic materials from sludge, and soils may represent a viable technique in future.	*Micrococcus sp. PHE9, T. tenuis* and *Mycobacterium sp. NJS-P* (Zhang et al., 2015).
Management of landfill leachate	Bioleaching, biodegradation	Leaching of heavy metals and hazardous substance, and it can significantly contribute to the remediation of landfill leachate.	*Pseudochrobactrum saccharolyticum, B. sphaeicus F6* (Wang et al., 2013, 2018).

(*continued on next page*)

TABLE 11.1 This table shows summary of the list of the action of EPS from various microorganism which enable the cleaning up of environmental pollutants—cont'd

Application	Process	Biological action of EPS	Bacteria involve
Soil cleaning and reclamation	Biosorbent, biodegradation, Bioaugmentation	EPS in enhancing soil adhesion, mitigating soil erosion, and decreasing the migration of metals from soils to aquatic environments.	*T. tenuis*, *Rhizobium pusense*, *Bacillus megaterium*, and *Lysinibacillus sphaericus* (Khumairah et al., 2022; Shabaan et al., 2022).
Cleaning the pipeline in petroleum sector	Bio-surfactants and bioemulsifiers	Microbial enhanced oil recovery	*Haloferax mediterranei*, *Enterobacter cloacae*, *O. intermedium* (Sun et al., 2011; Chien et al., 2017).
Removal of radioactive substances from environment	Biosorption of radionuclides	EPS used for uranium removal effectiveness	*Pseudomonas sp.* (Marqués et al., 1991), *O. intermedium*, *Shewanella oneidensis* (Li et al., 2016).

EPS, exopolysaccharides.

The petroleum sector employs xanthan gum derived from bacteria in diverse usages, including drilling for crude oil, the cracking process, and the maintenance of pipelines. Its inclusion in the drilling solution is a beneficial reason of exceptional ability to withstand salt and resist heat degradation (Chen and Stewart, 2002). Xanthan gum exerts a notable influence on sludge treatment and tertiary oil recovery while also playing a crucial role in enhancing drilling speed and inhibiting thawing. The remarkable ability to improve viscosity, thickening, and withstand salt and petroleum, xanthan gum serves as an exceptional additive for oilfield exploration earth, demonstrating a distinct technological advantage in oilfield development compared to polyacrylamide and carboxymethylcellulose, as well as a diverse kind of phyto-derived polysaccharides. Therefore, the natural bacteria that are chosen to make bio-surfactant as specialized forms of EPSs are integrated into drilling operations in order to get rid of benign by-products, such as sticky organic molecules or gases, that are helpful in the extraction of oil. This leads to a greater volume of oil that might be recovered. In the field of microbial enhanced oil recovery, genetically engineered *Enterobacter cloacae* are being exploited in an efficient manner (Sun et al., 2011).

11.4.7 Exopolysaccharides used to enhance the fertility and productivity of soil

When organic materials are mixed into the soil, the soil's enzymatic and microbiological activity increases. Through EPS along with other organic biocompounds, microorganisms have contributed in aggregation and durability of soil. Fungal hyphae combine soil components, forming structures that enhance aeration, water retention, and root penetration, so increasing soil productivity. Soil-based enzymes are mostly believed to originate from microorganisms, with some clearly linked to living cells; nonetheless, several enzymes can retain catalytic activity within cell detritus, in soil solution, or when complexed with clay or organic colloids.

Unlike soils that are heavily exploited, it is anticipated that soils that are carefully managed to improve quality, such as by implementing organic changes, crop rotations, and limited tillage, would exhibit a higher level of biological activity. This difference ought to be represented in an increased capacity to generate and secrete enzymes, along with an expanded capacity to stabilize and preserve extracellular enzymes through the development of soil-based enzyme complexes (De Caire et al., 2000). The research has confirmed that the implementation of EPS to the soil matrix may enhance soil microbial enzymatic activity, which further results in the higher productivity of crops due to efficient soil microbial interaction. A larger increase in enzymatic activity was seen as a consequence of EPS introduction, namely *T. tenuis*, in comparison to that of biomass elements. The disparity in increments resulting from EPS was partially because of the requirement for biomass to experience cellular lysis before the accidental discharge of intracellular enzymes, since the findings demonstrate minimal removal of extracellular enzymes. The EPS mixture consisted of proteins, some of which could demonstrate exoenzymatic behavior, along with additional compounds that may serve as stimulant compounds, encouraging the generation of exoenzymes by native soil microbes.

A handful of bacterial organisms, including *Rhizobium pusense, Bacillus megaterium*, and *Lysinibacillus sphaericus*, were identified as the most successful and promising cultures. This occurred because these cultures produced a significant amount of EPS, which were the sole material that was evaluated. Moreover, these cultures demonstrated superior performance in IAA and GA3 synthesis, phosphate solubilization, and potassium release, yielding corroborative results. Consequently, these cultures have been suggested for the augmented manufacturing of EPS and the way of efficient extraction, aimed at promoting sustainable plant development and agricultural yield while maintaining consistent productivity (Table 11.2).

With the activation of various enzymes to degrade the soil organic components, utilizing *rhizospheric* bacteria in the cultivation of shallots has the capacity to improve both the yield and quality of plants through the respective hormones that support their growth. *Rhizospheric* bacteria provide antagonistic effects against pathogens and enhance plant resilience to illnesses and environmental stressors when it comes to controlling pests and plant diseases. Biofertilizers will surely be more ecologically sustainable, therefore diminishing the reliance on detrimental chemical fertilizers and pesticides that may compromise environmental quality. EPS formed by microorganisms has the ability to boost the uptake of nitrogen, phosphorus, and potassium from the deepest layer of soil. PGPR in certain plants that are undergoing salt stress has been demonstrated to reduce the levels of sodium ions due to the synthesis and release of EPS by beneficial bacteria (Shabaan et al., 2022). EPS serve as immunomodulators that markedly enhance soil porosity, aggregation, and sodium ion retention. Bacteria osmotically adjust to salty conditions by intracellularly controlling ion density by the expulsion of Na^+/K^+ ions via antiporters or K^+/Na^+ ion exchangers (Khumairah et al., 2022). The persistence of rhizosphere bacteria enhances environmental quality by enhancing soil architecture and facilitating bioremediation and the breakdown of organic materials in the land (Ayu Lutfiani Amalia et al., 2025). Future research should concentrate on developing more efficient manufacturing procedures of these rhizosphere microbes, guaranteeing their sustainability and survivability throughout expanded period of preservation and processing. To preserve the proper functioning of the bacteria and guarantee their continued existing biological activity along with effectiveness when applied to crops, this modification is indeed essential.

TABLE 11.2 This table describes introduction of EPS to elevate the productivity of the crops.

Microbes sps.	Mode of action	Action	References
Steptomyces sp.	N-fixers, P, K, Fe, and Al solubilizers	Improved the development of shallot bulbs, and increased absorption of N, P, and K.	Haitami et al. (2024)
Pseudomonas putida	ACC deaminase producer	Root dry mass, leaf area, plant growth rate, total dry mass, plant height, bulb oven-dried mass, bulb size, harvest index, bulb fresh mass, bulb weight reduction, and sun-dried bulb mass.	Ahmad et al. (2024)
Rhodobacter spp. and *Rhodopseudo monas spp.*	Phosphate solubilizer and N- Fixer	Root size, concentration of chlorophyll, nitrate reductase activity, fresh and dry mass of roots and shoots, bulb count per clump, fresh and dry mass of bulbs per clump, and total productivity.	Ayu Lutfiani Amalia et al. (2025)
Streptomyces sp., *Bacillus sp.*, and *Corynebacterium sp.*	IAA and N fixers	Augmented root biomass and chlorophyll concentrations.	Haitami et al. (2024)

EPS, exopolysaccharides.

11.5 Discussion and conclusions

This chapter indicates that bacterial EPS may be an optimal polymer for use in the ecosystem. A comprehensive study is necessary before its implementation in the real world. The distinctive characteristics of EPS render it a valuable asset for cleanup techniques, necessitating more elucidation. The large volume of data about environmental uses of EPS is significantly restricted in comparison to the data accessible on ordinary polymers.

Because EPS possesses critical qualities that impede liquid flow, lower pore to restrict liquid migration, and lessen hydraulic conductivity, it is possible that the contribution of EPS formed by a diverse group of bacteria might be an efficient solution to the problem of dumpsite rehabilitation. All of these properties are crucial. Because of its relatively lower concentration and increased ability for humic acid elimination by 85%, the contribution of EPS of bacteria is more effective in the coagulation of chemicals. The Effect of EPS of bacteria could not be ignored, as it interferes with the decolorization efficacy of EPS due to the presence of various properties. The efficacy of decolorization is affected by the kind of dye, temperature, structural composition, pH, ionic strength, quantity of EPS, and duration of contact. Further analysis on the efficacy of dye removal efficiency/capacity of EPS generated by use of *Alteromonas sp.* showed 98.5%, and 97.9%, to Congo red, and straight black, respectively (Fig. 11.2).

The EPS secreted by bacterial consortia have considerable effectiveness in the leaching/cleanup of heavy metals. EPS of bacterial from hydrocarbon-enriched water exhibit absorption rates of 87.12% for cadmium (Cd), 19.82% for zinc (Zn), and 37.64% for copper (Cu) inside the matrix. Additionally, EPS formed from anaerobically induced sludge for the absorption of uranium in wastewater. Further, uranium removal facilitated by EPS is 98.3%. The natural bacterium *Enterobacter cloacae* is employed to release nonpoisonous by-products, which improve the flow of oil due to decreasing its viscosity. Genetically modified bacteria such as *Enterobacter cloacae* are efficiently employed in microbial enhanced oil recovery (MEOR) in an effective manner.

As a result, there exists a limited comprehension of the relevant features of EPS, the diverse aspects of its production and recovery, as well as its possible uses in preventing pollution. Moreover, the utilization of EPS in the processing of sewage solids flocculation, drainage, decolorizing agents, and disinfection remains still under investigation. A thorough examination is essential before considering its potential application in field processes. Microbial polysaccharides have gained relevance in the past decades, and efforts have intensified to broaden the applications of EPS. As a result, a more in-depth comprehension is required in order to shed light on the mechanisms, pathways, and genetic control of *Pseudomonas sp.* and other bacterial EPS. Exploring novel EPS with biotechnological relevance across various model species will clarify how different biomolecules are stabilized within the EPS under extreme conditions.

11.6 Future prospectives of the study

The usage of bacterial EPS-stabilized nanoparticles in wastewater cleanup, noted for their biogenic characteristics, has led to significant advancements. The bioremediation of wastewater and industrial effluent are two applications that are expected to benefit from the commercial availability of biopolymeric/metal nanoparticles in the future. When considering the future, it would be advantageous to combine of EPS with contemporary sewage treatment technologies such as ultra- and nanofiltration, reverse-osmosis electrodialysis, microbial fuel cells, and magnetic nanoparticles. It is anticipated that a more comprehensive understanding of functional interactions of bacterial extracellular polymers will pave the way for the development of novel approaches to remarkable biomaterials to the greatest extent of their potential.

Credit statements

All the mentioned writers have made substantial contributions to the development and composition of this article.

Conflict of interest

Authors declare no conflicts of interest.

Ethical statement

This article adheres to ethical standards and does not include any experiments involving human volunteers conducted by any of the writers.

Acknowledgments

The authors would like to thank the people who have worked with the authors in the laboratory. The authors express gratitude to Prof. Manabendra Dutta Choudhury, Vice Chancellor of Rabindra Nath Tagore University, Hojai, Assam, and Dr. Partha Palit of the Department of Pharmaceutical Sciences, Assam University, Silchar, Assam, for their substantial assistance in the drafting and structuring of this chapter. The perspectives and beliefs articulated in this article belong to the writers.

References

Abasolo-Pacheco, F., et al., 2016. Isolation and use of beneficial microbiota from the digestive tract of lions-paw scallop *Nodipecten subnodosus* and winged pearl oyster *Pteria sterna* in oyster aquaculture. Aquac. Res. 47 (10), 3042–3051. https://doi.org/10.1111/ARE.12754.

Abdel-Aziz, S. (2012) Acidic pH-shock induces the production of an exopolysaccharide by the fungus *Mucor rouxii*: utilization of beet-molasses. Available from: https://www.academia.edu/64878518/Acidic_pH_Shock_Induces_the_Production_of_an_Exopolysaccharide_by_the_Fungus_Mucor_rouxii_Utilization_of_Beet_Molasses (Accessed: 12 February 2025).

Ahmad, S., et al., 2024. Foliar application of plant growth regulators enhances drought tolerance by modulating growth and biochemical responses in sugarcane varieties. J. Agron. Crop Sci. 210 (6), e12784. https://doi.org/10.1111/JAC.12784.

Augimeri, R.V., Varley, A.J., Strap, J.L., 2015. Establishing a role for bacterial cellulose in environmental interactions: lessons learned from diverse biofilm-producing proteobacteria. Front. Microbiol. 6, 1282. https://doi.org/10.3389/FMICB.2015.01282.

Ayu Lutfiani Amalia, D., Hasnah Qurata, S.A., Nur Ardian, M., 2025. Potential of rhizospheric bacteria for increasing growth and yield of shallots (*Allium ascalonicum* L.) in saline soil. BIO Web Conf. 158, 03026. https://doi.org/10.1051/bioconf/202515803026.

Balíková, K., et al., 2022. Role of exopolysaccharides of *Pseudomonas* in heavy metal removal and other remediation strategies. Polymers 14 (20). https://doi.org/10.3390/POLYM14204253.

Bhaskar, P.V., Bhosle, N.B., 2006. Bacterial extracellular polymeric substance (EPS): a carrier of heavy metals in the marine food-chain. Env. Int 32 (2), 191–198. https://doi.org/10.1016/J.ENVINT.2005.08.010.

Biswas, J.K., et al., 2020. Exploration of an extracellular polymeric substance from earthworm gut bacterium (*Bacillus licheniformis*) for bioflocculation and heavy metal removal potential. Appl. Sci. 10 (1), 349. https://doi.org/10.3390/APP10010349.

Boggs, M.A., et al., 2016. Interactions of plutonium with *Pseudomonas sp.* strain EPS-1W and its extracellular polymeric substances. Appl. Env. Microbiol. 82 (24), 7093–7101. https://doi.org/10.1128/AEM.02572-16/SUPPL_FILE/ZAM999117552SO1.PDF.

Buthelezi, S.P., Olaniran, A.O., Pillay, B., 2012. Textile dye removal from wastewater effluents using bioflocculants produced by indigenous bacterial isolates. Molecules 17 (12), 14260–14274. https://doi.org/10.3390/MOLECULES171214260.

Cao, B., et al., 2011. Contribution of extracellular polymeric substances from *Shewanella sp.* HRCR-1 biofilms to U(VI) immobilization. Env. Sci. Technol. 45 (13), 5483–5490. https://doi.org/10.1021/ES200095J.

Caruso, C., et al., 2018. Production and biotechnological potential of extracellular polymeric substances from sponge-associated antarctic bacteria. Appl. Env. Microbiol. 84 (4), e01624-17. https://doi.org/10.1128/AEM.01624-17.

Chen, X., Stewart, P.S., 2002. Role of electrostatic interactions in cohesion of bacterial biofilms. Appl. Microbiol. Biotechnol. 59 (6), 718–720. https://doi.org/10.1007/S00253-002-1044-2.

Chen, Z., et al., 2017. Characterization of a novel bioflocculant from a marine bacterium and its application in dye wastewater treatment. BMC Biotech. 17 (1), 1–11. https://doi.org/10.1186/S12896-017-0404-Z/FIGURES/6.

Chien, C.Y., et al., 2017. Synthesis and characterization of regioselectively substituted curdlan hetero esters via an unexpected acyl migration. Carbohydr. Polym. 155, 440–447. https://doi.org/10.1016/J.CARBPOL.2016.08.067.

Chug, R., et al., 2016. Optimization of extracellular polymeric substances production using *Azotobacter beijreinckii* and *Bacillus* subtilis and its application in chromium (VI) removal. Bioresour. Technol. 214, 604–608. https://doi.org/10.1016/J.BIORTECH.2016.05.010.

Chug, R., et al., 2021. Maximizing EPS production from *Pseudomonas aeruginosa* and its application in Cr and Ni sequestration. Biochem. Biophys. Rep. 26. https://doi.org/10.1016/J.BBREP.2021.100972.

De Caire, G.Z., et al., 2000. Short communication: changes in soil enzyme activities following additions of cyanobacterial biomass and exopolysaccharide. Soil Biol. Biochem. 32 (13), 1985–1987. https://doi.org/10.1016/S0038-0717(00)00174-7.

Diels, L., van der Lelie, N., Bastiaens, L., 2002. New developments in treatment of heavy metal contaminated soils. Rev. Env. Sci. Biotechnol. 1 (1), 75–82. https://doi.org/10.1023/A:1015188708612/METRICS.

Donlan, R.M., 2002. Biofilms: microbial life on surfaces. Emerg. Infect. Dis. 8 (9), 881–890. https://doi.org/10.3201/EID0809.020063.

Elsholz, A.K.W., Wacker, S.A., Losick, R., 2014. Self-regulation of exopolysaccharide production in *Bacillus* subtilis by a tyrosine kinase. Genes Dev. 28 (15), 1710–1720. https://doi.org/10.1101/GAD.246397.114.

Fatima, T., Arora, N.K., 2021. *Pseudomonas entomophila* PE3 and its exopolysaccharides as biostimulants for enhancing growth, yield and tolerance responses of sunflower under saline conditions. Microbiol. Res. 244. https://doi.org/10.1016/J.MICRES.2020.126671.

Franklin, M.J., et al., 1994. *Pseudomonas aeruginosa* AlgG is a polymer level alginate C5-mannuronan epimerase. J. Bacteriol. 176 (7), 1821–1830. https://doi.org/10.1128/JB.176.7.1821-1830.1994.

Fratty, I.S., et al., 2022. The activity of BcsZ of *Salmonella Typhimurium* and its role in *Salmonella*-plants interactions. Front. Cell. Infect. Microbiol. 12, 967796. https://doi.org/10.3389/FCIMB.2022.967796/FULL.

Freitas, F., Alves, V.D., Reis, M.A.M., 2011. Advances in bacterial exopolysaccharides: from production to biotechnological applications. Trends Biotechnol. 29 (8), 388–398. https://doi.org/10.1016/J.TIBTECH.2011.03.008.

Guo, J., et al., 2017. Enhanced anaerobically digested swine wastewater treatment by the composite of polyaluminum chloride (PAC) and *Bacillus megatherium* G106 derived EPS. Sci. Rep. 7 (1), 1–10. https://doi.org/10.1038/s41598 017 09011 0.

Gutiérrez-Miceli, F.A., et al., 2017. Vermicomposting leachate as liquid fertilizer for the cultivation of sugarcane (*Saccharum sp.*). J. Plant Nutr. 40 (1), 40–49. https://doi.org/10.1080/01904167.2016.1193610.

Haitami, A., et al., 2024. Yield response and nutrient uptake of shallots by giving ameliorants and actinobacteria in water saturated cultivation in tidal land. IOP Conf. Ser.: Earth Environ. Sci. 1359 (1), 012006. https://doi.org/10.1088/1755-1315/1359/1/012006.

Hamadi, F., Latrache, H., 2008. Comparison of contact angle measurement and microbial adhesion to solvents for assaying electron donor–electron acceptor (acid–base) properties of bacterial surface. Colloids Surf. B 65 (1), 134–139. https://doi.org/10.1016/J.COLSURFB.2008.03.010.

Kazy, S.K., Sar, P., D'Souza, S.F, 2008. Studies on Uranium removal by the extracellular polysaccharide of a *Pseudomonas aeruginosa* strain. Biorem. J 12 (2), 47–57. https://doi.org/10.1080/10889860802052870.

Khumairah, F.H., et al., 2022. Halotolerant plant growth-promoting rhizobacteria isolated from saline soil improve nitrogen fixation and alleviate salt stress in rice plants. Front. Microbiol. 13. https://doi.org/10.3389/FMICB.2022.905210.

Kiliç, N., Dönmez, G., 2012. Remazol Blue Removal and EPS production by *Pseudomonas aeruginosa* and *Ochrobactrum sp*. Pol. J. Environ. Stud. 21(1),123-128.

Kiliç, N.K., et al., 2015. EPS production and bioremoval of heavy metals by mixed and pure bacterial cultures isolated from *Ankara* stream. Water Sci. Technol. 72 (9), 1488–1494. https://doi.org/10.2166/WST.2015.365.

Knoshaug, E.P., Ahlgren, J.A., Trempy, J.E., 2007. Exopolysaccharide expression in *Lactococcus lactis* subsp. cremoris Ropy352: evidence for novel gene organization. Appl. Env. Microbiol. 73 (3), 897–905. https://doi.org/10.1128/AEM.01945-06.

Kwaszewska, A.K., Brewczyñska, A., Szewczyk, E.M., 2006. Hydrophobicity and biofilm formation of lipophilic skin corynebacteria. Pol. J. Microbiol. 55 (3), 189–193.

Li, B., Logan, B.E., 2004. Bacterial adhesion to glass and metal-oxide surfaces. Colloids Surf. B 36 (2), 81–90. https://doi.org/10.1016/J.COLSURFB.2004.05.006.

Li, H., et al., 2016. Biological production of welan gum. Phys. Sci. Rev. 1 (10). https://doi.org/10.1515/PSR-2016-0049/MACHINEREADABLECITATION/RIS.

Li, J., McLandsborough, L.A., 1999. The effects of the surface charge and hydrophobicity of *Escherichia coli* on its adhesion to beef muscle. Int. J. Food Microbiol 53 (2–3), 185–193. https://doi.org/10.1016/S0168-1605(99)00159-2.

Li, O., et al., 2013. Optimization and characterization of polysaccharide-based bioflocculant produced by *Paenibacillus elgii* B69 and its application in wastewater treatment. Bioresour. Technol. 134, 87–93. https://doi.org/10.1016/J.BIORTECH.2013.02.013.

Li, S., et al., 2015. Adsorption of U (VI) by anaerobic sludge extracellular polymeric substances (EPS) in wastewater. In: 2015 3rd International Conference on Advances in Energy and Environmental Science, pp. 803–807.
Li, X., et al., 2015. Applicability of the welch method for examining self-noise level parameters for broadband seismometers. Geod. Geodyn. 6 (3), 233–239. https://doi.org/10.1016/J.GEOG.2015.04.002.
Long, D., et al., 2019. Cr(VI) reduction by an extracellular polymeric substance (EPS) produced from a strain of *Pseudochrobactrum saccharolyticum*. 3 Biotech 9 (3), 1–9. https://doi.org/10.1007/S13205-019-1641-8/METRICS.
Mahapatra, S., Banerjee, D., 2013. Optimization of a bioactive exopolysaccharide production from endophytic *Fusarium solani* SD5. Carbohydr. Polym. 97 (2), 627–634. https://doi.org/10.1016/J.CARBPOL.2013.05.039.
Mangwani, N., Kumari, S., Das, S., 2017. Marine bacterial biofilms in bioremediation of polycyclic aromatic hydrocarbons (PAHs) under terrestrial condition in a soil microcosm. Pedosphere 27 (3), 548–558. https://doi.org/10.1016/S1002-0160(17)60350-3.
Mao, Y., et al., 2011. Production of a novel biopolymer by culture of *Bacillus cereus* B-11 using molasses wastewater and its use for dye removal. Adv. Mater. Res. 230–232, 1119–1122. https://doi.org/10.4028/WWW.SCIENTIFIC.NET/AMR.230-232.1119.
Marín, M.L., et al., 1997. Lactic acid bacteria: hydrophobicity and strength of attachment to meat surfaces. Lett. Appl. Microbiol. 24 (1), 14–18. https://doi.org/10.1046/J.1472-765X.1997.00334.X.
Marqués, A.M., et al., 1991. Uranium accumulation by *Pseudomonas sp.* EPS-5028. Appl. Microbiol. Biotechnol. 35 (3), 406–410. https://doi.org/10.1007/BF00172734/METRICS.
Martins, P.S.D.O., De Almeida, N.F., Leite, S.G.F., 2008. Application of a bacterial extracellular polymeric substance in heavy metal adsorption in a co-contaminated aqueous system. Braz. J. Microbiol. 39 (4), 780–786. https://doi.org/10.1590/S1517-83822008000400034.
Mijakovic, I., Grangeasse, C., Turgay, K., 2016. Exploring the diversity of protein modifications: special bacterial phosphorylation systems. FEMS Microbiol. Rev. 40 (3), 398–417. https://doi.org/10.1093/FEMSRE/FUW003.
Moghannem, S.A.M., et al., 2018. Exopolysaccharide production from *Bacillus velezensis* KY471306 using statistical experimental design. Braz. J. Microbiol.: [Publication Braz. Soc. Microbiol. 49 (3), 452–462. https://doi.org/10.1016/J.BJM.2017.05.012.
Mohamed, S.S., et al., 2018. Characterization and applications of exopolysaccharide produced by marine *Bacillus altitudinis* MSH2014 from Ras Mohamed, Sinai, Egypt. Egypt. J. Basic Appl. Sci. 5 (3), 204–209. https://doi.org/10.1016/J.EJBAS.2018.05.009.
Moppert, X., et al., 2009. Investigations into the uptake of copper, iron and selenium by a highly sulphated bacterial exopolysaccharide isolated from microbial mats. J. Ind. Microbiol. Biotechnol. 36 (4), 599–604. https://doi.org/10.1007/S10295-009-0529-8.
More, T.T., et al., 2014. Extracellular polymeric substances of bacteria and their potential environmental applications. J. Env. Manage. 144, 1–25. https://doi.org/10.1016/J.JENVMAN.2014.05.010.
Naveen, B.P., et al., 2017. Physico-chemical and biological characterization of urban municipal landfill leachate. Env. Pollut. 220, 1–12. https://doi.org/10.1016/J.ENVPOL.2016.09.002.
Nichols, C.M., Bowman, J.P., Guezennec, J., 2005. Effects of incubation temperature on growth and production of exopolysaccharides by an antarctic sea ice bacterium grown in batch culture. Appl. Env. Microbiol. 71 (7), 3519–3523. https://doi.org/10.1128/AEM.71.7.3519-3523.2005.
Pal, A., Paul, A.K., 2008. Microbial extracellular polymeric substances: central elements in heavy metal bioremediation. Indian J. Microbiol. 48 (1), 49–64. https://doi.org/10.1007/S12088-008-0006-5.
Pan, X., Liu, J., Zhang, D., 2010. Binding of phenanthrene to extracellular polymeric substances (EPS) from aerobic activated sludge: a fluorescence study. Colloids Surf. B 80 (1), 103–106. https://doi.org/10.1016/J.COLSURFB.2010.05.002.
Patel, S. (2016) Review on biosorption of dyes by fungi. IJIRSET, 5(1), 2347–6710.
Poli, A. et al. (2011) Synthesis, production, and biotechnological applications of exopolysaccharides and polyhydroxyalkanoates by archaea. Archaea, 2011, 13. https://doi.org/10.1155/2011/693253.
Reid, A.N., Whitfield, C., 2005. functional analysis of conserved gene products involved in assembly of Escherichia coli capsules and exopolysaccharides: evidence for molecular recognition between Wza and Wzc for colanic acid biosynthesis. J. Bacteriol. 187 (15), 5470–5481. https://doi.org/10.1128/JB.187.15.5470-5481.2005.
Renou, S., et al., 2008. Landfill leachate treatment: review and opportunity. J. Hazard. Mater. 150 (3), 468–493. https://doi.org/10.1016/J.JHAZMAT.2007.09.077.

Salehizadeh, H., Shojaosadati, S.A., 2003. Removal of metal ions from aqueous solution by polysaccharide produced from *Bacillus firmus*. Water Res. 37 (17), 4231–4235. https://doi.org/10.1016/S0043-1354(03)00418-4.

Schmid, J., Sieber, V., Rehm, B., 2015. Bacterial exopolysaccharides: biosynthesis pathways and engineering strategies. Front. Microbiol. 6. https://doi.org/10.3389/FMICB.2015.00496.

Shabaan, M., et al., 2022. Salt-tolerant PGPR confer salt tolerance to maize through enhanced soil biological health, enzymatic activities, nutrient uptake and antioxidant defense. Front. Microbiol. 13, 901865. https://doi.org/10.3389/FMICB.2022.901865/BIBTEX.

Shukla, A., et al., 2020. Characterization of novel thorium tolerant *Ochrobactrum* intermedium AM7 in consort with assessing its EPS-Thorium binding. J. Hazard. Mater. 388, 122047. https://doi.org/10.1016/J.JHAZMAT.2020.122047.

Solís, M., et al., 2012. Microbial decolouration of azo dyes: a review. Process Biochem. 47 (12), 1723–1748. https://doi.org/10.1016/J.PROCBIO.2012.08.014.

Staudt, C., et al., 2004. Volumetric measurements of bacterial cells and extracellular polymeric substance glycoconjugates in biofilms. Biotechnol. Bioeng. 88 (5), 585–592. https://doi.org/10.1002/BIT.20241.

Sun, S., et al., 2011. Exopolysaccharide production by a genetically engineered *Enterobacter cloacae* strain for microbial enhanced oil recovery. Bioresour. Technol. 102 (10), 6153–6158. https://doi.org/10.1016/J.BIORTECH.2011.03.005.

Tewari, S., Arora, N.K., 2014. Multifunctional exopolysaccharides from *Pseudomonas aeruginosa* PF23 involved in plant growth stimulation, biocontrol and stress amelioration in sunflower under saline conditions. Curr. Microbiol. 69 (4), 484–494. https://doi.org/10.1007/S00284-014-0612-X.

Tewari, S., Sharma, S., 2020. Rhizobial exopolysaccharides as supplement for enhancing nodulation and growth attributes of *Cajanus cajan* under multi-stress conditions: a study from lab to field. Soil Tillage Res. 198, 104545. https://doi.org/10.1016/J.STILL.2019.104545.

Upadhyay, A., et al., 2017. Players over the surface: unraveling the role of exopolysaccharides in zinc biosorption by fluorescent *Pseudomonas* strain Psd. Front. Microbiol. 8 (FEB). https://doi.org/10.3389/FMICB.2017.00284.

Wang, L., et al., 2013. Cadmium removal from aqueous solution by compound bioflocculant: equilibrium, kinetics and reuse studies. Curr. Biotechnol. 2 (2), 117–124. https://doi.org/10.2174/22115501113029990011.

Wang, T., et al., 2018. Adsorption of heavy metals from aqueous solution by UV-mutant *Bacillus* subtilis loaded on biochars derived from different stock materials. Ecotoxicol. Env. Saf. 148, 285–292. https://doi.org/10.1016/J.ECOENV.2017.10.039.

Weigel, P.H., DeAngelis, P.L., 2007. Hyaluronan synthases: a decade-plus of novel glycosyltransferases. J. Biol. Chem. 282 (51), 36777–36781. https://doi.org/10.1074/JBC.R700036200.

Whitfield, C., Mainprize, I.L., 2010. TPR motifs: hallmarks of a new polysaccharide export scaffold. Struct. (Lond. Engl.: 1993) 18 (2), 151–153. https://doi.org/10.1016/J.STR.2010.01.006.

Xing, J., et al., 2013. Removal efficiency and mechanism of sulfamethoxazole in aqueous solution by bioflocculant MFX. J. Anal. Methods Chem. 2013 (1), 568614. https://doi.org/10.1155/2013/568614.

Yang, M., et al., 2017. Characterisation of an extracellular polysaccharide produced by *Bacillus mucilaginosus* MY6-2 and its application in metal biosorption. Chem. Ecol. 33 (7), 625–636. https://doi.org/10.1080/02757540.2017.1351554.

Zhang, Y., et al., 2015. Extracellular polymeric substances govern the development of biofilm and mass transfer of polycyclic aromatic hydrocarbons for improved biodegradation. Bioresour. Technol. 193, 274–280. https://doi.org/10.1016/J.BIORTECH.2015.06.110.

Zhang, Z., et al., 2009. A novel biosorbent for dye removal: extracellular polymeric substance (EPS) of *Proteus mirabilis* TJ-1. J. Hazard. Mater. 163 (1), 279–284. https://doi.org/10.1016/J.JHAZMAT.2008.06.096.

Zhang, Z.qiang, et al., 2007. Production and application of a novel bioflocculant by multiple-microorganism consortia using brewery wastewater as carbon source. J. Env. Sci. 19 (6), 667–673. https://doi.org/10.1016/S1001-0742(07)60112-0.

Zouboulis, A.I., Chai, X.L., Katsoyiannis, I.A., 2004. The application of bioflocculant for the removal of humic acids from stabilized landfill leachates. J. Env. Manage. 70 (1), 35–41. https://doi.org/10.1016/J.JENVMAN.2003.10.003.

CHAPTER

12

Recent progress and outlook of natural EPS for biomedical application

Onkar Nath Tiwari[a], *Dipankar Ghosh*[b] *and Kesavan Bhaskar Reddy*[c]

[a]Centre for Conservation and Utilisation of Blue Green Algae (CCUBGA), Division of Microbiology, ICAR-Indian Agricultural Research Institute (PUSA), New Delhi, India [b]Microbial Engineering and Algal Biotechnology Laboratory, Department of Biosciences, JIS University, Agarpara, Kolkata, West Bengal, India [c]Apollo Institute of Pharmaceutical Sciences, The Apollo University, Chittoor, Andhra Pradesh, India

12.1 Introduction

Fast developing, the fields of biotechnologies and biomedicine demand biological and biocompatible goods. These compounds are basic for applications like drug delivery, tissue engineering, and wound healing—where interactions with biological systems must be smooth and free of negative reactions—where interactions with biological systems must be smooth and free of negative reactions. Natural exopolysaccharides (EPS) are attracting more and more interest as potential candidates for these uses as demand for safe and effective biomedical tools rises. Including bacteria, fungi, and algae, numerous microorganisms produce EPS—high- molecular-weight polymers. These chemicals are adaptable elements in medical products since they can make hydrophilic environments, change viscosity, and create gels. EPS is a perfect candidate for various medical applications since recent studies have highlighted its great biocompatibility and bioactivity. Important processes in tissue engineering (Muthukumar et al., 2019), including cell differentiation, proliferation, and adhesion, have been shown to be helped by EPS. Among the noteworthy events involving sources of EPS is the start of manufacturing of a capsular polysaccharide from K. algae *glycosyltransferase gene* applied in food and pharmaceutical industries, thickening and stabilizing processes; pneumoniae employed for

Microbial Exopolysaccharides: An Integrated Approach to Green Applications

DOI: https://doi.org/10.1016/B978-0-443-36619-2.00012-7

their protective qualities. Every source of knowledge adds unique qualities that influence the probable applications of EPS. Because of its biocompatibility and ability to generate hydrogels under mild circumstances, brown seaweed derivative alginate—for example—has become a common material in cell encapsulation (Kouser et al., 2018). Among the typical characteristics of EPS, their biodegradability is quite high since it relates to the growing focus on waste reduction and sustainable practices in the medical sector. Moreover, their inherent bioactivities with antibacterial effects and immunological modulation make them useful therapeutic agents. These features not only provide practical materials but also possible collaborators for the development of fresh biomedical uses. At last, natural EPS highlight the need for biocompatible and bioactive materials in biomedicine. EPS offers a medical sector frontier of opportunity with many sources and good characteristics. As research keeps revealing and developing its applications, eminent chemicals such as these are likely to be essential in boosting the efficacy and safety of next biomedical developments. Microorganisms produce complex sugars known as EPS into their surrounding environment. Important for the biological activities of these species are cell protection, production of biofilm, and nutrient retention. One of their most essential characteristics is high flexibility, which enables many structural and functional elements supporting various uses, particularly in the biomedical field. Structural variability of EPS is explained by variations in monosaccharide content, glycosidic connections, and molecular weight. For instance, adjusting their functional characteristics and structure, EPS might be branching or linear. Functional groups, including carboxyl and sulfate, additionally help them to interact with proteins and other molecules, thereby improving their activity (Yang et al., 2019). EPS's remarkable biocompatibility is made possible by its structural adaptability, hence, they might be safely injected into biological systems without causing significant immune reactions. Another crucial issue for environmentally friendly medical applications is the biodegradability of EPS. Their natural source helps them to break down into harmless molecules easily absorbed by living things. Especially helpful in the manufacturing of biomedical items requiring later body clearance, like drug delivery systems, tissue scaffolds, and wound dressings, is this property (Negut et al., 2020). Apart from their biocompatibility and biodegradation, EPS also show bioactivity—they can interact elegantly with biological systems. Fundamental steps of tissue engineering rely on some EPS that stimulates cell adhesion, proliferation, and differentiation. For example, specific EPS have been found to support chondrocyte and osteoblast formation, hence they could be suitable candidates for usage in bone and cartilage regeneration treatments (Liu et al., 2019). Theoretically, EPS finds extensive biomedical applications. Compounds can be used in drug delivery systems for efficient encapsulation and release of medicinal molecules. Their ability to create hydrogels also makes them ideal for tissue engineering support manufacture. In immunotherapy, EPS can also be used as natural agents or sensors, therefore boosting the body's reaction to cancer cells or pathogens (Huang et al., 2021). At last, EPS's many structural and functional features make it a flexible biomaterial with tremendous promise for several medical uses. Scientists and doctors trying to develop innovative healthcare treatments find them interesting because of their biocompatibility, biodegradability, and bioactive qualities. To this end, current extensive summarization deals with the identification and characterization of natural EPS toward applications in drug delivery, tissue engineering, wound healing, skin regeneration, biocompatible coating, and implants, including exploration on antimicrobial and antiviral activities. Hence, natural EPS in biomedical implementation is promising with the potential to offer sustainable and effective remedies for diverse ranges of medical challenges (Fig. 12.1).

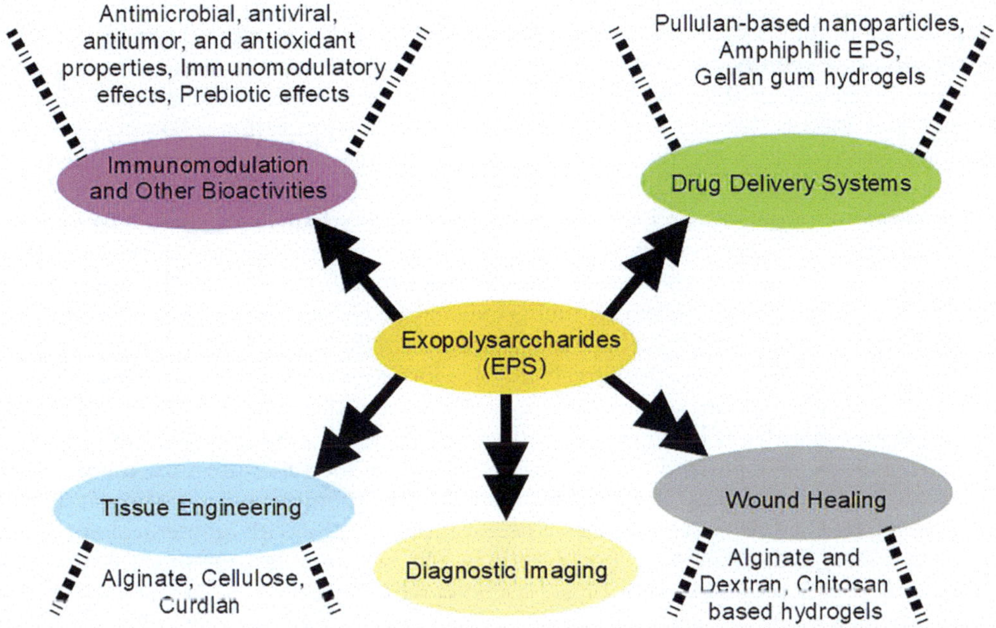

FIGURE 12.1 Recent progress and outlook of natural EPS for biomedical application to attain sustainability. *EPS*, Exopolysaccharides.

12.2 Sources and characterization of natural exopolysaccharides

The bio-complexes of carbohydrates produced by bacteria and fungi are microbial EPS. Along with key industrial needs like food, drugs, and biotechnological sectors, they serve several ecological roles, including creation of biofilm and desiccation prevention. The present essay addresses microbiological EPS sources and isolation, industrial production of EPS, and provides instances of previously investigated EPS, followed by recent results on genetic as well as metabolic engineering for optimization of production. Rich bio-producers of EPS are bacteria and fungi. EPS is mostly produced in bacteria by some gram-positive and gram-negative ones. For example, alginate produced by *Pseudomonas aeruginosa* is a major EPS that boosts biofilm stability (Powell et al., 2018). Likewise, fungi such as *Aureobasidium pullulans* are well-known for producing pullulan, which finds applications in the food and packaging sectors depending on their unique film-forming capabilities. With the help of several methods, including precipitation, centrifugation, and chromatography, which help to purify microbial EPS cultures, their separation from polysaccharide chains is achievable (Yadav et al., 2024). Among the EPS extensively studied include xanthan, hyaluronic acid, curdlan, and dextran. Xanthan gum, utilized extensively as a thickening agent in pharmaceuticals and foods, makes *Xanthomonas campestris*'s most popular polysaccharide, EPS manufactured commercially as xanthan gum (Chaturvedi et al., 2021). Mostly produced by *Streptococcus* species, hyaluronic acid is well-known for its skin moisturizing qualities and finds extensive application in both cosmetic and medical formulations. Made by the bacterium *Alcaligenes faecalis*, curdlan—which has a gelling property—is used in the food business. Made by *Leuconostoc mesenteroides*, dextran is of importance in the medical sector as a replacement for blood plasma (Díaz-Montes, 2021). EPS generation has been considerably changed by recent advances in metabolic and genetic engineering. By means of these approaches, researchers may control metabolic pathways to boost yield and modify the features of EPS to fit specific applications. For instance, it has been demonstrated that overexpression of genes in charge of polysaccharide biosynthesis greatly increases yield in several microbial hosts, including *Escherichia coli* (Mahmoud et al., 2021). Furthermore, under research are synthetic biology techniques to create novel EPS with particular purposes. All things considered, microbial EPS from fungi and bacteria have a lot of promise in several sectors. Improved manufacturing capabilities and creative applications of EPS have been possible thanks to developments in isolation techniques and genetic engineering. Understanding these polysaccharides and the techniques of their synthesis would not only boost industrial applications but also expose fresh chances for biotechnological developments.

Given their broad spectrum of therapeutic potential, the discovery of plant and algae extracellular polysaccharides (EPS) has attracted especially great attention. Naturally occurring polymers, EPS can be isolated and refined for application in food manufacture, medicines, and agriculture, among other fields. The extraction and purification techniques of plant and algal EPS are investigated in this essay, together with examples of EPS with biomedical applications and a discussion of environmental considerations and sustainable sourcing. Common techniques used in isolation of EPS from plants are enzymatic, alkaline, or hot water extraction. One such polymer found in fruit cell walls is pectin, a polysaccharide mostly derived from citrus peels or apple pomace by hot water extraction. The method aims to solubilize pectin with the least amount of degradation. Usually including dialysis for the removal of contaminants and precipitation techniques, purification. Alginate is also derived from brown algae by similar methods; the most often used extraction method is breaking down the cell walls of the algae using an alkaline solution. Many instances of plant and algal EPS with great biological promise abound. Research on pectin's possible use as a gelling agent and stabilizer in pharmaceutical formulations will help to enhance medication distribution systems (Freitas et al., 2021). Applied in wound healing and tissue engineering, alginate is widely known for its biocompatibility and hydrogel-forming capacity. Derived from red algae, carrageenan shows unique gelation properties and is employed in the food industry as a stabilizer and thickening agent; novel medicine formulation and bioactive component delivery find increasing relevance here. In addressing the environmental problem, eco-friendly procurement of EPS is becoming ever more important. The same holds true for plant-based EPS, where farming methods must balance production with ecological preservation. Overuse of algae resources may cause ecological problems that call for sustainable harvesting and aquaculture methods. Furthermore, by means of agricultural waste products, circular economy ideas can be applied to EPS manufacturing, so reducing the environmental impact of their procurement and guaranteeing resource economy (Pagliaccia et al., 2024). Finally, given their particular characteristics and uses, plant and algal EPS show great biopolymers in the biomedical context. Their applications depend much on the procedures of extraction and purification. However, the long-term supply of the biopolymers depends on sustainable sourcing methods; thus, further research and development in this field will surely open fresh doors for the use of plant and algal EPS. Also, this helps to lower environmental impact.

Derived from microbes, extracellular polymeric compounds (EPS) are multifunctional biomolecules having important roles in environmental and ecological systems. Understanding the functions of EPS, including biofilm development, nutrient cycling, and pollutant-polymer interaction, depends on knowing its characterization. High-tech analytical techniques, including chromatography, Fourier transform infrared spectroscopy (FTIR), and nuclear magnetic resonance (NMR), have become necessary instruments to clarify structural and functional aspects of EPS. One reliable method allowing the molecular structure of EPS to be found is NMR spectroscopy. It provides details on the atom distribution in the polymer, therefore helping to identify functional groups and chemical bond types found there. For example, it has been demonstrated how well NMR detects the polysaccharide structure of EPS from activated sludge, therefore providing important information on their monosaccharide composition and branching topologies. FTIR spectroscopy enhances NMR in providing functional group information on EPS. By measuring infrared absorption, FTIR generates specific vibrations of molecules at specific chemical bonds. Characterizing FTIR spectra helps one to identify the characteristic peaks corresponding to different chemical purposes, such as hydroxyl, carboxyl, and ether groups. This information is crucial in connecting the EPS structure to its biological purposes since the presence of specific functional groups may define the interaction of EPS with cells and with environmental pollutants. Moreover, extensively utilized to investigate the molecular weight and monosaccharide composition of EPS are chromatography techniques like size exclusion chromatography (SEC) and high-performance liquid chromatography (HPLC). Because it affects viscosity, stability, and the general polymer behavior during solution, molecular weight is a fundamental determinant of EPS characteristics. For biofilm growth and providing strength, for instance, higher molecular weight EPS are often superior to their low molecular weight equivalents (Fulaz et al., 2019). The biological purpose of EPS depends much on the branching patterns it exhibits. Within biofilms, branched EPS can build a more extensive network, which increases their mechanical stability and helps to separate embedded organisms from environmental pressures. Consequently, the structural complexity of EPS can help one to grasp their functions in both natural and manmade situations. Ultimately, structural and functional profiling of EPS (Yadav et al., 2014) depends much on advanced analytical instruments like NMR, FTIR, and chromatography. These methods allow one to investigate critical factors, including molecular weight, monosaccharide profile, and branching patterns. Incorporation of these analytical techniques will progressively help to clarify the complex link between EPS structure and their bioactivities as research develops, and support environmental biotechnology and other uses.

12.3 Recent progress in exopolysaccharides-based biomedical applications

12.3.1 Tissue engineering and regenerative medicine

Aimed at the repair or replacement of damaged tissues and organs, tissue engineering and regenerative medicine are fast-expanding domains of scientific research. Among the several materials used in this field, electrospun polymer scaffolds (EPS) have gained great attention due to their unique characteristics that support cell development and tissue regeneration (Fan et al., 2022). The use of EPS-based hydrogels in bone, cartilage, and vascular tissue engineering, as well as medication delivery and wound healing tools, is covered in this work. Made of biodegradable polymers capable of replicating mechanical and structural elements of natural tissue, EPS scaffolds have great surface area to effectively distribute nutrients and promote cellular adhesion and development by means of which according to present studies, these scaffolds can be made to release growth factors meant to enhance tissue healing systems. Apart from its application in bone tissue engineering, EPS hydrogels have also promised drug delivery systems hope. Drugs can be encapsulated by the hydrogels, which also stop their breakdown through regulated release (Lei et al., 2022). Formulating the release profiles of the medications increases their therapeutic efficacy and reduces side effects, so improving their usefulness. Furthermore, adding bioactive molecules will help to increase the bioactivity of the hydrogels, hence improving wound healing rates. Beyond bone and wound healing, EPS scaffolds find use in cartilage and vascular tissue engineering. EPS scaffolds provide a matrix in which chondrocytes may proliferate, therefore helping with cartilage trauma, which seldom heals spontaneously because cartilage is avascular. Research shows that EPS scaffolds can induce cartilage cell proliferation and the synthesis of extracellular matrix, both of which are required for the repair of cartilage function (Ahmadian et al., 2023). EPS materials improve endothelial cell development in vascular tissue engineering, hence producing vaseline and scaffold implant integration. Because they enable cell proliferation and tissue regeneration, EPS scaffolds are, all things considered, a major component of tissue engineering and regenerative medicine. Their importance in creating medical treatments is even more highlighted by their application in wound healing and drug delivery. More study will allow fresh ideas in the therapy of many tissue flaws and injuries to be revealed.

12.3.2 Drug delivery systems

Particularly with the development of novel carriers that increase the precision and efficiency of medication administration, drug delivery systems have considerably evolved. Application of EPS, or exopolymeric compounds, as drug carriers for controlled release of medications is one such exciting area. Particularly in fields like cancer treatment, gene therapy, and vaccination administration, EPS nanoparticles and microparticles have been investigated for their capacity to transport pharmaceuticals, especially microbial biopolymers, which are naturally occurring. As drug delivery vehicles, their unique qualities—biocompatibility, biodegradability, and hydrogel-forming capacity—make them interesting prospects. Therapeutic substances can be encapsulated by EPS-based nanoparticles and microparticles, therefore generating a protective niche that stabilizes the medications. Apart from enhancing medication solubility, encapsulation allows regulated rates of release, therefore preserving therapeutic effect over a long period. Particularly interesting in cancer treatment is the application of EPS-based drug delivery systems. Scientists have created EPS-based nanoparticles, for instance, to target cancer cells specifically without greatly compromising normal organs. Conventional chemotherapy has minimal negative effects when targeted selectively. The employment of EPS of *Lactobacillus plantarum* for the development of nanoparticles to deliver anticancer medicines has shown a considerable reduction of tumor growth in vitro and in vivo. Likewise, by means of the enhanced permeability and retention effect, EPS microparticles have also been found to help in medication concentration in tumor areas. Furthermore, beneficial in gene therapy is the use of EPS. In this regard, regulated release of genetic materials becomes crucial to effectively introduce therapeutic genes into target cells. Effective gene expression can be obtained via intracellular delivery of plasmid DNA or short interfering RNA by EPS-based nanoparticles. It has been shown, for example, that effective gene silencing in cancer cells by means of efficient RNA-based therapies delivered by EPS-based carriers (Liang et al., 2024). Vaccine candidates needing consistent release and continuous immune response are built using EPS-based platforms. Studies have shown that EPS can improve immunogenicity and effectiveness, therefore helping antigen-presenting cells to enable higher vaccination uptake. Recent research shows how EPS-based carriers may shield delicate RNA molecules during storage and injection, therefore implying their possible use in mRNA vaccination delivery (Saha and Datta, 2022). At last, a novel medication delivery method based on EPS-based nanoparticles and microparticles promises both targeted and regulated release. Their applications in vaccine distribution, gene therapy, and cancer treatment suggest their worth in optimizing therapeutic results. More investigation and development in this field promise to release the complete potential of EPS in transforming modern medicine.

12.3.3 Antimicrobial and antiviral applications

EPS have gained a prominent emphasis in recent years due to their inherent qualities that can be exploited in many spheres, particularly in medicine. Their antimicrobial and antiviral applications have become a key focus also. High-molecular-weight polysaccharides generated by microbially are EPS. These naturally occurring polymers contain several biologically active properties, including antibacterial and antiviral ones, which make them interesting candidates for application in medical equipment, specifically in order to avoid infection. Development of coatings for medical equipment is one of the most crucial domains for EPS application. Still, a persistent issue in the clinical environment is infections linked to implanted devices. Usually failing conventional infection control measures, they increase morbidity and medical expenses. Pathogen adherence could be much reduced by EPS's natural capacity to produce a protective and biocompatible layer on surfaces. For example, a study found that EPS generated from *Lactobacillus rhamnosus* was used as a catheter coating and showed a great decrease in microbial colonization, therefore lowering the risk of urinary tract infection (Khan et al., 2020). Advances such as these show EPS as a possible replacement to prevent infections in medical equipment without synthetic antibiotics. Moreover, EPS show promises as immunomodulating drugs. Their control of the immune response might strengthen the human resistance against infections. Important in immunological responses, certain EPS were discovered to increase the activity of macrophages and release cytokines. Some EPS from the bacterium *Streptococcus thermophilus*, for instance, were observed to raise immunoglobulin A synthesis, hence strengthening mucosal immunity (Mizuno et al., 2020). In the case of viral infections, in particular, this immunomodulating ability is very important since a good immune response can block viral replication and spread. Particularly in the medical sector, EPS's antibacterial and antiviral applications are overall rather important in the fight against infections. They are excellent candidates for the development of new preventive strategies since their natural origin and ability to provide defense against diseases help to strengthen immune systems. Medical environments will have a bright future for EPS application in making them safer, as studies keep looking at its full possibilities in several uses. Validation of these benefits and paving the path for general usage in medical practice would depend mostly on clinical trials and continuous research (Tisch et al., 2024).

12.3.4 Wound healing and skin regeneration

Essential reactions for the restoration of skin integrity following damage include wound healing and skin renewal. Improving these procedures is crucial to improving patient outcomes, particularly in diseases like the treatment of burns and diabetic wounds, where ideal and rapid healing can significantly affect quality of life. The use of EPS, microorganism-secreted compounds identified for their capacity to improve wound healing and reduce scar development, is one of the interesting approaches. Many of the characteristics of EPS make them appropriate for treating wounds (Shen et al., 2021). Crucially, components of the healing process, cellular proliferation and migration, have been shown to be stimulated by them. Furthermore, beneficial for cellular processes resulting in healing is EPS's biocompatible and moist environment. Studies have indicated that EPS dressings can both speed up healing and help to prevent scar tissue development. Research has shown, for instance, that EPS taken from some bacteria significantly promotes skin regeneration in animal models, therefore suggesting their clinical relevance (Xu et al., 2024). EPS is used in clinical settings as well as in laboratories, particularly in burn treatment and diabetic ulcer management. Typical consequences of burn injuries burns can help to reduce inflammation and act as a barrier against infection by means of EPS. By maintaining ideal moisture levels and improving new tissue growth, EPS-based topical treatments can help speed recovery. Furthermore, EPS has the ability to regenerate the wound bed and hence support the body's natural healing mechanisms in diabetic ulcers, whose repair is usually delayed by inadequate perfusion and concurrent diseases (Arslan et al., 2024). Moreover, under development and tested for efficacy are EPS-infused dressings. Throughout wound treatment, these dressings not only heal but also reduce pain and discomfort. Using natural and biocompatible materials like EPS offers a fresh and efficient substitute for traditional therapies and might completely change wound healing. All things considered, the utilization of EPS in wound healing is a creative development in modern medicine. EPS-based dressings and topical treatments represent a new horizon in burn and diabetic ulcer treatment since they help to minimize scar tissue development and enable faster closure. Underlining the need for more study and clinical use of EPS in the care of wounds, the range of natural products to improve patient outcomes keeps widening with advancing knowledge.

12.3.5 Biocompatible coatings and implants

Medical implant function and shelf life are improved by biocompatible coatings. Here, a significant advance is the use of ethylene propylene styrene (EPS) as a coating agent. Since they can improve biocompatibility and lower adverse immunological reactions, biomaterials based on EPS have shown promise, particularly in dentistry and orthopedic applications. Medical implants—including hip implants, dental crowns, and bone screws—tend to have trouble blending into the body. Foreign chemicals can be unfavorably reacted upon by the immune system, leading to inflammation and implant failure. In response to such difficulties, the compatibility of the implants with human tissues can be much enhanced by the use of biocompatible coatings such as EPS. EPS is an excellent candidate for coating uses because it has mechanical strength, flexibility, and chemical degradation resistance. EPS-based coatings can be used in dentistry applications on dental implants to promote tissue integration and help to control plaque development. Studies with coatings changed by bioactive chemicals such as hydroxyapatite have shown that the osteoconductivity of dental implants may be raised still more. As opposed to conventional materials, EPS has been shown to reduce the rate of inflammatory response and have less bone resorption but boost osseointegration.

Also improved are orthopedic applications with EPS coatings. For example, hip and knee implants covered in EPS have shown enhanced cellular adhesion and proliferation, hence increasing incorporation into surrounding bone. The hydrophobicity of EPS lowers inflammatory problems and fits the first healing stages. Reduced infection risk and a marked improvement in healing rate were displayed by EPS-coated orthopedic implants (Cao et al., 2019). One of the most significant characteristics of EPS coatings is also their ability to assist in lower undesired immunological reactions. EPS can improve the surroundings for healing and integration by means of a less immunogenic surface development. Including anti-inflammatory drugs into the EPS matrix will help to further enhance the regulation of the immune response, therefore enabling the body to more absorb the foreign material (Batool et al., 2021). Patients with a pre-existing autoimmune disease or those who are prone to inflammatory consequences especially depend on this capacity. All things considered, EPS-based coatings show great potential for helping medical implants to be biocompatible. Their use in dental and orthopedic settings has shown considerable promise in improving integration, lowering negative immune responses, and accelerating healing. The use of EPS and other biomaterials may alter the direction of implant technology and improve patient prognosis as technology advances (Braem et al., 2023).

12.4 Modifications and functionalization of exopolysaccharides

Because of their biocompatibility, biodegradability, and unusual chemical characteristics, EPS have attracted major interest in biological uses. Emphasizing their fundamental importance in many biomedical applications like drug delivery, tissue engineering, and wound healing, this essay investigates the alterations and functionalizing of EPS. Important considerations are new approaches for improving EPS characteristics, eminent scientists in the field, and possible future breakthroughs that would widen the range of influence of EPS in biomedicine. Bacteria secrete high-molecular-weight polymers into the external environment called EPS. EPS have been observed to exhibit different biological functions and activities with their different structures than the organism from which they are derived (Wang et al., 2023). EPS used to be derived from several sources, including the fermentation of fungus, yeasts, and bacteria. Early enthusiasm in EPS was mostly directed to their rheological characteristics and uses as food thickeners (Bibi et al., 2021). Still, studies have shown that they have great potential in biomedical uses, hence there has been a boom in interest in their modification and functionalization.

One of the main methods EPS is changed for biomedical applications is chemical alteration. This approach raises EPS's solubility, stability, and usability. Among the most often occurring chemical changes are methylation, acetylation, and sulfonation. For instance, higher anticoagulant action has shown great promise for sulfated polysaccharides. By means of sulfate groups, researchers have observed enhanced interaction with blood coagulation factors, thereby rendering them possibly valuable in uses in anticoagulant therapy (Zang et al., 2022). Another such is the derivatization of EPS utilizing hydrophobic groups, in which case, drug encapsulation efficiency can be much improved. One of the most extensively studied EPS, alginate, shows this, and hydrophobic alteration has resulted in increased loading of anticancer medicines for targeted therapy (Pan et al., 2024). Apart from chemical modification, physical techniques, including mixing and crosslinking, have also been applied to functionalize EPS. Strong candidates for tissue engineering scaffolds, blending with synthetic polymers, can offer better mechanical qualities and control of degradation rates. For tissue engineering uses, chitosan plus gelatin is one of the most significant pairings since it improves cell adherence and proliferation rates (Re et al., 2019). Moreover, crosslinking agents can create three-dimensional networks that provide EPS-based hydrogels with mechanical strength and adjustable porosity. In applications of tissue engineering, such characteristics enable enhanced cell migration and continuous drug release (Mohd Nadzir et al., 2021). One other crucial approach is the biological alteration of EPS. This method changes EPS structures by the use of enzymes or bacteria, generating polymers with higher biological activity. Lytic enzymes, for example, can generate oligosaccharides with more bioactivity by breaking certain glycosidic linkages in the EPS structure. Research on these bioactive oligosaccharides has shown that they have immunomodulating action, hence promoting tissue regeneration and wound healing (Sharma et al., 2022). This biological alteration potential is particularly interesting for use in regenerative medicine since cellular signaling pathways have to be strengthened. Natural sources of EPS have been investigated as the need for sustainable and bio-based materials attracts more attention. Among the valuable sources is the purple nonsulfur bacterium *Rhodobacter sphaeroides*, which produces a novel EPS with interesting applications in tissue engineering and medication delivery. From such fresh sources, scientists have recorded significant improvement in isolating, characterizing, and modifying EPS, producing materials with fresh characteristics and uses (Mouro et al., 2024). This focus on renewable energy complements contemporary green chemical and environmental trends. Furthermore, eminent experts such as Professor Dolors Montoya and Dr. Marina G. Gallardo have made significant contributions to the field by looking at the biological prospects of several forms of EPS. Combining biochemistry, materials science, and medicine to improve knowledge and application of EPS in the biomedical field, their studies have spurred multidisciplinary projects. Prospects for EPS in biomedical uses seem high going forward (Ahuja et al., 2023). Nanotechnology has brought nanoscale carriers formed from modified EPS under design for drug delivery systems that enhance therapeutic advantages and minimize negative effects. Furthermore, EPS combined with smart materials able to react to environmental cues (pH, temperature) creates interesting opportunities for systems of controlled medication release. All things considered, the functionalizing and modification of EPS have opened fresh avenues in biomedical uses. Scientists have made tremendous progress in enhancing the properties of EPS to be used in drug delivery, tissue engineering, and regenerative medicine by means of several chemical, physical, and biological approaches (Asgher et al., 2021). Future studies and multidisciplinary cooperation will retain EPS in its leadership position of biomedical innovation and release its potential.

12.5 Challenges and outlook on natural exopolysaccharides for biomedical applications

Natural EPS are already quite well-known in biomedical uses because of their unique properties, including biocompatibility, biodegradability, and adaptability. This article addresses all the difficulties and potential perspectives of employing natural EPS in biomedical contexts. Among the salient features are the consistent quality maintenance and manufacturing scale-up, cost-effective extraction methods, handling of regulatory concerns, and application of nanotechnology (Amer et al., 2025). Furthermore, included is the investigation of novel EPS sources and mixes, as well as the evolution of individualized medicine strategies. The first difficulty is the consistent quality of the EPS scaling-up procedure. Any EPS that finds usage in biological applications must be produced in great quantities. Most modern manufacturing methods rely on microbial systems based on fermentation processes, time-consuming and difficult to standardize. It has been underlined that EPS with a constant molecular weight and composition is rather important. Future developments should concentrate on bioreactor technology and process engineering if we are to raise yields and enable industrial scale-up. Developing low-cost and reasonably priced extraction techniques is another main difficulty (Mahata et al., 2021). For industrial-scale use, current extraction techniques can be financially unworkable and time-consuming. Usually requiring harmful solvents or conditions, both physical and chemical extraction techniques generate environmental problems. Green substitutes are desperately sought for, including enzymatic extraction, which not only reduces waste but also preserves EPS integrity additional recently, as shown by Adel et al. (2025), studies suggest additional innovative approaches using waste biomass from agricultural activities. These low-cost EPS might be produced, therefore increasing the accessibility of biomedical applications. Introduction of EPS-based products to the market faces significant challenges from regulatory and safety issues. As observed by Hasan et al. (2024), regional differences abound in the tight biocompatibility and safety regulations that regulatory authorities follow. Commercial success depends on establishing internationally acknowledged standards. Furthermore, researchers have to show by rigorous preclinical and clinical studies the success of EPS-based treatments. Establishing a comprehensive regulatory system will help to ease the shift through the complexity of implementing EPS-based medical goods. Regarding creativity, the identification of fresh EPS sources is exciting. Currently under investigation in tandem with unusual sources of marine creatures and plant exudates, conventional sources include bacterial and algal strains. Based on current research, some marine-sourced EPS may have unique properties allowing other therapeutic advantages (Khalifa et al., 2019). Investigating these fresh sources calls for multidisciplinary cooperation combining expertise in the domains of biomaterials science, marine biology, and microbiology to perhaps identify EPS with higher biological activity.

Furthermore, crucial is creativity in novel extraction techniques capable of regularly generating high-quality EPS with desired characteristics. High-quality EPS with minimum processing durations and bioactivity preservation has been produced by the combination of sophisticated technologies, including ultrasonic-assisted extraction and supercritical fluid extraction. Such advances could completely transform the way natural EPS is extracted and used in biomedical domains. Investigating synergistic actions of EPS combinations (Hosseini et al., 2024) piques still another area of interest. Certain combinations of EPS have cumulative effects more than the total of the individual polysaccharides, according to certain studies. Co-administration of many EPS, for instance, can increase cell adhesion, proliferation, and differentiation. Investigating interactions across several EPS can help to build more powerful therapeutic strategies for drug delivery or tissue engineering. EPS combined with nanotechnology presents interesting paths to maximize the action and delivery of medicinal medicines (Tiwari et al., 2024). Nanoparticles allow EPS to be contained in order to target drugs and reduce systemic toxicity. Furthermore, the coupling of EPS with nanostructures might increase the biocompatibility and stability of the nanoparticles, therefore providing a safer substitute for clinical uses. Recent advancements in 3D bioprinting also suggest a bright future whereby EPS can be combined with bioinks to create complex biological structures able to emulate tissue architecture (Muthukrishnan, 2021).

A developing area of interest in personalized medicine is customizing EPS-based therapy to the particular needs of individual patients. Knowing the physiology and disease condition of each individual patient will help to guide the customization of EPS-based treatments as the discipline develops. This trend advances more customized treatments from broad-brush techniques, therefore improving therapeutic results. Another area with significant promise is the building of diagnostic tools depending on EPS-based biosensors. Early detection and disease treatment will be improved by biosensors identifying some biomolecules at very low concentrations (Haleem et al., 2021). It has been described how modified EPS can be used in biosensors, and this opens directions for quick and reliable diagnosis. Cooperation between the sector and academics will help one to understand the complex regulatory environment for EPS-based medical devices. Working together, researchers and clinical developers can transform ideas resulting from the study into practical uses. Cooperation can help to direct the creation of protocols that meet legal criteria without sacrificing the efficacy of EPS-based products (Pereira et al., 2019). Furthermore, included in effective commercialization plans should be marketing elements, competitor analysis, and market needs assessment. In essence, even although natural EPS presents many difficulties for usage in biomedical applications, there is a great possibility for innovation and development. The potential for EPS in individualized medicine and creative therapeutic approaches can be exploited by means of the resolution of difficulties relating to production scalability, methods of extraction, regulatory criteria, and new technologies for integration. Multidisciplinary integration and an attention on sustainability help to greatly promise EPS for biomedical uses (De et al., 2023).

12.6 Conclusion

Based on EPS, the development of biomedical applications shows a great increase and a promising future in the field of biomedicine. Released by many bacteria, EPS are polymers with a great molecular weight that play vital roles in microbial and cell interactions. Recent research has underlined how well EPS might enhance tissue engineering, medication transport, and the synthesis of novel therapeutic agents. Emphasizing the need for constant research and development in order to bring future advancement, this essay is on the developments in EPS-based biomedical applications. EPS have also been underlined recently for their application in medicine delivery systems. Their biocompatibility and biodegradability make them suitable carriers of drugs. Research has shown, for instance, efficient encapsulation of anticancer medications inside EPS mats with a controlled release for focused therapy to lower adverse effects. Moreover, EPS helps nanoparticles to be stabilized, therefore improving the action of medications for the fight against diseases. The ability to create efficient, EPS-based drug delivery systems implies not just nanomedicine development but also opportunities in tailored patient demands in customized medicine. EPS have been investigated in tissue engineering to provide scaffolds able to support cell development and tissue regeneration. The special structural properties of EPS enable the reproduction of extracellular mats, the structures required for tissue regeneration and healing. Recent studies reveal that in wound healing processes, scaffolds mixed with EPS increase cell adhesion and proliferation and speed recovery timeframes. This reveals the possibilities of EPS in regenerative medicine, therefore influencing the therapy of degenerative diseases and injuries. Still, there are issues that draw attention to the need for more study and invention in this sector. First, the complexity and variation of EPS structures demand a better knowledge of their biological interactions. Second, the scalability of EPS manufacturing to commercial scales is still a vital area of research. Thirdly, it is obviously necessary to maximize their features for specific uses. Synthetic biology and bioprocessing techniques could offer answers to these problems, therefore enabling extensive use of EPS in industry. All things considered, the present developments in EPS-based biomedical applications mirror their transforming possibilities in biomedicine. From tissue engineering to bettering drug delivery methods, EPS offers fresh ideas to long-standing medical issues. To fully utilize their potential, though, research and invention are still absolutely vital. By means of ongoing research and development of EPS technologies, we could expose fresh therapeutic strategies with a great impact on the health results of patients.

Credit author statement

Dr. O.N. Tiwari and Dr. Dipankar Ghosh prepared the manuscript and figures. Dr. O. N. Tiwari, Dr. Dipankar Ghosh, and Dr. K. Bhaskar Reddy carried out the final editing and provided value addition to the chapter. All authors reviewed and approved the final version.

Conflict of interest

The authors declare that they have no conflict of interest.

Ethical statement

This chapter does not contain any studies involving human participants or animals performed by any of the authors.

Acknowledgment

Dr. O.N. Tiwari would like to thank CCUBGA and ICAR-IARI for their immense support and help. Dr. D. Ghosh appreciates the help and support of JIS University and JIS Group of Educational Initiatives.

References

Adel, R., Fahim, I.S., Bakhoum, E.S., Ahmed, A.M., AbdelSalam, S.S., 2025. Sustainable nanocellulose coating for EPS geofoam extracted from agricultural waste. Waste Manag. 191, 135–146.

Ahmadian, E., Eftekhari, A., Janas, D., Vahedi, P., 2023. Nanofiber scaffolds based on extracellular matrix for articular cartilage engineering: a perspective. Nanotheranostics 7 (1), 61.

Ahuja, V., Bhatt, A.K., Banu, J.R., Kumar, V., Kumar, G., Yang, Y.H., Bhatia, S.K., 2023. Microbial exopolysaccharide composites in biomedicine and healthcare: Trends and advances. Polymers 15 (7), 1801.

Amer, M.S., Barakat, K.M., Ibrahim, H.A., Matsuo, K., Ibrahim, M.I., 2025. An overview on marine bacterial exopolysaccharides and their industrial applications. J. Carbohydr. Chem. 1–38.

Arslan, N.P., Orak, T., Ozdemir, A., Altun, R., Esim, N., Eroglu, E., Karaagac, S.I., Aktas, C., Taskin, M., 2024. Polysaccharides and peptides with wound healing activity from bacteria and fungi. J. Basic Microbiol. 64 (12), e2400510.

Asgher, M., Qamar, S.A., Iqbal, H.M., 2021. Microbial exopolysaccharide-based nano-carriers with unique multifunctionalities for biomedical sectors. Biologia 76 (2), 673–685.

Batool, F., Özçelik, H., Stutz, C., Gegout, P.Y., Benkirane-Jessel, N., Petit, C., Huck, O., 2021. Modulation of immune-inflammatory responses through surface modifications of biomaterials to promote bone healing and regeneration. J. Tissue Eng. 12, 20417314211041428.

Bibi, A., Xiong, Y., Rajoka, M.S.R., Mehwish, H.M., Radicetti, E., Umair, M., Shoukat, M., Khan, M.K.I., Aadil, R.M., 2021. Recent advances in the production of exopolysaccharide (EPS) from *Lactobacillus spp.* and its application in the food industry: a review. Sustainability 13 (22), 12429.

Braem, A., Kamarudin, N.H.N., Bhaskar, N., Hadzhieva, Z., Mele, A., Soulié, J., Linklater, D.P., Bonilla-Gameros, L., Boccaccini, A.R., Roy, I., Drouet, C., 2023. Biomaterial strategies to combat implant infections: new perspectives to old challenges. Int. Mater. Rev. 68 (8), 1011–1049.

Cao, J., Zhao, Y., Liu, Y., Tian, S., Zheng, C., Liu, C., Zhai, Y., An, Y., Busscher, H.J., Shi, L., Liu, Y., 2019. Phosphorylcholine-based polymer encapsulated chitosan nanoparticles enhance the penetration of antimicrobials in a staphylococcal biofilm. ACS Macro Lett. 8 (6), 651–657.

Chaturvedi, S., Kulshrestha, S., Bhardwaj, K., Jangir, R., 2021. A review on properties and applications of xanthan gum. Microb. Polym.: Appl. Ecol. Perspect. 87–107.

De, K., Debnath, S., Ghosh, D., 2023. Elevating algal biomass generation toward sustainable utilization for high value added biomolecules generations. J. Appl. Biol. Biotechnol., 11 (1), 16–27.

Díaz-Montes, E., 2021. Dextran: Sources, structures, and properties. Polysaccharides 2 (3), 554–565.

Fan, J., Abedi-Dorcheh, K., Sadat Vaziri, A., Kazemi-Aghdam, F., Rafieyan, S., Sohrabinejad, M., Ghorbani, M., Rastegar Adib, F., Ghasemi, Z., Klavins, K., Jahed, V., 2022. A review of recent advances in natural polymer-based scaffolds for musculoskeletal tissue engineering. Polymers 14 (10), 2097.

Freitas, C.M.P., Coimbra, J.S.R., Souza, V.G.L., Sousa, R.C.S., 2021. Structure and applications of pectin in food, biomedical, and pharmaceutical industry: a review. Coatings 11 (8), 922.

Fulaz, S., Vitale, S., Quinn, L., Casey, E., 2019. Nanoparticle–biofilm interactions: the role of the EPS matrix. Trends Microbiol. 27 (11), 915–926.

Haleem, A., Javaid, M., Singh, R.P., Suman, R., Rab, S., 2021. Biosensors applications in medical field: a brief review. Sens. Int. 2, 100100.

Hasan, H.A., Rahim, N.F.M., Alias, J., Ahmad, J., Said, N.S.M., Ramli, N.N., Buhari, J., Abdullah, S.R.S., Othman, A.R., Jusoh, H.H.W., Juahir, H., 2024. A review on the roles of extracellular polymeric substances (EPSs) in wastewater treatment: source, mechanism study, bioproducts, limitations, and future challenges. Water 16 (19), 2812.

Hosseini, S.P., Mousavi, S.M., Jafari, A., 2024. Exploring biosynthesis strategies to boost the yield of exopolysaccharide-protein blend from *Bacillus arachidis* SY8 (T), an isolated native strain, as a potent adsorbent for heavy metals removal. Int. J. Biol. Macromol. 271, 132634.

Huang, X., Pan, J., Xu, F., Shao, B., Wang, Y., Guo, X., Zhou, S., 2021. Bacteria-based cancer immunotherapy. Adv. Sci. 8 (7), 2003572.

Khalifa, S.A., Elias, N., Farag, M.A., Chen, L., Saeed, A., Hegazy, M.E.F., Moustafa, M.S., Abd El-Wahed, A., Al-Mousawi, S.M., Musharraf, S.G., Chang, F.R., 2019. Marine natural products: a source of novel anticancer drugs. Mar. Drugs 17 (9), 491.

Khan, F., Tabassum, N., Kim, Y.M., 2020. A strategy to control colonization of pathogens: embedding of lactic acid bacteria on the surface of urinary catheter. Appl. Microbiol. Biotechnol. 104, 9053–9066.

Kouser, R., Vashist, A., Zafaryab, M., Rizvi, M.A., Ahmad, S., 2018. Biocompatible and mechanically robust nanocomposite hydrogels for potential applications in tissue engineering. Mater. Sci. Eng.: C 84, 168–179.

Lei, L., Bai, Y., Qin, X., Liu, J., Huang, W., Lv, Q., 2022. Current understanding of hydrogel for drug release and tissue engineering. Gels 8 (5), 301.

Liang, S., Wang, X., Li, C., Liu, L., 2024. Biological activity of lactic acid bacteria exopolysaccharides and their applications in the food and pharmaceutical industries. Foods 13 (11), 1621.

Liu, Y., Ma, Y., Zhang, J., Yuan, Y., Wang, J., 2019. Exosomes: A novel therapeutic agent for cartilage and bone tissue regeneration. Dose-Response 17 (4), 1559325819892702.

Mahata, C., Dhar, S., Ray, S., Das, D., 2021. Flocculation characteristics of anaerobic sludge driven-extracellular polymeric substance (EPS) extracted by different methods on microalgae harvesting for lipid utilization. Biochem. Eng. J. 167, 107898.

Mahmoud, Y.A.G., El-Naggar, M.E., Abdel-Megeed, A., El-Newehy, M., 2021. Recent advancements in microbial polysaccharides: synthesis and applications. Polymers 13 (23), 4136.

Mizuno, H., Tomotsune, K., Islam, M.A., Funabashi, R., Albarracin, L., Ikeda-Ohtsubo, W., Aso, H., Takahashi, H., Kimura, K., Villena, J., Sasaki, Y., 2020. Exopolysaccharides from *Streptococcus thermophilus* ST538 modulate the antiviral innate immune response in porcine intestinal epitheliocytes. Front. Microbiol. 11, 894.

Mohd Nadzir, M., Nurhayati, R.W., Idris, F.N., Nguyen, M.H., 2021. Biomedical applications of bacterial exopolysaccharides: a review. Polymers 13 (4), 530.

Mouro, C., Gomes, A.P., Gouveia, I.C., 2024. Microbial exopolysaccharides: Sructure, diversity, applications, and future frontiers in sustainable functional materials. Polysaccharides 5 (3), 241–287.

Muthukrishnan, L., 2021. Imminent antimicrobial bioink deploying cellulose, alginate, EPS and synthetic polymers for 3D bioprinting of tissue constructs. Carbohydr. Polym. 260, 117774.

Muthukumar, T., Song, J.E., Khang, G., 2019. Biological role of gellan gum in improving scaffold drug delivery, cell adhesion properties for tissue engineering applications. Molecules 24 (24), 4514.

Negut, I., Dorcioman, G., Grumezescu, V., 2020. Scaffolds for wound healing applications. Polymers 12 (9), 2010.

Pagliaccia, B., Campo, R., Carretti, E., Severi, M., Lubello, C., Lotti, T., 2024. Towards resource recovery-oriented solutions in agriculture exploiting structural extracellular polymeric substances (sEPS) extracted from aerobic granular sludge (AGS). Chem. Eng. J. 485, 149819.

Pan, W., Gu, F., Yan, X., Huang, J., Liao, H., Niu, F., 2024. Biomacromolecular carriers based hydrophobic natural products for potential cancer therapy. Int. J. Biol. Macromol., 132274.

Pereira, S.B., Sousa, A., Santos, M., Araújo, M., Serôdio, F., Granja, P., Tamagnini, P., 2019. Strategies to obtain designer polymers based on cyanobacterial extracellular polymeric substances (EPS). Int. J. Mol. Sci. 20 (22), 5693.

Powell, L.C., Pritchard, M.F., Ferguson, E.L., Powell, K.A., Patel, S.U., Rye, P.D., Sakellakou, S.M., Buurma, N.J., Brilliant, C.D., Copping, J.M., Menzies, G.E., 2018. Targeted disruption of the extracellular polymeric network of *Pseudomonas aeruginosa* biofilms by alginate oligosaccharides. npj Biofilms Microbiomes 4 (1), 13.

Re, F., Sartore, L., Moulisova, V., Cantini, M., Almici, C., Bianchetti, A., Chinello, C., Dey, K., Agnelli, S., Manferdini, C., Bernardi, S., 2019. 3D gelatin-chitosan hybrid hydrogels combined with human platelet lysate highly support human mesenchymal stem cell proliferation and osteogenic differentiation. J. Tissue Eng. 10, 2041731419845852.

Saha, I., Datta, S., 2022. Bacterial exopolysaccharides in drug delivery applications. J. Drug Deliv. Sci. Technol. 74, 103557.

Sharma, R., Kataria, A., Sharma, S., Singh, B., 2022. Structural characterisation, biological activities and pharmacological potential of glycosaminoglycans and oligosaccharides: A review. Int. J. Food Sci. Technol. 57 (1), 4–15.

Shen, S., Chen, X., Shen, Z., Chen, H., 2021. Marine polysaccharides for wound dressings application: an overview. Pharmaceutics 13 (10), 1666.

Tisch, M., Roháč, L., Reineke, T., Burkart, M., Kostev, K., 2024. Long-term benefits of EPs® 7630 in patients with acute sinusitis: A real-world cohort study. Front. Pharmacol. 15, 1358879.

Tiwari, O.N., Bobby, M.N., Kondi, V., Halder, G., Kargarzadeh, H., Ikbal, A.M.A., Bhunia, B., Thomas, S., Efferth, T., Chattopadhyay, D., Palit, P., 2024. Comprehensive review on recent trends and perspectives of natural exopolysaccharides: pioneering nano-biotechnological tools. Int. J. Biol. Macromol., 130747.

Wang, W., Ju, Y., Liu, N., Shi, S., Hao, L., 2023. Structural characteristics of microbial exopolysaccharides in association with their biological activities: a review. Chem. Biol. Technol. Agric. 10 (1), 137.

Xu, H., Li, Y., Song, J., Zhou, L., Wu, K., Lu, X., Zhai, X., Wan, Z., Gao, J., 2024. Highly active probiotic hydrogels matrixed on bacterial EPS accelerate wound healing via maintaining stable skin microbiota and reducing inflammation. Bioact. Mater. 35, 31–44.

Yadav, M.K., Song, J.H., Vasquez, R., Lee, J.S., Kim, I.H., Kang, D.K., 2024. Methods for detection, extraction, purification, and characterization of exopolysaccharides of lactic acid bacteria—a systematic review. Foods 13 (22), 3687.

Yang, X., Wan, Y., Zheng, Y., He, F., Yu, Z., Huang, J., Wang, H., Ok, Y.S., Jiang, Y., Gao, B., 2019. Surface functional groups of carbon-based adsorbents and their roles in the removal of heavy metals from aqueous solutions: A critical review. Chem. Eng. J. 366, 608–621.

Zang, L, Zhu, H, Wang, K, Liu, Y, Yu, F, Zhao, W., 2022. Not just anticoagulation—new and old applications of heparin. Molecules 27 (20), 6968.

CHAPTER

13

Potential application of natural EPS in pest control in organic farming

Dipankar Ghosh[a], *Indira Mikkili*[b] *and Onkar Nath Tiwari*[c]

[a]Microbial Engineering and Algal Biotechnology Laboratory, Department of Biosciences, JIS University, Agarpara, Kolkata, West Bengal, India [b]Department of Biotechnology, Vignan's Foundation for Science, Technology and Research, Vignan University, Vadlamudi, Guntur, Andhra Pradesh, India [c]Centre for Conservation and Utilisation of Blue Green Algae (CCUBGA), Division of Microbiology, ICAR-Indian Agricultural Research Institute (PUSA), New Delhi, India

13.1 Introduction

As consumers seek increasingly sustainable and pesticide-free food, organic farming has grown rather popular recently. The growing awareness of health issues resulting from chemical exposure and the necessity of environmental preservation drives mostly the popularity of this farming method. Organic farming appeals to a society more and more concerned about the environmental repercussions of mainstream farming methods since it emphasizes the use of natural processes to increase soil fertility and stimulate biodiversity (Gamage et al., 2023). One of the main ideas of organic farming, the avoidance of synthetic pesticides, has produced many consumers choosing organic products. A study by the soil association claims that UK organic food sales alone exceeded £2. 8 billion in 2020; consumer demand is rising (Bostan et al., 2019). This trend fits a larger worldwide movement toward sustainability as people pay a premium for goods that satisfy their values about the environment and health. Organic farmers have several difficulties, particularly in terms of pest control, notwithstanding their bright future. Organic producers have to employ various pest control strategies without petrochemical pesticides. These might call for biological control, for instance, bringing in natural enemies or using organic, permitted pesticides derived from plants. However, these strategies take more time than conventional techniques and are less effective; hence, harvests usually suffer

DOI: https://doi.org/10.1016/B978-0-443-36619-2.00013-9

(Hasan et al., 2020). Furthermore, organic farmers sometimes battle the mistaken belief that their method of production is less profitable than a conventional one. Though it is not always the case, this emphasizes the need for ongoing study and improvement in methods of organic pest control. Because organic methods often need more labor-intensive operations, which can further lower their competitiveness in the market, organic farmers can also have additional production expenses (Durham and Mizik, 2021). The requirement for environmentally acceptable biopesticide substitutes for man-made pesticides is especially crucial as organic farming is developing. Made from natural materials, new discoveries in biopesticides offer efficient means of pest management free from the negative impacts of conventional pesticides. Research indicates that applying biopesticides can help to maintain good control of pests with ecological balance assurance (Archana et al., 2022). These advances are absolutely essential for organic farmers to be able to sustainably generate food. All things considered, the rise in organic farming shows a growing trend in agricultural sustainability. Though problems such as pest control still exist, developments in environmentally friendly substitutes give organic farmers a bright future. Maintaining farming sustainably and free of chemicals, ongoing support for research and education into organic farming techniques will help to overcome these obstacles as the organic market keeps expanding (Soni et al., 2022).

Highly sought for in recent years for their sustainability and wide variety of uses are exopolysaccharides (EPS), biopolymers (Mouro et al., 2024). Made by bacteria, plants, and algae, their long-chain polysaccharides are polysaccharides that are vital in ecosystems. EPS are known for their ability to improve soil structure, hold water, and serve as a food source for microbes. Offering a good substitute for synthetic polymers, they are thus very important in agricultural settings and beyond. Produced by bacteria such as *Pseudomonas sp., Rhizobium sp.*, and *Lactobacillus sp.*, microbial EPS are particularly fascinating because of their great variety and functional properties (Bhagat et al., 2021). They can create biofilms that offer protection from abiotic stress and pathogens. For example, it is well known that the *Pseudomonas* fluorescens EPS increases plant resilience to abiotic stress and helps to acquire nutrients. This clarifies the importance of microbial EPS not only for maintaining the balance of ecosystems but also for raising agricultural output. Occurring in numerous species, including guar gum in guar beans and xanthan gum in *Xanthomonas* campestris, plant-based EPS play an equal part in this ecosystem. Food manufacturing makes great use of these biopolymers as thickeners and stabilizers. Their biodegradability also presents a better substitute for synthetic goods most often used nowadays. Like those of red algae, algal EPS shows traits that help to retain water and maintain soil quality, so improving plant growth (Tiwari et al., 2022). EPS's possible application as biocontrol agents and plant protectors is among the most well-known. EPS can help to lower plant diseases brought on by pathogens, so restricting the need for chemical pesticides. By restricting pathogenic access, EPS can stop the infection process by means of a protective film developed around plant roots (Abdalla et al., 2021). Studies on the use of EPS in agricultural surroundings show that it produces better crops with more yield and disease resistance. EPS are, all things considered, a workable answer for many agricultural issues. Their special qualities make them valuable biopolymers as soil conditioners for enhancing the health of soils, plant growth boosters, and natural protectants against infections; their sources range from microbial to algal and plant. The trend in using EPS for agriculture corresponds with worldwide tendencies in favor of environmental conservation and sustainability (Saha et al., 2020). More studies exposing the several applications of EPS will surely help to highlight their significance in creating environmentally friendly agricultural methods. Based on this current scenario, the

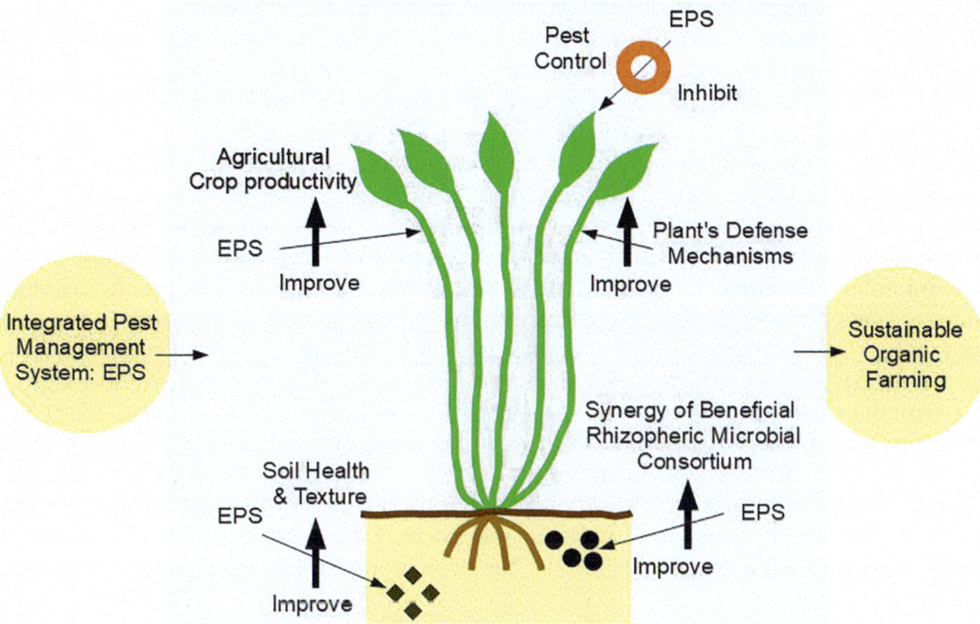

FIGURE 13.1 Potential application of natural EPS in pest control, plant growth-defense mechanism, soil health for effective organic farming to establish sustainable agriculture technology. *EPS*, Exopolysaccharides.

major goal of this extensive summarization is to depict diverse aspects of natural EPS (Fig. 13.1), considering its biochemical properties, sources, mode of action upon pest control, and soil health booster for establishing the integrated pest management to attain sustainable organic farming in the field of the agriculture sector.

13.2 Understanding natural exopolysaccharides and their properties

13.2.1 Sources of natural exopolysaccharides

Microbial EPS are microbe-secreted biopolymers that play essential functions in ecological and agricultural processes. This essay reviews the bacterial and fungal origins of EPS of agricultural interest, together with plant and algal EPS, highlighting their means of extraction and their known or prospective bioactivity. The review also addresses the influence of these biopolymers and the foremost drivers within the field, bearing in mind emerging trends. Bacterial EPS are produced by various bacterial species, both gram-positive and gram-negative bacteria (Li et al., 2021). Some examples include *Pseudomonas sp.*, *Bacillus sp.*, and *Rhizobium sp.* These bacteria produce EPS that are involved in biofilm development, soil stabilization, and promotion of plant growth. For example, the EPS of *Pseudomonas* fluorescens has been found to increase plant resistance against pathogenic fungi and facilitate nutrient uptake, thereby demonstrating significant agricultural importance (Mehmood et al., 2023). The process of this bioactivity is frequently based on the capacity of these EPS to function as a plant-protective cover, alleviate drought stress, and allow for beneficial plant-soil microorganism interactions. Fungal EPS is also an essential element of agricultural ecosystems. Fungi like *Trichoderma sp.* and mycorrhizal fungi produce EPS that increase soil health and plant growth (Devi et al., 2020). *Trichoderma harzianum* EPS, for instance, has been shown to enhance the resistance of crops against a number of stressors, such as biotic and abiotic stress. The EPS produced by these fungi tend to include functional groups capable of chelating minerals, water retention, and eliciting positive plant responses. This bifunctionality underscores their role in sustainable agriculture. Another aspect involves plant-derived EPS, for instance, in higher plants and algae. Plant EPS, which consists predominantly of pectins and hemicelluloses, provides cell wall stability and resilience to environmental pressures (Cosgrove, 2024). Agricultural purposes call for the isolation of such biopolymers for deciphering their function in crop survivability. The techniques used in isolation include alkaline or enzymatic treatment for releasing the EPS from plant biomass in a bioactive state. Algal EPS from species such as *Spirulina sp.* and *Chlorella sp.* also show significant promise with their own specific bioactivities, namely anti-inflammatory and antioxidant activity (Hamidi et al., 2020). Recent research indicates that algal EPS could be used to increase the fertility of soils and drive beneficial microbial activity. Extraction and structural elucidation of these polysaccharides are crucial to their use in agriculture. Traditional extraction methods like precipitation and hydrolysis have been the norm. However, innovation in methods of extraction, such as green extraction methods, has been on the rise in the recent past (Panda and Manickam, 2019).

Such processes reduce the ecological footprint as well as maintain bioactivity preservation. For example, enzymatic hydrolysis combined with membrane filtration methods provide effective and eco-friendly means of isolating and purifying polysaccharides (Kim et al., 2022). Such developments not only provide greater yields of EPS but also maintain the structural characteristics responsible for bioactivity. The bioactivity of EPS is supported by their intricate structures, and these may differ vastly depending on the sources. The polysaccharide matrix of EPS provides a gel-like scaffold that is capable of encapsulating other molecules and imparting protective or growth-stimulatory properties. New research on the structure-function interaction of these EPS has discovered their ability to serve as biocontrol agents, plant growth promotors, and soil conditioners, and has proven their multifunctional potential. For instance, EPS from marine algae have been shown to stimulate the growth of different terrestrial crops, as well as increase soil microbial diversity (Abinandan et al., 2019). Probing future developments for the microbial EPS field is upbeat. As global focus on green agriculture and ecological practices increases, demand for agricultural biopolymers that ensure increased crop outputs while reducing effects on the environment is increasing. Further research on microbial genetics as well as on biosynthesis systems, will enable the development of EPS types with tailored features for particular crop applications. Additionally, progress in synthetic biology will permit the development of microorganisms capable of synthesizing improved EPS variants with enhanced bioactivities. Their roles in sustaining soil health, growth promotion, and maintenance of desirable interactions within the agricultural environment indicate their relevance. The development of extraction techniques and a better understanding of their bioactivity will enable them to be used in actual agricultural applications (Bérard et al., 2020). With the agricultural industry moving toward sustainable practices, the incorporation of these biopolymers is bound to be an important factor in meeting the challenges of contemporary farming.

Microbe-secreted biopolymers with vital roles in ecological and agricultural systems are the sources of natural EPS microbial EPS. Coupled with plant and algal EPS, this essay discusses the bacterial and fungal sources of agricultural importance, coupled with their means of extraction and their known or predicted bioactivity. Considering new trends, the review also covers the impact of these biopolymers and the main drivers in the sector. Various bacterial species—both gram-positive and gram-negative—produce bacterial EPS (Li et al., 2021). Among some are *Rhizobium sp.*, *Bacillus sp.*, and *Pseudomonas sp.* These bacteria generate EPS implicated in soil stability, biofilm formation, and enhancement of plant development. For instance, the EPS of *Pseudomonas* fluorescens has been observed to boost plant resistance against pathogenic fungi and enable nutrient absorption, therefore proving great agricultural relevance (Mehmood et al., 2023). The capacity of these EPS to serve as a plant-protective cover, reduce drought stress, and enable favorable plant-soil microbial interactions often drives the mechanism of this bioactivity. Furthermore, a necessary component of agricultural ecosystems is fungal EPS. Fungi similar to *Trichoderma sp.* and mycorrhizal fungi generate EPS that improve plant development and soil condition (Devi et al., 2020). For example, *Trichoderma harzianum* EPS has been demonstrated to improve crop tolerance against several types of stress, including biotic and abiotic stress. These fungi generate functional groups able to chelate minerals, water retention, and induce beneficial plant responses. This bi-functionality emphasizes their part in environmentally friendly farming. Another element is plant-derived EPS, for example, found in higher plants and algae. Mostly composed of pectins and hemicelluloses, plant EPS give cell wall integrity and resilience against environmental stressors (Cosgrove, 2024). Agricultural uses demand the separation of such biopolymers for their interpretation in crop viability. The methods applied in isolation consist of alkaline or enzymatic treatment for EPS release from plant biomass in a bioactive form. With their own particular bioactivities—anti-inflammatory and antioxidant activity—algal EPS from species including *Spirulina sp.* and *Chlorella sp.* also show great promise (Hamidi et al., 2020). According to recent studies, algal EPS could boost soil fertility and stimulate good microbial activity. Their application in agriculture depends critically on their extraction and structural clarity. The usual techniques have been conventional extraction techniques, including hydrolysis and precipitation. On the other hand, recent innovations in ways of extraction, such as green extraction techniques, have been increasing (Panda and Manickam, 2019). These techniques preserve bioactivity and lower the environmental impact as well. For instance, effective enzymatic hydrolysis paired with membrane filtering techniques offers environmentally friendly methods of separating and cleaning polysaccharides (Kim et al., 2022). The complex structures of EPS help to explain their bioactivity. These could differ greatly based on the sources. The polysaccharide matrix of EPS offers a gel-like framework able to encapsulate other molecules and apply a protective or growth-promoting effect. motivating qualities. For example, EPS from sea algae have been found to have found to boost soil microbial variety as well as the growth of several terrestrial crops (Abinandan et al., 2019). Looking forward, developments for the microbial EPS field seem promising. Demand for agricultural products rises as attention on green farming and environmentally friendly methods spreads around the world, biopolymers guaranteeing higher crop yields while minimizing environmental impact is getting more. Additional investigation on biosynthetic systems and microbial genetics will help to evolution of EPS forms with specific characteristics for certain agricultural uses. Apart from that, advancement in synthetic biology will allow the synthesis of microbes fit for generating increased bioactivities and EPS variations. Their jobs in maintaining soil health, development encouragement, and preservation of desired interactions inside the agricultural surroundings highlight their relevance. The evolution of better extraction methods and knowing their bioactivity will help them to be applied practically in agriculture (Berard et al., 2020). The agricultural sector is headed toward sustainable methods, so the Using these biopolymers is definitely going to be crucial in overcoming the obstacles of modern cultivation.

13.2.2 Structural and functional properties

Exopolysaccharides are polymers of high molecular weight produced by many bacteria and fungi, among other microbes. Mostly, their chemical composition consists of monosaccharides joined together by glycosidic linkages. EPS has structural variations. These are resulting from variations in the monosaccharide units and the kinds of glycosidic bonds. Usually, EPS can comprise glucose, galactose, mannose, rhamnose, and others. The degree of polymerization influences molecular weight, hence influencing EPS characteristics in other uses (Babiak and Krzemińska, 2021). EPS has attracted more attention since its debut, several qualities relevant for agricultural systems' insect control. Viscosity is one of the amazing properties of EPS; it may be regulated depending on the chemical composition and molecular structure. This is a crucial characteristic since higher viscosity increases mucilage formation, therefore creating an exudate protective layer against several environmental pressures, including pest invasion. Small insect mobility can be reduced by viscous EPS solutions, therefore lowering their capacity to reach plant surfaces (Jha et al., 2022). The excellent film-forming capacity of EPS is another key quality. On plants, EPS can create a film covering the surface that guards against diseases and pests. The movie acts as a physical barrier preventing bug access and holding moisture. For instance, some EPS produced from microbial sources form stable films that effectively prevent bacterial and fungal diseases. Such a capacity to produce such films can help to reduce reliance on synthetic pesticides, hence promoting more ecologically friendly farming. Another appealing quality of EPS that helps to increase its application in pest management is its bioactivity. Antimicrobial action of some EPS helps them to stop the spread of various infections. The bioactive molecules could possibly interact with plant-defense chemicals, therefore strengthening the plant's natural resistance against pests (Jamiołkowska, 2020). A fresh approach in integrated pest control is presented by the interaction of EPS with plant chemicals. Research reveals a spectrum of interactions between EPS, plant surfaces, and insect species. For usage as a pest control, for instance, studies show that EPS can change the surface tension of water to maximize wetting of surfaces and increase coverage. Furthermore, EPS can stop the attachment of dangerous pests and help helpful microbes to stick to plant surfaces. Changing the dynamics of such interactions helps EPS to increase the health and resilience of plants, hence feeding into sustainable agricultural systems. Modern developments in the field of discipline accent highlight the need for EPS isolation and characterization techniques. Better techniques provide for a closer understanding of the molecular mechanisms behind their bioactivity. Furthermore, supporting the identification of microorganisms able to generate EPS with desired properties is an advancement in proteomics and genetics. These days, researchers want to use synthetic biology in engineering microbes to create custom-made EPS that would be useful for pest management. Historically, important turning points in microbial biotechnology have seen the study and application of EPS in agriculture. Our knowledge of microbes comes from the work of such pioneers as Robert Koch and Louis Pasteur (Céspedes et al., 2021). Their attempts to separate bacterial strains resulted in the creation of the contemporary biotechnological techniques, best shown by new studies on EPS's role in plant disease and pest resistance. Several studies over the last ten years have shown how well EPS controls pests in different crops. Research on EPS generated from *Lactobacillus* species, for example, shows considerable potential for controlling aphid infestation in lettuce (Zheng et al., 2022). Furthermore, underlining the potential of EPS in integrated pest control, another study found that EPS generated from *Pseudomonas* fluorescens will be able to stop the growth of fungus pathogens pathogenic to tomatoes. From fully pure agricultural uses to more significant environmental repercussions, the perspective on the use of EPS in pest control varies. While most people agree that natural substitutes like EPS should replace man-made pesticides, there are questions regarding the viability of increasing EPS production. To guarantee steady and cost-effective industrial-scale EPS production, optimization of this process must be guaranteed (Revin et al., 2023). The possibilities of EPS for control of pests seem to be growing. Research will be expected to create optimum EPS production via fermenting techniques, strain selection for high yields, and enhancement of their bioactive properties. Furthermore, demand for nontoxic pest control will rise as organic farming and cannabis production grow more frequently. Promising area of study with possible application in pest control is EPS (Tahir and Hasnain, 2020). Together, their molecular weight, chemical composition, and structural variations give physical characteristics, including viscosity, film-forming capacity, and bioactivity, that give plant defensive systems more strength. The application and knowledge of EPS in sustainable agriculture will continue to extend as further breakthroughs are made via study, therefore confirming its promise to reduce problems associated to pests.

13.2.3 Mechanisms of action

Integrated pest management (IPM) techniques use a variety of mechanisms of action to control insect populations in a practical manner, therefore enhancing agricultural output and sustainability. Among the methods of action, direct effects include feeding interference, toxicity, and repellence, are absolutely important. Management of pests also depends critically on indirect effects, including stimulation of plant defense systems and induction of beneficial bacteria. Repellence can help to show direct influence on pests by discouraging insects from actively invading plants. For instance, peppermint and citronella essential oils have been found to have strong repellent action (Coetzee et al., 2022). These compounds reduce the pests' capacity to locate their hosts, therefore preventing losses of crops. Conversely, toxins are those that, upon exposure, kill or incapacitate the pests. For instance, the biopesticide *Bacillus thuringiensis* produces toxins that especially target lepidopteran larvae, hence providing a selective and environmentally benign control (Devi et al., 2020; Vimala Devi et al., 2020). Another important tactic is disturbance of eating; some insecticides reduce feeding activity, influencing the behavior of development and reproduction among the insects. Apart from direct effects, indirect systems are also quite crucial for enhancing strategies of pest control. A main indirect result is the development of plant defense mechanisms. Researchers have shown that some bio-insecticides activate secondary metabolites in plants, acting as natural pesticides (Wani et al., 2020). This not only strengthens the plant's long-term ecological resilience but also helps it to fight against pest attacks at that specific period. Another vital element of indirect insect management is promoting good bacteria. Some bacteria, such as mycorrhizal fungi, can improve plant resilience against many pests and diseases as well as their general state. These beneficial bacteria have the ability to boost nutrient absorption, improving general plant strength and reducing the impact of insect outbreaks (Ouaarous et al., 2025). Finally, the development of films as a physical barrier serves as an adaptive strategy. These bio-degradable films could be placed on plant surfaces so that, at the same time, allowing air and water penetration, a barrier is created preventing insect entrance. This approach is becoming popular since it conforms to sustainable agriculture and reduces chemical reliance in farming (Chojnacka, 2024). Effective pest management is based on a multidimensional process including both direct and indirect effects. Sustainable agricultural methods can bloom by combining repellence, toxicity, feeding interference, and plant defense stimulation with support for beneficial bacteria and the use of physical barriers (Manjunath et al., 2023). This whole technique improves environmental health in addition to lowering pest hazards.

13.3 Potential applications of exopolysaccharides in pest control

13.3.1 Exopolysaccharides as biopesticides

Targeting certain agricultural pests like insect pests, nematodes, and fungal infections, EPS-based biopesticides have attracted a lot of research over the past several years. While being more ecologically benign and nontarget organism-friendly, EPS, produced by many microorganisms, can boost the efficiency of biopesticides. Insect pests are main hazards to agriculture, resulting in significant crop losses. This essay addresses the use of EPS-based biopesticides in pest control and provides some examples of EPS with proven pesticidal activity. Current studies indicate that aphids and other piercing-sucking insects can be effectively controlled by EPS from bacterial species, including *Bacillus* subtilis and *Pseudomonas* fluorescens. Like levans and xanthan, these EPS compounds have characteristics that affect insect or metabolic processes as well as eating behavior. The two-way benefit of *Bacillus* subtilis in environmentally friendly farming is shown by a recent study (Bastas et al., 2023), which showed that EPS from the bacterium not only slowed aphid development but also encouraged plant development. Still another type of pests attacking crops are nematodes. Recent research shows EPS can also fight these small roundworms. It has been found that several fungi, including *Trichoderma harzianum*, have nematicidal EPS. It acts by giving nematodes an adverse environment and simultaneously promotes the multiplication of helpful microorganisms capable of ruling harmful species. It has been noted (Nafady et al., 2022) that EPS from *Trichoderma harzianum* showed great nematicidal activity against *Meloidogyne incognita*, offering a potential method of nematode infestation control. Another immediate concern for crop well-being is fungal infections. Additionally, acting as a defense layer preventing fungal invasion is EPS. For instance, using EPS from *Lactobacillus plantarum* has been shown to be quite successful against the pathogenic fungus *Botrytis cinerea*, which compromises various fruits and vegetables (Romero et al., 2022). Increased resistance and yield came from the EPS's stimulation of the plant's defense system and monitoring of the fungus's development. All things considered, the creation of EPS-based biopesticides offers a hopeful method of environmentally friendly pest management. These biopesticides offer a less dangerous choice to synthetic pesticides since they may destroy fungal infections, nematode parasites, and insect pests. The particularity of some EPS, notably those of *Bacillus sp.* and *Trichoderma sp.*, also shows their possible value as effective agents in reducing agricultural pests and improving plant condition.

13.3.2 Exopolysaccharides as plant protectants

Due to their capacity to increase plant resistance to viruses and insects, polysaccharides—also known as EPS—have attracted especially attention in the last few years as plant protectants. The defensive mechanisms of the plant depend mostly on these microbially generated biopolymers. A vital response that prepares plants to fight would-be pathogens (Sarkar et al., 2025), EPS can induce systemic acquired resistance (SAR). Therefore, the use of EPS is being more and more appreciated as a technique by which plant vigor and health can be promoted in agriculture, especially with the world seeking sustainable approaches to fend off crop loss. Plants under biotic stress—that is, from insects or pathogens—often set off a series of defense mechanisms. Among the crucial reactions is systematic acquired resistance. Important in initiating defense mechanisms in the remote areas of the plant, EPS has been reported to cause the synthesis of signaling molecules like salicylic acid and jasmonic acid (Li et al., 2019). Using EPS from particular bacterial isolates clearly raised the amount of expression of defense genes in tomato plants, thus generating more resistance against fungal diseases (Drira et al., 2021). EPS capacity to boost plant health goes beyond mere resistance against diseases. By means of better nutrient acquisition and higher soil conditions, EPS increases plant vigor. They could improve soil structure, increase water-holding capacity, and help good microorganisms proliferate. For crops like wheat and maize, for example, EPS have been shown to boost root development and biomass, which finally results in higher yield (Dar et al., 2021). EPS's two-way properties—that of a protective layer and growth stimulation—offer a whole crop management plan. Furthermore, given the continuous emphasis on sustainable farming, EPS presents a green choice over synthetic pesticides. Application of EPS improves soil condition and biodiversity in addition to reducing chemical inputs dependency. Recent research show that EPS application can be matched with agro-ecosystems to sustain higher output at the same time as a greener environment is promoted. All things considered, EPS coatings show a good path for increasing plant health and vitality as well as plant immunity against pests and diseases. Their approaches of inducing systematic acquired resistance and improving soil fertility make them assets in sustainable farming operations (Velmourougne et al., 2023). Given continuous research showing the whole potential of EPS, including these biopolymers into agricultural methods could have a major effect on ensuring food production under rising environmental pressure.

13.3.3 Exopolysaccharides in soil health and pest suppression

Important in maintaining soil structure and increasing microbial diversity, EPS in soil health and pest suppression. EPS are microbial polymers with high molecular weights produced by a wide spectrum of microorganisms. These substances improve the aggregate stability of soil, hence improving water retention and lowering of erosion. By giving a stable environment for soil microbes, EPS supports soil health and nutrient cycling, therefore promoting plant development. The development of soil aggregates is one of the most significant ways EPS influences soils. Groups of soil particles called aggregates let air and water flow through the ground, hence improving drainage and root penetration. In their study, Naveen and Balachandar (2025) found that by binding soil particles together, EPS from soil bacteria, including *Pseudomonas*, might enhance the structure of soil. This concentration allows a diversified microbial community to be produced since varied microorganisms can coexist here in these nourishing surroundings. Furthermore, useful soil bacteria encouraged by EPS help control pests. For instance, some species of bacteria and fungi can create naturally occurring insecticides. They accomplish this by either breaking down harmful chemicals in the soil or by vying with poisonous insects. Applications of EPS-generating microbial inoculants can essentially reduce insect numbers in agricultural environments, so producing healthier crops free from the need for chemical pesticides, claims Saha et al. (2020). EPS helps microorganisms to break down organic debris, therefore releasing nutrients into the soil. Plants find such nutrients more easily, which increases their development and output. Research showing plants grown on soils with added EPS show enhanced nutrient absorption led to higher biomass and yield (Zhou et al., 2024). This improved nutrient availability supports sustainable agriculture in addition to plant health. Generally speaking, soil health depends critically on EPS's function in microbial diversity and soil structure. EPS has a multifarious function in supporting plant development by improving soil aggregate formation, supporting beneficial microbes, and raising nutrient availability. With an eye toward its sustainability in farming, researchers are always looking at EPS applications in agriculture (Redmile-Gordon et al., 2020). Particularly in view of rising needs for food production in an environmentally responsible manner, the utilization of EPS-producing organisms for soil management holds enormous possibilities for agricultural operations of the future.

13.3.4 Exopolysaccharides in integrated pest management

Effective techniques of pest management are much advanced by combining environmentally friendly activities, such as EPS-based solutions, with other organic approaches of pest control. EPS, are biopolymers in nature and can boost plant growth and help resist pests in the environment. Usually involving the use of beneficial insects and biocontrol agents, organic control methods are fit for integration with EPS measures to create a more integrated and sustainable IPM program. Harmony between the combined approach and beneficial insects is important for their success. For example, some EPS helps plants grow, thereby making them more resistant and decreasing the chances of pests entering (Morcillo and Manzanera, 2021). Attracting helpful insects, such as ladybugs and lacewings, which are known to reduce aphid numbers, this quality can be especially helpful. Apart from improving plant vitality, EPS makes the surroundings suitable for the survival of these helpful insects. EPS can also be a nutritional supply for biocontrol agents, therefore enhancing their effectiveness. For example, using plant extracts high in EPS has been demonstrated to increase predatory insect populations, so allowing them to more effectively help to manage pests. Such synergy between EPS and biocontrol chemicals shows the great possibility of such a mix in creating successful pest control programs. Establishing viable IPM projects based on EPS calls for multifarious strategies that balance the use of biocontrol agents and EPS formulations with other organic methods. Management of these habitats, rotation of crops, and targets of specific pest control operations may be needed. These initiatives aim to reduce the usage of chemical pesticides by means of efficient insect control, hence safeguarding crops. These techniques used together encourage a stronger agricultural ecosystem capable of resisting pest pressure without too harming the surroundings. Furthermore, ongoing investigation on optimal EPS formulations will surely provide fresh data optimizing compatibility with present biocontrol methods. Knowledge on plant-microbe interactions could lead to fresh applications enhancing the effectiveness of projects aimed at controlling pests even more. All things considered, EPS-based strategies taken in concert with other natural pest management methods have considerable promise for producing sustainable IPM programs (Singh et al., 2022). In addition to enhancing agricultural productivity, these combination tactics do not simply benefit the environment, but they also increase compatibility with beneficial insects and other biocontrol agents. The development of strong and reliable pest control systems depends on constant research and the use of these strategies.

13.4 Formulation and delivery systems

The agricultural sector is upon waste-effective "gentle on the environment" methods of pest management. Among these activities, reaching good target pest control largely relies on the production of effective environmental pest control (EPS) products such as sprays, coatings, and granules. The formulations have to be adjusted for effective delivery methods that enhance performance with minimum impact on nontarget species and the surroundings. Great promise exists in new developments in EPS product formulation. For instance, sprays are usually used since they are easy to use and cover large areas. But adding surfactants and stabilizers helps to improve adhesion and lower drift; so, their formulation has to be changed (Katz et al., 2022). The right mix of these elements guarantees that the active agents reach the intended pests with efficiency and lowers the product loss by environmental dispersion or evaporation. Likewise, coatings have proven to be a consistent method to guard crops since they can offer active compounds gradually and act as barriers against insects, therefore improving plant health. Conversely, granules provide regulated release and convenience of application, therefore addressing certain pest infestations at different phases of development. Using nanotechnology in EPS compositions is a major breakthrough in insect control. Active compounds can become more powerful by targeted distribution, surface area, and solubility of nanoparticles. For example, active component release from nano carriers can be delayed postencapsulation, thereby reducing the environmental impact and overall treatment, amongst other benefits to the environment. Moreover, said to be able by nanotechnology to increase the absorption and permanency of EPS products, which is their efficacy against pests (Kumar et al., 2022). Moreover, improved distribution system optimization for these formulations will lead to greater results in insect control. The deployment of technology such as drone-enabled supported spraying and precision agriculture techniques made it possible to apply it in a much more targeted way, which results in reducing the runoff and also the demand for supplies. They serve to lay a very fundamental-based role to ensure that environmentally sound farming is carried out, and they are the site-specific pest control. All things considered, the continuous inventiveness in successful EPS formulations, along with the application of nanotechnology and enhanced delivery technologies, is propelling pest control toward a more efficient, concentrated, and ecologically friendly future (Mondéjar-López et al., 2024). As the study progresses, these technologies will be increasingly accountable for the creation of sustainable agriculture by means of better pest control strategies without compromising ecosystem integrity.

13.5 Synergistic effects with other biocontrol agents

Given the effects of chemical pesticides on the surroundings, the hunt for sustainable pest management methods has been more important throughout the last few years. Integration of extracellular polymeric compounds (EPS) with various natural chemicals, such as plant extracts and microbial agents (Lahiri et al., 2019), is one of the most fascinating paths of research in this field. Apart from enhancing plant resilience, this combined approach raises the effectiveness of pest control, hence reducing crop vulnerability to invasion. Microbial natural polymers called EPS find application in many different fields of agriculture. Soil quality regulate plant tolerance against biotic stresses, and promotes plant growth enhance by these. EPS is combined with plant extracts with bioactive elements; its benefits can be significantly strengthened. For example, studies have shown that if EPS is mixed with neem oil, the neem plants derived from the seeds of the neem tree, that mixing can produce more effective ways of pest management. Among the active components in neem oil, azadirachtin enhances EPS to disrupt insect life cycles, therefore lowering their population densities (Fabreag et al., 2019). Apart from plant extracts, EPS can be blended with microbial components, such as beneficial fungi and bacteria, to increase pest resistance. Among bacterial species, *Bacillus* thuringiensis is one that generates poisons specifically targeting and killing problematic insects. Researchers can better distribute and preserve the bacteria in the agricultural environment by means of EPS matrix entrapping, therefore improving their pest control efficiency. For example, research demonstrates that EPS enhances the efficacy and survival rate of microbiological agents to such an extent that pest control is improved. Combining EPS with other natural chemicals also provides a whole approach to insect control. Minimized is this primary flaw of conventional pesticides, which increases the possibility of resistance by bug populations. By supplementing this method with biocontrol agents (parasitoids or natural predator insects), nature's way of controlling by natural ecological processes, will result in the formation of a balanced system (the system with FPS set in EPS as a base). Reduced chemical use also fits environmental sustainability and is helpful for the preservation of agricultural output. All things considered, combining EPS with microbial agents and plant extracts offers a respectable approach to raising the efficacy of pest management (Riseh et al., 2022). Sustainable agricultural technologies are based on this cooperative approach since it not only improves the efficiency of pest management but also raises plant resilience to pests. More investigation in this area is required to generate more sophisticated, ecologically friendly pest control plans fit for the goals of world sustainability.

13.6 Challenges in exopolysaccharides production and application for pest control in organic farming

Using EPS (exopolysaccharides) as pest management in organic farming presents some significant difficulties for manufacture and use. Made by different microorganisms, EPS, which is produced, has drawn attention because of its possible sustainability for farming. Though EPS shows enormous potential as a biopesticide, its actual use is hampered. Variability in yield and quality of EPS is one of the main obstacles in its manufacturing process. Different environmental conditions and specific strains of the microorganisms themselves used in their manufacturing help to contribute to such variety. Furthermore, necessary for organic agricultural systems with an eye on biodiversity to have consistent, predictable EPS yields under varying conditions. To enhance production and quality, researchers keep working to maximize culture conditions, including environmental parameters and nutritional supplies (Jyoti et al., 2024). Still, it is difficult to produce a homogeneous product that satisfies the rigorous organic certification standards. Apart from production problems, the efficacy and compatibility of EPS as an organic pest control agent impede its application. EPS may have antifungal and antibacterial properties, however its effectiveness against a range of pests may be rather different. According to current research, EPS may interact in concert with other organic pest control strategies—such as insecticidal soaps or plant extracts to improve general pest control practices (Holka and Kowalska, 2023). Still, further research is required to define efficient application methods since the mechanisms of interaction between EPS and other biopesticides are in their infancy. Furthermore, influencing the acceptance of EPS in organic farming are customer opinions and the legal surroundings.

Although EPS is expected to offer advantages, consumers are mostly ignorant of biopesticides, which can postpone their acceptance. Farmers might not use EPS products without particular direction and assurances of safety and effectiveness. Regulatory authorities have to create carefully defined guidelines for the use of EPS in organic systems if they want to encourage more acceptance. In essence, even if EPS shows great promise for organic farming in terms of pest control, obstacles have to be solved (Baker et al., 2020). Major obstacles to be solved are variations in EPS production, efficacy against a spectrum of pests, and customer acceptance. Constant research and cooperation among scientists, authorities, and farmers will be needed to totally realize EPS in sustainable agriculture.

13.7 Conclusion

Given the rising need for environmentally friendly farming, organic pest management has become increasingly more important recently. Out of all the several techniques under research, EPS is among the most fascinating ones. High-molecular-weight carbohydrates made by bacteria, EPS, have significant potential to support environmental sustainability and economic viability as well as to help to lower insect concerns. Mostly applied in organic pest control, EPS is a biocontrol agent. Studies show that EPS can raise the effectiveness of parasite species aimed at agricultural pests as well as natural predators. For instance, past studies have already shown that certain bacterial EPS can enhance the survival and reproductive rates of beneficial insects, thereby raising their capacity to naturally reduce numbers of their own pests. It is a biocontrol method of suppressing synthetic pesticide use and reducing the chemical residue in food crops, which therefore removes harmful effects on nontarget species such as pollinators and beneficial insects. Strongly also against plant diseases, EPS possesses antifungal and antibacterial qualities. Following organic agricultural guidelines lets farmers protect their crops against several kinds of diseases by using EPS as an antibacterial agent. In addition to enhancing agricultural output, this helps the environment by reducing the chemical runoff into the water bodies. Using EPS for pest control seems to provide very clear environmental benefits (Friuli et al., 2023). Farmers could assist in preserving biodiversity and lowering soil and water contamination by using fewer chemical pesticides. Moreover, EPS may retain water and improve the condition of the soil, so promoting more stable agricultural systems. Better microbial activity connected with EPS utilization and more soil organic matter helps the agroecosystem to be generally healthy. For farmers financially, using EPS in organic pest control would save a lot of money (Durham and Mizik, 2021). Farmers may greatly lower their cost of production by lowering crop losses from diseases and pests and relying less on expensive chemical inputs. Using sustainable agriculture with EPS can even assist in creating markets since customers are drawn to organic items generated with the least potential influence on the environment. All things considered, EPS gives some economic and environmental benefits as well as a quite useful tool for organic pest management. This is why the possibilities of biocontrol and disease control underline how effective and adaptable they are. In other words, scientists find their powers, and EPS could start to enter an important role in defining the probable directions of sustainable development.

EPS-based pest management in organic agriculture has bright future potential since EPS compounds are recognized to help to support sustainable development. Natural pest control mechanism (Reddy et al., 2022), microbial biopolymers called EPS can enhance soil quality, stimulate plant development, and naturally occur pest control mechanism. Since organic farming focuses on pollutants and environmental responsibility, EPS's inclusion into pest control plans is a novel idea for reducing the use of chemical pesticides, which is currently a need. Recent research shows how effectively EPS increases crop resilience against invasion by pests. Some EPS-producing bacteria, for example, have been found to cause systemic resistance in plants, hence enhancing their natural protection against phytopathogens and insects such as aphids and spider mites. These techniques not only minimize the ecological effect of traditional approaches of pest control but also diminish the loss of crops by means of pest attacks. Adoption of EPS-based solutions corresponds with the underlying ideas of organic farming, meant to preserve biodiversity, soil, and plant health as the knowledge of the negative consequences of chemical pesticides rises. Still, further study is required to identify which EPS-producing microbe strains would be most suited under different agricultural environments before EPS may take front stage in organic pest control. The study of the interactions between plant crop types and different EPS compounds is one essential field of inquiry (Chen et al., 2024). Understanding these links could thus help to enable optimal usage of EPS focused on particular crops and regional settings, so improving both effectiveness and farmers' acceptance of EPS. Research and development will also have to focus on creating user-friendly EPS products for farmers, combining EPS with companion technologies, including biopesticides or other organic additions. Adoption of EPS among farmers will depend on developing appropriate rules and best practices for its application in pest control systems being adopted (Sabbahi et al., 2022). All things considered, EPS-based pest control technologies have significant potential to support organic farming methods aiming at increased sustainability. Not only could they help to reduce the usage of synthetic chemicals, but they also have the ecologically friendly ability to improve plant condition. Emphasizing useful applications and farmer involvement will be crucial as research goes on so that the benefits of EPS are fully realized in organic farming surroundings.

Credit author statement

Dr. Dipankar Ghosh prepared the manuscript and figures. Dr. Dipankar Ghosh, Dr. O. N. Tiwari, and Dr. M. Indira carried out the final editing and provided value addition to the chapter. All authors reviewed and approved the final version.

Conflict of interest

The authors declare that they have no conflict of interest in this chapter.

Ethical statement

The authors confirm that this chapter includes no studies involving human participants or animals.

Acknowledgment

Dr. D. Ghosh appreciates the help and support of JIS University and JIS Group of Educational Initiatives. Dr. O.N. Tiwari would like to thank CCUBGA and ICAR-IARI for their immense support and help.

References

Abdalla, A.K., Ayyash, M.M., Olaimat, A.N., Osaili, T.M., Al-Nabulsi, A.A., Shah, N.P., Holley, R., 2021. Exopolysaccharides as antimicrobial agents: mechanism and spectrum of activity. Front. Microbiol 12. doi:10.3389/fmicb.2021.664395.

Abinandan, S., Subashchandrabose, S.R., Venkateswarlu, K., Megharaj, M., 2019. Soil microalgae and cyanobacteria: The biotechnological potential in the maintenance of soil fertility and health. Crit. Rev. Biotechnol 39 (8), 981–998. doi:10.1080/07388551.2019.1654972.

Archana, H.R., Darshan, K., Amrutha Lakshmi, M., Ghoshal, T., Bashayal, B.M., Aggarwal, R., 2022. Biopesticides: a key player in agro-environmental sustainability. In: Trends of Applied Microbiology for Sustainable Economy. Elsevier, India, pp. 613–653. doi:10.1016/B978-0-323-91595-3.00021-5.

Babiak, W., Krzemińska, I., 2021. Extracellular polymeric substances (EPS) as microalgal bioproducts: a review of factors affecting EPS synthesis and application in flocculation processes. Energies 14 (13), 4007. doi:10.3390/en14134007.

Baker, B.P., Green, T.A., Loker, A.J., 2020. Biological control and integrated pest management in organic and conventional systems. Biol. Control 140. doi:10.1016/j.biocontrol.2019.104095.

Bastas, K.K, Kumar, A., Sivakumar, U., 2023. Microbial biocontrol: molecular perspective in plant disease management. Microorg. Sustain. 49. doi:10.1007/978-981-99-3947-3.

Bérard, A., Clavel, T., Bourvellec, C.L., Davoine, A., Gall, S.Le, Doussan, C., Bureau, S., 2020. Exopolysaccharides in the rhizosphere: a comparative study of extraction methods. Application to their quantification in Mediterranean soils. Soil Biol. Biochem. 149, 107961. doi:10.1016/j.soilbio.2020.107961.

Bhagat, N., Raghav, M., Dubey, S., Bedi, N., 2021. Bacterial exopolysaccharides: insight into their role in plant abiotic stress tolerance. J. Microbiol. Biotechnol. 31 (8), 1045–1059. doi:10.4014/jmb.2105.05009.

Bostan, I., Onofrei, M., Gavriluţă (Vatamanu), A.F., Toderaşcu, C., Lazăr, C.M., 2019. An integrated approach to current trends in organic food in the EU. Foods 8 (5), 144. doi:10.3390/foods8050144.

Céspedes, L.D., Fonseca, Y.M.C., 2021. The relationship between microbiology as a science and the microbiology subject. Chem. Biomol. Eng. 6 (3), 43–48.

Chen, Q., Song, Y., An, Y., Lu, Y., Zhong, G., 2024. Mechanisms and impact of rhizosphere microbial metabolites on crop health, traits, functional components: a comprehensive review. Molecules 29 (24), 5922. doi:10.3390/molecules29245922.

Chojnacka, K., 2024. Sustainable chemistry in adaptive agriculture: a review. Curr. Opin. Green Sustain. Chem. 46, 100898. doi:10.1016/j.cogsc.2024.100898.

Coetzee, D., Militky, J., Venkataraman, M., 2022. Functional coatings by natural and synthetic agents for insect control and their applications. Coatings 12 (4), 476. doi:10.3390/coatings12040476.

Cosgrove, D.J., 2024. Structure and growth of plant cell walls. Nat. Rev. Mol. Cell Biol 25 (5), 340–358. doi:10.1038/s41580-023-00691-y.

Dar, A., Zahir, Z.A., Iqbal, M., Mehmood, A., Javed, A., Hussain, A., Bushra, M.A., 2021. Efficacy of rhizobacterial exopolysaccharides in improving plant growth, physiology, and soil properties. Env. Monit. Assess 193 (8). doi:10.1007/s10661-021-09286-6.

Devi, R., Kaur, T., Kour, D., Rana, K.L., Yadav, A., Yadav, A.N., 2020. Beneficial fungal communities from different habitats and their roles in plant growth promotion and soil health. Microb. Biosyst. 5 (1), 21–24. doi:10.21608/MB.2020.32802.1016.

Drira, M., Elleuch, J., Hlima, H.B, Hentati, F., Gardarin, C., Rihouey, C., Cerf, D.Le, Michaud, P., Abdelkafi, S., Fendri, I., 2021. Optimization of exopolysaccharides production by porphyridium sordidum and their potential to induce defense responses in *Arabidopsis thaliana* against fusarium oxysporum. Biomolecules 11 (2), 282. doi:10.3390/biom11020282.

Durham, T.C., Mizik, T., 2021. Comparative economics of conventional, organic, and alternative agricultural production systems. Economies 9 (2), 64. doi:10.3390/economies9020064.

Fabreag, M.A.C., Familara, J.A., 2019. Biodegradation of expanded polystyrene (EPS)(Styrofoam) block as feedstock to *Tribolium castaneum* (red flour beetle) imago: a promising plastic-degrading process. World News Nat. Sci. (24) 145–156.

Friuli, M., Pellegrino, R., Lamanna, L., Nitti, P., Madaghiele, M., Demitri, C., 2023. Materials engineering to help pest control: a narrative overview of biopolymer-based entomopathogenic fungi formulations. J. Fungi 9 (9). doi:10.3390/jof9090918.

Gamage, A., Gangahagedara, R., Gamage, J., Jayasinghe, N., Kodikara, N., Suraweera, P., Merah, O., 2023. Role of organic farming for achieving sustainability in agriculture. Farming Syst. 1 (1), 100005. doi:10.1016/j.farsys.2023.100005.

Hamidi, M., Safarzadeh Kozani, P., Safarzadeh Kozani, P., Pierre, G., Michaud, P., Delattre, C., 2020. Marine bacteria versus microalgae: who is the best for biotechnological production of bioactive compounds with antioxidant properties and other biological applications? Mar. Drugs 18 (1), 28. doi:10.3390/md18010028.

Hasan, M.K., Tanaka, T.S.T., Alam, M.M., Ali, M.R., Saha, C.K., 2020. Impact of modern rice harvesting practices over traditional ones. Rev. Agric. Sci., 8, 89–108. doi:10.7831/ras.8.0_89.

Holka, M., Kowalska, J., 2023. The potential of adjuvants used with microbiological control of insect pests with emphasis on organic farming. Agriculture 13 (9), 1659. doi:10.3390/agriculture13091659.

Jamiołkowska, A., 2020. Natural compounds as elicitors of plant resistance against diseases and new biocontrol strategies. Agronomy 10 (2), 173. doi:10.3390/agronomy10020173.

Jha, S., Bhadani, N.K., Kumar, A., Sengupta, T.K., 2022. Glucose-induced biofilm formation in *Bacillus thuringiensis* KPWP1 is associated with increased cell surface hydrophobicity and increased production of exopolymeric substances. Curr. Microbiol. 79 (1). doi:10.1007/s00284-021-02699-z.

Jyoti, K., Soni, K., Chandra, R., 2024. Optimization of the production of exopolysaccharide (EPS) from biofilm-forming bacterial consortium using different parameters. Microbe 4, 100117. doi:10.1016/j.microb.2024.100117.

Katz, J.S., Chou, D.K., Christian, T.R., Das, T.K., Patel, M., Singh, S.N., Wen, Y., 2022. Emerging challenges and innovations in surfactant-mediated stabilization of biologic formulations. J. Pharm. Sci. 111 (4), 919–932. doi:10.1016/j.xphs.2021.12.002.

Kim, D., Yu, J.H., Hong, K.S., Jung, C.D., Kim, H., Kim, J., Myung, S., 2022. Green production of low-molecular-weight xylooligosaccharides from oil palm empty fruit bunch via integrated enzymatic polymerization and membrane separation for purification. Sep. Purif. Technol. 293. doi:10.1016/j.seppur.2022.121084.

Kumar, R., Kumar, N., Rajput, V.D., Mandzhieva, S., Minkina, T., Saharan, B.S., Kumar, D., Sadh, P.K., Duhan, J.S., 2022. Advances in biopolymeric nanopesticides: a new eco-friendly/eco-protective perspective in precision agriculture. Nanomaterials 12 (22). doi:10.3390/nano12223964.

Lahiri, D., Dash, S., Dutta, R., Nag, M., 2019. Elucidating the effect of anti-biofilm activity of bioactive compounds extracted from plants. J. Biosci. 44 (2). doi:10.1007/s12038-019-9868-4.

Li, N., Han, X., Feng, D., Yuan, D., Huang, L.i-J., 2019. Signaling crosstalk between salicylic acid and ethylene/jasmonate in plant defense: do we understand what they are whispering? Int. J. Mol. Sci. 20 (3), 671. doi:10.3390/ijms20030671.

Li, W., Liu, M., Saboor Siddique, M., Graham, N., Yu, W., 2021. Contribution of bacterial extracellular polymeric substances (EPS) in surface water purification. Env. Pollut. 280, 116998. doi:10.1016/j.envpol.2021.116998.

Manjunath, M.S., Kumar Koshariya, A., Sharma, N., Rajput, A., Kumar Pandey, S., Singh, S., Kumar, R., Singh, B.V., 2023. Exploring the use of aromatic compounds in crop growth and protection. Int. J. Plant Soil Sci. 35 (21), 78–89. doi:10.9734/ijpss/2023/v35i213948.

Mehmood, N., Saeed, M., Zafarullah, S., Hyder, S., Rizvi, Z.F., Gondal, A.S., Jamil, N., Iqbal, R., Ali, B., Ercisli, S., Kupe, M., 2023. Multifaceted impacts of plant-beneficial *Pseudomonas spp.* in managing various plant diseases and crop yield improvement. ACS Omega 8 (25), 22296–22315. doi:10.1021/acsomega.3c00870.

Mondéjar-López, M., Paz García-Simarro, M., Navarro-Simarro, P., Gómez-Gómez, L., Ahrazem, O., Niza, E., 2024. A review on the encapsulation of "eco-friendly" compounds in natural polymer-based nanoparticles as next generation nano-agrochemicals for sustainable agriculture and crop management. Int. J. Biol. Macromol. 280, 136030. doi:10.1016/j.ijbiomac.2024.136030.

Morcillo, R., Manzanera, M., 2021. The effects of plant-associated bacterial exopolysaccharides on plant abiotic stress tolerance. Metabolites 11 (6), 337. doi:10.3390/metabo11060337.

Mouro, C., Gomes, A.P., Gouveia, I.C., 2024. Microbial exopolysaccharides: structure, diversity, applications, and future frontiers in sustainable functional materials. Polysaccharides 5 (3), 241–287. doi:10.3390/polysaccharides5030018.

Nafady, N.A., Sultan, R., El-Zawahry, A.M., Mostafa, Y.S., Alamri, S., Mostafa, R.G., Hashem, M., Hassan, E.A., 2022. Effective and promising strategy in management of tomato root-knot nematodes by *Trichoderma harzianum* and *Arbuscular Mycorrhizae*. Agronomy 12 (2). doi:10.3390/agronomy12020315.

Naveen, S., Balachandar, D., 2025. Extracellular polymeric substances of plant-growth-promoting rhizobacteria modulate the positive plant-soil feedback in maize via soil conditioning. Sci. Total Env. 975, 179256. doi:10.1016/j.scitotenv.2025.179256.

Ouaarous, M., El Fakhouri, K., Taarji, N., Baouchi, A., Amri, M., Ramdani, C., Sobeh, M., Mesfioui, A., El Bouhssini, M., 2025. Impact of field insect pests on seed and nutritional quality of some important crops: a comprehensive review. ACS Omega 10 (9), 8779–8792. doi:10.1021/acsomega.4c08982.

Panda, D., Manickam, S., 2019. Cavitation technology—the future of greener extraction method: a review on the extraction of natural products and process intensification mechanism and perspectives. Appl. Sci. 9 (4), 766. doi:10.3390/app9040766.

Reddy, M.S.P., Pandey, V., Bai, A.T., Pandey, R., Nadaf, H.A., Kancharlapalli, S.J., 2022. Soil microbes as biopolymers to enhance soil mechanical properties. In: Microbial Resource Technologies for Sustainable Development. Elsevier, India, pp. 47–76. doi:10.1016/B978-0-323-90590-9.00003-1.

Redmile-Gordon, M., Gregory, A.S., White, R.P., Watts, C.W., 2020. Soil organic carbon, extracellular polymeric substances (EPS), and soil structural stability as affected by previous and current land-use. Geoderma 363, 114143. doi:10.1016/j.geoderma.2019.114143.

Revin, V.V., Liyaskina, E.V., Parchaykina, M.V., Kurgaeva, I.V., Efremova, K.V., Novokuptsev, N.V., 2023. Production of bacterial exopolysaccharides: xanthan and bacterial cellulose. Int. J. Mol. Sci. 24 (19). doi:10.3390/ijms241914608.

Riseh, R.S., Tamanadar, E., Pour, M.M., Thakur, V.K., 2022. Novel approaches for encapsulation of plant probiotic bacteria with sustainable polymer gums: application in the management of pests and diseases. Adv. Polym. Tech. 2022. doi:10.1155/2022/4419409.

Romero, J., Albertos, I., Díez-Méndez, A., Poveda, J., 2022. Control of postharvest diseases in berries through edible coatings and bacterial probiotics. Sci. Hortic. 304, 111326. doi:10.1016/j.scienta.2022.111326.

Sabbahi, R., Hock, V., Azzaoui, K., Saoiabi, S., Hammouti, B., 2022. A global perspective of entomopathogens as microbial biocontrol agents of insect pests. J. Agric. Food Res. 10, 100376. doi:10.1016/j.jafr.2022.100376.

Saha, I., Datta, S., Biswas, D., 2020. Exploring the role of bacterial extracellular polymeric substances for sustainable development in agriculture. Curr. Microbiol. 77 (11), 3224 3239. doi:10.1007/s00284 020 02169 y.

Sarkar, R., Sarkar, M., Roychoudhury, A., 2025. Rhizosphere microbiome signalling and biotic stress tolerance in plants. J. Stress Physiol. Biochem. 21 (2), 57–83.

Singh, R.P., Manchanda, G., Bhattacharjee, K., Panosyan, H., 2022. Microbial Syntrophy-Mediated Eco-enterprising. Elsevier, India, pp. 1–346. doi:10.1016/C2021-0-00471-5.

Soni, R., Gupta, R., Agarwal, P., Mishra, R., 2022. Organic farming: a sustainable agricultural practice. Vantage: J. Themat. Anal. 21–44. doi:10.52253/vjta.2022.v03i01.03.

Tahir, S., Hasnain, S., 2020. Potential uses of lipopolysaccharide and exopolysaccharide isolated from *Agrobacterium spp*. J. Microbiol. Mol. Genet. 1 (1). doi:10.52700/jmmg.v1i1.6.

Tiwari, O.N., Ghosh, D., Debnath, S., Sahu, M., Vanitha, K., 2022. Multifaceted utilization of microalgal biomass towards industrial applications. In: Biotechnology for Waste Biomass Utilization. Apple Academic Press, Oakville, ON, Canada, pp. 79–127.

Velmourougane., K., Thapa, S., Prasanna, R., 2023. Prospecting microbial biofilms as climate smart strategies for improving plant and soil health: a review. Pedosphere 33 (1), 129–152.

Vimala Devi, P.S., Duraimurugan, P., Poorna Chandrika, K.S.V., Vineela, V., Hari, P.P., 2020. Novel formulations of *Bacillus thuringiensis* var. *kurstaki*: An eco-friendly approach for management of lepidopteran pests. World J. Microbiol. Biotechnol. 36 (5). doi:10.1007/s11274-020-02849-8.

Wani, J.A., Wali, A.F., Majid, S., Rasool, S., Rehman, M.U., Rashid, S.M., Ali, S., Farooq, S., Rasool, S., Ahmad, A., Qamar, W., 2020. Bio-pesticides: application and possible mechanism of action. In: Bioremediation and Biotechnology, Vol 2: Degradation of Pesticides and Heavy Metals, 2. Springer International, pp. 97–119. doi:10.1007/978-3-030-40333-1_6.

Zheng, Z., Zhao, M., Zhang, Z., Hu, X., Xu, Y., Wei, C., He, H., 2022. Lactic acid bacteria are prevalent in the infrabuccal pockets and crops of ants that prefer aphid honeydew. Front. Microbiol. 12. doi:10.3389/fmicb.2021.785016.

Zhou, Y., Cui, X., Wu, B., Wang, Z., Liu, Y., Ren, T., Xia, S., Rittmann, B.E., 2024. Microalgal extracellular polymeric substances (EPS) and their roles in cultivation, biomass harvesting, and bioproducts extraction. Bioresour. Technol. 406, 131054. doi:10.1016/j.biortech.2024.131054.

CHAPTER

14

Microbial exopolysaccharides (EPS) as a potential nano-theragnostic agent: Current perspectives and scope

Nilimanka Das[a], *Rishabh Rana*[b,c], *Subodh Kumar*[c,d] *and Deepak Kumar*[c]

[a]Regional Institute of Pharmaceutical Science and Technology (RIPSAT), Agartala, Tripura, India [b]Faculty of Biotechnology and Applied Sciences, Shoolini University, Bajhol, Himachal Pradesh, India [c]Department of Pharmaceutical Chemistry, School of Pharmaceutical Sciences, Shoolini University, Bajhol, Himachal Pradesh, India [d]Department of Pharmaceutics, School of Pharmaceutical Sciences, Shoolini University, Bajhol, Himachal Pradesh, India

14.1 Introduction

Microbial exopolysaccharides (EPS) are complex, high molecular weight polysaccharides released by microorganisms such as bacteria, fungus, yeast, and algae in the environment (Paul et al., 2023; Ahuja et al., 2024). The discovery of an EPS in wine in the 19th century opened a new era of bacterial EPS, which later came to be known as dextran and *Leuconostoc mesenteriodes* prokaryote was found to be the source of it (Linker and Jones, 1966; Rehm, 2010; Nwodo et al., 2012). These naturally occurring biopolymers exhibit wide range of applications in industrial, environmental, biofuel and biomedical fields owing to their unique composition, structure, and characteristics (Waoo et al., 2023). The EPS are the polysaccharides that bacteria secrete into their growing media which act as the organism's defense system (Kaur and Dey, 2023). The constituent unit of these biopolymers can vary widely, containing repeating units of amino sugars, uronic acids and polysaccharides (such as glucose, galactose and mannose). Functional groups like phosphate, acetyl or sulfate are often found to be present in the linear or branching structure of EPS which adds to its adaptability (Williams, 2019; Khan et al., 2022).

DOI: https://doi.org/10.1016/B978-0-443-36619-2.00014-0

Food processing, medicines, biotechnology, and environmental management are just a few of the industries that use EPS. They are crucial in food and cosmetic compositions because of their functional qualities, which include thickening, gelling, emulsifying, retaining water and film-forming capabilities (Korcz and Varga, 2021). Because of their biocompatibility, nontoxicity and capacity to boost immunological responses, they are used in biomedical applications including tissue engineering, wound healing and drug delivery systems (Laubach et al., 2021; Mansour et al., 2023). Microbial EPS are being investigated in wide scale for their potential in nano-theragnostic application where they can be employed simultaneously as drug delivery device for carrying therapeutic moieties along with its diagnostic imaging ability (Sachin and Karn, 2021). They are becoming essential in the development of bio-based sustainable and environment friendly green technologies as there is a growing demand for eco-friendly, biodegradable materials (Nguyen et al., 2024; Tiwari et al., 2024).

One possible cause of bacteria in the catheter insertion site is the formation of biofilm. As a complex organic material, they produce an extracellular mucopolysaccharide compound that is colonized by bacteria. Calcium and magnesium ions, along with the Tamm–Horsfall protein, are integrated into the biofilm (Trautner and Darouiche, 2004; Majumder et al., 2018). Following catheter insertion, biofilm formation began when organisms adhered to the host's conditioning layer that formed beside the catheter surface. The biofilm protects the organisms from the effects of host defense and antibiotic resistance (Costerton et al., 1999; Sutherland, 2001; Sharma et al., 2023). Following the implantation of the indwelling catheter during the first episode of bacteriuria, a single species is discovered. When a catheter remains in place for a few days, mature biofilm may form. For those who have long-term indwelling catheters, three to five organisms are often isolated (Hooton et al., 2010; Pelling et al., 2019; Gunardi et al., 2021) (Fig. 14.1).

Bacterial extracellular polymeric substances (EPSs) are classified based on their attachment to the cell. As capsular polysaccharides (CPS), they tightly bind to the cell surface, whereas as free slime polysaccharides, they loosely attach or are completely released into the extracellular environment (Mahmoud et al., 2021; Xiao et al., 2022; Netrusov et al., 2023). They have garnered significant interest in both academic and industrial settings owing to their unparalleled physiochemical properties and wide-ranging bioactivities (Zannini et al., 2016; Stevanović and Filipović, 2024). These are types of extrinsic fluids that cells can readily discharge into the surrounding environment. As a result, the procedures for removing them from the cell-free supernatant are fairly easy and economical. Numerous physiological and biochemical characteristics, as well as metabolic flexibility, are characteristics of bacteria, which proliferate quickly. It is feasible to modify the yield as well as the structural and functional characteristics of bacterial EPSs by genetic and metabolic engineering techniques, as well as growth condition optimization (Schmid et al., 2015; Pereira et al., 2019; Xiao et al., 2022). EPS are utilized in multiple sectors, notably the food, pharmaceutical, cosmetic, agriculture, and biotechnology industries, where they act as thickening, stabilizing, gelling, emulsifying, and film-forming agents (Netrusov et al., 2023). Such polymers are typically produced as CPS or selected EPS, which contribute not only to microbial survival but also to ecological functions such as biofilm formation, desiccation tolerance, and protection from environmental stressors (Whitfield et al., 2015; Strieth et al., 2021; Nguyen et al., 2024). Furthermore, the continuous exploration of EPS for its feasibility in nano-theragnostic platforms underscores their biocompatibility, nontoxicity, and low immunogenicity (Kong et al., 2023). Microbial EPS, such as xanthan, dextran, levan, gellan, pullulan, etc., produced by the industries are used in various formulations demonstrating their

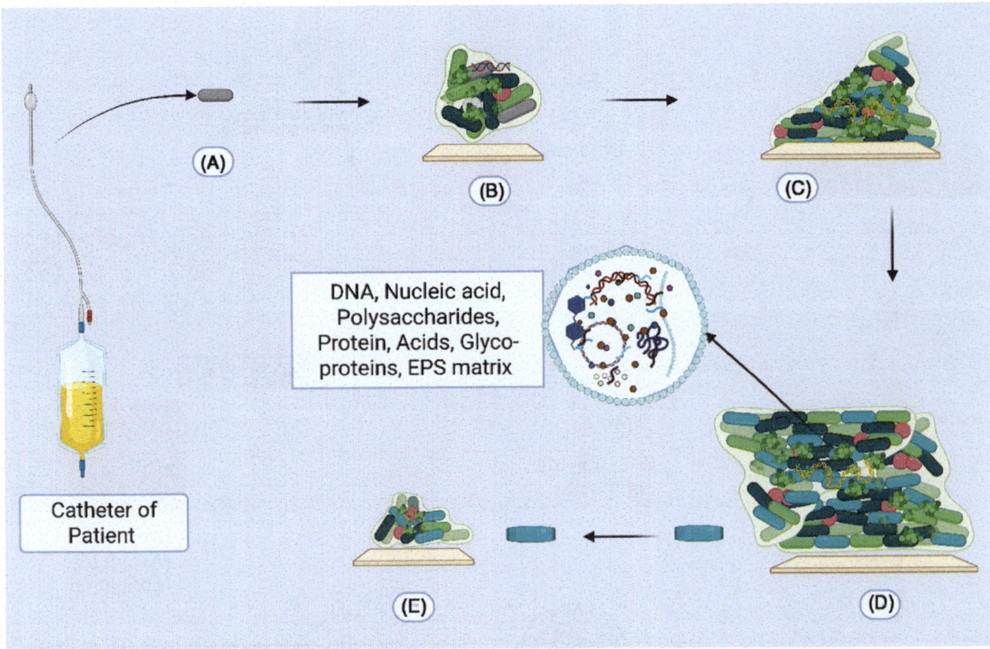

FIGURE 14.1 Formation of bacterial extracellular polymeric substances from biofilm. (A) Bacteria float freely. (B) Bacterial stickiness or attachment to the surface. (C) Development of extracellular polymeric substances (EPSs) and monolayer formation. (D) Dispersal of biofilms: Planktonic cell separation and reversal. (E) Resume of the biofilm's life cycle.

scalability and commercial viability (Duru and Toksoy Oner, 2017; Waoo et al., 2023). EPS have also shown promising biological properties such as antioxidant, anti-inflammatory, antitumor, immuno-modulatory, hypocholesterolemia, and antimicrobial activities, making them attractive for use in functional foods, nutraceuticals, and as biopharmaceuticals. Their role in modulating the human gut microbiota and serving as prebiotics has also drawn attention in the context of digestive health, a depersonalized nutrition (Costerton et al., 1999; Sutherland, 2001; Sharma et al., 2023). EPS based biodegradable films, bioplastics, and composites are being studied as potential replacements for petroleum-based polymers in packaging and materials science. In environmental applications, EPS play a significant role in heavy metal sequestration, oil spill remediation, and wastewater treatment due to their excellent metal-binding and flocculating capabilities (Kumari et al., 2023; Sharma et al., 2024; Sulaiman and Hassan, 2024) (Fig. 14.2).

Bacterial EPS are gaining significant momentum in nanotechnology and biomedicine due to their innate biocompatibility, biodegradability, and nanoparticle-stabilizing capabilities (Sathiyanarayanan et al., 2017; Escárcega-González et al., 2018; Dey et al., 2023; Netrusov et al., 2023). In green synthesis of nanoparticles, especially metal nanoparticles like silver, EPS acts as a reducing agent as well as a stabilizer, which outperforms conventional chemical approaches (Singh et al., 2018; Bahrulolum et al., 2021; Chugh et al., 2021; Dey et al., 2023; Singh et al., 2023; Zeinivand et al., 2023). The resulting EPS-mediated nanoparticles exhibit potent antimicrobial activity that effectively targets bacteria, fungi, and biofilms and showed effective results that

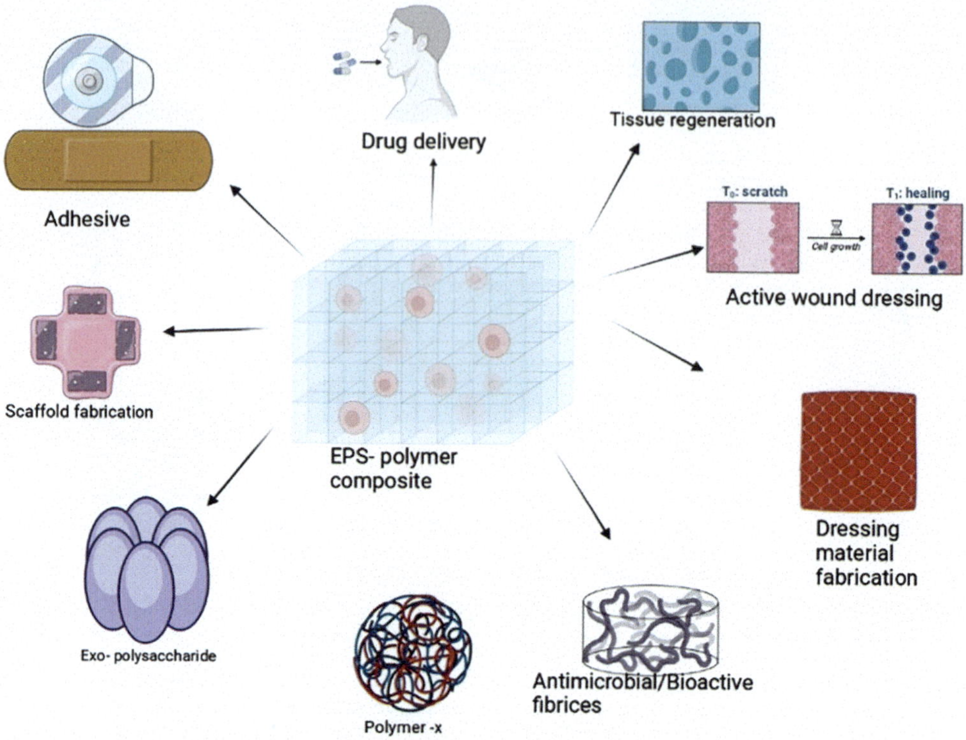

FIGURE 14.2 Current activities of EPS (exopolysaccharides).

promise to enhance antibacterial coatings, wound dressings, and medical devices (Chugh et al., 2021; Thambirajoo et al., 2021; Pang et al., 2023). In drug delivery and anticancer therapy, EPS enhances nanoparticle biocompatibility and targeting, improving therapeutic delivery while minimizing side effects (Raveendran et al., 2013; Abdelnasser et al., 2017; Song et al., 2024). Additionally, EPS-stabilized nanoparticles hold potential as contrast agents in medical imaging and as labels for tracking in bioimaging studies, enhancing diagnostic and monitoring capabilities. From tissue engineering to diagnostics, EPS serves as the foundation of scaffolds, biosensors, and antimicrobial materials, promoting tissue regeneration, disrupting biofilms, and safeguarding against infection (Gargouch et al., 2021; Duarte Mouro and Gouveia, 2024; Ghareeb et al., 2024). By merging green synthesis with multifunctional biomedical applications (Santiesteban et al., 2016; Nandhini et al., 2024; Zhang et al., 2024), EPS-based nanomaterials offer sustainable, versatile, and high-performance platforms poised to redefine healthcare technologies (Kaur and Dey, 2023; Ma et al., 2024).

Since it emphasizes the adaptability and potential of EPS in developing contemporary technologies, especially in the domains of nanotechnology and biomedicine, the chapter is significant in a number of important ways. Microbial EPS have emerged as multifunctional biopolymers with tremendous promise across biomedical and environmental technologies (Sajna et al., 2021; Kiran et al., 2024). In drug delivery, EPS-based systems offer regulated release, enhanced targeting, and reduced side effects, collectively improving therapeutic efficacy. Their utility extends into nanoparticle synthesis, where EPS serve as eco-friendly reducing and stabilizing agents, enabling the sustainable production of metal and semiconductor nanoparticles without toxic chemicals.

Crucially, EPS-coated nanoparticles exhibit exceptional biocompatibility and safety, displaying minimal toxicity or immune response, which is vital for applications such as wound healing, biosensors, and implant coatings (Saha and Datta, 2022; Adegbolagun et al., 2023; Cheng et al., 2023; Nag et al., 2024). The versatility of EPS allows them to function not only as therapeutic carriers but also as bioadhesives, scaffolds, and sensor platforms, underlining their broad capacity to enhance health outcomes. Additionally, the biodegradability, nontoxicity, and renewable sourcing of EPS align with the growing demand for sustainable materials, positioning them at the forefront of green nanotechnology and biomedicine (Huang et al., 2015; Singh et al., 2018; Bahrulolum et al., 2021; Sharma et al., 2021).

14.2 Biological and physicochemical properties of exopolysaccharides

A variety of biological and physicochemical characteristics of microbial EPS support their widespread and growing use in biotechnology, nanotechnology, pharmaceuticals, and biomedicine. The molecular weight, monomer composition, functional group, and three-dimensional conformation of EPS are critical parameters that determine their functionality and influence their behavior in different environments (Zannini et al., 2016; Paul et al., 2023). EPS can typically be separated into two main categories: Heteropolysaccharides, which contain two or more units of distinct monosaccharides (glucose, fructose, galactose, mannose, rhamnose, fucose, N-acetylglucosamine, and uronic acids), and homopolysaccharides, which are either unbranched or branched and made up of single-type monosaccharides like glucose and fructose connected by glycosidic bonds (Sutherland, 1972; Mohd Nadzir et al., 2021; Netrusov et al., 2023). Depending on the producing microorganism and culture conditions, EPS can vary significantly in branching, charge density, and hydrophilic-hydrophobic balance, allowing for tailored applications in diverse sectors (Krsmanovic et al., 2021). EPS is extremely valuable for a variety of applications due to its integrated biological and physicochemical characteristics. Their solubility, viscosity, and thermal stability provide practical benefits in industrial processes, controlled drug release, food processing, and tissue engineering, while their antioxidant, antibacterial, anti-inflammatory, and anticancer properties establish them as powerful and versatile therapeutic agents (Kaur and Dey, 2023; Mouro et al., 2024). Additionally, EPS have shown immuno-modulatory effects, making them potential candidates for vaccine adjuvants and immune-boosting nutraceuticals (Nazeam and Singab, 2022). The rheological behavior of EPS, including shear thinning, gel formation, and emulsion-stabilizing capacity is another key feature that adds value in food and cosmetic formulations, as well as injectable biomedical hydrogels (Romero et al., 2013). In pharmaceutical applications, EPS can enhance bioavailability by protecting sensitive drugs from degradation and facilitating targeted delivery through surface modification or co-encapsulation strategies (Saha and Datta, 2022).

Moreover, the ability of EPS to bind metals and other charged molecules opens the door for their use in biosorption, environmental detoxification, and smart sensor development. These diverse capabilities are directly linked to the structural composition of the EPS, which dictates how the molecules interact with other biological or chemical entities (Gupta and Diwan, 2017;

Jaroszuk-Ściseł et al., 2025). Therefore, an in-depth understanding of the structure-functional relationship is essential for advancing the rational design of EPS-based systems tailored for specific applications in healthcare, industry, and environmental science (Mohd Nadzir et al., 2021; Kaur and Dey, 2023). Continued research on the correlation between EPS composition and its performance under varying physiological and process conditions will be fundamental for future innovations. Advanced analytical techniques such as NMR spectroscopy, FTIR, SEM, and rheometry, along with molecular modeling and simulations, are increasingly being employed to decipher these complex relationships and guide functional optimization of EPS formulations (Kiran et al., 2024; Nguyen et al., 2024).

14.2.1 Structure of microbial exopolysaccharides and their actions

Microbial EPS are high molecular weight carbohydrates released into the extracellular environment by bacteria, fungus, and algae (Mouro et al., 2024). In microbial physiology, these macromolecules are essential for biofilm formation, desiccation resistance, nutrition retention, and stress tolerance. Structural organization is the primary determinant of EPS's physicochemical and biological characteristics, and it can vary greatly according on the generating microbe, culture conditions, nutrient availability, and environmental stresses (Sathiyanarayanan et al., 2017; Sharma et al., 2024). In terms of structure, microbial EPS can be broadly divided into two categories: Heteropolysaccharides, which are made up of two or more distinct monosaccharide units arranged in repeating sequences, and homopolysaccharides, which are made up of repeating units of a single type of monosaccharide (such as glucose in dextran or fructose in levan). With repeating units of three to seven distinct sugars, heteropolysaccharides usually have more structural complexity and functional diversity. They frequently comprise glucose, galactose, rhamnose, mannose, arabinose, and uronic acids (Schmid et al., 2015; Bhattacharya and Samanta, 2025; Wu et al., 2024). The variety of EPS structures is also greatly influenced by the kind and location of glycosidic bonds (e.g., α-1,6; β-1,4; α-1,3), the degree of branching, molecular weight, and the presence of substituent groups like acetyl, sulfate, phosphate, or pyruvate groups (Kaur and Dey, 2023). The biological activities and industrial utility of EPS are determined by these structural traits, which also affect important aspects, including solubility, viscosity, gelling ability, charge density, and interaction with metals or proteins (Daba et al., 2021; Kaur and Dey, 2023; Wu et al., 2024). Noncarbohydrate components such as peptides, lipids, or nucleic acids can also be present in advanced structural variations in EPS to form glycoconjugates or glycolipids, which increase their potential for use in drug delivery systems, immunomodulatory treatments, and biosensors (Harvey, 2015). The glucuronic acid and galacturonic acid units found in *Klebsiella pneumonia's* capsular extracellular polymer substances (EPSs) are known to confer strong negative charge and biofilm resilience. Similar to this, *Xanthomonas campestris* generates xanthan gum, which has a high viscosity even at low concentrations. It has a cellulose backbone and trisaccharide side chains (Benincasa et al., 2016; Khadka et al., 2023).

Crucially, the correlation between EPS's structure, properties, and functions is essential to their use in medicinal, biotechnological, and nanotechnological applications. The rational design of EPS-based hydrogels, nanoparticles, coatings, and scaffolds is made possible by an understanding of how particular structural features contribute to desired qualities, such as mucoadhesiveness, immunogenicity, or biocompatibility (Hasan et al., 2018; Dey et al., 2023; Patel et al., 2024). Furthermore, there are ways to customize particular structural features to satisfy the requirements of cutting-edge industrial and biological applications through genetic

TABLE 14.1 Structural characteristics of common microbial exopolysaccharides (EPS).

EPS	Main monomers	Glycosidic bonds	Type	References
Dextran	Glucose	α-1,6	Homopolysaccharide	(Amspacher and Curreri, 1953)
Levan	Fructose	β-2,6	Homopolysaccharide	(Domżał-Kędzia et al., 2023)
Xanthan	Glucose, mannose, glucuronic acid	β-1,4, α-1,3	Heteropolysaccharide	(Berezina et al., 2024)
Alginate	Mannuronic acid, guluronic acid	β-1,4	Heteropolysaccharide	(Stender et al., 2019)
Pullulan	Maltotriose (glucose units)	α-1,4 and α-1,6	Homopolysaccharide	(Prajapati et al., 2013)

engineering and metabolic pathway modification in EPS-producing microorganisms (Romero et al., 2013; Stender et al., 2019; Kaur and Dey, 2023) (Table14.1).

14.2.2 Physicochemical characteristics

Microbial EPS have imbued a wide range of physicochemical parameters that determine their applicability for various industrial, biological, medicinal, and environmental applications. The degree of branching, type and location of glycosidic connections, the molecular weight, the monomer composition, and the presence of functional groups like sulfate, phosphate, acetyl, and uronic acid residues, all have a significant impact on these characteristics (Schmid et al., 2015; Ahuja et al., 2024). Food thickeners, cosmetic emulsifiers, and biomedical hydrogels depend on EPS's rheological expressions like viscosity, gel-forming ability, shear thinning capabilities, etc. EPS's molecular weight can vary from a few thousand to several million Daltons. The formulations for higher structural integrity or extended-release profiles warrant involvement of higher molecular weight EPS, which in turn shows higher viscosity and improved mechanical strength (Brummer, 2006; da Silva and Rao, 2007). Solubility is another crucial characteristic that changes based on EPS structure and environmental pH and ionic strength. Many EPS are soluble in water, but by incorporating hydrophobic or charged side chains, their solubility can be changed, allowing the formation of colloidal suspensions, films or nanoparticles (Umut, 2020; Netrusov et al., 2023; Zhao et al., 2023). In a pharmaceutical delivery system, regulated solubility is crucial for drug disposition and bioavailability (Zhang et al., 2018; Kumari et al., 2023). Processing conditions and product shelf life are also significantly influenced by the thermal and pH stability of EPS. High-heat-resistant EPS can be utilized in pasteurized or sterilized products, whereas pH-responsive EPS is best suited for stimuli-responsive biomedical devices or controlled drug release in specific gastrointestinal tract regions (Mende et al., 2015; Suryawanshi et al., 2022).

The way EPS interacts with metal ions, proteins or other polymers is determined by its electric charge, neutral, anionic, or cationic. This has an effect on EPS performance in bioremediation, wastewater treatment, and biosorption. Anionic EPS containing carboxyl or sulfate functional groups have considerable metal-binding ability and are therefore useful in eliminating radionuclides or heavy metals from contaminated settings (Gupta and Diwan, 2017;

Kondakindi et al., 2024; Singh et al., 2024). Additionally, viscoelastic qualities are essential when creating EPS-based hydrogels for tissue engineering, regenerative medicine, and wound healing, where the mechanical characteristics must resemble those of natural tissues. Better retention and absorption of encapsulated medications on mucosal surfaces is made possible by their viscoelastic behaviors, which also strengthen their role as mucoadhesive agents. Certain EPS's capacity to form films is used to create medication delivery membranes, edible coatings, and biodegradable packaging. Because of their physicochemical stability, active substances (such as enzymes, antioxidants, or antimicrobials) can be included with EPS to create multipurpose delivery systems (Díaz-Montes and Castro-Muñoz, 2021; Wu et al., 2022; Muñoz-Tebar et al., 2023; Walait et al., 2023; Parvin et al., 2024; Xu et al., 2025). Amphiphilic EPS's surface activity and emulsifying ability are very useful in the food, cosmetics, and nanotechnology industries since they allow for the creation of micro- and nano-emulsions as well as the stabilization of oil-water combinations (Azmi et al., 2019; Díaz-Montes and Castro-Muñoz, 2021; Daba et al., 2021) (Table 14.2).

14.2.3 Functional properties and biocompatibility

Naturally occurring polymers, namely microbial EPS, are produced by a variety of microorganisms such as yeast, fungi, bacteria, algae, etc. They combine remarkable biocompatibility with great performance. These water-soluble, high-molecular-weight biopolymers exhibit a variety of useful qualities, including gelling, emulsifying, water retention, and drug encapsulation, which make them indispensable in various fields, including pharmaceuticals, food, and cosmetics, and many more (Osemwegie et al., 2020; Abbot et al., 2022; Bolouri et al., 2022; Guo et al., 2024; Salehi and Rashidinejad, 2025). EPS are perfect materials for drug delivery systems, wound dressings, tissue-engineering scaffolds, and controlled release platforms because of their nontoxicity, low immunogenicity, biodegradability, and capacity to promote cell growth and tissue regeneration in biomedical contexts (Moscovici, 2015; Mehta et al., 2021; Mansour et al., 2023; Waoo et al., 2023). Stance, EPS such as alginate, pullulan, dextran, and hyaluronic acid have been successfully formulated into hydrogels and injectable scaffolds for sustained ocular drug delivery and cardiac repair, while bacterial cellulose mimics the structure and hydration properties of collagen to enable moist, enzyme-resistant wound dressings (Jacob et al., 2021; Teixeira et al., 2023). Furthermore, several EPS, including chitosan, levan, and curdlan, have bioactivity in the form of immuno-modulatory, anticancer, antioxidant, and antibacterial properties. Improvements in microbial engineering and EPS-based composites are allowing for both increased strength and customized functionality, even though there are still issues with optimizing large-scale production, mechanical robustness, and regulatory compliance. EPS are sustainable, affordable biopolymers that have the potential to completely transform a variety of biological and therapeutic applications (Mohd Nadzir et al., 2021; Alaneme et al., 2023; Kiran et al., 2024; N S et al., 2024).

14.3 Current nano-theragnostic applications of exopolysaccharides

Microbial EPS are widely acknowledged as sustainable and efficient agents for theragnostic applications, which unite two separate properties, therapeutic and diagnostic, into a single

TABLE 14.2 Physicochemical properties of microbial exopolysaccharides and their functional examples.

S.no.	Property	Description	Example	References
1	Solubility	Ability to dissolve in water	Xanthan gum (soluble), alginate (pH-dependent)	(Ergün and Ergün, 2024)
2	Molecular weight	Affects viscosity and bioactivity	High-molecular-weight dextran	(Zarour et al., 2017)
3	Viscosity	Resistance to flow in solution	Xanthan gum, pullulan	(Brunchi et al., 2023)
4	Thermal stability	Resistance to heat	Xanthan gum (high stability)	(Lambert and Rinaudo, 1985)
5	Rheological behavior	Flow properties under stress	Xanthan gum (shear-thinning)	(Reinoso et al., 2019)
6	Charge	Ionic nature	Alginate (anionic), chitosan (cationic)	(Alsamman and Sánchez, 2022)
7	Film-forming capacity	Ability to create thin films	Pullulan	(Farris et al., 2014)
8	pH Sensitivity	Behavior changes with pH	Alginate (gel stability at neutral pH)	(Qin et al., 2021)
9	Functionalization	Presence of modifiable chemical groups	Sulfated or acetylated EPS	(Gupta and Diwan, 2017)

system. They are perfect for a range of biomedical applications due to their distinct biological characteristics and structural adaptability. Here, an attempt has been made to explain the applications of EPS in nano-theragnostic systems. Microbial EPS provide a great deal of flexibility for creating nano-theragnostic platforms that can be used for gene delivery, cancer treatment, antimicrobial therapies, and bio-sensing (Manivasagan and Oh, 2016; Zhao et al., 2021; Tiwari et al., 2024). EPS is positioned as a game-changing material in next-generation personalized medicine due to its biocompatibility, ease of functionalization, and capacity to combine therapeutic and diagnostic roles. Their scope and commercial feasibility will be expanded by more research and technical developments (Moscovici, 2015; Nguyen et al., 2024).

14.3.1 Role of exopolysaccharides in nanoparticle synthesis and mode of action

Microbial EPS play a fundamental role in nanoparticle synthesis credits to their unique structural and chemical features (Sathiyanarayanan et al., 2017; Dey et al., 2023). These natural polymers, composed of diverse sugar residues rich in hydroxyl, carboxyl, and amino groups, exhibit strong interactions with metal ions through chelation, electrostatic binding, and electron donation (Escárcega-González et al., 2018; Mehta et al., 2021). For such interactions, EPS can function as an efficient bio-reducing agent, transforming metal ions like Ag^{+} and Au^{3+} into zero-valent nanoparticles, while simultaneously acting as stabilizing or capping agents to prevent particle aggregation by forming a protective polysaccharide shell. This dual functionalization leads to the creation of stable, uniform nanoparticles under gentle, eco-friendly conditions fully embodying the ideals of green chemistry (Dong et al., 2020; Javed et al., 2020; Zhu et al., 2022). Moreover, the specific composition and functionalization of EPS (for example, through sulfation or carboxylation) enable precise control over nanoparticle size, shape, surface charges, and morphology, making synthesis highly customized and scalable (Menichetti et al., 2023; Salehirozveh et al., 2024). Integrating EPS into nanoparticle production not only ensures environmentally sustainable processes but also delivers finely tuned nanomaterials suited for antimicrobial coatings, catalysis, biosensing, environmental remediation, and a wide array of biomedical applications (Raveendran et al., 2013; Parvin et al., 2024; Xu et al., 2025).

14.3.2 Exopolysaccharides-based drug delivery systems

The distinct structural and chemical features of microbial EPS are an essential tool for the synthesis and stabilization of nanoparticles (Sathiyanarayanan et al., 2017). Microorganisms secrete natural polymers called EPS, which are made up of different sugar residues with functional groups like hydroxyl, carboxyl, and amino acids that allow them to interact with metal ions and other substances. Because of these interactions, EPS is very useful for the synthesis of environmentally benign, sustainable, and green nanoparticles (Yadav et al., 2022). When nanoparticles are developed, EPS serve as stabilizing, capping, and reducing agent and at the same time, they provide biocompatibility and functional flexibility to the system. Their nontoxic and environmentally benign properties make them perfect for green chemistry approaches in nanotechnology, which have uses in industry, environmental remediation, and medicine. The potential for producing scalable and customized nanoparticles will be increased by further developments in EPS property customization (Sharma et al., 2019; Alqarni et al., 2022).

14.3.3 Diagnostic applications in imaging and biosensors

The use of microbial EPS as functional materials in diagnostic applications has grown, especially in biosensors and medical imaging. Their importance in creating sophisticated diagnostic instruments is bolstered by their biocompatibility, diversity of functional groups, and capacity to establish stable interactions with metals and polymers (Saha and Datta, 2022). Microbial EPS offer a strong foundation for improving diagnostic technologies by using functionalized, sustainable nanoparticle systems. Their contributions to the development of biosensors and imaging demonstrate their revolutionary potential in environmental diagnostics and personalized medicine (Ramesh et al., 2022; Saha and Datta, 2022).

14.3.4 Pharmacological properties of exopolysaccharides

Microbial EPS exhibit a rich array of pharmacological properties that position them as promising agents in the development of novel therapeutics. These carbohydrate polymers, produced by bacteria, fungi, algae, etc., possess functional groups such as hydroxyl, carboxyl, amino, sulfate, and phosphate, which contribute to their bioactivity (Ibrahim et al., 2022). EPS have been shown to exhibit potent anti-inflammatory, antioxidant, antibacterial, and anticancer effects, making them strong candidates for use in drug development. For example, compounds like curdlan can stimulate macrophage activity to combat infections, while lactobacilli-derived EPS boost immune response and neutralize free radicals(Li et al., 2015; Deepak et al., 2016; Abdelnasser et al., 2017; Andrew and Jayaraman, 2020; Abdalla et al., 2021; Abdel-Wahab et al., 2022; El Awady et al., 2024; Hu et al., 2025). Their natural origin, biocompatibility, biodegradability, and low immunogenicity further support their applicability in drug delivery and regenerative medicines, where they can act as carriers, scaffolds, or adjuvants with minimal side effects. Although EPS hold significant therapeutic potential, further research is needed to enhance their production at an industrial scale, functional customization, and clinical efficacy to fully realize their promise as sustainable pharmacological agents (Lambert and Rinaudo, 1985; Reinoso et al., 2019).

14.4 Advantages of exopolysaccharides in nano-theragnostic

Microbial EPS are a promising for combining therapeutic and diagnostic functions into a single nano-platform, translating them into potential nano-theragnostic agents (Paul et al., 2023). The following are the main benefits that demonstrate their potential in nanomedicine:

14.4.1 Innate biocompatibility and biodegradability

Being naturally occurring biopolymers, microbial EPS have remarkable intrinsic biocompatibility and biodegradability, which is much required for nano-theragnostic applications. These exo-biopolymers are nontoxic and well-tolerated by biological systems and hence can be incorporated into the body without causing immunological rejection or severe inflammatory reactions. The biodegradability of EPS based nanoparticles can break them down into innocuous byproducts through metabolism and removed from the biological system once they deliver the therapeutic or diagnostic payload (Yildiz and Karatas, 2018; Ahuja et al., 2023;

Kaur and Dey, 2023; Mouro et al., 2024). This lowers the possibility of toxicity buildup. When creating theragnostic devices that integrate immunological regulation, drug administration, and imaging, especially for in vivo applications, the intrinsic safety profile is essential (Janib et al., 2010; Saha and Datta, 2022). A regulatory advantage is also provided by the fact that certain EPS, including dextran, alginate, and xanthan, are already accepted as safe or authorized for use in food and medicine. They are more appealing for biomedical technology because of their environmentally friendly, sustainable production method, which uses microbial fermentation instead of harsh chemicals (Ahmad et al., 2015; Łętocha et al., 2022; Revin et al., 2023). A class of nanomaterials that balance safety and effectiveness is the end result, making it perfect for next-generation precision medicine platforms that need to be multifunctional without sacrificing biological compatibility (Malik et al., 2023; Kurul et al., 2025).

14.4.2 Multimodal functionality

When microbial EPS are applied to nanoparticles, EPS provide potent multimodal activity that allows them to function as immunomodulators, drug transporters, and imaging agents all at once, all on the same platform (Bayda et al., 2019; Bansal et al., 2020; Kurul et al., 2025). Superparamagnetic iron oxide nanoparticles (SPIONs) coated with dextran, for instance, are a fundamental platform utilized in biosensing, medicinal administration, and magnetic resonance imaging (MRI) (Adams and Barbante, 2013; Grauer et al., 2019). In addition to preventing aggregation and improving biocompatibility, the dextran coating offers a large number of surface locations where therapeutic medicines or targeted ligands can be attached (Swami et al., 2025). Precise imaging and therapy monitoring are made possible by these nanoparticles' great MRI contrast (T_2 relativity) and ability to be targeted (for example, with folate or Herceptin) to certain tumor cells (Chen et al., 2009). Beyond imaging, dextran-SPIONs can be used for hyperthermia therapy, which uses an alternating magnetic field to deliver medications like doxorubicin or curcumin for synergistic treatment benefits while producing localized heat that harms tumor cells (Farzin et al., 2020; Vangijzegem et al., 2023). Additionally, EPS-based nanogels with iron oxide cores can be customized for MRI-guided, pH-sensitive drug release, offering a clever theragnostic method. This EPS-functionalized nanoplatform, which combines immunomodulation, targeted delivery, controlled release, and diagnostic imaging, represents the cutting edge of personalized, precision nanomedicine and provides safe, efficient, and adaptable treatments for cancer and other complicated illnesses (Damiri et al., 2023; Kong et al., 2023; Rodríguez et al., 2025).

14.4.3 Controlled, stimuli-responsive release

Treatment specificity and efficacy can be significantly increased by engineering microbial EPS into regulated, stimuli-responsive delivery systems that release therapeutic moieties precisely at target areas in response to physiological or environmental stimuli (Srivastava and Choudhury, 2022; Zhang et al., 2022). EPS-based carriers undergo structural changes (swelling, degradation, or phase transition) that trigger drug release precisely where needed by incorporating functional moieties that react to internal triggers, such as an acidic pH in tumor tissues or within endosomes, elevated temperatures at inflamed or hyperthermic tumor sites, or overexpressed enzymes like Matrix Metalloproteinases (MMPs) and proteases (Lin et al., 2021; Samec et al., 2022; Sousa et al., 2023). For instance, enzyme-responsive systems break in the presence of disease-associated enzymes to release their cargo, while EPS-conjugated nanoparticles with pH-sensitive connections collapse or inflate in acidic microenvironments, allowing selective drug discharge within tumors (Gao et al., 2010; Hu et al., 2014; Javan Nikkhah and Vandichel, 2022). Under mild hyperthermia, phase change is induced in thermo-responsive EPS hydrogels or nanogels, allowing on-demand medication release. For more complex spatiotemporal regulation of payload delivery, EPS systems can be made to respond to several stimuli in addition to internal ones. These stimuli could be pH, temperature, enzymes, redox or even external signals like light, magnetic field or ultrasound (Gao et al., 2010; Kaur et al., 2016; Song et al., 2017; Granton et al., 2024). These technologies provide in situ triggered release that is self-regulated, minimal off-target effects, and improved therapeutic efficacy within precise therapeutic windows.

14.4.4 Immune system interactions

Microbial EPS, particularly β-glucans, are powerful adjuvants in nano-theragnostic systems and are remarkable in regulating immune responses. Complement receptor 3 (CR3), Toll-like receptor (TLRs), scavenger receptors on neutrophils, NK cells, macrophages, dendritic cells (DCs), etc., are immunological receptors involved in the innate immune response that recognize these polysaccharides, which imitate pathogen-associated molecular patterns (PAMPs) (Goodridge et al., 2009; Hajishengallis and Lambris, 2011; Taban et al., 2022). Strong immune signaling cascades are set off by this interaction, which can strengthen adaptive immunity by boosting cytokine secretion, improving antigen presentation, and activating innate defenses (Oerlemans et al., 2021). In cancer treatments, β-glucan-coated nanoparticles can increase DC and T-cell activation in the tumor microenvironment, "reprogram" tumor-associated macrophages toward a proinflammatory M1 phenotype, and enable targeted destruction of cancerous cells. They have been demonstrated to increase $CD4^+$ and $CD8^+$ T-cell tumor infiltration, boost NK cell activity and antibody-dependent cytotoxicity, and work in concert with immuno therapies such as checkpoint inhibitors to produce better results (Van Dalen et al., 2019; He et al., 2021; Liu et al., 2023; Qi et al., 2025). Furthermore, in vaccine models, *Lactobacilli*-derived EPS have successfully increased T-cell proliferation, antibody generation, and defense against bacterial infections (Lei et al., 2018; Zhang et al., 2021, 2024).

Additionally, β-glucan particles and nanoparticles greatly enhance DC maturation and antigen absorption, laying a solid basis for vaccine development. Long-term protection against recurring infections or tumors is provided by their epigenetic reprogramming, which activates trained immunity and stimulates innate immunological memory (Geller et al., 2019; Paris et al., 2020; Kulkarni and Nanjappa, 2024; Dowdall and Hoare, 2025). EPS-based adjuvants provide a flexible, biocompatible, and potent way to combine immunological activation and therapeutic efficacy in cutting-edge nanomedicine platforms because of their strong safety profiles, which demonstrate minimal toxicity even at large dosages, and their chemical structures that can be modified for additional functional modification (Nafee et al., 2014; Perera et al., 2021; Díaz-Dinamarca et al., 2022; Facciolà et al., 2022; Kim et al., 2022; Rohokale and Guo, 2023).

14.4.5 Enhanced antimicrobial action

Silver nanoparticles produced from microbial EPS provide a potent dual-action antibacterial and imaging platform that is perfect for theragnostic applications. By transforming silver ions into evenly distributed nanoparticles that are naturally covered in a protective polysaccharide coating, EPS serves as a stabilizing and reducing agent in these systems (Saravanan et al., 2017; Ribeiro et al., 2022). Particle stability and low toxicity are guaranteed by this biocapping, which also improves biocompatibility and makes them suitable for use in biomedical settings. These EPS-silver nanoparticles have strong antibacterial properties and can kill bacteria, fungus, and even disrupt biofilms by damaging membranes, producing reactive oxygen species, and interfering with DNA replication (Gherasim et al., 2020; More et al., 2023; Rodrigues et al., 2024). They can also be used as contrasting agents in imaging owing to their special optical and plasmonic characteristics which allows both detection and therapy on a single platform. Additionally, by avoiding harmful chemicals, the green synthesis approach promotes environmentally friendly, sustainable production. A next generation theragnostic solution with great potential in areas ranging from infection control and wound care to diagnostic imaging and beyond, EPS-derived silver nanoparticles seamlessly combine imaging capabilities with strong antimicrobial efficacy and environmental friendliness (Vu et al., 2009; Costa et al., 2018; Hsu et al., 2023; Jiang et al., 2023; Kurul et al., 2025; Sidhic et al., 2025).

14.4.6 Customizable surface chemistry

The remarkable surface chemistry of microbial EPS provides a flexible platform for accurate targeting in nano-theragnostic applications. They provide reactive anchor points for conjugating different targeted ligands like antibodies, peptides, aptamers or small molecules. Thanks to their plentiful functional groups which include hydroxyls, carboxyl's, amines, sulfates and phosphates (Sivakumar et al., 2014; Ma et al., 2021; Yaraki et al., 2022). For example, covalently attaching peptides such as human epidermal growth factor receptor 2 (HER2)-targeting sequences or RGD motifs to EPS-coated nanoparticles allows for high-affinity binding to overexpressed tumor receptors, greatly increasing tumor accumulation in comparison to passive delivery (Blanco et al., 2012). By directing nanoparticles toward particular cell types, such as EGFR- or HER2-positive cancer cells, antibodies or antibody fragments attached to EPS surfaces improve selective uptake and lessen off-target effects (Ma et al., 2021; Yaraki et al., 2022; Javid et al., 2024; Thongpon et al., 2025).

Furthermore, EPS facilitates robust ligand display under physiological settings by enabling stable peptide attachment using chemistries such as maleimide-thiol or EDC/NHS. Targeting, imaging (MRI, fluorescence, PET) and therapeutic delivery (chemotherapy, heat) may all be done simultaneously with this modular surface functionalization, all while preserving biocompatibility and reducing immunogenicity. Personalized nano-theragnostic is essentially made possible by EPS surface customization, opening the door to safe, effective, and focused treatments in oncology and other disease areas (Hui et al., 2014; Glasgow and Chougule, 2015; Rial-Hermida et al., 2021; Peng et al., 2025).

14.4.7 Regulatory-friendly and scalable

Many people consider microbial EPS like dextran, alginate, xanthan and gellan to be highly scalable, regulatory-friendly and ideal for nano-theragnostic applications (Huu Thanh et al., 2024; Nguyen et al., 2024). Strong safety and regulatory precedents have been established by the long history of safe use of many of these EPS in food, pharmaceutical, and medical products, including pullulan in medical implants, xanthan gum and alginate as food additives and pharmaceutical excipients, gellan in ophthalmic and sustained-release formulations, and dextran as a plasma expander since the 1950s (U.S.Congress, 1993; Morris and Harding, 2009; Vuddanda et al., 2016; Khan et al., 2022; Zustiak and Jain, 2023). Regulatory approval for new therapeutic or diagnostic uses is streamlined by their GRAS status or inclusion in pharmacopoeias around the world.

In terms of manufacturing, the manufacture of EPS makes use of well-established microbial fermentation techniques that are economical and ecologically friendly. They show high-yield, reproducible under GMP conditions in bioprocess engineering, durable solid-state fermentation, and stirred-tank and scale-down bioreactors (Seviour et al., 2010; Teworte et al., 2021; Vaishnav et al., 2021; Netrusov et al., 2023; Jyoti et al., 2024; Thakur and Yadav, 2024). Microbial cell factories are further optimized by genetic and metabolic engineering, which lowers production costs and environmental impacts while providing exact control over EPS yield, molecular weight, and functionality (Cho et al., 2022). In comparison to new synthetic nanomaterials, EPS-derived nano-theragnostic platforms offer a more efficient development pathway and are well-positioned for clinical translation and commercial deployment due to their acknowledged safety, regulatory acceptance and scalable green production (Saravanan et al., 2017; Cho et al., 2022; Dey et al., 2023; Harun-Ur-Rashid et al., 2023).

14.5 Challenges and limitations

Microbial EPS have great potential for usage in biomedical applications as nano-theragnostic agents, combining therapeutic and diagnostic properties into one (Asgher et al., 2020). However, its clinical translation and widespread use are restricted due to several obstacles that need to be overcome (Austin, 2021). Microbial EPS has enormous promise as nano-theragnostic agents, but it is imperative to overcome obstacles related to production, stability, functionalization, and safety (Nguyen et al., 2024). Overcoming these impediments is essential to explore the potential of EPS in cutting-edge medical domains, innovations in biotechnology, nanotechnology and regulatory science (Wang et al., 2024). Its nonbiodegradability is one such issue, as it contributes to plastic pollution, especially in marine settings where it can linger for generations. Although EPS is recyclable, but its low density and bulkiness make the process challenging. This makes it expensive and inefficient to process and transport, particularly when contaminated with food waste (Daba et al., 2021; Moshood et al., 2022; Zhao et al., 2022). Furthermore, flame retardants can lessen the risk of fire caused by combustible EPS, but they may also pose additional health or environmental hazards. Because of its restricted biocompatibility, EPS cannot be used in direct-contact applications in the medical and biological fields, such as implants or devices that come into touch with living tissues. Its hazardous ingredients, like stabilizers or plasticizers, could leak into the environment and cause injury (Dowbysz et al., 2022; Hu et al., 2025). Its sustainability is further complicated by the fact that EPS's large carbon footprint is a result of its energy-intensive production method, which mostly uses fossil fuels. Sustainable production practices, enhanced recycling infrastructure, and environmentally friendly additives are required to address these issues, lessen the environmental impact of EPS, and expand its use in longer-term, more responsible applications (Dowbysz et al., 2022; Sattar et al., 2025).

14.6 Future perspectives and ongoing research

EPS are poised to revolutionize diverse fields through their multifunctional capabilities and green credentials (Rühmann et al., 2015; Duarte Mouro and Gouveia, 2024). In nanotechnology and nano-therapeutics, EPS-derived nanoparticles offer eco-friendly drug delivery and diagnostic platforms with enhanced targeting and reduced side effects (Raveendran et al., 2013; Subhaswaraj et al., 2020; Dey et al., 2023). Their exceptional biocompatibility supports innovation in wound healing, tissue engineering, and scaffold fabrication, positioning EPS as a sustainable alternative to synthetic polymers (Xu et al., 2023). By engineering stimuli-responsive or pH-sensitive EPS, next-generation drug delivery systems can release therapeutics precisely when and where needed. Functionalizing EPS with targeting ligands or chemical groups also opens the door to personalized medicine tailored to individual patient profiles (Karimi et al., 2016; Bachta et al., 2025). In food and agriculture, EPS-producing microbes can generate biofertilizers, soil conditioners, and nanoencapsulation systems that boost productivity while protecting the environment (Ates, 2015; Kaur and Dey, 2023). Meanwhile, industrial biotechnology gains from synthetic biology and genetic engineering advancements that maximize EPS yield, reduce costs, and create novel polymers with bespoke properties. Collectively, these advances underscore EPS's potential to drive sustainable innovation across health, agriculture, and industry (Benhadda et al., 2023; de Souza and Gupta, 2024; Kiran et al., 2024).

14.7 Conclusions

Plant protein-based encapsulation technologies and microbial EPS will be used in the food, health, and biomedical industries. With their structural diversity, capacity to synthesize green nanoparticles, and potent bioactivities, EPS have the potential to completely transform nanotheragnostic, regenerative medicine, and targeted drug delivery. Customized EPS systems for exact therapeutic, diagnostic, and environmental applications will be made possible by advancements in metabolic engineering and functional characterization. At the same time, plant protein matrices provide sustainable, scalable encapsulation solutions that safeguard and regulate the release of enzymes and probiotics. This method not only improves digestive support and gut health, but it also produces innovative, multipurpose items like smart enzyme supplements and enriched plant-based yoghurts. These biopolymeric platforms promise to support the next wave of personalized nutrition, green nanomedicine, and environmentally friendly bio-manufacturing as research moves toward clinical trials, industrial-scale production, and regulatory approval. This represents a significant advancement in fusing sustainability, safety, and efficacy.

Acknowledgment

The authors would like to thank their respective institutes for providing support and necessary facilities.

References

Abbot, V., Paliwal, D., Sharma, A., Sharma, P., 2022. A review on the physicochemical and biological applications of biosurfactants in biotechnology and pharmaceuticals. Heliyon 8 (8), e10149. doi:10.1016/j.heliyon.2022.e10149.

Abdalla, A.K., Ayyash, M.M., Olaimat, A.N., Osaili, T.M., Al-Nabulsi, A.A., Shah, N.P., Holley, R., 2021. Exopolysaccharides as antimicrobial agents: Mechanism and spectrum of activity. Front. Microbiol. 12, 664395. doi:10.3389/fmicb.2021.664395.

Abdel-Wahab, B.A., F. Abd El-Kareem, H., Alzamami, A., A. Fahmy, C., H. Elesawy, B., Mostafa Mahmoud, M., Ghareeb, A., El Askary, A., H. Abo Nahas, H., G. M. Attallah, N., Altwaijry, N., M. Saied, E., 2022. Novel exopolysaccharide from marine *Bacillus* subtilis with broad potential biological activities: Insights into antioxidant, anti-inflammatory, cytotoxicity, and anti-Alzheimer activity. Metabolites, 12. MDPI AG, Basel, Switzerland doi:10.3390/metabo12080715.

Abdelnasser, S.M., Yahya, M., S., M., Mohamed, W.F., Asker, M.M.S., Shady, A., H., M., Mahmoud, M.G., Gadallah, M.A., 2017. Antitumor exopolysaccharides derived from novel marine *Bacillus*: Isolation, characterization aspect and biological activity. Asian Pac. J. Cancer Prev. : APJCP. 18 (7), 1847–1854. doi:10.22034/APJCP.2017.18.7.1847.

Adams, F.C., Barbante, C., 2013. Nanoscience, nanotechnology and spectrometry. Spectrochim. Acta B 86, 3–13.

Adegbolagun, T.I., Odeniyi, O.A., Odeniyi, M.A., 2023. Drug delivery applications and future prospects of microbial exopolysaccharides. Polim. Med. 53 (2), 117–127. doi:10.17219/pim/176590.

Ahmad, M., Man, C., 2015. Microbial polysaccharides and their modification approaches: A review. Int. J. Food Prop. 18. doi:10.1080/10942912.2012.693561.

Ahuja, V., Bhatt, A.K., Banu, J.R., Kumar, V., Kumar, G., Yang, Y.-H., Bhatia, S.K., 2023. Microbial exopolysaccharide composites in biomedicine and healthcare: Trends and advances. Polymers, 15. MDPI AG (Basel, Switzerland. https://doi.org/10.3390/polym15071801.

Ahuja, V., Chauhan, S., Dasgupta, D., Wadhwa, P., Raj, T., Yang, Y.-H., Bhatia, S.K., 2024. Microbial exopolysaccharide composites with inorganic materials and their biomedical applications: A review. Carbohydr. Polym. Technol. Appl. 7, 100482. doi:10.1016/j.carpta.2024.100482.

Alaneme, K.K., Anaele, J.U., Oke, T.M., Kareem, S.A., Adediran, M., Ajibuwa, O.A., Anabaranze, Y.O., 2023. Mycelium based composites: A review of their bio-fabrication procedures, material properties and potential for green building and construction applications. Alex. Eng. J. 83, 234–250. doi:10.1016/j.aej.2023.10.012.

Alqarni, L.S., Alghamdi, M.D., Alshahrani, A.A., Nassar, A.M., 2022. Green nanotechnology: Recent research on bioresource-based nanoparticle synthesis and applications. J. Chem. 2022 (1), 4030999. doi:10.1155/2022/4030999.

AlSamman, M.T., Sánchez, J., 2022. Chitosan- and alginate based hydrogels for the adsorption of anionic and cationic dyes from water. Polymers, 14 (8). doi:10.3390/polym14081498.

Amspacher, W.H., Curreri, A.R., 1953. Use of dextran in control of shock resulting from war wounds. A.M.A. Arch. Surg. 66 (6), 730–740. doi:10.1001/archsurg.1953.01260030750004.

Andrew, M., Jayaraman, G., 2020. Structural features of microbial exopolysaccharides in relation to their antioxidant activity. Carbohydr. Res. 487, 107881. doi:10.1016/j.carres.2019.107881.

Asgher, M., Qamar, S.A., Iqbal, H., 2020. Microbial exopolysaccharide-based nano-carriers with unique multi-functionalities for biomedical sectors. Biologia 76. doi:10.2478/s11756-020-00588-7.

Ates, O., 2015. Systems biology of microbial exopolysaccharides production. Front. Bioeng. Biotechnol. 3. Available from: https://www.frontiersin.org/journals/bioengineering-and-biotechnology/articles/10.3389/fbioe.2015.00200 .

Austin, C.P., 2021. Opportunities and challenges in translational science. Clin. Transl. Sci. 14 (5), 1629–1647. doi:10.1111/cts.13055.

Azmi, N.A., Elgharbawy, A.A.M., Motlagh, S.R., Samsudin, N., Salleh, H.M., 2019. Nanoemulsions: Factory for food, pharmaceutical and cosmetics. Processes, 7. MDPI AG (Basel, Switzerland. https://doi.org/10.3390/pr7090617.

Bachta, P., Jakhmola, V., Nidhi, N., Joshi, P., Bahuguna, R., Chaudhary, M., Kumar, K., Singh, A., Kumar, P., 2025. A comprehensive study on pH-sensitive nanoparticles for the efficient delivery of drugs. J. Adv. Biotechnol. Exp. Ther. 8, 200. doi:10.5455/jabet.2025.17.

Bahrulolum, H., Nooraei, S., Javanshir, N., Tarrahimofrad, H., Mirbagheri, V.S., Easton, A.J., Ahmadian, G., 2021. Green synthesis of metal nanoparticles using microorganisms and their application in the agrifood sector. J. Nanobiotechnol. 19 (1), 86. doi:10.1186/s12951-021-00834-3.

Bansal, S.A., Kumar, V., Karimi, J., Singh, A.P., Kumar, S., 2020. Role of gold nanoparticles in advanced biomedical applications. Nanoscale Adv. 2 (9), 3764–3787.

Bayda, S., Adeel, M., Tuccinardi, T., Cordani, M., Rizzolio, F., 2019. The history of nanoscience and nanotechnology: from chemical–physical applications to nanomedicine. Molecules 25 (1), 112.

Benhadda, F., Zykwinska, A., Colliec-Jouault, S., Sinquin, C., Thollas, B., Courtois, A., Fuzzati, N., Toribio, A., Delbarre-Ladrat, C., 2023. Marine versus non-marine bacterial exopolysaccharides and their skincare applications. Mar. Drugs 21 (11). doi:10.3390/md21110582.

Benincasa, M., Lagatolla, C., Dolzani, L., Milan, A., Pacor, S., Liut, G., Tossi, A., Cescutti, P., Rizzo, R., 2016. Biofilms from *Klebsiella pneumoniae*: Matrix polysaccharide structure and interactions with antimicrobial peptides. Microorganisms 4 (3). doi:10.3390/microorganisms4030026.

Berezina, O.V., Rykov, S.V., Schwarz, W.H., Liebl, W., 2024. Xanthan: Enzymatic degradation and novel perspectives of applications. Appl. Microbiol. Biotechnol. 108 (1), 227. doi:10.1007/s00253-024-13016-6.

Bhattacharya, M., Samanta, D., 2025. Microbial polysaccharides and their classification: A brief study on fundamental biochemical properties. J. Med. Pharm. Allied Sci. 14, 55–67. doi:10.55522/jmpas.V14I2.6851.

Blanco, M.D., Teijón, C., Olmo, R.M., Teijón, J.M., 2012. Targeted nanoparticles for cancer therapy. In: Sezer, A.D. (Ed.), Recent advances in novel drug carrier systems. IntechOpen, London, United Kingdom. https://doi.org/10.5772/51382.

Bolouri, P., Salami, R., Kouhi, S., Kordi, M., Asgari Lajayer, B., Hadian, J., Astatkie, T., 2022. Applications of essential oils and plant extracts in different industries. Mol. (Basel Switz.) 27 (24). doi:10.3390/molecules27248999.

Brummer, R., 2006. Rheology Essentials of Cosmetic and Food Emulsions. Springer, Berlin, Germany doi:10.1007/3-540-29087-7.

Brunchi, C.-E., Morariu, S., Iftime, M.-M., Stoica, I., 2023. Xanthan gum in solution and solid-like state: Effect of temperature and polymer concentration. J. Mol, Liq. 387, 122600. doi:10.1016/j.molliq.2023.122600.

Chen, T.-J., Cheng, T.-H., Chen, C.-Y., Hsu, S.C.N., Cheng, T.-L., Liu, G.-C., Wang, Y.-M., 2009. Targeted herceptin-dextran iron oxide nanoparticles for noninvasive imaging of HER2/neu receptors using MRI. J. Biol. Inorg. Chem. : JBIC : Publ. Soc. Biol. Inorg. Chem. 14 (2), 253–260. doi:10.1007/s00775-008-0445-9.

Cheng, X., Xie, Q., Sun, Y., 2023. Advances in nanomaterial-based targeted drug delivery systems. Front. Bioeng. Biotechnol. 11, 1177151. doi:10.3389/fbioe.2023.1177151.

Cho, J.S., Kim, G.B., Eun, H., Moon, C.W., Lee, S.Y., 2022. Designing microbial cell factories for the production of chemicals. JACS Au. 2 (8), 1781–1799. doi:10.1021/jacsau.2c00344.

Chugh, D., Viswamalya, V.S., Das, B., 2021. Green synthesis of silver nanoparticles with algae and the importance of capping agents in the process. J. Genet., Eng., Biotechnol. 19 (1), 126. doi:10.1186/s43141-021-00228-w.

Costa, O.Y.A., Raaijmakers, J.M., Kuramae, E.E., 2018. Microbial extracellular polymeric substances: Ecological function and impact on soil aggregation. Front. Microbiol.,, 9. https://www.frontiersin.org/journals/microbiology/articles/10.3389/fmicb.2018.01636.

Costerton, J.W., Stewart, P.S., Greenberg, E.P., 1999. Bacterial biofilms: A common cause of persistent infections. Sci. (N. Y. N.Y.) 284 (5418), 1318–1322. doi:10.1126/science.284.5418.1318.

da Silva, J., Rao, M., 2007. Rheological behavior of food gels. Food Engineering Series. Springer US, New York, USA, pp. 339–401. doi:10.1007/978-0-387-70930-7_6.

Daba, G.M., Elnahas, M.O., Elkhateeb, W.A., 2021. Contributions of exopolysaccharides from lactic acid bacteria as biotechnological tools in food, pharmaceutical, and medical applications. Int, J. Biol, Macromol. 173, 79–89. doi:10.1016/j.ijbiomac.2021.01.110.

Damiri, F., Fatimi, A., Paiva-Santos, A.C., Varma, R., Berrada, M., 2023. Smart stimuli-responsive polysaccharide nanohydrogels for drug delivery: A review. J. Mater. Chem. B 11. doi:10.1039/D3TB01712E.

de Souza, F.M., Gupta, R.K., 2024. Bacteria for bioplastics: Progress, applications, and challenges. ACS Omega 9 (8), 8666–8686. doi:10.1021/acsomega.3c07372.

Deepak, V., Ramachandran, S., Balahmar, R.M., Pandian, S.R.K., Sivasubramaniam, S.D., Nellaiah, H., Sundar, K., 2016. In vitro evaluation of anticancer properties of exopolysaccharides from Lactobacillus acidophilus in colon cancer cell lines. In Vitro Cell. Dev. Biol. - Anim. 52 (2), 163–173. doi:10.1007/s11626-015-9970-3.

Dey, G., Patil, M.P., Banerjee, A., Sharma, R.K., Banerjee, P., Maity, J.P., Singha, S., Taharia, M., Shaw, A.K., Huang, H.-B., Kim, G.-D., Chen, C.-Y.,, 2023. The role of bacterial exopolysaccharides (EPS) in the synthesis of antimicrobial silver nanomaterials: A state-of-the-art review. J. Microbiol. Methods 212, 106809. doi:10.1016/j.mimet.2023.106809.

Díaz-Dinamarca, D.A., Salazar, M.L., Castillo, B.N., Manubens, A., Vasquez, A.E., Salazar, F., Becker, M.I., 2022. Protein-based adjuvants for vaccines as immunomodulators of the innate and adaptive immune response: Current knowledge, challenges, and future opportunities. Pharmaceutics 14 (8). doi:10.3390/pharmaceutics14081671.

Díaz-Montes, E., Castro-Muñoz, R., 2021. Edible films and coatings as food-quality preservers: An overview. Foods (Basel Switz.) 10 (2). doi:10.3390/foods10020249.

Domżał-Kędzia, M., Ostrowska, M., Lewińska, A., Łukaszewicz, M., 2023. Recent developments and applications of microbial levan, a versatile polysaccharide-based biopolymer. Mol. (Basel Switz.) 28 (14). doi:10.3390/molecules28145407.

Dong, B., Liu, G., Zhou, J., Cai, L., Wang, J., Jin, R., 2020. Roles of molecular weight-fractionated extracellular polymeric substance in transformation of Au(III) to Au nanoparticles in aqueous environments. Sci, Total Env. 728, 138889. doi:10.1016/j.scitotenv.2020.138889.

Dowbysz, A., Samsonowicz, M., Kukfisz, B.,, 2022. Recent advances in bio-based additive flame retardants for thermosetting resins. Int, J. Env., Res, Public Health 19 (8). doi:10.3390/ijerph19084828.

Dowdall, N., Hoare, T., 2025. β-1,3 Glucan microparticles and nanoparticles: Fabrication methods and applications in immunomodulation and targeted drug delivery. Adv, Healthc. Mater. 14 (14), 2501006. doi:10.1002/adhm.202501006.

Duarte Mouro, C.F., Gouveia, I., 2024. Microbial exopolysaccharides: Structure, diversity, applications, and future frontiers in sustainable functional materials. Polysaccharides 5. doi:10.3390/polysaccharides5030018.

Duru, O.A., Toksoy Oner, E., 2017. Microbial xanthan, levan, gellan, and curdlan as food additives. Microbial biopolymer-based materials: Science and engineering. John Wiley & Sons, Inc, Hoboken, New Jersey, US, pp. 149–173. https://doi.org/10.1002/9781119048961.ch7.

El awady, M.E., Mohamed, S.S., Abo Elsoud, M.M., Mahmoud, M.G., Anwar, M.M., Ahmed, M.M., Eltaher, A., Magdeldin, S., Attallah, A., Elhagry, A.E., Abdelhamid, S.A., 2024. Insight into antioxidant and anti-inflammatory effects of marine bacterial natural exopolysaccharide (EPSSM) using carrageenan-induced paw edema in rats. Sci, Rep. 14 (1), 5113. doi:10.1038/s41598-024-53502-5.

Ergün, H., Ergün, M.E., 2024. Modeling xanthan gum foam's material properties using machine learning methods. Polymers 16 (6). doi:10.3390/polym16060740.

Escárcega-González, C.E., Garza-Cervantes, J.A., Vázquez-Rodríguez, A., Morones-Ramírez, J.R., 2018. Bacterial exopolysaccharides as reducing and/or stabilizing agents during synthesis of metal nanoparticles with biomedical applications. Int. J. Polym. Sci. 2018 (1), 7045852. doi:10.1155/2018/7045852.

Facciolà, A., Visalli, G., Laganà, A., Di Pietro, A., 2022. An overview of vaccine adjuvants: Current evidence and future perspectives. Vaccines 10 (5). doi:10.3390/vaccines10050819.

Farris, S., Unalan, I.U., Introzzi, L., Fuentes-Alventosa, J.M., Cozzolino, C.A., 2014. Pullulan-based films and coatings for food packaging: Present applications, emerging opportunities, and future challenges. J. Appl, Polym, Sci. 131 (13). doi:10.1002/app.40539.

Farzin, A., Etesami, S.A., Quint, J., Memic, A., Tamayol, A., 2020. Magnetic Nanoparticles in Cancer Therapy and Diagnosis. Adv., Healthc. Mater. 9 (9), e1901058. doi:10.1002/adhm.201901058.

Gao, W., Chan, J.M., Farokhzad, O.C., 2010. pH-responsive nanoparticles for drug delivery. Mol. Pharm. 7 (6), 1913–1920. doi:10.1021/mp100253e.

Gargouch, N., Elleuch, F., Karkouch, I., Tabbene, O., Pichon, C., Gardarin, C., Rihouey, C., Picton, L., Abdelkafi, S., Fendri, I., Laroche, C., 2021. Potential of exopolysaccharide from *Porphyridium marinum* to contend with bacterial proliferation, biofilm formation, and breast cancer. Marine Drugs, 19. MPDI AG, Switzerland doi:10.3390/md19020066.

Geller, A., Shrestha, R., Yan, J., 2019. Yeast-derived β-glucan in cancer: Novel uses of a traditional therapeutic. Int. J. Mol. Sci. 20 (15). doi:10.3390/ijms20153618.

Ghareeb, A., Fouda, A., Kishk, R.M., El Kazzaz, W.M., 2024. Unlocking the therapeutic potential of bioactive exopolysaccharide produced by marine actinobacterium *Streptomyces vinaceusdrappus* AMG31: A novel approach to drug development. Int. J. Biol. Macromol. 276, 133861. doi:10.1016/j.ijbiomac.2024.133861.

Gherasim, O., Puiu, R.A., Bîrcă, A.C., Burduşel, A.-C., Grumezescu, A.M., 2020. An updated review on silver nanoparticles in biomedicine. Nanomaterials, 10 (11). doi:10.3390/nano10112318.

Glasgow, M., Chougule, M., 2015. Recent developments in active tumor targeted multifunctional nanoparticles for combination chemotherapy in cancer treatment and imaging. J. Biomed, Nanotechnol. 11. doi:10.1166/jbn.2015.2145.

Goodridge, H., Wolf, A., Underhill, D., 2009. Glucan recognition by the innate immune system. Immunol. Rev. 230, 38–50. doi:10.1111/j.1600-065X.2009.00793.x.

Granton, E., Brown, L., Defaye, M., Moazen, P., Almblad, H., Randall, T.E., Rich, J.D., Geppert, A., Abdullah, N.S., Hassanabad, M.F., Hiroki, C.H., Farias, R., Nguyen, A.P., Schubert, C., Lou, Y., Andonegui, G., Iftinca, M., Raju, D., Vargas, M.A., … Yipp, B.G., 2024. Biofilm exopolysaccharides alter sensory-neuron-mediated sickness during lung infection. Cell 187 (8), 1874–1888.e14. doi:10.1016/j.cell.2024.03.001.

Grauer, O., Jaber, M., Hess, K., Weckesser, M., Schwindt, W., Maring, S., Wölfer, J., Stummer, W., 2019. Combined intracavitary thermotherapy with iron oxide nanoparticles and radiotherapy as local treatment modality in recurrent glioblastoma patients. J. Neurooncol. 141, 83–94.

Gunardi, W.D., Karuniawati, A., Umbas, R., Bardosono, S., Lydia, A., Soebandrio, A., Safari, D., 2021. Biofilm-producing bacteria and risk factors (gender and duration of catheterization) characterized as catheter-associated biofilm formation. Int. J. Microbiol. 2021, 8869275. doi:10.1155/2021/8869275.

Guo, Y., Qiao, D., Zhao, S., Liu, P., Xie, F., Zhang, B., 2024. Biofunctional chitosan–biopolymer composites for biomedical applications. Mater. Sci. Eng.: R: Rep. 159, 100775. doi:10.1016/j.mser.2024.100775.

Gupta, P., Diwan, B., 2017. Bacterial Exopolysaccharide mediated heavy metal removal: A Review on biosynthesis, mechanism and remediation strategies. Biotechnol. Rep. 13, 58–71. doi:10.1016/j.btre.2016.12.006.

Hajishengallis, G., Lambris, J.D., 2011. Microbial manipulation of receptor crosstalk in innate immunity. Nat. Rev., Immunol. 11 (3), 187–200. doi:10.1038/nri2918.

Harun-Ur-Rashid, M., Jahan, I., Foyez, T., Imran, A.B., 2023. Bio-inspired nanomaterials for micro/nanodevices: A new era in biomedical applications. Micromachines, 14. MDPI AG, Basel, Switzerland doi:10.3390/mi14091786.

Harvey, D.J., 2015. Analysis of carbohydrates and glycoconjugates by matrix-assisted laser desorption/ionization mass spectrometry: An update for 2009–2010. Mass Spectrom. Rev. 34 (3), 268–422. doi:10.1002/mas.21411.

Hasan, A., Morshed, M., Memic, A., Hassan, S., Webster, T.J., Marei, H.E.-S., 2018. Nanoparticles in tissue engineering: applications, challenges and prospects. Int, J. Nanomed. 13, 5637–5655. doi:10.2147/IJN.S153758.

He, Y., de Araújo Júnior, R.F., Cruz, L.J., Eich, C., 2021. Functionalized nanoparticles targeting tumor-associated macrophages as cancer therapy. Pharmaceutics 13 (10). doi:10.3390/pharmaceutics13101670.

Hooton, T.M., Bradley, S.F., Cardenas, D.D., Colgan, R., Geerlings, S.E., Rice, J.C., Saint, S., Schaeffer, A.J., Tambayh, P.A., Tenke, P., Nicolle, L.E., 2010. Diagnosis, prevention, and treatment of catheter-associated urinary tract infection in adults: 2009 international clinical practice guidelines from the infectious diseases society of America. Clin. Infect. Dis. 50 (5), 625–663. doi:10.1086/650482.

Hsu, J.C., Tang, Z., Eremina, O.E., Sofias, A.M., Lammers, T., Lovell, J.F., Zavaleta, C., Cai, W., Cormode, D.P., 2023. Nanomaterial-based contrast agents. Nat. Rev., Methods Primers. 3. doi:10.1038/s43586-023-00211-4.

Hu, Q., Katti, P.S., Gu, Z., 2014. Enzyme-responsive nanomaterials for controlled drug delivery. Nanoscale 6 (21), 12273–12286. doi:10.1039/c4nr04249b.

Hu, X., Fang, X., Zhao, Y., Shao, W., Hou, J., Li, X., Liu, J., Zhai, M., Tian, F., Yan, Y., Lu, Y., 2025. Gel foam loaded O2-consuming microbial community and the stratification structure in preventing coal spontaneous combustion. ACS Appl. Mater. Interfaces 17 (9), 14085–14096. doi:10.1021/acsami.4c22558.

Huang, Y., Wang, Y.-J., Wang, Y., Yi, S., Fan, Z., Sun, L., Lin, D., Anreddy, N., Zhu, H., Schmidt, M., Chen, Z.-S., Zhang, M., 2015. Exploring naturally occurring ivy nanoparticles as an alternative biomaterial. Acta Biomater doi:10.1016/j.actbio.2015.07.035.

Hui, J., Al Zaki, A., Cheng, Z., Popik, V., Zhang, H., Luning Prak, N., Tsourkas, A., 2014. Facile method for the site-specific, covalent attachment of full-length IgG onto nanoparticles. Small 10. doi:10.1002/smll.201303629.

Huu Thanh, N., Trang, P., Nguyen, P.-T., Le Buanec, H., Rabetafika, H., Razafindralambo, H., 2024. Advances in microbial exopolysaccharides: Present and future applications. Biomolecules 14, 1162. doi:10.3390/biom14091162.

Ibrahim, H.A.H., Abou Elhassayeb, H.E., El-Sayed, W.M.M., 2022. Potential functions and applications of diverse microbial exopolysaccharides in marine environments. J. Genet. Eng. Biotechnol. 20 (1), 151. doi:10.1186/s43141-022-00432-2.

Jacob, S., Nair, A.B., Shah, J., Sreeharsha, N., Gupta, S., Shinu, P., 2021. Emerging role of hydrogels in drug delivery systems, tissue engineering and wound management. Pharmaceutics 13 (3). doi:10.3390/pharmaceutics13030357.

Janib, S.M., Moses, A.S., MacKay, J.A., 2010. Imaging and drug delivery using theranostic nanoparticles. Adv. Drug Deliv. Rev. 62 (11), 1052–1063. doi:10.1016/j.addr.2010.08.004.

Jaroszuk-Ściseł, J., Nowak, A., Pac-Sosińska, M., Kołodyńska, D., Komaniecka, I., 2025. Heavy metal biosorption ability of EPS obtained from cultures of *Fusarium culmorum* strains with different effects on cereals. Sustainability, 17 doi:10.3390/su17083744.

Javan Nikkhah, S., Vandichel, M., 2022. Modeling polyzwitterion-based drug delivery platforms: A perspective of the current State-of-the-Art and beyond. ACS Eng. Au. 2 (4), 274–294. doi:10.1021/acsengineeringau.2c00008.

Javed, R., Zia, M., Naz, S., Aisida, S.O., ul Ain, N., Ao, Q., 2020. Role of capping agents in the application of nanoparticles in biomedicine and environmental remediation: recent trends and future prospects. J. Nanobiotechnol. 18 (1), 172. doi:10.1186/s12951-020-00704-4.

Javid, H., Oryani, M.A., Rezagholinejad, N., Esparham, A., Tajaldini, M., Karimi-Shahri, M., 2024. RGD peptide in cancer targeting: Benefits, challenges, solutions, and possible integrin-RGD interactions. Cancer Med. 13 (2), e6800. doi:10.1002/cam4.6800.

Jiang, Z., Zhang, M., Li, P., Wang, Y., Fu, Q., 2023. Nanomaterial-based CT contrast agents and their applications in image-guided therapy. Theranostics 13 (2), 483–509. doi:10.7150/thno.79625.

Jyoti, K., Soni, K., Chandra, R., 2024. Optimization of the production of exopolysaccharide (EPS) from biofilm-forming bacterial consortium using different parameters. Microbe 4, 100117. doi:10.1016/j.microb.2024.100117.

Karimi, M., Eslami, M., Sahandi-Zangabad, P., Mirab, F., Farajisafiloo, N., Shafaei, Z., Ghosh, D., Bozorgomid, M., Dashkhaneh, F., Hamblin, M.R., 2016. pH-Sensitive stimulus-responsive nanocarriers for targeted delivery of therapeutic agents. Wiley Interdiscip. Rev., Nanomed. Nanobiotechnology 8 (5), 696–716. doi:10.1002/wnan.1389.

Kaur, N., Dey, P., 2023. Bacterial exopolysaccharides as emerging bioactive macromolecules: from fundamentals to applications. Res. Microbiol. 174 (4), 104024. doi:10.1016/j.resmic.2022.104024.

Kaur, P., Aliru, M.L., Chadha, A.S., Asea, A., Krishnan, S., 2016. Hyperthermia using nanoparticles–Promises and pitfalls. Int. J. Hyperth. : Off. J. Eur. Soc. Hyperthermic Oncol. N. Am. Hyperth. Group 32 (1), 76–88. doi:10.3109/02656736.2015.1120889.

Khadka, S., Ring, B.E., Pariseau, D.A., Mike, L.A., 2023. Characterization of *Klebsiella pneumoniae* extracellular polysaccharides. Curr. Protoc. 3 (11), e937. doi:10.1002/cpz1.937.

Khan, R., Shah, M.D., Shah, L., Lee, P.-C., Khan, I., 2022. Bacterial polysaccharides-A big source for prebiotics and therapeutics. Front. Nutr. 9, 1031935. doi:10.3389/fnut.2022.1031935.

Kim, Y., Hu, Y., Jeong, J., Jung, S., 2022. Injectable, self-healable and adhesive hydrogels using oxidized Succinoglycan/chitosan for pH-responsive drug delivery. Carbohydr. Polym. 284, 119195. doi:10.1016/j.carbpol.2022.119195.

Kiran, N.S., Yashaswini, C., Singh, S., Prajapati, B.G., 2024. Revisiting microbial exopolysaccharides: A biocompatible and sustainable polymeric material for multifaceted biomedical applications. 3 Biotech 14 (4), 95. doi:10.1007/s13205-024-03946-3.

Kondakindi, V.R., Pabbati, R., Erukulla, P., Maddela, N.R., Prasad, R., 2024. Bioremediation of heavy metals-contaminated sites by microbial extracellular polymeric substances – A critical view. Environ. Chem. Ecotoxicol. 6, 408–421. doi:10.1016/j.enceco.2024.05.002.

Kong, X., Gao, P., Wang, J., Fang, Y., Hwang, K.C., 2023. Advances of medical nanorobots for future cancer treatments. J. Hematol, Oncol. 16 (1), 74. doi:10.1186/s13045-023-01463-z.

Korcz, E., Varga, L., 2021. Exopolysaccharides from lactic acid bacteria: Techno-functional application in the food industry. Trends Food Sci. Technol. 110, 375–384. doi:10.1016/j.tifs.2021.02.014.

Krsmanovic, M., Biswas, D., Ali, H., Kumar, A., Ghosh, R., Dickerson, A.K., 2021. Hydrodynamics and surface properties influence biofilm proliferation. Adv. Colloid Interface Sci. 288, 102336. doi:10.1016/j.cis.2020.102336.

Kulkarni, N.A., Nanjappa, S.G., 2024. Advances in dendritic-cell-based vaccines against respiratory fungal infections. Vaccines., 12. MDPI, Basel, Switzerland doi:10.3390/vaccines12090981.

Kumari, L., Choudhari, Y., Patel, P., Gupta, G.D., Singh, D., Rosenholm, J.M., Bansal, K.K., Kurmi, B.D., 2023. Advancement in solubilization approaches: A step towards bioavailability enhancement of poorly soluble drugs. Life (Basel Switz.) 13 (5). doi:10.3390/life13051099.

Kumari, S., Rao, A., Kaur, M., Dhania, G. (2023). Petroleum-based plastics versus bio-based plastics: A review. Nat. Environ. Pollut. Technol., 22(3), 1111–1124.

Kurul, F., Turkmen, H., Cetin, A.E., Topkaya, S.N., 2025. Nanomedicine: How nanomaterials are transforming drug delivery, bio-imaging, and diagnosis. Next Nanotechnol. 7, 100129. doi:10.1016/j.nxnano.2024.100129.

Lambert, F., Rinaudo, M., 1985. On the thermal stability of xanthan gum. Polymer 26 (10), 1549–1553. doi:10.1016/0032-3861(85)90092-8.

Laubach, J., Joseph, M., Brenza, T., Gadhamshetty, V., Sani, R.K., 2021. Exopolysaccharide and biopolymer-derived films as tools for transdermal drug delivery. J. Control. Release 329, 971–987. doi:10.1016/j.jconrel.2020.10.027.

Lei, X., Zhang, H., Hu, Z., Liang, Y., Guo, S., Yang, M., Du, R., Wang, X., 2018. Immunostimulatory activity of exopolysaccharides from probiotic *Lactobacillus casei* WXD030 strain as a novel adjuvant in vitro and in vivo. Food Agric. Immunol. 29, 1086–1105. doi:10.1080/09540105.2018.1513994.

Łętocha, A., Miastkowska, M., Sikora, E., 2022. Preparation and characteristics of alginate microparticles for food, pharmaceutical and cosmetic applications. Polymers, 14 doi:10.3390/polym14183834.

Li, W., Xia, X., Tang, W., Ji, J., Rui, X., Chen, X., Jiang, M., Zhou, J., Zhang, Q., Dong, M., 2015. Structural characterization and anticancer activity of cell-bound exopolysaccharide from *Lactobacillus helveticus* MB2-1. J. Agric, Food Chem. 63 (13), 3454–3463. doi:10.1021/acs.jafc.5b01086.

Lin, L., Chi, J., Yan, Y., Luo, R., Feng, X., Zheng, Y., Xian, D., Li, X., Quan, G., Liu, D., Wu, C., Lu, C., Pan, X., 2021. Membrane-disruptive peptides/peptidomimetics-based therapeutics: Promising systems to combat bacteria and cancer in the drug-resistant era. Acta Pharm. Sin. B 11 (9), 2609–2644. doi:10.1016/j.apsb.2021.07.014.

Linker, A., Jones, R.S., 1966. A new polysaccharide resembling alginic acid isolated from pseudomonads. J. Biol, Chem. 241 (16), 3845–3851.

Liu, S., Wei, W., Wang, J., Chen, T., 2023. Theranostic applications of selenium nanomedicines against lung cancer. J. Nanobiotechnol 21 (1), 96. doi:10.1186/s12951-023-01825-2.

Ma, W., Zhan, Y., Zhang, Y., Mao, C., Xie, X., Lin, Y., 2021. The biological applications of DNA nanomaterials: Current challenges and future directions. Signal Transduct. Target. Ther. 6 (1), 351. doi:10.1038/s41392-021-00727-9.

Ma, X., Tian, Y., Yang, R., Wang, H., Allahou, L.W., Chang, J., Williams, G., Knowles, J.C., Poma, A., 2024. Nanotechnology in healthcare, and its safety and environmental risks. J. Nanobiotechnol. 22 (1), 715. doi:10.1186/s12951-024-02901-x.

Mahmoud, Y.A.-G., El-Naggar, M.E., Abdel-Megeed, A., El-Newehy, M., 2021. Recent advancements in microbial polysaccharides: Synthesis and applications. Polymers 13 (23). doi:10.3390/polym13234136.

Majumder, M.M.I., Ahmed, T., Khan, A.R., Ahmed, S., 2018. Microbiology of Catheter Associated Urinary Tract Infection. IntechOpen, London, United Kingdom doi:10.5772/intechopen.80080.

Malik, S., Muhammad, K., Waheed, Y., 2023. Nanotechnology: A revolution in modern industry. Molecules 28 (2), 661.

Manivasagan, P., Oh, J., 2016. Marine polysaccharide-based nanomaterials as a novel source of nanobiotechnological applications. Int, J. Biol, Macromol. 82, 315–327. doi:10.1016/j.ijbiomac.2015.10.081.

Mansour, A., Romani, M., Acharya, A.B., Rahman, B., Verron, E., Badran, Z., 2023. Drug delivery systems in regenerative medicine: An updated review. Pharmaceutics 15 (2). doi:10.3390/pharmaceutics15020695.

Mehta, K., Shukla, A., Saraf, M., 2021. Articulating the exuberant intricacies of bacterial exopolysaccharides to purge environmental pollutants. Heliyon 7 (11), e08446. doi:10.1016/j.heliyon.2021.e08446.

Mende, S., Rohm, H., Jaros, D., 2015. Influence of exopolysaccharides on the structure, texture, stability and sensory properties of yoghurt and related products. Int. Dairy J. 52. doi:10.1016/j.idairyj.2015.08.002.

Menichetti, A., Mavridi-Printezi, A., Mordini, D., Montalti, M., 2023. Effect of size, shape and surface functionalization on the antibacterial activity of silver nanoparticles. J. Funct. Biomater., 14 (5). doi:10.3390/jfb14050244.

Mohd Nadzir, M., Nurhayati, R.W., Idris, F.N., Nguyen, M.H., 2021. Biomedical applications of bacterial exopolysaccharides: A review. Polymers 13 (4). doi:10.3390/polym13040530.

More, P.R., Pandit, S., Filippis, A.D.e, Franci, G., Mijakovic, I., Galdiero, M., 2023. Silver nanoparticles: Bactericidal and mechanistic approach against drug resistant pathogens. Microorganisms 11 (2). doi:10.3390/microorganisms11020369.

Morris, G., Harding, S., 2009. Polysaccharides, microbial. Encyclopedia of Microbiology. Academic press, Amsterdam, Netherlands, pp. 482–494. doi:10.1016/B978-012373944-5.00135-8.

Moscovici, M., 2015. Present and future medical applications of microbial exopolysaccharides. Front. Microbiol. 6, 1012. doi:10.3389/fmicb.2015.01012.

Moshood, T.D., Nawanir, G., Mahmud, F., Mohamad, F., Ahmad, M.H., AbdulGhani, A., 2022. Sustainability of biodegradable plastics: New problem or solution to solve the global plastic pollution? Curr. Res. Green Sustain. Chem. 5, 100273. doi:10.1016/j.crgsc.2022.100273.

Mouro, C., Gomes, A.P., Gouveia, I.C., 2024. Microbial exopolysaccharides: Structure, diversity, applications, and future frontiers in sustainable functional materials. Polysaccharides., 5. MDPI AG, Basel, Switzerland, pp. 241–287. doi:10.3390/polysaccharides5030018.

Muñoz-Tebar, N., Pérez-Álvarez, J.A., Fernández-López, J., Viuda-Martos, M., 2023. Chitosan edible films and coatings with added bioactive compounds: Antibacterial and antioxidant properties and their application to food products: A review. Polymers, 15 doi:10.3390/polym15020396.

N S, K., Chandrashekar, Y., Singh, S., Prajapati, B., 2024. Revisiting microbial exopolysaccharides: A biocompatible and sustainable polymeric material for multifaceted biomedical applications. 3 Biotech 14, 1–24. doi:10.1007/s13205-024-03946-3.

Nafee, N., Husari, A., Maurer, C., Lu, C., De Rossi, C., Steinbach, A., Hartmann, R., Lehr, C.-M., Schneider, M., 2014. Antibiotic-free nanotherapeutics: Ultra-small, mucus-penetrating solid lipid nanoparticles enhance the pulmonary delivery and anti-virulence efficacy of novel quorum sensing inhibitors. J. Control. Release : Off. J. Control. Release Soc. 192. doi:10.1016/j.jconrel.2014.06.055.

Nag, S., Kar, S., Mishra, S., Stany, B., Seelan, A., Mohanto, S., Haryini, S., S., Kamaraj, C., Subramaniyan, V., 2024. Unveiling green synthesis and biomedical theranostic paradigms of selenium nanoparticles (SeNPs) - A State-of-the-Art comprehensive update. Int, J. Pharm. 662, 124535. doi:10.1016/j.ijpharm.2024.124535.

Nandhini, J., Karthikeyan, E., Elizabeth Rani, E., Karthikha, V.S., Sakthi Sanjana, D., Jeevitha, H., Rajeshkumar, S., Venugopal, V., Priyadharshan, A., 2024. Advancing engineered approaches for sustainable wound regeneration and repair: Harnessing the potential of green synthesized silver nanoparticles. Eng. Regen. 5 (3), 306–325. doi:10.1016/j.engreg.2024.06.004.

Nazeam, J.A., Singab, A.N.B., 2022. Immunostimulant plant proteins: Potential candidates as vaccine adjuvants. Phytother. Res.: PTR 36 (12), 4345–4360. doi:10.1002/ptr.7624.

Netrusov, A.I., Liyaskina, E.V., Kurgaeva, I.V., Liyaskina, A.U., Yang, G., Revin, V.V., 2023. Exopolysaccharides producing bacteria: A review. Microorganisms 11 (6). doi:10.3390/microorganisms11061541.

Nguyen, H.-T., Pham, T.-T., Nguyen, P.-T., Le-Buanec, H., Rabetafika, H.N., Razafindralambo, H.L., 2024. Advances in microbial exopolysaccharides: Present and future applications. Biomolecules, 14 doi:10.3390/biom14091162.

Nwodo, U.U., Green, E., Okoh, A.I., 2012. Bacterial exopolysaccharides: Functionality and prospects. Int, J. Mol, Sci 13 (11), 14002–14015. doi:10.3390/ijms131114002.

Oerlemans, M.M.P., Akkerman, R., Ferrari, M., Walvoort, M.T.C., de Vos, P., 2021. Benefits of bacteria-derived exopolysaccharides on gastrointestinal microbiota, immunity and health. J. Funct. Foods 76, 104289. doi:10.1016/j.jff.2020.104289.

Osemwegie, O.O., Adetunji, C.O., Ayeni, E.A., Adejobi, O.I., Arise, R.O., Nwonuma, C.O., Oghenekaro, A.O., 2020. Exopolysaccharides from bacteria and fungi: Current status and perspectives in Africa. Heliyon 6 (6), e04205. doi:10.1016/j.heliyon.2020.e04205.

Pang, Q., Jiang, Z., Wu, K., Hou, R., Zhu, Y., 2023. Nanomaterials-based wound dressing for advanced management of infected wound. Antibiot. (Basel Switz,) 12 (2). doi:10.3390/antibiotics12020351.

Paris, S., Chapat, L., Martin-Cagnon, N., Durand, P.-Y., Piney, L., Cariou, C., Bergamo, P., Bonnet, J.-M., Poulet, H., Freyburger, L., De Luca, K., 2020. β-Glucan as trained immunity-based adjuvants for rabies vaccines in dogs. Front. Immunol. 11. https://www.frontiersin.org/journals/immunology/articles/10.3389/fimmu.2020.564497. Available from:.

Parvin, N., Kumar, V., Joo, S.W., Mandal, T.K., 2024. Cutting-edge hydrogel technologies in tissue engineering and biosensing: An updated review. Mater. (Basel Switz.) 17 (19). doi:10.3390/ma17194792.

Patel, B., Darji, P., Ivan, P., Fnu, J., Nalla, S., Khatri, V., Parikh, S. (2024). A comprehensive review and insight into the latest advancements in nanotechnology. Biosci. Biotechnol. Res. Asia, 21(September), 985–1000.

Paul, P., Nair, R., Mahajan, S., Gupta, U., Aalhate, M., Maji, I., Singh, P.K., 2023. Traversing the diverse avenues of exopolysaccharides-based nanocarriers in the management of cancer. Carbohydr. Polym. 312, 120821. doi:10.1016/j.carbpol.2023.120821.

Pelling, H., Nzakizwanayo, J., Milo, S., Denham, E.L., MacFarlane, W.M., Bock, L.J., Sutton, J.M., Jones, B.V., 2019. Bacterial biofilm formation on indwelling urethral catheters. Lett. Appl. Microbiol. 68 (4), 277–293. doi:10.1111/lam.13144.

Peng, H., Chen, I.A., Qimron, U., 2025. Engineering phages to fight multidrug-resistant bacteria. Chem. Rev. 125 (2), 933–971. doi:10.1021/acs.chemrev.4c00681.

Pereira, S.B., Sousa, A., Santos, M., Araújo, M., Serôdio, F., Granja, P., Tamagnini, P., 2019. Strategies to obtain designer polymers based on cyanobacterial extracellular polymeric substances (EPS). Int. J. Mol. Sci. 20 (22). doi:10.3390/ijms20225693.

Perera, S.R., Sokaribo, A.S., White, A.P., 2021. Polysaccharide vaccines: A perspective on non-typhoidal *Salmonella*. Polysaccharides., 2, pp. 691–714. doi:10.3390/polysaccharides2030042.

Prajapati, V., Jani, G., Khanda, S., 2013. Pullulan: An exopolysaccharide and its various applications. Carbohydr. Polym. 95, 540–549. doi:10.1016/j.carbpol.2013.02.082.

Qi, Y., Yan, J., Huang, X., Jiang, X., Li, R., Wan, J., Li, Y., Miao, Z., Song, Z., Liu, Y., Zhang, L., Zhang, Z., 2025. Targeting tumor-associated macrophage polarization with traditional Chinese medicine active ingredients: Dual reversal of chemoresistance and immunosuppression in tumor microenvironment. Pharmacol. Res. 216, 107788. doi:10.1016/j.phrs.2025.107788.

Qin, X.-S., Luo, Z.-G., Li, X.-L., 2021. An enhanced pH-sensitive carrier based on alginate-Ca-EDTA in a set-type W1/O/W2 double emulsion model stabilized with WPI-EGCG covalent conjugates for probiotics colon-targeted release. Food Hydrocoll. 113, 106460. doi:10.1016/j.foodhyd.2020.106460.

Ramesh, M., Janani, R., Deepa, C., Rajeshkumar, L., 2022. Nanotechnology-enabled biosensors: A review of fundamentals, design principles, materials, and applications. Biosensors 13 (1). doi:10.3390/bios13010040.

Raveendran, S., Poulose, A.C., Yoshida, Y., Maekawa, T., Kumar, D.S., 2013. Bacterial exopolysaccharide-based nanoparticles for sustained drug delivery, cancer chemotherapy and bioimaging. Carbohydr. Polym. 91 (1), 22–32. doi:10.1016/j.carbpol.2012.07.079.

Rehm, B.H.A., 2010. Bacterial polymers: Biosynthesis, modifications and applications. Nat. Rev. Microbiol. 8 (8), 578–592. doi:10.1038/nrmicro2354.

Reinoso, D., Martín-Alfonso, M.J., Luckham, P.F., Martínez-Boza, F.J., 2019. Rheological characterisation of xanthan gum in brine solutions at high temperature. Carbohydr. Polym. 203, 103–109. doi:10.1016/j.carbpol.2018.09.034.

Revin, V.V., Liyaskina, E.V., Parchaykina, M.V., Kurgaeva, I.V., Efremova, K.V., Novokuptsev, N.V., 2023. Production of bacterial exopolysaccharides: Xanthan and bacterial cellulose. Int. J. Mol. Sci. 24 (19). doi:10.3390/ijms241914608.

Rial-Hermida, M.I., Rey-Rico, A., Blanco-Fernandez, B., Carballo-Pedrares, N., Byrne, E.M., Mano, J.F., 2021. Recent progress on polysaccharide-based hydrogels for controlled delivery of therapeutic biomolecules. ACS Biomater. Sci. Eng. 7 (9), 4102–4127. doi:10.1021/acsbiomaterials.0c01784.

Ribeiro, A.I., Dias, A.M., Zille, A., 2022. Synergistic effects between metal nanoparticles and commercial antimicrobial agents: A review. ACS Appl., Nano Mater. 5 (3), 3030–3064. doi:10.1021/acsanm.1c03891.

Rodrigues, A.S., Batista, J.G.S., Rodrigues, M.Á.V., Thipe, V.C., Minarini, L.A.R., Lopes, P.S., Lugão, A.B., 2024. Advances in silver nanoparticles: A comprehensive review on their potential as antimicrobial agents and their mechanisms of action elucidated by proteomics. Front. Microbiol. 15, 1440065. doi:10.3389/fmicb.2024.1440065.

Rodríguez, C.F., Paula, G.-S., Alan, S.-B., Coryna, R., Paula, V., Luis, H.R., Cruz, J.C., 2025. Magnetoliposomes for nanomedicine: Synthesis, characterization, and applications in drug, gene, and peptide delivery. Expert Opin. Drug Deliv. 1–30. doi:10.1080/17425247.2025.2506829.

Rohokale, R., Guo, Z., 2023. Development in the concept of bacterial polysaccharide repeating unit-based antibacterial conjugate vaccines. ACS Infect. Dis. 9 (2), 178–212. doi:10.1021/acsinfecdis.2c00559.

Romero, N., Muñoz, J., Cox, P., Heuer, A., Guerrero, A., 2013. Influence of chitosan concentration on the stability, microstructure and rheological properties of O/W emulsions formulated with high-oleic sunflower oil and potato protein. Food Hydrocoll. 30, 152–162. doi:10.1016/j.foodhyd.2012.05.004.

Rühmann, B., Schmid, J., Sieber, V., 2015. Methods to identify the unexplored diversity of microbial exopolysaccharides. Front. Microbiol., 6. https://www.frontiersin.org/journals/microbiology/articles/10.3389/fmicb.2015.00565.

Sachin, K., Karn, S.K., 2021. Microbial fabricated nanosystems: Applications in drug delivery and targeting. Front. Chem. 9. Available from: https://www.frontiersin.org/journals/chemistry/articles/10.3389/fchem.2021.617353.

Saha, I., Datta, S., 2022. Bacterial exopolysaccharides in drug delivery applications. J. Drug Deliv. Sci. Technol. 74, 103557. doi:10.1016/j.jddst.2022.103557.

Sajna, K., Sharma, S., Nadda, A., 2021. Microbial Exopolysaccharides: An Introduction. Springer International Publishing, Cham, Switzerland, pp. 1–18. doi:10.1007/978-3-030-75289-7_1.

Salehi, M., Rashidinejad, A., 2025. Multifaceted roles of plant-derived bioactive polysaccharides: A review of their biological functions, delivery, bioavailability, and applications within the food and pharmaceutical sectors. Int. J. Biol. Macromol. 290, 138855. doi:10.1016/j.ijbiomac.2024.138855.

Salehirozveh, M., Dehghani, P., Mijakovic, I., 2024. Synthesis, functionalization, and biomedical applications of iron oxide nanoparticles (IONPs). J. Funct. Biomater.,, 15 doi:10.3390/jfb15110340.

Samec, T., Boulos, J., Gilmore, S., Hazelton, A., Alexander-Bryant, A., 2022. Peptide-based delivery of therapeutics in cancer treatment. Mater. Today Bio. 14, 100248. doi:10.1016/j.mtbio.2022.100248.

Santiesteban, D.Y., Kubelick, K., Dhada, K.S., Dumani, D., Suggs, L., Emelianov, S., 2016. Monitoring/imaging and regenerative agents for enhancing tissue engineering characterization and therapies. Ann. Biomed. Eng. 44 (3), 750–772. doi:10.1007/s10439-015-1509-y.

Saravanan, C., Rajesh, R., Kaviarasan, T., Muthukumar, K., Kavitake, D., Shetty, P.H., 2017. Synthesis of silver nanoparticles using bacterial exopolysaccharide and its application for degradation of azo-dyes. Biotechnol. Rep. (Amst. Neth.) 15, 33–40. doi:10.1016/j.btre.2017.02.006.

Sathiyanarayanan, G., Dineshkumar, K., Yang, Y.-H., 2017. Microbial exopolysaccharide-mediated synthesis and stabilization of metal nanoparticles. Crit. Rev. Microbiol 43 (6), 731–752. doi:10.1080/1040841X.2017.1306689.

Sattar, A., Hafeez, S., Hedar, M., Saeed, M., Hussain, T., Intisar, A., 2025. Polymer/POSS based robust and emerging flame retardant nanocomposites: A comprehensive review. Nano-Struct. Nano-Objects 41, 101427. doi:10.1016/j.nanoso.2024.101427.

Schmid, J., Sieber, V., Rehm, B., 2015. Bacterial exopolysaccharides: Biosynthesis pathways and engineering strategies. Front. Microbiol. 6, 496. doi:10.3389/fmicb.2015.00496.

Seviour, R., Mcneil, B., Fazenda, M., Harvey, L., 2010. Operating bioreactors for microbial exopolysaccharide production. Crit. Rev. Biotechnol. 31, 170–185. doi:10.3109/07388551.2010.505909.

Sharma, D., Kanchi, S., Bisetty, K., 2019. Biogenic synthesis of nanoparticles: A review. Arab. J. Chem. 12 (8), 3576–3600. doi:10.1016/j.arabjc.2015.11.002.

Sharma, M., Tellili, N., Kacem, I., Rouissi, T., 2024. Microbial biopolymers: From production to environmental applications—A review. Applied Sci.,, 14 doi:10.3390/app14125081.

Sharma, S., Mohler, J., Mahajan, S.D., Schwartz, S.A., Bruggemann, L., Aalinkeel, R., 2023. Microbial biofilm: A review on formation, infection, antibiotic resistance, control measures, and innovative treatment. Microorganisms 11 (6). doi:10.3390/microorganisms11061614.

Sharma, S., Sudhakara, P., Singh, J., Ilyas, R.A., Asyraf, M.R.M., Razman, M.R., 2021. Critical review of biodegradable and bioactive polymer composites for bone tissue engineering and drug delivery applications. Polymers, 13 (16). doi:10.3390/polym13162623.

Sidhic, J., Aswathi, M.K., Prasad, A., Tom, A., Mohan, P., Sarbadhikary, P., Narayanankutty, A., George, S., Abrahamse, H., George, B.P., 2025. Advancements in metal and metal oxide nanoparticles for targeted cancer therapy and imaging: Mechanisms, applications, and safety concerns. J. Drug Deliv. Sci. Technol. 105, 106622. doi:10.1016/j.jddst.2025.106622.

Singh, A., Ahmed, M., Iqbal, U., Srivastava, U., Saini, P., 2024. Simultaneous biosorption of chromium and zinc from aqueous solution by *Lactobacillus plantarum* exopolysaccharides and biofilm characterized by structural and functional analysis. Food Saf. Health 2 (4), 497–508. doi:10.1002/fsh3.12065.

Singh, J., Dutta, T., Kim, K.-H., Rawat, M., Samddar, P., Kumar, P., 2018. Green' synthesis of metals and their oxide nanoparticles: Applications for environmental remediation. J. Nanobiotechnol. 16 (1), 84. doi:10.1186/s12951-018-0408-4.

Singh, N.A., Narang, J., Garg, D., Jain, V., Payasi, D., Suleman, S., Swami, R.K., 2023. Nanoparticles synthesis via microorganisms and their prospective applications in agriculture. Plant Nano Biol. 5, 100047. doi:10.1016/j.plana.2023.100047.

Sivakumar, B., Aswathy, R.G., Sreejith, R., Nagaoka, Y., Iwai, S., Suzuki, M., Fukuda, T., Hasumura, T., Yoshida, Y., Maekawa, T., Sakthikumar, D.N., 2014. Bacterial exopolysaccharide based magnetic nanoparticles: A versatile nanotool for cancer cell imaging, targeted drug delivery and synergistic effect of drug and hyperthermia mediated cancer therapy. J. Biomed. Nanotechnol. 10 (6), 885 899. doi:10.1166/jbn.2014.1820.

Song, J., Wang, X., Huang, Y., Qu, Y., Teng, L., Wang, D., Meng, Z., 2017. Antidepressant-like effects of *Marasmius androsaceus* metabolic exopolysaccharides on chronic unpredictable mild stress-induced rat model. Mol. Med. Rep. 16 (4), 5043–5049. doi:10.3892/mmr.2017.7145.

Song, M., Aipire, A., Dilxat, E., Li, J., Xia, G., Jiang, Z., Fan, Z., Li, J., 2024. Research progress of polysaccharide-gold nanocomplexes in drug delivery. Pharmaceutics 16 (1). doi:10.3390/pharmaceutics16010088.

Sousa, A., Phung, A.N., Škalko-Basnet, N., Obuobi, S., 2023. Smart delivery systems for microbial biofilm therapy: Dissecting design, drug release and toxicological features. J. Control. Release 354, 394–416. doi:10.1016/j.jconrel.2023.01.003.

Srivastava, N., Choudhury, A.R., 2022. Stimuli-responsive polysaccharide-based smart hydrogels and their emerging applications. Ind. Eng. Chem. Res. 62 (2), 841–866.

Stender, E.G.P., Dybdahl Andersen, C., Fredslund, F., Holck, J., Solberg, A., Teze, D., Peters, G.H.J., Christensen, B.E., Aachmann, F.L., Welner, D.H., Svensson, B., 2019. Structural and functional aspects of mannuronic acid-specific PL6 alginate lyase from the human gut microbe Bacteroides cellulosilyticus. J. Biol. Chem. 294 (47), 17915–17930. doi:10.1074/jbc.RA119.010206.

Stevanović, M., Filipović, N., 2024. A Review of recent developments in biopolymer nano-based drug delivery systems with antioxidative properties: insights into the last five years. Pharmaceutics, 16. MDPI, Basel, Switzerland doi:10.3390/pharmaceutics16050670.

Strieth, D., Schwarz, A., Stiefelmaier, J., Erdmann, N., Muffler, K., Ulber, R., 2021. New procedure for separation and analysis of the main components of cyanobacterial EPS. J. Biotechnol. 328, 78–86. doi:10.1016/j.jbiotec.2021.01.007.

Subhaswaraj, P., Syed, A., Siddhardha, B., 2020. Novel nanotherapeutics as next-generation anti-infective agents: Current trends and future prospectives. Curr. Drug Discov. Technol. 17 (4), 457–468.

Sulaiman, S., Hassan, N., 2024. Application of biodegradable plastic and their environmental impacts: A review. World J. Adv. Res. Rev. 21, 2139–2148. doi:10.30574/wjarr.2024.21.1.0155.

Suryawanshi, N., Naik, S., Jujjawarapu, S., 2022. Exopolysaccharides and their applications in food processing industries. Food Sci. Appl. Biotechnol. 5, 22. doi:10.30721/fsab2022.v5.i1.165.

Sutherland, I., 2001. Biofilm exopolysaccharides: A strong and sticky framework. Microbiol. (Read. Engl.) 147 (Pt 1), 3–9. doi:10.1099/00221287-147-1-3.

Sutherland, I.W., 1972. Bacterial exopolysaccharides. Adv. Microb. Physiol. 8, 143–213. doi:10.1016/s0065-2911(08)60190-3.

Swami, R., Popli, P., Sal, K., Challa, R.R., Vallamkonda, B., Garg, M., Dora, C.P., 2025. A review on biomacromolecular ligand-directed nanoparticles: New era in macrophage targeting. Int. J. Biol. Macromol. 306, 141740. doi:10.1016/j.ijbiomac.2025.141740.

Taban, Q., Mumtaz, P.T., Masoodi, K.Z., Haq, E., Ahmad, S.M., 2022. Scavenger receptors in host defense: From functional aspects to mode of action. Cell Commun., Signal 20 (1), 2. doi:10.1186/s12964-021-00812-0.

Teixeira, M.O., Marinho, E., Silva, C., Antunes, J.C., Felgueiras, H.P., 2023. Pullulan hydrogels as drug release platforms in biomedicine. J. Drug Deliv. Sci. Technol. 89, 105066. doi:10.1016/j.jddst.2023.105066.

Teworte, S., Malcı, K., Walls, L., Halim, M., Rios-Solis, L., 2021. Recent advances in fed-batch microscale bioreactor design. Biotechnol. Adv. 55, 107888. doi:10.1016/j.biotechadv.2021.107888.

Thakur, R., Yadav, S., 2024. Biofilm forming, exopolysaccharide producing and halotolerant, bacterial consortium mitigates salinity stress in Triticum aestivum. Int. J. Biol. Macromol. 262, 130049. doi:10.1016/j.ijbiomac.2024.130049.

Thambirajoo, M., Maarof, M., Lokanathan, Y., Katas, H., Ghazalli, N.F., Tabata, Y., Fauzi, M.B., 2021. Potential of nanoparticles integrated with antibacterial properties in preventing biofilm and antibiotic resistance. Antibiot. (Basel Switz,) 10 (11). doi:10.3390/antibiotics10111338.

Thongpon, P., Tang, M., Cong, Z., 2025. Peptide-based nanoparticle for tumor therapy. Biomedicines,, 13 doi:10.3390/biomedicines13061415.

Tiwari, O.N., Bobby, M.N., Kondi, V., Halder, G., Kargarzadeh, H., Ikbal, A.M.A., Bhunia, B., Thomas, S., Efferth, T., Chattopadhyay, D., Palit, P., 2024. Comprehensive review on recent trends and perspectives of natural exo-polysaccharides: Pioneering nano-biotechnological tools. Int, J. Biol, Macromol. 265, 130747. doi:10.1016/j.ijbiomac.2024.130747.

Trautner, B.W., Darouiche, R.O., 2004. Role of biofilm in catheter-associated urinary tract infection. Am. J. Infect. Control 32 (3), 177–183. doi:10.1016/j.ajic.2003.08.005.

U.S.Congress, O.T.A., 1993. Biopolymers: Making materials nature's way September 1993. Off. Technol. Assess. Sept. 1–80.

Umut, E., 2020. Surface Properties of Polysaccharides BT - Polysaccharides of Microbial Origin: Biomedical Applications. Springer International Publishing, Cham, Switzerland, pp. 1–26. doi:10.1007/978-3-030-35734-4_31-1.

Vaishnav, A., Upadhyay, K., Tipre, D., Dave, S., 2021. Bio-prospecting of Fruits Waste for Exopolysaccharide Production by Bacteria. 353–371. Springer Nature, Singapore doi:10.1007/978-981-16-1955-7_15.

Van Dalen, F.J., Van Stevendaal, M.H.M.E., Fennemann, F.L., Verdoes, M., Ilina, O., 2019. Molecular repolarisation of tumour-associated macrophages. Molecules,, 24 doi:10.3390/molecules24010009.

Vangijzegem, T., Lecomte, V., Ternad, I., Van Leuven, L., Muller, R.N., Stanicki, D., Laurent, S., 2023. Superparamagnetic iron oxide nanoparticles (SPION): From fundamentals to State-of-the-Art innovative applications for cancer therapy. Pharmaceutics 15 (1). doi:10.3390/pharmaceutics15010236.

Vu, B., Chen, M., Crawford, R.J., Ivanova, E.P., 2009. Bacterial extracellular polysaccharides involved in biofilm formation. Mol. (Basel Switz.) 14 (7), 2535–2554. doi:10.3390/molecules14072535.

Vuddanda, P.R., Montenegro-Nicolini, M., Morales, J., Velaga, S., 2016. Effect of plasticizers on the physico-mechanical properties of pullulan based pharmaceutical oral films. Eur. J. Pharm. Sci. 96. doi:10.1016/j.ejps.2016.09.011.

Walait, M., Mir, H.R., Anees, K., 2023. Edible biofilms and coatings; its characterization and advanced industrial applications. Nat. Resour. Hum. Health 3 (1), 28–37. doi:10.53365/nrfhh/149622.

Wang, T., Lai, L., Huang, Y., Su, E., 2024. Nanozyme: An emerging tool for food packaging. Food Control 155, 110104. doi:10.1016/j.foodcont.2023.110104.

Waoo, A.A., Singh, S., Pandey, A., Kant, G., Choure, K., Amesho, K.T.T., Srivastava, S., 2023. Microbial exopolysaccharides in the biomedical and pharmaceutical industries. Heliyon 9 (8), e18613. doi:10.1016/j.heliyon.2023.e18613.

Whitfield, G.B., Marmont, L.S., Howell, P.L., 2015. Enzymatic modifications of exopolysaccharides enhance bacterial persistence. Front. Microbiol. 6, 471. doi:10.3389/fmicb.2015.00471.

Williams, D.F. (2019). Hydrogels in Regenerative Medicine. A. Atala, R. Lanza, A. G. Mikos, (Eds.), 627–650). Academic Press, London, United Kingdom. doi:10.1016/B978-0-12-809880-6.00036-9.

Wu, D.T., Jeffreys, N., Diba, M., Mooney, D.J., 2022. Viscoelastic biomaterials for tissue regeneration. Tissue Eng. C Methods 28 (7), 289–300. doi:10.1089/ten.TEC.2022.0040.

Wu, M., Pakroo, S., Nadai, C., Molinelli, Z., Speciale, I., De Castro, C., Tarrah, A., Yang, J., Giacomini, A., Corich, V., 2024. Genomic and functional evaluation of exopolysaccharide produced by *Liquorilactobacillus mali* t6-52: Technological implications. Microb. Cell Fact 23 (1), 158. doi:10.1186/s12934-024-02431-z.

Xiao, M., Ren, X., Yu, Y., Gao, W., Zhu, C., Sun, H., Kong, Q., Fu, X., Mou, H., 2022. Fucose-containing bacterial exopolysaccharides: Sources, biological activities, and food applications. Food Chem.: X 13, 100233. doi:10.1016/j.fochx.2022.100233.

Xu, L., Zhang, J., Luo, J., Cui, Y., Chen, J., Zeng, B., Deng, Z., Shao, L., 2025. Double-sided protector" Janus hydrogels for skin and mucosal wound repair: applications, mechanisms, and prospects. J. Nanobiotechnol. 23 (1), 387. doi:10.1186/s12951-025-03438-3.

Xu, R., Fang, Y., Zhang, Z., Cao, Y., Yan, Y., Gan, L., Xu, J., Zhou, G., 2023. Recent advances in biodegradable and biocompatible synthetic polymers used in skin wound healing. Mater. (Basel Switz.) 16 (15). doi:10.3390/ma16155459.

Yadav, S., Chandra, S., Yadav, A., Kumar, A., Awasthi, M., 2022. Green Synthesis of Silver Nano-Particle from Cyanobacteria and Effect on Microalgal Growth and Production of Exopolysaccharide (EPS). In: Tiwari, A. (Ed.), IntechOpen, United Kingdom doi:10.5772/intechopen.106039.

Yaraki, M.T., Zahed Nasab, S., Zare, I., Dahri, M., Moein Sadeghi, M., Koohi, M., Tan, Y.N., 2022. Biomimetic metallic nanostructures for biomedical applications, catalysis, and beyond. Ind. Eng. Chem. Res. 61 (22), 7547–7593. doi:10.1021/acs.iecr.2c00285.

Yildiz, H., Karatas, N., 2018. Microbial exopolysaccharides: Resources and bioactive properties. Process Biochem. 72. doi:10.1016/j.procbio.2018.06.009.

Zannini, E., Waters, D., Coffey, A., Arendt, E., 2016. Production, properties, and industrial food application of lactic acid bacteria-derived exopolysaccharides. Appl. Microbiol. Biotechnol. 100. doi:10.1007/s00253-015-7172-2.

Zarour, K., Llamas-Arriba, M., Prieto, A., Ruas-Madiedo, P., Dueñas, M., Palencia, P., Aznar, R., Mebrouk, K., López, P., 2017. Rheology and bioactivity of high molecular weight dextrans synthesised by lactic acid bacteria. Carbohydr. Polym. 174. doi:10.1016/j.carbpol.2017.06.113.

Zeinivand, M., Aghaei, S.S., Zargar, M., Ghasemzadeh, M.A., 2023. Exopolysaccharide-mediated silver nanoparticles synthesized from *Lactobacillus paracasei* with antimicrobial, antibiofilm and antioxidant activities. Arch. Microbiol. 205 (5), 210. doi:10.1007/s00203-023-03497-w.

Zhang, H., Pan, N., Ma, C., Liu, B., Xiu, L., Tong, H., Sheng, S., Liang, Y., Li, H., Ma, F., Bao, X., Hu, W., Wang, X., 2021. Vaccine composition formulated with a novel lactobacillus-derived exopolysaccharides adjuvant provided high protection against *Staphylococcus aureus*. Vaccines 9 (7). doi:10.3390/vaccines9070775.

Zhang, H., Sheng, S., Li, C., Bao, X., Zhao, L., Chen, J., Guan, P., Li, X., Pan, N., Liang, Y., Wang, X., Sun, J., Wang, X., 2024. Mucosal immunization with the lung *Lactobacillus*-derived amphiphilic exopolysaccharide adjuvanted recombinant vaccine improved protection against *P. aeruginosa* infection. PLoS Pathog 20 (11), e1012696. doi:10.1371/journal.ppat.1012696.

Zhang, Q., Wang, Z., Shen, S., Wang, J., Cao, J., Deng, Y., Meng, H., Ma, L., 2024. Integrating enzyme-nanoparticles bring new prospects for the diagnosis and treatment of immune dysregulation in periodontitis. Front. Cell. Infect. Microbiol. 14, 1494651. doi:10.3389/fcimb.2024.1494651.

Zhang, X., Xing, H., Zhao, Y., Ma, Z., 2018. Pharmaceutical dispersion techniques for dissolution and bioavailability enhancement of poorly water-soluble drugs. Pharmaceutics 10 (3). doi:10.3390/pharmaceutics10030074.

Zhang, Y., Dong, L., Liu, L., Wu, Z., Pan, D., Liu, L., 2022. Recent advances of stimuli-responsive polysaccharide hydrogels in delivery systems: A review. J. Agric. Food Chem. 70 (21), 6300–6316.

Zhao, A., Sun, J., Liu, Y., 2023. Understanding bacterial biofilms: From definition to treatment strategies. Front. Cell. Infect. Microbiol. 13, 1137947. doi:10.3389/fcimb.2023.1137947.

Zhao, D., Jiang, J., Liu, L., Wang, S., Ping, W., Ge, J., 2021. Characterization of exopolysaccharides produced by *Weissella confusa* XG-3 and their potential biotechnological applications. Int. J. Biol. Macromol. 178, 306–315. doi:10.1016/j.ijbiomac.2021.02.182.

Zhao, J.R., Zheng, R., Tang, J., Sun, H.J., Wang, J., 2022. A mini-review on building insulation materials from perspective of plastic pollution: Current issues and natural fibres as a possible solution. J. Hazard. Mater. 438, 129449. doi:10.1016/j.jhazmat.2022.129449.

Zhu, Q., Chua, M.H., Ong, P.J., Cheng Lee, J.J., Le Osmund Chin, K., Wang, S., Kai, D., Ji, R., Kong, J., Dong, Z., Xu, J., Loh, X.J., 2022. Recent advances in nanotechnology-based functional coatings for the built environment. Mater. Today Adv. 15, 100270. doi:10.1016/j.mtadv.2022.100270.

Zustiak, S.P., Jain, E., 2023. Feature reviews in pharmaceutical technology. Pharmaceuticals,, 16 doi:10.3390/ph16101336.

CHAPTER

15

EPS: A promising antioxidant and antibiofilm agent

Laxmipriya Koijam[a], *Avijeet Singh Oinam*[b], *Bijaya Thongataba*[c], *Tabitha Langhu*[a], *Md. Nazneen Bobby*[d] *and Onkar Nath Tiwari*[e]

[a]Department of Biotechnology, T.S. Paul Manipur Women's College, Mongsangei, Manipur, India [b]Department of Botany, Regional College, Imphal, Manipur, India [c]Department of Zoology, T.S. Paul Manipur Women's College, Mongsangei, Manipur, India [d]Department of Biotechnology, Vignan's Foundation for Science, Technology and Research, Vignan University, Vadlamudi, Guntur, Andhra Pradesh, India [e]Centre for Conservation and Utilisation of Blue Green Algae (CCUBGA), Division of Microbiology, ICAR-Indian Agricultural Research Institute (PUSA), New Delhi, India

15.1 Introduction

Microbes grow primarily in two forms: As free-floating or as sessile forms within biofilms (Bjarnsholt et al., 2011). Biofilms are organized microbial communities enclosed in a self-produced extracellular polymeric matrix, which is mostly water along with exopolysaccharide such as polysaccharides, proteins, and extracellular DNA (eDNA), which may include both prokaryotic and eukaryotic cells (Rumbaugh and Sauer, 2020). The increasing prevalence of multidrug-resistant (MDR) bacteria, driven in part by excessive antibiotic use and prolonged hospital stays, poses a significant challenge to effective patient treatment. Bacterial biofilms are now recognized as major contributors to global health concerns due to their inherent resistance to antibiotics, evasion of the host immune system, and resilience against various environmental stresses (Schulze et al., 2021). Geesey (1982) clarifies "exopolysaccharide of biological origin that participate in the formation of microbial aggregates." exopolysaccharide can be produced by a broad range of living organisms such as plants, animals, fungi, and bacteria (Moscovici, 2015). Most biological activity to produce exopolysaccharide in response to the stress stimuli from natural environments is typical of these micro-organisms, both prokaryotic (Bacteria and

DOI: https://doi.org/10.1016/B978-0-443-36619-2.00015-2

Archaea) and eukaryotic (algae and fungi) (Sivaperumal et al., 2014; Uzma and Chowdappa, 2017). This may be in the form of a tightly bound capsule around their cell or as more loosely associated extracellular slime. EPS secretion is universal among diverse groups of microorganisms as a metabolic by-product, and it plays an important role in biofilm formation, environmental resilience, and multiple ecological functions (Ziadi et al., 2018).

The formidable extracellular polymeric substance plays unique roles within diverse biological niches. One of their key roles is **biofilm formation, where exopolysaccharides (EPS) facilitate microbial attachment to surfaces and enhance survival during epiphytic or saprophytic life stages (Tampakaki et al., 2009) of the**rapeutic potential; for example, their antioxidant activity makes them valuable in the development of functional foods and applications in other industrial sectors (Badel et al., 2011). Biofilms were first described in the seventeenth century by Anton van Leeuwenhoek, who observed microbial aggregates on dental plaque using his primitive microscope. The term "biofilm" was later coined in 1978 by Canadian microbiologist Bill Costerton, who defined it as "a structured community of bacterial cells enclosed in a self-produced polymeric matrix and adherent to an inert or living surface." The structural integrity and functionality of the biofilm were supported by EPS (Chandki et al., 2011; Yasir et al., 2018; Gupta et al., 2019).

Free radicals are widely recognized as harmful to living organisms. To diminish the damage caused by these reactive species, both synthetic and natural antioxidants have been employed. However, synthetic antioxidants are associated with adverse effects, including hepatotoxicity and carcinogenicity. Consequently, there is a growing need to develop safe, natural antioxidants to protect human health. Reactive oxygen species (ROS) such as superoxide anions, hydrogen peroxide, and hydroxyl radicals are etiologies of rheumatoid arthritis, atherosclerosis, cancer, and neurodegenerative disorders (Kondo et al., 2023). These ROS attack vital biomolecules such as DNA and proteins, leading to cellular dysfunction and damage. Identifying effective, nontoxic natural antioxidants remains a scientific challenge. Notably, exopolysaccharides (EPS) derived from bacteria, fungi, and various medicinal plants have demonstrated promising antioxidant activities (Krupanidhi et al., 2022).

Cells within biofilms are able to survive under harsh environmental conditions due to their encapsulation in a matrix of high-molecular-weight exopolysaccharide, which facilitate cell attachment and cohesion (Branda et al., 2005; Flemming and Wingender, 2010). Studies show that microbes in biofilms can resist ultraviolet radiation, toxic metals, acidic conditions, desiccation, and pH variations (Hall-Stoodley et al., 2004). However, biofilms can also have deteriorative effects, such as corrosion in industrial pipelines. Biofilm can be deteriorated by disintegrating the components of the extracellular polymeric substance matrix. Key strategies for biofilm dissolution include the use of enzymes such as alginate lyase, DNase, and various hydrolases, which target specific constituents of the EPS. This review focuses on the antioxidant properties of EPS, along with novel strategies for controlling biofilm formation by targeting the extracellular polymeric substance matrix.

15.2 Structural and functional aspects of exopolysaccharides

EPS are organic polymers derived from microbes, which facilitate cell-to-cell binding and particles (cohesion) and to the substratum (adhesion). Polysaccharides or glycans are rich in nature and show diverse chemical composition, physical features, and biological functions

(Wingender et al., 1999; Muhamad et al., 2019). Most bacterial cells are surrounded by a polysaccharide membrane called the glycocalyx. The EPS matrix supports cell-cell adhesion, surface attachment, and aggregation (Flemming and Wingender, 2010).

Exopolysaccharide are complex matrices composed of proteins, lipids, polysaccharides, and eDNA, and they play a crucial role in the pathogenesis of numerous microbial infections (Ch'ng et al., 2019). The generation of biofilm is facilitated through the presence of an extracellular polymeric substance matrix that yields to high cell density, cellular proximity, and long-term spatial organization within the biofilm structure (Flemming and Wingender, 2010). EPS are involved in different biological functions both at the level of the cell and at a multicellular or complex metaorganism. They protect them from death by desiccation and phagocytosis, enable nutrient concentration, and reduce toxic compound exposure. Biofilm formation offers an advantage for attachment to surfaces and survival in the environment for long-term during epiphytic or saprophytic life stages, is also mediated by EPSs. Bacterial species are additionally covered by a polysaccharide layer called the glycocalyx in several. The extracellular polymeric substance matrix is believed to be necessary for cell-cell adhesion, surface attachment, and microcolony formation (Flemming and Wingender, 2010).

EPS are generally categorized into homo-exopolysaccharides (HoEPSs), which contain only one type of monosugar in their repeat units, and hetero-exopolysaccharides (HeEPSs), which include more than two different types of monosugars in the repeating unit (Zhou et al., 2019). EPSs play several crucial roles, such as facilitating the interactions of bacteria with their environment and providing protection against hostile conditions. These protections include resistance to toxic compounds such as bile salts, hydrolyzing enzymes (e.g., lysozyme, gastric, and pancreatic enzymes), metal ions, and antibiotics. A secondary role of EPSs is to protect the bacteria against different environmental strains like pH adaptations, temperature changes, and osmolarity variations.

Additionally, they contribute to evasion of host immune responses and protection against bacteriophage (phage) attacks (Castro-Bravo et al., 2017; Lynch et al., 2018). Furthermore, polysaccharides are being explored as potential targets for the inhibition of biofilm (BF) formation, which is crucial for controlling persistent infections.

15.3 Antioxidant potential of exopolysaccharides

Free radicals are harmful to living organisms. They can cause significant cellular damage (Mahapatra and Benerjee, 2013). To reduce these effects, both synthetic and natural antioxidants are used. Concerns have been raised about synthetic antioxidants because they may be linked to liver toxicity and cancer (Liu et al., 2009). As a result, there is a growing focus on creating safe, natural, and nontoxic antioxidants to protect human health from free radical damage. Antioxidants are substances that prevent oxidation by neutralizing free radicals. This process reduces oxidative stress and promotes tissue regeneration through cellular repair mechanisms.

The antioxidant activity of EPS depends on several factors, such as glycosidic bond types, branching patterns, and the composition and ratio of monosaccharides (Zhao et al., 2018). Natural polysaccharides are more likely to biodegrade and are safer than synthetic polymers (Wang et al., 2014; Ngan et al., 2014). ROS govern various biological processes such as apoptosis, cell signaling, and ion transportation (Barnham et al., 2004). Excessive accumulation of ROS and other free radicals triggers oxidative stress, which contributes to the emergence and development of various diseases (Prakash, 2023). Oxidative stress is linked to a range of diseases, including cardiovascular disease (Cinato et al., 2024), steatotic liver disease (Gensluckner et al., 2024), and diabetes (Bai et al., 2024). Lactic acid bacterial EPSs can strengthen cellular defense mechanisms through antioxidant activity that minimizes the oxidative damage caused by ROS and free radicals (Xu et al., 2020).

Recent studies revealed the benefits of EPS regarding the health and other industrial potential applications. In the food industry, EPSs are seen as a promising choice to plant-based polysaccharides as they can improve consistency, flavor, and stability in fermented products. Additionally, EPSs have important functional activities, especially antioxidant aspects, which make them useful for functional foods and various other industrial applications. The EPS exhibits strong antioxidant properties, scavenges hydroxyl and superoxide anion radicals, and significantly reduces malondialdehyde (MDA) levels (Pan and Mei, 2010).

15.4 Antioxidant mechanism of microbial exopolysaccharides

Cyanobacteria-derived EPS have demonstrated notable antioxidant properties. These biopolymers can influence oxidative reactions by enhancing the formation of ROS and enhancing the function of antioxidant enzymes. Moreover, cyanobacterial EPS are known for their ability to stimulate the immune system (Costa et al., 2010). Exopolymers extracted from three *Anabaena* strains and *Tolypothrix tenuis* have displayed antioxidant effects against several ROS, including superoxide anion ($O_2^{\bullet -}$), hydrogen peroxide (H_2O_2), hydroxyl radicals ($^{\bullet}OH$), and nitric oxide ($NO^{\bullet}$) (Parwani et al., 2014). Cyanobacterial EPS act as potent antioxidants that may alleviate oxidative stress. Given the known link between oxidative stress and inflammation, multiple studies indicate that cyanobacterial EPSs might mitigate inflammatory responses by reducing oxidative damage. Various cyanobacterial EPSs have been utilized as protective films for drug encapsulation and therapeutic delivery due to their compatibility and stability (Bellini et al., 2018). These EPSs have shown beneficial effects in models of chronic conditions like rheumatoid arthritis and inflammatory bowel disease (IBD). Gas chromatography-mass spectrometry (GC-MS) analysis verified the presence of sulfated polysaccharides with glucose subunits in cyanobacterial samples (Bouyahya et al., 2024). Importantly, these sulfated polysaccharides exhibited significantly greater antioxidant activity, ranging from 65% to 75%, in comparison to their nonsulfated variants. Furthermore, the radical scavenging efficiency increased proportionally with both the concentration of sulfated polysaccharides (SPs) and the presence of sulfate groups, elevating the role of sulfation in enhancing antioxidant properties (Zou et al., 2010).

Marine SPs show notable antioxidant activity by increasing the function of key antioxidant enzymes and reducing lipid peroxidation byproducts. This contributes to a strengthened antioxidant defense system, particularly through the upregulation of catalase activity. The antioxidant potential of polysaccharides extracted from the cyanobacterium *Nostoc commune* has been evaluated using various in vitro assays, with the DPPH assay showing the most effective results (EC_{50} = 2719.1 mg/L). Similarly, polysaccharides derived from *Phormidium versicolor* demonstrated strong in vitro antioxidant capacity, including efficient free radical scavenging, inhibition of β-carotene bleaching, and significant reducing power. Sacran, a SP isolated from *Aphanothece sacrum*, has shown both antioxidant and anti-inflammatory properties. In vivo studies reveal that sacran enhances the activities of major antioxidant enzymes-superoxide dismutase (SOD), serum catalase, and hepatic glutathione S-transferase, while significantly reducing serum MDA levels and monoamine oxidase activity. These effects highlight its potent antioxidant and antiaging potential (Motoyama et al., 2016).

During fermentation, lactic acid bacteria (LAB) synthesize and release complex carbohydrate polymers, i.e., EPS (Yang et al., 2023). The characteristics of EPS produced by *Lactobacillus plantarum* are influenced by various factors, including the bacterial strain, cultivation conditions, and the types of sugars involved in their synthesis (Xiong et al., 2023). EPS extracted from *L. plantarum* VAL6 has shown notable antioxidant abilities, as demonstrated by DPPH free radical scavenging tests and total antioxidant capacity evaluations. Several studies have noted similar antioxidant properties of *L. plantarum* EPS (Wang et al., 2014; Xu et al., 2019); their effectiveness is mainly linked to the presence of hydroxyl and other reactive functional groups. Additionally, the antioxidant potential of polysaccharides from *Boletus aereus* depends on glucose content (Zheng J-Q et al., 2014). This indicates that the composition of monosaccharides can significantly affect antioxidant properties. In this context, EPS from *L. plantarum* that is rich in mannose has shown particularly high antioxidant activity (Bomfin et al., 2020). The EPS extracted from *Weissella cibaria* SJ14, which has 75.9% mannose, has exhibited excellent activity in scavenging hydroxyl radicals, likely due to its high mannose concentration (Zhu et al., 2018). In a similar manner, high-galactose heteropolysaccharides from *W. confusa* demonstrated significant scavenging abilities. They effectively reduced DPPH and hydrogen peroxide radicals (Adebayo-Tayo et al., 2018). Furthermore, two EPS fractions from *Lactobacillus delbrueckii* ssp. *bulgaricus* SRFM-1, referred to as r-EPS1 and r-EPS2, contained high levels of galactose and showed strong antioxidant activity (Tang et al., 2017). The antioxidant effectiveness of EPS appears to be influenced by the types of glycosidic bonds and their branching patterns (Tang et al., 2017). It has been suggested that α-(1→2) and α-(1→6) glycosidic bonds allow for greater molecular flexibility compared to β-(1→3) and β-(1→4) bonds (Delbarre-Ladrat et al., 2014; Plazinski et al., 2016; Peesapati et al., 2021). For example, EPS from *Bacillus amyloliquefaciens*, which is an α-glucan made of glucose units linked by two α-(1→3) and one α-(1→6) bonds, shows a strong ability to scavenge superoxide anions (Zhao et al., 2018). The antioxidant activity is seen to improve when there is a significant level of branching of EPS. Yang et al. (2022) reported that two polysaccharide fractions (THPS-1 and THPS-2) from *Tetragenococcus halophilus* SNTH-8 had a significant degree of branching and exhibited powerful antioxidant and emulsifying activities.

15.5 Antibiofilm activity of exopolysaccharides

As per Flemming et al. (2007), biofilm is compared with the "cities of microbes" while the extracellular polymeric substance matrix serves as the "house of biofilm." Thomas Neu and John Lawrence first compiled the diverse functions of EPS in their review (Neu and Lawrence, 2009).

The adaptation of biofilm on the surfaces requires structural and physiological changes, as it is reflected in the viscoelastic properties of the biofilm matrix. Persat et al. (2015) illustrated that the eDNA affects the viscoelastic behavior of biofilms. Biofilm are highly complex structure embedded in a self-generated extracellular polymeric substance matrix in association with bacteria, cyanobacteria, and fungi. Biofilms usually form on both living and nonliving surfaces (Okuda et al., 2018; Silvia et al., 2020). The Extracellular polymeric substance (EPS) matrix acts as a physical barrier, protecting embedded microorganisms from host immune responses and antimicrobial agents, thereby affecting treatment in any type of infection and leading to relapse of extracellular polymeric substance matrix acts as a physical barrier and protects the embedded microorganisms from any host immune response or from antimicrobial agents, thus affecting the treatment in any type of infection, which further leads to relapses of the therapy. Biofilms are considered a significant cause of chronic bacterial infections. Formation of biofilm on medical devices such as surgical instruments or any hospital equipment are associated with the hospital-acquired (noscomial) infection (Nwodo et al., 2012; Soto, 2013).

The establishment of biofilm is a natural and intricate method that develops when the bacterial colonies attain a critical density, resulting in the growth of numerous microcolonies. It is a successive activity concerning the initial attachment, irreversible attachment, early biofilm growth, maturation, and dispersion (Sauer et al., 2002). Their adhesion becomes irreversible once they start producing their EPS. According to Melandera et al. (2020) and Danquah et al. (2022), bacteria, fungi, and algae are the source for natural antibiofilm agents having a broad spectrum. These microbial compounds can suppress the formation of biofilm through multiple mechanisms by acting either independently or synergistically (Raj and Thomas, 2021). These include suppression of initial cell adhesion and attachment; disruption of extracellular polymeric substance matrix synthesis; degradation of biofilm components; inhibition of cell division and viability; prevention of biofilm maturation; substrate limitation; and downregulation of quorum-sensing (QS) molecules, ultimately blocking the QS networks essential for biofilm development (Lu et al., 2019). Rendueles et al. (2013) stated that the ability of the EPS to suppress the development of biofilm is through different mechanism such as modulating gene expression in intercellular, modifying the physicochemical properties of living (biotic) and nonliving (abiotic) surfaces, and disrupting the integrity of biofilm by exploiting their biosurfactant properties. Wu et al. (2020) demonstrated that EPS273 secreted by the marine bacterium *Pseudomonas stutzeri* have shown to suppress the biofilm establishment by influencing the PhoP-PhoQ system and impacting QS pathways. Likewise, the distribution of *Pseudomonas aeruginosa* biofilms by EPS A101, derived from *Vibrio* strain, occurs through the EPS–bacterial surface interactions (Jiang et al. 2011).

Several studies have emphasized that EPS can hinder the formation of microbial biofilm and their adhesion, attributable to their biological or physicochemical features (Bernal and Llamas, 2012; Rendueles et al., 2013). Champion et al. (2022) reported that EPS MO245 not only blocked the adhesion of *P. aeruginosa* and *Vibrio harveyi* but also suppressed the formation of biofilm and degraded established biofilms, without showing toxicity, a characteristic which is noted in other EPSs. Zanna et al. (2022) also reported that EPS NCIMB2021from marine bacterium *Pseudoalteromonas* can exhibit antiadhesive features by absorbing onto a chromium surface, thus inhibiting the biofilm formation. Moreover, EPS obtained from *Bacillus licheniformis* showed the disrupting of biofilm in dose dependent against four bacterial strain including *P. aeruginosa* (Vishwakarma and Sinha, 2020). These observations suggest that nontoxic EPS can be considered as a promising food-grade additive to regulate the formation of infective biofilm in the food industry.

In the cyanobacterium *Synechocystis*, the production of EPS led to a significant decrease in surface hydrophobicity, along with notable antibiofilm properties. Hydrophobicity assays revealed a substantial 90% reduction in surface hydrophobicity in the EPS-producing strain (Li et al., 2021). This reduction in hydrophobicity is particularly important, as microbial antibiofilm activity is often linked to the prevention of initial colonization and subsequent biofilm development on surfaces.

Antibiofilm agents generally operate through three primary strategies, i.e., targeting the cellular membrane, disrupting the extracellular matrix, and interfering with intracellular signaling pathways. One key approach involves the regulation or inhibition of matrix synthesis, often mediated through QS systems, which coordinate bacterial communication and biofilm development. Additionally, direct enzymatic degradation or digestion of the extracellular polymeric substance matrix can effectively compromise the structural integrity of biofilms, thereby prevent their formation or facilitating their disruption (Akbarian et al., 2022).

15.6 Mechanisms of antibiofilm action

Main approach employed to combat biofilms are:

- Biofilm inhibition, which aims to prevent the initial formation of biofilms, and
- Biofilm dispersal, which aims at disposing on well-established biofilms (Kim et al., 2015).

Disrupting the EPS polymers is an essential part in order to stimulate the detachment of biofilm, as the polymers retain the structure of the biofilm and also activate the bacterial cell for self-dispersal. As per Li and Lee (2017), disrupting the component of the biofilm matrix, suppressing the QS, and hindering the non-QS regulatory pathways are some of the novel antibiofilm strategies for targeting early-stage biofilm formation or mature, established biofilms. The enzymatic degradation brings about the disruption of the biofilm matrix, which is essentially made up of EPS, eDNA, and proteins. Deoxyribonucleases (DNases) that target eDNA, polysaccharide-degrading enzymes that break down EPS, and proteases that degrade protein components are some of the precise enzymes that have been used as agents for antibiofilm.

In general, antibiofilm agents act through multiple mechanisms, which include:

- Inhibition of initial bacterial adhesion,
- Degradation and removal of mature biofilms,
- Disruption of bacterial communication via QS, and
- Inhibition of non-QS regulatory (Flemming and Wingender, 2010; Mann and Wozniak, 2012).

15.7 Plant-derived compounds as natural antibiofilm agents

According to Soliman et al. (2018) and DeFrancesco et al. (2017), the crude or the fractionated extracts from plants and their secondary metabolites have emerged as a potential source of antibiofilm due to their broad range of bioactivities, which may include the capacity to suppress the formation of biofilm as well as disrupt the well-established biofilm.

For example, a crude extract of *Adiantum philippense* L. was found to significantly reduce the constituent of biofilm. It interfered with biofilm development at multiple levels by focusing the adhesion proteins, deforming preformed biofilms, and inhibiting extracellular polymeric substance production-thereby restraining biofilm formation at the initial stages (Lopes et al., 2018). Antibiofilm substances of plant origin may also act by preventing the polymerization and functionalization of proteins, which are crucial for the structural and regulatory components of biofilms. By disrupting these protein-based processes, such agents effectively inhibit the establishment and maintenance of the biofilm matrix.

15.8 Role of exopolysaccharides-degrading enzymes in biofilm control

Numerous enzymes can operate on the constituents of exopolysaccharide in the disintegration of the biofilm matrix (Overhage et al., 2008).

α-amylase, Dispersin B (DspB), and DNase I are enzymes chiefly used for antibiofilm approaches. These enzymes are promising agents that can be used in antimicrobial therapies. In nature, the extracellular polymeric substance matrix in *Actinobacillus pleuropneumoniae* can be degraded by:

Dispersin B (DspB) thus promotes the scattering of biofilm (Kaplan et al., 2004b). Likewise, the biofilm matrix of *Staphylococcus epidermidis* is interfered with by Dispersin B (DspB), which is isolated from *Aggregatibacter actinomycetemcomitans* (Kaplan et al., 2004a). Poly-β-1,6-N-acetylglucosamine (PNAG) is a crucial exopolysaccharide component of the extracellular polymeric substance , and DspB can hydrolyse the β-1,6-glycosidic linkages, resulting in the interrupting and disintegration of the biofilm structure (Izano et al., 2007).

Alginate is a polysaccharide derived from algae and is a major ingredient of the extracellular polymeric substance matrix for certain bacteria. An enzyme, i.e., Alginate lyase, which has the capacity to disintegrate alginate, can break down and decompose the biofilm produced by the alginate-producing bacteria. (Germoni et al., 2016). As per Ethica et al. (2021), Alginate lyases have been obtained from various organisms, such as algae, bacteria, fungi, as well as from viruses too. Marine bacteria are found to be a promising organism to yield and show a vast diversity of alginate lyases (Barzkar et al., 2022). Alginate lyase (AlyP1400) is a prominent enzyme that is developed from the marine bacterium *Pseudoalteromonas* sp. 1400 and is proved to efficiently disintegrate the biofilm of *Pseudomonas aeruginosa by* disintegrating the component of alginate within the extracellular polysaccharide matrix (Yang et al., 2017).

Amylases are one among the enzymes that have the capacity to hydrolyze the structure of polysaccharides within the extracellular polymeric substance matrix, thus altering biofilm formation. When amalyse is used along with amyloglucosidase, it can improve the disintegration of extracellular polymeric substance , ruining the integrity of biofilm. Currently, α-amylase is found to be an efficient antibiofilm agent against the pathogenic bacteria such as *Vibrio cholerae, Staphylococcus aureus*, and *Pseudomonas aeruginosa* (Lahiri et al., 2021). *Bacillus amyloliquefaciens, Bacillus licheniformis, Bacillus stearothermophilus, Bacillus subtilis*, and *Bacillus cereus* are the major sources for α-amylase production (Vaikundamoorthy et al., 2018).

Proteases have been shown to have a notable potential for antibiofilm. Mukherji et al. (2015) demonstrated the effect of proteolytic enzymes on biofilm. Subtilisin from *Bacillus* sp., lysostaphin from *Staphylococcus*, and bacteriophage-derived lysins are some of the efficient proteases that are being used for inhibiting the biofilm formation (Thallinger et al., 2013). As a new technique, the formation and adhesion of biofilm of *Candida albicans* can be inhibited when proteases are immobilized on a polypropylene surface, and this technique is sustainable and nontoxic in nature for the surface development of biofilm.

Biofilm properties of both gram-positive and gram-negative bacteria can be destroyed by a well-known prominent enzyme, i.e., Deoxyribonuclease (DNase) (Tetz et al., 2009). These enzymes can cleave the phosphodiester bonds in the backbone of eDNA, which is part of the structural component of biofilm that supports the initial attachment of microbes and the aggregation of extracellular polymeric substance. DNase disrupts the biofilm integrity and longevity by degrading the eDNA. According to Deng et al. (2022), DNase I, which is an endonuclease, can target the phosphodiester bonds near pyrimidine bases, resulting in the production of polynucleotides with 5′-phosphate termini. eDNA is an essential part required during the formation, structural stability, and management of biofilm (Das et al., 2010; Periasamy et al., 2012; Chagnot et al., 2013). DNase I can inhibit the biofilm by disintegrating eDNA within the matrix, deteriorating the structural integrity. DNase I has been proven to be a promising enzyme that can efficiently disrupt the biofilms of *Listeria monocytogenes* and *Campylobacter* spp., indicating as an antibiofilm agent. According to Quan et al. (2024), a combination of two enzymes, such as DNase I and lysostaphin, proved to disrupt the biofilm rather than using these enzymes alone. Its significant observation encourages the combination therapies of enzymes for controlling the biofilm. As per Rasouli et al. (2020), Cellobiose dehydrogenase (CDH) isolated from *Aspergillus niger* proved to have potential antibiofilm activity against *Staphylococcus epidermidis* and *Pseudomonas aeruginosa.* These enzymes destabilized the biofilm matrix through the hydrolysis of polysaccharides, resulting in the elevation of terminal-reducing sugars concentration. Thus, CDH proved to be a promising candidate for controlling biofilm formation and also for its therapeutic applications.

15.9 Exopolysaccharides quorum sensing quenching

Quorum sensing (QS) is a mechanism of communicating intercellularly, allowing them to adapt their behavior in connection to the modification in cell population density, pH, signaling molecule and their association with the microbial community (Malendera et al., 2020). QS plays a vital role in managing various social and physiological behaviors such as motility, bioluminescence, virulence, sporulation, antibiotic production, and genetic competence of microbial communities (Reuter et al., 2016; Barzegari et al., 2020). Disrupting the QS signaling pathway is an effective strategy for controlling and terminating the formation of biofilm by microbes. By inhibiting QS, it becomes possible to interfere with early bacterial adhesion, thereby preventing the development of mature biofilm structures (Paluch et al., 2020).

The idea of QS was first described in two bioluminescent marine bacterial species, *Vibrio fischeri* and *Vibrio harveyi*, where QS-governed enzymes were found to be responsible for light production (Bassler et al., 2006). Since then, QS systems have been recognized as key regulators of biofilm formation, in part by modulating the expression of efflux pump genes that contribute to biofilm resilience. Many EPS have demonstrated biological activity by interfering with QS systems and virulence factors, often by modulating gene expression or competitively inhibiting bacterial adhesion processes (Rendueles et al., 2013; Abdel-Aziz and Aeron, 2014). For instance, *Streptococcus thermophilus* produces an EPS capable of inhibiting the formation of violacein, a QS-regulated pigment in the biosensor strain *Chromobacterium violaceum*, as well as the synthesis of the C6-HSL signal molecule. These findings emphasize the capacity of EPSc to alter the display of QS and virulence-associated genes in infective microorganisms (Karadeniz et al., 2021).

Lactobacillus crustorum ZHG 2-1 has been proven to be a new quorum-quenching (QQ) bacterium that have the ability to degrade the signaling molecules such as N-3-oxododecanoyl-DL-homoserine lactone (3-oxo-C12-HSL) and N-butyryl-DL-homoserine lactone (C4-HSL). This QQ allow to serve as a potent antibiofilm agent against *Pseudomonas aeruginosa* (Cui et al., 2020).

Various enzymes and the compounds of the QQ have been found to ruin the communication pathway of bacteria, and acylases are one among the enzymes.

These enzymes degrade N-acyl-L-homoserine lactones (AHLs), key signal molecules in gram-negative bacterial QS systems, thereby inhibiting activities such as biofilm development and virulence expression. In addition to microbial enzymes, a wide array of plant-based natural products has been identified as potential quorum-sensing inhibitors (QSIs) (Zhong et al., 2020). These compounds are being explored for their role in future biofilm-targeting strategies due to their efficacy and eco-friendly nature.

The mechanisms by which antibiofilm agents interfere with QS are mainly twofold:

1. Inhibition or enzymatic degradation of QS signal molecules, such as AHLs or autoinducing peptides (AIPs), and
2. Mimicry of signal molecules, which prevents their binding to QS receptors, effectively blocking downstream gene activation. (Kalia, 2013)

Furthermore, various bacterial enzymes and secondary metabolites have been shown to interfere with QS pathways in pathogenic bacteria, ultimately disrupting biofilm development and maintenance (Khan et al., 2019).

15.10 Conclusion

This study examines the dual role of EPS as both a key component of microbial exopolysaccharide based biofilms and a promising therapeutic agent. Biofilms, organized microbial communities within an exopolysaccharide matrix, are major contributors to chronic infections due to immunity to antibiotics, immune clearance, and environmental stress. Biofilm formation and the rising incidence of antibiotic resistance in bacteria are major concerns within the scientific community. Preventing biofilm development may represent one of the most effective strategies to hinder bacteria from establishing a protective environment that promotes their proliferation, shields them from external agents, and enhances their resistance to antibiotics. A major functional aspect of EPS is also its antioxidant capacity, influenced by structural factors such as monosaccharide content, branching, and glycosidic linkage type. Cyanobacterial and lactic acid bacterial EPS, particularly those enriched in mannose, galactose, or sulfate groups, demonstrate high free radical scavenging activity and potential to mitigate oxidative stress-related diseases. The review explores the antibiofilm potential of EPS and the importance of certain enzymes that can serve as an acceptable substitute for conventional chemical sanitizers, reducing environmental hazards while efficiently controlling biofilms. Enzymes prove to be eco-friendly compared with chemical sanitizers. Thus, overall, EPS represents both a challenge in biofilm-associated infections and an opportunity as a bioactive compound for antioxidant, antibiofilm, and therapeutic applications, especially when sourced from safe and natural origins of microbes subjected to scaled-up production.

Credit statement

All authors have shared their work in data collections and drafting the manuscript.

Conflict of interest

The authors declared that there is no conflict of interest.

Ethical statement

This study does not involve any human participants or animals and therefore did not need ethical approval.

Acknowledgment

The authors would like to acknowledge our Respective Departments of, T.S. Paul Manipur Women's College; Regional College, Manipur, India and Vignan University, Guntur, A.P., and CCUBGA, IARI, New Delhi-12, India for their support.

References

Abdel-Aziz, S.M., Aeron, A., 2014. Bacterial biofilm: Dispersal and inhibition strategies. SAJ Biotechnol. 1, 105. https://doi.10.18875/2375-6713.1.105.

Adebayo-Tayo, B., Ishola, R., Oyewunmi, T., 2018. Characterization, antioxidant and immunomodulatory potential of exopolysaccharide produced by wild type and mutant *Weissella confusa* strains. Biotechnol. Rep. 19, e00271. https://doi.10.1016/j.btre.2018.e00271.

Akbarian, M., Chen, S.H., Kianpour, M., Farjadian, F., Tayebi, L., Uversky, V.N., 2022. A review on biofilms and the currently available antibiofilm approaches: matrix-destabilizing hydrolases and anti-bacterial peptides as promising candidates for the food industries. Int. J. Biol. Macromol. 219, 1163–1179. https://doi.org/10.1016/j.ijbiomac.2022.08.192.

Badel, S., Bernardi, T., Michaud, P., 2011. New perspectives for *Lactobacilli* exopolysaccharides. Biotechnol. Adv. 29 (1), 54–66. https://doi.org/10.1016/j.biotechadv.2010.08.011.

Bai, M., Lu, W., Tan, J., Mei, X., 2024. HINT2 may be one clinical significance target for patients with diabetes mellitus and reduced ROS-induced oxidative stress and ferroptosis by MCU. Horm. Metab. Res. 56 (10), 670–678. https://doi.org/10.1055/a-2251-7654.

Barnham, K.J., Masters, C.L., Bush, A.I., 2004. Neurodegenerative diseases and oxidative stress. Nat. Rev. Drug Discov. 3 (3), 205–214. https://doi.10.1038/nrd1330.

Barzkar, N, Sheng, R, Sohail, M, Jahromi, ST, Babich, O, Sukhikh, S, et al., 2022. Alginate lyases from marine bacteria: An enzyme ocean for sustainable future. Molecules 27 (11), 3375. https://doi.org/10.3390/molecules27113375.

Barzegari, A., Kheyrolahzadeh, K., Khatibi, S.M.H., Sharifi, S., Memar, M.Y., Vahed, S.Z., 2020. The battle of probiotics and their derivatives against biofilms. Infect. Drug Resist. 13, 659–672. https://doi.org/10.2147/IDR.S232982.

Bassler, B.L, Losick, R., 2006. Bacterially Speaking. Cell 125 (2), 237–246. https://doi.org/10.1016/j.cell.2006.04.001.

Bellini, E., Ciocci, M., Savio, S., Antonaroli, S., Seliktar, D., Melino, S., Congestri, R, 2018. *Trichormus variabilis* (Cyanobacteria) biomass: From the nutraceutical products to novel EPS-cell/protein carrier systems. Drugs 16, 298.

Bernal, P., Llamas, M.A., 2012. Promising biotechnological applications of antibiofilm exopolysaccharides. Microb. Biotechnol. 5, 670–673. https://doi.org/10.1111/j.1751-7915.2012.00359.x.

Bomfim, V.B., Casarotti, S.N., Silva, R., Penna, A.L.B., 2020. Partial characterization and antioxidant activity of exopolysaccharides produced by *Lactobacillus plantarum* CNPC003. *LWT*. Food Sci. Technol. 127, 109349. https://doi.org/10.1016/j.lwt.2020.109349.

Bouyahya, A., Bakrim, S., Chamkhi, I., Tarik, A., 2024. Bioactive substances of cyanobacteria and microalgae: Sources, metabolism, and anticancer mechanism insight. Biomed. Pharmacother. 158, 113974. https://doi.org/10.1016/j.biopha.2023.113974.

Branda, S.S., Vik, A., Friedman, L., Kolter, R., 2005. Biofilms: The matrix revisited. Trends Microbiol. 13 (1), 20–26. https://doi.org/10.1016/j.tim.2004.11.006.

Bjarnsholt, T., Moser, C., Jensen, P Ø, Høiby, N., 2011. Biofilm Infections. Springer, New York.

Castro-Bravo, N., Hidalgo-Cantabrana, C., Rodriguez-Carvajal, M.A., Ruas-Madiedo, P., Margolles, A., 2017. Gene replacement and fluorescent labeling to study the functional role of exopolysaccharides in *Bifidobacterium*. Front. Microbiol. 8, 1405. https://doi.org/10.3389/fmicb.2017.

Chagnot, C., Zorgani, M.A., Astruc, T., Desvaux, M., 2013. Proteinaceous determinants of surface colonization in bacteria: Bacterial adhesion and biofilm formation from a protein secretion perspective. Front. Microbiol. 4, 303. https://doi.10.3389/fmicb.2013.00303.

Champion, M., Portier, E., Vallée-Réhel, K., Linossier, I, 2022. Anti-biofilm activity of a hyaluronan-like exopolysaccharide from the marine vibrio MO245 against pathogenic bacteria. Mar. Drugs 20, 728. https://doi.org/10.3390/md20110728.

Chandki, R., Banthia, P., Banthia, R., 2011. Biofilms: A microbial home. J. Indian Soc. Periodontol. 15 (2), 111. https://doi.org/10.4103/0972-124X.84377.

Ch'ng, J.-H, Chong, K.K., Lam, L.N., Wong, J.J., Kline, K.A., 2019. Biofilm-associated infection by enterococci. Nat. Rev. Microbiol. 17, 82–94. https://doi.10.1038/s41579-018-0107-z.

Cinato, M., Andersson, L., Miljanovic, A., Laudette, et al., 2024. Role of perilipins in oxidative stress: Implications for cardiovascular disease. Antioxidants 13 (2), 209. https://doi.org/10.3390/antiox13020209.

Costa, L.S., Fidelis, G.P., Cordeiro, S.L., Oliveira, R.M., Sabry, D.D., Câmara, et al., 2010. Biological activities of sulfated polysaccharides from tropical seaweeds. Biomed. Pharmacother. 64 (1), 21–28. https://doi.org/10.1016/j.biopha.2009.03.005.

Cui, T., Bai, F., Sun, M., Lv, X., Li, X., Zhang, D, et al., 2020. *Lactobacillus crustorum* ZHG 2-1 as novel quorum-quenching bacteria reducing virulence factors and biofilms formation of *Pseudomonas aeruginosa*. LWT, 117,, 108696. https://doi.10.1016/j.lwt.2019.108696.

Das, T., Sharma, P.K., Busscher, H.J., Van Der Mei, H.C., Krom, B.P., 2010. Role of extracellular DNA in initial bacterial adhesion and surface aggregation. Appl. Env. Microbiol. 76 (10), 3405–3408. https://doi.10.1128/aem.03119-09.

Danquah, C.A., Minkah, P.A.B., Agana, T.A., Moyo, P., Tetteh, M., Junior, I.O.D, et al., 2022. Natural Products as Antibiofilm Agents. https://doi.org/10.5772/intechopen.104434.

DeFrancesco, A.S., Masloboeva, N., Syed, A.K., DeLoughery, et al., 2017. Genome-wide screen for genes involved in eDNA release during biofilm formation by *Staphylococcus aureus*. Proc. Nat. Acad. Sci. U.S.A 114 (29), E5969–E5978. https://doi.org/10.1073/pnas.1704544114.

Delbarre-Ladrat, C., Sinquin, C., Lebellenger, L., Zykwinska, A., Colliec-Jouault, S., 2014. Exopolysaccharides produced by marine bacteria and their applications as glycosaminoglycan-like molecules. Front. Chem. 2, 85. https://doi.org/10.3389/fchem.2014.00085.

Deng, W., Lei, Y., Tang, X., Li, D., Liang, J., et al., 2022. DNase inhibits early biofilm formation in *Pseudomonas aeruginosa* or *Staphylococcus aureus* induced empyema models. Front. Cell. Infect. Microbiol. 12, 917038. http://doi.org/10.3389/fcimb.2022.917038.

Ethica, S.N., Zilda, D.S., Oedjijono, O., Nurgayah, W., Muhtadi, M., 2021. Bioprospection of alginate lyase from bacteria associated with brown algae *Hydroclathrus* sp. as antibiofilm agent: A review. AACL Bioflux. 14 (4), 1974–1989. https://doi.org/10.2139/ssrn.5027680.

Flemming, H.C., Neu, T.R., Wozniak, D., 2007. The EPS matrix: The house of biofilm cells. J. Bacteriol. 189 (2), 7945–7947. https://doi.10.1128/JB.00858-07.

Flemming, H.-C., Wingender, J., 2010. The biofilm matrix. Nat. Rev. Microbiol. 8 (9), 623–633. https://doi.org/10.1038/nrmicro2415.

Geesey, G.G., 1982. Microbial exopolymers: Ecological and economic considerations. ASM News 48, 9–14.

Gensluckner, S., Wernly, B., Datz, C., Aigner, E., 2024. Iron, oxidative stress, and metabolic dysfunction-associated steatotic liver disease. Antioxidants 13 (2), 208. https://doi.org/10.3390/antiox13020208.

Gupta, P., Pruthi, P.A., Pruthi, V., 2019. Role of exopolysaccharides in biofilm formation in introduction to biofilm engineering. Am. Chem. Soc. 17–57. https://doi.org/10.1021/bk-2019-1323.ch002.

Germoni, L., Bremer, P., Lamont, I., 2016. The effect of alginate lyase on the gentamicin resistance of *Pseudomonas aeruginosa* in mucoid biofilms. J. Appl. Microbiol. 121 (1), 126–135. https://doi.org/10.1111/jam.13153.

Hall-Stoodley, L., Costerton, J.W., Stoodley, P., 2004. Bacterial biofilms: From the natural environment to infectious diseases. Nat. Rev. Microbiol. 2 (2), 95–108. https://doi.org/10.1038/nrmicro821.

Izano, E.A., Wang, H., Ragunath, C., Ramasubbu, N., Kaplan, J.B., 2007. Detachment and killing of *Aggregatibacter actinomycetemcomitans* biofilms by dispersin B and SDS. J. Dent. Res. 86 (7), 618–622. https://doi.10.1177/154405910708600707.

Jiang, P., Li, J., Han, F., Duan, G., Lu, X., Gu, Y., Yu, W., 2011. Antibiofilm activity of an exopolysaccharide from marine bacterium *Vibrio* sp. QY101. PLoS One 6 (4), e18514. https://doi.org/10.1371/journal.pone.0018514.

Kalia, V.C., 2013. Quorum sensing inhibitors: An overview. Biotechnol. Adv. 31, 224–245. https://doi.org/10.1016/j.biotechadv.2012.10.004.

Kaplan, J.B., Ragunath, C., Velliyagounder, K., Fine, D.H., Ramasubbu, N., 2004a. Enzymatic detachment of *Staphylococcus epidermidis* biofilms. Antimicrob. Agents Chemother. 48 (7), 2633–2636. https://doi.org/10.1128/AAC.48.7.2633-2636.2004.

Kaplan, J.B., Velliyagounder, K., Ragunath, C., Rohde, H., Mack, D., Knobloch, J.K.M., Ramasubbu, N., 2004b. Genes involved in the synthesis and degradation of matrix polysaccharide in *Actinobacillus actinomycetemcomitans* and *Actinobacillus pleuropneumoniae* biofilms. J. Bacteriol. 186 (24), 8213–8220. https://doi.org/10.1128/JB.186.24.8213-8220.

Karadeniz, D.G., Kaskatepe, B., Kiymaci, M.E., Tok, K.C., Gumustas, M., Karaaslan, C., 2021. Microbial exopolysaccharide production of *Streptococcus thermophilus* and its antiquorum sensing activity. Arch. Microbiol. 203, 3331–3339. https://doi.org/10.1007/s00203-021-02313-7.

Khan, F., Oloketuyi, S.F., Kim, Y.M., 2019. Diversity of bacteriaand bacterial products as anti-biofilm and antiquorum sensing drugs against pathogenic bacteria. Curr. Drug Targets 20, 1156–1179. https://doi.org/10.2174/1389450120666190423161249.

Kim, S., Jung, U.T., Kim, S.K., Lee, J.H., Choi, H.S., Kim, C.S., Jeong, M.Y., 2015. Nanostructured multifunctional surface with antireflective and antimicrobial characteristics. ACS Appl. Mater. Interfaces 7 (1), 326–331. https://doi.org/10.1021/am507503c.

Kondo, N., Kanai, T., Okada, M., 2023. Rheumatoid arthritis and reactive oxygen species: A review. Curr. Issues Mol. Biol. 45, 3000–3015. https://doi.org/10.3390/cimb45040197.

Krupanidhi, A.M., Dabadi, P., Anusha, M.M., Deepika, B.V., et al., 2022. Antioxidant properties of medicinal plants: A systematic review. Syst. Rev. Pharm. 13 (7), 457–463. https://doi.org/10.31838/srp.2022.7.67.

Lahiri, D., Nag, M., Banerjee, R., Mukherjee, D., Garai, S., Sarkar, T., et al., 2021. Amylases: Biofilm inducer or biofilm inhibitor? Front. Cell. Infect. Microbiol. 11, 660048. https://doi.org/10.3389/fcimb.2021.660048.

Li, J., Fang, D., Ye, R., Zhou, C., Li, P., 2021. The Released polysaccharide inhibits cell aggregation and biofilm formation in the cyanobacterium *Synechocystis* sp. PCC 6803. Eur. J. Phycol. 56, 119–128. https://doi.org/10.1080/09670262.2020.1777469.

Li, X.H., Lee, J.H., 2017. Antibiofilm agents: A new perspective for antimicrobial strategy. J. Microbiol. 55 (10), 753–766. https://doi.org/10.1007/s12275-017-7084-2.

Liu, J., Luo, J., Ye, H., Sun, Y., Lu, Z., Zeng, X. (2009), Production, characterization and antioxidant activities in vitro of exopolysaccharides from endophytic bacterium *Paenibacillus polymyxa* EJS-3" carbohydrate polymers, 78 (2), 275–281. https://doi.org/10.1016/j.carbpol.2009.04.039

Lopes, S.P., Azevedo, N.F., Pereira, M.O., 2018. Quantitative assessment of individual populations within polymicrobial biofilms. Sci. Rep. 8, 9494. https://doi.org/10.1038/s41598-018-27734-3.

Lu, L., Hu, W., Tian, Z., Yuan, D., Yi, G., Zhou, Y., et al., 2019. Developing natural products aspotential anti-biofilm agents. Chin, Med. 14 (11), 1–17. https://doi.org/10.1186/s13020-019-0232-2.

Lynch, K.M., Zannini, E., Coffey, A., Arendt, E.K., 2018. Lactic acid bacteria exopolysaccharides in foods and beverages: Isolation, properties, characterization, and health benefits. Annu. Rev. Food Sci. Technol. 9, 155–176. https://doi.org/10.1146/annurev-food-030117-012329.

Mahapatra, S., Banerjee, D. (2013), Evaluation of in vitro antioxidant potency of exopolysaccharide from endophytic *Fusarium solani* SD5. Int. J. Biol. Macromol., 53 62–66. https://doi.org/10.1016/j.ijbiomac.2012.11.003.

Mann, E.E., Wozniak, D.J., 2012. *Pseudomonas* biofilm matrix composition and niche biology. FEMS Microbiol. Rev. 36, 893–916. https://doi.org/10.1111/j.1574-6976.2011.00322.x.

Melandera, R.J., Basaka, A.K., Melandera, C., 2020. Natural products as inspiration for the development of bacterial antibiofilm agents. Nat. Prod. Rep. 37 (11), 1454–1477. https://doi.org/10.1039/d0np00022a.

Moscovici, M., 2015. Present and future medical applications of microbial exopolysaccharides. Front. Microbiol. 6, 1012. https://doi.org/10.3389/fmicb.2015.01012.

Motoyama, K., Tanida, Y., Hata, K., Hayashi, T., Hashim, I.I.A., Higashi, T., Ishitsuka, Y., Kondo, Y., Irie, T., Kaneko, S., et al., 2016. Anti-inflammatory effects of novel polysaccharide sacran extracted from cyanobacterium Aphanothece sacrum in various inflammatory animal models. Biol. Pharm. Bull 39 (7), 1172–1178. https://doi.org/10.1248/bpb.b16-00208.

Muhamad, I.I., Lazim, N.A.M., Selvakumaran, S., 2019. Natural polysaccharide-based composites for drug delivery and biomedical applications. In: Hasnain, M.S., Nayak, A.K. (Eds.), Natural Polysaccharides in Drug Delivery and Biomedical Applications. Academic Press, Cambridge, pp. 419–440. https://doi.org/10.1016/B978-0-12-817055-7.00018-2.

Mukherji, R., Patil, A., Prabhune, A., 2015. Role of extracellular proteases in biofilm disruption of gram-positive bacteria with special emphasis on *Staphylococcus aureus* biofilms. Enzyme Eng. 4, 126. https://doi.org/10.4172/2329-6674.1000126.

Neu, T.R., Lawrence, J.R., 2009. Microbial Glycobiology: Structures, Relevance and Applications. Elsevier, San Diego, CA, USA.

Ngan, L.T.K., Wang, S.L., Hiep, I.M., Luong, P.M., Vui, N.T., et al., 2014. Preparation of chitosan nanoparticles by spray drying and their antibacterial activity. Res. Chem. Intermed. 40, 2165–2175. https://doi.org/10.1007/s11164-013-1064-7.

Nwodo, U.U., Green, E., Okoh, A.I., 2012. Bacterial exopolysaccharides: Functionality and prospects. Int. J. Mol. Sci. 13 (11), 14002–14015. https://doi.org/10.3390/ijms131114002.

Okuda, K-I., Nagahori, R., Yamada, S., Sugimoto, S., Sato, C., Sato, M., et al., 2018. Thecomposition and structure of biofilms developed by *Propionibacterium acnes* isolated from cardiac pacemaker devices. Front. Microbiol. 9 (182), 1–12. https://doi.org/10.3389/fmicb.2018.00182.

Overhage, J., Campisano, A., Bains, M., Torfs, E.C.W., Rehm, B.H.A., Hancock, R.E.W., 2008. Human host defense peptide LL-37 prevents bacterial biofilm formation. Infect. Immun. 76 (9), 4176–4182. https://doi.org/10.1128/IAI.00319-08.

Paluch, E., Rewak-Soroczyñska, J., Jêdrusik, I., Mazurkiewicz, E., Jermakow, K., 2020. Prevention of biofilm formation by quorum quenching. Appl. Microbiol. Biotechnol. 104, 1871–1881. https://doi.org/10.1007/s00253-020-10349-w.

Pan, D., Mei, X., 2010. Antioxidant activity of an exopolysaccharide purified from *Lactococcus lactis* subsp. *lactis* 12. Carbohydr. Polym. 80 (3), 908–914. https://doi.org/10.1016/j.carbpol.2010.01.001.

Parwani, L., Bhatnagar, M., Bhatnagar, A., Sharma, V., 2014. Antioxidant and iron-chelating activities of cyanobacterial exopolymers with potential for wound healing. J. Appl. Phycol 26 (3), 1473–1482. https://doi.org/10.1007/s10811-013-0180-7.

Peesapati, S., Sajeevan, K.A., Patel, S.K., Roy, D., 2021. Relation between glycosidic linkage, structure and dynamics of α- and β-glucans in water. Biopolymers 112 (6), e23423. https://doi.org/10.1002/bip.23423.

Periasamy, S., Joo, H.S., Duong, A.C., Bach, T.H.L., Tan, V.Y., Chatterjee, S.S., 2012. How *Staphylococcus aureus* biofilms develop their characteristic structure. Proc. Nat. Acad. Sci. U.S.A 109 (4), 1281–1286. https://doi.org/10.1073/pnas.1115006109.

Persat, A., Nadell, C.D., Kim, M.K., Ingremeau, F., Siryaporn, A., et al., 2015. The mechanical world of bacteria. Cell 161, 988–997. https://doi.org/10.1016/j.cell.2015.05.005.

Plazinski, W., Drach, M., Plazinska, A., 2016. Ring inversion properties of 1→2, 1→3 and 1→6-linked hexopyranoses and their correlation with the conformation of glycosidic linkages. Carbohydr. Res. 423, 43–48. https://doi.org/10.1016/j.carres.2016.03.004.

Prakash Reddy, V., 2023. Oxidative stress in health and disease. Biomedicines 11 (10), 2925. https://doi.org/10.3390/biomedicines11102925.

Quan, Lin, Sheng, M, Tian, Y, Li, B, Kang, Z, Yang, Y, Xu, Z, Soteyome, T, Guo, L, Sun, H, Gao, Y, Yu, L, Pan, Q, Tan, Y, 2024. Antibiofilm Activity and Synergistic Effects of DNase I and Lysostaphin Against Staphylococcus Aureus Biofilms, Food Quality and Safety. https://doi.org/10.1093/fqsafe/fyae024.

Raj, S.S., Thomas, N.G., 2021. Anti biofilm agents - nature the cure. Int. Dent. J. Stud. Res. 9 (1), 25–32. https://doi.org/10.18231/j.idjsr.2021.006.

Rasouli, R., Navidinia, M., Ghahfarokhi, M.S., Roudsari, R.V., Adabian, S., Baghestani, A.R., 2020. Antibiofilm activity of cellobiose dehydrogenase enzyme (CDH) isolated from *Aspergillus niger* on biofilm of clinical *Staphylococcus epidermidis* and *Pseudomonas aeruginosa* isolates. Arch. Clin. Infect. Dis. 15 (1), 1–7. https://doi.org/10.5812/archcid.100960.

Rendueles, O., Kaplan, J.B., Ghigo, J.-M., 2013. Antibiofilm polysaccharides: Bacterial antibiofilm polysaccharides. Env. Microbiol. 15, 334–346. https://doi.org/10.1111/j.1462-2920.2012.02810.x.

Reuter, K., Steinbach, A., Helms, V., 2016. Interfering with bacterial quorum Sensing. Perspect. Med. Chem. 8, 1–15. https://doi.org/10.4137/PMC.S13209.

Rumbaugh, K.P., Sauer, K., 2020. Biofilm dispersion. Nat. Rev. Microbiol. 18 (10), 571–586. https://doi.org/10.1038/s41579-020-0385-0.

Sauer, K., Webb, J.S., Davies, D.G., Greenberg, E.P., Kinney, K., Costerton, J.W., Stewart, P.S., 2002. Biofilm life-cycle: Reversible attachment, irreversible attachment, microcolony formation, maturation, and dispersal. J. Bacteriol. 184 (16), 4477–4482. https://doi.org/10.1128/jb.184.4.1140-1154.2002.

Schulze, A., Mitterer, F., Pombo, J.P., Schild, S., 2021. Biofilms by bacterial humanpathogens: Clinical relevance-development, composition and regulation-therapeutical strategies. Microb. Cell 8 (2), 28–56. https://doi.org/10.15698/mic2021.02.741.

Silva, N.B.S., Marques, L.A., Roder, D.D.B., 2020. Antibiofilm activity of natural products: promising strategies for combating microbial biofilms. Ann. Pub. Health Rep. 4 (1), 92–99. https://doi.org/10.36959/856/502.

Sivaperumal, P., Kamala, K., Rajaram, R., 2015. Bioactive DOPA melanin isolated and characterised from a marine actinobacterium *Streptomyces* sp. MVCS6 from Versova coast. Nat. Prod. Res. 29, 2117–2121. https://doi.org/10.1080/14786419.2014.988712.

Soliman, C., Walduck, A.K., Yuriev, E., Richards, J.S., et al., 2018. Structural basis for antibody targeting of the broadly expressed microbial polysaccharide poly-N-acetylglucosamine. J. Biol, Chem. 293 (14), 5079–5089. https://doi.org/10.1074/jbc.RA117.001148.

Soto, S.M., 2013. Role of efflux pumps in the antibiotic resistance of bacteria embedded in a biofilm. Virulence 4 (3), 223–229. https://doi.org/10.4161/viru.23724.

Tang, W., Dong, M., Wang, W., Han, S., Rui, X., Chen, X., et al., 2017. Structural characterization and antioxidant property of released exopolysaccharides from *Lactobacillus delbrueckii* ssp. *bulgaricus* SRFM-1. Carbohydr. Polym. 173, 654–664. https://doi.org/10.1016/j.carbpol.2017.06.030.

Tetz, G.V., Artemenko, N.K., Tetz, V.V., 2009. Effect of DNase and antibiotics on biofilm characteristics. Antimicrob. Agents Chemother 53, 1204–1209. https://doi.org/10.1128/aac.00471-08.

Thallinger, B., Prasetyo, E.N., Nyanhongo, G.S., Guebitz, G.M., 2013. Antimicrobial enzymes: An emerging strategy to fight microbes and microbial biofilms. Biotechnol. J. 8 (1), 97–109. https://doi.org/10.1002/biot.201200313.

Uzma, F., Chowdappa, S., 2017. Antimicrobial and antioxidant potential of endophytic fungi isolated from ethnomedicinal plants of Western Ghats, Karnataka. J. Pure Appl. Microbiol. 11, 1009–1025. https://doi.org/10.22207/jpam.11.2.43.

Vaikundamoorthy, R., Rajendran, R., Selvaraju, A., Moorthy, K., Perumal, S., 2018. Development of thermostable amylase enzyme from *Bacillus cereus* for potential antibiofilm activity. Bioorg. Chem. 77, 494–506. http://doi.org/10.1016/j.bioorg.2018.02.014.

Vishwakarma, J., Sinha, V.L., 2020. Unraveling the anti-biofilm potential of green algal sulfated polysaccharides against *Salmonella enteric* and *Vibrio harveyi*. Appl. Microbiol. Biotechnol. 104, 6299–6314. https://doi.org/10.1007/s00253-020-10653-5.

Wang, K., Li, W., Rui, X., Chen, X., Jiang, M., Dong, M., 2014. Structural characterization and bioactivity of released exopolysaccharides from *Lactobacillus plantarum* 70810. Int. J. Biol. Macromol. 67, 71–78. https://doi.org/10.1016/j.ijbiomac.2014.02.056.

Wingender, J., Neu, T., Flemming, H.-C., 1999. Microbial Extracellular Polymer Substances. Springer, Heidelberg, Germany.

Wu, Z., Zheng, R., Zhang, J., Wu, S., 2020. Transcriptional profiling of *Pseudomonas aeruginosa* PAO1 in response to anti-biofilm and anti-infection agent exopolysaccharide EPS273. J. Appl. Microbiol. 130, 265–277. https://doi.org/10.1111/jam.14764.

Xiong, J., Liu, D.M., Huang, Y.Y., 2023. Exopolysaccharides from *Lactiplantibacillus plantarum*: Isolation, purification, structure–function relationship, and application. Eur. Food Res. Technol. 249, 1431–1448. https://doi.org/10.1007/s00217-023-04237-6.

Xu, Y., Cui, Y., Yue, F., Liu, L., Shan, Y., Liu, B., Zhou, Y., Lü, X., 2019. Purification, characterization and bioactivity of exopolysaccharides produced by *Lactobacillus plantarum* KX041. Int. J. Biol. Macromol. 128, 480–492. https://doi.org/10.1016/j.ijbiomac.2019.01.117.

Xu, X., Peng, Q., Zhang, Y., Tian, D., Zhang, P., et al., 2020. A novel exopolysaccharide produced by *Lactobacillus coryniformis* NA-3 exhibits antioxidant and biofilm-inhibiting properties in vitro. Food Nutr. Res. 64, 3. https://doi.org/10.29219/fnr.v64.3694.

Yang, A., Tang, W.S., Si, T., Tang, J.X., 2017. Influence of physical effects on the swarming motility of *Pseudomonas aeruginosa*. Biophys. J. 112 (7), 1462–1471. https://doi.org/10.1016/j.bpj.2017.02.018.

Yang, X., Wu, J., An, F., Xu, J., Bat-Ochir, M., Wei, L., et al., 2022. Structure characterization, antioxidant and emulsifying capacities of exopolysaccharide derived from *Tetragenococcus halophilus* SNTH-8. Int. J. Biol. Macromol. 208, 288–298. https://doi.org/10.1016/j.ijbiomac.2022.03.049.

Yang, Y., Jiang, G., Tian, Y., 2023. Biological activities and applications of exopolysaccharides produced by lactic acid bacteria: A mini-review. World J. Microbiol. Biotechnol. 39, 155. https://doi.org/10.1007/s11274-023-03610-7.

Yasir, M., Willcox, M.D.P., Dutta, D., 2018. Action of antimicrobial peptides against bacterial biofilms. Materials 11 (12), 2468. https://doi.org/10.3390/ma11122468.

Zanna, S., Mercier, D., Gardin, E., Allion-Maurer, A., Marcus, P., 2022. EPS for bacterial anti-adhesive properties investigated on a model metal surface. Colloids Surf. B Biointerfaces 213, 112413. https://doi.org/10.1016/j.colsurfb.2022.112413.

Zhao, W., Zhang, J., Jiang, Y.Y., Zhao, X., Hao, X.N., Li, L., et al., 2018. Characterization and antioxidant activity of the exopolysaccharide produced by *Bacillus amyloliquefaciens* GSBa-1. J. Microbiol. Biotechnol. 28 (8), 1282–1292. https://doi.org/10.4014/jmb.1803.03016.

Zheng, J-Q., Wang, J-Z., Shi, C-W., Mao, D-B., He, P-X., Xu, C-P., 2014. Characterization and antioxidant activity for exopolysaccharide from submerged culture of *Boletus aereus*. Process Biochem 49, 1047–1053. https://doi.org/10.1016/j.procb/io.2014.03.009.

Zhong, L., Ravichandran, V., Zhang, N., Wang, H., Bian, X., Zhang, Y., Li, A., 2020. Attenuation of *Pseudomonas aeruginosa* Quorum Sensing by Natural Products: Virtual Screening, Evaluation and Biomolecular Interactions. Int. J. Mol. Sci. 21 (6), 2190. https://doi.org/10.3390/ijms21062190.

Zhou, Y., Cui, Y., Qu, X., 2019. Exopolysaccharides of lactic acid bacteria: Structure, bioactivity and associations: A review. Carbohydr. Polym 207, 317–332. https://doi.org/10.1016/j.carbpol.2018.11.093.

Zhu, Y., Wang, C., Jia, S., Wang, B., Zhou, K., Chen, S., et al., 2018. Purification, characterization and antioxidant activity of the exopolysaccharide from *Weissella cibaria* SJ14 isolated from Sichuan paocai. Int. J. Biol. Macromol. 115, 820–828. https://doi.org/10.1016/j.ijbiomac.2018.04.055.

Ziadi, M., Bouzaiene, T., M'Hir, S., Zaafouri, K., Mokhtar, F., Hamdi, M., Boisset-Helbert, C., 2018. Evaluation of the efficiency of ethanol precipitation and ultrafiltration on the purification and characteristics of exopolysaccharides produced by three lactic acid bacteria. BioMed. Res. Int. 11. https://doi.org/10.1155/2018/1896240.

Zou, C., Du, Y., Li, Y., Yang, J., Zhang, L., 2010. Preparation and in vitro antioxidant activity of lacquer polysaccharides with low molecular weights and their sulfated derivatives. Int. J. Biol. Macromol 46 (2), 140–144. https://doi.org/10.1016/j.ijbiomac.2009.11.010.

CHAPTER

16

EPS: A potential anti-inflammatory agent

Paramita Paul[a,b], Sucheta Sarkar[c], Mrinmoy Pratim Chetia[d], Tapan Kumar Shaw[e], Tarun Kumar Dua[a], Ranabir Sahu[a] and Gouranga Nandi[a]

[a]Department of Pharmaceutical Technology, University of North Bengal, Darjeeling, West Bengal, India [b]Institute of Pharmacy, Bankura, West Bengal, India [c]Guru Nanak Institute of Pharmaceutical Science & Technology, Panihati, Kolkata, West Bengal, India [d]School of Pharmacy, Arunachal University of Studies, Namsai, Arunachal Pradesh, India [e]Department of Pharmaceutical Technology, JIS University, Agarpara, Kolkata, West Bengal, India

16.1 Introduction

Microorganisms with a variety of morphologies produce a varied type of high molecular weight polysaccharides called exopolysaccharides (EPSs) (Wang et al., 2023b). EPSs are obtained from lactic acid bacteria, and microalgae are receiving more and more attention in contemporary healthcare and biomedicine. It should be noted that there is not many studies that have looked into these medicinal activities in cyanobacteria. It has recently been noted that certain EPSs may help reduce inflammation by preventing oxidative stress; as a result, they may be a viable option for treating a number of medical issues (Prete et al., 2024). However, a variety of parameters, such as molecular weight, chemical composition, sugar units, producer strain, surface features, overall structure, etc., frequently affect how effective they are. It was also discovered that the functional groups that sugars contain determine their biological activity. To increase antioxidant activity, for instance, hydroxyl groups in monosaccharide units can transfer electrons or hydrogen to free radicals' neutralization (Andrew and Jayaraman, 2020). The unique physicochemical characteristics of EPSs include their nontoxic nature, biocompatibility, and biodegradability. These rendered EPSs are appropriate for use in clinical and industrial settings. Certain EPSs derived from particular strains, their monosaccharide composition, and their structural features may be utilized for their potential as anti-inflammatory and antioxidant agents.

DOI: https://doi.org/10.1016/B978-0-443-36619-2.00016-4

The body uses inflammation as an important physiological defense mechanism to shield itself against harmful substances and various stimuli, such as injuries and infections. Inflammatory cells mediate inflammation and have a tendency to reinstate tissue homeostasis. Pro-inflammatory mediators, or cytokines, are released by inflammatory cells in response to inflammatory stimuli that initiate and exacerbate the acute phase of the reaction (Abdulkhaleq et al., 2018). Long term inflammation can exacerbate diseases such as excessive tissue damage. Additionally, it may contribute to the onset of conditions including cancer, lupus erythematosus, asthma, psoriasis, rheumatic arthritis, cardiovascular diseases, and atherosclerosis. Examining the molecular changes in inflammatory markers, EPS was found to have anti-inflammatory characteristics. Recent studies have demonstrated that EPS suppresses the development of an inducible nitric oxide synthase that results in oxidative stress and inhibits TNF-α, COX-2, IL-1β, and IL-6. Furthermore, by hindering the interaction between LPS and Toll-like receptor 4 (TLR4), as confirmed by the well-known TLR4 inhibitor TAK-242, EPS has demonstrated an anti-inflammatory effect (Kwon et al., 2020).

16.2 Exopolysaccharides: Source, uses, applications

An extensive range of microorganisms, particularly bacteria, are accomplished of producing a variety of EPSs, such as gellan gum and welan gum, which are widely used in food and medicine. According to the conditions of fermentation, fungi produced EPSs such as pullulan, scleroglucan, pleuran, etc. The marine microorganisms that thrive in severe conditions, including high salinity, low temperature, and high pressure, generate distinct EPSs such as HYD 657, levan, EPS-S3, and others (Qi et al., 2022). Some of the EPS were identified, and their source, structure, along with the industrial applications are compiled in Table 16.1.

EPSs' unique properties are being used in a wide range of industries. They are described below:

Food industries: EPSs are widely used in the food manufacturing as thickeners, stabilizers, and food packaging because of their immunomodulatory, viscosity, emulsifying, binding, texture development, gelling ability, and other qualities (Jindal and Khattar, 2018; Prasad and Purohit, 2023). They are also employed in both fermented and nonfermented food products (Korcz and Varga, 2021). For instance, gellan gum, derived from *Sphingomonas paucimobilis*, is broadly used as a gelling agent, thickening agent, or emulsifying agent in foodstuffs (Prajapati et al., 2013).

Environmental application: EPSs have a key role in environmental remediation. They have shown excellent sorption capabilities for heavy metal adsorption, thereby mitigating industrial pollution concerns (Prasad and Purohit, 2023). Zinc and nickel can be effectively removed from wastewater using EPS-alginate composites containing urea and biuret (Benettayeb et al., 2023), while *Bacillus cereus* VK1 EPS effectively adsorbs mercury (Kalpana et al., 2018). Moreover, in wastewater treatment, EPS-based silver nanoparticles have been used to break down harmful azo-dyes (Saravanan et al., 2017). Additionally, a novel EPS derived from lactic acid bacteria demonstrated high adsorption capacity for methylene blue (Li et al., 2020).

Cosmetics and pharmaceuticals industries: The unique properties of EPS, like structural complexity, biological activities, biodegradability, and biocompatibility, have drawn attention in the field of cosmetics and pharmaceuticals (Radchenkova et al., 2013). For example, HYD

TABLE 16.1 Summary of marketed microbial EPS with their source, structure, industrial applications, and potential clinical benefits.

Name of EPS	Microbial source	Structure	Clinical uses/industrial applications	References
Dextran	*Leuconostoc mesenteroids*	An α-1,6-glucan structure with α-1,3-glycosidic branch linkages	Plasma extender, wound healer, antithrombotic agent, iron deficiency treatment, and organ preservation. Other uses include food packaging, photography, cell culture techniques, and as a cryoprotectant.	Côté (2007), Díaz-Montes (2021)
Gellan gum	*Sphingomonas elodea*	Composed of one β-D-glucuronic acid, two β-D-glucose units, and single α-L-rhamnose molecule	Stabilizing thickening, and gelling ingredient in food sector.	Abdl Aali and Al-Sahlany (2024), Lin et al. (2024)
Welan gum	*Sphingomonas* sp. ATCC 31555	Tetrasaccharide repeats featuring single side chains of L-mannose or L-rhamnose	Pharmaceutical applications- thickener, stabilizer, binding, emulsifier, and suspending agent.	Wang et al. (2024)
Pullulan	*Aureobasidium pullulans*	A succession of α-(1→6)-linked α-(1→4) α-d triglucosides	Clinical applications – In treating several respiratory or lungs disorders. Pharmaceutical industries- as coatings for tablets, pills, and granules.	Ponnusami and Gunasekar (2015), Thakur et al. (2023)
Scleroglucan	*Sclerotium rolfsii*	A β-d-(1-3)-glucopyranosyl and β-d-(1-6)-glucopyranosyl group linear chain	Clinical applications – reduce cholesterolemia. Cosmetic industries for the formulation of hair sprays, skin care, and bath products.	Mehta and Jadhav (2023), Rahman et al. (2021)
Pleuran	*Pleurotus ostreatus*	A glucose polymer with β-(1→3) linkages and occasional β-(1→6) branches	Clinically used to lessen the time of diarrhoea and severity of children's acute gastroenteritis symptoms.	Karácsonyi and Kuniak (1994), Wzorek-Łyczko et al. (2021)

(*continued on next page*)

TABLE 16.1 Summary of marketed microbial EPS with their source, structure, industrial applications, and potential clinical benefits—cont'd

Name of EPS	Microbial source	Structure	Clinical uses/industrial applications	References
Ulvan	Isolated from green algae genus Ulva, commonly known as sea lettuce	Composed of repeating disaccharide units, with iduronic acid, glucuronic acid, xylose, and rhamnose as the main building blocks	Clinically used for its antioxidant, anti-inflammatory, anticancer, antiviral, and immunomodulatory properties. Cosmetic industry- for moisturizing, collagen synthesis, and emulsifying capabilities.	Kidgell et al. (2019)
Hyd 657	Obtained from marine bacterium *Alteromonas macleodii* subsp. Deepsane	Glucose, galacturonic acid, galactose, rhamnose, glucuronic acid, fucose, mannose, 3-0-(1 carboxyethyl)-d-glucuronic acid	Cosmetic industry- It soothes and reduces irritation of sensitive skin and shows protective effect on keratinocytes.	Casillo et al. (2018)
Levan	*Zymomonas mobilis, Aspergillus sydowii, Erwinia herbicola, B. subtilis, Aspergillus versicolor*	β-D-fructose units linked through β-(2→6) glycosidic bonds	Clinically used to reduce fat, hyperlipidemia, and body weight cosmetic industry—preparation of skin creams, moisturizers.	Domżał-Kędzia et al. (2023), Hundschell et al. (2020), Srikanth et al. (2015)
Eps-S3	*Pantoea* sp.	D-galactose and L-rhamnose at a proportion of 2:1	Clinically used for its wound healing applications.	Qi et al. (2022)
Hyaluronan	*Streptococcus* sp.	Two repeating units of saccharides	diagnostic indicator for liver disorders, rheumatoid arthritis, and cancer	Netrusov et al. (2023)
Curdlan	*Alcaligenes faecalis, Rhizobium radiobacter*, and *Agrobacterium* sp.	Linear (1,3)-β-glucan or polysaccharide made of glucose molecules linked by β-1,3-glycosidic bonds	Used in encapsulation of drugs like salbutamol sulfate, indomethacin, etc.	Kogan et al. (2007), Netrusov et al. (2023)
Xanthan	*Xanthomonas*	A polymer of D-mannose, D-glucose, and D-glucuronic acid	Pharmaceutical applications- utilized as a stabilizer, emulsifier, thickener, and suspending agent.	Netrusov et al. (2023)

EPS, exopolysaccharides.

657 has been used to soothe and reduce irritations on sensitive skin (Le Costaouëc et al., 2012). Levan from *H. smyrnensis* AAD6T demonstrated high compatibility and anticoagulant properties (Erginer et al., 2016). Another novel EPS, EPS-S3 derived from *Pantoea* sp., has demonstrated potential in promoting wound healing (Sahana and Rekha, 2020; Qi et al., 2022). Dextran is utilized as a moisturizer for its thickening, viscous, and emulsifier properties (Rana and Upadhyay, 2020).
Agriculture applications: EPS removes heavy metals and toxins from soil, increases the soil's fertility, increases water retention ability, prevents soil erosion, promotes plant-microbe interactions, supports nutrient exchange, inhibits pathogens, and acts as an environmentally friendly substitute in agrochemical formulations. All of these contribute to sustainable agriculture (Arora et al., 2015). Gellan and Curdlan are more frequently used as soil conditioners, which increase the water retention ability of the soil (Rana and Upadhyay, 2020).

16.3 Exopolysaccharide: Advantages and limitations

Compared to polysaccharides originating from plants and animals, microbial EPSs offer more advantages due to their structure reproducibility, optimizing production potential, and multifunctionality. Regardless of the environment, microbial EPSs maintain a consistent structure and composition, unlike polysaccharides originating from plants and animals. Because of their reliability, they are useful in the food, medicine, and cosmetics industries (Radchenkova et al., 2013; Casillo et al., 2018). Moreover, alterations to growth conditions and advancements in synthetic biology can enhance the production of microbial EPS, leading to higher yields. On the other hand, their functional properties make them useful as emulsifiers, stabilizers, and texture enhancers in food items (Elnahas et al., 2017), while their bioactive properties offer prospective health benefits like antioxidant, anti-inflammatory, anticancer, and hypolipidemic effects.

Despite their potential applications, EPSs have several limitations. One major barrier is our incomplete knowledge of potential EPS-producing species, as not all microbes are capable of producing EPSs and are not efficient at the same time. Another significant disadvantage is the complex extraction process, which involves multiple steps and the use of chemicals, including formaldehyde, ethanol, sodium hydroxide, and hydrochloric acid. The high cost, complicated equipment, and processing time needed for typical extraction techniques, including membrane filtration and electro-separation, make large-scale production inefficient. Furthermore, mass production of EPSs is still a major concern, as the majority of research is limited to laboratory-scale processes (Hasan et al., 2024). More research is needed to identify new choices with enhanced capacity because different microbial species produce varying amounts of EPS.

16.4 About inflammation

Inflammation is a complex body's defensive system, triggered in response to harmful stimuli (El Awady et al., 2024). It is essential for removing harmful substances and re-establishing tissue homeostasis (Kumar et al., 2024). The vertebrate immune system is shaped by microorganisms, and less exposure to microorganisms as a result of urbanization has led to inflammatory diseases such as multiple sclerosis, lupus, and allergies. Probiotics are a kind of microorganism known for health benefits, though their exact mechanisms are still unknown (Zamora-Pineda et al., 2023).

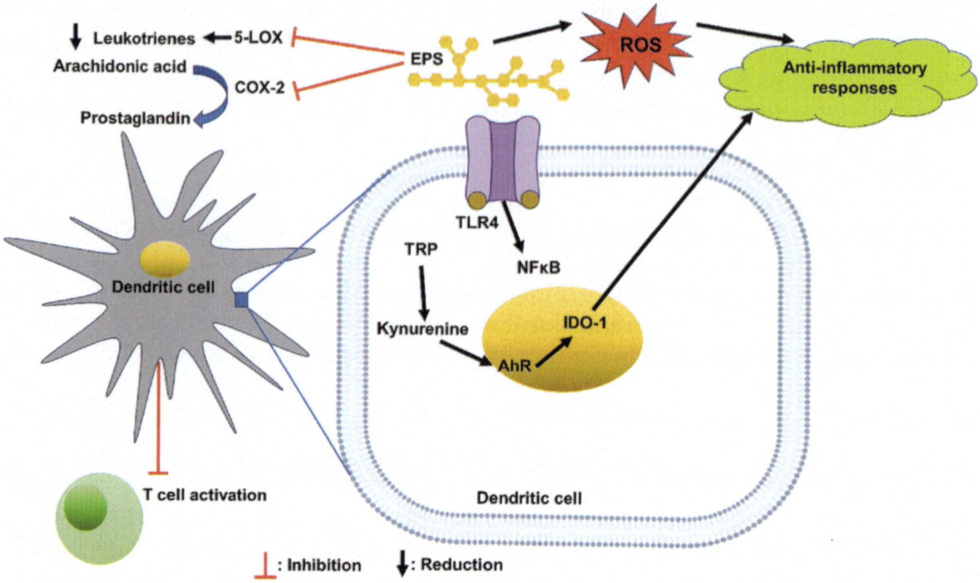

FIGURE 16.1 Probable anti-inflammatory mechanism of EPS. *EPS*, Exopolysaccharides.

There are two types of inflammation: controlled and uncontrolled. Immune system function against pathogens, wound healing, and homeostasis are all controlled by inflammation. On the other hand, cytokine storms and chronic inflammation are uncontrolled inflammation that leads to complex disorders like cancer, metabolic syndrome, cardiovascular disease, neurodegenerative diseases, and autoimmune diseases. Immune and nonimmune cells that actively participate in the inflammatory process are epithelial cells, muscle cells, fibroblasts, myeloid cells, endothelial cells, B and T lymphocytes, and adipocytes are among them. These cells use growth factors, cytokines, chemokines, matrix metalloproteases, and membrane molecules to communicate. IL-6, IL-1β, TNF-α, etc., are important pro-inflammatory cytokines; IL-6 has a significant role in cancer, autoimmune illnesses, cytokine storms, and chronic inflammatory diseases (Hirano, 2021).

16.5 Anti-inflammatory mechanisms of exopolysaccharides

The probable anti-inflammatory mechanism of EPS by modulating various pathways involved in the inflammatory response is illustrated in Fig. 16.1. EPS inhibit COX-2 and 5-LOX enzymes, having a vital role in the inflammatory process. 5-LOX is linked to the production of leukotrienes, while COX-2 is an enzyme involved in the synthesis of prostaglandins (Abdel-Wahab et al., 2022). Additionally, it promoted the release of anti-inflammatory cytokines like IL-10 while suppressing the generation of pro-inflammatory cytokine mediators, including TNF-α, IL-1β, and IL-6. Research has demonstrated that EPS inhibits the activation of intracellular COX-2 signaling and the NF-κB/TLR4 signal transduction pathway (Kwon et al., 2020). Furthermore, some EPS can directly interact with TLR4, preventing its activation by inflammatory stimuli like LPS. Antigen-specific CD4+ T cell activation and dendritic cell maturation can both be inhibited by EPS derived from *Bifidobacterium* breve (Hickey et al., 2021).

16.6 Research status

Over the past few years, several works have been done in cell lines or in animal models to confirm the anti-inflammatory activity of EPS from different sources. Most of the work has been done with either RAW264.7 cells or THP-1 dual cell lines. Some work has been assessed using Caco-2 cells, UVB-HaCaT cell or red blood cells of rats. Very few in vivo assessments have been done in experimental animals or animal inflammation models. They are summarized in Table 16.2 below.

16.6.1 Challenges and future directions

Challenges and Prospects for Further Study depicted in Fig. 16.2, a number of issues and potential research avenues need to be resolved in order to guarantee the effective deployment and broad acceptance of EPSs, despite their promising uses and features.

It is essential to make efforts to overcome the obstacles in the manufacture of EPS on a big scale. Therefore, it is important to standardize manufacturing and purifying techniques in order to improve industrial scalability and reproducibility. Additionally, by better understanding the biosynthesis routes of EPSs from various bacterial strains, genetic alterations can be optimized, potentially leading to higher EPS yields and the ability to customize the functionality of materials. Understanding the distinct biosynthetic routes and sugar residues produced by yeasts, fungi, and microalgae may also help to broaden the spectrum of EPS applications and functionality across different industries.

Current research has concentrated on tackling the complexity and variability of EPS generation techniques. In order to improve EPS production and functionality, Bibi et al. looked into approaches to standardize the generation and purification processes S a focus on genetic changes and biosynthesis pathway optimization (Bibi et al., 2021). Similar to this, Xiong et al. investigated the functional and structural connections between EPSs produced by lactic acid bacteria, emphasizing the possibility of creating valuable biopolymers (Bibi et al., 2021; Xiong et al., 2023).

However, there are a number of obstacles to putting these developments into practice. Because of the intricacy of biosynthetic pathways and the varying needs of various bacterial strains, it is still difficult to produce EPSs consistently on an industrial scale. Significant obstacles are also presented by technical difficulties in expanding production processes and guaranteeing cost-effectiveness with novel approaches (Díaz-Cornejo et al., 2023).

Ongoing research should focus on removing the legal and technical barriers that prevent the industrial use of EPS production techniques in order to address these issues. Furthermore, investigating different strategies like innovative fermentation techniques or sophisticated bioreactor designs may offer more effective and scalable production routes.

Furthermore, a key concern in the production of EPS is still economic viability. Numerous studies have looked into this topic and found that using inexpensive industrial and agricultural

TABLE 16.2 Anti-inflammatory activity of some exopolysaccharides and possible mechanism.

Exopolysaccharide source	Cell line used in in vitro study	In vivo model used	Anti-inflammatory mechanism	References
Pleurotus sajor-caju mushroom	-	Swiss male mice of 25–30g	Reduces edema and nociception.	Silveira et al. (2015)
L.plantarum R301 bacteria	RAW 264.7 Cell line	-	Reduces the expression of Cox-2 and the synthesis of the pro-inflammatory cytokine IL-6.	Wang et al. (2023a)
L.fermentum, L.plantarum, L.helveticus, L.rhamnosus, and *L.salivarius*	LPS-stimulated RAW264.7 macrophages	-	Reduce pro-inflammatory factors, neutrophil infiltration, and regulate intestinal flora.	Zhang et al. (2024)
Marine *Kocuria sp.* Bacterial EPS	-	Albino male rats of 160±20 g	Reduce carrageenan-produced paw edema by suppressing oxidative and inflammatory mediators such as MDA, ROS, NO, IL-6, and COX-2.	El Awady et al. (2024)
Marine *Bacillus subtilis*	Lipoxygenase (5-LOX), Cyclooxygenase (COX2) Inhibition	-	Demonstrates strong inhibitory effects on 5-LOX compared to many anti-inflammatory drugs (Ibuprofen, celecoxib).	Abdel-Wahab et al. (2022)
Salinity-induced *Auxenochlorella protothecoides* algae	RAW 264.7 cells	-	Prevents inflammation caused by LPS by inhibiting inflammatory mediators such as ROS, iNOS, TNF-α, and IL-6.	Zhou et al. (2023)
Cordyceps sinensis Cs-HK1	THP-1 dual cell lines	-	Inhibits the proinflammatory reactions of THP-1 cells triggered by LPS *in vitro*, including NF-κB inflammatory activation, NO synthesis, TNF-α release, and IL 8.	Li et al. (2021)
Bacillus thuringiensis	RAW 264.7	-	Inhibit the generation of several pro-inflammatory cytokines and nitric oxide generated by lipopolysaccharide, and they also had impacts on proliferation.	Gao et al. (2022)

(*continued on next page*)

Pediococcus pentosaceus 4412	RAW264.7	-	Suppresses the STAT-1/NF-κB pathway and reduces the levels of COX-2, iNOS, and another proinflammatory.	Angelin and Kavitha (2025)
Lactiplantibacillus plantarum Z-1	RAW264.7		Effectively reduce the activation of the intracellular COX-2 signaling cascade, the NF27 κB/TLR4 signaling pathway, and elevate the release of IL-10 and lessen the production of inflammatory mediators.	Xu et al. (2024)
Cordyceps sinensis Cs-HK1	THP-1 dual cell lines	Male 204 C57BL/6 mice (20–25 g)	Inhibits the generation of inflammatory cytokines and reduces inflammation by preventing the NF-κB pathway from being activated.	Li et al. (2020b)
Lactobacillus plantarum L-14	RAW 264.7	-	Block the interaction between LPS and TLR4 to reduce inflammation.	Kwon et al. (2020)
Lactobacillus paraplantarum BGCG11	-	Male Wistar rats	When neutrophil infiltration stays constant, it decreases the level of pro-inflammatory mediators like IL-1, TNF-α, and iNOS without influencing the elevated levels of anti-inflammatory cytokines like IL-10 and IL-6.	Dinić et al. (2018)
Phormidium sp. ETS05	-	Zebrafish embryos and larvae	Downregulates the expression of some inflammatory mediators.	Zampieri et al. (2020)
Schizophyllum commune	RAW264.7	-	Inhibits lipopolysaccharides-induced iNOS expression. Effective in NO-related disorders.	Du et al. (2017a)
Bacillus subtilis	-	C57BL/6, OTII, *Tlr4*$^{-/-}$, and *Ido*$^{-/-}$ mice	Promoting anti-inflammatory M2 macrophages or dendritic cells helps suppress excessive T cell responses.	Zamora-Pineda et al. (2023)

(*continued on next page*)

TABLE 16.2 Anti-inflammatory activity of some exopolysaccharides and possible mechanism—cont'd

Exopolysaccharide source	Cell line used in in vitro study	In vivo model used	Anti-inflammatory mechanism	References
Bacillus axarquiensison	-	Albino male rats of 160±20 g	Prevents the synthesis of pro-inflammatory mediators, such as TNF-α, IL-6, COX-2, IL-12, and iNOS, during inflammation.	Ibrahim et al. (2019)
Schizophyllum commune	-	Male C57BL/6 mice (7–8 weeks-old) (18–22g)	By reducing macrophage infiltration in inflammatory tissues and alleviating DSS-induced colitis, medium (936 kDa) and high molecular weight (1437 kDa) exopolysaccharides showed intestinal anti-inflammatory effects.	Du et al. (2017b)
Lactiplantibacillus plantarum B7	Red blood cells of rats	-	The prevention of RBC membrane lysis under hypotonic and heat stress suggests the anti-inflammatory potential of EPSs.	Ageeli and Mohamed (2025)
Bifidobacterium longum infantis	RAW264.7 cells, splenic lymphocytes, and NK cells of the mouse	-.	Decreases inflammatory mediators, downregulates NF-κB signaling, increases NK cell function and splenic lymphocyte proliferation, and improves the formation and phagocytic activity of RAW264.7 macrophages.	Yue et al. (2023)

(*continued on next page*)

The SS-01 strain of *Lacticaseibacillus paracasei*	UVB-HaCaT cell model	-	UVB irradiation significantly increases the release of inflammatory cytokines from HaCaT cells; however, LP-EPS efficiently inhibits this release.	Su et al. (2022)
Mycoplasma mycoides subsp. mycoides	Cattle macrophages and lymphocytes	-	It increases the secretion of IL-10 from bovine macrophages and decreases TNF-α and IL-12p40 produced by LPS.	Totté et al. (2015)
Streptococcus thermophilus	Caco-2 cells	Male BALB/c mice (6–8 weeks old)	Reduce intestinal inflammation by reducing pro-inflammatory cytokine levels, which also counteract decreased tight junction protein expression. It also maintains the intestinal barrier integrity against lipopolysaccharide-induced disruption in Caco-2 monolayers, by increasing tight protein junctions and reducing the pro-inflammatory response.	Chen et al. (2019)

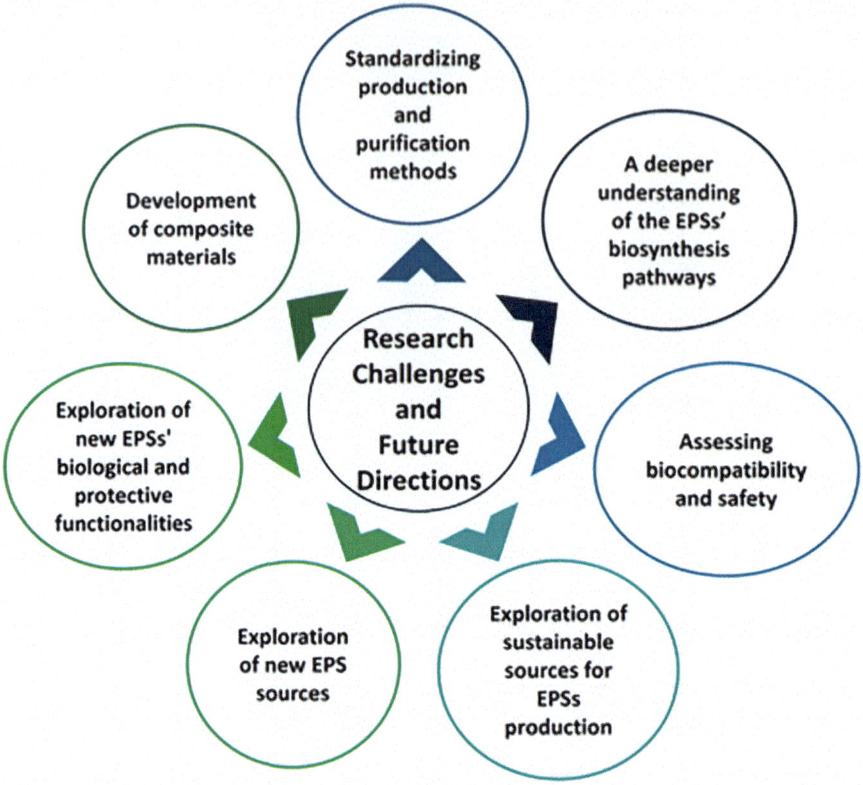

FIGURE 16.2 An illustration of the research issues and potential paths for the creation of functional EPS materials (Mouro et al., 2024). *EPS*, Exopolysaccharides.

wastes as substitute raw materials can significantly lower costs while improving sustainability. Incorporating these alternative raw materials promotes the utilization of renewable resources, which helps to address economic issues while also supporting environmental goals (Chavan et al., 2021; Liu et al., 2021; Mehariya et al., 2023; Możejko-Ciesielska et al., 2023a, 2023b; Chacón et al., 2024).

Apart from economic factors, the popularity of EPSs depends on their safety and biocompatibility, especially in vital areas like the food and medicinal sectors. In order to fulfill industrial standards, recent research has placed a strong emphasis on thorough evaluations of EPSs' toxicity profiles and regulatory compliance (Guerrero et al., 2024).

Future studies should keep looking into extremophilic bacteria to identify EPSs with special qualities like radiation protection and harsh condition resistance. Examining their pharmacological properties, such as their anti-inflammatory, antibacterial, antioxidant, and anticancer effects, EPSs could find more applications. To do this, investigating cutting-edge manufacturing processes like 3D printing and electro-spinning provides creative approaches to integrating EPSs into functional materials, allowing for the modification of material properties and extending their uses in a variety of industries.

Combining EPSs with other materials to build composite materials with better properties may also enable a new generation of useful EPS materials.

Therefore, even though there are still difficulties in producing and using EPS, new developments in technology and continuous study offer encouraging answers. Resolving these issues will increase EPSs' industrial viability and broaden their potential uses in a variety of industries.

16.7 Conclusion

EPSs derived from microorganisms in a harsh environmental condition are a type of natural carbohydrate biopolymer. They are eco-friendly and biodegradable. They are useful in areas like food, cosmetics, medicine, and other industries. Moreover, EPSs' exceptional biocompatibility and bioactive qualities enhance their efficacy as encapsulating agents, stabilizers, and perform many other functions due to their capacity to assume a range of topologies. Prominent microbial extracellular polymers, such as bacterial cellulose, dextran, levan, curdlan, pullulan, hyaluronic acid, xanthan gum, and gellan, have had important functional characteristics and several different uses in the past. Recent research, however, shows that EPSs made from extremophile bacteria have promising potential and unique defensive qualities like radiation protection and environmental tolerance. Their pharmacological properties, such as anti-inflammatory, antibacterial, antioxidant, and anticancer qualities, should be further studied in order to reveal more advantages and expand their usefulness in a variety of domains.

Our main focus was anti-inflammatory effect of the EPSs. EPS can be synthesized from various microorganisms and is effective against several inflammation of the body. It generally acts by inhibiting pro-inflammatory mediators, including IL-6, TNF-α, IL-12, and COX-2, and stimulates the generation of IL-10, an anti-inflammatory cytokine. Many in vitro and in vivo investigations confirmed that EPS, derived from microorganisms, has an anti-inflammatory effect.

Acknowledgments

The authors appreciate the e-resources and necessary facilities provided by University of North Bengal for completing this chapter.

References

Abdel-Wahab, B.A., F Abd El-Kareem, H., Alzamami, A., A Fahmy, C., H Elesawy, B., Mostafa Mahmoud, M., Ghareeb, A., El Askary, A., H Abo Nahas, H., G M Attallah, N., Altwaijry, N., M Saied, E., 2022. Novel exopolysaccharide from Marine *Bacillus* subtilis with broad potential biological activities: insights into antioxidant, anti-inflammatory, cytotoxicity, and anti-Alzheimer activity. Metabolites 12 (8), 715.

Abdl Aali, R.A.K., Al-Sahlany, S.T.G., 2024. Gellan gum as a unique microbial polysaccharide: its characteristics, synthesis, and current application trends. Gels 10 (3), 183.

Abdulkhaleq, L.A., Assi, M.A., Abdullah, R., Zamri-Saad, M., Taufiq-Yap, Y.H., Hezmee, M.N.M., 2018. The crucial roles of inflammatory mediators in inflammation: a review. Vet World 11 (5), 627–635.

Ageeli, A.A., Mohamed, S.F., 2025. Extraction, purification and characterization of exopolysaccharide from *Lactiplantibacillus plantarum* B7 with potential antioxidant, antitumor and anti-inflammatory activities. Processes 13 (4), 935.

Andrew, M., Jayaraman, G., 2020. Structural features of microbial exopolysaccharides in relation to their antioxidant activity. Carbohydr. Res. 487, 107881.

Angelin, J., Kavitha, M., 2025. Structural characterization and in vitro anti-inflammatory activity of exopolysaccharide produced by *Pediococcus pentosaceus* 4412. Int. Immunopharmacol. 150, 114301.

Arora, A., Sharma, P., Katti, D.S., 2015. Pullulan-based composite scaffolds for bone tissue engineering: improved osteoconductivity by pore wall mineralization. Carbohydr. Polym. 123, 180–189.

Benettayeb, A., Guibal, E., Bhatnagar, A., Morsli, A., Kessas, R., 2023. Effective removal of nickel (II) and zinc (II) in mono-compound and binary systems from aqueous solutions by application of alginate-based materials. Int. J. Env. Anal. Chem. 103 (9), 2016–2037.

Bibi, A., Xiong, Y., Rajoka, M.S.R., Mehwish, H.M., Radicetti, E., Umair, M., et al., 2021. Recent advances in the production of exopolysaccharide (EPS) from *Lactobacillus spp.* and its application in the food industry: a review. Sustainability 13 (22), 12429.

Casillo, A., Lanzetta, R., Parrilli, M., Corsaro, M.M., 2018. Exopolysaccharides from marine and marine extremophilic bacteria: structures, properties, ecological roles and applications. Mar. Drugs 16 (2), 69.

Chacón, M., Wongsirichot, P., Winterburn, J., Dixon, N., 2024. Genetic and process engineering for polyhydroxyalkanoate production from pre-and post-consumer food waste. Curr. Opin. Biotechnol. 85, 103024.

Chavan, S., Yadav, B., Tyagi, R.D., Drogui, P., 2021. A review on production of polyhydroxyalkanoate (PHA) biopolyesters by *Thermophilic* microbes using waste feedstocks. Bioresour. Technol. 341, 125900.

Chen, Y., Zhang, M., Ren, F., 2019. A role of exopolysaccharide produced by *Streptococcus thermophilus* in the intestinal inflammation and mucosal barrier in caco-2 monolayer and dextran sulphate sodium-induced experimental murine colitis. Molecules 24 (3), 513.

Côté, G.L., 2007. Flavorings and other value-added products from sucrose. Novel Enzyme Technology for Food Applications. Elsevier, pp. 243–269. doi:10.1533/9781845693718.2.243.

Díaz-Cornejo, S., Otero, M.C., Banerjee, A., Gordillo-Fuenzalida, F., 2023. Biological properties of exopolysaccharides produced by *Bacillus spp.* Microbiol. Res. 268, 127276.

Díaz-Montes, E., 2021. Dextran: Sources, structures, and properties. Polysaccharides 2 (3), 554–565.

Dinić, M., Pecikoza, U., Djokić, J., Stepanović-Petrović, R., Milenković, M., Stevanović, M., et al., 2018. Exopolysaccharide produced by probiotic strain *Lactobacillus paraplantarum* BGCG11 reduces inflammatory hyperalgesia in rats. Front. Pharmacol. 9, 1.

Domżał-Kędzia, M., Ostrowska, M., Lewińska, A., Łukaszewicz, M., 2023. Recent developments and applications of microbial levan, a versatile polysaccharide-based biopolymer. Molecules 28 (14), 5407.

Du, B., Yang, Y., Bian, Z., Xu, B., 2017a. Characterization and anti-inflammatory potential of an exopolysaccharide from submerged mycelial culture of *Schizophyllum* commune. Front. Pharmacol. 8, 252.

Du, B., Yang, Y., Bian, Z., Xu, B., 2017b. Molecular weight and helix conformation determine intestinal anti-inflammatory effects of exopolysaccharide from *Schizophyllum* commune. Carbohydr. Polym. 172, 68–77.

El Awady, M.E., Mohamed, S.S., Abo Elsoud, M.M., Mahmoud, M.G., Anwar, M.M., Ahmed, M.M., et al., 2024. Insight into antioxidant and anti-inflammatory effects of marine bacterial natural exopolysaccharide (EPSSM) using carrageenan-induced paw edema in rats. Sci. Rep. 14 (1), 5113.

Elnahas, M.O., Amin, M.A., Hussein, M.M., Shanbhag, V.C., Ali, A.E., Wall, J.D., 2017. Isolation, characterization and bioactivities of an extracellular polysaccharide produced from *Streptomyces sp.* MOE6. Molecules 22 (9), 1396.

Erginer, M., Akcay, A., Coskunkan, B., Morova, T., Rende, D., Bucak, S., et al., 2016. Sulfated levan from *Halomonas smyrnensis* as a bioactive, heparin-mimetic glycan for cardiac tissue engineering applications. Carbohydr. Polym. 149, 289–296.

Gao, Z., Wu, C., Wu, J., Zhu, L., Gao, M., Wang, Z., et al., 2022. Antioxidant and anti-inflammatory properties of an aminoglycan-rich exopolysaccharide from the submerged fermentation of *Bacillus thuringiensis.* Int. J. Biol. Macromol. 220, 1010–1020.

Guerrero, B.G., Santos, K.d.L., Kamimura, E.S., de Oliveira, C.A.F., 2024. Application of microbial exopolysaccharides in packaging films for the food industry: a review. Int. J. Food Sci. Technol. 59 (1), 17–29.

Hasan, H.A., Rahim, N.F.M., Alias, J., Ahmad, J., Said, N.S.M., Ramli, N.N., et al., 2024. Review on the roles of extracellular polymeric substances (EPSs) in wastewater treatment: source, mechanism study, bioproducts, limitations, and future challenges. Water 16 (19), 2812.

Hickey, A., Stamou, P., Udayan, S., Ramón-Vázquez, A., Esteban-Torres, M., Bottacini, F., et al., 2021. Bifidobacterium breve exopolysaccharide blocks dendritic cell maturation and activation of CD4+ T cells. Front. Microbiol. 12, 653587.

Hirano, T., 2021. IL-6 in inflammation, autoimmunity and cancer. Int. Immunol. 33 (3), 127–148.

Hundschell, C.S., Jakob, F., Wagemans, A.M., 2020. Molecular weight dependent structure of the exopolysaccharide levan. Int. J. Biol. Macromol. 161, 398–405.

Ibrahim, G.S., Mahmoud, M.G., EL Awady, M.E., Ahmed, M.M., Fathy, S.A., 2019. Characterization and anti-inflammatory activity effect of exopolysaccharide from *Bacillus axarquiensison* paw rats carrageenan model. J. Biosci. Appl. Res. 5 (4), 501–518.

Jindal, N., Khattar, J.S., 2018. Microbial polysaccharides in food industry. Biopolymers for Food Design. Elsevier, pp. 95–123. doi:10.1016/B978-0-12-811449-0.00004-9.

Kalpana, R., Angelaalincy, M.J., Kamatchirajan, B.V., Vasantha, V.S., Ashokkumar, B., Ganesh, V., et al., 2018. Exopolysaccharide from *Bacillus cereus* VK1: enhancement, characterization and its potential application in heavy metal removal. Colloids Surf. B 171, 327–334.

Karácsonyi, Š., Kuniak, Ľ., 1994. Polysaccharides of *Pleurotus ostreatus*: Isolation and structure of pleuran, an alkali-insoluble β-D-glucan. Carbohydr. Polym. 24 (2), 107–111.

Kidgell, J.T., Magnusson, M., de Nys, R., Glasson, C.R.K., 2019. Ulvan: a systematic review of extraction, composition and function. Algal Res. 39, 101422.

Kogan, G., Šoltés, L., Stern, R., Gemeiner, P., 2007. Hyaluronic acid: a natural biopolymer with a broad range of biomedical and industrial applications. Biotechnol. Lett. 29, 17–25.

Korcz, E., Varga, L., 2021. Exopolysaccharides from lactic acid bacteria: techno-functional application in the food industry. Trends Food Sci. Technol. 110, 375–384.

Kumar, S., Kaushik, D., Sharma, S.K., 2024. Autoimmune disorders: types, symptoms, and risk factors. Artificial Intelligence and Autoimmune Diseases. Studies in Computational Intelligence. Springer, Singapore, pp. 3–31. doi:10.1007/978-981-99-9029-0_1.

Kwon, M., Lee, J., Park, S., Kwon, O.-H., Seo, J., Roh, S., 2020. Exopolysaccharide isolated from *Lactobacillus plantarum* L-14 has anti-inflammatory effects via the toll-like receptor 4 pathway in LPS-induced RAW 264.7 cells. Int, J. Mol, Sci. 21 (23), 9283.

Le Costaouëc, T., Cérantola, S., Ropartz, D., Ratiskol, J., Sinquin, C., Colliec-Jouault, S., et al., 2012. Structural data on a bacterial exopolysaccharide produced by a deep-sea *Alteromonas macleodii* strain. Carbohydr. Polym. 90 (1), 49–59.

Li, C., Chen, D., Ding, J., Shi, Z., 2020. A novel hetero-exopolysaccharide for the adsorption of methylene blue from aqueous solutions: isotherm, kinetic, and mechanism studies. J. Clean. Prod. 265, 121800.

Li, L.-Q., Song, A.-X., Wong, W.-T., Wu, J.-Y., 2021. Modification and enhanced anti-inflammatory activity by Bifidobacterial fermentation of an exopolysaccharide from a medicinal fungus Cs-HK1. Int, J. Biol, Macromol. 188, 586–594.

Li, L.-Q., Song, A.-X., Yin, J.-Y., Siu, K.-C., Wong, W.-T., Wu, J.-Y., 2020. Anti-inflammation activity of exopolysaccharides produced by a medicinal fungus Cordyceps sinensis Cs-HK1 in cell and animal models. Int. J. Biol. Macromol. 149, 1042–1050.

Lin, Q., Si, Y., Zhou, F., Hao, W., Zhang, P., Jiang, P., et al., 2024. Advances in polysaccharides for probiotic delivery: properties, methods, and applications. Carbohydr. Polym. 323, 121414.

Liu, H., Kumar, V., Jia, L., Sarsaiya, S., Kumar, D., Juneja, A., et al., 2021. Biopolymer poly-hydroxyalkanoates (PHA) production from apple industrial waste residues: a review. Chemosphere 284, 131427.

Mehariya, S., Plöhn, M., Jablonski, P., Stagge, S., Jönsson, L.J., Funk, C., 2023. Biopolymer production from biomass produced by Nordic microalgae grown in wastewater. Bioresour. Technol. 376, 128901.

Mehta, P.P., Jadhav, P.D., 2023. Green composites materials as a carrier for pulmonary drug delivery. Green Sustainable Process for Chemical and Environmental Engineering and Science. Elsevier, pp. 131–186. doi:10.1016/B978-0-323-95169-2.00014-6.

Mouro, C., Gomes, A.P., Gouveia, I.C., 2024. Microbial exopolysaccharides: structure, diversity, applications, and future frontiers in sustainable functional materials. Polysaccharides 5 (3), 241–287.

Możejko-Ciesielska, J., Moraczewski, K., Czaplicki, S., Singh, V., 2023. Production and characterization of polyhydroxyalkanoates by *Halomonas alkaliantarctica* utilizing dairy waste as feedstock. Sci. Rep. 13 (1), 22289.

Możejko-Ciesielska, J., Ray, S., Sankhyan, S., 2023. Recent challenges and trends of polyhydroxyalkanoate production by extremophilic bacteria using renewable feedstocks. Polymers 15 (22), 4385.

Netrusov, A.I., Liyaskina, E.V., Kurgaeva, I.V., Liyaskina, A.U., Yang, G., Revin, V.V., 2023. Exopolysaccharides producing Bacteria: a review. Microorganisms 11 (6), 1541.

Ponnusami, V., Gunasekar, V., 2015. Production of pullulan by microbial fermentation. Polysaccharides. Springer, Cham, pp. 581–596. doi:10.1007/978-3-319-16298-0_58.

Prajapati, V.D., Jani, G.K., Zala, B.S., Khutliwala, T.A., 2013. An insight into the emerging exopolysaccharide gellan gum as a novel polymer. Carbohydr. Polym. 93 (2), 670–678.

Prasad, S., Purohit, S.R., 2023. Microbial exopolysaccharide: sources, stress conditions, properties and application in food and environment: a comprehensive review. Int, J. Biol, Macromol. 242, 124925.

Prete, R., Dell'Orco, F., Sabatini, G., Montagano, F., Battista, N., Corsetti, A., 2024. Improving the antioxidant and anti-inflammatory activity of fermented milks with exopolysaccharides-producing *Lactiplantibacillus plantarum* strains. Foods 13 (11), 1663.

Qi, M., Zheng, C., Wu, W., Yu, G., Wang, P., 2022. Exopolysaccharides from marine microbes: source, structure and application. Mar. Drugs 20 (8), 512.

Radchenkova, N., Vassilev, S., Panchev, I., Anzelmo, G., Tomova, I., Nicolaus, B., et al., 2013. Production and properties of two novel exopolysaccharides synthesized by a thermophilic bacterium *Aeribacillus pallidus* 418. Appl. Biochem. Biotechnol. 171, 31–43.

Rahman, H.U., Asghar, W., Khalid, N., 2021. *Scleroglucan* and *Schizophyllan*: microbial polysaccharides of functional importance. Polysaccharides of Microbial Origin: Biomedical Applications. Springer, pp. 1–28. doi:10.1007/978-3-030-35734-4_16-1.

Rana, S., Upadhyay, L.S.B., 2020. Microbial exopolysaccharides: synthesis pathways, types and their commercial applications. Int. J. Biol. Macromol. 157, 577–583.

Sahana, T., Rekha, P., 2020. A novel exopolysaccharide from marine bacterium Pantoea sp. YU16-S3 accelerates cutaneous wound healing through Wnt/β-catenin pathway. Carbohydr. Polym. 238, 116191.

Saravanan, C., Rajesh, R., Kaviarasan, T., Muthukumar, K., Kavitake, D., Shetty, P.H., 2017. Synthesis of silver nanoparticles using bacterial exopolysaccharide and its application for degradation of azo-dyes. Biotechnol. Rep. 15, 33–40.

Silveira, M.L., Smiderle, F.R., Agostini, F., Pereira, E.M., Bonatti-Chaves, M., Wisbeck, E., et al., 2015. Exopolysaccharide produced by Pleurotus sajor-caju: its chemical structure and anti-inflammatory activity. Int. J. Biol. Macromol. 75, 90–96.

Srikanth, R., Reddy, C.H.S., Siddartha, G., Ramaiah, M.J., Uppuluri, K.B., 2015. Review on production, characterization and applications of microbial levan. Carbohydr. Polym. 120, 102–114.

Su, Y., Zhang, Y., Fu, H., Yao, F., Liu, P., Mo, Q., et al., 2022. Physicochemical and anti-UVB-induced skin inflammatory properties of *Lacticaseibacillus paracasei subsp. paracasei* SS-01 strain exopolysaccharide. Fermentation 8 (5), 198.

Thakur, A., Sharma, S., Naman, S., Baldi, A., 2023. Pullulan based polymeric novel drug delivery systems: a review on current state of art and prospects. J. Drug Deliv. Sci. Technol. 90, 105117.

Totté, P., Puech, C., Rodrigues, V., Bertin, C., Manso-Silvan, L., Thiaucourt, F., 2015. Free exopolysaccharide from *Mycoplasma mycoides* subsp. mycoides possesses anti-inflammatory properties. Vet. Res. 46 (1), 122.

Wang, J., Zhang, J., Guo, H., Cheng, Q., Abbas, Z., Tong, Y., et al., 2023. Optimization of exopolysaccharide produced by *Lactobacillus plantarum* R301 and its antioxidant and anti-inflammatory activities. Foods 12 (13), 2481.

Wang, W., Ju, Y., Liu, N., Shi, S., Hao, L., 2023. Structural characteristics of microbial exopolysaccharides in association with their biological activities: A review. Chem. Biol. Technol. Agric. 10 (1), 137.

Wang, Y., Zhang, T., Zhu, L., Li, R., Jiang, Y., Li, Z., et al., 2024. Optimization of welan gum extraction and purification using lysozyme and alkaline protease. Appl. Microbiol. Biotechnol. 108 (1), 70.

Wzorek-Łyczko, K., Piwowarczyk, A., Kuchar, E., 2021. Protocol of the study: the effectiveness of pleuran in the treatment of acute gastroenteritis in children—a randomised, placebo-controlled, double-blind trial (EPTAGE). BMJ Open 11 (3), e042370.

Xiong, J., Liu, D., Huang, Y., 2023. Exopolysaccharides from *Lactiplantibacillus plantarum*: Isolation, purification, structure–function relationship, and application. Eur. Food Res. Technol. 249 (6), 1431–1448.

Xu, Q., Wang, M.-M., Li, X., Ding, Y.-R., Wei, X.-Y., Zhou, T., 2024. Antioxidant and anti-inflammatory activities and action mechanisms of exopolysaccharides from *Lactiplantibacillus plantarum* Z-1. Food Biosci. 62, 105247.

Yue, Y., Wang, Y., Yang, R., Liu, D., Cheng, Y., Li, S., et al., 2023. Exopolysaccharides of *Bifidobacterium longum* subsp. infantis E4 on the immune and anti-inflammatory effects in vitro. J. Funct. Foods 107, 105699.

Zamora-Pineda, J., Kalinina, O., Sperling, A.I., Knight, K.L., 2023. Mechanism of TLR4-mediated anti-inflammatory response induced by exopolysaccharide from the probiotic *Bacillus* subtilis. J. Immunol. 211 (8), 1232–1239.

Zampieri, R.M., Adessi, A., Caldara, F., Codato, A., Furlan, M., Rampazzo, C., et al., 2020. Anti-inflammatory activity of exopolysaccharides from *Phormidium sp.* ETS05, the most abundant cyanobacterium of the therapeutic Euganean thermal muds, using the zebrafish model. Biomolecules 10 (4), 582.

Zhang, J., Zhang, H., Xiao, Y., Wang, H., Zhang, H., Lu, W., 2024. Interspecific differences and mechanisms of *Lactobacillus*-derived anti-inflammatory exopolysaccharides. Int. J. Biol. Macromol. 263, 130313.

Zhou, R., Qian, Y., Lei, Z., Tang, Y., Li, Y., 2023. Production and characterization of exopolysaccharides from salinity-induced *Auxenochlorella protothecoides* and the analysis of anti-inflammatory activity. Int. J. Biol. Macromol. 240, 124217.

CHAPTER

17

EPS: A potential application in bone tissue engineering for repairing and regeneration

Subhajit Mandal[a,b], *Tanmay Jit*[c], *Ranabir Sahu*[b], *Gouranga Nandi*[b] and *Partha Palit*[c]

[a]Gitanjali College of Pharmacy, Birbhum, West Bengal, India [b]Department of Pharmaceutical Technology, University of North Bengal, Darjeeling, West Bengal, India [c]Department of Pharmaceutical Sciences, Assam University (A Central University), Silchar, Assam, India

17.1 Introduction

17.1.1 Overview of bone tissue engineering

The dynamic equilibrium between the interests of osteoblasts, which create new bone, and osteoclast-mediated bone absorption is necessary for the physiological process of bone remodeling (Bauso et al., 2024). Small skeletal drawbacks are typically fixed independently, as bone is a self-healing tissue; however, severe bone lesions hinder bone regeneration. About 1.71 billion people worldwide suffer from several bone defects brought on by trauma, osteoporosis, ageing, degenerative diseases, autoimmune diseases, or tumor removal, etc. These flaws are a primary cause of disability (Cieza et al., 2020; Wu et al., 2021). Since pathological fragility shattering is often confused and causes severe morbidity or death, it is generally acknowledged that the situation is a real public health crisis due to the significant economic and civil expenses involved, especially in treating these injuries (Sözen et al., 2016). Finding novel therapeutic policies that might enhance patients' quality of life while averting drawbacks, including donor site morbidity, pain, rejection, disease mediation, and high costs, is thus crucial. The tissue engineering (TE) shelter could be the biggest practical replacement to bone rebuilding if these disadvantages are removed and clinical necessities are met (Delpierre et al., 2023). Bone tissue engineering (BTE), an emerging and attractive therapy approach for bone abnormalities, is based on the development of biomaterials that encourage tissue regeneration. But even though a lot of effort

DOI: https://doi.org/10.1016/B978-0-443-36619-2.00017-6

has been put into developing biomaterials that can satisfy all therapeutic needs in recent years, however the ideal material has not yet been discovered (Bauso et al., 2024).

17.1.2 Role of biomaterials in bone repair and regeneration

Replicating natural tissue with biomaterials and bioactive ingredients may maintain, improve, or restore tissue function. In order to provide the new bone structure and the structural support it requires, BTE targets to build three-dimensional scaffolds.

Typically, adding growth hormones and cells speeds up the induction of osteogenic lineage. When intact, bone tissue serves vital purposes for the human body. Because it gives the body mechanical support, it is intimately tied to movement. Furthermore, bone oversees preserving mineral homeostasis and serves as the primary barrier of protection for important organs due to its stiffness. Being a dynamic, multipurpose tissue, bones are prone to damage that might impair their functionality. Orthopedic and dental bone abnormalities are common problems that may be caused by neoplasms, infections, trauma, congenital diseases, or just ageing. Grafts are thus necessary to repair damaged tissue and ensure that the implant and host bone form a strong relationship. The optimum possible constructional environment for the cell engaged in bone regeneration must be provided by the materials used in this phenomenon.

Autogenous bone is appraising the "gold standard" for bone development for its osteogenic, osteoconductive, and osteoinductive properties. Its disadvantages, such as its reliance on bone supply and the potential for vascular-nervous lesions and noxious in the recipient bed, restrict its utilization. Less encroaching alternatives include xenogeneic, alloplastic, and allogeneic bone transplants (Girón et al., 2021). The restriction of xenogenous grafts comprises the possibility of rejection or disease transmission, their limited ability to completely integrate with the local bone, and their prolonged presence in the implanted location. Reduced osteoinductive qualities are shown by allogeneic bone, and immunoreactions and infection transmission are possible with such transplants. Alloplastic biomaterials, which include metals, ceramics, and polymers that may be linked to cells or growth factors, have been expanded for research in TE. Because of its benefits, like its ease of handling, wide range of sizes and forms, and plenty of materials for production, this kind of graft has been studied. Additionally, the risk of surgical morbidity, disease transmission, and lack of antigenicity are reduced when alloplastic biomaterials are used (Dimitriou et al., 2011; Rolvien et al., 2017)

Because of the biocompatibility, biodegradability, and capacity to replicate the extracellular matrix (ECM) microbial exopolysaccharide (EPS), which are high molecular weight polysaccharides released by microbes, have drawn a lot of interest as potential alternative biomaterials for bone repair and regeneration. In addition to supporting osteoblast adhesion, proliferation, and differentiation, EPS may function as three-dimensional scaffolds. Mechanical strength and osteconductivity can be improved by mixing it with other materials, such as gellen gum or hydroxyapatite (HA) (Waoo et al., 2023; Mouro et al., 2024).

17.1.3 Exopolysaccharides as emerging biomaterials

EPSs are a heterogeneous set of biopolymers produced by microbes in reaction to environmental stressors, including light intensity, temperature, pH, and salt. They provide eco-friendly and biodegradable substitutes for synthetic polymers. EPS are intriguing biopolymers synthesized by many microorganisms, such as yeast, fungi, bacteria, and microalgae. Furthermore, because of their supreme qualities such as biodegradability, biocompatibility, and the potential to form fibers, films, and membrane natural polymers like chitosan and chitin, which are developed in the exoskeletons of arthropods as well as the cell walls of some yeast and fungi, make a substantial contribution to the field of natural polysaccharides. Chitin and chitosan are mainly obtained from marine sources and have important uses in biomedical, food, cosmetic, and pharmaceutical applications. This is in contrast to EPS, which microorganisms produce and expel in relation to a variety of environmental stressors like salinity, chemical substances, temperature changes, pH variations, and radiation (Rinaudo, 2006; Mouro et al., 2024). Natural polymers, such as EPSs, are essential for the survival and adaptation of microbes in challenging environments. Besides their exceptional capacity to flourish in adverse conditions, EPSs have other benefits compared to synthetic polymers. Their organic nature renders them biodegradable and ecologically sustainable, in contrast to petroleum-derived polymers that may last in the environment for extended periods. This quality promotes EPSs as ecological and sustainable options across many research and industrial applications (Salimi and Farrokh, 2023; Waoo et al., 2023).

Moreover, EPSs may assume many shapes, such as capsular and grin, each serving distinct roles in cellular defense. Capsular structures provide a defensive barrier over the cell. On the other hand, slime EPS can be released into their surroundings as a thick, sticky substance that helps with sticking to surfaces, forming colonies, and shielding against environmental challenges. This adaptability enables EPSs to be fabricated into many shapes, including hydrogels, sponges, biodegradable films, and nanocapsules for the regulated fluency of chemicals, over other uses (Angelin and Kavitha, 2020; Saha and Datta, 2022).

Future research should focus on optimizing EPS-based biomaterials for enhanced osteoconductivity, mechanical strength, and biocompatibility, enabling sustainable, effective, and clinically translatable bone TE applications (Fig.17.1).

17.2 Extracellular polysaccharides (EPS): Structure and properties

17.2.1 Classification of exopolysaccharides

Extracellular polymer (EPS) are typically divided into two categories: Heteropolysaccharides, which are composed of two units of distinct monosaccharides, such as fructose, glucose, fucose, galactose, rhamnose, mannose, uronic acids, acetylglucosamine, and homopolysaccharides, which can be split and consist of a single kind of monosaccharide, like fructose or glucose, joined by glycosidic bonds. β-D-glucans (bacterial cellulose), α-D-glucans (reuteran, dextran, and alternan), fructans (inulin and levan), and polygalactans are the distinct types of homopolysaccharides. Polymers like hyaluronic acid, alginate, and xanthan are a variety of heteropolysaccharides. The phylogenetic classification of EPS-producing bacteria includes Gram-negative bacteria belonging to the *Alphaproteobacteria* class, which includes the genera *Acetobacter, Gluconacetobacter, Gluconobacter, Komagataeibacter, Neoasaia, Kozakia, Agrobacterium, Zymomonas,* and *Rhizobium*; the *Betaproteobacteria* class, which includes the genera *Achromobacter,* and *Alcaligenes*; and the *Gammaproteobacteria* class, which includes the genera *Pseudomonas, Azotobacter, Xanthomonas, Enterobacter, Alteromonas, Halomonas, Pseudoalteromonas, Vibrio, Erwinia,* and *Klebsiella. Bacilli* (e.g., *Lactobacillus, Bacillus, Paenibacillus, Streptococcus,* and *Leuconostoc*

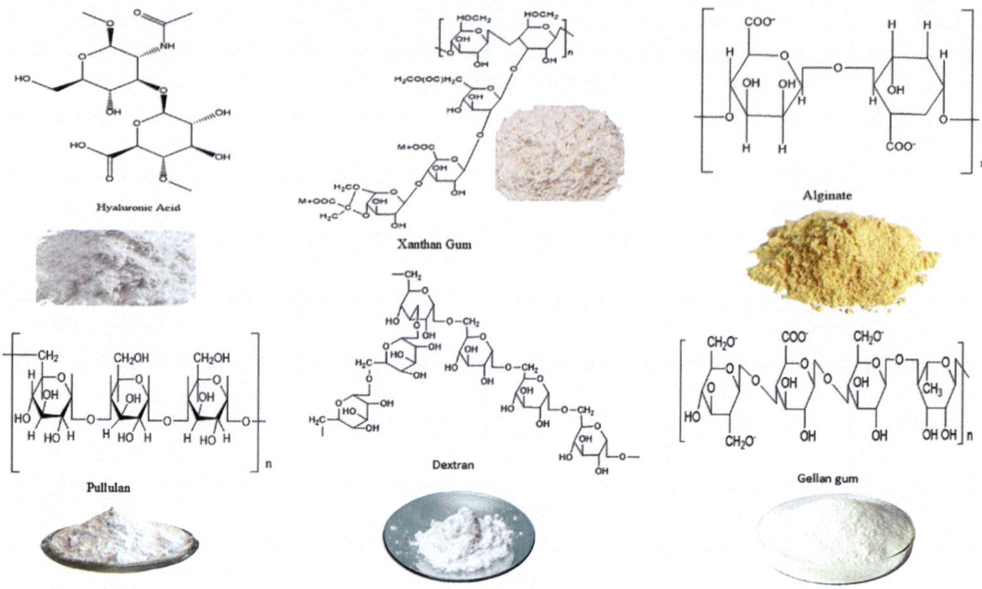

FIGURE 17.1 Chemical compositions of the most prevalent exopolysaccharides.

genera), *Actinomycetia* (e.g., *Rhodococcus genera, Bifidobacterium*), and *Clostridium* (e.g., *Sarcina sp.*) are among the gram-positive bacteria that make up their scaffolds (Netrusov et al., 2023; Kumari et al., 2025).

17.2.2 Physicochemical and biological properties relevant to bone tissue engineering

Long-chain polysaccharides include branching and repeating units of sugars or their derivatives, including galactose, glucose, rhamnose, and fructose in varying amounts, and are known as EPS, which are metabolic byproducts of microorganisms. Extreme conditions such as antibiotics, osmotic stress, phage attacks, dehydration, and toxic chemicals are all prevented by EPS for microbial cells. Additionally, they take part in the microbial cells' communication process. EPS vary in terms of source, molecule size, structure, and charge. For EPS production, a range of sources was investigated, including eukaryotes and prokaryotes, including fungi, eubacteria, algae, and archaebacteria. EPS may be extracted and purified utilizing a variety of methods, such as various chromatography systems, ultrafiltration, salting out, and ultracentrifugation. The degree of molecular weight, polymerization, and structural composition of EPS are physicochemical properties that govern its technical characteristics, which include gelling, stabilization, thickening, and water binding. As stabilizers, emulsifiers, gelling agents, viscosity enhancers, and water-binding agents, EPS improves texture, consistency, and mouthfeel while decreasing syneresis in a variety of culinary products, such as beverages, jams, yoghurt, jellies, and bread. Because of its special qualities and health-promoting benefits, EPS is widely utilized in different industrial areas, like wastewater treatment, food additives, cosmetics, medicines, textiles, adhesives, etc. EPS have antioxidant, antibacterial, cholesterol-lowering, anticancer, anticoagulant, and immunoregulatory-boosting properties (Ali et al., 2023).

17.2.3 Biocompatibility, biodegradability, and functionalization potential

Since these natural polysaccharides are more accessible, biodegradable, nontoxic, and environmentally benign upon synthetic polymers, they have great attention in tissue engineering and the creation of biomaterials. Because of their broad biocompatibility and advantageous mechanical properties, eco-friendly polysaccharide-based biomaterials are used in the production of artificial grafts, tissue scaffolds, implantation, drug delivery, wound closure, bone filler materials, etc. Crucially, they are smoothly absorbed by the tissues and cells involved in the human body. In the search for efficient biocompatible eco-friendly polysaccharides, bacterial EPS have been the focus of research due to their sustainability, speed of synthesis, and environmental friendliness. High-molecular-weight organic biomacromolecules known as extracellular polymers (EPSs) are released into the environment by the host organism. They have a role in defense, phagocytosis, symbiosis, signaling, cell growth via nutrient trapping, and desiccation prevention. In general, uronic acids, a variety of monosaccharide compositions, many functional groups, and a few noncarbohydrate components can make up EPSs (Andrew and Jayaraman, 2022).

Future studies should enhance EPS functionalization, tailoring their physicochemical properties for improved biocompatibility, biodegradability, and mechanical performance to advance their application in sustainable bone TE and regenerative medicine.

17.3 Challenges in bone tissue engineering

17.3.1 Limitations of current synthetic and natural polymers

Collagen, chitosan, alginate, gelatin, silk fibroin, hyaluronic acid, and cellulose are instances of natural polymers that have outstanding biodegradability, biocompatibility, and bioactivity. Nevertheless, they have several significant drawbacks, including inflated cost and processing difficulties, uncontrolled degradation, contamination, and immunogenicity risks, poor mechanical strength, and stability, etc. Like natural polymers, synthetic polymers (PLA, PCL, PGA, PEG, PLGA, and PEEK) have additional unique problems. These include poor long-term durability, adverse degradation byproducts, balancing porosity and mechanical strength, low bioactivity, and lack of cellular signals (Donnaloja et al., 2020; Alkaron et al., 2024).

17.3.2 Need for bioactive and bioinspired materials

Collagen (organic phase) and HA (inorganic phase) makes up most bone, a dynamic, hierarchical composite tissue with exceptional self-healing properties. The biological signals necessary for integration with host tissue or for promoting bone development are often absent from conventional synthetic materials used in bone transplants, such as metals or inert ceramics. Thus, for a number of reasons, bioactive and bioinspired materials have become essential parts of contemporary BTE: Better mechanical compatibility, increased osteointegration and osteoinduction, stimulation of cellular activity, biodegradability, remodeling, and mimicking the architecture of natural bones (Stevens, 2008; Lee et al., 2014).

Future research should focus on developing cost-effective, bioactive, and bioinspired materials that replicate bone's hierarchical structure, enhance osteointegration, and balance biodegradability with mechanical strength for advanced regenerative applications.

17.4 Exopolysaccharides in bone tissue engineering applications

17.4.1 Scaffold fabrication using exopolysaccharides

Tissue scaffolds obtain a structural foundation to which cells may cling when they are introduced to the matrix, eventually filling it. An ideal scaffold may store growth factors, cytokines, and nutrients for cell division. The tissue scaffold's architecture is an important part of its design as it influences cell attachment, migration, the management of cell phases that have mechanical and biological effects, and the modification of the diffusion of vital cell nourishment. Additionally, by improving mechanical contact among the implanted biomaterial and neighboring native bone, a porous surface raises mechanical stability at this critical point (Thakur et al., 2024).

17.4.2 Exopolysaccharides-based hydrogels and nanocomposites

Hydrogels have special qualities like high-water content, mechanical qualities, and biocompatibility that may be changed, and hydrogels have created an interest in TE. Because the three-dimensional structure of cross-linked polymer chains closely mimics the ECM of actual tissues, they provide the perfect environment for tissue regeneration and cell proliferation.

Among the distinctive features of nanoclays like montmorillonite, loponite, and halloysite are their enormous surface area, intrinsic biocompatibility, and high aspect ratio. Nanoparticles may be dispersed across hydrogel networks for the enhancement of mechanical properties, increase bioactivity, and improve swelling behavior, making them an attractive option for bone TE applications.

There are several benefits of adding nanoclay to hydrogels for bone repair. Hydrogels' mechanical qualities might be greatly improved by nanoclay reinforcement, reaching a mechanical strength that is on par with that of real bone tissue (Hwang and Lee, 2024; Tiwari et al., 2024; Chowdhury et al., 2025).

17.4.3 Cell adhesion, proliferation, and differentiation support

When included in scaffold materials, extracellular proteins like collagen type I, laminin, vitronectin, and fibronectin improve the adherence of human mesenchymal stem cells (hMSCs) generated since bone marrow via integrin-mediated signaling, which is essential for efficient bone regeneration (Mathews et al., 2012). It has been shown that scaffolds covered with ECM-like elements, such as decellularized bone matrix incorporated into PLGA scaffolds, support the ongoing proliferation of hESCs, with an increase in cell numbers over time. Mesenchymal stem cells (MSCs) lose some of their ability to proliferate in composite scaffolds with mineralized ECM deposited by MSCs under flow perfusion, but they exhibit markedly increased early markers

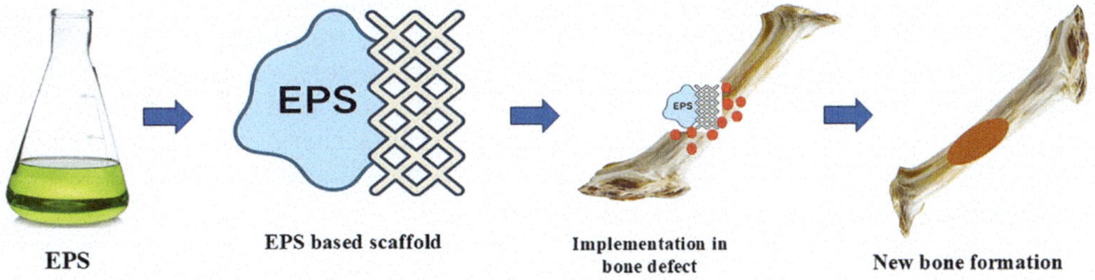

FIGURE 17.2 Regeneration of new bone tissues using EPS. *EPS*, Exopolysaccharides.

of differentiation (such as alkaline phosphatase activity), illustrating the trade-off between proliferation and lineage commitment facilitated by the existence of ECM (Liao et al., 2010; Rutledge et al., 2014).

Future studies should optimize EPS-based hydrogels and nanocomposite scaffolds to enhance mechanical strength, bioactivity, and cell differentiation, advancing their clinical translation for efficient bone regeneration and TE applications (Fig. 17.2).

17.5 Mechanism of exopolysaccharides mediated bone regeneration

17.5.1 Osteoinduction and osteoconduction

The technique by which bone develops from existing bone into a deficit using a scaffold that acts as a passive framework is known as osteoconduction. The porosity, connectivity, and surface topography of the scaffold are critical for promoting osteoblast migration, vascularization, and nutrient flow. EPS-based scaffolds may copy the ECM to guide the formation of bone tissue while promoting cell proliferation, cell adhesion, and differentiation. An example, EPS hydrogels and composite matrices with interconnected pores have been shown to have enhanced osteoconductive properties. Osteoinduction involves attracting MSCs and forcing them to develop into osteoblasts using growth factors like TGF-β, Wnt signaling proteins, and bone morphogenic proteins. Because EPS may serve as bioactive carriers and bind, protect, and release these growth factors, they can have a long-lasting osteogenic impact. Additionally, certain EPS have the capacity to modify the immune milieu, which reduces inflammation and promotes bone formation (Hutmacher, 2000; Wang et al., 2022b).

17.5.2 Interaction with cellular pathway

Bone morphogenetic proteins (BMPs), namely BMP-4, BMP-2, and BMP-7, bind to serine/threonine kinase receptors, activating Smad1/5/8 proteins. By encouraging the production of osteogenic genes, they go to the nucleus and boost osteoblast formation (Runx2, Osterix). EPS may function as sustained-release carriers for BMPs, halting their rapid degradation and increasing their local concentration at the bone defect site. When Wnt proteins bind to the Frizzled/LRP5/6 receptors, osteogenic transcription programs are triggered, and β-catenin accumulates. This pathway not only inhibits bone resorption but also stimulates bone formation.

EPS scaffolds with ECM-like surfaces may enhance interactions between cells and the matrix by indirectly activating Wnt signaling in MSCs. Integrins are transmembrane receptors that bind to ECM proteins. This binding sets off downstream processes that regulate cell adhesion, migration, and survival, including MAPK, PI3K/Akt, and focal adhesion kinase (FAK). Functional groups or RGD-like motifs found in natural polysaccharides that mimic EPS, such as chitosan and hyaluronic acid, interact with integrins to enhance osteoblast adhesion and spreading (Baron and Kneissel, 2013; Yuan et al., 2015).

17.5.3 Mineralization and extracellular matrix mimicking

Bone mineralization is the process by which hydroxyapatite crystals are formed within the collagen matrix, giving bone its stiffness and strength. Osteocalcin, osteopontin, and bone sialoprotein are examples of ECM proteins that control this process, which is mediated by osteoblasts. The hydroxyl, carboxyl, and phosphate functional groups of EPS allow it to act as a nucleation site for HA crystals. The capacity of phosphorylated and sulfated EPS to bind calcium ions may facilitate calcium-phosphate precipitation on the scaffold surface. Certain EPS, such as bacterial cellulose and alginate, may mineralize in vitro to form bioactive, bone-like composites with enhanced osteoconductive properties. The ECM provides a three-dimensional environment that stimulates cell adhesion, migration, proliferation, and differentiation via biochemical cues and mechanical properties. Since EPS is like naturally occurring ECM polysaccharides (such as hyaluronic acid, glycosaminoglycans, and chondroitin sulphate), it may interact with integrins and cell-surface receptors to change cellular function. Porous or nanofibrous features that resemble the ECM's topography may be included in EPS scaffolds to enhance cell–matrix interactions. By mimicking ECM rigidity and ligand presentation, EPS may influence osteogenesis and regulate mechanotransduction pathways, such as YAP/TAZ signaling (Rosellini et al., 2018).

Future research should explore EPS-mediated modulation of osteogenic pathways and ECM mimicry to enhance mineralization, osteoinduction, and osteoconduction, enabling next-generation bioactive scaffolds for efficient bone regeneration.

17.6 Examples of exopolysaccharides-based materials in bone repair

17.6.1 Alginate-based scaffolds

Alginate is a hydrophilic, anionic polymer that occurs naturally, mostly made from bacteria and brown seaweed, and is one of the most prevalent biosynthesized compounds. Blocks of α-L-guluronic acid (G) and (1 - 4)-linked β-D-mannuronic acid (M) monomers are found in alginate (Sun and Tan, 2013) (Fig.17.3).

One of the biggest problems in reconstructive surgery is bone regeneration. Lack of bone tissue may result from several causes, including trauma and tumor excision. Inducing osteogenesis in situ is a desired method of bone tissue healing. One method for creating bone tissue is to seed stem cells that can differentiate into bone tissue into an injectable scaffold. As a consequence, several studies have looked at the application of injectable scaffolds based on alginate for bone regeneration. When MSCs and alginate were used as a scaffold, enough bone tissue development was seen. Thus, alginate may be used to create tissue in gels and exhibits both angiogenic and osteogenic qualities (Xia et al., 2012). Numerous studies have

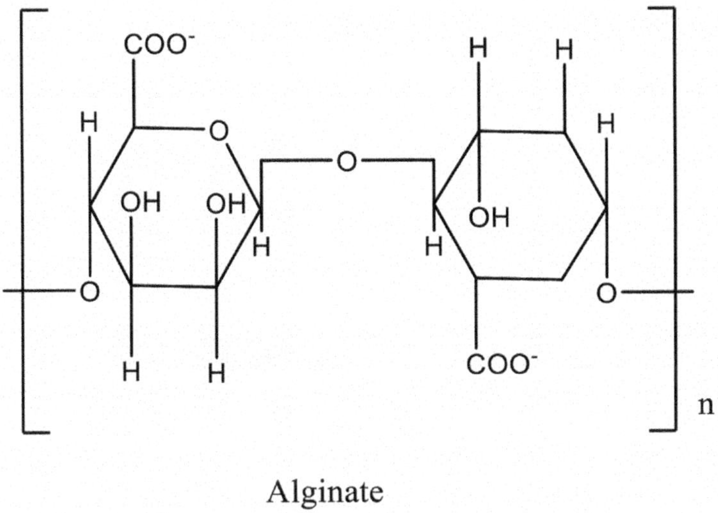

FIGURE 17.3 Chemical structure of alginate (Sun and Tan, 2013).

shown bone regeneration using injectable scaffolds that combine undifferentiated MSCs or ASCs with hydrogels or microspheres based on alginate. These findings showed that BMP and TGF-β administration may cause MSCs and ASCs to undergo osteogenic differentiation into mature osteocytes (Florczyk et al., 2012; Xia et al., 2012).

17.6.2 Chitosan and its derivatives

A deacetylated derivative of chitin, chitosan (CS), is mostly derived from crab exoskeletons. This linear polymer is made up of units that are dispersed at random, specifically: NAG, which is 2-acetamido-2-deoxy-β-D-glucan or (1 - 4). N-acetyl-D-glucosamine, and (1 - 4) 2-amino-2- 1, dideoxy-β-D-glucan (D-glucosamine) is connected by β (1 - 4) links.

The molar ratio of the D-glucosamine units to the sum of the NAG and D-glucosamine units indicates the degree of deacetylation. CS may be depolymerized by enzymes such as lysozyme, lipases, and glucosaminidases. Chitosan has structural similarities with glycosaminoglycans (GAG), one of the modules of the ECM that interacts with collagen fibers and is essential for cell-cell adhesion.

After depolymerization, chitosan produces bioactive chito-oligosaccharides with strong antimicrobial qualities, as well as its monomeric products, such as glucosamine, which the body metabolizes or eliminates. As a result, CS is highly biocompatible with almost every tissue in the body and biodegradable. Significant osteoconductivity but little osteoinductive activity has been shown by chitosan. It causes mesenchymal and osteoblast cells to proliferate, as well as in vivo neovascularization. CS is manufactured in a variety of geometries, including sponges, fibers, films, and other intricate structures, for orthopedic purposes (LogithKumar et al., 2016).

17.6.3 Pullulan, gellan gum, and dextran-based systems

To straight bone regeneration, a nanogel membrane composed of acryloyl and cholesteryl binding pullulan was assessed. In a rat, calvarial defect, this nanogel membrane and a commercially available collagen membrane were tested, and both promoted bone repair in comparison to the control, where no membrane was used. The pullulan-associated nanogel membrane greatly improved bone regeneration earlier, and after two weeks of surgery, newly produced bone was less mature (Miyahara et al., 2012).

Systems based on gellan gum are showing promise as materials for bone regeneration and repair because of their biodegradability, biocompatibility, and even adaptability to particular uses. These systems, which are often made as hydrogels, may imitate the ECM and encourage osteoblasts (bone-forming cells) to proliferate, adhere, and differentiate. In order to further promote bone repair and treat any complications like infection, they may also be made to supply growth factors and antibiotics (Xu et al., 2020).

Several studies mixed this polysaccharide with synthetic or natural polymers to assess the bone regeneration potential of biomaterials produced from dextran. For instance, dextran has been applied to polyvinyl formal (PVF) sponges to improve the synthetic polymer's low osteogenic potential because of its fibrous structure. A rat femoral defect's bone formation was greatly enhanced by a dextran-coated PVF sponge along with a huge water capability, according to an investigation of the scaffold's capacity to store water (Togami et al., 2015).

17.6.4 Exopolysaccharides-based composite materials with ceramics

Because of their great strength, low wear rates, and biocompatibility, ceramics like alumina and zirconia are utilized in orthopedic implants, such as hip and knee replacements. Bioactive ceramics: Used in dental implants, bone transplants, and other procedures requiring bone regeneration, bioactive ceramics (tricalcium phosphate and HA) may form a direct link with bone tissue. Porous ceramics: Because porous ceramics may support cell development and encourage tissue regeneration, they are employed in TE approaches such as medication delivery systems and scaffolds for cell growth (Shanmugam and Sahadevan, 2018).

Future research should focus on optimizing EPS-based and polysaccharide composite scaffolds with ceramics to enhance osteoinduction, mechanical stability, and controlled bioactivity for advanced, sustainable bone regeneration therapies.

Preclinical and clinical studies (Table 17.1–17.2).

17.7 Advantages and limitations of exopolysaccharides in bone tissue engineering

17.7.1 Biological performance advantages

For scaffolds that integrate with host tissue and gradually resorb as new bone forms, EPS-based materials' exceptional biocompatibility and controlled biodegradation are essential (Mohd Nadzir et al., 2021). Because they are produced cheaply and on an industrial scale, many EPS are economically feasible alternatives for BTE as compared to more expensive biomaterials (Bagnol, Grijpma et al., 2022). In comparison to native EPS, EPS composites, which blend EPS with other synthetic or natural polymers, enhance biological and structural performance in ways such as maintaining biodegradability and biocompatibility and having a higher mechanical strength. When the appropriate components are added, there may be antibacterial or anti-inflammatory effects (Ahuja et al., 2023). Since EPS are natural polymers, they enable the

TABLE 17.1 In vitro studies on EPS-based scaffolds.

Study	EPS type/scaffold composition	Fabrication method	In vitro findings	References
Bone tissue engineering	Chitosan + poly (butylene succinate)	Solvent casting and salt leaching	Enhanced osteogenic differentiation and cell adhesion	Costa-Pinto et al. (2014)
Cartilage regeneration	Chitosan - gelatin scaffold	Freeze-drying with glutaraldehyde crosslinking	Promoted chondrocyte proliferation and ECM deposition	Whu et al. (2013)
Cancer inhibition / immunomodulation	Marine microalgal EPS (Thraustochytriidae sp.)	EPS isolated and purified	Inhibited proliferation of cancer cells; modulated immune responses (e.g., IL-6, IFN-γ)	Park et al. (2017)
Cartilage scaffold	Collagen and gum arabic EPS	Crosslinking with EDC/NHS	High biocompatibility and mechanical stability; supported cartilage cell culture	Rekulapally et al. (2021)
Skin tissue engineering	Microbial EPS and PCL/gelatin nanofibers	Electrospinning	High porosity; supported fibroblast adhesion, proliferation, and drug release	Hivechi et al. (2021)

EPS, exopolysaccharides; *PCL*, polycaprolactone.

production of hydrogels or porous scaffolds. For instance, gellan-based hydrogels and HA may stimulate the development of adipose stem cells of human in vitro for 21 days, indicating that they may be used to create bone tissue (Mohd Nadzir et al., 2021).

17.7.2 Manufacturing and scalability issues

Scaling up for clinically important dimensions is made more difficult by the need to balance many manufacturing criteria for producing large, customized scaffolds, including those that use EPS. These needs include cell integration, structural stability, and nutrition flow (Forrestal et al., 2017). Conventional scaffold construction methods, such as freeze-drying, may be labor-intensive and have poor repeatability. Even while throughput and consistency are increased by innovative manufacturing techniques like 3D printing and electrospinning, and new drying methods like microwave vacuum drying, including EPS into these scalable processes is still technically challenging (Ko et al., 2012).

TABLE 17.2 In vivo bone healing with EPS-based scaffolds

Study focus	EPS type/scaffold composition	Animal model/defect site	Key findings	References
Osteogenic differentiated bone marrow mesenchymal stem cells OBMSC-exo-loaded acellular fish-scale scaffold.	Exosome-loaded decellularized fish scale (DE-FS) scaffold	Mouse calvarial critical-sized defect	Achieved ~80% bone healing versus ~6% in controls; improved trabecular architecture and histology	Wang et al. (2022c)
(Exosome-enriched but not classic EPS) polydopamine-coated scaffold with exosomes.	Silk fibroin–chitosan–nanohydroxyapatite scaffold with PDA and BMSC-derived exosomes	Rabbit critical-sized radial defect.	In vivo: enhanced bone repair visualized via 3D-CT, histology confirmed superior regeneration in the treated group	Zhou et al. (2023)
Gellan gum-based injectable scaffold along with calcitonin.	Gellan gum hydrogel and salmon calcitonin-loaded silsesquioxane NPs, hydroxyapatite and PRP	Rat femoral defect (2 × 5 mm)	Greater bone formation (~75% defect fill), increased ALP, calcium, and osteogenic marker gene expression	Ahmadipour et al. (2022)

EPS, exopolysaccharides.

A multimodal method may greatly improve the stability and manufacturing of EPS for bone TE. The use of EPS in bone healing scaffolds depends on the optimal yield and repeatability of industrial-scale extraction and purification processes (Bagnol et al., 2022). By combining EPS with complementary polymers (such as polycaprolactone) and using biocompatible cross-linkers like genipin, structural flaws and fast degradation problems may be resolved, significantly enhancing scaffold integrity and mechanical strength (Uma Thanu Krishnan Neela et al., 2025). The stability, solubility, and bioactivity of EPS may be further improved by chemical changes such as sulfation, phosphorylation, carboxymethylation, and acetylation (Shah et al., 2024). Furthermore, it has been demonstrated that formulation techniques such as the addition of stabilizing agents, careful blending with hydrophilic polymers, and meticulous management of storage parameters (such as moisture, heat exposure, and sterilization procedures) can reduce molecular deterioration and maintain EPS functionality over time (Szymańska and Winnicka, 2015).

17.7.3 Regulatory and safety concerns

Reactive chemical groups are often used to improve mechanical strength or scaffold crosslinking. Aldehyde-based chemicals, such as Schiff base connections, may be dangerous, for example, because certain aldehydes have been linked to neurological side effects, inflammation, or cancer. Therefore, it is necessary to precisely assess their kind, concentration, and safety profile. To maintain biosafety, residual compounds such as EDC/NHS coupling agents or methacrylation reagents that are introduced by processing techniques and polymer modifications must be reduced and identified. Since EPS is microbially produced, contaminants such as endotoxins, leftover solvents, and crosslinkers must be carefully eliminated. Particularly when EPS are chemically altered (e.g., with aldehyde or methacrylate groups) to improve material qualities, regulatory agencies need verified techniques to detect and remove such impurities. If not properly regulated, these substances may cause toxicity, inflammatory reactions, or immunogenicity (Wang et al., 2022a).

Future research should prioritize scalable, safe EPS production through green synthesis, biocompatible crosslinking, and regulatory standardization to ensure consistent quality, mechanical integrity, and biosafety for clinical bone TE applications.

17.8 Future perspectives and research directions

17.8.1 Biofunctionalization and smart exopolysaccharides-based biomaterials

EPS may be customized for certain TE purposes, as shown by the potential biocompatibility and structural support of EPS-derived scaffolds like kefiran-based porous constructions or gellan-HA hydrogels (for bone). Examples of EPS materials that have been transformed into hydrogels and pH- and thermo-responsive nanoparticles are pullulan and gellan. These materials enable the controlled, stimuli-triggered release of medicines, which is ideal for more precise and effective treatment methods.

New biomaterials enable dynamic functionality in bone healing by reacting to environmental stimuli like mechanical stress, light, temperature, magnetic fields, and enzymatic activity to coordinate scaffold behavior or drug distribution. The nanostructure and porosity of EPS-derived aerogels may be maintained by supercritical CO_2-assisted drying (SC-CO_2), producing highly porous, mechanically resistant scaffolds appropriate for regenerative applications.

17.8.2 3D bioprinting and advanced fabrication techniques

The use of a gellan gum microgel bath regulated with trisodium citrate allowed for the maximum resolution 3D printing of collagen structures, such as hands, ears, and heart forms. This technique enhanced build resolution and print accuracy, providing a scalable approach to soft tissue architecture fabrication. The development of gellan gum-based inks for 3D printing has advanced, according to a recent analysis. To advance tissue-like 3D structures for regenerative applications, researchers are adjusting compositions and using innovative fabrication techniques to increase mechanical integrity, printability, and structural complexity.

Researchers produced bioceramic scaffolds with hollow strut shapes using extrusion-based 3D printing, attaining both customized porosity and mechanical strength. These patterns facilitate cellular ingrowth, osseointegration, and effective nutrition flow, all of which are necessary for bone healing. At the forefront of TE, 4D-fabricated systems scaffolds that may change over time in response to external stimuli are envisioned. These dynamic materials, which include hydrogels or shape-memory polymers, have the potential to create scaffolds for adaptive bone regeneration that change as the body heals.

17.8.3 Integration with stem cells and gene therapy approaches

Because of its structural similarity and biocompatibility, polysaccharide-based systems are being investigated increasingly for directing MSCs toward angiogenic and osteogenic development. By giving stem cells mechanical support and signaling signals, these materials have the potential to aid in bone repair. To generate new hybrid scaffolds, biodegradable polymers like PCL and HA have been mixed with stem cells, such as human dental pulp-derived mesenchymal stem/stromal cells.

These constructs, which are made by additive manufacturing or bioink printing, have great in vivo biocompatibility, promote osteogenic differentiation, and have the capacity to regenerate bone. They also show high cell survival. MSC attachment, proliferation, and expression of bone-specific markers (such as osteopontin) are all markedly improved by additively built polycaprolactone (PCL) scaffolds adorned with ECM formed from MSCs, which promotes osteogenesis. To increase bone production directly and via paracrine stimulation of host tissue, stem cells may be engineered to express important osteogenic or angiogenic genes, including BMP2, VEGF, Cbfa1, and Osterix. This approach may lessen the requirement for more implanted cells and promote stronger, more effective bone healing.

17.9 Conclusion

Since EPS are naturally biocompatible, biodegradable, and may promote cell adhesion and proliferation, they are a promising family of natural polymers, along with significant promise in BTE. Because of their structural adaptability, scaffolds that closely resemble the ECM of bone may be designed via functional changes and composite synthesis. Recent developments in 3D bioprinting, nanocomposites, and EPS-based hydrogel systems have shown increased osteoconductivity and better mechanical performance, especially when paired with MSCs, osteoinductive chemicals, or bioactive ceramics. Notwithstanding these benefits, several issues need to be resolved before EPS may be completely implemented in clinical settings. These include the requirement for long-term safety and effectiveness evidence, batch-to-batch variability in microbial production, regulatory obstacles, and the low mechanical strength of native forms. The development of sophisticated manufacturing techniques like 4D printing and EPS-based bioinks for personalized medicine, as well as integration with smart materials and gene therapy, must be the fundamental areas of future study. To sum up, EPS-based biomaterials provide a very flexible and long-lasting framework for bone growth. EPS have capability to develop into next-generation biomaterials that may overcome the present constraints in bone repair and regenerative treatments with more multidisciplinary study and technical innovation.

Credit authorship contribution statement

Subhajit Mandal: Writing, formal analysis; Tanmay Jit: Formal analysis, software; Ranabir Sahu: Visualization; validation; Gouranga Nandi: Analysis, visualization; Partha Palit: Visualization, validation, and conceptualization, supervision; write-up- review and editing, methodology, project administration, resources, visualization, supervision.

Declaration of generative AI and AI-assisted technologies in the writing process of chapter

During the end process after drafting of the chapter, the author used Grammarly, an AI-assisted editing tool for language correction, specifically to improve spelling, grammar, word selection, readability, and clarity. After using this tool, the authors re-examined and revised the text as required.

Conflict of interest

The authors declare no conflict of interest.

Ethical statement

Not applicable.

Acknowledgments

Dr. Partha Palit is highly grateful to the authorities of Assam University, Silchar for their continuous encouragement, support, and facilities for preparing the chapters.

References

Ahmadipour, S., Varshosaz, J., Hashemibeni, B., Manshaei, M., Safaeian, L., 2022. In vivo assessment of bone repair by an injectable nanocomposite scaffold for local co-delivery of autologous platelet-rich plasma and calcitonin in a rat model. Drug Dev. Ind. Pharm. 48 (3), 98–108. doi:10.1080/03639045.2022.2087080.

Ahuja, V., Bhatt, A.K., Banu, J.R., Kumar, V., Kumar, G., Yang, Y.-H., Bhatia, S.K., 2023. Microbial exopolysaccharide composites in biomedicine and healthcare: Trends and advances. Polymers 15 (7), 1801. doi:10.3390/polym15071801.

Ali, A.H., Bamigbade, G., Tarique, M., Esposito, G., Obaid, R., Abu-Jdayil, B., Ayyash, M., 2023. Physicochemical, rheological, and bioactive properties of exopolysaccharide produced by a potential probiotic *Enterococcus faecalis* 84B. Int. J. Biol. Macromol. 240, 124425. doi:10.1016/j.ijbiomac.2023.124425.

Alkaron, W., Almansoori, A., Balázsi, C., Balázsi, K., 2024. A critical review of natural and synthetic polymer-based biological apatite composites for bone tissue engineering. J. Compos. Sci. 8 (12), 523. doi:10.3390/jcs8120523.

Andrew, M., Jayaraman, G., 2022. Molecular characterization and biocompatibility of exopolysaccharide produced by moderately *Halophilic Bacterium Virgibacillus dokdonensis* from the Saltern of Kumta coast. Polymers 14 (19), 3986. doi:10.3390/polym14193986.

Angelin, J., Kavitha, M., 2020. Exopolysaccharides from probiotic bacteria and their health potential. Int. J. Biol. Macromol. 162, 853–865. doi:10.1016/j.ijbiomac.2020.06.190.

Bagnol, R., Grijpma, D., Eglin, D., Moriarty, T.F., 2022. The production and application of bacterial exopolysaccharides as biomaterials for bone regeneration. Carbohydr. Polym. 291, 119550. doi:10.1016/j.carbpol.2022.119550.

Baron, R., Kneissel, M., 2013. WNT signaling in bone homeostasis and disease: From human mutations to treatments. Nat. Med. 19 (2), 179–192. doi:10.1038/nm.3074.

Bauso, L.V., La Fauci, V., Longo, C., Calabrese, G., 2024. Bone tissue engineering and nanotechnology: A promising combination for bone regeneration. Biology 13 (4), 237. doi:10.3390/biology13040237.

Chowdhury, S., Kumari, G., Jit, T., 2025. Nano-formulation of phytochemicals to combat neurodegenerative diseases: Current and thorough strategies. Pharmacol. Res.-Nat. Prod., 100249 doi:10.1016/j.prenap.2025.100249.

Cieza, A., Causey, K., Kamenov, K., Hanson, S.W., Chatterji, S., Vos, T., 2020. Global estimates of the need for rehabilitation based on the global burden of disease study 2019: A systematic analysis for the global burden of disease study 2019. Lancet 396 (10267), 2006–2017. doi:10.1016/S0140-6736(20)32340-0.

Costa-Pinto, A.R., Vargel, I., Tuzlakoglu, K., Correlo, V., Sol, P., Faria, S., Neves, N.M., 2014. Influence of scaffold composition over in vitro osteogenic differentiation of hBMSCs and in vivo inflammatory response. J. Biomater. Appl. 28 (9), 1430–1442. doi:10.1177/0885328213512385.

Delpierre, A., Savard, G., Renaud, M., Rochefort, G.Y., 2023. Tissue engineering strategies applied in bone regeneration and bone repair. Bioengineering 10 (6), 644. doi:10.3390/bioengineering10060644.

Dimitriou, R., Jones, E., McGonagle, D., Giannoudis, P.V., 2011. Bone regeneration: current concepts and future directions. BMC Med. 9 (1), 66. doi:10.1186/1741-7015-9-66.

Donnaloja, F., Jacchetti, E., Soncini, M., Raimondi, M.T., 2020. Natural and synthetic polymers for bone scaffolds optimization. Polymers 12 (4), 905. doi:10.3390/polym12040905.

Florczyk, S.J., Leung, M., Jana, S., Li, Z., Bhattarai, N., Huang, J.I., Zhang, M., 2012. Enhanced bone tissue formation by alginate gel-assisted cell seeding in porous ceramic scaffolds and sustained release of growth factor. J. Biomed. Mater. Res. A 100 (12), 3408–3415. doi:10.1002/jbm.a.34288.

Forrestal, D.P., Klein, T.J., Woodruff, M.A., 2017. Challenges in engineering large customized bone constructs. Biotechnol. Bioeng. 114 (6), 1129–1139. doi:10.1002/bit.26222.

Girón, J., Kerstner, E., Medeiros, T., Oliveira, L., Machado, G., Malfatti, C., Pranke, P., 2021. Biomaterials for bone regeneration: An orthopedic and dentistry overview. Braz. J. Med. Biol. Res. 54, e11055. doi:10.1590/1414-431X2021e11055.

Hivechi, A., Milan, P.B., Modabberi, K., Amoupour, M., Ebrahimzadeh, K., Gholipour, A.R., Delattre, C., 2021. Synthesis and characterization of exopolysaccharide encapsulated PCL/gelatin skin substitute for full-thickness wound regeneration. Polym. (Basel) 13 (6). doi:10.3390/polym13060854.

Hutmacher, D.W., 2000. Scaffolds in tissue engineering bone and cartilage. Biomaterials 21 (24), 2529–2543. doi:10.1016/S0142-9612(00)00121-6.

Hwang, H.S., Lee, C.-S., 2024. Nanoclay-composite hydrogels for bone tissue engineering. Gels 10 (8), 513. doi:10.3390/gels10080513.

Ko, J., Kolehmainen, K., Ahmed, F., Jun, M.B., Willerth, S.M., 2012. Towards high throughput tissue engineering: development of chitosan-calcium phosphate scaffolds for engineering bone tissue from embryonic stem cells. Am. J. Stem. Cells 1 (1), 81–89.

Kumari, J., Kumawat, R., Prasanna, R., Jothieswari, D., Debnath, R., Ikbal, A.M.A., Tiwari, O.N., 2025. Microbial exopolysaccharides: Classification, biosynthetic pathway, industrial extraction, and commercial production to unveil its bioprospection: A comprehensive review. Int. J. Biol. Macromol., 139917 doi:10.1016/j.ijbiomac.2025.139917.

Lee, E.J., Kasper, F.K., Mikos, A.G., 2014. Biomaterials for tissue engineering. Ann. Biomed. Eng. 42 (2), 323–337. doi:10.1007/s10439-013-0859-6.

Liao, J., Guo, X., Nelson, D., Kasper, F.K., Mikos, A.G., 2010. Modulation of osteogenic properties of biodegradable polymer/extracellular matrix scaffolds generated with a flow perfusion bioreactor. Acta Biomater. 6 (7), 2386–2393. doi:10.1016/j.actbio.2010.01.011.

LogithKumar, R., KeshavNarayan, A., Dhivya, S., Chawla, A., Saravanan, S., Selvamurugan, N., 2016. A review of chitosan and its derivatives in bone tissue engineering. Carbohydr. Polym. 151, 172–188. doi:10.1016/j.carbpol.2016.05.049.

Mathews, S., Bhonde, R., Gupta, P.K., Totey, S., 2012. Extracellular matrix protein mediated regulation of the osteoblast differentiation of bone marrow derived human mesenchymal stem cells. Differentiation 84 (2), 185–192. doi:10.1016/j.diff.2012.05.001.

Miyahara, T., Nyan, M., Shimoda, A., Yamamoto, Y., Kuroda, S., Shiota, M., Kasugai, S., 2012. Exploitation of a novel polysaccharide nanogel cross-linking membrane for guided bone regeneration (GBR). J. Tissue Eng. Regen. Med. 6 (8), 666–672. doi:10.1002/term.475.

Mohd Nadzir, M., Nurhayati, R.W., Idris, F.N., Nguyen, M.H., 2021. Biomedical applications of bacterial exopolysaccharides: A review. Polymers 13 (4), 530. doi:10.3390/polym13040530.

Mouro, C., Gomes, A.P., Gouveia, I.C., 2024. Microbial exopolysaccharides: Structure, diversity, applications, and future frontiers in sustainable functional materials. Polysaccharides 5 (3), 241–287. doi:10.3390/polysaccharides5030018.

Netrusov, A.I., Liyaskina, E.V., Kurgaeva, I.V., Liyaskina, A.U., Yang, G., Revin, V.V., 2023. Exopolysaccharides producing bacteria: A review. Microorganisms 11 (6), 1541. doi:10.3390/microorganisms11061541.

Park, G.-T., Go, R.-E., Lee, H.-M., Lee, G.-A., Kim, C.-W., Seo, J.-W., Hwang, K.-A., 2017. Potential anti-proliferative and immunomodulatory effects of marine microalgal exopolysaccharide on various human cancer cells and lymphocytes in vitro. Mar. Biotechnol. 19 (2), 136–146. doi:10.1007/s10126-017-9735-y.

Rekulapally, R., Udayachandrika, K., Hamlipur, S., Sasidharan Nair, A., Pal, B., Singh, S., 2021. Tissue engineering of collagen scaffolds crosslinked with plant based polysaccharides. Prog. Biomater. 10 (1), 29–41. doi:10.1007/s40204-021-00149-4.

Rinaudo, M., 2006. Chitin and chitosan: Properties and applications. Prog. Polym. Sci. 31 (7), 603–632. doi:10.1016/j.progpolymsci.2006.06.001.

Rolvien, T., Barvencik, F., Klatte, T.O., Busse, B., Hahn, M., Rueger, J.M., Rupprecht, M., 2017. ß-TCP bone substitutes in tibial plateau depression fractures. Knee 24 (5), 1138–1145. doi:10.1016/j.knee.2017.06.010.

Rosellini, E., Zhang, Y.S., Migliori, B., Barbani, N., Lazzeri, L., Shin, S.R., Cascone, M.G., 2018. Protein/polysaccharide-based scaffolds mimicking native extracellular matrix for cardiac tissue engineering applications. J. Biomed. Mater. Res. A 106 (3), 769–781. doi:10.1002/jbm.a.36272.

Rutledge, K., Cheng, Q., Pryzhkova, M., Harris, G.M., Jabbarzadeh, E., 2014. Enhanced differentiation of human embryonic stem cells on extracellular matrix-containing osteomimetic scaffolds for bone tissue engineering. Tissue Eng C Methods 20 (11), 865–874. doi:10.1089/ten.tec.2013.0411.

Saha, I., Datta, S., 2022. Bacterial exopolysaccharides in drug delivery applications. J. Drug Deliv. Sci, Technol. 74, 103557. doi:10.1016/j.jddst.2022.103557.

Salimi, F., Farrokh, P., 2023. Recent advances in the biological activities of microbial exopolysaccharides. World J. Microbiol. Biotechnol. 39 (8), 213. doi:10.1007/s11274-023-03660-x.

Shah, I.A., Kavitake, D., Tiwari, S., Devi, P.B., Reddy, G.B., Jaiswal, K.K., Shetty, P.H., 2024. Chemical modification of bacterial exopolysaccharides: Antioxidant properties and health potentials. Curr. Res. Food Sci. 9, 100824. doi:10.1016/j.crfs.2024.100824.

Shanmugam, K., Sahadevan, R., 2018. Bioceramics—An introductory overview. Fundam. Biomater.: Ceram. 1–46.

Sözen, T., Özışık, L., Başaran, N.Ç., 2016. An overview and management of osteoporosis. Eur. J. Rheumatol. 4 (1), 46. doi:10.5152/eurjrheum.2016.048.

Stevens, M.M., 2008. Biomaterials for bone tissue engineering. Mater. Today 11 (5), 18–25. doi:10.1016/S1369-7021(08)70086-5.

Sun, J., Tan, H., 2013. Alginate-based biomaterials for regenerative medicine applications. Materials 6 (4), 1285–1309. doi:10.3390/ma6041285.

Szymańska, E., Winnicka, K., 2015. Stability of chitosan—a challenge for pharmaceutical and biomedical applications. Mar. Drugs 13 (4), 1819–1846. doi:10.3390/md13041819.

Thakur, S., HPS, A.K., Bairwan, R.D., Yahya, E.B., Jha, K., Adnan, A.S., Khan, M.R., 2024. Potential application of polysaccharide-based aerogel scaffolds for bone tissue engineering. Adv. Mater. Lett. 15 (2), 1746–2402. doi:10.5185/amlett.2024.021746.

Tiwari, O.N., Bobby, M.N., Kondi, V., Halder, G., Kargarzadeh, H., Ikbal, A.M.A., Chattopadhyay, D., 2024. Comprehensive review on recent trends and perspectives of natural exo-polysaccharides: pioneering nano-biotechnological tools. Int. J. Biol. Macromol. 265, 130747. doi:10.1016/j.ijbiomac.2024.130747.

Togami, W., Sei, A., Okada, T., Taniwaki, T., Fujimoto, T., Tahata, S., Mizuta, H., 2015. Effects of the water-holding capability of polyvinyl formal sponges on osteogenic ability in in vivo experiments. J. Biomed. Mater. Res. B 103 (1), 188–194. doi:10.1002/jbm.b.33200.

Uma Thanu Krishnan Neela, N., Szewczyk, P.K., Karbowniczek, J.E., Polak, M., Knapczyk-Korczak, J., Stachewicz, U., 2025. Improving stability and mechanical strength of electrospun chitosan-polycaprolactone scaffolds using genipin cross-linking for biomedical applications. Macromol. Rapid Commun. 46 (13), e2400869. doi:10.1002/marc.202400869.

Wang, L., Guo, X., Chen, J., Zhen, Z., Cao, B., Wan, W., Ge, Z., 2022. Key considerations on the development of biodegradable biomaterials for clinical translation of medical devices: With cartilage repair products as an example. Bioact. Mater. 9, 332–342. doi:10.1016/j.bioactmat.2021.07.031.

Wang, Y., Kong, B., Chen, X., Liu, R., Zhao, Y., Gu, Z., Jiang, Q., 2022. BMSC exosome-enriched acellular fish scale scaffolds promote bone regeneration. J. Nanobiotechnol. 20 (1), 444. doi:10.1186/s12951-022-01646-9.

Wang, Y., Zhang, H., Hu, Y., Jing, Y., Geng, Z., Su, J., 2022. Bone repair biomaterials: A perspective from immunomodulation. Adv. Funct. Mater. 32 (51), 2208639. doi:10.1002/adfm.202208639.

Waoo, A.A., Singh, S., Pandey, A., Kant, G., Choure, K., Amesho, K.T., Srivastava, S., 2023. Microbial exopolysaccharides in the biomedical and pharmaceutical industries. Heliyon 9 (8). doi:10.1016/j.heliyon.2023.e18613.

Whu, S.W., Hung, K.-C., Hsieh, K.-H., Chen, C.-H., Tsai, C.-L., Hsu, S.-H., 2013. In vitro and in vivo evaluation of chitosan–gelatin scaffolds for cartilage tissue engineering. Mater. Sci. Eng. C 33 (5), 2855–2863. doi:10.1016/j.msec.2013.03.003.

Wu, A.-M., Bisignano, C., James, S.L., Abady, G.G., Abedi, A., Abu-Gharbieh, E., Asaad, M., 2021. Global, regional, and national burden of bone fractures in 204 countries and territories, 1990–2019: A systematic analysis from the global burden of disease study 2019. Lancet Healthy Longev. 2 (9), e580–e592. doi:10.1016/S2666-7568(21)00172-0.

Xia, Y., Mei, F., Duan, Y., Gao, Y., Xiong, Z., Zhang, T., Zhang, H., 2012. Bone tissue engineering using bone marrow stromal cells and an injectable sodium alginate/gelatin scaffold. J. Biomed. Mater. Res. A 100 (4), 1044–1050. doi:10.1002/jbm.a.33232.

Xu, L., Bai, X., Yang, J., Li, J., Xing, J., Yuan, H., Li, J., 2020. Preparation and characterisation of a gellan gum-based hydrogel enabling osteogenesis and inhibiting Enterococcus faecalis. Int. J. Biol. Macromol. 165, 2964–2973. doi:10.1016/j.ijbiomac.2020.10.083.

Yuan, N., Rezzadeh, K.S., Lee, J.C., 2015. Biomimetic scaffolds for osteogenesis. Recept. Clin. Investig. 2 (3), 898.

Zhou, Y., Deng, G., She, H., Bai, F., Xiang, B., Zhou, J., Zhang, S., 2023. Polydopamine-coated biomimetic bone scaffolds loaded with exosomes promote osteogenic differentiation of BMSC and bone regeneration. Regen. Ther. 23, 25–36. doi:10.1016/j.reth.2023.03.005.

CHAPTER

18

EPS: A potential agent as wound healer

Amlan Bishal[a], *Tapan Kumar Shaw*[b], *Bratati Bandyopadhyay*[a,b], *Mohini Mondal*[a,b] *and* *Paramita Paul*[c]

[a]Bharat Technology, Uluberia, Howrah, West Bengal, India [b]Department of Pharmaceutical Technology, JIS University, Agarpara, Kolkata, West Bengal, India [c]Department of Pharmaceutical Technology, University of North Bengal, Darjeeling, West Bengal, India

18.1 Introduction

Wounds are disruptions in the structural integrity of tissues beneath the skin, which are caused by trauma, surgical intervention, burns, or pathological processes like diabetes. The kind of wound is usually determined primarily by the time taken to heal such a wound and the pattern of healing. Acute wounds are those wounds that result in surgical cuts or minor cuts and usually take a predictable time to heal as they go through the normal stages of healing in a smooth way. Conversely, the types of wounds that last long and do not progress well through the normal healing process are termed chronic wounds, e.g., pressure ulcers, venous leg ulcers, and diabetic foot ulcers. The process of wound healing is very complex. It is dependent on the coordinated expression of various cellular and molecular pathways as it goes through phases of healing hemostasis, then inflammation, then proliferation, and lastly, remodeling (Herman et al., 2025). Conventional wound care practices involve various forms of treatment with the view to healing wounds and preventing infection. This involves the use of dressings and bandages, particularly those that can support moist wound healing, such as hydrogel, alginates, and hydrocolloids. Antiseptics, growth hormones, and antibiotics are common topical medicines that are used to regulate the microbial load and enhance the healing of the tissues. New methods such as scaffolding of bioactive, nanotechnology, tissue engineering, and stem cell therapy have been a possible factor in tissue regeneration. Polymers such as chitosan, alginate, hyaluronic acid, including different exopolysaccharides (EPS), are widely applied in wound healing due to their biocompatibility and regenerative potential (Kolimi et al., 2022).

DOI: https://doi.org/10.1016/B978-0-443-36619-2.00018-8

Microorganisms such as bacteria, fungi, and algae produce the high-molecular-weight EPS. These biopolymers possess their bioactive properties such as anti-inflammatory, antioxidant, antibacterial, and immunomodulatory activity (Netrusov et al., 2023). Capsular polysaccharides are an EPS, loosely attached to the cell wall of microbes, and the mucous complex, released to the area surrounding the microbes. EPS are unique in structure and normally consist of monosaccharides, namely glucose, galactose, and fructose, and other functional groups like sulphate, phosphate, and carboxyl that provide them with increased bioactivity (Tiwari et al., 2020).

EPS are highly sought after as prospective wound healing agents because they are highly biocompatible, biodegradable and have a wide range of biological activities. Their antibacterial activity inhibits the formation of biofilm and infection, and their anti-inflammatory activity controls the inflammatory response and promotes faster tissue regeneration (Ahuja et al., 2024). EPS also holds wounds in the most suitable condition for healing, moisturized and hydrated (Ousey etal.,2016). They are naturally biodegradable and thus can be degraded safely without toxic byproducts, and are also immunomodulatory, that is, can be used to stimulate the immune cell activity to aid in healing (Parvin et al., 2025). Although EPS have several advantages, they also have several disadvantages, such as high production of these polymers inhibiting further production, the structural complexity of the molecules making them hard to be consistent and reproducible, and a lack of clinical translation because a lot of in-vivo studies and clinical trials are needed to ascertain their safety and efficacy in human use (Kulchar et al., 2023).

18.2 Mechanisms of wound healing by exopolysaccharides

18.2.1 Hemostasis and clot formation

Following tissue injury, adenosine diphosphate of the damaged tissue facilitates the binding of platelets to exposed type I collagen. When activated, they secrete glycoproteins promotes platelet aggregation (Meyer et al., 1981). Moreover, the fundamental clotting cascade is activated by products released by platelets that lead to the production of thrombin that converts fibrinogen to fibrin. Platelet plug is reinforced by fibrin mesh that forms a robust hemostatic seal (LaPelusa and Dave, 2025). Other than clotting, platelets are also essential in wound healing since they secrete growth factors like TGF-beta, platelet-derived growth factors (PDGF), epidermal growth factor (EGF), and fibroblast growth factor (FGF) that are necessary in the recruitment of monocytes and neutrophils, the inflammatory stage of the wound-healing process begins (Scopelliti et al., 2022). During the initial phase of wound healing, the EPS enhances the process of clotting by interacting with platelets and coagulation factors. Some EPS exhibit a hemostatic ability to enhance the formation of fibrin network and to bind platelets, other than clot formation (Opneja et al., 2019).

18.2.2 Inflammatory phase

Inflammatory response starts after about 6–8 hours of injury, and vascular permeability is increased, permitting extravasation of plasma and neutrophils into the tissues around the injury site (Hannoodee and Nasuruddin, 2024). The primary protection against infection is the phagocytosis of detritus and bacteria by neutrophils. When bacteria and detritus are digested, neutrophils die and liberate cell enzymes into the extracellular space, which in turn degrade tissue further. Neutrophils are substituted by macrophages, and thus they digest as well as destroy cell wastes and other foreign matter (Rosales, 2018). Moreover, macrophages stimulate the growth of endothelium and smooth muscle cell development (Osorio-Valencia and Zhou, 2024). EPS also stimulates immune response by activating macrophages and neutrophils, and leads to the release of interleukins such as IL-6, TNF-alpha, and IL-10. Other EPS are also anti-inflammatory in nature and can help in controlling over-inflammation and preventing the development of chronic wounds. Besides, EPS has been shown to have antioxidant activity, which helps reduce oxidative stress in the wound environment and assists in healing (Delano and Ward, 2016).

18.2.3 Proliferation and tissue regeneration

The proliferation phase begins normally after four days of injury and lasts up to twenty first day, in the case of acute wounds, depending on the size of the wound and the health of the patient. Fibroblasts and endothelial cells also play a critical role in wound healing by coordinating the angiogenesis process, collagen formation, and development of scar tissue, wound contraction, and re-epithelialization (Schultz et al., 2011). EPS stimulate fibroblast proliferation and collagen synthesis, while upregulating VEGF to promote angiogenesis, which are essential in the regeneration of tissues. By upregulating VEGF, they increase the growth of new vasculature and angiogenesis

18.2.4 Extracellular matrix remodeling

The wound healing process is associated with the remodeling of collagen to regain rupture strength similar to that of normal, healthy skin. Due to this process, the cell and capillary density also decrease. The fibroblast cells play a vital role in this process (Mathew-Steiner et al., 2021). Epithelial tissues composed of rapidly growing keratinocytes undergo terminal differentiation to become the definitive barrier-forming system, as evidenced by the presence of characteristic protein indicators such as cytokeratin K1 and K10, involucrin, and loricrin (Bragulla and Homberger, 2009). EPS supports the structural integrity of wounds by interacting with key ECM components, including hyaluronic acid, collagen, and fibronectin. Further, a few EPS mimic hyaluronic acid and act as natural scaffolds for tissue regeneration (Fig.18.1). They help in regulating matrix metalloproteinase (MMP) activity, preventing excessive ECM degradation (Diller and Tabor, 2022).

18.2.5 Antimicrobial and biofilm disruption

Some EPS have a possible antibacterial property to counter wound infections by breaking the biofilm of the organism. These attached bacteria make complex bacterial consortia covered by a protective sheath rich in EPS (Zhao et al., 2023). This matrix forms structural reinforcement that defends the bacteria against the effects of antimicrobial compounds, hence rendering biofilms extremely difficult to treat using traditional tropics (Almatroudi, 2025). Disruption of the EPS matrix is a potential way of treating biofilm-related diseases. EPS can also have an antibiofilm capability that can reduce colonization by bacteria and chances of disease (Pinto et al., 2020).

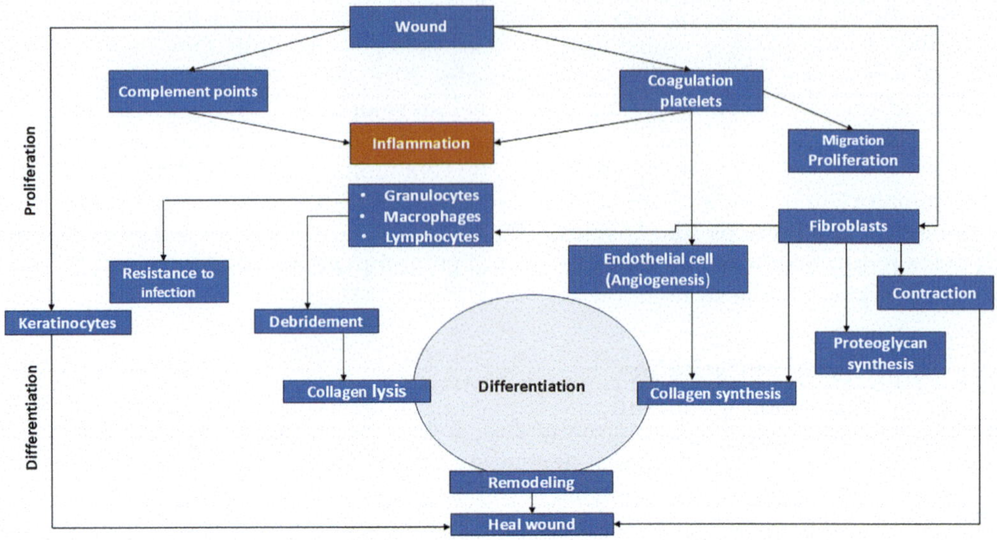

FIGURE 18.1 Flowchart illustrating the complex biological wound healing process. The diagram outlines the sequential and overlapping phases, including inflammation, proliferation, and differentiation. These processes collectively contribute to tissue regeneration and healed wound formation.

18.2.6 Epithelialization and wound closure

EPS promote keratinocyte adhesion and growth, facilitating complete wound closure (Fig 18.2). They assist in the preservation of the optimum level of moisture in the wound, decreasing dryness and the possibility of scarring (Kaur and Dey, 2023).

18.3 Research status of exopolysaccharides as a wound healer

Nowadays, studies related to EPS in wound healing have made considerable progress in the past decade and offer valuable insights and developments. Some of these are depicted here to understand the fact of EPS as a wound healer.

A study determined the antioxidant, antibacterial, and wound healing properties of a new exopolysaccharide (EPS-Ca6) produced by *Lactobacillus sp.* Ca6. In an excision rat wound model, EPS-Ca6 significantly accelerates wound healing and achieves complete closure within 14 days. Histological analysis proves that EPS-Ca6 promoted full re-epithelialization and complete epidermal regeneration (Trabelsi et al., 2017). A notable decrease in inflammation and an improvement in wound healing is found in *Lactobacillus casei* and its exopolysaccharide-treated rats (Majid et al., 2017). Another study highlights the role of bacterial biofilms in chronic wounds, where they promote antibiotic resistance, blocking immune responses, and causing persistent inflammation (Mendoza et al., 2019).

The EPS from *Papiliotrema terrestris* PT22AV is a mannose-rich, biocompatible polymer with strong antibacterial activity and dose-dependent wound healing effects, making it a potent candidate for skin wound healing processes (Hivechi et al., 2021).

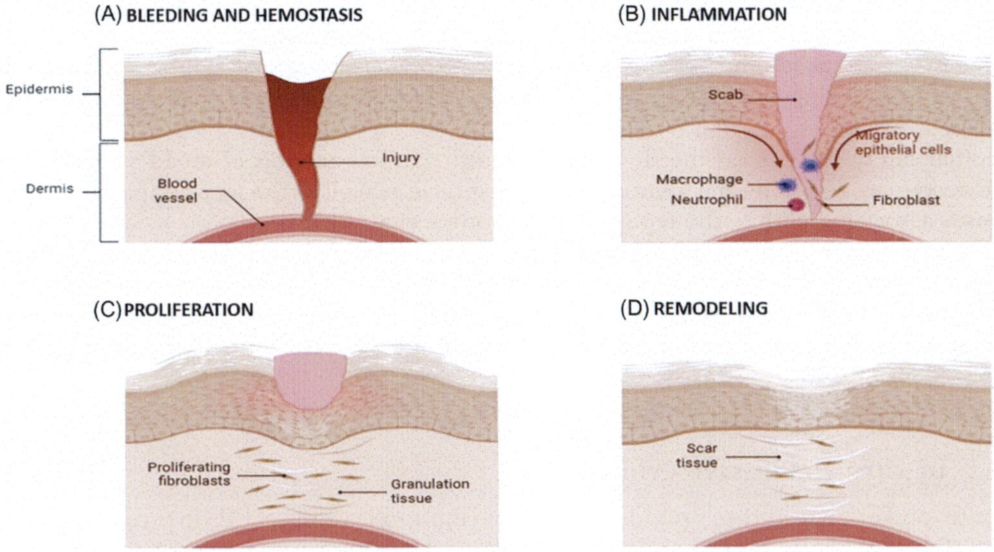

FIGURE 18.2 (A) Hemostasis involves clot formation; (B) inflammation causes immune cell infiltration; (C) proliferation includes fibroblast activity, angiogenesis, and re-epithelialization; (D) remodeling involves tissue maturation and scar formation.

Researchers found the wound healing activity of an EPS produced by cold-adapted yeast, *Rhodotorula mucilaginosa* GUMS16, on full-thickness wounds in a rat model. They extracted and lyophilized the EPS, then incorporated it with polycaprolactone (PCL) and gelatin polymer to create a nanocomplex using electrospinning methods. The 2% EPS-nanofibers complex showed the maximum cell viability. In-vivo results demonstrated that the EPS-nanofibers complex significantly accelerated wound healing, indicating strong efficacy for skin tissue engineering applications (Hivechi et al., 2021).

Microbial EPSs extracted from *Lactiplantibacillus plantarum* RO30 provide good healing efficacy on burn wound models in Wistar rats, which shows 0.5% (w/w) concentration stimulates the healing process in burn wound areas (Elmansy et al., 2022).

Microbial EPS are promising biopolymers for wound healing owing to their biocompatibility and multifunctional bioactivities. Microbial EPS are produced by bacteria, yeasts, and fungi, which promote wound repair by enhancing fibroblast proliferation, collagen synthesis, and re-epithelialization (Salimi and Farrokh, 2023).

A review study determines polysaccharides and peptides found in marine organisms, their in-vitro wound healing effects, and their different mechanisms of action. The study concluded that lyophilized exopolysaccharide (L-EPS) of *Lactiplantibacillus plantarum* GD2 has wound healing activity by increasing collagen production, activation of TGF- o/Smad Pathway, decreased inflammation, proliferation of fibroblast, and aids in angiogenesis. It demonstrates the fact that L-EPS is a promising agent in the advanced wound healing treatment methodology (Demir and Aslim, 2024).

Researchers highlight those bacterial sources such as *Actinobacteria*, *Firmicutes*, and *Proteobacteria*, as well as fungal sources like yeasts and mushrooms, produce EPS, bacteriocins, and lipopeptides that induce wound repair by re-epithelialization and angiogenesis, thus providing medicinal importance of natural compounds for wound care (Arslan et al., 2024).

Other results of the wound healing potential of EPS of *Weissella cibaria* exhibit a possible influence on wound sealing and skin production via the DNA damage repair pathway (Kahraman-Ilıkkan et al., 2024).

Another study focuses on the development and characterization of an EPS isolated from marine resources, revealing the wound healing response as well as the focus impacts on cellular responses associated with tissue repair (Abdel-monem et al., 2025).

EPS derived from lactic acid bacteria (LAB) has multiple health as well as industrial applications, and structural features of EPSs affect their bioactivity. It has antioxidant, immunomodulatory, and antibiofilm properties, as well as drug delivery applications. Sophisticated characterization and genetic engineering methods are also improving the knowledge and production of EPS (Liu et al., 2025).

18.3.1 In-vitro and in-vivo evidence

In-vitro and in-vivo studies provide evidence on the applications of EPS in promoting wound healing. Table 18.1 highlights the role of EPS in wound healing based on current studies.

18.4 Challenges and future direction

The use of EPS as wound healing agents has demonstrated prospective therapeutic value in the recent past based on the antimicrobial, anti-inflammatory, and regenerative effects of the EPS. Nevertheless, there are a number of challenges that must be clarified in order to utilize their potential to the fullest extent (Kiran et al., 2024). A key challenge is that EPS production is highly variable, limiting standardization and reproducibility because the biochemical composition of EPS can vary widely based on the microbial strain and growth conditions. This inconsistency makes it difficult to translate research findings in the lab to a uniform clinical practice (Netrusov et al., 2023). Also, the process of scaling up production to meet the demand without compromising quality and bioactivity is a serious issue, and the regulatory aspects of ensuring patient safety and product efficacy are also a problem. The future studies will probably be devoted to the embedding of such advanced bioengineering strategies as nanotechnology-based delivery systems and hydrogel formulations to increase the stability and target delivery of EPS in wound sites (Ma et al., 2024). In addition, the signaling pathways manipulated by EPS can be better understood at the molecular level, which will lead to the development of combination therapeutics, thus enhancing the overall healing process. To address these issues, it will be vital to have interdisciplinary cooperation between microbiologists, bioengineers, and clinical researchers, which will lead to the commercialization of EPS-based wound healing therapies (Deo et al., 2025). EPS produced by *Lactobacillus plantarum* was found to have a potential positive effect on wound healing and collagen deposition in diabetic rat models with pro- and anti-inflammatory cytokine regulation, but the differences in yields and purity proved problematic to reproducibility (Wang et al., 2024). Antibiofilm and fibroblast-stimulating EPS were also produced by marine bacteria, including *Pseudoalteromonas sp.*, but the strain-specific needs and labor-intensive purification steps also limited their scalability (Artini et al., 2023). In

TABLE 18.1 Current evidence on the role of EPS in wound healing in different animal models.

Year	EPS source/type	In-vitro evidence	In-vivo/animal model evidence	Key findings/outcomes	References
2020	Marine bacterium (*Pantoea* sp.)	Enhanced cell adhesion, proliferation, and migration in fibroblasts and keratinocytes.	Rat model: Accelerated re-epithelialization, upregulation of HB-EGF, FGF, E-cadherin, β-catenin.	Modulates healing via the Wnt/β-catenin pathway.	Sahana and Rekha (2020)
2021	Yeast EPS (with PCL/gelatin nanofiber)	Highest cell activity in 2% EPS-loaded nanofiber.	Full-thickness wound in rats; faster healing with EPS nanofiber dressing.	The EPS nanofiber enhances tissue regeneration.	Hivechi et al. (2021)
2022	Marine LAB (*Lactiplantibacillus plantarum* EI6) EPS	Scratch assay: 77% wound closure at 48 h versus 60% in control; promoted fibroblast migration and proliferation.	Not specified	Safe, bioactive polymer with significant in-vitro wound healing.	Zaghloul and Ibrahim (2022)
2022	Probiotic LAB EPS	Antioxidant activity; promoted fibroblast migration.	Burn wound in rats; greater contraction and tissue regeneration versus control.	EPS supports burn healing and fibroblast activity.	Elmansy et al. (2022)

(*continued on next page*)

TABLE 18.1 Current evidence on the role of EPS in wound healing in different animal models—cont'd

Year	EPS source/type	In-vitro evidence	In-vivo/animal model evidence	Key findings/outcomes	References
2023	Yeast (*Papiliotrema terrestris*) EPS	Not specified	Full-thickness wound model in rats; dose-dependent wound closures up to 99.2% after 14 days.	EPS at 10 mg/mL is comparable to phenytoin cream; enhanced re-epithelialization and connective tissue ordering.	Hamidi et al. (2023)
2024	Bacterial/fungal polysaccharides and peptides	Both in vitro and in vivo wound healing potency is reported.	Various animal models	Broad-spectrum wound healing efficacy.	Arslan et al. (2024)
2025	Bacterial (sulfated EPS, BAP-2 from *Bacillus altitudinis*)	Downregulation of inflammatory cytokines in macrophage and monocytic cell lines; increased fibroblast proliferation.	Rat wound model; ~98% wound closure and re-epithelialization.	Strong anti-inflammatory, pro-proliferative, and healing effects.	Asharaf et al. (2025)

EPS, exopolysaccharides; *LAB*, lactic acid bacteria.

TABLE 18.2 Discussion of wound healing applications of EPS regulatory and safety.

EPS type/source	Usage in wound healing	Regulatory considerations	Safety tests carried out	Clinical significance	References
Alginate (*Brown algae*)	Healing wound moist dressings, hemostats.	Products bearing the CE mark in the EU; FDA-approved alginate dressings.	Biocompatibility, sterility, pyrogenicity, and in-vivo wound healing model.	Commercially available to a great extent; biocompatible and absorbable.	Aderibigbe and Buyana (2018)
Pullulan (*Aureobasidium pullulans*)	Wound healing antimicrobial film.	Approved as generally recognized as safe by the FDA for food application.	In-vitro cytotoxicity, human patch testing, and irritation test.	Biodegradabl; good oxygen barrier and biocompatibility.	Rai et al. (2021)
Levan (*Bacillus subtilis*)	Chronic wound scaffold and gel.	No drug approved, requires safety data to use in the clinic.	MTT assay, acute dermal toxicity, ELISA, wound closure assays.	Has good regenerative potential and no regulatory clearance.	Al-Dhuayan et al. (2021)
Curdlan (*Alcaligenes faecalis*)	Injectable gel, scaffold material.	Not yet approved as drugs; food-grade in some countries.	Biocompatibility, sub-chronic toxicity, and determination of inflammatory response.	Studied the wound healing of burns; requires additional clinical testing.	Nurzynska et al. (2021)

(*continued on next page*)

TABLE 18.2 Discussion of wound healing applications of EPS regulatory and safety—cont'd

EPS type/source	Usage in wound healing	Regulatory considerations	Safety tests carried out	Clinical significance	References
Dextran (*Leuconostoc* spp.)	Hydrogel wound dressing.	Approved by FDA as an excipient; GRAS in certain forms.	Cytotoxicity (MTT) test, dermal irritation test, sensitization, and biodegradability.	Applied in hydrogel and scaffold systems; safe to use in topical products.	Zhao and Jalili (2022)
Gellan gum (*Sphingomonas elodea*)	Wound healing film-forming material.	US FDA approved as a food ingredient (GRAS); drug not approved.	Cytotoxicity (ISO 10993-5), Hemocompatibility, Subacute dermal toxicity.	Good moisture-retainer film-former.	Feketshane et al. (2022)
Xanthan gum (*Xanthomonas campestris*)	Thickening and gelation agent in wound formulations.	FDA GRAS status; accepted to be used in foods and cosmetics.	Dermal irritation, Biodegradability studies, Sensitization, In-vivo skin irritation tests.	Topical use is safe; bioadhesive films and hydrogels are used.	Furtado et al. (2022)
Hyaluronic acid (EPS-like, from bacteria)	Moisturizer, repair skin, wound healing.	FDA approved (topical and injectable) CE marked.	Cytotoxicity, irritation, sensitization, and subchronic toxicity testing based on ISO 10993.	It is used in clinics in the regeneration of skin and the management of chronic wounds.	Antoszewska et al. (2024)

EPS, exopolysaccharides.

a different research, recombinant EPS nanoparticles of *E. coli* were used to deliver curcumin, and it demonstrated prolonged release and enhanced wound healing in infected wounds, but there were regulatory and safety issues related to the use of genetically modified organisms (Lan et al., 2025). Moreover, dextran, like EPS, produced by *Leuconostoc mesenteroides*, was found to be highly biocompatible and wound healing in preclinical studies, and was progressed to commercialization, but regulatory issues regarding consistency and documentation slowed its entry to the market.

18.5 Regulatory and safety aspects for exopolysaccharides in wound healing applications

Regulatory and safety issues of EPS use in wound healing products are of utmost importance to the safe transition of research findings to clinical trials. EPS is a biopolymer of microbial origin, and thus it undergoes intensive safety tests prior to human use. As discussed in the Table. 18 2, regulatory bodies like the United States Food and Drug Administration (FDA), the European Medicines Agency (EMA), demand a large amount of preclinical testing, including cytotoxicity, genotoxicity, immunogenicity, and biocompatibility testing, to ensure EPS do not elicit adverse biological activity (Nguyen et al., 2024). Besides, the process of purification of the EPS and any chemical treatment, e.g., sulfation or carboxymethylation to improve the healing properties, should be provided and evaluated in terms of its toxicity in good laboratory practices (GLP). Regulatory approval also involves documenting manufacturing practices under good manufacturing practices (GMP) in a manner that is sterile and reproducible (Yadav et al., 2024). Also, EPS may be used in modern wound care products, like hydrogel dressings or nanofiber scaffolds, which could require classification as a medical device or combination product, which would attract additional regulatory attention. Thus, it is evident that although EPS have a potential of bioactivity against the growth of microbes and anti- inflammatory properties, their safe use in wound healing is contingent on a high level of regulatory compliance, clear safety profiles, and disclosure of manufacturing practices (Norahan et al., 2023).

18.6 Conclusion

EPS are valuable biopolymers with significant potential in wound therapy based on their broad biological properties like antimicrobial, anti-inflammatory, and tissue-regeneration. The fact that they are biocompatible, biodegradable, and can regulate a wound microenvironment makes them appealing solutions in the integration with the advanced wound care items like hydrogels, films, or nanofibrous scaffolds. Nonetheless, in order to achieve their highest potential as therapeutic agents and to accomplish safe clinical translation, critical regulatory and safety hurdles must be resolved. It involves extensive preclinical tests, GLP/GMP compliance, and conformance with the health regulatory standards by global healthcare watchdogs. The potential of EPS as a part of the wound healing technologies presents itself as very bright in the light of innovative, effective, and safer substitutes of conventional wound healing methods, as future research would keep finding new sources and techniques of EPS and perfecting the qualities of functional properties.

Credit author statement

Amlan Bishal: Writing–original draft and editing, Tapan Kumar Shaw: Formal analysis, Bratati Bandyopadhyay: Data curation, Mohini Mondal: Writing review and editing, Paramita Paul: Software.

Conflict of interest

The authors declared that they have not any potential conflict of interest with any third party or any other organization whatsoever.

Ethical statement

Ethics approval and consent to participate: Not applicable

Acknowledgments

The authors are thankful to their institutes for providing the necessary facilities to complete the review.

List of abbreviations

BAP-2	Sulfated polyglucurono-galactopyranan
CE	(European conformity)
DNA	Deoxyribonucleic acid
E-cadherin	Epithelial cadherin
ECM	Extracellular matrix
EGF	Epidermal growth factor
ELISA	Enzyme-linked immunosorbent assay
EMA	European medicines agency
EPS	Exopolysaccharides
FDA	Food and drug administration
FGF	Fibroblast growth factor
GLP	Good laboratory practices
GMP	Good manufacturing practices
GRAS	Generally recognized as safe
HB-EGF	Heparin-binding epidermal growth factor
IL	Interleukin
ISO	International organization for standardization
LAB	Lactic acid bacteria
MMP	Matrix metalloproteinase
PCL	Polycaprolactone
PDGF	Platelet-derived growth factor
TGF-beta	Transforming growth factor beta
TNF-alpha	Tumor necrosis factor alpha
VEGF	Vascular endothelial growth factor
β-catenin	Beta-catenin

References

Abdel-monem, D.A., Sabry, S.A., Ghozlan, H.A., Zaghloul, E.H., 2025. Preparation of novel marine *Enterococcus faecium* MSD8 exopolysaccharide ointment and in vivo evaluation of its impact on cutaneous wound healing in male Albino Rats. Probiotics & Antimicrob. Prot. 17 (3), 963–975. https://doi.org/10.1007/s12602-024-10334-z.

Aderibigbe, B., Buyana, B., 2018. Alginate in wound dressings. Pharmaceutics 10 (2), 42. https://doi.org/10.3390/pharmaceutics10020042.

Ahuja, V., Chauhan, S., Dasgupta, D., Wadhwa, P., Raj, T., Yang, Y.-H., Bhatia, S.K., 2024. Microbial exopolysaccharide composites with inorganic materials and their biomedical applications: a review. Carbohydr. Polym. Technol. Appl. 7, 100482. https://doi.org/10.1016/j.carpta.2024.100482.

Al-Dhuayan, I., Kotb, E., Alqosaibi, A., Mahmoud, A., 2021. Histological studies on a newly isolated *Bacillus* subtilis D10 protease in the debridement of burn wound eschars using mouse model. Pharmaceutics 13 (7), 923. https://doi.org/10.3390/pharmaceutics13070923.

Almatroudi, A., 2025. Biofilm resilience: molecular mechanisms driving antibiotic resistance in clinical contexts. Biology 14 (2), 165. https://doi.org/10.3390/biology14020165.

Antoszewska, M., Sokolewicz, E.M., Barańska-Rybak, W., 2024. Wide use of hyaluronic acid in the process of wound healing—a rapid review. Sci. Pharm. 92 (2), 23. https://doi.org/10.3390/scipharm92020023.

Arslan, N.P., Orak, T., Ozdemir, A., Altun, R., Esim, N., Eroglu, E., Karaagac, S.I., Aktas, C., Taskin, M., 2024. Polysaccharides and peptides with wound healing activity from bacteria and fungi. J. Basic Microbiol. 64 (12). https://doi.org/10.1002/jobm.202400510.

Artini, M., Papa, R., Vrenna, G., Trecca, M., Paris, I., D'Angelo, C., Tutino, M.L., Parrilli, E., Selan, L., 2023. Antarctic marine bacteria as a source of anti-biofilm molecules to combat ESKAPE pathogens. Antibiotics 12 (10), 1556. https://doi.org/10.3390/antibiotics12101556.

Asharaf, S., Chakraborty, K., Paulose, S.K., Dhara, S., Chakraborty, R.D., Varghese, C., 2025. A sulfated exopolysaccharide from *Bacillus altitudinis* MTCC13046 accelerates cutaneous wound healing via dermal fibroblast migration: insights into an in vivo wound re-epithelialization. Int. J. Biol. Macromol. 305, 141001. https://doi.org/10.1016/j.ijbiomac.2025.141001.

Bragulla, H.H., Homberger, D.G., 2009. Structure and functions of keratin proteins in simple, stratified, keratinized and cornified epithelia. J. Anat. 214 (4), 516–559. https://doi.org/10.1111/j.1469-7580.2009.01066.x.

Delano, M.J., Ward, P.A., 2016. The immune system's role in sepsis progression, resolution, and long-term outcome. Immunol. Rev. 274 (1), 330–353. https://doi.org/10.1111/imr.12499.

Demir, A., Aslim, B. (2024). Investigation of multifaceted wound healing effect of exopolysaccharide (EPS) produced from probiotic strain *Lactiplantibacillus plantarum* GD2 as in vitro and in ovo. https://doi.org/10.21203/rs.3.rs-4058404/v1

Deo, R., Lakra, U., Ojha, M., Nigam, V.K., Sharma, S.R., 2025. Exopolysaccharides in microbial interactions: signalling, quorum sensing, and community dynamics. Nat. Prod. Res. 39 (11), 3224–3239. https://doi.org/10.1080/14786419.2024.2405867.

Diller, R.B., Tabor, A.J., 2022. The role of the extracellular matrix (ECM) in wound healing: a review. Biomimetics (Basel, Switzerland) 7 (3), 87. https://doi.org/10.3390/biomimetics7030087.

Elmansy, E.A., Elkady, E.M., Asker, M.S., Abdou, A.M., Abdallah, N.A., Amer, S.K., 2022. Exopolysaccharide produced by *Lactiplantibacillus plantarum* RO30 isolated from Romi cheese: characterization, antioxidant and burn healing activity. World J. Microbiol. Biotechnol. 38 (12). https://doi.org/10.1007/s11274-022-03439-6.

Feketshane, Z., Alven, S., Aderibigbe, B.A., 2022. Gellan gum in wound dressing scaffolds. Polymers 14 (19), 4098. https://doi.org/10.3390/polym14194098.

Furtado, I.F.S.P.C., Sydney, E.B., Rodrigues, S.A., Sydney, A.C.N., 2022. Xanthan gum: applications, challenges, and advantages of this asset of biotechnological origin. Biotechnol. Res. Innov. 6 (1), e202204. https://doi.org/10.4322/biori.202205.

Hamidi, M., Okoro, O.V., Ianiri, G., Jafari, H., Rashidi, K., Ghasemi, S., Castoria, R., Palmieri, D., Delattre, C., Pierre, G., Mirzaei, M., Nie, L., Samadian, H., Shavandi, A., 2023. Exopolysaccharide from the yeast *Papiliotrema terrestris* PT22AV for skin wound healing. StatPearls Publishing. J. Adv. Res. 46, 61–74 doi:10.1016/j.jare.2022.06.012.

Hannoodee, S., Nasuruddin, D.N., 2024. Acute inflammatory response. StatPearls. StatPearls Publishing.

Herman, T.F., Popowicz, P., Bordoni, B., 2025. Wound classification. StatPearls. StatPearls Publishing. http://www.ncbi.nlm.nih.gov/books/NBK554456/.

Hivechi, A., Milan, P.B., Modabberi, K., Amoupour, M., Ebrahimzadeh, K., Gholipour, A.R., Sedighi, F., Amini, N., Bahrami, S.H., Rezapour, A., Hamidi, M., Delattre, C., 2021. Synthesis and characterization of exopolysaccharide encapsulated PCL/gelatin skin substitute for full-thickness wound regeneration. Polymers 13 (6), 854. https://doi.org/10.3390/polym13060854.

Kahraman-Ilıkkan, Ö., Akad Dinçer, S., Bağdat, E.Ş., Çağlar, I., Dertli, E., İspirli, H., Şahin, F.İ., 2024. The wound healing effect and DNA damage repair of exopolysaccharide extracted from *Weissella cibaria* MED17. Bioac. Carbohydr. Diet. Fibre 31, 100400. https://doi.org/10.1016/j.bcdf.2023.100400.

Kaur, N., Dey, P., 2023. Bacterial exopolysaccharides as emerging bioactive macromolecules: from fundamentals to applications. Res. Microbiol. 174 (4), 104024. https://doi.org/10.1016/j.resmic.2022.104024.

Kiran, N.S., Yashaswini, C., Singh, S., Prajapati, B.G., 2024. Revisiting microbial exopolysaccharides: a biocompatible and sustainable polymeric material for multifaceted biomedical applications. 3 Biotech 14 (4), 95. https://doi.org/10.1007/s13205-024-03946-3.

Kolimi, P., Narala, S., Nyavanandi, D., Youssef, A.A.A., Dudhipala, N., 2022. Innovative treatment strategies to accelerate wound healing: trajectory and recent advancements. Cells 11 (15), 2439. https://doi.org/10.3390/cells11152439.

Kulchar, R.J., Singh, R., Ding, S., Alexander, E., Leong, K.W., Daniell, H., 2023. Delivery of biologics: topical administration. Biomaterials 302, 122312. https://doi.org/10.1016/j.biomaterials.2023.122312.

Lan, J., Zou, J., Xin, H., Sun, J., Han, T., Sun, M., Niu, M., 2025. Nanomedicines as disruptors or inhibitors of biofilms: opportunities in addressing antimicrobial resistance. J. Control. Release, StatPearls, StatPearls Publishing, 381, 113589. https://doi.org/10.1016/j.jconrel.2025.113589.

LaPelusa, A., Dave, H.D., 2025. Physiology, Hemostasis. StatPearls. StatPearls Publishing. http://www.ncbi.nlm.nih.gov/books/NBK545263/.

Liu, W., Wei, Y., Xiang, R., Dong, B., Yang, X., 2025. Lactic acid bacteria exopolysaccharides unveiling multifaceted insights from structure to application in foods and health promotion. Foods 14 (5), 823. https://doi.org/10.3390/foods14050823.

Ma, X., Tian, Y., Yang, R., Wang, H., Allahou, L.W., Chang, J., Williams, G., Knowles, J.C., Poma, A., 2024. Nanotechnology in healthcare, and its safety and environmental risks. J. Nanobiotechnol. 22 (1), 715. https://doi.org/10.1186/s12951-024-02901-x.

Majid, S.A., Khorasgani, M.R., Emami, S.H., Talebi, A., Gholami, A.M., 2017. Study of Diabetic cutaneous wound healing in rats treated with *Lactobacillus Casei* and its exopolysaccharide. BioIT 8 (4), 1236–1245.

Mathew-Steiner, S.S., Roy, S., Sen, C.K., 2021. Collagen in wound healing. Bioeng 8 (5), 63. https://doi.org/10.3390/bioengineering8050063.

Mendoza, R.A., Hsieh, J.-C., D Galiano, R., 2019. The impact of biofilm formation on wound healing. In: Dogan, K.H. (Ed.), Wound Healing—Current Perspectives. Intech Open. https://doi.org/10.5772/intechopen.85020.

Meyer, F.A., Weisman, Z., Frojmovic, M.M., 1981. Platelet adhesion to collagen is inhibited by adenosine diphosphate but unaffected by cell shape. Thromb. Haemostasis 46 (2), 515–520.

Netrusov, A.I., Liyaskina, E.V., Kurgaeva, I.V., Liyaskina, A.U., Yang, G., Revin, V.V., 2023. Exopolysaccharides producing bacteria: a review. Microorganisms 11 (6), 1541. https://doi.org/10.3390/microorganisms11061541.

Nguyen, H.-T., Pham, T.-T., Nguyen, P.-T., Le-Buanec, H., Rabetafika, H.N., Razafindralambo, H.L., 2024. Advances in microbial exopolysaccharides: present and future applications. Biomolecules 14 (9), 1162. https://doi.org/10.3390/biom14091162.

Norahan, M.H., Pedroza-González, S.C., Sánchez-Salazar, M.G., Álvarez, M.M., Trujillo De Santiago, G., 2023. Structural and biological engineering of 3D hydrogels for wound healing. Bioactive Mat. 24, 197–235. https://doi.org/10.1016/j.bioactmat.2022.11.019.

Nurzynska, A., Klimek, K., Palka, K., Szajnecki, Ł., Ginalska, G., 2021. Curdlan-based hydrogels for potential application as dressings for promotion of skin wound healing—preliminary in vitro studies. Materials 14 (9), 2344. https://doi.org/10.3390/ma14092344.

Opneja, A., Kapoor, S., Stavrou, E.X., 2019. Contribution of platelets, the coagulation and fibrinolytic systems to cutaneous wound healing. Thromb. Res. 179, 56–63. https://doi.org/10.1016/j.thromres.2019.05.001.

Osorio-Valencia, S., Zhou, B., 2024. Roles of macrophages and endothelial cells and their crosstalk in acute lung injury. Biomedicines 12 (3), 632. https://doi.org/10.3390/biomedicines12030632.

Ousey, K., Cutting, K.F., Rogers, A.A., Rippon, M.G., 2016. The importance of hydration in wound healing: reinvigorating the clinical perspective. J. Wound Care 25 (3), 122–130. https://doi.org/10.12968/jowc.2016.25.3.122.

Parvin, N., Joo, S.W., Mandal, T.K., 2025. Biodegradable and stimuli-responsive nanomaterials for targeted drug delivery in autoimmune diseases. J. Funct. Biomat. 16 (1), 24. https://doi.org/10.3390/jfb16010024.

Pinto, R.M., Soares, F.A., Reis, S., Nunes, C., Van Dijck, P., 2020. Innovative strategies toward the disassembly of the EPS matrix in bacterial biofilms. Front. Microbiol. 11, 952. https://doi.org/10.3389/fmicb.2020.00952.

Rai, M., Wypij, M., Ingle, A.P., Trzcińska-Wencel, J., Golińska, P., 2021. Emerging trends in pullulan-based antimicrobial systems for various applications. Int. J. Mol. Sci. 22 (24), 13596. https://doi.org/10.3390/ijms222413596.

Rosales, C., 2018. Neutrophil: a cell with many roles in inflammation or several cell types? Front.Physiol. 9, 113. https://doi.org/10.3389/fphys.2018.00113.

Sahana, T.G., Rekha, P.D., 2020. A novel exopolysaccharide from marine bacterium Pantoea sp. YU16-S3 accelerates cutaneous wound healing through Wnt/β-catenin pathway. Carbohydr. Polym. 238, 116191. https://doi.org/10.1016/j.carbpol.2020.116191.

Salimi, F., Farrokh, P., 2023. Recent advances in the biological activities of microbial exopolysaccharides. World J. Microbiol. Biotechnol. 39 (8), 213. https://doi.org/10.1007/s11274-023-03660-x.

Schultz, G.S., Chin, G.A., Moldawer, L., Diegelmann, R.F., 2011. Principles of Wound Healing. In: Fitridge, R., Thompson, M. (Eds.), Mechanisms of Vascular Disease: A Reference Book for Vascular Specialists. University of Adelaide Press. http://www.ncbi.nlm.nih.gov/books/NBK534261/.

Scopelliti, F., Cattani, C., Dimartino, V., Mirisola, C., Cavani, A., 2022. Platelet derivatives and the immunomodulation of wound healing. Int. J. Mol. Sci. 23 (15), 8370. https://doi.org/10.3390/ijms23158370.

Tiwari, O.N., Sasmal, S., Kataria, A.K., Devi, I., 2020. Application of microbial extracellular carbohydrate polymeric substances in food and allied industries. 3 Biotech 10 (5), 221. https://doi.org/10.1007/s13205-020-02200-w.

Trabelsi, I., Ktari, N., Ben Slima, S., Triki, M., Bardaa, S., Mnif, H., Ben Salah, R., 2017. Evaluation of dermal wound healing activity and in vitro antibacterial and antioxidant activities of a new exopolysaccharide produced by *Lactobacillus sp.* Ca 6. Int. J. Biol. Macromol. 103, 194–201. https://doi.org/10.1016/j.ijbiomac.2017.05.017.

Wang, X., Li, X., Liu, J., Tao, Y., Wang, T., Li, L., 2024. *Lactobacillus Plantarum* promotes wound healing by inhibiting the NLRP3 inflammasome and pyroptosis activation in diabetic foot wounds. J. Inflamm. Res. 17, 1707–1720. https://doi.org/10.2147/JIR.S449565.

Yadav, M.K., Song, J.H., Vasquez, R., Lee, J.S., Kim, I.H., Kang, D.-K., 2024. Methods for detection, extraction, purification, and characterization of exopolysaccharides of lactic acid bacteria—a systematic review. Foods 13 (22), 3687. https://doi.org/10.3390/foods13223687.

Zaghloul, E.H., Ibrahim, M.I.A., 2022. Production and characterization of exopolysaccharide from newly isolated marine probiotic *Lactiplantibacillus plantarum* EI6 with in vitro wound healing activity. Front. Microbiol. 13. https://doi.org/10.3389/fmicb.2022.903363.

Zhao, A., Sun, J., Liu, Y., 2023. Understanding bacterial biofilms: from definition to treatment strategies. Front. Cell. Infect. Microbiol. 13, 1137947. https://doi.org/10.3389/fcimb.2023.1137947.

Zhao, Y., Jalili, S., 2022. Dextran, as a biological macromolecule for the development of bioactive wound dressing materials: a review of recent progress and future perspectives. Int. J. Biol. Macromol. 207, 666–682. https://doi.org/10.1016/j.ijbiomac.2022.02.114.

CHAPTER

19

The significant role of exopolysaccharides in biodegradable, renewable, and sustainable bioplastic production technology

Vadde Venkatesh[a], *Gundu Shravani*[b], *Prasanna Kumar Desu*[c] *and Vanitha Kondi*[a]

[a]Department of Pharmaceutics, Vishnu Institute of Pharmaceutical Education and Research, Narsapur, Medak, Telangana, India [b]Department of Pharmaceutical Chemistry, Vishnu Institute of Pharmaceutical Education and Research, Narsapur, Medak, Telangana, India [c]Principal & Professor, Dr. Samuel George Institute of Pharmaceutical Sciences, Markapur, Andhra Pradesh, India

19.1 Introduction

To reduce plastic pollution and fossil resource depletion, bioplastics have become a promising alternative to petroleum-based plastics (Ibrahim et al., 2021). As the global demand for sustainable materials increases, bioplastics derived from renewable biological resources have garnered significant attention due to their potential to reduce the ecological footprint of plastic products. This overview examines the types, advantages, challenges, and scientific implications of bioplastics in the context of material science, environmental sustainability, and industrial applications (Altamira-Algarra et al., 2023).

19.1.1 Overview of bioplastics

Bioplastics consist of many polymers that can be divided into two categories:

DOI: https://doi.org/10.1016/B978-0-443-36619-2.00019-X

19.1.1.1 Bio-based plastics

Plastics made from plants, algae, or microbes are renewable. Plastics made from biomaterials may degrade (Dilshad et al., 2021). Examples: Polylactic acid (PLA), polyhydroxyalkanoates (PHA), and polyamide 11 (PA11).

19.1.1.2 Biodegradable plastics

These plastics are designed to degrade under natural environmental conditions, either through microbial action or through chemical processes. Some biodegradable plastics are also bio-based, while others are synthetic (Mouro et al., 2024).

Examples: Polybutylene succinate (PBS), polycaprolactone (PCL)

19.1.2 Advantages

1. Sustainability: Derived from renewable resources like plants, algae, and microorganisms, reducing dependence on fossil fuels.
2. Biodegradability: Many bioplastics, such as PLA and PHA, are biodegradable, breaking down naturally over time and reducing plastic pollution.
3. Lower carbon footprint: Bioplastics may reduce greenhouse gas emissions because feedstocks absorb CO_2 during growth.
4. Nontoxic: Bioplastics are safer for food packaging and medical devices since they lack BPA, phthalates, and PVC.
5. Reduced waste: Biodegradable polymers disintegrate faster than regular plastics, minimizing landfill and ocean plastic waste.
6. Versatility: Bioplastics are utilized in packaging, textiles, medical devices, agricultural, and automotive.
7. Reduced dependency on petroleum: Bioplastics use renewable resources to replace petroleum-based plastics.
8. Compostability: Industrial composting of some bioplastics replenishes soil and reduces landfill waste.

19.1.3 Disadvantages

1. Higher production costs: Currently, the production of bioplastics is more expensive than conventional plastics due to the cost of raw materials (e.g., crops, fermentation) and manufacturing processes.
2. Limited performance: Bioplastics may not match the durability, heat resistance, or mechanical properties of petroleum-based plastics, limiting their applications in some industries.
3. Land use and food security: The use of food crops like corn and sugarcane for bioplastics production competes with food production, raising concerns about food security and land availability.
4. Waste management issues: Bioplastics require specific environmental conditions (such as industrial composting facilities) for proper degradation. Without proper waste management infrastructure, biodegradability may not occur as intended.

5. Environmental impact of raw materials: The agricultural production of raw materials for bioplastics can lead to deforestation, water use, pesticide use, and land degradation, which can offset some of the environmental benefits.
6. Limited recycling options: Some bioplastics, like PLA, are not easily recyclable in existing plastic recycling systems, leading to challenges in developing a circular economy for these materials.
7. Inadequate biodegradation in natural environments: While many bioplastics are biodegradable, some require specific conditions (e.g., industrial composting environments) to break down, and may not degrade effectively in natural environments like oceans or landfills.
8. Potential for greenwashing: Some companies may market bioplastics as fully sustainable without addressing the broader environmental impacts (such as land use or water consumption), leading to potential "green washing" of products.

19.1.4 Applications of bioplastics in various sectors

1. Packaging: Bioplastics, particularly PLA and PHA, have become prominent in food packaging, consumer goods, and beverage containers. Their biodegradability makes them attractive for single-use packaging solutions, which are a significant contributor to plastic waste (Gamage et al., 2024).
2. Medical and pharmaceutical: Bioplastics are used in medical devices, surgical sutures, drug delivery systems, and other healthcare applications. Their biocompatibility and biodegradability make them ideal for temporary implants and controlled-release systems.
3. Agriculture: Mulch films, crop protection, and slow-release fertilizers use biodegradable films and coatings made from bioplastics like PHA and PBS. These materials mitigate farming's plastic waste.
4. Textiles: Bioplastics are also employed in the textile industry for the production of fibers used in clothing, upholstery, and home goods. PLA-based fibers are already used in various fabrics, and ongoing research is focused on improving their performance characteristics.
5. Automotive and electronics: Bioplastics are being integrated into the production of automotive parts and electronic devices. Their potential to reduce vehicle weight and contribute to more sustainable manufacturing processes makes them appealing to industries seeking greener alternatives (Gamage et al., 2024).

19.1.5 The need for sustainable alternatives

The increasing environmental burden posed by conventional petroleum-based plastics has necessitated the urgent development of sustainable alternatives. Conventional plastics are characterized by their nonbiodegradability, excessive carbon footprint, and contribution to global plastic pollution (Mouro et al., 2024). With the escalating concerns surrounding climate change, ecosystem degradation, and resource depletion, there is a growing demand for innovative materials that can offer functionality while minimizing environmental impact. Sustainable alternatives, particularly biodegradable and bio-based polymers, have emerged as promising candidates to address these challenges.

1. Environmental and economic challenges of conventional plastics.
   a. Persistence in the environment

   b. High carbon emissions and resource depletion
   c. Inefficiencies in waste management
2. Criteria for sustainable alternatives.
   a. Bio-based composition
   b. Biodegradability and composability
   c. Lower carbon footprint
   d. Functional Equivalence
   e. Economic Viability
3. Emerging sustainable alternatives
   a. Bioplastics
   b. Recyclable and renewable alternatives
4. Challenges and future directions.
   a. Cost and scalability
   b. Performance limitations
   c. Waste management infrastructure

19.2 Exopolysaccharides: Structure and function

Microorganisms release high-molecular-weight carbohydrate polymers called exopolysaccharides (EPS). These biopolymers are essential for microbial physiology, biofilm development, and environmental interactions (Benalaya et al., 2024). EPS exhibit diverse physicochemical properties and structural variations, making them valuable in biotechnological applications such as bioplastics, pharmaceuticals, and the food industry. Understanding their structure and function is essential for harnessing their potential in sustainable material development (Nadzir et al., 2021).

EPS are structurally diverse biopolymers that play essential roles in microbial ecology, biofilm formation, and industrial applications. Their functional properties, including biodegradability, biocompatibility, and hydrogel-forming abilities, make them promising candidates for sustainable bio-material innovations (Mu et al., 2021). Ongoing research in microbial biotechnology and polymer science continues to explore novel EPS-based materials for applications in medicine, food, and bioplastics (Fig.19.1). Future advancements in genetic engineering and metabolic pathway optimization will further enhance the production and functionalization of EPS for eco-friendly material development (Laroche, 2022).

19.3 Exopolysaccharides as a base for bioplastic production

The biodegradable plastics are being considered as an alternative to petroleum-based plastics due to global demand for sustainable products. Bioplastics made from microorganism-secreted EPS, high-molecular-weight polysaccharides, seem a promising base for the development of bioplastics (Moradi et al., 2021). These biopolymers are biodegradable and have flexible functional features like high-water retention, biocompatibility, and biodegradability, making them great candidates for ecologically friendly and renewable bioplastics. EPS is a potential material for biodegradable, renewable, and multifunctional bioplastics. With continued research and

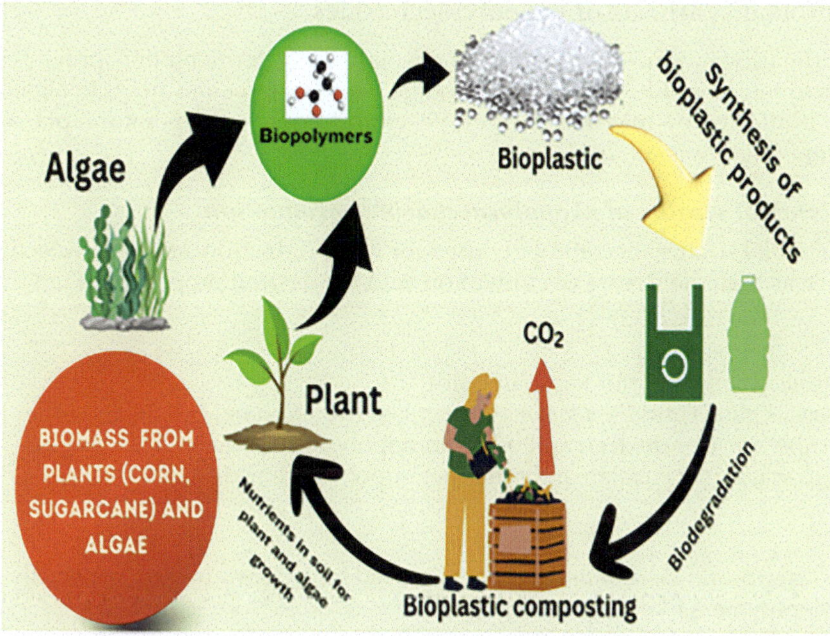

FIGURE 19.1 EPS as a base for bioplastic production. *EPS*, Exopolysaccharides.

technology breakthroughs, EPS-based bioplastics can reduce plastic pollution and petroleum dependence (Altamira-Algarra et al., 2024a). The future of EPS in bioplastic production hinges on improving scalability, optimizing production processes, and enhancing the material properties to meet the demands of various industrial applications.

1. Renewable feedstock: EPS are produced by microorganisms that can be cultivated on renewable resources, such as agricultural waste, carbon dioxide, or industrial by-products, making the production process sustainable and less reliant on fossil fuels.
2. Biodegradability: Many EPS biodegrade naturally, solving the plastic waste problem in landfills and oceans.
3. Versatility: The chemical structure of EPS can be modified to tailor specific properties required for bioplastic applications, such as flexibility, strength, and moisture resistance.

19.4 Production technology

The production of EPS for bioplastic applications involve several technological steps, ranging from microbial fermentation to polymer extraction and processing. The optimization of production techniques and the integration of EPS into bioplastic manufacturing is critical for achieving sustainable, large-scale production. Below is a detailed overview of the key aspects of EPS production technology, microbial synthesis, optimization strategies, and integration into bioplastic manufacturing (Altamira-Algarra et al., 2024b).

19.4.1 Microbial synthesis of exopolysaccharides

EPS are primarily produced by microorganisms during fermentation processes. The microorganisms involved in EPS synthesis can range from bacteria and fungi to algae. Microbial fermentation is affected by microorganism type, carbon source, temperature, pH, and oxygen levels (Altamira-Algarra et al., 2025).

19.4.1.1 Microbial strains of exopolysaccharides production

Several microbial strains are commonly used for EPS production, and their selection is based on factors such as high yield, ease of cultivation, and the desired properties of the EPS.

- Bacteria:
  - *Xanthomonas campestris* (for xanthan gum)
  - *Pseudomonas putida* and *Pseudomonas aeruginosa* (for alginate, gellan, and other EPS)
  - *Lactobacillus spp.* (for dextran and other homopolysaccharides)
  - *Enterobacter spp.* (for curdlan and other heteropolysaccharides)
- Fungi:
  - *Aspergillus* spp. and *Penicillium* spp. are also used for EPS production, especially for fungal-based biopolymers like mannan and glucan.
- Algae:
  - Brown algae (e.g., *Laminaria, Ascophyllum*) for alginates and carrageenan production.

19.4.1.2 Fermentation process

- Batch fermentation: In batch fermentation, bacteria create EPS for metabolism in a fixed volume of nutrient-rich media. This is the most prevalent EPS manufacturing method, although time and nutrient exhaustion limit it (Muneer et al., 2021).
- Continuous fermentation: This method involves the continuous addition of nutrients and the removal of culture broth, allowing for more efficient EPS production and a continuous supply of bioplastic feedstock. This method offers higher yields and is more scalable for industrial production.
- Fed-batch fermentation: This hybrid approach involves periodic nutrient additions to the culture medium during fermentation to maintain optimal growth conditions. It is commonly used to enhance EPS production efficiency (Okoro et al., 2021).

19.4.1.3 Key parameters in microbial fermentation

The key parameters that influence EPS production are:

- Carbon source: Sugars (glucose, sucrose, lactose), waste streams (glycerol, molasses), and even CO_2 can serve as carbon sources. The choice of carbon source directly affects the yield and composition of EPS.
- Nitrogen source: Organic nitrogen sources like peptone or inorganic nitrogen salts (ammonium nitrate) are critical for cellular growth and EPS biosynthesis.

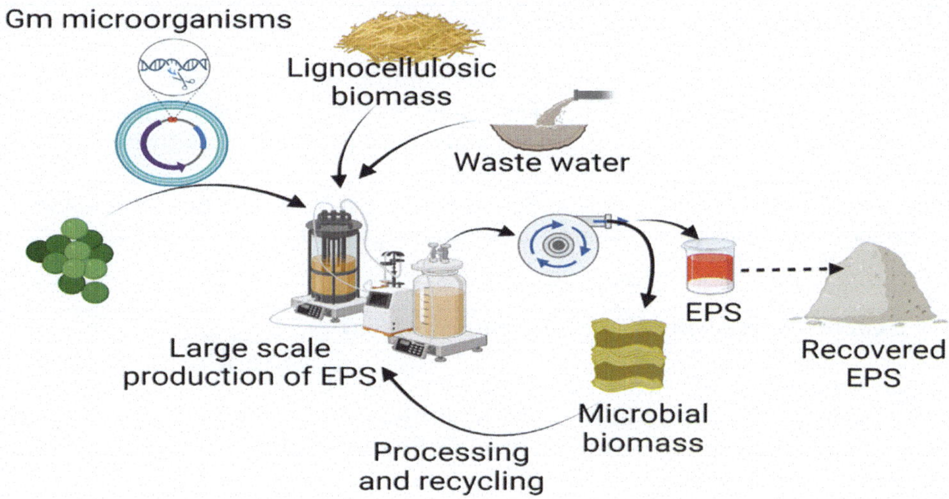

FIGURE 19.2 Microbial exopolysaccharides.

- pH and temperature: These environmental factors must be optimized for the specific microbial strain to maximize EPS production. For example, *Xanthomonas campestris* typically produces high yields of xanthan gum at pH 7–8 and 28°C–30°C.
- Oxygen and agitation: Aerobic fermentation is often required, with adequate oxygen supply to support the growth of the microorganism and EPS synthesis (Fig. 19.2). Agitation speeds are also crucial for maintaining uniform growth conditions (Okoro et al., 2021).

19.4.2 Optimization of exopolysaccharides production

Optimization solutions are essential to increase EPS yield and quality and lower microbial fermentation costs. These strategies focus on optimizing the fermentation conditions, media composition, and microbial strain improvement.

19.4.2.1 Nutritional optimization

The media composition significantly affects the rate of EPS production. The following strategies are used for optimizing the nutrient components:

- Carbon and nitrogen sources: Tailoring the carbon and nitrogen sources can enhance EPS yield. For example, using inexpensive waste products like agricultural residues (e.g., corn stalks, sugarcane bagasse) or by-products (e.g., glycerol from biodiesel production) can lower production costs.
- Trace elements and growth factors: Some microbial strains require specific trace elements (e.g., calcium, magnesium) or vitamins (e.g., biotin, thiamine) for optimal growth and EPS production (Dirpan et al., 2023).

19.4.2.2 *Environmental factors optimization*

- Temperature and pH control: Monitoring and controlling temperature and pH in the fermentation process can maximize the efficiency of EPS synthesis. For instance, some bacterial strains may require a slightly acidic environment, while others may thrive under neutral conditions.
- Oxygen transfer rate: Ensuring adequate oxygen levels during fermentation is critical for aerobic microorganisms. Using aerators or optimizing agitation speeds can improve oxygen transfer efficiency.

19.4.2.3 *Genetic engineering and strain improvement*

- Metabolic engineering: By genetically modifying microbial strains to enhance metabolic pathways that lead to higher EPS yields, researchers can optimize production efficiency. For instance, overexpressing genes involved in EPS biosynthesis or introducing synthetic pathways can increase EPS synthesis.
- Mutagenesis: Random mutagenesis followed by screening for EPS-producing strains with improved yields is another common approach.
- CRISPR/Cas and synthetic biology: Advanced genetic tools such as CRISPR can be used to precisely edit genes involved in EPS production, enhancing microbial productivity and EPS properties (Dirpan et al., 2023).

19.4.2.4 *Bioreactor design and process control*

- Bioreactor design: The choice of bioreactor (stirred-tank, airlift, or bubble-column) affects the efficiency of EPS production. The design should ensure proper aeration, mixing, and nutrient distribution.
- Process monitoring: Online sensors for monitoring key fermentation parameters (e.g., pH, dissolved oxygen, temperature, biomass concentration) allow for precise control and optimization of EPS production.

19.4.3 Integration into bioplastic manufacturing

Once EPS are synthesized, they need to be isolated, purified, and processed into bioplastics. The integration of EPS into bioplastic manufacturing involves several steps:

19.4.3.1 *Exopolysaccharides isolation and purification*

After microbial fermentation, EPS are typically isolated from the culture medium using various methods, such as:

- Precipitation: EPS can be precipitated from the fermentation broth using alcohol (e.g., ethanol or isopropanol), which helps concentrate and purify the polymer.
- Filtration: Techniques such as ultra-filtration or dialysis may be used to remove small molecules and other impurities from the EPS.
- Centrifugation: High-speed centrifugation can separate EPS from the microbial cells and other suspended solids.

19.4.3.2 Modification of exopolysaccharides properties

To enhance the material properties of EPS-based bioplastics, chemical modifications or blending with other biopolymers (such as PLA, PHA, or starch) may be necessary. This can improve properties such as:

- Mechanical strength (through cross-linking or reinforcement)
- Thermal stability (by adding plasticizers or stabilizers)
- Moisture resistance (by incorporating hydrophobic agents)

19.4.3.3 Bioplastic processing techniques

- Film casting: EPS can be dissolved in water or other solvents, and the solution can be cast into thin films for packaging applications.
- Injection molding: EPS-based bioplastics can be processed using traditional polymer molding techniques such as extrusion and injection molding to produce solid objects and containers.
- Blending and compounding: Blending EPS with other biodegradable polymers (e.g., PHA, PLA) improves the mechanical properties and expands the range of applications for bioplastics (Narayanan et al., 2023).

19.5 Exopolysaccharides as a base for bioplastic production in scientific writing

The growing global concern over environmental pollution, particularly plastic waste accumulation, has intensified the need for sustainable alternatives to conventional petroleum-based plastics. EPS, biopolymers produced by microorganisms, have emerged as an important class of materials with significant potential for addressing environmental sustainability issues. EPS are biodegradable, renewable, and environmentally friendly, making them promising candidates for bioplastic production. In this context, understanding the sustainability and environmental benefits of EPS is essential for evaluating their role in reducing the environmental impact of plastic materials (Varghese et al., 2022).

19.5.1 Biodegradability and environmental impact

An important environmental benefit of EPS-based bioplastics is biodegradability. EPS breaks down naturally in soil, water, or compost, unlike typical plastics that last for centuries.

19.5.1.1 Biodegradability in various environments

- Soil: EPS-based materials, such as xanthan gum, gellan gum, and alginate, undergo rapid biodegradation in soil, reducing their long-term environmental impact. Environmental elements like moisture, temperature, and microbial population affect deterioration.
- Water: EPS are typically water-soluble and can be degraded by aquatic microorganisms, contributing to a reduction in plastic pollution in water bodies. Some EPS, like alginate, are known to degrade effectively in marine environments.

- Composting: Many EPS-based bioplastics can be broken down efficiently in composting environments, where microorganisms metabolize them into nontoxic byproducts such as carbon dioxide, water, and humus.

19.5.1.2 *Comparison with petroleum-based plastics*

Polyethylene and polypropylene, on the other hand, biodegrade slowly and can last decades. Plastic trash accumulates in landfills, oceans, and ecosystems. EPS-based bioplastics are biodegradable and reduce long-term waste and contamination (Samadhiya et al., 2022).

19.5.2 Renewable resources and reduced carbon footprint

The sustainability of EPS as bioplastics is closely tied to the use of renewable resources and their relatively low carbon footprint. The production of EPS involves microbial fermentation using renewable raw materials, such as agricultural waste, glycerol, or carbon dioxide, as carbon sources. This makes EPS production more sustainable compared to the petroleum-based raw materials used for conventional plastics (Samadhiya et al., 2022).

19.5.2.1 *Comparison with petroleum-based plastics*

- Agricultural by-products: Sugars (from sugarcane, corn, or molasses), glycerol (from biodiesel manufacturing), and starch can be used to culture EPS-producing microbes. These by-products minimize agricultural waste and production expenses.
- CO_2 utilization: EPS-producing bacteria like *Pseudomonas* and *Corynebacterium* can use CO as a carbon source, decreasing the environmental impact and contributing to carbon sequestration (Samadhiya et al., 2022).

19.5.2.2 *Low carbon emissions*

The carbon footprint of EPS-based bioplastics is significantly lower than that of conventional plastics. The manufacture of petroleum-based polymers is energy-intensive and releases significant amounts of CO into the atmosphere. However, renewable feedstocks and energy-efficient fermentation reduce greenhouse gas emissions in EPS synthesis. Because EPS are biodegradable, their end-of-life disposal has less environmental impact than plastics (Samadhiya et al., 2022).

19.5.2.3 *Life cycle assessments*

Studies that conduct lifecycle assessments (LCA) of EPS-based bioplastics often show that their environmental impact—from raw material extraction to disposal—is lower than that of petroleum-based plastics. These assessments measure EPS's environmental advantages by assessing raw material source, production, transportation, use, and disposal.

19.5.3 Reduction of plastic pollution and marine impact

The oceans receive millions of tons of plastic waste each year, making it a global problem. EPS-based bioplastics reduce plastic pollution's environmental impact (Luo et al., 2022).

19.5.3.1 *Biodegradation in aquatic environments*

EPS are often water-soluble and can be readily degraded by marine microorganisms. This makes EPS-based bioplastics an attractive option for marine applications, such as biodegradable packaging, fishing nets, and agricultural films, where their ability to degrade in seawater prevents the long-term accumulation of waste in oceans. In contrast, traditional plastics persist in marine habitats for decades, harming marine life through ingestion, entanglement, and chemical leaching (Vijayanand et al., 2023).

19.5.3.2 *Preventing ocean micro plastic pollution*

One of the major environmental challenges of conventional plastics is the formation of microplastics—small plastic particles that are less than 5 mm in diameter. These microplastics are often ingested by marine organisms and enter the food chain. EPS-based bioplastics, being biodegradable, are less likely to contribute to microplastic pollution. As they degrade, they break down into nontoxic substances that do not persist in the environment or pose a threat to aquatic organisms (Onyena et al., 2021).

19.5.4 Reduction of toxicity and nontoxic degradation products

EPS-based bioplastics are generally nontoxic, both to humans and the environment. As they degrade, the byproducts are typically harmless, consisting mainly of water, carbon dioxide, and organic matter that can be assimilated by microorganisms. This stands in contrast to petroleum-based plastics, which can release harmful chemicals such as bisphenol A (BPA), phthalates, and styrene upon degradation (Panigrahy et al., 2022).

19.5.4.1 *Nontoxicity ecosystems*

EPS-based materials, such as xanthan gum and alginate, are recognized for their biocompatibility and nontoxic nature. They have been used in food products, pharmaceuticals, and medical devices without posing health risks. Their degradation products, including organic acids and sugars, are naturally occurring compounds that do not cause harm to ecosystems when released into the environment (Panigrahy et al., 2022).

19.5.4.2 *Safe disposal and composting*

The regulated breakdown of EPS-based bioplastics into nutrient-rich organic matter in composting facilities is safe. Compost enriches soil, promoting sustainable agriculture and lowering synthetic fertilizer consumption (Panigrahy et al., 2022).

19.5.5 Potential for circular economy

Circular economies reduce waste and maximize resources. EPS-based bioplastics are renewable and biodegradable, supporting circular economy concepts (Panigrahy et al., 2022).

19.5.5.1 *Closed loop cycling*

EPS-based materials can be recycled in a closed-loop system, where the waste generated at the end of a product's life is reintegrated into the production process. This contrasts with conventional plastics, which are often downcycled or discarded, leading to the accumulation of nondegradable waste in landfills (Mousavi et al., 2024).

19.5.5.2 Biorefinery models

EPS production could be integrated into bio-refinery models, where waste biomass from agriculture, forestry, or food processing industries is converted into EPS and other valuable bio-based products. This decreases fossil fuel use and diverts organic waste from landfills, promoting sustainability (Mousavi et al., 2024).

19.6 Challenges and future directions

EPS holds great promise as a sustainable alternative to petroleum-based plastics; there are still significant challenges that must be addressed to fully realize its potential in bioplastic applications. These challenges range from technological barriers in the production process to scalability issues (Alias et al., 2022).

19.6.1 Technological barriers

19.6.1.1 Limited strain selection and optimization

One of the primary barriers to large-scale EPS production is the limited availability of high-yield microbial strains that can consistently produce EPS in sufficient quantities and at a low cost. Not all microorganisms are equally efficient in synthesizing EPS, and many of the strains used for EPS production require further genetic or metabolic optimization (Findrik and Meixner, 2023).

1. Genetic engineering: The use of genetic engineering to modify microorganisms can enhance their EPS production capabilities, but the application of genetic modification techniques is still limited, especially in nonmodel organisms. Furthermore, the regulatory concerns around genetically modified organisms (GMOs) may limit the widespread adoption of engineered strains.
2. Microbial diversity: There is a need to explore more microbial diversity (including bacteria, fungi, and algae) to discover novel strains that can produce EPS with specific properties suitable for different bioplastic applications. This requires extensive screening and research, which can be resource-intensive (Findrik and Meixner, 2023).

19.6.1.2 Process control and optimization

Fermentation processes for EPS production are often complex and highly sensitive to environmental variables. Maintaining optimal conditions in large-scale fermentation systems can be difficult (Rabiya and Sen, 2022).

- Strain performance variability: Even under controlled laboratory conditions, the performance of microbial strains can vary, which presents a challenge for consistency in EPS production.

- Complexity of fermentation: The microbial fermentation process can be slow and energy-intensive, requiring careful control of variables to maximize yields. In large-scale applications, this is even more challenging, as minor fluctuations can lead to significant losses in EPS yield and quality.

19.6.1.3 Exopolysaccharides purification and processing challenges

Once EPS are synthesized, they need to be extracted, purified, and processed into usable bioplastic forms. The purification is complex, labor-intensive, and expensive. Common challenges include:

- Separation of EPS from culture media: EPS are often produced in aqueous fermentation broths and need to be separated from microbial cells and other residual materials. Methods like centrifugation, filtration, and precipitation are typically used but may not be efficient for large-scale production.
- Solubility and viscosity issues: Some EPS, like xanthan gum, are highly viscous, which can make processing difficult. High viscosity can complicate the blending or extrusion processes required to produce bioplastic films and other products.

19.6.2 Enhancing production scalability of exopolysaccharides

The scalability of EPS production remains one of the most significant challenges to its commercial viability. For EPS-based bioplastics to be widely adopted as sustainable materials, production processes must be both cost-effective and efficient at a large scale. Several strategies can be explored to address these scalability challenges.

19.6.2.1 Exopolysaccharides purification and processing challenges

Laboratory-scale fermentation to industrial-scale production needs major process control, equipment, and infrastructure improvements. Scalability-boosting methods include:

- Bioreactor design: The design of the bioreactor used for microbial fermentation must be optimized for high-efficiency EPS production. This includes ensuring proper mixing, aeration, and temperature control. More advanced bioreactor designs, such as membrane bioreactors or airlift reactors, may help reduce energy consumption and improve overall yield.
- Automated process control: The integration of automated monitoring and control systems can help maintain consistent fermentation conditions, ensuring that microbial growth and EPS synthesis are optimized in large-scale bioreactors (Rabiya and Sen, 2022).
- Feedstock availability and cost: The availability of affordable and renewable feedstocks is crucial for scaling up EPS production. Research into alternative raw materials, such as agricultural waste, carbon dioxide, or industrial by-products, can help lower costs and increase the sustainability of EPS production. Utilizing waste products from other industries could create a closed-loop system, improving the overall economics of EPS production.

19.6.2.2 Economic feasibility and cost reduction

One of the most significant barriers to the large-scale commercialization of EPS-based bioplastics is the cost of production. The cost of raw materials, fermentation, and purification makes EPS production expensive. Strategies to cut costs include:

- Waste valorization: Using waste products such as glycerol, agricultural residues, or CO_2 as raw materials for EPS production could reduce feedstock costs and enhance process sustainability.
- Improved strain engineering: Genetic and metabolic engineering of microbial strains to boost EPS yield per input material can minimize production costs. This would reduce the amount of feedstock required and improve the economic feasibility of EPS-based bioplastic production (Rabiya and Sen, 2022).
- Scale-up of purification processes: Advances in filtration technologies, such as ultra-filtration and membrane filtration, may offer more efficient methods for purifying EPS and reducing associated costs. The development of continuous flow systems for EPS extraction and purification could also streamline the process.

19.6.2.3 Integration with other biopolymers

While EPS are biodegradable, their mechanical properties (such as strength, flexibility, and water resistance) can sometimes be inadequate for certain bioplastic applications. To overcome these limitations and improve the performance of EPS-based bioplastics, they can be blended with other biopolymers like PLA, PHA, or starch.

- Blending and compounding: Mixing EPS with other biodegradable polymers can enhance the material properties of bioplastics, such as tensile strength, thermal stability, and moisture resistance. This also helps reduce the overall reliance on EPS alone and makes the production of bioplastics more scalable.
- Co-processing techniques: Co-processing EPS with other bioplastics in a single manufacturing process can help simplify the production line and reduce costs. This could involve simultaneous processing steps, such as extrusion or injection molding, to produce more diverse types of bioplastic products.

19.6.2.4 Government policy and industry support

Government policies and industry support play a crucial role in the advancement of EPS-based bioplastics. Research funding, subsidies for sustainable materials, and incentives for bio-based innovation can help address many of the challenges associated with scalability. Additionally, partnerships between academia, industry, and government will facilitate the commercialization of EPS-based bioplastics by fostering collaboration and reducing the risks associated with scaling up.

19.6.3 Future trends

To address the challenges of EPS-based bioplastic production, future research and technological advancements should focus on the following areas:

19.6.3.1 Enhanced microbial strains and genetic engineering

Develop and utilize genetic modification techniques to enhance EPS production in microorganisms, such as CRISPR-Cas9 or synthetic biology approaches. This can lead to higher yields, more consistent quality, and lower production costs. Explore microbial consortia or co-cultures that utilize multiple microbial species to produce diverse EPS types with tailored properties, improving the range of potential applications.

19.6.3.2 Integrated bio-refinery models

Implement bio-refinery models that integrate EPS production with other bio-based products, such as bio-fuels, bio-energy, and bio-chemicals. This approach maximizes the utilization of feedstock and minimizes waste, making EPS production more efficient and economically viable.

19.6.3.3 Circular economy and waste valorization

Promote circular economy principles, where EPS-based bioplastics are produced using waste materials and can be fully recycled or composted at the end of their lifecycle. This reduces the need for virgin resources and minimizes environmental pollution.

19.6.3.4 Commercialization and market penetration

Work toward lowering production costs and improving the performance of EPS-based bioplastics to compete with conventional plastics. The development of cost-effective, scalable production methods will facilitate the widespread adoption of EPS in various industries, including packaging, agriculture, and medicine.

19.7 Summary and conclusion

Microorganisms develop EPS, biopolymers having bioplastic potential. The rising plastic pollution problem can be solved by these biodegradable, renewable, and sustainable polymers. EPS-based bioplastics can be made from agricultural byproducts and industrial waste, making them more environmentally friendly. EPS has potential, but microbial strain optimization, fermentation process scalability, and cost-effective purification remain obstacles.

To enable large-scale commercial manufacturing, technological constraints, including fermentation system efficiency and EPS yield, must be overcome. Research continues to improve EPS's strength, flexibility, and moisture resistance by combining it with other biopolymers. In conclusion, while obstacles remain, EPS-based bioplastics are a potential step toward sustainable materials that can reduce plastic pollution and advance circular economy concepts. Microbial biotechnology, process optimization, and product development must continue to evolve to maximize EPS's bioplastic innovation potential.

Credit author statement

1. **V Venkatesh:** Conceptualization, methodology, investigation, data curation, writing—original draft, visualization.
2. **G Shravani:** Methodology, formal analysis, validation, writing—review and editing.
3. **Prasanna Kumar Desu:** Supervision, resources, project administration, writing—review and editing.
4. **Vanitha Kondi:** Conceptualization, supervision, funding acquisition, critical revision of the manuscript, writing—review and editing.

Conflict of interest

The authors declare that there is no conflict of interest regarding the publication of this manuscript.

Ethical statement

The authors declare that this book chapter entitled "The significant role of exopolysaccharides in biodegradable, renewable, and sustainable bioplastic production technology" is an original work and has been prepared in accordance with ethical standards of academic research and publication. This chapter does not involve any studies with human participants or animals performed by any of the authors. All data, concepts, and interpretations are derived from previously published literature, which has been appropriately cited and acknowledged. The authors confirm that there is no plagiarism, data fabrication, or falsification, and that all sources of information have been duly referenced.

Acknowledgment

The authors sincerely acknowledge the invaluable contributions of researchers and scientists worldwide whose pioneering studies and published literature on exopolysaccharides, microbial biopolymers, and sustainable bioplastic technologies formed the scientific foundation of this review. Proper acknowledgment of the work of others is essential, and the authors have made every effort to cite original sources wherever applicable.

The authors also express their gratitude to the management and faculty of Vishnu Institute of Pharmaceutical Education and Research, Narsapur, Telangana, and Dr. Samuel George Institute of Pharmaceutical Sciences, Markapur, Andhra Pradesh, for providing academic support, research facilities, and a conducive environment for scholarly work.

References

Alias, N.H., Abdullah, N., Othman, N.H., Marpani, F., Zainol, M.M., Shayuti, M.S.M. (2022). Sustainability challenges and future perspectives of biopolymer. In: Nadda, A.K., Sharma, S., Bhat, R. (Eds), Biopolymers. Springer Series on Polymer and Composite Materials. Springer, Cham. 373–389. https://doi.org/10.1007/978-3-030-98392-5_17

Altamira-Algarra, B., Garcia, J., Gonzalez-Flo, E., 2025. Cyanobacteria microbiomes for bioplastic production: critical review of key factors and challenges in scaling from laboratory to industry set-ups. Bioresour. Technol. 422, 132231. https://doi.org/10.1016/J.BIORTECH.2025.132231.

Altamira-Algarra, B., Lage, A., García, J., Gonzalez-Flo, E., 2024. Photosynthetic species composition determines bioplastics production in microbiomes: strategy to enrich cyanobacteria PHB-producers. Algal Res. 79, 103469. https://doi.org/10.1016/J.ALGAL.2024.103469.

Altamira-Algarra, B., Lage, A., Meléndez, A.L., Arnau, M., Gonzalez-Flo, E., García, J., 2024. Bioplastic production by harnessing cyanobacteria-rich microbiomes for long-term synthesis. Sci. Total Environ. 954, 176136. https://doi.org/10.1016/J.SCITOTENV.2024.176136.

Altamira-Algarra, B., Rueda, E., Lage, A., San León, D., Martínez-Blanch, J.F., Nogales, J., García, J., Gonzalez-Flo, E., 2023. New strategy for bioplastic and exopolysaccharides production: enrichment of field microbiomes with cyanobacteria. New Biotechnol. 78, 141–149. https://doi.org/10.1016/J.NBT.2023.10.008.

Benalaya, I., Alves, G., Lopes, J., Silva, L.R., 2024. A review of natural polysaccharides: sources, characteristics, properties, food, and pharmaceutical applications. Int. J. Mol. Sci., 25 (2), 1322. https://doi.org/10.3390/IJMS25021322.

Dilshad, E., Waheed, H., Ali, U., Amin, A., Ahmed, I., 2021. General structure and classification of bioplastics and biodegradable plastics. Bioplastics for Sustainable Development. Springer, Singapore, pp. 61–82. https://doi.org/10.1007/978-981-16-1823-9_2.

Dirpan, A., Ainani, A.F., Djalal, M., 2023. A review on biopolymer-based biodegradable film for food packaging: trends over the last decade and future research. Polymers 15 (13), 2781. https://doi.org/10.3390/POLYM15132781.

Findrik, E., Meixner, O., 2023. Drivers and barriers for consumers purchasing bioplastics – a systematic literature review. J. Cleaner Prod. 410, 137311. https://doi.org/10.1016/J.JCLEPRO.2023.137311.

Gamage, A., Thiviya, P., Liyanapathiranage, A., Wasana, M.L.D., Jayakodi, Y., Bandara, A., Manamperi, A., Dassanayake, R.S., Evon, P., Merah, O., Madhujith, T., 2024. Polysaccharide-based bioplastics: Eco-friendly and sustainable solutions for packaging. J. Comp. Sci. 8 (10), 413. https://doi.org/10.3390/JCS8100413.

Ibrahim, N.I., Shahar, F.S., Hameed Sultan, M.T., Md Shah, A.U., AzrieSafri, S.N., Mat Yazik, M.H., 2021. Overview of bioplastic introduction and its applications in product packaging. Coatings (11) 11. https://doi.org/10.3390/COATINGS11111423.

Laroche, C., 2022. Exopolysaccharides from microalgae and cyanobacteria: diversity of strains, production strategies, and applications. Mar. Drugs 20 (5). https://doi.org/10.3390/MD20050336.

Luo, Y., Douglas, J., Pahl, S., Zhao, J., 2022. Reducing plastic waste by visualizing marine consequences. Environ. Behav. 54 (4), 809–832. https://doi.org/10.1177/00139165221090154.

Moradi, M., Guimarães, J.T., Sahin, S., 2021. Current applications of exopolysaccharides from lactic acid bacteria in the development of food active edible packaging. Curr. Opin. Food Sci. 40, 33–39. https://doi.org/10.1016/j.cofs.2020.06.001.

Mouro, C., Gomes, A.P., Gouveia, I.C., 2024. Microbial exopolysaccharides: structure, diversity, applications, and future frontiers in sustainable functional materials. Polysaccharides 5 (3), 241–287. https://doi.org/10.3390/POLYSACCHARIDES5030018.

Mousavi, M., Kowsari, E., Gheibi, M., Cheshmeh, Z.A., Teymoorian, T., Ramakrishna, S., 2024. Assessing bioplastics' economic, commercial, political, and energy potential with circular economy modeling: a sustainable solution to plastic waste management. Mater. Circ. Econ. 6 (1), 1–36. https://doi.org/10.1007/S42824-023-00098-2.

Mu, Y., Zeng, H., Chen, W., 2021. Quercetin inhibits biofilm formation by decreasing the production of EPS and altering the composition of EPS in *Staphylococcus epidermidis*. Front. Microbiol. 12, 631058. https://doi.org/10.3389/FMICB.2021.631058/BIBTEX.

Muneer, F., Nadeem, H., Arif, A., Zaheer, W., 2021. Bioplastics from biopolymers: an eco-friendly and sustainable solution of plastic pollution. Polymer Science - Series C 63 (1), 47–63. https://doi.org/10.1134/S1811238221010057/METRICS.

Nadzir, M.M., Nurhayati, R.W., Idris, F.N., Nguyen, M.H., 2021. Biomedical applications of bacterial exopolysaccharides: a Review. Polymers 13 (4), 530. https://doi.org/10.3390/POLYM13040530.

Narayanan, M., Lee, J., Barathi, S., Kandasamy, S., 2023. Marine bioresources are a boon for bioplastic production as an alternative to conventional plastics—a review. Biomass Conv. Bioref. 1–17. https://doi.org/10.1007/S13399-023-04783-9.

Okoro, O.V., Gholipour, A.R., Sedighi, F., Shavandi, A., Hamidi, M., 2021. Optimization of exopolysaccharide (EPS) production by *Rhodotorula mucilaginosa* sp. GUMS16. ChemEngineering 5 (3), 39. https://doi.org/10.3390/CHEMENGINEERING5030039.

Onyena, A.P., Aniche, D.C., Ogbolu, B.O., Rakib, Md.R.J., Uddin, J., Walker, T.R., 2021. Governance strategies for mitigating microplastic pollution in the marine environment: a review. Microplastics 1 (1), 15–46. https://doi.org/10.3390/MICROPLASTICS1010003.

Panigrahy, N., Priyadarshini, A., Sahoo, M.M., Verma, A.K., Daverey, A., Sahoo, N.K., 2022. A comprehensive review on eco-toxicity and biodegradation of phenolics: recent progress and future outlook. Environ. Technol. Innov. 27, 102423. https://doi.org/10.1016/J.ETI.2022.102423.

Rabiya, R., Sen, R., 2022. Artificial intelligence driven advanced optimization strategy vis-à-vis response surface optimization of production medium: bacterial exopolysaccharide production as a case-study. Biochem. Eng. J. 178, 108271. https://doi.org/10.1016/J.BEJ.2021.108271.

Samadhiya, K., Sangtani, R., Nogueira, R., Bala, K., 2022. Insightful advancement and opportunities for microbial bioplastic production. Front. Microbiol. 12, 674864. https://doi.org/10.3389/FMICB.2021.674864/XML/NLM.

Varghese, S., Dhanraj, N.D., Rebello, S., Sindhu, R., Binod, P., Pandey, A., Jisha, M.S., Awasthi, M.K., 2022. Leads and hurdles to sustainable microbial bioplastic production. Chemosphere 305, 135390. https://doi.org/10.1016/J.CHEMOSPHERE.2022.135390.

Vijayanand, M., Ramakrishnan, A., Subramanian, R., Issac, P.K., Nasr, M., Khoo, K.S., Rajagopal, R., Greff, B., Wan Azelee, N.I., Jeon, B.H., Chang, S.W., Ravindran, B., 2023. Polyaromatic hydrocarbons (PAHs) in the water environment: a review on toxicity, microbial biodegradation, systematic biological advancements, and environmental fate. Environ. Res. 227, 115716. https://doi.org/10.1016/J.ENVRES.2023.115716.

CHAPTER

20

Chemico-structural features and respective biomedical utility co-relationships of different categorized exopolysaccharides

Akshayaa Sankar[a], *Ramachandran Venkataraman*[a], *Ramya Devi Durai*[a], *Sivanjineyulu Veluri*[b], *Marek Brzeziński*[b] *and Vedha Hari B Narayanan*[a]

[a]Pharmaceutical Technology Laboratory, School of Chemical & Biotechnology, SASTRA Deemed to be University, Thanjavur, Tamil Nadu, India [b]Centre of Molecular and Macromolecular Studies, Polish Academy of Sciences, Lodz, Poland

20.1 Introduction

Exopolysaccharides (EPS) are a group of polysaccharides that are produced during the metabolic process of fungi, bacteria, etc. Microbial EPS are extracellular carbohydrate-based polysaccharides produced exclusively by microorganisms, including bacteria, yeast, fungi, and microalgae (Benalaya et al., 2024). These produced EPS are of high molecular weight polymers. A wide variety of useful, structural EPS are produced by microbial organisms or microbial cells, which can be either prokaryotic or eukaryotic, and also homopolymer or heteropolymer in their structural makeup. Microbial EPS are widely used in ecological, environmental, food, biomedical, wound healing, agricultural, and pharmaceutical fields, largely because of their unique structural and functional properties. These compounds, produced by microbes, are naturally biodegradable, biocompatible, nontoxic, and environmentally friendly (Mouro et al., 2024). The structural diversity and composition of EPS are tremendous in nature. These EPS fall into two basic types: Homopolysaccharides and heteropolysaccharides (HePs). Homopolysaccharides are primarily composed of repeating units of a single type of sugar, such as glucose or fructose. In contrast, HePs consist of a variety of monosaccharides, including D-glucose,

DOI: https://doi.org/10.1016/B978-0-443-36619-2.00020-6

D-lactose, rhamnose, and others, which are organized in intricate, branched structures (Vashisth and Pruthi, 2016). Due to their distinctive structural and functional characteristics, EPS hold great potential as a sustainable resource in bioremediation (Prajapati et al., 2013). Additionally, they are widely utilized as stabilizers, emulsifiers, thickeners, and gelling agents in various industrial applications (Ullah et al., 2019). In recent times, microbial EPS have exceptional biological macromolecules generated from marine sources of microalgae, such as chitin/chitosan, got an attention in biomedical applications, pharmaceutical formulation developments, and excipients in drug delivery systems. One such example of a marine microorganism, such as an extremophile, produces EPS-S3 from *Pantoea* species, which are robust in nature, with high biocompatibility and high biodegradability, and hence, it is extensively applied in wound healing within tissue engineering, primarily through the Wnt/β-catenin signaling pathway, demonstrating significant potential for skin regeneration. Furthermore, they are also widely employed and play a crucial role in biofilm formation. The structure of EPS helps to form a structural matrix, which provides protection and makes the cell resilient by forming three-dimensional microcolonies. Apart from food, pharmaceutical, and microbiological functions, EPS have demonstrated their utility in bioremediation, wastewater treatment, and heavy metal removal (Nadzir et al., 2021). Certain EPS, such as xanthan gum, exhibit exceptional rheological characteristics, and because of this nature, they are extensively utilized in the oil and petroleum industry for various purposes, including enhanced oil recovery, pipeline cleaning, and other related applications. These EPS outperform many synthetic polymers (Schmid, 2018). The major advantage of using microbial EPS in contrast to plant-derived or any synthetic polysaccharides is mainly because of three important factors: Sustainability, customizable characteristics, and consistent production. Fermentation allows controlled and stable yields irrespective of climatic conditions. The microbial exopolysaccharide can be tailored genetically or metabolically engineered to provide various structural properties. The microbial EPS are highly sustainable because it has a biodegradable, nontoxic, and biocompatible nature in case of biomedical and tissue engineering applications (Mohammed et al., 2021). Thus, EPS have several potential applications in many different fields (Nguyen et al., 2024). But there are certain challenges associated with EPS, which include regulatory hurdles, scale-up limitations because of huge production costs, and comprehensive structure, property, or functional relationships that are still evolving and limiting the rational design of EPS (Bajpai et al., 2016). However, EPS are versatile, eco-friendly biopolymers with exceptional structural diversity and functional properties, as schematically depicted in Fig. 20.1.

The overall literature summarizes that EPS from various sources open up a wide range of platforms to production, purification, characterization, and exploration for various biomedical, tissue engineering and drug delivery applications.

20.2 Sources of exopolysaccharides

EPS are higher molecular weight polymers made up of sugar bases such as fructose or rhamnose, which are prominently produced or secreted by microorganisms into the extracellular environment (Abarquero et al., 2022). Since these polysaccharides are produced by microbes such as bacteria, fungi, algae, cyanobacteria, and marine sources, it has diverse applications, and their structural property makes them a valuable, sustainable source for pharmaceutical, tissue engineering, industrial, and biomedical applications. Among various sources, each source

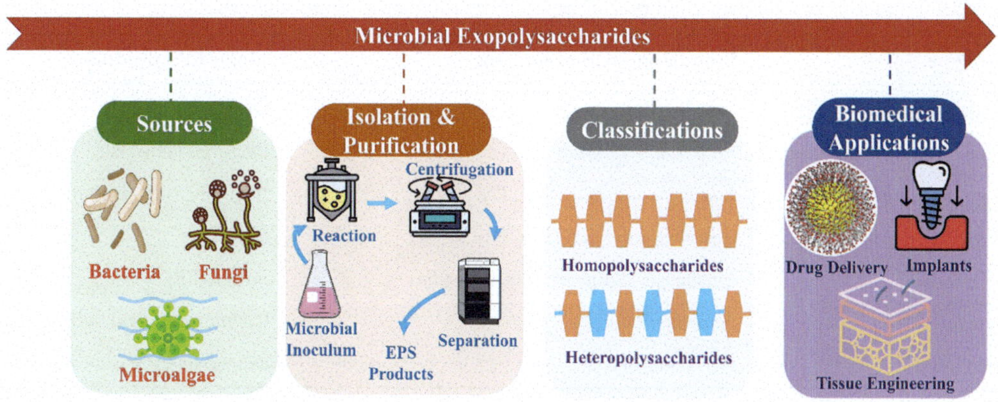

FIGURE 20.1 Overall schematic representation of exopolysaccharides with application, production, classification, and sources.

has its own structural property, and based on modification, the biological activity or functional property of the biomolecule is altered (Donot et al., 2012).

Among all the other mentioned sources, bacterial EPS produced by bacteria are considered predominant in terms of production, with a high amount of yield because of their major population in the environment, and also for their ability to grow in inexpensive environments and substrates. The major examples of bacterial EPS include dextran, levan, xanthan, alginate (ALG), bacterial cellulose, and extremophiles. Each exopolysaccharide is produced by a unique species of bacteria, which in turn provides unique biological functions. The bacterial species commonly involved in producing EPS are lactic acid bacteria, pathogenic bacteria, and cellulose-producing bacteria (Osemwegie et al., 2020). Homopolysaccharides and heteropolysaccharides, such as dextran and levan, are synthesized by various bacterial species, including *Lactobacillus lactis* and *Streptococcus thermophilus*. These have their potential application in gut microbiological modification, food, and pharmaceutical applications. Xanthan, also known as xanthan gum, is produced by *Xanthomonas campestris* and is widely applied in the oil industry, where it serves as an effective stabilizer and thickening agent. *Pseudomonas aeruginosa* synthesizes ALG, which is highly useful and a potential drug carrier in drug delivery applications. Bacterial cellulose is produced by *Acetobacter Xylinum,* which has vast applications in biomedical scaffolds, wound healing, and biofilm formation. The sources of EPS, along with the producing microorganisms with suitable examples, have been listed in Table 20.1.

Apart from bacterial sources, fungi, algae, cyanobacteria, and actinomycetes are also involved in synthesizing EPS. The fungal species include filamentous and yeast as subcategories. The filamentous fungi of species *Aspergillus niger* and *Penicillium* produce β-β-glucans, which are used as immunomodulators, and *Aureobasidium pullulans*, Cryptococcus produce pullulans, which are neutral polysaccharides that are a potential drug delivery carrier in encapsulating drugs or biomolecules to give a sustained release system. Algae and cyanobacteria of species such as *Microalgae porphyridium, Chlorella,* and *Spirulina*, Nostoc produce sulfated EPS and uronic acids, which are mainly applied as antioxidant, antiviral, and nutraceuticals. Actinomycetes such as *Streptomyces spp* produce branched EPS with a functional property mainly of antibiotic production, biofilm formation, etc. (Poli et al., 2010). Numerous sources, such as bacterial and

TABLE 20.1 Sources of EPS with example, along with the producing microorganism.

Source type	Producing microorganisms	Major EPS produced
Lactic acid bacteria	*Lactobacillus, Streptococcus thermophilus*	Dextran, levan
Cellulose-producing bacteria	*Acetobacter xylinum*	Bacterial cellulose
Pathogenic bacteria	*Xanthomonas campestris, Pseudomonas aeruginosa*	Alginate and xanthan gum
Fungi	*Aureobasidium pullulans*	Pullulans
Algae	*Microalgae (Porphyridium, Chlorella)*	Sulfated EPS
Cyanobacteria	*Spirulina, Nostoc*	Uronic acids and sulfated EPS
Actinomycetes	*Streptomyces spp.*	Branched EPS

EPS, exopolysaccharides.

other microbial organisms, have been involved in the production of EPS with varying yields and properties. Therefore, suggesting the need for an appropriate purification and characterization technique to confirm the high-quality EPS.

20.3 Production, isolation, and purification of exopolysaccharides

The synthesis of EPS by microorganisms involves a complex series of steps that involves majorly of four main pathways at both intracellular and extracellular levels. The major steps involved in synthesis basically include precursor synthesis, polymerization, and activation of sugar nucleotides and export from the microbial cell (Bagnol et al., 2022). The principal mechanisms involved in these pathways in microbial biosynthesis of EPS are as follows:

i) Wzx/Wzy dependent pathway
ii) ABC transport-dependent pathway
iii) Synthase-dependent pathway
iv) Synthesis of protein using the Extracellular pathway

Homopolysaccharides are typically synthesized through a synthase-based pathway as well as an extracellular production pathway. In contrast, HePs are produced via the transporter-ATP binding cassette (ABC)-dependent and Wzx/Wzy-dependent pathways (Nadzir et al., 2021).

20.3.1 Precursor synthesis

The monomers of EPS were initially synthesized or originate from central metabolic pathways such as glycolysis, pentose phosphate pathway, and tricarboxylic acid (TCA). The simple sugars in the pathway, such as glucose, lactose, and sucrose into nucleotides such as uridine diphosphate glucose, Guanosine diphosphoglucose, and thymidine diphosphate rhamnose. These nucleotides act as donors to activate the glucose transferase enzymes (Vashisth and Pruthi, 2016).

20.3.2 Activation of sugar nucleotides and polymerization

The sugar bases into activated nucleotides are sequentially attached to lipid carriers, which are attached in the bacterial cell membrane. This conversion and attachment were catalyzed using a glycosyltransferase enzyme, which in turn converts the monomeric sequence into branched structures. Once after branching, the polymerization reaction occurs through the above-mentioned 3 pathways, viz, ABC transporter dependent pathway, Wzx/Wzy dependent pathway, and synthase dependent pathway (Sonawdekar and Gupte, 2016).

20.3.2.1 *Wzx/Wzy dependent pathway*

The Wzx/Wzy-dependent pathway occurs in three main stages: Activation of nucleotides, attachment to lipid carriers, and polymerization. In the initial stage, sugar residues are actively transported into the cell and transformed into various activated monomeric forms. These monomers are then attached to an undecaprenyl phosphate (C55) lipid carrier located within the inner membrane. During the second stage, glycosyltransferases catalyze the stepwise addition of sugar units to assemble repeating oligosaccharide structures, which are subsequently translocated across the cytoplasmic membrane by the Wzx flippase. In the final stage, the translocated oligosaccharide units may undergo enzymatic modifications, such as methylation or acetylation, before being polymerized into long polysaccharide chains by the Wzy polymerase (Mouro et al., 2024). Polysaccharides generated through this pathway are HePs, characterized by a broad diversity of sugar components. Once fully assembled, these polysaccharides are transported to the cell surface with the assistance of ABC transporters. In probiotic bacteria, important EPS such as gellan, xanthan, and kefiran are known to be synthesized through this Wzx/Wzy-dependent mechanism (Srivastava et al., 2022).

20.3.2.2 *Transporter-ATP binding cassette transporter dependent pathway*

In the Wzx/Wzy- and ABC transport-dependent pathway, *Myxococcus Xanthus*, which produces the exopolysaccharide xanthan gum, utilizes this mechanism for EPS biosynthesis. Full-length polysaccharides are assembled by specific glycosyltransferases located in the inner membrane of the cytoplasm. This pathway is primarily employed for the synthesis of capsular polysaccharides (Xu et al., 2019a).

20.3.2.3 *Synthase-dependent pathway*

Homopolysaccharides, composed of repeating units of a single sugar such as ALG, are produced through the synthase-dependent pathway. In this mechanism, the activated sugar nucleotide, such as UDP-sugar, is processed by a membrane-bound synthase enzyme, which facilitates the polymerization of the sugar units. For example, in bacterial cellulose synthesis, the synthase enzyme embedded in the inner membrane acts as a transporter, guiding the elongating polysaccharide chain across the membrane to the extracellular space (Liang and Wang, 2015). This pathway allows the direct production of high-molecular-weight Homopolysaccharides with consistent structural properties, which are crucial for applications in food, pharmaceuticals, and biomedical industries (Lynch et al., 2018).

20.3.2.4 *Export of exopolysaccharides*

Once the polysaccharide is fully assembled, it is transported across the cell membrane and released into the extracellular environment, where it can exist as a capsule, as part of a biofilm, or as free-floating polymer chains. Environmental factors such as temperature, pH, oxygen concentration, and osmotic conditions play a significant role in controlling both the secretion process and the final molecular weight of the exopolysaccharide (Tallon et al., 2003).

20.3.3 Isolation of crude exopolysaccharides

After the production of EPS using various biosynthetic pathways, the selection of a strain, for example, for bacteria such as *Lactobacillus* for dextran, *Xanthomonas* for xanthan, and *Pseudomonas* for ALG, as well as for fungi and algae, allows for the production of unique EPS with distinct biological functions. These strain selection and fermentation are carried out using continuous fermentation, batch fermentation, and fed-batch fermentation methods. After the fermentation procedure, the EPS are removed from the microbial cell, protein, and other metabolites (Srivastava et al., 2022). The isolation process of EPS includes steps such as precipitation, cell removal, and recovery. The cells can be removed from the microbial biomass using centrifugation at high rpm of 12,000 and ultra-high filtration for large-scale production. After the precipitation process, heat treatment has to be given to remove enzymes that degrade EPS. After the removal of the cell, EPS can be precipitated by adjusting solubility, pH, salt, and alcohol precipitation by adding ethanol or methanol to the culture supernatant, which will precipitate the EPS, which are further dried (Xu et al., 2019b).

20.3.4 Purification and characterization of exopolysaccharides

The purification process of EPS includes the removal of proteins, nucleic acids, debris, pigments, and any slats to provide high-quality analytical-grade EPS. EPS can be purified using several techniques, such as dialysis or ultrahigh filtration, in which dialysis using a semipermeable membrane with a cut-off of 10k–100k Dalton removes salt and other substances (Wang et al., 2019). Chromatographic purification provides high-purity EPS. These techniques include size-exclusion chromatography, which separates EPS molecules according to their molecular weight; ion-exchange chromatography, which fractionates them based on their electrical charge; and affinity chromatography, which is employed to isolate EPS according to their specific binding interactions or affinities. The purified EPS can be dried using spray drying or freeze drying to get a desirable dried EPS that will retain EPS structure and bioactivity. The obtained EPS will be characterized for molecular weight using gel permeation chromatography, functional group and chemical composition analysis using FTIR analysis, and immunomodulatory, antioxidant, and antimicrobial activity can be evaluated using bioactivity assays (Raveendran et al., 2013). Thus, various transporter pathways are involved in the production and isolation of different types of EPS with varying molecular weights and properties, which paves way for utilizing them in various fields such as pharmaceuticals and the food industry.

20.4 Types of exopolysaccharides

EPS are polymers with high molecular weight that are released outside the cells by microorganisms into the surrounding environment, either by biofilm or capsular layers. The EPS are broadly classified into two types. Homopolysaccharides, which are made up of monomeric sugar units, and heteropolysaccharides, which have different types of monosaccharides arranged as repeating units or structure (Kumari et al., 2025). In detail, EPS are classified more broadly into various classifications based on biological origin, sugar composition, structure and linkage, function, solubility, and rheology.

20.4.1 Classification of exopolysaccharides

EPS can be categorized according to their sugar composition into homopolysaccharides, which consist of a single type of monosaccharide, such as glucose, fructose, or galactose. The homopolysaccharides include EPS, which are of dextran, levan, and cellulose. Heteropoly saccharides are composed of two or more different monosaccharide structures, which are arranged in repeating units. Examples for HePs are xanthan, ALG, and gellan gum. Based on biological origin, EPS can be classified based on the production using different strains, such as bacterial EPS, fungi, algae, cyanobacteria, and actinomycetes. Based on the structural linkage, EPS can be classified into linear, branched, charged, and neutral EPS. EPS based on function is classified into capsular, slime, and surface EPS. Capsular EPS are firmly attached to the microbial cell surface and form a capsular protective structure (Wang et al., 2023). Structural EPS helps to form a biofilm matrix and helps in microbial adhesion. Slime-based EPS are loosely attached and form biofilm and slime layers on them. Based on the viscosity, high-viscosity EPS and low-viscosity EPS are classified. Xanthan, gellan, and ALG are highly viscous and have gel-forming ability, whereas pullulan and dextran possess low viscosity.

20.4.2 Structural and chemical properties

The structural and chemical composition of EPS is diverse and varies based on the produced organism, variation in growth conditions, and based on biological synthesis and pathways. These structural variations result in functional alteration in biological, pharmaceutical, industrial, and other applications. The most common structural properties of EPS include molecular weight, structure of the polymer, charge distribution, and morphology. In general, EPS are of high molecular weight in the range of 10^4 to $10^{7,}$ which contributes to gel-forming ability, high viscosity, and mechanical stability. Charge distribution of EPS, such as neutral charge (pullulan), anionic charge (uronic acid), can influence the EPS to interact with metal ions and proteins to form biofilms or flocculation. Structure and morphology determine the industrial applications of EPS, such as cellulose, which provides a rigid structure that can be used in microfibril, whereas xanthan gum is amorphous in nature and can be used in gel-like structures (Waoo et al., 2023). The chemical composition of EPS includes monosaccharide units, hydrophilicity, glycosidic linkage, functional groups, and noncarbohydrate substituents. Most of the EPS are hydrophilic in nature and retain a huge amount of water, enhancing biofilm formation. The chemical functional groups, such as carboxyl, amino, and hydroxy groups, crosslink with cations to form gels, so that they are widely used in pharmaceutical applications. The noncarbohydrate substituents present in EPS include acetyl, pyruvyl, succinyl, sulfate, and phosphate groups, which show chemical diversity and alteration in bioactivity (Panchal et al., 2022). Based on the sugar moieties attached to the backbone of the polymer and structural linkages, the EPS has been classified as homo, hetero, capsular, neutral, leaner, and branched polymers.

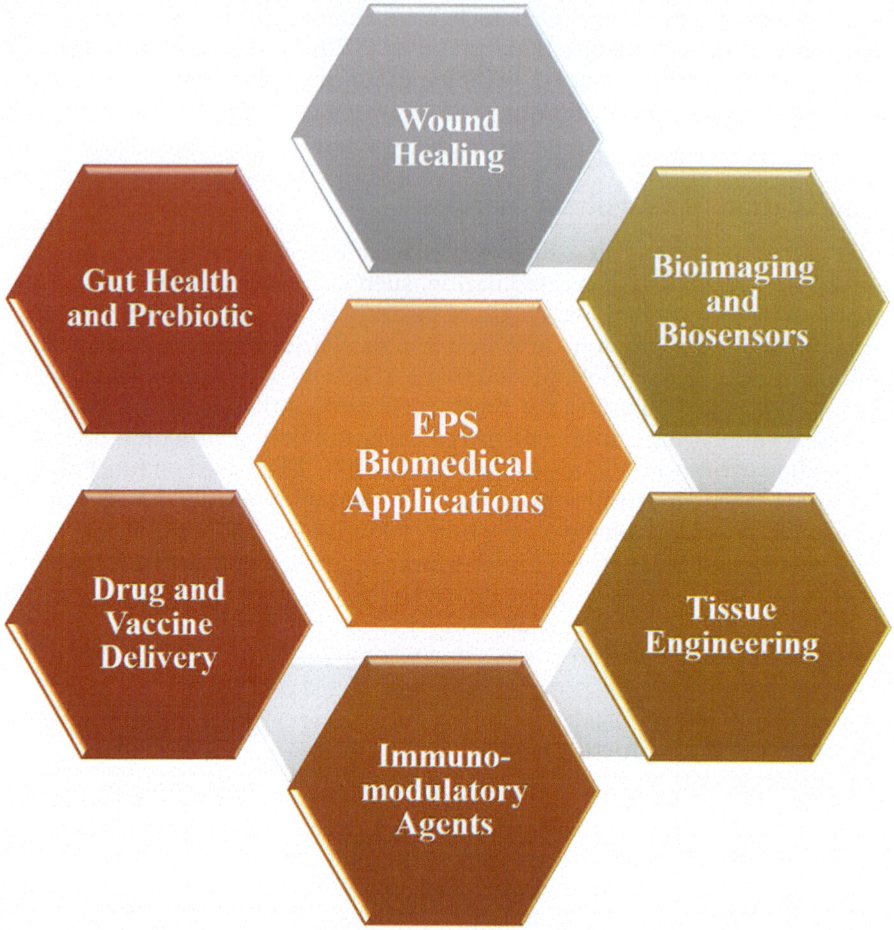

FIGURE 20.2 Overall application of EPS. *EPS*, Exopolysaccharides.

20.5 Biomedical application of exopolysaccharides

Microbial EPS are considered extracellular biopolymers that are secreted by bacteria, fungi, algae, and other microbes (Ahuja et al., 2023). This EPS has potential application in the pharma sector, tissue Engineering, as well as wound healing process, vaccine and gene delivery, and other potential biomedical applications, only because of its biodegradability, biocompatibility, nontoxicity, and structural diversity. Most of the EPS utilized in biomedical applications include dextran, levan, pullulan, xanthan, hyaluronic acid, and ALG (Hamidi et al., 2022). EPS have broad biomedical applications, which have been mentioned in Fig. 20.2, including their use in drug delivery systems, wound healing, tissue engineering, regenerative medicine, as well as functioning as antioxidant and anti-inflammatory agents and immunomodulation, prebiotics and gut health, bioadhesives and biofilm for medical device implants (Rahbar Saadat et al., 2019; Kumar et al., 2021).

20.5.1 Nano drug delivery system

EPS has a desirable quality of being hydrophilic, which in turn makes pseudoplastic solutions in water, in addition to biodegradable and biocompatible properties (Moscovici, 2015). This property makes EPS a widely used polymer in drug delivery and as a drug carrier system. Grafting of hydrophobic segments into a hydrophilic chain forms an amphiphilic polymer that forms stable nanogel structures (Waoo et al., 2023). One such example for this type is cholesteryl-based pullulan forms stable hydrogel nanoparticles as a drug carrier for a drug complex (Severo et al., 2022). In that case, ALG blend crosslinks with divalent cations to form a hydrogel, a crosslinked three-dimensional structure which remains insoluble yet retains a large amount of water, which enables them to form scaffolds and deliver drugs in the form of beads. Therefore, nanoparticles within this gel matrix are capable of encapsulating drugs or biomolecules within their internal structure, facilitating penetration into cells or tissues, and delivering therapeutic effects directly to the target site. The EPS-based hydrogel provides sustained release (Markovic et al., 2016). EPS such as xanthan, dextran, and levan are widely used to form nanoparticles to encapsulate drugs. Since EPS can be modified chemically in its structure, it can be modified to get attached to site-specific targeting. EPS plays an important and diverse role in drug carriers, drug targeting, and sustained release systems, mainly in vaccine adjuvants, immunomodulators, and mucoadhesive films (Tabernero and Cardea, 2020). The biomedical applications of EPS in drug delivery systems, tissue engineering, and biomedical implants and devices have been schematically represented in Figs. 20.3 and 20.4.

20.5.2 Tissue engineering

EPS has the structural ability to form scaffolds, hydrogels, and bioactive materials that exactly mimic the extracellular matrices (ECM), in turn promoting cell growth, proliferation, and tissue or bone regeneration (Hivechi et al., 2021). Apart from wound healing, scaffold engineering for skin repair or regeneration, EPS plays a vital role in neural or cardiac tissue engineering by modifying its structure to mimic the soft conductive properties of neural and cardiac tissues. This domain aims to design a structure that can repair, mimic, support, or enhance the function affected by any infractions in the case of the heart, myocardial infarctions. Hyaluronic acid methacrylic anhydride mimics the extracellular matrix in heart valve functions. Dextran-based conductive hydrogels are widely used in nerve tissue engineering by combining with graphene. Bacterial cellulose produced by bacterial fermentation with selective strains and cultivation parameters yields a controllable 3-D nanofibril network structure that mimics the exact collagen-like nature and possesses excellent biocompatibility in bone, cartilage, or skin tissue regeneration, compared to effective cartilage regeneration in plant cellulose (Chauhan et al., 2024). Bacterial cellulose shows superior results in acute or chronic wound healing compared to plant cellulose because of its high-water absorption capacity and resistance to biodegradation. ALG-based materials find extensive applications in tissue engineering, including the culture of bone, cartilage, vascular, and liver cells. ALG scaffolds can effectively create chondrocyte culture stem cells, which can effectively reduce or combat osteoarthritis and orthoplasty by effective cartilage regeneration (Zhao et al., 2021).

20.5.3 Medical implants/devices

Recent advances in medical technology have led to a significant improvement in devices such as biosensors and prosthetic implants. These devices are commonly fabricated from

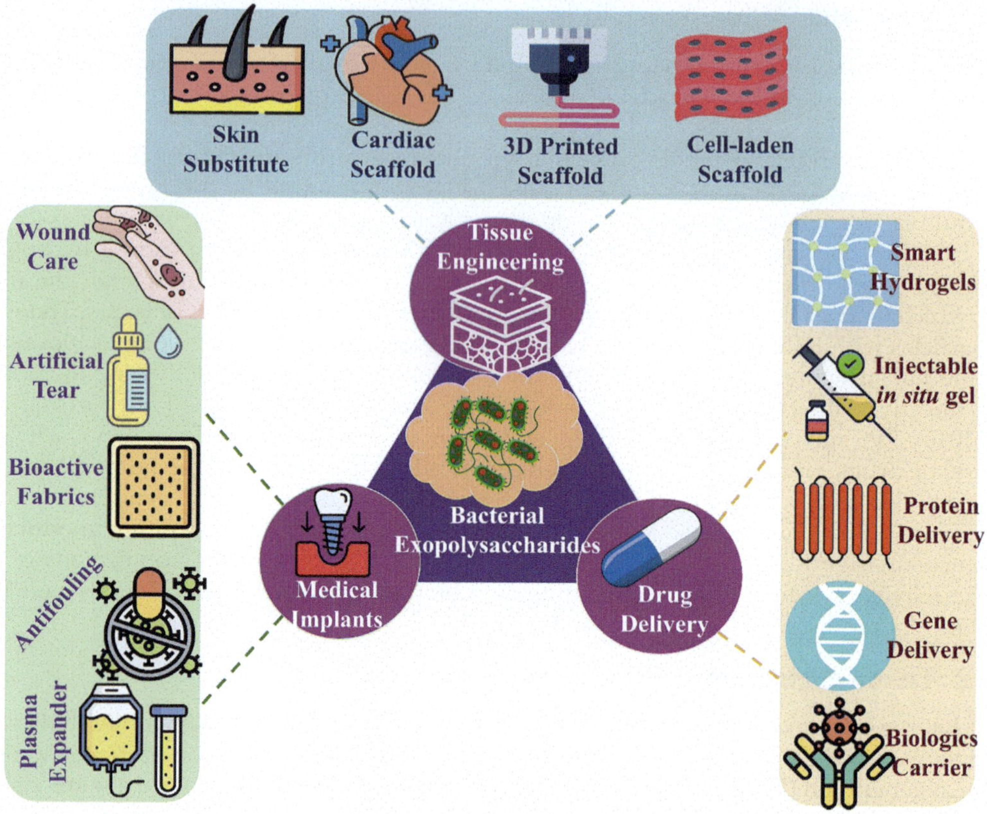

FIGURE 20.3 Biomedical applications of exopolysaccharides.

mechanically robust polymers or metals to enhance longevity and minimize the need for revision surgeries. However, the considerable rigidity of these long-lasting materials can trigger negative immune responses, which in severe instances may lead to significant health hazards (Salave et al., 2022). Dextran, one of the EPS, has been used as a coating material for improving the biocompatibility of medical devices, and the study demonstrates the possibility of utilizing dextran as a coating material for neural probes. The molecular weight and coating thickness of dextran can be adjusted to control the stiffness and degradation time. This study reported that scar tissue was infrequently observed in Wistar rats following four months of implantation with a dextran-coated neural probe (Saveleva et al., 2019). Despite its widespread use in bone implants, the corrosion behavior of metals continues to be a challenge. Another study indicated that coating a titanium-based implant with organic polymers, such as dextran, can enhance its resistance to corrosion when exposed to simulated body fluid. The EPS that are utilized in various biomedical applications have been listed in Table 20.2.

20.5.3.1 Surgical sealant

Surgical sealants are specialized materials designed to control bleeding during surgical procedures. Fixative substances like fibrin, chitosan, and dextran are promising candidates for

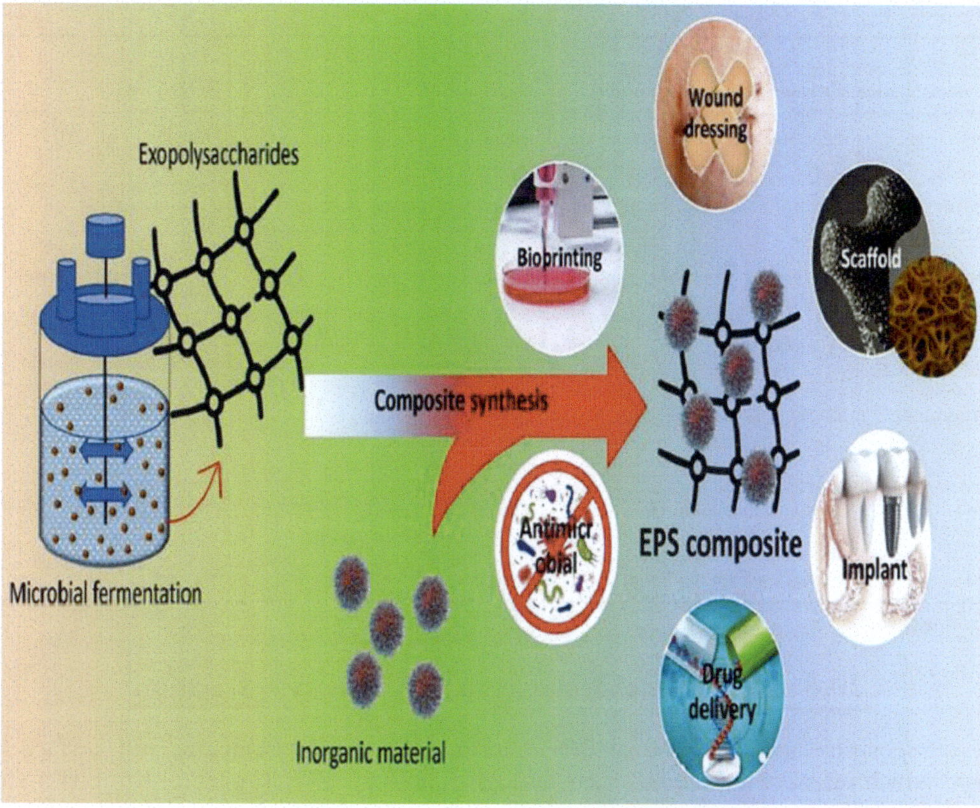

FIGURE 20.4 Biomedical application of EPS in medical device implants and devices. *EPS*, Exopolysaccharides.

making these sealants. It has been demonstrated that dextran–chitosan in situ gels exhibited strong adhesive characteristics while maintaining low cytotoxicity and limited swelling. In their experiments with rabbits that had liver injuries, the dextran–chitosan composite worked well as an adhesive glue and showed little cell toxicity (Nadzir et al., 2021). Among bacterial EPS, sulfated levan has emerged as a promising candidate for cardiac tissue engineering owing to its excellent biocompatibility and anticoagulant properties. One study demonstrated the creation of adhesive, free-standing multilayer films composed of sulfated levan combined with ALG and chitosan. The inclusion of sulfated levan significantly improved both the mechanical strength and adhesive performance of these films. In vitro testing using the C2C12 myoblast cell line showed that the multilayer films were cytocompatible and supported myoconduction. These results underscore the potential of sulfated levan–ALG–chitosan membranes for use in cardiac tissue engineering applications (Fig. 20.4). Isolated and purified microbial EPS possess safe, biocompatible, and biodegradable characteristics with wide applications in tissue engineering, wound repair, nanodrug delivery, and surgical sealants.

TABLE 20.2 EPS and biomedical applications.

Exopolysaccharides	Producing microorganisms	Potential biomedical applications	References
Dextran	*Leuconostoc mesenteroides*	Plasma expander, wound healing, drug carrier	Hussain et al. (2017)
Gellan gum	*Sphingomonas paucimobilis*	Sustained release, gelling agent	Zia et al. (2018)
Alginate	*Azotobacter vinelandii, Pseudomonas aeruginosa*	Controlled drug delivery, encapsulation of drugs	Tabernero and Cardea (2020)
Cellulose	Acetobacter	Wound dressings, tissue scaffolds	Wan et al. (2020)
Hyaluronic acid	*Bacillus subtilis*	Bone and cartilage regeneration, scaffold formation	De Lucia et al. (2020)
Xanthan	*Xanthomonas campestris*	Tissue engineering scaffold, ophthalmic drug carrier	Petri (2015)
Pullulan	*Aureobasidium pullulans*	Oral drug delivery, biofilm	Shingel (2004)

EPS, exopolysaccharides.

20.6 Exopolysaccharide versus bio activity correlation

The composition and structure of EPS determine their microstructure and surface morphology, which affect or determine their biological activity. EPS in general exhibits a strong correlation by altering various parameters such as chemical composition, polysaccharide chain, molecular weight, presence of functional group, charge distribution, degree of polymerization, and monosaccharide composition to produce variation in biological function or activity (Shingel, 2004). The chemical structure modifications, such as phosphorylation, acetylation, sulfonation, and carboxymethylation, strongly influence changes or alterations in biological activities (Wang et al., 2023). Many of the biological activities include antioxidant, antitumor, immunomodulatory, gut microbial modulating, antimicrobial activity, etc. In the aspects of molecular weight, low molecular weight EPS in general exhibit antioxidant and antimicrobial effects due to their higher solubility in nature and interact better with biological targets. In the case of higher molecular weight EPS, it possesses the ability to form gels and scaffolds, which are mainly utilized in drug delivery and carrier systems, and tissue engineering applications (Zhou et al., 2021). Likewise, EPS are rich in sugar bases such as fructose or rhamnose, which provide biological functions such as immunomodulation and anti-inflammatory effects by activating macrophage or cytokine release. Charge distribution is another factor that influences bioactivity. Negatively charged EPS generally contains uronic acid or sulfate groups, which show antioxidant and antimicrobial properties because of their ability to chelate metal ions and interact with biological molecules (Casillo et al., 2017). The chemically modified EPS, such as phosphorylated or carboxymethylated derivatives, shows significant anticancer and antiviral activity, which clearly indicates that modification in structure or chemical modification enhances biological activity. On the other hand, the degree of acetylation in EPS shows good gut microbiota modulation and biofilm-forming ability (Sayem et al., 2011).

20.7 Role of genetic engineering and regulatory requirements

Many studies were conducted to genetically modify the EPS-producing microorganisms to increase the production yield and also to develop new polymer varieties. In controlling microbial activity, various enzymes play a crucial role in EPS production by participating in multiple metabolic pathways. These enzymes control the nucleotide sugar metabolite formation, determination of chain length, polymerization, and excretion of polymers. Recently, extensive studies have been performed by concentrating on the mechanisms behind the production of EPS through the overexpression of controlling genetic codons. The recent advancements in bioinformatic tools boost the research on variations in sugar supplements, the structure of proteins, genetic modifications, and their effect on EPS production. The ultimate aim of the genetic manipulation is to improve the overall yield of EPS production with minimal substrates and low-cost processing and reactors (Salave et al., 2022). A strategy emphasizes optimizing the gene responsible for transcriptional regulation of EPS production. In this study involving *Sphingomonas* strains, the promoter gene was inserted with changeable strengths upstream of EPS production genes, leading to wider biosynthesis stages and, consequently, differences in EPS yield. Notably, the production of the EPS was significantly improved by manipulating the strong promoter genes, stressing the importance of transcriptional control in metabolic engineering. Furthermore, the targeted inhibition of genes related to specific glycosyltransferase expression. The gene GumI was deactivated in *X. campestris*, leading to the increased production of different xanthan polymers with flexible viscoelasticity properties, highlighting the importance of genetic engineering of microbes in the development of polymers with required properties. Additionally, systems biology drastically changed the entire view of genetic manipulation and its benefits by connecting the data from omics and developing genetic-level metabolic models, supporting the identification of the important regulatory genes and metabolic steps in EPS production. In summary, genetic engineering strategies, encompassing gene overexpression, transcriptional regulation, structural modification, and systems biology, have significantly advanced the production and functional diversification of EPSs. These approaches offer promising avenues for the development of EPSs with tailored properties for various industrial applications. Due to the wider application of EPS, such as pharmaceuticals, food, and biotechnology, the products containing EPS are regulated by different regulatory agencies around the world. In the United States, the EPS applicable for food product development is considered as generally recognized as safe (GRAS) or regulated through the food additive applications process. In the case of pharmaceutical products containing EPS, they must comply with current good manufacturing practices (cGMP) as recommended by the United States-Food and Drug Administration (US-FDA). However, according to the European Union (EU), the safety assessment data need to be produced before the marketing of EPS /EPS containing products to be used in European countries. With respect to the safety of EPS, the materials should be tested for heavy metals, microbial pathogens, and endotoxins (Bhunia et al., 2018).

20.8 Conclusion

Microbial EPS are multipurpose biomaterials with important properties for transversely different applications related to food, biomedical, and pharmaceutical fields. Their variety of structures, dextran (homo) and xanthan, and gellan (hetero) promise unique characteristics applicable for drug delivery, tissue regeneration, and wound dressing. The production of EPS is primarily achieved through fermentation processes involving microorganisms like bacteria, yeast, and fungi, which are favored due to their rapid growth rates, ease of cultivation, and ability to produce high yields of EPS using cost-effective substrates. Advancements in genetic engineering have further enhanced EPS production by optimizing precursor metabolite pathways and modifying gene expression to increase yield and tailor properties for specific applications. Despite their promising applications, challenges remain in large-scale production, purification, and regulatory approval of EPS-based products. Ongoing research aims to address these issues, focusing on improving production efficiency and ensuring compliance with safety standards. In conclusion, microbial EPS constitute a valuable class of biopolymers with a wide range of potential applications in the biomedical sector. Continued research and development are essential to realize their potential and overcome existing challenges fully. Their biocompatibility, biodegradability, and versatility make them ideal candidates for a wide range of therapeutic and pharmaceutical applications.

Credit statement

AS, RV, SV, and VH B.N. contributed to the design and development of the manuscript. AS, SV collected literature and prepared the draft, and RV created images, while DRD, MB, and VH B.N. reviewed the draft. All authors participated in writing the manuscript and critically revised and approved it.

Acknowledgments and funding information

All the authors thank SASTRA Deemed University, Thanjavur, India, and CBMM-PAN, Lodz, Poland, for the infrastructural support.

List of abbreviations

ABC	Transporter-ATP binding cassette
ALG	Alginate
cGMP	Current good manufacturing practice
ECM	Extracellular matrices
EPS	Exopolysaccharide
EU	European Union
GRAS	Generally recognized as safe
HePs	Hetero polysaccharide
US-FDA	United States-Food and Drug Administration

References

Abarquero, D., Renes, E., Fresno, J.M., Tornadijo, M.E., 2022. Study of exopolysaccharides from lactic acid bacteria and their industrial applications: a review. Int. J. Food Sci. Technol. 57 (1), 16–26. https://doi.org/10.1111/ijfs.15227.

Ahuja, V., Bhatt, A.K., Banu, J.R., Kumar, V., Kumar, G., Yang, Y.H., Bhatia, S.K., 2023. Microbial exopolysaccharide composites in biomedicine and healthcare: trends and advances. Polymers (7) 15. https://doi.org/10.3390/polym15071801.

Bagnol, R., Grijpma, D., Eglin, D., Moriarty, T.F., 2022. The production and application of bacterial exopolysaccharides as biomaterials for bone regeneration. Carbohydr. Polym. 291, 119550. https://doi.org/10.1016/j.carbpol.2022.119550.

Bajpai, V.K., Rather, I.A., Majumder, R., Shukla, S., Aeron, A., Kim, K., Kang, S.C., Dubey, R.C., Maheshwari, D.K., Lim, J., Park, Y.H., 2016. Exopolysaccharide and lactic acid bacteria: perception, functionality and prospects. Bangladesh J. Pharm. 11 (1), 1–23. https://doi.org/10.3329/bjp.v11i1.23819.

Benalaya, I., Alves, G., Lopes, J., Silva, L.R., 2024. A review of natural polysaccharides: sources, characteristics, properties, food, and pharmaceutical applications. Int. J. Mol. Sci. 25 (2). https://doi.org/10.3390/ijms25021322.

Bhunia, B., Prasad Uday, U.S., Oinam, G., Mondal, A., Bandyopadhyay, T.K., Tiwari, O.N., 2018. Characterization, genetic regulation and production of cyanobacterial exopolysaccharides and its applicability for heavy metal removal. Carbohydr. Polym. 179, 228–243. https://doi.org/10.1016/j.carbpol.2017.09.091.

Casillo, A., Parrilli, E., Sannino, F., Mitchell, D.E., Gibson, M.I., Marino, G., Lanzetta, R., Parrilli, M., Cosconati, S., Novellino, E., Randazzo, A., Tutino, M.L., Corsaro, M.M., 2017. Structure-activity relationship of the exopolysaccharide from a *Psychrophilic bacterium*: a strategy for cryoprotection. Carbohydr. Polym. 156, 364–371. https://doi.org/10.1016/j.carbpol.2016.09.037.

Chauhan, R., Rajput, V.D., Minkina, T., Prasad, R., Zhu, Y., Varma, A., Goel, A., 2024. Deciphering the role of nanoparticles induced microbial exopolysaccharides in soil amelioration and plant health. Clean Technol. Environ. Policy. https://doi.org/10.1007/s10098-024-02858-3.

De Lucia, O., Murgo, A., Pregnolato, F., Pontikaki, I., De Souza, M., Sinelli, A., Cimaz, R., Caporali, R., 2020. Hyaluronic acid injections in the treatment of osteoarthritis secondary to primary inflammatory rheumatic diseases: a systematic review and qualitative synthesis. Adv. Ther. 37 (4), 1347–1359. https://doi.org/10.1007/s12325-020-01256-7.

Donot, F., Fontana, A., Baccou, J.C., Schorr-Galindo, S., 2012. Microbial exopolysaccharides: main examples of synthesis, excretion, genetics and extraction. Carbohydr. Polym. 87 (2), 951–962. https://doi.org/10.1016/j.carbpol.2011.08.083.

Hamidi, M., Okoro, O.V., Milan, P.B., Khalili, M.R., Samadian, H., Nie, L., Shavandi, A., 2022. Fungal exopolysaccharides: properties, sources, modifications, and biomedical applications. Carbohydr. Polym. 284, 119152. https://doi.org/10.1016/j.carbpol.2022.119152.

Hivechi, A., Milan, P.B., Modabberi, K., Amoupour, M., Ebrahimzadeh, K., Gholipour, A.R., Sedighi, F., Amini, N., Hajir Bahrami, S., Rezapour, A., Hamidi, M., Delattre, C., 2021. Synthesis and characterization of exopolysaccharide encapsulated PCL/gelatin skin substitute for full-thickness wound regeneration. Polymers 13 (6), 1–16. https://doi.org/10.3390/polym13060854.

Hussain, A., Zia, K.M., Tabasum, S., Noreen, A., Ali, M., Iqbal, R., Zuber, M., 2017. Blends and composites of exopolysaccharides; properties and applications: a review. Int. J. Biol. Macromol. 94, 10–27. https://doi.org/10.1016/j.ijbiomac.2016.09.104.

Kumar, N., Hasnain, M.S., 2021. Biomedical composites: Perspectives and applications (Materials Horizons: From Nature to Nanomaterials). Springer Nature, Singapore. https://doi.org/10.1007/978-981-33-4753-3.

Kumari, J., Kumawat, R., Prasanna, R., Jothieswari, D., Debnath, R., Ikbal, A.M.A., Palit, P., Rawat, R., Gopikrishna, K., Tiwari, O.N., 2025. Microbial exopolysaccharides: classification, biosynthetic pathway, industrial extraction and commercial production to unveil its bioprospection: a comprehensive review. Int. J. Biol. Macromol. 297, 139917. https://doi.org/10.1016/j.ijbiomac.2025.139917.

Liang, T.W., Wang, S.L., 2015. Recent advances in exopolysaccharides from *Paenibacillus spp.*: production, isolation, structure, and bioactivities. Mar. Drugs 13 (4), 1847–1863. https://doi.org/10.3390/md13041847.

Lynch, K.M., Zannini, E., Coffey, A., Arendt, E.K., 2018. Lactic acid bacteria exopolysaccharides in foods and beverages: isolation, properties, characterization, and health benefits. Annu. Rev. Food Sci. Technol. 9, 155–176. https://doi.org/10.1146/annurev-food-030117-012537.

Markovic, D., Zarubica, A., Stojkovic, N., Vasic, M., Cakic, M., Nikolic, G., 2016. Alginates and similar exopolysaccharides in biomedical application and pharmacy: controled delivery of drugs. Adv. Technol. 5 (1), 39–52. https://doi.org/10.5937/savteh1601039m.

Mohammed, A.S.A., Naveed, M., Jost, N., 2021. Polysaccharides; classification, chemical properties, and future perspective applications in fields of pharmacology and biological medicine (a review of current applications and upcoming potentialities). J. Polym. Environ. 29 (8), 2359–2371. https://doi.org/10.1007/s10924-021-02052-2.

Moscovici, M., 2015. Present and future medical applications of microbial exopolysaccharides. Front. Microbiol. 6, 1–11. https://doi.org/10.3389/fmicb.2015.01012.

Mouro, C., Gomes, A.P., Gouveia, I.C., 2024. Microbial exopolysaccharides: structure, diversity, applications, and future frontiers in sustainable functional materials. Polysaccharides 5 (3), 241–287. https://doi.org/10.3390/polysaccharides5030018.

Nadzir, M.M., Nurhayati, R.W., Idris, F.N., Nguyen, M.H., 2021. Biomedical applications of bacterial exopolysaccharides: a review. Polymers 13 (4), 1–25. https://doi.org/10.3390/polym13040530.

Nguyen, H.T., Pham, T.T., Nguyen, P.T., Le-Buanec, H., Rabetafika, H.N., Razafindralambo, H.L., 2024. Advances in microbial exopolysaccharides: present and future applications. Biomolecules 14 (9). https://doi.org/10.3390/biom14091162.

Osemwegie, O.O., Adetunji, C.O., Ayeni, E.A., Adejobi, O.I., Arise, R.O., Nwonuma, C.O., Oghenekaro, A.O., 2020. Exopolysaccharides from bacteria and fungi: current status and perspectives in Africa. Heliyon 6 (6), e04205. https://doi.org/10.1016/j.heliyon.2020.e04205.

Panchal, R., Prajapati, K., Prajapati, M., Sharma, S., Saraf, M., 2022. Bacterial exopolysaccharides: types, its biosynthesis and their application in different fields. Acta Sci. Biotechnol. 3 (2), 3–11.

Petri, D.F.S, 2015. Xanthan gum: a versatile biopolymer for biomedical and technological applications. J. Appl. Polym. Sci. (23) 132. https://doi.org/10.1002/app.42035.

Poli, A., Anzelmo, G., Nicolaus, B., 2010. Bacterial exopolysaccharides from extreme marine habitats: production, characterization and biological activities. Mar. Drugs 8 (6), 1779–1802. https://doi.org/10.3390/md8061779.

Prajapati, V.D., Jani, G.K., Khanda, S.M., 2013. Pullulan: an exopolysaccharide and its various applications. Carbohydr. Polym. 95 (1), 540–549. https://doi.org/10.1016/j.carbpol.2013.02.082.

Rahbar Saadat, Y., Yari Khosroushahi, A., Pourghassem Gargari, B., 2019. A comprehensive review of anticancer, immunomodulatory and health beneficial effects of the lactic acid bacteria exopolysaccharides. Carbohydr. Polym. 217, 79–89. https://doi.org/10.1016/j.carbpol.2019.04.025.

Raveendran, S., Poulose, A.C., Yoshida, Y., Maekawa, T., Kumar, D.S., 2013. Bacterial exopolysaccharide based nanoparticles for sustained drug delivery, cancer chemotherapy and bioimaging. Carbohydr. Polym. 91 (1), 22–32. https://doi.org/10.1016/j.carbpol.2012.07.079.

Salave, S., Rana, D., Sharma, A., Bharathi, K., Gupta, R., Khode, S., Benival, D., Kommineni, N., 2022. Polysaccharide based implantable drug delivery: development strategies, regulatory requirements, and future perspectives. Polysaccharides 3 (3), 625–654. https://doi.org/10.3390/polysaccharides3030037.

Saveleva, M., Vladescu, A., Cotrut, C., Van Der Meeren, L., Surmeneva, M., Surmenev, R., Parakhonskiy, B., Skirtach, A.G., 2019. The effect of hybrid coatings based on hydrogel, biopolymer and inorganic components on the corrosion behavior of titanium bone implants. J. Mater. Chem. B 7 (43), 6778–6788. https://doi.org/10.1039/c9tb01287g.

Sayem, S.M.A., Manzo, E., Ciavatta, L., Tramice, A., Cordone, A., Zanfardino, A., De Felice, M., Varcamonti, M., 2011. Anti-biofilm activity of an exopolysaccharide from a sponge-associated strain of *Bacillus licheniformis*. Microb. Cell Fact. 10, 1–12. https://doi.org/10.1186/1475-2859-10-74.

Schmid, J., 2018. Recent insights in microbial exopolysaccharide biosynthesis and engineering strategies. Curr. Opin. Biotechnol. 53, 130–136. https://doi.org/10.1016/j.copbio.2018.01.005.

Severo, I.A., Dias, R.R., do Nascimento, T.C., Deprá, M.C., Maroneze, M.M., Zepka, L.Q., Jacob-Lopes, E., 2022. Microalgae-derived polysaccharides: potential building blocks for biomedical applications. World J. Microbiol. Biotechnol. 38 (9), 1–17. https://doi.org/10.1007/s11274-022-03342-0.

Shingel, K.I., 2004. Current knowledge on biosynthesis, biological activity, and chemical modification of the exopolysaccharide, pullulan. Carbohydr. Res. 339 (3), 447–460. https://doi.org/10.1016/j.carres.2003.10.034.

Sonawdekar, S., Gupte, A., 2016. Production and characterization of exopolysaccharide produced by oil emulsifying bacteria. Int. J. Curr. Microbiol. Appl. Sci. 5 (2), 254–262. https://doi.org/10.20546/ijcmas.2016.502.028.

Srivastava, N., Kumari, S., Kurmi, S., Pinnaka, A.K., Choudhury, A.R., 2022. Isolation, purification, and characterization of a novel exopolysaccharide isolated from marine bacteria *Brevibacillus borstelensis* M42. Arch. Microbiol. 204 (7), 1–16. https://doi.org/10.1007/s00203-022-02993-9.

Tabernero, A., Cardea, S., 2020. Microbial exopolysaccharides as drug carriers. Polymers (9) 12. https://doi.org/10.3390/POLYM12092142.

Tallon, R., Bressollier, P., Urdaci, M.C., 2003. Isolation and characterization of two exopolysaccharides produced by *Lactobacillus plantarum* EP56. Res. Microbiol. 154 (10), 705–712. https://doi.org/10.1016/j.resmic.2003.09.006.

Ullah, S., Khalil, A.A., Shaukat, F., Song, Y., 2019. Sources, extraction and biomedical properties of polysaccharides. Foods 8 (8), 304. https://doi.org/10.3390/foods8080304.

Vashisth, P., Pruthi, V., 2016. Synthesis and characterization of crosslinked gellan/PVA nanofibers for tissue engineering application. Mater. Sci. Eng. C 67, 304–312. https://doi.org/10.1016/j.msec.2016.05.049.

Wan, Y., Yang, S., Wang, J., Gan, D., Gama, M., Yang, Z., Zhu, Y., Yao, F., Luo, H., 2020. Scalable synthesis of robust and stretchable composite wound dressings by dispersing silver nanowires in continuous bacterial cellulose. Composites, Part B 199, 108259. https://doi.org/10.1016/j.compositesb.2020.108259.

Wang, B., Song, Q., Zhao, F., Xiao, H., Zhou, Z., Han, Y., 2019. Purification and characterization of dextran produced by *Leuconostoc pseudomesenteroides* PC as a potential exopolysaccharide suitable for food applications. Process Biochem. 87, 187–195. https://doi.org/10.1016/j.procbio.2019.08.020.

Wang, W., Ju, Y., Liu, N., Shi, S., Hao, L., 2023. Structural characteristics of microbial exopolysaccharides in association with their biological activities: a review. Chem. Biol. Technol. Agric. 10 (1), 1–16. https://doi.org/10.1186/s40538-023-00515-3.

Waoo, A.A., Singh, S., Pandey, A., Kant, G., Choure, K., Amesho, K.T.T., Srivastava, S., 2023. Microbial exopolysaccharides in the biomedical and pharmaceutical industries. Heliyon 9 (8), e18613. https://doi.org/10.1016/j.heliyon.2023.e18613.

Xu, Y., Cui, Y., Wang, X., Yue, F., Shan, Y., Liu, B., Zhou, Y., Yi, Y., Lü, X., 2019. Purification, characterization and bioactivity of exopolysaccharides produced by *Lactobacillus plantarum* KX041. Int. J. Biol. Macromol. 128, 480–492. https://doi.org/10.1016/j.ijbiomac.2019.01.117.

Xu, Z., Chen, G., Xue, L., Zhang, H., Wang, J., Xiang, H., Li, J., Zheng, K., 2019. Isolation, structural characterizations and bioactivities of exopolysaccharides produced by *Bacillus licheniformis*. Int. J. Biol. Macromol. 141, 298–306. https://doi.org/10.1016/j.ijbiomac.2019.08.217.

Zhao, X.H., Niu, Y.N., Mi, C.H., Gong, H.L., Yang, X.Y., Cheng, J.S.Y., Zhou, Z.Q., Liu, J.X., Peng, X.L., Wei, D.X., 2021. Electrospinning nanofibers of microbial polyhydroxyalkanoates for applications in medical tissue engineering. J. Polym. Sci. 59 (18), 1994–2013. https://doi.org/10.1002/pol.20210418.

Zhou, Y., Cui, Y., Suo, C., Wang, Q., Qu, X., 2021. Structure, physicochemical characterization, and antioxidant activity of the highly arabinose-branched exopolysaccharide EPS-M2 from *Streptococcus thermophilus* CS6. Int. J. Biol. Macromol. 192, 716–727. https://doi.org/10.1016/j.ijbiomac.2021.10.047.

Zia, K.M., Tabasum, S., Khan, M.F., Akram, N., Akhter, N., Noreen, A., Zuber, M., 2018. Recent trends on gellan gum blends with natural and synthetic polymers: a review. Int. J. Biol. Macromol. 109, 1068–1087. https://doi.org/10.1016/j.ijbiomac.2017.11.099.

CHAPTER

21

Current status and advancements of natural exopolysaccharides as potential excipients and adjuvants in the pharmaceutical field

Sanjukta Sen[a,b], *Piu Jana*[a], *Esha Das*[a] and *Gouranga Nandi*[a]

[a]Division of Pharmaceutics, Department of Pharmaceutical Technology, University of North Bengal, Darjeeling, West Bengal, India [b]Gitanjali College of Pharmacy, Birbhum, West Bengal, India

21.1 Introduction

Recently, natural exopolysaccharides (EPSs) have played a promising role due to their unique structural design and characteristics, and also for their potential uses in technology, industry, research, and medicine (Rana and Upadhyay, 2020; Mohd Nadzir et al., 2021; Kaur & Dey, 2023). High-molecular-weight carbohydrate biopolymers derived from bacteria were initially referred to as EPS by Sutherland in 1972 (Osemwegie et al., 2020). The microbes, like bacteria, yeasts, fungi, and microalgae, produce EPSs, which are essentially carbohydrate-based biopolymers (Costa et al., 2021; Rahbar Saadat et al., 2021; Hamidi et al., 2022; Sørensen et al., 2022). Microbial EPSs are of interest in many industrial areas because of their exceptional material characteristics, as well as their composition, which is less hazardous, completely biodegradable, and biocompatible. Microbial EPSs have recently been found to be an excellent model for the rapid production of nanoparticles of metallic material, and EPS-generated metal mitigation methods are gaining popularity as easy, safe, and eco-friendly green chemistry techniques (Sathiyanarayanan et al., 2017). Since microbial polymers first appeared commercially in the 1960s, their use has expanded significantly (Quesada et al., 1993). Microbial EPSs are essential for cellular interactions, nourishment, and the macro- and microenvironments. The rise of the synthetic biology field has made it possible to manipulate bacteria to produce certain EPSs with

DOI: https://doi.org/10.1016/B978-0-443-36619-2.00021-8

desired characteristics. By adding or deleting genes from the BGC or using artificial promoters to control the expression of these genes, this technique typically involves changing the metabolic pathways of the microorganisms (Wagner and Alper, 2016). The application of synthetic biology to make EPSs offers multiple benefits over conventional techniques. For example, it enables the making of EPSs with certain properties such as increased stability, improved solubility, and increased utility. It also enables the manufacture of EPS on a large scale, which is essential for commercial applications (Sathiyanarayanan et al., 2017).

Recently, Lactic acid bacteria (LABs), one kind of EPS-producing organism, have received a lot of attention from both the academic and industrial sectors (Moscovici, 2015; Sathiyanarayanan et al., 2017; Xu et al., 2019). It has been demonstrated that over 40 LAB species, or bacteria with a qualified presumption of safety (QPS) or generally regarded as safe (GRAS), may produce an array of EPSs with no endangering human health (Xu et al., 2019). These days, microbial EPSs are utilized for their hydrocolloid techno-activities as thickening agents, gelling agents, water-binding agents, stabilizers, and emulsifiers in pharmaceutical product formulations (such as excipients and drug delivery agents) and in fermented items (e.g., cheese and yogurt) as thickeners, stabilizers, and emulsifiers. They have, however, also been demonstrated to offer several health advantages to people due to their varied physiological and biological roles (Moscovici, 2015; Torino et al., 2015; Andrew and Jayaraman, 2020; Rana and Upadhyay, 2020). Some of the current examples recorded in the literature include antioxidant, antihypertensive, hypocholesterolemia, antiviral, anticancer, and immunomodulating activities (Moscovici, 2015; Torino et al., 2015; Xu et al., 2019; Andrew and Jayaraman, 2020). The potential use of LAB-derived EPSs as immunobiotic additives and innovative drug delivery vehicles in vaccine manufacturing and theranostic techniques, respectively, is a more novel and relevant uses that deserve much more attention both now and in the future.

Furthermore, many of these polysaccharides are already commercially available, while others are still being developed. In the pharmaceutical industry, EPSs also aid in the sustained, controlled release of medication into the human body (Moscovici, 2015; Tabernero and Cardea, 2020). EPS are anticipated to become more important in novel medication formulation techniques as studies into their therapeutic value expand.

21.2 Role of exopolysaccharides as pharmaceutical excipients

21.2.1 Functionality of excipients in drug formulations

Generally, drugs are always prepared in a formulated dosage form instead of their pure form. These may differ from a comparatively simple solution to a complex dosage form by using suitable excipients in the formulation. EPS are biopolymers synthesized by different kinds of microorganisms and have drawn significant interest in pharmaceutical formulation owing to their multifaceted properties. Because of their distinctive properties, they can be used as excipients to optimize drug delivery systems by various mechanisms (Manjanna et al., 2009).

21.2.1.1 *Sustained-release dosage form*

A variety of EPS forms gel or matrices that permit sustained drug release. EPS, such as scleroglucan and pullulan, are utilized to manufacture sustained-release tablets. These EPS releases are gradually released over an extended period. This is especially beneficial for maintaining therapeutic levels of7 drugs in the bloodstream and reducing fluctuations that might cause side effects (Sheth and Lachman, 1967; Hu et al., 2024). It is crucial to be able to modify the release profile according to various dosage forms, such as injectable formulations and oral tablets.

21.2.1.2 *Thickeners and stabilizers*

In several pharmaceutical applications, EPS play an important role as thickeners and stabilizers because of their renowned ability to alter the viscosity of the formulation, which is important for obtaining an appropriate texture and flow characteristics. By increasing the viscosity, EPS can produce a more stable formulation that can withstand changes over time (Tabernero and Cardea, 2020). Recent achievements in exudate gum modification, such as gum karaya, gum tragacanth, gum arabica, and k-carrageenan, have enhanced their beneficial characteristics and applications across various industries due to their emulsification, stabilization, and thickening behavior.

21.2.1.3 *Bioavailability enhancers*

Certain EPS effectively improve the solubility and absorption of poorly soluble drugs. This enhancement occurs by solubilization and modifying the release kinetic mechanism. In solubilization, EPS binds with poorly water-soluble drugs and forms complexes, thus increasing the solubility as well as absorption. Then, in the release kinetic mechanism, EPS forms a gel or matrix in which drugs are encapsulated, and release can be extended over time, enabling a prolonged therapeutic effect. Scleroglucan has been observed to improve the solubility and absorption of various poorly soluble drugs by forming hydrophilic gels that improve bioavailability by enhancing drug dissolution in the GIT (Werning et al., 2022).

21.2.2 Case studies

21.2.2.1 *Xanthan gum*

Xanthan is an extracellular polymer consisting of a homopolysaccharide D-glucose backbone and a trisaccharide side chain that alternates between two mannose residues and glucuronic acid and is produced by *Xanthomonas campestris* (Nguyen et al., 2024). Xanthan gum is commonly used in the food industry as a thickener and stabilizer. In pharmaceutical formulation, it is used as a binder because of its capacity to form a viscous solution that increases the drug release profile, tablet mechanical strength, and improves the stability of suspension. Xanthan gum is a desirable choice for many applications due to its exceptional rheological properties, which help to maintain stability under different conditions (Ng et al., 2020; Mouro et al., 2024).

21.2.2.2 *Dextran*

EPS dextran is produced by the lactic acid bacteria *Leuconostoc mesenteroides* in the presence of sucrose. Due to its viscosity, solubility, thermal, and rheological properties, dextran has commercial significance in food, research, and pharmaceutical applications. It has been used in drug delivery systems because it is biocompatible and can form a hydrogel. As a binder, it increases tablet integrity and improves the controlled release of active ingredients. Additionally, dextran has been used as a plasma expander and stabilizer in various pharmaceutical formulations (Díaz-Montes, 2021).

21.2.2.3 *Bacterial cellulose*

Bacterial cellulose is an EPS produced by a variety of bacterial species. Among them, Gluconacetobacter species are well recognized for producing cellulose effectively. It has been used proficiently in wound dressings, tissue engineering, and controlled drug release systems due to its high purity and mechanical strength. In pharmaceutical formulation, it acts as a binder that shows water-holding capacity, mechanical, and stabilizing properties without impacting its chemical structure, making it suitable for controlled drug delivery systems (Netrusov et al., 2023; Mouro et al., 2024).

21.2.2.4 *Alginate*

Alginate is a natural EPS of glucuronic acid and mannuronic acid obtained from brown algae. At pH > 6, alginate instantly produces a gel when it comes into contact with divalent cations like Ca2+, Ba2+, or Zn2+ through the ionotropic gelation method. Conversely, hydration of alginic acid forms a high-viscosity gel, "acid gel," at low pH. These attributes make the polymer exceptional and allow it to be used in various applications (Laurienzo, 2010). Alginate acts as a disintegrating agent, enhancing the breakdown of the dosage form by swelling in the presence of water and releasing the active medicament (Tønnesen and Karlsen, 2002).

21.2.2.5 *Pullulan*

Pullulan is an EPS originating from the microbial cell surface, mainly from the yeast-like fungus *Aureobasidium pullulans*. It is a linear glucan made up of maltotrose units linked by α-1,6-glycosidic bonds. Pullulan has been widely used in food, pharmaceutical, and biomedical applications due to its stabilizing, binding, thickening, texture-improving, food packaging, and disintegrating properties (Pacheco et al., 2021). Sharma et al. (2016) investigated the plasticizing effect of sorbitol, glycerol, and liquid glucose on Pullulan/tamarind pectin-based oral disintegrating film.

21.2.2.6 Exopolysaccharides-BMS

EPS-BMS is an exopolysaccharide produced from *Leuconostoc citreum*. Its efficacy as a stabilizer is considerably influenced by its high molecular weight. Compared to commercial stabilizers such as xanthan gum and gum arabica, EPS-BMS has better emulsifying ability. Due to its thickening properties, this EPS can efficiently stabilize the oil-in-water emulsion by the formation of a macromolecular barrier in the aqueous region, which increases the stability of the emulsion for extended periods of time in both acidic and basic environments (Bibi et al., 2021).

21.2.2.7 *Exopolysaccharides-S11*

EPS-S11, obtained from *Lactobacillus paracasei* H9, has been thoroughly investigated for its ability to improve the stability of sodium caseinate solution. EPS-S11 interacts with casein molecules and forms a network-like structure that enhances the elasticity and water-holding capacity. By these mechanisms, EPS-S11 substantially prevents the aggregation of casein by increasing the electrostatic repulsion between particles and decreases the turbidity of the solution, which contributes to preserving the texture, preventing syneresis, and maintaining the stability of yogurt and other dairy products (Jurášková et al., 2024).

21.3 Natural exopolysaccharides as adjuvants in drug delivery

Natural EPS have drawn attention to their promising use as adjuvants in immunotherapy and vaccine development to alter immune responses. The primary mechanisms, such as Toll-like receptor (TLR) engagement, cytokine modulation, antibody formation, and immune cell activation by natural EPS, improve drug delivery.

21.3.1 Mechanism of adjuvant activity

21.3.1.1 *Immune cell activation*

EPS can activate various immune cells, including dendritic cells (DC), macrophages, and natural killer cells (NK). EPS derived from *Lactobacillus casei*, particularly the WXD030 strain, serves as a prominent example of how natural polysaccharides can modulate immune responses: In-vitro study L-EPS shown the significant stimulation of cytokines (NO, TNF-α, IL-1β, and IL-6) while also increasing the proliferation and phagocytic activity of immune cells. On the other hand, in an in vivo application, L-EPS significantly increases T-cell proliferation and specific antibody production when given as adjuvants with foot-and-mouth disease vaccines (Xiu et al., 2018). In another research, researchers developed *Lactobacillus* EPS nanoparticle adjuvants that were prepared by sulfation-modified EPS and quaternization modified chitosan for the development of an efficient vaccine for severe acute respiratory syndrome coronavirus 2 (SARS-CoV-2) to provide potent and long-lasting immunity (Zhang et al., 2024).

21.3.1.2 *Entrapment of bioactive compounds*

EPS has the ability to incorporate bioactive ingredients, preventing degradation and improving therapeutic efficacy by forming a protective layer around the bioactive compounds. This encapsulation enables controlled drug release, extended periods of therapeutic action, and reduces the dose frequency. Researchers used 3%w/v EPS obtained from *Lactobacillus plantarum* ATCC 8014 to formulate a stable o/w emulsion with a 40:60 oil-to-water ratio. This finding confirmed that EPS successfully incorporated eugenol, maintaining its stability over 90 days under critical conditions like temperature, pH, and salinity (Balyan et al., 2024).

21.3.1.3 *Antigen presentation*

EPS from probiotics facilitates the interaction between host immune cells and probiotics. This interaction is important for the induction of antigen-presenting cells, like DC and macrophages, which process and present antigens to T cells, resulting in an effective adaptive immune response. EPS from *Lactobacillus kiferi* increased antibody titer against some specific antigens and stimulated a mixed Th1, Th2, and Th17 cell response. This improves the immune system and makes it more effective against several kinds of pathogens (Xiu et al., 2020).

21.3.2 Applications in vaccine formulations and immune therapies

The most affordable approach to preventing the wide range of infectious diseases is vaccination. There is growing evidence that EPS functions as an adjuvant to improve the immunogenicity of antigens, stimulate the production of cytokines, and activate various immune cells, like DC and macrophages. In order to boost immune response and reduce the antigen dose, adjuvants serve as essential components in vaccinations (Liu et al., 2011). Wang and colleagues discovered the adjuvant potentials of sulfated β-glucan of *S. cerevisiae*, significantly stimulating the release of IL-2 and interferon-gamma (IFN-γ) and markedly increasing lymphocyte proliferation. In vivo experiments demonstrated enhanced generation of CD4+ T cells, antibody titer, IL-2, INF-γ, IL-4, and IL-6 production, and also decreased the number of CD8+ T cells. Consequently, sulfated EPS showed a potent adjuvant effect on the influenza vaccination (Wang et al., 2016). Zhang et al. (2021) formulated a novel vaccine (rMntC-EPS) with *S. aureus* antigen MntC and *Lactobacillus casei* EPS. The rMntC-EPS vaccine induced γδ T cells and IL-17A; it could possibly be crucial for anti-*S. aureus* immunity and provide up to 90% protection against pulmonary infections with *S. aureus*.

21.4 Recent advancements in EPS-based pharmaceutical formulation

Amphiphilic Exopolysaccharides (EPS) have extensive applications in the field of drug delivery. Modifying EPS required to protect the incorporated drug molecule from enzymatic degradation inside the body is an effortless, economical process, prolonging the duration of action and reducing the side effects and shelf life of the drugs (Abbasi et al., 2022).

21.4.1 Nano- and micro-delivery systems

Levan is a homopolysaccharide that stabilizes selenium, cobalt, and iron-containing nanoparticles. Levan-coated nanoparticles can be used as a cascade of carbon source and eliminate the toxicity-related problem of microelements (Bondarenko et al., 2016). In their research, the authors demonstrated the biocompatibility of eps for casing magnetic nanoparticles and managing cancer with the synergistic effect of EPS on the drug. After being subjected to the drug-loaded magnetic nanoparticles, the authors reported that the nanoparticles remained biocompatible, had negligible cytotoxicity, and substantially lowered the number and lifespan of cancer cells (Sivakumar et al., 2014). Regarding the purpose of quick identification of disease, diagnosis, and health condition determination, super-magnetic iron oxide dextran coating nanoparticles permit simultaneous detection of molecular phenotype in vivo by targeted magnetic resonance and hybrid modality imaging (Tassa et al., 2011).

21.4.2 Transdermal drug delivery systems

Friedman et al. fabricated chitosan-alginate nanoparticles to produce antimicrobial activity against *Propionibacterium acnes*. They exert bactericidal action by disrupting the cell membrane. Additionally, nanoparticles reduced the secretion of cytokines from human eukaryotic cells (Friedman et al., 2013). The authors prepared a sodium alginate and trehalose encapsulated Bacillus Calmette–Guérin (BCG) microneedle patch, where sodium alginate serves as a penetration enhancer. Once the patch is applied, it gets dissolved and minimizes the chance of an allergy depending on the degree of moisture retention (Arshad et al., 2020). The mechanical attribute of cellulose scaffolds makes them suitable for tissue rejuvenation. Elasticity, toughness, and porosity of the cellulose scaffold are important for tissue regeneration and bone repair

(Shelke et al., 2014). The cytocompatible nature of chitosan hydrogels established them as an efficient vehicle to promote embryonic stem cell growth (Nair et al., 2007).

21.4.3 Enhancing bioavailability and targeted drug delivery

Nataraj and coresearchers reported for the first time the antianemic effect of the prepared exopolysaccharide EPSKAR-1 iron complex and its ability to alter the bioavailability of iron. Their research demonstrated that EPSK-1 has the ability to increase blood hemoglobin levels in anemic rats, which in turn facilitates the absorption, ingestion, and utilization of iron, thereby boosting iron bioavailability (Nataraj et al., 2023). Morifuji et al. observed how EPS from fermented milk by lactic acid altered carotenoid bioavailability. Through the diffusion process, EPS alter the physical characteristics of milk and influence the absorption of carotenoids. These EPS affect the adhesion of bacterial cell membranes to the human intestinal surface and enhance the penetration of carotene in the intestinal surface (Morifuji et al., 2020).

21.4.4 Sustainability and biocompatibility of exopolysaccharides

On controlling the amount of reactive oxygen species (ROS), the cell membrane repolarization process, and caspase enzyme function, *Bacillus sp.* fermented EPS is biodegradable, biocompatible, and efficient in shielding bodies from harmful UV rays (Li et al., 2022). In order to reduce considerable substantial manufacturing expenses of EPS, investigating the waste material sustainable strategy that supports both sustainable development and profitability (Mouro et al., 2024).

21.5 Challenges and limitations in the use of natural exopolysaccharides

It is difficult to synthesize EPS using lactic acid bacteria because of morphological diversity, certain ecological constraints, and strain on products. It takes a lot of evaluation to select a highly strain product, which is a very laborious, expensive, and time-consuming process. The fermentation process is a sophisticated and complex procedure that has to be handled very gently. During production for the same standard product, a thorough evaluation is required to avoid variation in the function and composition of EPS (Zhou et al., 2023). Bacteria's growth condition determines the standard, effectiveness, and amount of synthesized EPS. In contrast to nondairy bacteria, dairy bacteria produce EPS in a considerably lower amount (Papagianni, 2012).

21.6 Discussion and conclusions

EPS have established themselves as versatile biopolymers with significant applications in the pharmaceutical field, characterized by their unique physicochemical properties such as biocompatibility, biodegradability, and nontoxicity. These attributes make EPS ideal candidates for various biomedical applications, including drug delivery systems, vaccine formulations, and excipients in pharmaceuticals. The future of EPS in the pharmaceutical field is bright, with promising directions emerging from advances in technology that enable the modification of EPS to improve their drug encapsulation abilities and release profiles. The integration of EPS into advanced drug delivery platforms, such as microencapsulation and controlled release systems, will likely expand their applications in personalized medicine. As personalized medicine continues to gain traction, EPS can play a pivotal role by serving as biocompatible carriers for targeted therapies tailored to individual patient profiles. This evolution will lead to enhanced therapeutic strategies that could revolutionize patient care in the coming years.

Credit statement

Sanjukta Sen: Literature review, writing original draft; Piu Jana: Literature review, writing original draft; Esha Das: Literature review, writing original draft; Gouranga Nandi: Supervision, editing.

Conflict of interest

The authors declare that there is no conflict of interest with this book chapter and its submission.

Ethical statement

Not applicable

Acknowledgment

The authors are thankful to the authority of the University of North Bengal for providing the digital library facility, and also thankful to all the authors whose works are cited in this chapter.

References

Abbasi, A., Saadat, T.R., Saadat, Y.R., 2022. Microbial exopolysaccharides–β-glucans–as promising postbiotic candidates in vaccine adjuvants. Int. J. Biol. Macromol. 223, 346–361.

Andrew, M., Jayaraman, G., 2020. Structural features of microbial exopolysaccharides in relation to their antioxidant activity. Carbohydr. Res. 487, 107881.

Arshad, M.S., Fatima, S., Nazari, K., Ali, R., Farhan, M., Muhammad, S.A., Chang, M.-W., et al., 2020. Engineering and characterisation of BCG-loaded polymeric microneedles. J. Drug Targeting 28 (5), 525–532.

Balyan, S., Dadwal, V., Patil, B.S., 2024. *Lactobacillus*-isolated exopolysaccharide as emulsifier ensure extended stability of eugenol encapsulation, potent anti-microbial activity, and application on fresh produce. Food Biosci. 61, 104632.

Bibi, A., Xiong, Y., Rajoka, M.S.R., Mehwish, H.M., Radicetti, E., Umair, M., Aadil, R.M., et al., 2021. Recent advances in the production of exopolysaccharide (EPS) from *Lactobacillus spp.* and its application in the food industry: a review. Sustainability 13 (22), 12429.

Bondarenko, O.M., Ivask, A., Kahru, A., Vija, H., Titma, T., Visnapuu, M., Visnapuu, T., et al., 2016. Bacterial polysaccharide levan as stabilizing, non-toxic and functional coating material for microelement-nanoparticles. Carbohydr. Polym. 136, 710–720.

Costa, J.A.V., Lucas, B.F., Alvarenga, A.G.P., Moreira, J.B., de Morais, M.G., 2021. Microalgae polysaccharides: an overview of production, characterization, and potential applications. Polysaccharides 2((4), 759–772.

Díaz-Montes, E., 2021. Dextran: sources, structures, and properties. Polysaccharides 2((3), 554–565.

Friedman, A.J., Phan, J., Schairer, D.O., Champer, J., Qin, M., Pirouz, A., Modlin, R.L., et al., 2013. Antimicrobial and anti-inflammatory activity of chitosan–alginate nanoparticles: a targeted therapy for cutaneous pathogens. J. Invest. Dermatol. 133((5), 1231–1239.

Hamidi, M., Okoro, O.V., Milan, P.B., Khalili, M.R., Samadian, H., Nie, L., Shavandi, A., 2022. Fungal exopolysaccharides: properties, sources, modifications, and biomedical applications. Carbohydr. Polym. 284, 119152.

Hu, S., Xu, C., Zhang, Y., Du, Y., Tang, J., Chen, L., 2024. Preparation of enteric capsules with pulsatile drug delivery potential using pullulan and polyacrylic acid resin III. Arabian J. Chem. 17, 105691.

Jurášková, D., Ribeiro, S.C., Bastos, R., Coelho, E., Coimbra, M.A., Silva, C.C., 2024. Exopolysaccharide (EPS) produced by *Leuconostoc mesenteroides* SJC113: characterization of functional and technological properties and application in fat-free cheese. Macromol 4 (3), 680–696.

Kaur, N., Dey, P., 2023. Bacterial exopolysaccharides as emerging bioactive macromolecules: from fundamentals to applications. Research in microbiology 174 (4), 104024.

Laurienzo, P., 2010. Marine polysaccharides in pharmaceutical applications: an overview. Mar. Drugs 8 (9), 2435–2465.

Li, F., Hu, X., Qin, L., Li, H., Yang, Y., Zhang, X., Bao, M., et al., 2022. Characterization and protective effect against ultraviolet radiation of a novel exopolysaccharide from *Bacillus marcorestinctum* QDR3-1. Int. J. Biol. Macromol. 221, 1373–1383.

Liu, C.F., Tseng, K.C., Chiang, S.S., Lee, B.H., Hsu, W.H., Pan, T.M., 2011. Immunomodulatory and antioxidant potential of *Lactobacillus* exopolysaccharides. J. Sci. Food Agric. 91 (12), 2284–2291.

Manjanna, K., Shivakumar, B., Pramodkumar, T., 2009. Natural exopolysaccharides as novel excipients in drug delivery: a review. Arch. Appl. Sci. Res. 1 (2), 230–253.

Mohd Nadzir, M., Nurhayati, R.W., Idris, F.N., Nguyen, M.H., 2021. Biomedical applications of bacterial exopolysaccharides: A review. Polymers 13 (4), 530.

Morifuji, M., Ichikawa, S., Kitade, M., Fukasawa, T., Asami, Y., Manabe, Y., Sugawara, T., 2020. Exopolysaccharides from milk fermented by lactic acid bacteria enhance dietary carotenoid bioavailability in humans in a randomized crossover trial and in rats. Am. J. Clin. Nutr. 111 (4), 903–914.

Moscovici, M., 2015. Present and future medical applications of microbial exopolysaccharides. Front. Microbiol. 6, 1012.

Mouro, C., Gomes, A.P., Gouveia, I.C., 2024. Microbial exopolysaccharides: structure, diversity, applications, and future frontiers in sustainable functional materials. Polysaccharides 5 (3), 241–287.

Nair, L.S., Starnes, T., Ko, J.-W.K., Laurencin, C.T., 2007. Development of injectable thermogelling chitosan–inorganic phosphate solutions for biomedical applications. Biomacromolecules 8 (12), 3779–3785.

Nataraj, B.H., Gowda, B.S., Kapila, S., Arora, S., Puniya, A.K., Nagpal, R., Behare, P.V., 2023. Influence of exopolysaccharide EPSKar1–iron complexation on iron bioavailability and alleviating iron deficiency anaemia in Wistar rats. Food Funct. 14 (10), 4931–4947.

Netrusov, A.I., Liyaskina, E.V., Kurgaeva, I.V., Liyaskina, A.U., Yang, G., Revin, V.V., 2023. Exopolysaccharides producing bacteria: a review. Microorganisms 11 (6), 1541.

Ng, J.Y., Obuobi, S., Chua, M.L., Zhang, C., Hong, S., Kumar, Y., Ee, P.L.R., et al., 2020. Biomimicry of microbial polysaccharide hydrogels for tissue engineering and regenerative medicine–a review. Carbohydr. Polym. 241, 116345.

Nguyen, H.-T., Pham, T.-T., Nguyen, P.-T., Le-Buanec, H., Rabetafika, H.N., Razafindralambo, H.L., 2024. Advances in microbial exopolysaccharides: present and future applications. Biomolecules 14 (9), 1162.

Osemwegie, O.O., Adetunji, C.O., Ayeni, E.A., Adejobi, O.I., Arise, R.O., Nwonuma, C.O., Oghenekaro, A.O., 2020. Exopolysaccharides from bacteria and fungi: current status and perspectives in Africa. Heliyon 6 (6).

Pacheco, M.S., Barbieri, D., da Silva, C.F., de Moraes, M.A., 2021. A review on orally disintegrating films (ODFs) made from natural polymers such as pullulan, maltodextrin, starch, and others. Int. J. Biol. Macromol. 178, 504–513.

Papagianni, M., 2012. Metabolic engineering of lactic acid bacteria for the production of industrially important compounds. Comput. Struct. Biotechnol. J. 3 (4), e201210003.

Quesada, E., Béjar, V., Calvo, C., 1993. Exopolysaccharide production by *Volcaniella eurihalina*. Experientia 49, 1037–1041.

Rahbar Saadat, Y., Yari Khosroushahi, A., Pourghassem Gargari, B., 2021. Yeast exopolysaccharides and their physiological functions. Folia Microbiol. 66 (2), 171–182.

Rana, S., Upadhyay, L.S.B., 2020. Microbial exopolysaccharides: synthesis pathways, types and their commercial applications. Int. J. Biol. Macromol. 157, 577–583.

Sathiyanarayanan, G., Dineshkumar, K., Yang, Y.-H., 2017. Microbial exopolysaccharide-mediated synthesis and stabilization of metal nanoparticles. Crit. Rev. Microbiol. 43 (6), 731–752.

Sharma, R., Kamboj, S., Singh, G., Rana, V., 2016. Development of aprepitant loaded orally disintegrating films for enhanced pharmacokinetic performance. Eur. J. Pharm. Sci. 84, 55–69.

Shelke, N.B., James, R., Laurencin, C.T., Kumbar, S.G., 2014. Polysaccharide biomaterials for drug delivery and regenerative engineering. Polym. Adv. Technol. 25 (5), 448–460.

Sheth, P., Lachman, L., 1967. The coating of tablets. Fr. Patent 1 (480), 874.

Sivakumar, B., Aswathy, R.G., Sreejith, R., Nagaoka, Y., Iwai, S., Suzuki, M., Maekawa, T., et al., 2014. Bacterial exopolysaccharide based magnetic nanoparticles: a versatile nanotool for cancer cell imaging, targeted drug delivery and synergistic effect of drug and hyperthermia mediated cancer therapy. J. Biomed. Nanotechnol. 10 (6), 885–899.

Sørensen, H.M., Rochfort, K.D., Maye, S., MacLeod, G., Brabazon, D., Loscher, C., Freeland, B., 2022. Exopolysaccharides of lactic acid bacteria: production, purification and health benefits towards functional food. Nutrients 14 (14), 2938.

Tabernero, A., Cardea, S., 2020. Microbial exopolysaccharides as drug carriers. Polymers 12 (9), 2142.

Tassa, C., Shaw, S.Y., Weissleder, R., 2011. Dextran-coated iron oxide nanoparticles: a versatile platform for targeted molecular imaging, molecular diagnostics, and therapy. Acc. Chem. Res. 44 (10), 842–852.

Tønnesen, H.H., Karlsen, J., 2002. Alginate in drug delivery systems. Drug Dev. Ind. Pharm. 28 (6), 621–630.

Torino, M.I., Font de Valdez, G., Mozzi, F., 2015. Biopolymers from lactic acid bacteria. Novel applications in foods and beverages. Front. Microbiol. 6, 834.

Wagner, J.M., Alper, H.S., 2016. Synthetic biology and molecular genetics in non-conventional yeasts: current tools and future advances. Fungal Genet. Biol. 89, 126–136.

Wang, M., Zhang, L., Yang, R., Fei, C., Wang, X., Zhang, K., Xue, F., et al., 2016. Improvement of immune responses to influenza vaccine (H5N1) by sulfated yeast beta-glucan. Int. J. Biol. Macromol. 93, 203–207.

Werning, M.L., Hernández-Alcántara, A.M., Ruiz, M.J., Soto, L.P., Dueñas, M.T., López, P., Frizzo, L.S., 2022. Biological functions of exopolysaccharides from lactic acid bacteria and their potential benefits for humans and farmed animals. Foods 11 (9), 1284.

Xiu, L., Sheng, S., Hu, Z., Liu, Y., Li, J., Zhang, H., Wang, X., et al., 2020. Exopolysaccharides from *Lactobacillus kiferi* as adjuvant enhanced the immuno-protective against *Staphylococcus aureus* infection. Int. J. Biol. Macromol. 161, 10–23.

Xiu, L., Zhang, H., Hu, Z., Liang, Y., Guo, S., Yang, M., Wang, X., et al., 2018. Immunostimulatory activity of exopolysaccharides from probiotic *Lactobacillus casei* WXD030 strain as a novel adjuvant in vitro and in vivo. Food Agric. Immunol. 29 (1), 1086–1105.

Xu, Y., Cui, Y., Yue, F., Liu, L., Shan, Y., Liu, B., Lü, X., et al., 2019. Exopolysaccharides produced by lactic acid bacteria and *Bifidobacteria*: structures, physiochemical functions and applications in the food industry. Food Hydrocolloids 94, 475–499.

Zhang, H., Pan, N., Ma, C., Liu, B., Xiu, L., Tong, H., Ma, F., et al., 2021. Vaccine composition formulated with a novel *Lactobacillus*-derived exopolysaccharides adjuvant provided high protection against *Staphylococcus aureus*. Vaccines 9((7), 775.

Zhang, S., Fan, W., Ding, C., Zhang, M., Liu, S., Liu, W., Song, S., et al., 2024. Self-assembling sulfated lactobacillus exopolysaccharide nanoparticles as adjuvants for SARS-CoV-2 subunit vaccine elicit potent humoral and cellular immune responses. ACS Appl. Mater. Interfaces 16 (15), 18591–18607. doi:10.1021/acsami.4c01384.

Zhou, Z., Zeng, X., Wu, Z., Guo, Y., Pan, D., 2023. Relationship of gene-structure–antioxidant ability of exopolysaccharides derived from lactic acid bacteria: a review. J. Agric. Food Chem. 71 (24), 9187–9200.

CHAPTER

22

Latest status and advancements of EPS in the food industry and sustainable eco-friendly packaging

Dilip Kumar Roy[a], *Partha Palit*[b], *Aparna Beena Unni*[c] *and Rajesh Kumar Mukherjee*[d]

[a]Department of Pharmaceutical Technology, JIS University, Agarpara, Kolkata, West Bengal, India [b]Department of Pharmaceutical Sciences, Assam University (A Central University), Silchar, Assam, India [c]Institute of Physics, University of Silesia, Chorzów, Poland [d]Department of Pharmaceutical Technology, Brainware University, Kolkata, West Bengal, India

22.1 Introduction

Exopolysaccharides (EPS) extend endurable packaging solutions and food, demanding further research on biopolymers and microbial production alteration. The food industry greatly favors EPS, which are having the high molecular weight polysaccharides produced by bacteria, due to their beneficial and health-promoting qualities (Rahbar Saadat et al., 2021; Jurášková et al., 2022; Yang et al., 2023; Castro et al., 2024). Both traditional food handling and environmentally friendly packaging can benefit from their uses. Packaging performance compared to such synthetic plastics, EPS-based films often do not let water vapor through very well, are not very strong, and do not block oxygen very well, which makes them less useful for business (Chen et al., 2023a, 2023b; Eze et al., 2022; Giavasis, 2013; Harutoshi, 2013; Kırtel et al., 2017; Yates et al., 2021). Research shows that mixing EPS with other biopolymers or nanoparticles can make it more durable. However, prototypes have trouble with scalability and consistency in bioactive or edible coatings. Improving the performance of packaging when mixed with chitosan (which increases tensile strength by 40%) or nano-cellulose (which lowers WVTR by 60%), EPS films' barriers get better. Electrospinning is now used in scalable prototypes to make bioactive coatings that are uniform. This was shown to work in pilot trials that extended the shelf life of dairy products by six months. These hybrids work just as well as PET, but they can also be composted (Moradi et al., 2021).

DOI: https://doi.org/10.1016/B978-0-443-36619-2.00022-X

The global scenarios signify the EPS in the sustainable packaging and food sector, the EPS in the food sector are involved in as indicated in the following items:

22.1.1 Summary of the exopolysaccharides

EPS can strengthen sustainable and functional food innovations via boosting structural and microbial enhancement. EPS are high molecular weight polysaccharides produced by microbes, which have attracted the considerable attention in food business owing to their enormous uses (Prajapati et al., 2013). The EPS functions as thickening agents, gelling agents, stabilizers, and emulsifiers, enhancing the texture and the quality of several food products, particularly in the fermentation of goods such as yogurt and cheese. EPS yields are still low and changeable because they depend on microbial strains, fermentation conditions, and expensive substrates like sugars, which can make up to 40% of production costs. Recent reviews have pointed out that it is hard to move from (pilot)lab-scale bioreactors to industrial bioreactors because of problems with oxygen transfer, heat transfer, optimizing the medium, and finding cheaper alternatives to agro-waste. Strain selection (for example, high-producing *Sphingobium* sp.) and cheap substrates like molasses or whey help with low yields and variability. This cuts costs by 30%–50%. New impeller designs and CFD modeling help solve oxygen and heat transfer problems that come up when scaling up, getting 10–15 g/L EPS in 1000 L bioreactors. Metabolic engineering makes output even more stable by turning up the Uridine Diphosphate (UDP)-glucose pathways (Chen et al., 2023a; Guevara Guerrero et al., 2024).

22.1.2 Health benefits

EPS are important for sustainable packaging and functional foods, involving research on microbial refining and eco-friendly uses. Beyond its huge technological advancements, the EPS also confers several health benefits. They function as the prebiotics that promote intestinal health by supporting beneficial microbes in the gut. Furthermore, many EPS have antioxidant properties and may aid in lowering cholesterol levels, making them important for the development of functional meals. EPS produced by lactic acid bacteria (LAB) have been shown the effectives in reducing oxidative stress and enhancing immune responses.

22.1.3 Market growth

The demand for EPS is increasing due to a shift toward natural ingredients in food production. This trend is driven by customer preferences for healthier options and the growing awareness of the possible health risks associated with synthetic substances (Prajapati et al., 2013; Rahbar Saadat et al., 2019). As a result, the EPS is being investigated as a substitute for artificial stabilizers and emulsifiers in diverse food compositions and blends. Barriers in the market and with regulations, nondairy EPS must follow strict food safety rules, especially when it comes to labeling and getting approval for dairy or global markets. This slows down adoption, even though some, like *Xanthan*, are GRAS. Life-cycle assessments that show true sustainability over plastics are not being done as often as they should be, and consumers are not being taught about safety and benefits. Taking care of market and regulatory problems like FDA novel food approvals, like the EU 2023 listings, makes it possible for nondairy EPS like gellan to get

GRAS status. LCA studies show that LCA has a carbon footprint that is 70% smaller than LDPE. Apps that teach consumers about safety (no migration above 10 µg/kg) show this. Labeling compliance uses blockchain to make it easier to trace products, which speeds up market entry in India and the EU.

22.1.4 Sustainable packaging solutions trends in reusable packaging

The food business is seeing a notable transition toward sustainable packaging alternatives. This trend is mostly driven by environmental apprehensions and related to plastic waste, that necessarily for the sustainable materials. Companies are progressively embracing compostable and biodegradable packaging alternatives that reduce the environmental impact (Baptista et al., 2022; Serra et al., 2023; Chen et al., 2024). EPS yields are still low and changeable because they depend on microbial strains, fermentation conditions, and expensive substrates like sugars, which can make up to 40% of production costs. Recent reviews have pointed out that it is hard to move from (pilot)lab-scale bioreactors to industrial bioreactors because of problems with oxygen transfer, heat transfer, optimizing the medium, and finding cheaper alternatives to agro-waste. Strain selection (for example, high-producing *Sphingobium* sp.) and cheap substrates like molasses or whey help with low yields and variability. This cuts costs by 30%–50% new impeller designs and Computational Fluid Dynamics (CFD) modeling help solve oxygen and heat transfer problems that come up when scaling up, getting 10–15 g/L EPS in 1000 L bioreactors. Metabolic engineering makes output even more stable by turning up the UDP-glucose pathways (Guevara Guerrero et al., 2024). There has been more research on EPS in food and eco-friendly packaging, but there are still big gaps in scaling up production, making the film better in manufacturing, and following the rules to get a consistent product. Increasing production and consistency in QA/QC in the recent progress has been made it easier to scale up by improving fed-batch fermentation and using waste from agriculture and industry for example, using CRISPR-Cas9 to genetically modify *Lactobacillus plantarum* strains has increased EPS yields by 2–3 times and lowered sugar costs to less than 20%. Standardized ISO 22000 protocols and spectroscopy for real-time monitoring of microbial purity and consistency during audits fill in the gaps between QA and QC. This makes sure that everything is the same from lab to industrial bioreactors.

22.1.5 Role of the exopolysaccharides in packaging

EPS enhance biodegradable packaging, and research on their barrier and film-forming properties. EPS is employed not only in food items but also in applications related to packaging. Their film-forming qualities render them the most appropriate for the production of biodegradable coatings that prolong the shelf life and reduce dependency on traditional polymers. These coatings can safeguard the food from moisture and oxygen, therefore improving preservation while maintaining the preservation of the environment (Jurášková et al., 2022; Panchal et al., 2022; Benhadda et al., 2023; Vanin et al., 2023). Novel research trend needed more advanced microbial engineering to make customized modified EPS structures, more in-depth structural and functional studies using tools like Nuclear Magnetic Resonance (NMR), and interdisciplinary trials for active packaging with antimicrobials. There are also gaps in microalgae EPS because it is hard to grow, and because there are claims that it is good for your health in vivo as a prebiotic or antioxidant. Synthetic biology and microbial engineering make custom EPS, like

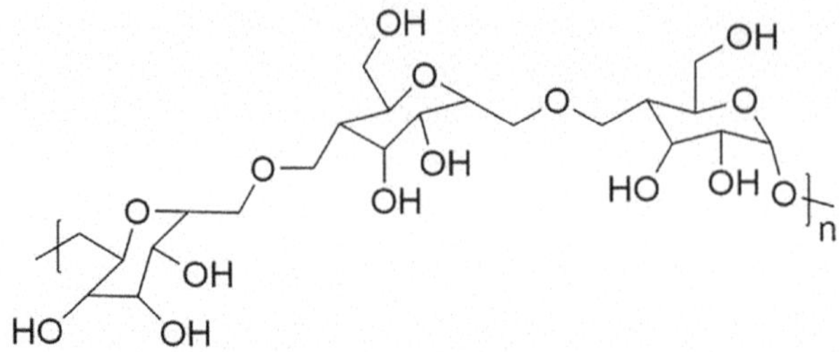

FIGURE 22.1 Structure of pullulan. Adapted from (https://www.mdpi.com/2079-4983/11/1/20) under CC YY open access.

sulfated variants for antimicrobials, by using NMR/Fourier Transform Infrared (Spectroscopy) (FTIR) to map structure to function. Microalgae EPS (for example, from *Spirulina*) yields rise five times with photobioreactors. In vivo trials show that it has prebiotic effects (modulating gut microbiota) and antioxidant activity (ORAC > 500 μmol TE/g). Interdisciplinary antimicrobial packaging trials combine EPS with essential oils, resulting in a 90% reduction in pathogens (Kiran et al., 2024).

22.1.6 Consumers' demand and the regulatory assistance

EPS may boost sustainable food packaging with their natural antimicrobial properties and enhance food quality and stability. Laws are intended to reduce the use of plastic supports, driving the adoption of environmentally friendly, safe packaging. Consumers are increasingly preferring companies that demonstrate a commitment to sustainability, thereby driving innovation in environmentally friendly packaging solutions that use natural materials such as EPS. In this regard, the following Fig. 22.1 depicts various bioresources for waste valorization that may be used as raw materials for the large-scale production of EPS for food packaging. Approaches to valorizing waste reduce the environmental burden and can create high-value EPS biopolymers for abundant applications in food resources and food packaging.

Whereas the EPS is utilized in the food industry in the following context:

A. Texture and stability of the food: EPS organically solidifies and stabilizes the food in a way that provides a wide variety of foods that are better and more uniform. Soured dairy products, such as cheese and yogurt frequently associated with enhancing their sensory appeal and improving their emulsification in the oral cavity. EPS, such as *Xanthan* gum and gellan (gel-like property), have the ability to form a gel and can serve as an alternative to synthetic chemicals.

B. Health benefits: Some of the EPS have prebiotic (beneficial bacteria) properties that improve gut health. In addition, they may function as antioxidants that aid in the reduction of reactive stress responses within the gut mobility. Because EPS can do two things, it is useful for both technological uses and possible health benefits.

C. Emulsification: EPS are effective in adding additives for stability to oil-in-water emulsions in foods like stews and sauces for salads. They are better for making emulsions more stable than certain conventional emulsifiers, which makes them a popular choice in labeled clean goods.

TABLE 22.1 Differences in biodegradable and exopolysaccharides packaging.

S. no.	Feature	Biodegradable packaging	Exopolysaccharides packaging
1	Definition	Decomposes naturally via microorganisms	Polysaccharides produced by microbes
2	Materials used	Plant-based materials (e.g., Polylactic Acid (PLA))	Microbial-derived polysaccharides
3	Environmental impact	Reduces plastic waste	Biodegradable with potential for sustainable production
4	Cost	Generally, more expensive	May have higher production costs
5	Decomposition conditions	May require specific composting conditions	Naturally breaks down but depends on microbial activity
6	Applications	Food packaging, disposable items	Food preservation, pharmaceuticals

D. Sustainable packaging solutions: The use of EPS in packaging materials is tremendously growing, a trend that is beneficial for the environment.

i. "Biodegradable films": This can add EPS to compostable films that cover food. These layers can make food preserve better by preventing air from getting in to and stopping it from going rotten. These enhancements are in line with the growing need for long-lasting packing options that are better for human nature.

ii. Packaging at some point: Combining the antimicrobials with the EPS-based films makes it possible for these materials to actively stop the growth of microbes on the food surfaces, which makes food safer without using artificial stabilizers. The use of EPS as a natural alternative to manufactured preservatives and stabilizers improves the labels and encourages eco-friendly practices in food packaging (Table 22.1).

22.2 Origin and advancements of exopolysaccharides

EPS have biodegradable possibilities in biotechnology and food, that enhances the research on microbial sources and fermentation. EPS are synthesized by several microorganisms, including bacteria and fungi, which possess high molecular weight polysaccharides. These biopolymers are surrounded by an environment that serves numerous purposes, including protecting microbial cells, enhancing food texture, and acting as natural preservatives. EPS has gathered information and triggered attention in the food industry owing to being able to improve the rheological properties of food products, making them valuable as emulsifiers, thickeners, and gelling agents (Donot et al., 2012; Moscovici, 2015; Freitas et al., 2017; Wünsche and Schmid, 2023).

22.2.1 Sources of exopolysaccharides

EPS have more potential for industrial and pharmaceutical uses, requiring their biosynthesis and microbial sources.

a. Bacterial origins: Bacteria synthesize diverse extracellular polymeric compounds during metabolic processes, generally excreting these polysaccharides into their environment, which facilitates the development of protective biofilms and improves nutrient retention and surface adhesion. Environmental conditions remarkably impacted deep to the production of exopolysaccharide (EPS) in the bacteria. These processes, which are mainly involved in the multifaceted metabolic pathways that employ nucleotide as sugar metabolites, that to be as an intermediate product. Particularly, the marine type of bacteria—especially those are inhabiting extreme frozen environments such as hydrothermal chilling vents—demonstrate their rare EPS profiles. These specific structured EPS have shown varied optimistic potentials those are in the field of industrial applications (Hayashi et al., 1996; Laroche, 2022; Huang et al., 2022; Zhang et al., 2022; Shao et al., 2023; Benhadda et al., 2023; Visentin et al., 2024; Tao et al., 2024).
b. Fungal origins: Fungi are the most common and crucial source of EPS, producing a wide, varied range of substances with diverse structural and notable functional characteristics effects. Some of the fungal EPS are often known for their primarily important physical and pharmacological properties, including anticancer and antioxidant feasible activities. Their adequate extraction processes are often more practical than those from any of the plant sources, enhancing their values in industries such as agro-food processing and in pharmaceuticals. Due to their prosperous bioactive profiling, the fungal EPS are commonly thoroughly scrutinized for biotechnological use.
c. Marine microorganisms as EPS producers: Marine microorganisms, especially those from extreme or polar habitats, are recognized for producing structurally distinctive EPS. These molecules are being explored for their potential therapeutic and industrial applications due to their unique chemical composition. Aquatic-ascertained EPS are particularly the most promising as the novel bio-catalysts and medicinal agents, that are offering the most innovative solutions in drug process development and in the biotechnological studies.
d. Synthesis and metabolic pathways: EPS development commonly involves various mediate the processes and most likely the metabolic mechanisms, which may differ between homo-polysaccharides and hetero-polysaccharides. Homo-polysaccharides are usually synthesized with the help of extracellular enzymes, whereas the hetero-polysaccharides are generally produced by the intracellularly and then secreted toward the cell from the outer end. In process research that mostly focused on associating the specific genes and enzymes that are generally involved in these pathways to better understand them and to influence the EPS biosynthesis (Fig. 22.2).

22.2.2 Biological functions and composition of exopolysaccharides

EPS are promising for deeper analysis and applications due to their assorted features and roles in microbial systems.

Chemical composition: They are mixed and mostly indulge nature of polymers primarily composed of monomers of sugar, mostly like galactose, rhamnose, and glucose, although

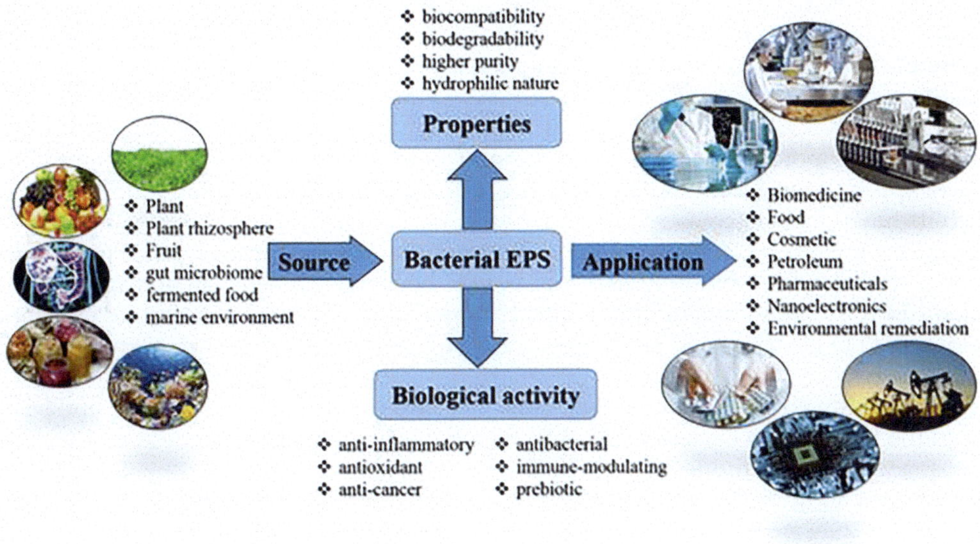

FIGURE 22.2 Showing bacterial EPS properties, biological activity, application, and sources. Adapted with open access permission CC BY (https://www.mdpi.com/2076-2607/11/6/1541). *EPS*, Exopolysaccharides.

their proportions and combinations widely vary. Additional functional groups like pyruvate, phosphate, and acetate can also be present, providing better functionality and diversity (Schmid et al., 2015; Zeidan et al., 2017; Jeong et al., 2022; Wünsche and Schmid, 2023; Chen et al., 2024).

Key biological functions:

a. Formation of biofilms: It provides biofilms to help in the structural support, facilitating the microbial adhesion to surfaces, and protection from environmental stresses.
b. Cell-to-cell communication: It helps in the communication of microbial environments that commonly influence the nature of pathogenicity and cell proliferation.
c. Protection: EPS offer defense against desiccation, predation, and antimicrobial agents, enhancing microbial resilience.

These functions underline the importance of EPS in microbial survival and environmental adaptation.

22.2.3 Recent researches on exopolysaccharides

EPS show promising prospects for future development, necessitating further research on their modification and synthesis. Recent studies have underscored the broad utility and biological activity of EPS across multiple domains, including food, pharmaceuticals, and environmental science (Branda et al., 2005; Torino et al., 2015).

a. *Pseudomonas alcaligenes:* A thermo-tolerant strain that is commonly isolated from a hot spring in the region of Chile and was found to be a novel with high stability in thermal and structurally modified and effective emulsification properties with an antioxidative activity.

These characteristics make it a strong candidate for use in food processing. Genetic analysis revealed insights into the biosynthetic pathways of this EPS, highlighting the potential of extremophilic bacteria as valuable EPS sources.

b. *Saccharomyces cerevisiae:* This EPS is a yeast-derived that promises the potential immunomodulatory and antioxidant effects, distinctly in macrophage activation. Due to its probiotic and thermo-stability potential, it carries and upholds promise for food supplement and pharmaceutical applications. Structural analysis suggested that the key sulfation modifications that enrich its functional effects.

c. *Bifidobacterium strains:* EPS secreted by *Bifidobacterium* species illustrate notable immunomodulatory activities. These bio-polymers are about to boost their potency in managing immune-related disorder likely causing inflammatory bowel disease. Mono-saccharide examination has identified D-galactose and D-glucose as key constituents, contributing to their bioactivity and complexity.

d. Chemical modifications: Recent studies show advances in the chemical reaction and have enhanced the functions of bacterial EPS, which includes methods like sulfation and acetylation to boost antioxidant activities. These modifications are fundamental for the development of more efficient and functionally superior biopolymers for industrial use.

e. Composites: They are enhancing their interest in a combination of natural EPS with inorganic chemicals to improve the mechanical and chemical properties. These fusions of materials are being thoroughly studied for their varied and advanced applications in biodegradable films, drug delivery systems, and in technologies such as pollution remediation and water purification.

22.2.4 Recent advancements and modifications in its production and applications

EPS hold enormous promise in biotechnology and pharmaceuticals, requiring continued exploration, improved production, and modification.

a. Novel production techniques and sources like:

i. Salt-tolerant bacteria: Researchers have identified that such bacteria are the emerging sources with improved unique modest properties that are suited to environments that overcome from severity to normal levels. Their enriched structural features make them highly useful for specialized industrial usage. However, the unique features such as low production yields and high costs, remain a hurdle to commercialization. Ongoing efforts are engrossed on upgradation of optimizing genetic strategies and fermentation conditions to boost efficiency in its production.

ii. Thermotolerant strains: A new EPS was identified from the thermotolerant strain *Pseudomonas alcaligin*'s, isolated from a hot spring in Chile. Recent studies have focused on the remarkable thermal stability of certain specific EPS, along with their significant antioxidant and emulsification properties. These features indicate strong potential for use as food additives and in other industrial processes where stability and bioactivity are critical (Rahbar Saadat et al., 2019; Santra and Banerjee, 2021; Vadnerker et al., 2022).

b. Bioactive functionalities: EPS derived from various microbial sources have demonstrated promising bioactivities, including antioxidant, antibacterial, antifungal, and antiviral effects. For example, EPS produced by Bifidobacterium strains show immunomodulatory properties relevant to human health. EPS are being examined for managing their therapeutic potential on colorectal cancer and inflammatory bowel diseases (Coviello et al., 2005; Andrew and Jayaraman, 2020).

c. Drug delivery development of nanoparticle: Emulsions have been employed to develop the nanoparticles for targeted drug delivery approaches. These nanoform formulations are modifying the drug stability, bioavailability, and solubility, which exhibits the flexibility of EPS as a drug carrier in biomedical utilization.
d. Modifications in chemical and its structural engineering: Advances in engineering have focused on modifying their structure through chemical techniques such as phosphorylation and acetylation of EPS. These alterations aim to focus on improving the functional parameters and their properties such as their biological activity, thermal resistance, and solubility. This facilitates to improve its development of EPS for specific medical and industrial applications.
e. Advanced techniques used in the structural and characterization: To understand the structural and functional relationships better, researchers are conducting various analytical tools and techniques that include FTIR, nuclear high-performance liquid chromatography (HPLC), and Magnetic resonance (NMR). These techniques are quite essential for improving its production strategies, functional properties, and ensuring high-quality control standards.

22.2.5 Applications in the food industry

EPS have auspicious applications in food and biomedicine, involving further research into multifunctional and eco-friendly solutions. EPS is implicated to enhance the crucial role in progressively improving the food industry that distinctly used in the development of food products and acceptable tangible packaging solutions. Their unique bioactive and nutritional properties support a range of applications, such as:

i. Enhancement of texture: EPS, such as gellan and xanthan gums, are broadly used to improve the fruitiness flavor and enhance the texture of foods like sauces, cheese, and yogurt. They furnish the consistency and overall increase its shelf life (Babitskaya et al., 2005; Vadnerker et al., 2022; Serra et al., 2023; Visentin et al., 2024).
ii. Nutritional functions that benefit human health: It contributes to a healthy gut due to their prebiotic effects, fostering the growth of beneficial microbes. They show proof that helps to reduce oxidative stress.
iii. Rational packaging: In the food sectors, it is progressively checking and searching for alternatives, to adopting environment friendly packaging systems, and also offers a promising option for sustainable packaging due to their:
    a. Biodegradability: There are specific types of engineered EPS that to reduce synthetic plastics dependency, biodegradable on minimizing environmental waste.
    b. Pharmaceutical and biomedical approach: Bioactivity and biocompatibility, microbes are perfect prospects for drug development, particularly in therapies like anticancer. Their criteria also enhance controlled release via drug encapsulation, raising evaluation limits across various pharmaceutical operations.
    c. Environmental improvements: It shows the capacity to encounter the environmental clean-up strains, such as the isolation of degradation and heavy metals of environmental impurities. These functions also rectify their utility in wastewater and biological enrichment systems.

d. Biomedical innovations: EPS are broadly investigated in biomedical arrangements for operations such as delivery systems and tissue structure. Their structural stretch, disintegrable, and immuno-regulatory possibilities make them suitable for designing advanced biomaterials for revitalizing medicine.

22.2.6 Enhancing biotechnological utilization in the food industry sectors

EPS are promising natural additives as an alternative in food biotechnology, requiring research for optimization and microbial engineering. It maintains to acquire attention in the food industry due to its versatility in health-promoting properties.

Key areas of transformation include:

a. Natural gelling agents: It utilizes to improve and impact the rheological properties of dairy applications, particularly in food products. Enhancing stability and viscosity, it also enhances the replacement of synthetic supplements with natural substitutes. In gluten-unchained baked products and other possible compositions, it helps to maintain the structure and the characteristics of gluten, enhancing the baking performance and good sensory appeal.

b. Feasible foods in health promotions: They mostly act as prebiotics, that having a promising healthy gut microbiome that maintains the healthy modulations in the gut homeostasis. Their intake can advance to overcome digestive health and prevent gastrointestinal disorders, making them beneficial components in functional foods.

c. Bioactive composition: Many EPS have acquired an additional health-boosting property, including cholesterol-lowering effects, anti-inflammatory, and antioxidant. These bioactivities make EPS suitable for formulating health-supportive food products, especially in the nutraceuticals.

d. Raise the shelf-life and formation of edible films: They are being produced into coatings that help extend the shelf life of food by retaining moisture, reducing oxygen permeability, and biodegradable films. These textures provide eco-friendly packaging solutions that furnish the plastic waste and reduce food decomposition.

e. Improvement in the safety of food: It possesses antimicrobial effects that prohibit the enrichment of harmful pathogens such as *Listeria monocytogenes* and *Escherichia coli*. These antimicrobial EPS can lead to be applied to reduce the risk of microbial contamination during storage and preservation to improve safety.

f. Fat replacement: EPS, like inulin, can be used as a fat substitute in reduced-fat or low-calorie food products. This application is increasingly important as consumers seek healthier options without sacrificing taste or texture.

g. Innovative functional beverages: EPS are increasingly being incorporated into functional beverages due to their ability to enhance mouthfeel and offer health benefits. Acting as natural thickeners, EPS improve the texture of drinks while also contributing prebiotic effects. These dual benefits enhance the beverages that seem to appeal to consumers those seeking both sensory satisfaction and wellness.

h. EPS in meat products applications: In meat screening, it enhances both product quality by improving texture and water-holding capacity. Their integration leads to enhanced flavor retention and juicier meat products.

22.2.7 Role of exopolysaccharides in bakery products

EPS improve gluten-free baking by ameliorating texture, shelf life, and nutrition, demanding further research on their fermentation amalgamation. The absence of gluten in baked foods that mostly results in reduced deprived texture, moisture retention, and shortening of shelf life. It also offers as a natural solution to these hassles through the following applications:

a. Improved natural texturizers: This natural texturizer effectively mimics some of the gluten's structural roles. That their incorporation enhances softness, provides a more compatible dough structure, and leads to be the most desirable mouthfeel in gluten-free products such as cakes, bread and muffins.
b. Moisture retention: They are hydrophilic in nature and bind water efficiently to help in retaining moisture in baked food items. This reduces contraction (liquid separation), ensures a softer fragment to help in maintaining freshness during storage.
c. Enhance shelf-life: It inhibits retrogradation of starch, a major cause of hinderance in gluten-free breads. By preventing and preserving moisture levels and starch crystallization, it extends the shelf life and improve the sensory quality of gluten-free products.
d. Improved handling and structure of bread: When add-mixture to gluten-free dough, it enhances its proofing due to its gas-holding capacity that leads to made fermentation results better and improved its loaf volume. This is especially benefitted in the wagging of gluten, which normally improvising its structure and elasticity.
e. Natural alternatives and clean labeling: It produced by lactic acid bacteria that offer alternatives to commonly used natural to synthetic additives in gluten-free baking. This aligns with consumer demands for simple, minimally processed ingredient lists and supports clean-label trends.
f. Sourdough fermentation: Using mother dough during fermentation with lactic acid bacteria that produce the EPS which shown to improve the overall grade of gluten-free bread. The fermentation process enriches its flavor, complexity and increases its yield, further contributing to improved stability, texture, and product appeal.
g. Prebiotic nutritional properties: Those are produced by lactic acid bacteria strains, that have been identified its prebiotic efficacy by enhancing its growth of beneficial gut bacteria. This counts nutritional value to gluten-unbound baked foods and enrich their appeal among health-conscious consumers.

22.2.8 Exopolysaccharides physicochemical properties

EPS upholds promising impact on acting as stabilizers and natural bio-thickeners in food, requiring additional research for enhancing their functionality.

i. Enhancement in viscosity: It contributes a stringent role in the fermented dairy products where viscosity matters, such as in yogurts. Increase in viscosity leads to improve texture, creaminess and overall sensory qualities, for example, strains like *Streptococcus thermophilus* have been shown to produce considerably thickened final products due to that it identified high EPS yield.
ii. Water-holding capacity: It possesses robust hydrophilic characteristics that allowing to retain and absorb efficient water. These characteristics is beneficial in the wide varieties of applications, which includes maintaining the moisture of bakery, dairy and meat products

throughout its processing and storage. High water-holding capacity enhances its texture, that prevents from drying, and improves it product stability. This property helps to stabilize the product by preventing excessive water loss, which is important for maintaining texture during consumption and storage. The ability of EPS to bind water also reduces contraction (the separation of whey, gels), resulting in a smoother texture.

iii. Gel formation and its structural modification: It contributes in the formation of a gel-like matrix structure in yogurt and other dairy products. It improves the gel stability and strength, leading to a smooth texture. This modification is essential for achieving the desired consistency in its structured product like yogurt.

iv. syneresis minimizing whey reduction and separation: The existence of EPS has been merged to low-grade levels in syneresis fermentation industries. For example, yogurt made with high EPS-producing significantly reduced contraction and separation that leads to strains showed a relative commercial starter culture. This is attributed to the water-holding capacity of EPS, which helps maintain product integrity.

v. Sensory attributes improved mouthfeel: The textural improvements provided by EPS also enhance sensory attribute like smoothness, which are crucial for consumer acceptance. Products with higher EPS content are often referred due to their desirable mouthfeel, sensory acceptance with closely resembling full-fat varieties even when lower fat content is utilized.

vi. Functional properties of stabilizing factors: It acts as a natural stabilizer, reducing the requirement of synthetic additives in dairy compositions. This ranges with consumer trends demand for favoring clean-label products, which leads to low cost with savings in production quality by minimizing stabilizers.

22.2.9 Environmental benefits in food packaging industries

EPS potentially enhance the promises for viable biodegradable and antimicrobial food packaging.

a. Biodegradable packaging: They are produced from natural sources, which makes them biodegradable, especially when used in packaging; they decompose easily in the environment compared to traditional plastic materials, which reduces waste and pollution free that can be associated with synthetic packaging materials, i.e., natural decomposers.

b. Barrier effects (moisture control): It can form films that can interact as an effective barrier that counters moisture. This enriches the shelf life of consumable food products by preventing damage due to oxidation, thus controlling and checking food waste.

c. Control of clean labeling and synthetic additives: Minimize or eliminate synthetic additives and preservatives from production. This can be ordered with consumer inclination for contributes and natural ingredients to a healthier ecosystem by reducing the harmful chemicals in the environment.

d. antimicrobial effects: This can inhibit the growth of denatured pathogens and microorganisms. This maintains food safety without relying on synthetic antimicrobial agents, further enhancing environmental health (Hayashi et al., 1996; Huang et al., 2022; Zhang et al., 2022; Laroche, 2022; Benhadda et al., 2023; Shao et al., 2023; Tao et al., 2024; Visentin et al., 2024).

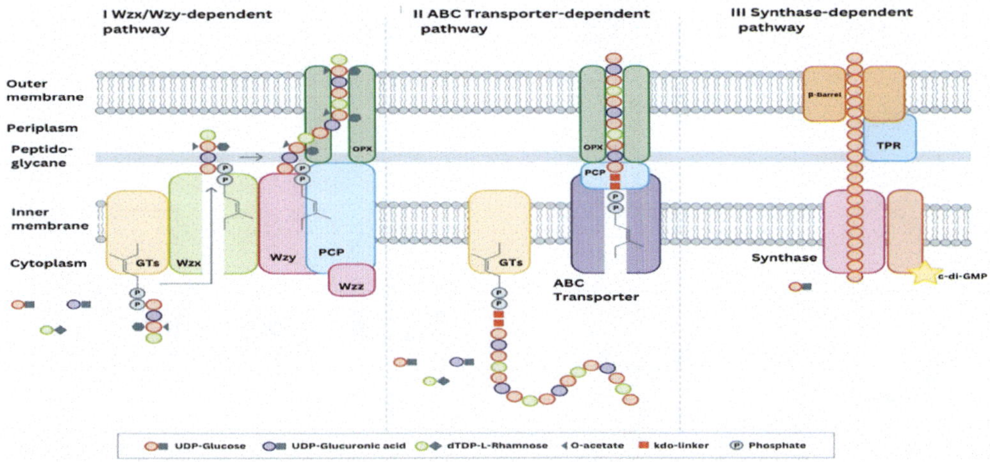

FIGURE 22.3 Schematic presentation of the three major pathways for" intracellular biosynthesis" of EPS. (I) Wzx/Wzy-dependent pathway: Repeating units assembled at C55-lipid linker by glycosyltransferases (GTs), translocated by Wzx flippase, polymerized via Wzy polymerase and PCP. Transport facilitated by PCP and OPX proteins; (II) ABC transporter-dependent pathway: Polysaccharide chain assembly on poly-do-linker, transported across membranes and cell wall with involvement of PCP and OPX proteins. ABC transporter ensures effective synthesis and transfer; (III) synthase-dependent pathway: Polymerization and transport via synthase complex spanning the cell envelope, including TPR proteins. TPR proteins coordinate the synthesis and transport of intracellular exopolysaccharides. Adapted under the terms of the Creative Commons Attribution License (CC BY)].

e. Sustainable sources with resource efficiency: It can be generated from renewable resources through fermentation via microbial processes or by utilizing agricultural by-products.
f. Material efficiency and lightweight design: Those films are conventional plastic films, which can lead to lower consumption of energy during transportation and lower carbon discharge associated with logistics (Rahbar Saadat et al., 2019; Santra and Banerjee, 2021; Vadnerker et al., 2022).
g. Active packaging systems: This can be combined with other natural combinations to produce the active packaging systems that not only secure food but also enrich and improve its quality through antioxidant properties, which can induce product shelf life and reduce unwanted wastage (Freitas et al., 2017; Chaudhari et al., 2021; Chen J. et al., 2024).

Impressive physico-chemical and biological properties of Food packaging materials are depicted in Fig. 22.3 above to expedite the EPS for better understanding as raw materials.

22.3 Applications of exopolysaccharide in the biosynthesis in the eco-friendly packaging and food industries

Exopolysaccharide biosynthesis offers sustainable production for food industries and a biopolymer eco-friendly packaging system.

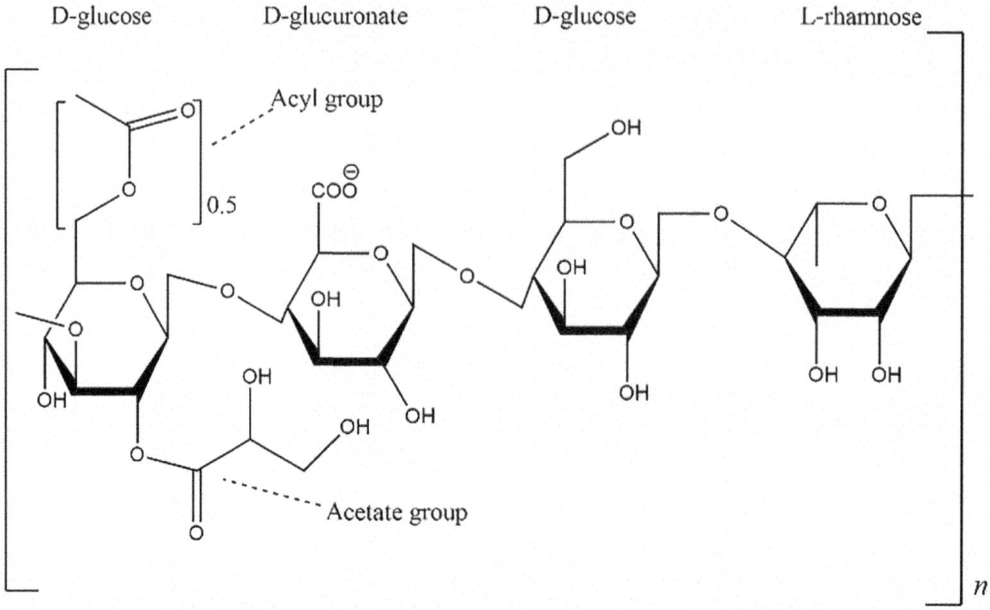

(A)

FIGURE 22.4 Chemical structure of gellan gum; natural high-acyl gellan (https://www.mdpi.com/2310-2861/10/3/183). Adapted under CC YY open access Repeating units assembled at C55-lipid linker by glycosyltransferases (GTs), translocated by Wzx flippase, polymerized via Wzy polymerase and Planar Cell Polarity (PCP). Transport facilitated by PCP and Outer Membrane Polysaccharide (OPX) proteins; (B) ABC transporter-dependent pathway: Polysaccharide chain assembly on poly-do-linker, transported across membranes and cell wall with involvement of PCP and OPX proteins. ABC transporter ensures effective synthesis and transfer; (C) synthase-dependent pathway: Polymerization and transport via synthase complex spanning the cell envelope, including TPR proteins. TPR proteins coordinate the synthesis and transport of intracellular exopolysaccharides. Adapted under the terms of the Creative Commons Attribution License (CC BY) (https://www.frontiersin.org/journals/microbiology/about#about-open). *EPS*, Exopolysaccharides.

The process involved in the process of bio-synthesis of EPS having various pathways that vary and turn on the growing micro-organism. This system governs the secretion, polymerization, and transportation of sugar units that form various EPS structures used across the packaging and food industries (Xu et al., 2019; Amiri et al., 2019; Amiri et al., 2019; Xu et al., 2019; Castro et al., 2024).

Key biosynthesis pathways involve like:

i. Wzx/wzy-dependent pathway: This pathway is often linked with the synthesis of heteropolysaccharides. It involves a sequence of proteins that transport sugar precursors that beyond the bacterial covering, where they are eventually polymerized and makeup into complex chains (Prajapati et al., 2013; Giavasis, 2013).

ii. ABC-based pathway: ATP-binding cassette transporters, this process enables the export of polysaccharides from the cytoplasm to the cell exterior. It takes part in an implicit crucial role in regulating the amount and timing of EPS secretion (Fig. 22.4).

iii. Synthase-based pathway: Initially responsible for the manufacturing of homo-polysaccharides, this requires the action of enzyme-specific synthase that catalyzes the polymerization of sugar monomers—such as glucose—into long chain structures.

iv. Extracellular synthesis via sucrase enzymes: Some EPS are synthesized extracellularly from sucrose or related sugars. Sucrase enzymes cleave glycosidic bonds and facilitate the formation of polysaccharide chains directly outside the cell, contributing to biofilm formation and extracellular matrix development.

22.3.1 Functional properties of exopolysaccharides in food systems

EPS are natural additives for improving the potential in the functional foods. The structural diversity of EPS leads to a range of functional properties that enhance the quality, stability, and health benefits of food products.

a. Rheological developments: It improves the viscosity and texture of various food commodities, distinctly in dairy products like milk, cheese, and yogurt. As natural bio-thickeners, they enhance product stability without using any synthetic additives (Yamanaka et al., 2013; Kırtel et al., 2017).

b. Emulsification: It can maintain and stabilize the emulsions and foams by a phase separation step checked. This is valuable in products such as sauces, salad dressings, and whipped dairy products, where a maintained layered texture and appearance are important (Jurášková et al., 2022; Baptista et al., 2022; Panchal et al., 2022).

c. Health benefits: It exhibits the effects of prebiotic, that support gut health by enhancing the growth of beneficial microbiota. Additionally, some have antioxidant properties and provide to the reduction of cholesterol, making them crucial in the developmental stages of functional foods processing that mainly aimed to the targeted consumers' health-focused is the primary objective.

22.3.2 Applications in biodegradable packaging

EPS enhance sustainable food packaging through improved biodegradability and performance. The advantages are widely approachable in food, sustainable packaging that contributes to environmental bio-conservation efforts in the food industry (Rahbar Saadat et al., 2019; Santra and Banerjee, 2021; Vadnerker et al., 2022).

i. Biodegradable packaging: It can be utilized in biodegradable coatings that are effective to control moisture. These coatings play an environmental beneficial and are alternatives to petroleum-based plastics, which helps to reduce pollution and toxic-waste.

ii. Natural coating in food: A huge portion of edible protective coatings is made from EPS, which can extend the shelf life of our delicate foods (Xu et al., 2019; Amiri et al., 2019). These natural coatings minimize moisture and delay the decomposition process without requiring any artificial preservatives into it and that aligns with clean-label prospectives (Jurášková et al., 2022; Baptista et al., 2022; Panchal et al., 2022).

iii. Sustainable replacements to synthetic materials: The combination of EPS into packaging reflects an upscaling industries momentum on using and retaining sustainability. Their

TABLE 22.2 Functional benefits of EPS-based films in food packaging.

Functional role	Example application	Benefit
Natural barrier	Fruit and vegetable coatings	Reduces moisture and oxygen ingress, extends shelf life
Antioxidant/antimicrobial	Meat or dairy films	Helps inhibit spoilage and pathogenic microbes
Texture and emulsion stabilizer	Yogurt, sauces	Improves texture, consistency, and product stability
Edible/active packaging	Coatings, dissolvable wraps	Adds nutritional value, can incorporate actives
Environmental sustainability	Replacement for plastics	Biodegradable, reduces plastic pollution

EPS, exopolysaccharides.

functional performance, renewable origin, and biodegradability make substitute for conventional plastics, offering both convenience to the environment and product protection (Table 22.2).

22.4 Recent research on exopolysaccharides

22.4.1 Current trends in exopolysaccharides in the eco-friendly food and packaging industry

EPS are favorable, sustainable biopolymers for eco-friendly food and packaging, requiring research on microbial production and its properties. EPS are the biopolymers that are made by metabolites and biofilm residues of microorganisms, especially lactic acid bacteria (LAB), and they have become quite important future in the food sector. The unique properties of this make them suitable for various applications, including as natural stabilizers, emulsifiers and thickeners. As consumer preferences a huge shift toward clean-label products, those which are free from artificial additives, EPS are increasingly recognized as valuable alternatives (Rahbar Saadat et al., 2021, 2019; Santra and Banerjee, 2021; Mohd Nadzir et al., 2021; Yang et al., 2023).

22.4.2 Current trends in the food industry

Research is required for the extraction and formulation of EPS as natural food additives to enhance food quality.

a. Natural additives: There is a high growing demand for natural ingredients in the food products industry. EPS are seen to be as effective replacements for synthetic additives, aligning with consumer desires and fruitful needs for healthier options without compromising the quality. They make food much more stable, last longer, and have a better texture (Coviello et al., 2005; Moscovici, 2015; Mohd Nadzir et al., 2021; Chaudhari et al., 2021). This makes them have very beneficial properties in dairy products, sauces, and fermented meals.

b. Health benefits: Studies show that the EPS has high health-promoting qualities, such as prebiotic actions that help in gut health and antioxidant activities. These benefits are making them a part of functional foods that are meant to improve people's health and give them better options (Tiwari et al., 2019).

c. Different uses: The EPS are used in a lot of different food products, like cheese, yogurt, plant-based drinks, and meat products. Their capacity to work with other such ingredients, i.e., EPS product, makes them work better by making it thicker and more pleasant to the palate (Freitas et al., 2017; Wünsche and Schmid, 2023; Chen J. et al., 2024).

22.4.3 Innovative ideas for eco-friendly packaging

EPS are important to developing biodegradable, eco-friendly packaging through fermentation research its characterization.

a. Biodegradable films: The EPS are being utilized much more nowadays to make biodegradable films and coatings for food packaging. These films are good for the environment and keep moisture and oxygen cut out from this making it sustainable and stable. This trend is part of a broader movement toward sustainable packaging solutions that reduce the reliability on traditional plastics and overcome from this (Freitas et al., 2017; Wünsche and Schmid, 2023; Chen et al., 2024).

b. Health benefits: In those studies, it shows that EPS has huge health benefits, such as prebiotic activity that helps intestinal health and possible antioxidant effects. These benefits are essential to make them a part of functional foods that are meant to make people's health a way better.

c. Future perspectives: The future of EPS in the food business market seems good, and research is still going on to learn more about their structure and function.

d. New ideas in microbial fermentation: Methods that could identify and lead to the production of new types of EPS with better qualities that are more suited for certain application processes. The rules and regulations about food safety and sustainability change, EPS will probably be vital in meeting those standards while also meeting customers' needs for products with clear labels (Babiak and Krzemińska, 2021; Baptista et al., 2022; Vadnerker et al., 2022).

22.5 Problems and possible solutions for exopolysaccharides in the food business and eco-friendly packaging

22.5.1 Optimizing the generation and yield of exopolysaccharides

Optimizing the production of exopolysaccharide is crucial for sustainable packaging and improving food quality. It is one of the biggest problems with using them on a large scale. The techniques involved for the extraction and isolation of EPS significantly impact their final yield and toward better quality. Alteration in fermentation conditions and downstream the processing techniques which can lead to variability in EPS production. These inconsistencies pose challenges in achieving the desired purity and functional properties required for foods, particularly in verifying consistency across individual production batches (Jurášková et al., 2022; Castro et al., 2024).

22.5.2 Regulatory aspects

EPS can uphold the promises that withstand industrial standard safety regulations. This goes through a rigorous regulatory control parameter, especially within the dairy sector. These regulatory barriers can sometimes confuse product development to marketing aspects, which require careful compliance with food safety and labeling claim standards. This restriction makes it hard for some markets to use EPS from nondairy sources, therefore manufacturers have to look for other options that meet the rules (Rahbar Saadat et al., 2021; Castro et al., 2024).

22.5.3 Technological limitations

EPS intends to enhance extraction methods that combine automation and enhance large-scale use while maintaining bioactivity and quality products. The existing ways of isolation and characterization of EPS can be more work-intensive and expensive. There are a lot of extraction methods that can change or damage the polysaccharide structure, which can influence their existing qualities and the way they work. For more uses, it is important to find ways to extract the EPS that are efficient, scalable, and keep their integrity (Huang et al., 2022; Baptista et al., 2022; Chen et al., 2024).

22.5.4 Market reception

Authentic communication and consumer awareness are the primary keys to enhancing EPS acceptance and safety. More and more people are choosing natural ingredients, however not everyone is okay with EPS as a food additive. To improve and increase customer trust and acceptance of food goods, people need to learn about the benefits and safety of the EPS.

22.5.5 Biotechnology developments that potentially happen in the future

Biotechnology improves the exopolysaccharide production via its modified genetic engineering techniques for high-quality biopolymers. Biotechnology could help make EPS production methods more efficient. Genetic modification can show some promising chances that enhance both the yield and functional properties of microorganisms by modulating metabolic pathways, it is easy to increase production quality and characteristics to meet specific nutritional and industrial standards (Giavasis, 2013; Schmid et al., 2015; Freitas et al., 2017; Zeidan et al., 2017; Yates et al., 2021).

22.5.6 Rational packaging systems

EPS improve sustainable, biodegradable food packaging, that imperative to conduct research. Sustainability is highly crucial in the food and packaging environment; researchers prioritize the production of biodegradable packaging that helps EPS as an important component. This upholds the material's natural properties like moisture while dealing remarkably by reducing environmental results compared to conventional plastic-based packaging.

22.5.7 Functional food development

EPS can meet consumer requirements and improve gut health for natural functional foods. Their antioxidant and prebiotic properties enhance their bioactive functions, suggesting a functional food aimed at digestive health and reducing oxidative stress. On formulating food products that uphold with specific EPS types to meet consumer demand that's growing and foremost important and linked with health-promoting dietary options.

22.5.8 Working together on research

Coordination of academia, industry, and regulators is important for safe EPS adoption in food. To deal with the problems that come with using EPS, it will be important for academia, industry, and regulatory organizations to work together. Working together can help create standard procedures for manufacturing and testing, which will make it easier for EPS to be used in a wider range of food applications.

22.6 Significance of exopolysaccharides research scope in food packaging

EPS show promise for enhancing sustainable food packaging, require essential research on formulation, and must uphold scalability. For their wide range of applications in the food industry, they can be applied in the development of sustainable, eco-friendly packaging solutions. However, there are still some areas that need to be fixed by undertaking more research in order to make more potential use and solve problems that are existing.

22.6.1 Important future scope and research directions

EPS uphold promises for sustainable food packaging and require more research on processing and microbial enhancements.

a. Optimization of manufacturing processes strain selection: Research should focus on finding and describing novel microbial strains that have higher EPS yields and other useful traits. This includes looking into less common LAB species that might make distinct EPS with certain functions (Torino et al., 2015; Moscovici, 2015; Kırtel et al., 2017; Mohd Nadzir et al., 2021; Castro et al., 2024).

b. Fermentation conditions: Finding the best fermentation conditions (temperature, pH, and nutrient availability) can make EPS generation more efficient. Changing these settings for different types of microbes could make EPS more productive and of better quality.

22.6.2 Understanding how structure and function relate to each other

Future EPS research should target structural characterization for optimized food and packaging applications.

a. Chemical characterization: We need to do more in-depth research on the chemical structure of different EPS to figure out how certain structural aspects affect how they work. This knowledge will make it easier to create EPS that are made for certain uses in food and packaging.
b. Functional mechanisms: Researchers should look into how EPS work to improve health, for by acting as prebiotics or antioxidants. This could mean looking into how they interact with gut bacteria and how they affect human health in a physical way.

22.6.3 Additional uses like

EPS show immense potential for biodegradable packaging and functional foods, which showcase the need for further improvement and research.

a. Functional foods: We need to look into how EPS can be included in functional foods that offer health advantages beyond just basic nutrition. This includes making items that help with specific health problems like diabetes, obesity, or gut health.
b. New ways to package things: Future research should focus on making packaging materials that break down naturally and use EPS as a key part. Finding out how they block things and how they interact with food can lead to new ideas that make food last longer and are good for the environment.

22.6.4 Regulatory regimes and how well people accept them

Future EPS investigations focus on consumer education and safety regulations for trust in sustainable food systems.

a. Regulatory compliance: As EPS are used more and more in food, it's important to know what the rules are in different places. The goal of research should be to come up with rules on how to safely employ EPS in food.
b. Teaching consumers: To get people to embrace EPS in the market, they need to know more about its benefits and safety. Campaigns to teach people about these biopolymers can help clear up any confusion and encourage their use in products with clean labels.

22.6.5 Sustainability evaluations

EPS tenders sustainable research on their potential environmental impact and production.

a. Life cycle analysis: Doing full life cycle evaluations (LCA) of EPS production and use can assist figure out how their effects on the environment compare to those of traditional components and packaging materials.
b. Circular economy models: Looking into how EPS can fit into circular economy models that reduce waste and reuse resources can help us learn more about sustainable practices in the food business.

22.6.6 Working together across fields

Interdisciplinary collaboration is crucial for advancing EPS research in food and packaging. Partnerships across sectors, working together with businesses, schools, and government agencies, will be very important for moving EPS research forward. Combining knowledge from microbiology, food science, packaging technology, and consumer behavior can help bring forward new ideas (Freitas et al., 2017; Wünsche and Schmid, 2023; Chen et al., 2024).

22.7 Exopolysaccharides that are currently used in the eco-friendly food business and packaging

EPS can improve food materials quality and help in biodegradable packaging, which requires enhanced research on production efficiency.

It is because they have several uses; EPS are becoming more common in the food business. They improve the texture, stability, and nutritional value of food. They are also very important for eco-friendly packaging solutions. Here are some of the most important EPS that are now available, and what they can be used as follows:

A. EPS that are often used:

i. *Xanthan gum* is made by the bacteria *Xanthomonas campestris*. It is commonly used as a thickener and stabilizer in salad dressings, sauces, and gluten-free baked goods. Because it acts as a pseudoplastic, it can alter viscosity when shear force is applied, which makes it great for many food uses.

ii. *Leuconostoc mesenteroides* makes this polysaccharide, which is recognized for its solutions that are very soluble and have a low viscosity. People often use dextran in baked goods to make them bigger and better in texture.

iii. *Gellan gum* comes from *Sphingomonas elodea* and is used in dairy goods, sweets, and as a vegan substitute for gelatin since it can gel. It makes the gel very strong and stable, even when the temperature changes (Fig. 22.5).

iv. *Pullulan* is a polysaccharide made by *Aureobasidium pullulans*. It is recognized for its ability to create films, which makes it good for edible films and coatings that can help keep food fresh and break down naturally (Fig. 22.6).

v. *Curdlan:* Curdlan is a linear glucan made by *Alcaligenes faecalis*. It is mostly utilized to make low-fat dairy products and as a gelling agent in many foods since it can gel at different temperatures.

vi. *Inulin* is a type of fructo-oligosaccharide that is present in many plants. It is a prebiotic fiber that helps keep the gut healthy. It is used in low-fat foods to replace fat and make dairy products taste better (Badel et al., 2011; Amiri et al., 2019; Costa et al., 2021; Baptista et al., 2022; Panchal et al., 2022; Vanin et al., 2023; Benhadda et al., 2023).

B. Uses in food processing: EPS are used in many parts of the food business because they are useful in:

i. Texture enhancement in EPS that makes yogurt, cheese, sauces, and condiments feel better in the mouth and thicker. They are natural substitutes for manufactured thickeners and stabilizers (Baptista et al., 2022; Huang et al., 2022; Serra et al., 2023; Chen J. et al., 2024; Visentin et al., 2024).

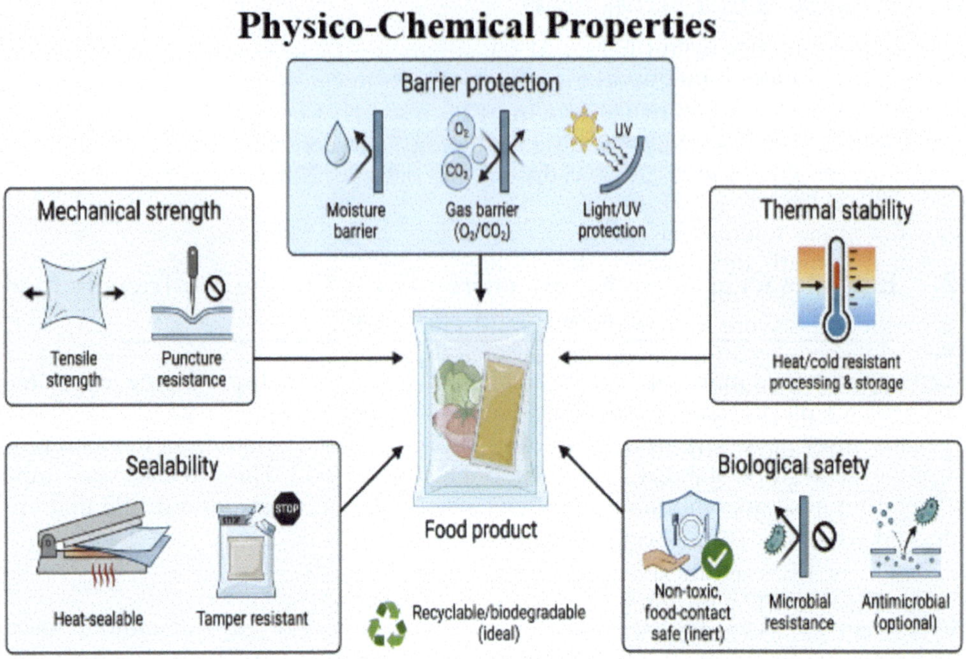

FIGURE 22.5 Physico-Chemical Properties. (Created in BioRender. Mukherjee, R. K. 2026, https://BioRender.com/p8qbn2n).

ii. Health benefits in many EPS have prebiotic effects, which means they help gut health by encouraging good bacteria to grow. They also have antioxidant qualities that can help lower oxidative stress (Prajapati et al., 2013; Xu et al., 2019; Jurášková et al., 2022; Mummaleti et al., 2022).

iii. *Xanthan gum* and gellan gum are the best examples of emulsifiers that work well. They keep oil-in-water emulsions stable, which are often seen in salad dressings and sauces (Table 22.3).

22.8 Exopolysaccharides face some problems and have some possible future uses in the food industry and eco-friendly packaging

EPS in the sustainable packaging and food industries, having huge prospective in requiring consumer awareness for market growth, research on production and regulations.

A. Problems with production efficiency: Getting EPS out, separating them, and cleaning them up can be time-consuming and expensive. Different strains of microbes and fermentation conditions can cause EPS yields and quality to vary. It is very important to make these processes as efficient as possible in order to increase production to fulfil industrial needs.

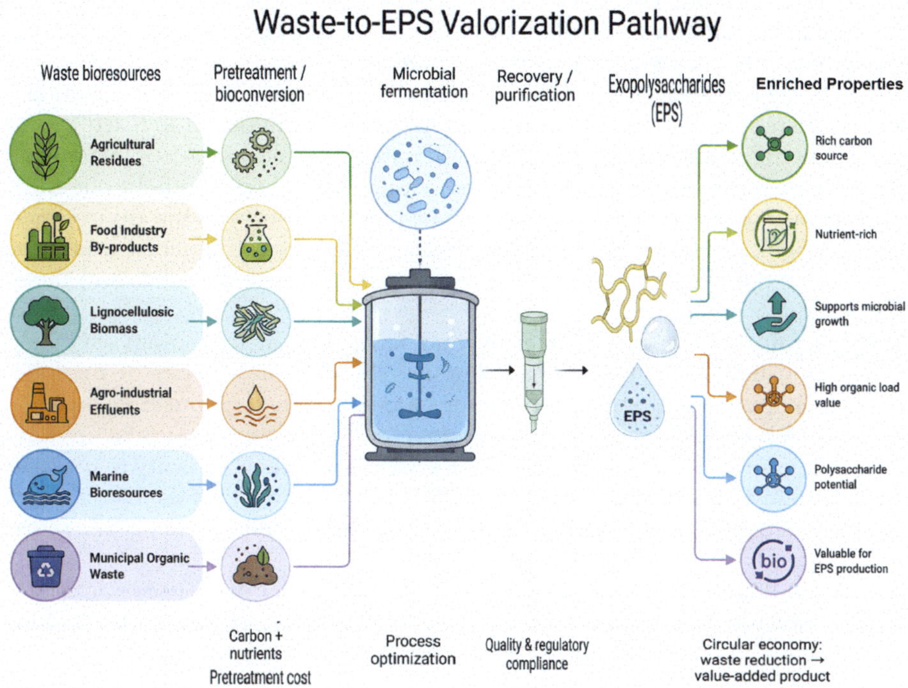

FIGURE 22.6 Waste to ESP Valorization Pathway (Created in BioRender. Mukherjee, R. K. 2026, https://BioRender.com/p8qbn2n).

TABLE 22.3 Key features of major polysaccharides can be explored in edible packaging.

Polysaccharide	Main source	Key properties	Advantages	Limitations
Cellulose	Plant cell walls	High strength, inert	Biodegradable, renewable	Poor solubility
Starch	Cereals, tubers	Film-forming	Abundant, edible	Brittle, sensitive to water
Chitosan	Fungal/bacterial	Antimicrobial	Biocompatible, bioactive	Limited solubility in water
Exopolysaccharides	Lactic acid bacteria, fungi	Viscosity, gelling	Prebiotic, functional, biodegradable	Film-forming can require modifications

B. Regulatory barriers: EPS from nondairy sources have trouble with regulations, especially in dairy applications, where labeling restrictions limit the use of nondairy polysaccharides. Because of this constraint, we need to focus on EPS from lactic acid bacteria, which are known to be harmless. However, getting these polysaccharides accepted in different markets is still a problem (Moscovici, 2015; Zeidan et al., 2017; Santra and Banerjee, 2021; Mohd Nadzir et al., 2021; Vadnerker et al., 2022; Huang et al., 2022).

C. Functional characterization: The varied structures of EPS generated by different bacteria cause their functional qualities to differ. To make EPS work better for certain foods, you need to know how these connections work. It's hard to guess how EPS will behave in food systems because the ways we currently use to describe them are not always accurate.
D. Consumer acceptance: To get more people to adopt EPS, it's important to teach them about its safety and benefits as the market moves toward natural ingredients. People may not want to use EPS as a substitute because they have the wrong ideas about synthetic chemicals.
E. Concerns about sustainability: EPS may have some positive effects on the environment, but we also need to look at how they are made to see if they are sustainable. It is important to carefully look at the effects on the environment of getting raw materials and the energy needed for production.
F. Biotechnology's future: Genetic engineering and fermentation technology can help EPS produce more and better functional characteristics. Finding new types of microbes and improving fermentation conditions can make production methods more efficient.
G. More uses: Research should focus on making novel food products that take advantage of the special qualities of EPS, namely their prebiotic effects and their capacity to make low-fat foods taste better. Also, looking into their application in functional meals that target certain health benefits could lead to new industries (Giavasis, 2013; Freitas et al., 2017; Yates et al., 2021).
H. Better extraction methods: It's important to come up with better ways to extract and purify that cause less damage and get more of the desired substance. This includes looking into nondestructive methods that keep the EPS intact while it is being processed.

22.9 Conclusion

EPS have immense potential for sustainable food packaging due to their antimicrobial and biodegradable properties.

EPS are increasingly touted as a viable answer to sustainable food packaging due to their biodegradable, antimicrobial, and functional properties as the industry trends toward the use of eco-friendly materials. Setting the agenda in the food industry is concerns over growth sustainability brought by consumers; thus, the trend is packaged sustainably. Among such innovations, microbial EPS have attracted attention due to their exceptional properties. EPS has great potential in sustainable packaging; EPS is a biopolymer produced by microorganisms that are nontoxic, biodegradable, and functional. EPS are of particular interest in the food packaging arena as they possess antioxidant, antimicrobial, and prebiotic functions, enhance shelf life, and increase safety. EPS films are being developed as an alternative to petroleum-based plastics and are more sustainable. A very promising sector with respect to circularity is the food packaging sector, which up to now has been predicated on a linear model of production and consumption—take, make, and dispose-to one that factors in designs for reuse, recycling, and composting. Polysaccharides derived product, such as edible films from seaweed, self-composting containers, and biodegradable trays and cups, have entered India under the leadership of Ecopack. These innovations herald a larger movement toward the amalgamation of tradition with innovation, matching consumer demands against sustainability goals. A broad allusion is made to the integration of EPS in packaging to fulfil global sustainability targets while presenting a scientifically substantiated and scalable solution to plastic waste. Constant research and commercialization of EPS-based materials could revolutionize how food products are stored, transported, and consumed. In essence, EPS stands out as a modern advancement in sustainable packaging. Multifaceted EPS holds multifunctional benefits and fits more than convenience offerings in terms of an eco-conscious practice.

Credit statement

Dilip Kumar Roy: Conceptualization, supervision, writing—original draft and editing; Partha Palit: Investigation, data curation, writing—editing and reviewing; Aparna Beena Unni: Visualization, formal analysis, visualization, writing—review; Rajesh Kumar Mukherjee: Writing—original draft and editing.

Conflict of interest

The authors declare no conflicts of interest regarding this manuscript.

Ethical statement

In this study, we did not go for any preclinical or clinical studies, hence not require any approval.

Acknowledgments

Authors are highly thankful to JIS University, Kolkata, University of Silacea, Poland, and Assam UNIVERSITY for providing support, infrastructure, and platform to frame the chapter successfully.

Declaration of generative AI and AI-assisted technologies in the writing process

During the write-up of this manuscript, the author used Grammarly, an AI-assisted editing tool for language correction, specifically to improve spelling, grammar, word selection, readability, and clarity. After using this tool, the authors re-examined and revised the text as required.

References

Andrew, M., et al., 2020. Structural features of microbial exopolysaccharides in relation to their antioxidant activity. Carbohydr. Res. 487, 107881. doi:10.1016/j.carres.2019.107881.

Amiri, S., et al., 2019. Exopolysaccharides production by *Lactobacillus acidophilus* LA5 and *Bifidobacterium animalis* subsp. lactis BB12: Optimization of fermentation variables and characterization of structure and bioactivities. Int. J. Biol. Macromol. 123, 752–765. doi:10.1016/j.ijbiomac.2018.11.084.

Babiak, W., et al., 2021. Extracellular polymeric substances (EPS) as microalgal bioproducts: A review of factors affecting EPS synthesis and application in flocculation processes. Energies 14, 4007. doi:10.3390/en14134007.

Babitskaya, V.G., et al., 2005. Polysaccharides of *Ganoderma lucidum*: Factors affecting their production. Appl. Biochem. Microbiol. 41, 169–173. doi:10.1007/s10438-005-0029-1.

Badel, S., et al., 2011. New perspectives for *Lactobacilli* exopolysaccharides. Biotechnol. Adv. 29, 54–66. doi:10.1016/j.biotechadv.2010.08.011.

Baptista, S.,, et al., 2022. Extraction of the bacterial extracellular polysaccharide FucoPol by membrane-based methods: Efficiency and impact on biopolymer properties. Polymers 14, 390. doi:10.3390/polym14030390.

Benhadda, F.,, et al., 2023. Marine versus non-marine bacterial exopolysaccharides and their skincare applications. Mar. Drugs 21, 582. doi:10.3390/md21110582.

Branda, S.S.,, et al., 2005. Biofilms: The matrix revisited. Trends Microbiol. 13, 20–26. doi:10.1016/j.tim.2004.11.006.

Castro, S.,, et al., 2024. Exploring the potential of probiotics in dentistry: A literature review. Odovtos - Int. J. Dent. Sci. 26, 24–36. doi:10.15517/ijds.2024.59138.

Chaudhari, V.,, et al., 2021. Therapeutic and industrial applications of curdlan with overview on its recent patents. Front. Nutr. 8, 646988. doi:10.3389/fnut.2021.646988.

Chen, C.C., et al., 2023a. High-molecular-weight exopolysaccharides production from *Tuber brochii* cultivated by submerged fermentation. Int. J. Mol. Sci. 24, 4875. doi:10.3390/ijms24054875.

Chen, G.,, et al., 2023b. Extraction, characterization, and biological activities of exopolysaccharides from plant root soil fungus *Fusarium merismoides* A6. Braz. J. Microbiol. 54, 199–211. doi:10.1007/s42770-022-00842-x.

Chen, J.,, et al., 2024. Characteristic analysis and fermentation optimization of a novel *Aureobasidium pullulans* RM1603 with high pullulan yield. J. Biosci. Bioeng. 137, 335–343. doi:10.1016/j.jbiosc.2023.12.018.

Costa, J.A.V.,, et al., 2021. Microalgae polysaccharides: An overview of production, characterization, and potential applications. Polysaccharides 2, 759–772. doi:10.3390/polysaccharides2040046.

Coviello, T.,, et al., 2005. *Scleroglucan*: A versatile polysaccharide for modified drug delivery. Molecules 10, 6–33. doi:10.3390/10010006.

Donot, F.,, et al., 2012. Microbial exopolysaccharides: Main examples of synthesis, excretion, genetics and extraction. Carbohydr. Polym. 87, 951–962. doi:10.1016/j.carbpol.2011.08.083.

Eze, C.O.,, et al., 2022. *Lactobacillus*—A multifunctional genus. *Lactobacillus* Exopolysaccharide: An Untapped Biopolymer. IntechOpen.

Freitas, F.,, et al., 2017. Engineering aspects of microbial exopolysaccharide production. Bioresour. Technol. 245, 1674–1683. doi:10.1016/j.biortech.2017.05.092.

Giavasis, I., 2013. Production of microbial polysaccharides for use in food. In: McNeil, B. (Ed.), Microbial Production of Food Ingredients, Enzymes and Nutraceuticals. Elsevier, pp. 413–468.

Guevara Guerrero, B.,, et al., 2024. Application of microbial exopolysaccharides in packaging films for the food industry: A review. Int. J. Food Sci. Technol. 59 (1), 17–29. doi:10.1111/ijfs.16789.

Harutoshi, T., 2013. Exopolysaccharides of lactic acid bacteria for food and colon health applications. Lactic Acid Bacteria—R&D for Food, Health and Livestock Purposes. InTech.

Hayashi, T.,, et al., 1996. Calcium spirulan, an inhibitor of enveloped virus replication, from a blue-green alga *Spirulina platensis*. J. Nat. Prod. 59, 83–87. doi:10.1021/np960017o.

Huang, H.,, et al., 2022. Biopolymers produced by *Sphingomonas* strains and their potential applications in petroleum production. Polymers 14, 1920. doi:10.3390/polym14091920.

Jeong, J.,, et al., 2022. Bacterial succinoglycans: Structure, physical properties, and applications. Polymers 14, 276. doi:10.3390/polym14020276.

Jurášková, D.,, et al., 2022. Exopolysaccharides produced by lactic acid bacteria: From biosynthesis to health-promoting properties. Foods 11, 156. doi:10.3390/foods11020156.

Kırtel, O.,, et al., 2017. Microbial polysaccharides as food ingredients. Microbial Production of Food Ingredients and Additives. Elsevier, pp. 347–383.

Kiran, N.S.,, et al., 2024. Revisiting microbial exopolysaccharides: A biocompatible and sustainable polymeric material for multifaceted biomedical applications. 3 Biotech 14 (4), 95. doi:10.1007/s13205-024-03946-3.

Laroche, C., 2022. Exopolysaccharides from microalgae and cyanobacteria: Diversity of strains, production strategies, and applications. Mar. Drugs 20, 336. doi:10.3390/md20050336.

Mohd Nadzir, M.,, et al., 2021. Biomedical applications of bacterial exopolysaccharides: A review. Polymers 13, 530. doi:10.3390/polym13040530.

Moradi, M.,, et al., 2021. Current applications of exopolysaccharides from lactic acid bacteria in the development of food active edible packaging. Curr. Opin. Food Sci. 40, 33–39. doi:10.1016/j.cofs.2020.06.001.

Moscovici, M., 2015. Present and future medical applications of microbial exopolysaccharides. Front. Microbiol. 6, 1012. doi:10.3389/fmicb.2015.01012.

Mummaleti, G.,, et al., 2022. Characterization of levan produced from coconut inflorescence sap using *Bacillus* subtilis and its application as a sweetener. LWT 154, 112697. doi:10.1016/j.lwt.2021.112697.

Panchal, R.,, et al., 2022. Bacterial exopolysaccharides: Types, biosynthesis and applications in different fields. Acta Sci. Biotechnol. 3, 1–9.

Prajapati, V.D.,, et al., 2013. Pullulan: An exopolysaccharide and its various applications. Carbohydr. Polym. 95, 540–549. doi:10.1016/j.carbpol.2013.02.082.

Rahbar Saadat, Y.,, et al., 2021. Yeast exopolysaccharides and their physiological functions. Folia Microbiol. 66, 171–182. doi:10.1007/s12223-021-00856-2.

Rahbar Saadat, Y.,, et al., 2019. A comprehensive review of anticancer, immunomodulatory and health beneficial effects of the lactic acid bacteria exopolysaccharides. Carbohydr. Polym. 217, 79–89. doi:10.1016/j.carbpol.2019.04.025.

Santra, H.K.,, et al., 2021. Production, optimization, characterization and drought stress resistance by β-glucan-rich heteropolysaccharide from an endophytic fungi *Colletotrichum alatae* LCS1. Front. Fungal Biol. 2, 796010. doi:10.3389/ffunb.2021.796010.

Schmid, J.,, et al., 2015. Bacterial exopolysaccharides: Biosynthesis pathways and engineering strategies. Front. Microbiol. 6, 496. doi:10.3389/fmicb.2015.00496.

Serra, M.,, et al., 2023. Microbial hyaluronic acid production: A review. Molecules 28, 2084. doi:10.3390/molecules28052084.

Shao, Z.,, et al., 2023. Anti-diabetic activity of a novel exopolysaccharide produced by the mangrove endophytic fungus *Penicillium janthinellum* N29. Mar. Drugs 21, 270. doi:10.3390/md21050270.

Tao, X.,, et al., 2024. Characterization and antioxidant properties of three exopolysaccharides produced by the *Cyclocarya paliurus* endophytic fungus. Int. J. Biol. Macromol. 271, 132110. doi:10.1016/j.ijbiomac.2024.132110.

Tiwari, O.N.,, et al., 2019. Purification, characterization and biotechnological potential of new exopolysaccharide polymers produced by cyanobacterium *Anabaena* sp. CCC 745. Polymer 178, 121695. doi:10.1016/j.polymer.2019.121695.

Torino, M.I.,, et al., 2015. Biopolymers from lactic acid bacteria: Novel applications in foods and beverages. Front. Microbiol. 6, 834. doi:10.3389/fmicb.2015.00834.

Vadnerker, P.,, et al., 2022. Characterization of exopolysaccharide produced by *Ganoderma* sp. TP and its immunomodulatory properties. Rom. Biotechnol. Lett. 28, 3527–3535. doi:10.25083/rbl/27.3/3527.3535.

Vanin, A.P.,, et al., 2023. β-(1→3) (1→6) glucan from *Schizophyllum* commune 227E.32: High yield production via glucose/xylose co-metabolization. Carbohydr. Polym. 320, 121176. doi:10.1016/j.carbpol.2023.121176.

Visentin, T.G.,, et al., 2024. Effects of temperature, pH and C/N ratio of sugarcane waste water processing (vinasse) on *Phormidium autumnale* heterotrophic cultivation. Algal Res. 77, 103349. doi:10.1016/j.algal.2023.103349.

Wünsche, J.,, et al., 2023. *Acetobacteraceae* as exopolysaccharide producers: Current state of knowledge and further perspectives. Front. Bioeng. Biotechnol. 11, 1166618. doi:10.3389/fbioe.2023.1166618.

Xu, et al., 2019. Exopolysaccharides produced by lactic acid bacteria and bifidobacteria: Structures, physiochemical functions and applications in the food industry. Food Hydrocoll. 94, 475–499. doi:10.1016/j.foodhyd.2019.03.032.

Yamanaka, D.,, et al., 2013. Royal sun medicinal mushroom, *Agaricus brasiliensis* Ka21 (higher *Basidiomycetes*), as a functional food in humans. Int. J. Med. Mushrooms 15, 335–343. doi:10.1615/IntJMedMushr.v15.i4.10.

Yang, Y.,, et al., 2023. Biological activities and applications of exopolysaccharides produced by lactic acid bacteria: A mini-review. World J. Microbiol. Biotechnol. 39, 155. doi:10.1007/s11274-023-03610-7.

Yates, L.E.,, et al., 2021. Bacterial glycoengineering as a biosynthetic route to customized glycomolecules. In: Rapp, E., Reichl, U. (Eds.), Advances in Glycobiotechnology. Springer, pp. 167–200. doi:10.1016/j.apmr.2020.12.020.

Zeidan, A.A.,, et al., 2017. Polysaccharide production by lactic acid bacteria: From genes to industrial applications. FEMS Microbiol. Rev. 41 (S1), S168–S200. doi:10.1093/femsre/fux017.

Zhang, P.,, et al., 2022. Alginate production of *Pseudomonas* strains and its application in preparation of alginate-biomass hydrogel for heavy metal adsorption. Int. J. Biol. Macromol. 222, 1511–1521. doi:10.1016/j.ijbiomac.2022.09.252.

CHAPTER

23

Integrated application of exopolysacharides as pesticide and herbicide for green cultivation

Prasanna Sathiadevan Prathibha[a], *Vinayaka Hegde*[b] *and Hiba Kanhirala*[c]

[a]Senior Scientist, Entomology, Crop Protection Division, ICAR-Central Plantation Crops Research Institute, Kasaragod, Kerala, India [b]Head, Crop Protection Division, ICAR-Central Plantation Crops Research Institute, Kasaragod, Kerala, India [c]Young Professional, Entomology Section, ICAR-Central Plantation Crops Research Institute, Kasaragod, Kerala, India

23.1 Introduction

Sustainable and eco-friendly farming methods are inevitable to conserve natural resources, maintain the diversity of living being and guarantee long-lasting food security and safety. Rather than depending exclusively on synthetic pesticides, exploiting the potential of natural microbial antagonists to resist pests and diseases is known as microbial control. Utilization of microbial bio-control agents (MBCAs) is found to be very efficient against an array of phyto-pathogens and insect pests. These are greatly integrated into pest/disease management programmes these days. As far as biopesticides are concerned, significant developments have occurred, particularly the use of new strains for enhanced efficacy, as well as formulating stable and slow-release formulations with the help of nanotechnology. MBCAs are effective when they possess key traits like antagonistic potential, rhizosphere competence, production of lytic enzymes, antibiotic effects andproduction of toxins to fight against pests and diseases, and high adaptability to various environmental conditions. According to Lahlali et al. (2022), microbial biocontrol agents (MBCA), viz., *Bacillus* spp., *Burkholderia* spp., *Pseudomonas* spp., and *Trichoderma* sp., are effective in controlling pathogenic fungi like *Fusarium* spp., *Pythium* spp., *Rhizoctonia solani, Agrobacterium*

DOI: https://doi.org/10.1016/B978-0-443-36619-2.00023-1

radiobacter var. *radiobacter, Phytophthora* spp., and *Erwinia* spp. that induce soil-borne and foliar diseases. Either these microbial control agents precisely aim soil-inhabiting pathogens directly, or trigger resistance/tolerance response in plants. MBCAs mediated pathogen suppression mechanisms include hyperparasitism, antibiosis, competition for the substrate or nutrients, interruption of quorum-sensing signals, synthesis of siderophores, production of cell wall-lytic proteins, and the induction of systemic resistance in plants. MBCAs used to manage pests and diseases often produce exopolysaccharides (EPS), which are sugar-based polymers secreted that play crucial roles in their survival, stress tolerance, and interaction with plants and other organisms. In the context of insect pest management, EPS can play a role in various ways. Some of the EPS producing micro organisms are exploited as bioagents in pest management because of their potential to interact with the insect exoskeleton and disrupt development or cause mortality. Certain EPS produced by cyanobacteria can be incorporated into nano-formulations with lipids to target insect pests like the Mediterranean fruit fly. Additionally, EPS-producing bacteria like *Agrobacterium* can be used with their lipo-polysaccharides (LPS), which has cidal activity against stored grain pests. It can act as the synergist of insecticide/biocontrol agents, some EPS forming microorganism can provide a physical barrier against root-feeding insects by forming biofilm, some EPS exhibit antifeedant property, which deters herbivorous insects from feeding on a plant, or by disrupting oviposition, or by disrupting the insect microbiome symbiosis in the gut. Thus, EPS producing microorganisms have immense potential in inducing tolerance/resistance against pests and diseases in plants. Thus, the mechanism of action of EPSs in reducing biotic stress in plants and its use as pesticides and herbicides in green cultivation are discussed in detail in this chapter

23.2 Microbial exopolysaccharides in alleviating biotic stress in crops

As far as plants are concerned, biotic stress refers to the deleterious effects produced by several biotic factors, viz., phytopathogens, phytophagous insects, other living organisms, and weeds. Biotic stresses influence the normal physiology and development of the plant, which eventually cause reduction in productivity and yield. In a number of ways, MBCAs shield plants from pests and disease incursions. viz., production of lipopeptides, hydrolytic enzymes, antifungal substances, or antibiotics (Choudhary et al., 2016). They can interact directly or indirectly with pathogens and suffocate plant diseases by using one or more resistance, tolerance, or evasion mechanisms. It has been discovered that EPS generated by fungus, algae, and plant growth-promoting rhizobacteria (PGPR) has a more beneficial impact on crop growth and physiology. They can flourish in a variety of abiotic stress situations because of PGPR. The following mechanisms are involved.

1) Through the synthesis of siderophores, which chelate iron and make it accessible to plant-roots,
2) through nitrogen fixing, that convert free nitrogen in the atmosphere to an available form,
3) production of phytohormones and the uptake of nitrogen and phosphorus. PGPR enhances the absorption of particular nutrients and minerals because ofraise in the movement of particular ions in the rhizosphere created by the bacterium. Most of the PGPR produce EPSs that protect its own and their particular plant hosts from the deleterious phytopathogens in the rhizosphere.

Microbial EPS hold striking potential for insect pest management in agriculture, primarily through their role in biopesticides, and to enhance plant defense mechanisms. As per Kumar et al. (2019), organic pesticides encompass ≈5% only in Indian pesticide market. It includes 970 microbial formulations and 15 microbial species that are registered with the Central Insecticides Board and Registration Committee (CIBRC). In order to combat phytophagous insects and nematodes, approximately two hundred formulations based on insect (Entomo) pathogenic fungi (EPF)—such as *Beauveria bassiana, B. brongniartii, Metarhizium anisopliae, Lecanicillium lecanii,* and *Hirsutella thompsonii*—as well as nematophagous fungi—such as *Purpureocillium lilacinum* and *Pochonia chlamydosporia*—were included by 2017. To date, 45 bacterial formulations have been registered, including 30 formulations of *Bacillus thuringiensis (Bt)* subsp. Kurstaki that are effective in controlling many lepidopteran pests, twelve formulations of *Bt* subsp. *israelensis*, and three formulations of *Bt* subsp. *Sphaericus* that are effective in controlling mosquitoes. Two viruses, viz., *Helicoverpa armigera* nucleo-polyhedrovirus (NPV) (twenty-two formulations) and *Spodoptera litura* NPV (five formulations), are registered for the management of pests of cotton and other crops.

23.3 Mechanism of exopolysaccharides in mitigating plant diseases

Pathogen attacks in plants are categorized into three, viz., necrotrophy, biotrophy, and hemibiotrophy. In plants, the defense signaling mechanism varies based on the nature of infection (Barna, 2022). In order to fight against infections and herbivores, plants employ a sophisticated system of defense mechanisms. Induced structural defense, which includes the development of an abscission layer and a robust cell wall that includes cork, and biochemical defense, which includes the activation of defense-related genes such as pathogenesis-related (PR) genes, the production of a series of enzymes and receptor-like kinases (RLKs), as well as triggering molecules and plant hormones like ethylene (ET), salicylic acid (SA), and jasmonic acid (JA). These hormones and genes work together to trigger various defense responses, often in a coordinated manner through signaling pathways and crosstalk. There are two types of biochemical-mediated resistance mechanisms in plants, viz., induced systemic resistance (ISR) and systemic acquired resistance (SAR). Their signaling pathways and stimulants are different from one another. ISR is activated by beneficial microbes and the EPSs they generate, but SAR is normally turned on by pathogens and relies on SA, ET, and JA pathways, which are primarily engaged

23.3.1 Disease suppression by induced systemic resistance in plants

Beneficial bacteria and fungi strengthen the plant's defense mechanisms against a range of pathogens. In plants, this process is referred to as ISR. The plant can react to future disease or pest attacks more rapidly and powerfully due to this priming effect. SAR, which is brought on by a pathogen attack, is not the same as ISR. ISR entails "priming" the defenses of the plant. This means the plant is not directly resistant to the pathogen, but its defenses are "on alert" and ready to be activated rapidly. EPS produced by certain bacteria, particularly rhizosphere bacteria, can act as messenger molecules that trigger defense mechanisms in the plant, both locally and systemically. It will enhance the plant's defense against herbivores/pathogens. This induced resistance is a form of plant immunity, and the plant becomes more resistant to a range of invaders. This ISR process prepares the plant for a more robust defense response rather than directly killing or inhibiting the pathogen or insect. A plant's defense systems, such as the synthesis of defense-related proteins and other chemicals, can be triggered when it comes into contact with an infection. A reasonably well-known antagonistic fungus that is frequently used to biologically control plant diseases is *Trichoderma* sp. Besides disease fungal suppression by myco-parasitism, antibiosis, and competition, EPS produced by this fungus inhabit the outer skin of the root but do not reach the conducting tissues due to the intervention of the plant defense system through SA-related genes. Therefore, it serves as a systemic pathogen checking the incidence of pests and diseases (Poveda, 2021). Purified EPS from the bacteria *Burkholderia gladioli* IN26 can ISR in cucumber against *Colletotrichum orbiculare,* which causes anthracnose disease. When the seeds of cucumber were soaked in 200 ppm of the EPS generated by the bacteria *Burkholderia gladioli* IN26 isolate, they could achieve a maximum level of disease protection (Kyungseok et al., 2008). *Bacillus*-mediated ISR is shown in a variety of plants, viz., *Nicotiana tabacum, Lycopersicon esculentum, Arabidopsis thaliana,* and others (Hyakumachi et al., 2013; Jiang et al., 2016; Wu et al., 2018). Certain strains of *B. thuringiensis* generate biochemicals like fengycin, chitinase, bacteriocins, and other volatile organic compounds that are hostile to plant diseases are produced (Kim et al., 2004; Zheng et al., 2013). By causing systemic resistance, *Bt* strain 199, serovar sotto RG1-6, and *fukuokaensis* B88-82 have been shown to be successful in preventing tomato wilt disease brought on by *Fusarium oxysporum* and *Ralstonia solanacearum* (Akram et al., 2013). Similarly, Wang et al. (2023) noticed that EPS produced by *Bt* strain-4F5 could induce ISR in mustard to the stem rot pathogen (*Sclerotinia sclerotiorum*), which is marked by elevated level of defense-related enzymes in the mustard leaves pretreated with *Bt*-4F5 EPSs. The *Bt* serovar *kenyae* strain 4F5 may cause systemic resistance in *Brassica campestris* through signaling pathways that are dependent on salicylic acid, ethylene, and jasmolic acid. From treated tissues to untreated areas, the systemic signals may spread. However, it's unclear which substances caused systemic resistance. Furthermore, the results of qPCR study showed that the defense-related genes HEL and PDF1.2 were upregulated, signifying the enhancement of JA/ET signaling pathways by *Bt* 4F5 EPS. Chen et al. (2013) reported a reduction in biocontrol potential of *B. subtilis* against *Ralstonia solanacearum* when EPS was removed from *B. subtilis*

Endophytic EPSs are able to create an independent microclimate inside, which supports the bacteria to thrive within the plant. For instance, the gene gumD controls the production of EPSs in *Gluconacetobacter diazotrophicus*, and the mutants that are devoid of EPSs were failed to colonize paddy plants (Meneses et al., 2011). According to Kelly et al. (2013), *Mesorhizobium loti's* EPSs were crucial in signaling nodulation in *Lotus* spp. Relying on the SA signaling pathways and mitogen-activated kinases (MAPK) in *Arabidopsis*, the EPSs generated by *B. cereus* AR156 stimulate ISR to *Pseudomonas syringae* pv. Tomato DC3000 (Jiang et al., 2016). Xanthan gum, an exopolysaccharide produced by the bacterium *Xanthomonas,* which was responsible for inducing susceptibility to *Xanthomonas campestris* pv *campestris* (a pathogen of crucifers and cause "black rot disease") in compatible interactions, was found to elicit defense mechanisms in nonhost species of *X. campestris,* such as barley, coffee, and wheat. Luiz et al. (2016) demonstrated that the disease severity was reduced by about 74.67% in xanthan gum-treated cauliflower plants inoculated with the pathogen, *X. campestris* pv. *campestris.* In addition, a significant reduction in the photosynthetic rate, stomatal conductance, and transpiration when plants were treated with xanthan was observed. It indicates the closure of natural openings (i.e., entry point of the pathogen) such as stomata and hydathodes, thus hampering the entry of *Xanthomonas.*

According to Blainski et al. (2018), EPS produced by *Lactobacillus plantarum* can be used against the *Xanthomonas gardneri* pathogen that causes "bacterial spot" disease on tomato. EPS treatment reduced bacterial spot disease by 72.0% compared to the untreated control. Tomato plants treated with EPS showed increased defense responses, including total phenolic compounds and flavonoid content, phenylalanine ammonia-lyase (PAL), glutathione reductase (GR), and lipoxygenase (LOX) activities. Under in-vitro conditions, the EPS did not show much inhibition of the test pathogen *X. gardneri,* and thus it was accomplished that EPS might have provided protection against the disease by triggering resistance mechanisms in tomato plants. This kind of resistance was seen even when there was only a 3-day gap between the administration of EPS and the pathogen's plant inoculation. When EPS was applied seven days prior to the inoculation, there was no bacterial spot at all.

23.3.2 Disease suppression by antibiosis

Some EPS exhibit antimicrobial activity, potentially inhibiting the growth of certain pathogens, which could indirectly influence in reducing pathogen load. EPS formed by lactic acid bacteria (LAB) was found to be antibiotic to plant pathogenic microorganisms like *Xanthomonas campestris, X. vesicatoria,* and *Erwinia carotovora* (Trias et al., 2008). Some of the EPSs produced by microorganisms have potential antiviral properties, which opens the way for their use in plant protection against viral diseases of crop plants. For instance, lipopolysaccharides produced by plant pathogenic bacteria *Pantoea agglomerans* strains (Bulyhina et al., 2021), and EPS produced by *Serratia* sp. Gsm01 (Ipper et al., 2008) and *P. chlororaphis* subsp. *Aureofaciens* (Balko et al., 2010). According to Garmasheva et al. (2024), 25 EPS produced by LAB had an inhibitory effect on bacterial plant pathogens, viz., *Pseudomonas fluorescens, P. syringae, Pectobacterium carotovorum, Clavibacter michiganensis, Xanthomonas campestris,* and *X. vesicatorium*. The GC–MS analysis of EPS revealed glucose as the major component (85.85%–97.98%) in it, and the molecular masses ranged between 3.8 and 5.0×10^4 Da. It caused considerable hindrance in the growth of plant pathogenic bacteria after 3–6 hour of cultivation, as it possessed hydroxyl radical scavenging property. He claimed that EPSs produced by LAB showed antibiofilm action to *P. carotovorum, P. syringae,* and *P. fluorescens,* which reduces the accumulation of biofilm by pathogenic bacteria. About 8%–29% of *X. campestris*, 3%–32% of *X. vesicatoria*, and 1%–41% of *Clavibactor michiganensis* formed biofilms as a result of it. The EPSs were found to be antiviral in action against tobacco mosaic virus (TMV). It reduced the infectivity potential of the virus around 61%–92% at 100 μg/mL concentration, thus substantiating LAB EPS's in vitro antiviral, antibacterial and antibiofilm potential against TMV and other phytopathogenic bacteria.

The EPS that was isolated from the acid-tolerant PGPR *Rhodopseudomonas palustris* KTSSR54 showed potent antifungal activity against the paddy crop fungal diseases *Magnaporthe oryzae* PTRC63, *Curvularia lunata* SPB0627, and *Bipolaris oryzae* NPT0508. Minimum inhibitory concentration (MIC) ranged from 3.75 to 30 mg/mL. The presence of living KTSSR54 cells and their EPS caused abnormal mycelial growth, such as shrinking and hyphal tip swelling in all the three test pathogens, indicating disruption of fungal growth. In addition to its antagonistic activity, both KTSSR54 cells and its EPS significantly enhanced rice seed germination, providing a potential benefit for crop establishment. In the subsequent soil plot experiments, it was observed that inoculation of rice seedlings with the KTSSR54 strain, followed by EPS treatment, reduced fungal infection in rice and improved crop yield, with grain weight losses 2.5 times less than those in uninoculated control groups (Nookongbut et al., 2020). Strains EPS817 and EPS894 of *Pseudomonas fluorescens* were discovered to protect strawberry plants against *Phytophthora cactorum*-caused root rot disease. Different bioactive metabolites produced by *P. fluorescens* strains were treated in combination to increase the degree of disease suppression that each strain achieved when used alone. In terms of their capacity for biocontrol, the strains exhibit considerable physiological differences. While EPS894 created pyoluteorin and phenazine-1-carboxylic acid, EPS817 produced 2, 4-diacetyl-phloroglucinol and HCN. Interaction investigations between bacterial strains and *P. cactorum* under laboratory conditions showed a notable inhibition of cyst germination. According to Augusti et al. (2011), scanning electron microscopy demonstrated that every strain prevented *P. cactorum* cysts from germinating on the surface of strawberry roots. EPSs and microbes that have antimicrobial properties have been recently utilized in the synthesis of edible coatings (EC) on fruits and vegetables. It serves as a biological covering on it and prevents the development of deleterious microorganisms that spoil the fruit/vegetable. It is an environmentally friendly tactic to extend the shelf life by avoiding postharvest deterioration and decay due to microbial growth. Moreover, it prevents the moisture loss, retains firmness, controls respiration rate, heals the physical damage, and it improves the overall appearance of the horticultural produce (Alvarez et al., 2021).

As LABs possess antibacterial activity those are widely utilized in EC for extending the shelf life of fresh fruits and vegetables. (Marin et al., 2019; Agriopoulou et al., 2020). *Lactiplantibacillus plantarum* A6 strain is endowed with antifungal property against the pathogens that cause postharvest loses in cherry tomato, *Solanum lycopersicum* var. *cerasiforme* (Manjarres Melo et al., 2021). The postharvest microbial deterioration in *Lactiplantibacillus plantarum* A6 strain-treated cherry tomatoes by *Aspergillus niger, Fusarium sp*., and *Rhizopus stolonifer* was significantly less when compared to uncoated fruits. The edible coat was prepared by incorporating the LAB strain *L. plantarum* A6 with EPS produced by *Weissella confusa* JCA4.

23.3.3 Disease suppression by biofilm formation

Bacterial biofilms are complex communities of bacteria adhered to their surfaces that are held together by self-made polymer matrixes generally by extracellular DNA, secreted proteins, and EPS. EPS biofilms play a significant role in plant disease suppression by providing a protective matrix for beneficial microbes, enhancing plant resilience, and inhibiting pathogen growth through various mechanisms. Biofilm acts as a physical shield from external stresses like desiccation, antibiotics, etc. However, for its development and structure, microbial communities rely upon an array of factors, viz., species, temperature, pH, and the ionic concentration, etc. These EPSs are vital for cell adhesion, cell aggregation, and water holding, in addition to creating a protective barrier. It functions as a storehouse of essential nutrients while providing a protective coating. Additionally, the proteins in the matrix provide electron donors and acceptors, binding enzymes (adhesins), and aid in the absorption of vital nutrients. For bacteria, EPS amino acids store carbon and nitrogen, whereas nucleic acids store nutrients and genetic information (Velmourougane et al., 2017). According to More et al., (2014), humic substances provide electron donors and acceptors and maintain adhesion. Normally, bacteria generate two types of EPSs, viz., released polysaccharides (RPS-EPS) and capsuled EPS (CPS). The bacterial cell releases RPS-EPS into the media, and CPS is affixed to the cell surface. When EPS transforms from a soluble to a filamentous matrix, its extracellular enzymes help with food uptake and distribution (Laspidou et al., 2002). In order to create a stable population of bacteria with a synergistic interaction, EPS promotes bacterial cohesiveness, biofilm attachment to the root surface, and the interchange of metabolic signals between bacterial cells (Drogue, 2013). Biofilms can physically impede pathogen access to plant tissues, especially in the root environment. Biofilms produced by PGPR, *Streptomyces, Pseudomonas, Serratia marcescens*, and *B. subtilis* can be employed as biological control agents (Bais et al., 2004; Vejan et al., 2016; Arrebola et al., 2019). EPSs present in these biofilms assist in the development of soil aggregates, improving soil structure, texture, and water holding capacity of the soil. This may adversely affect the surrounding pathogen and create a microclimate that is favorable for beneficial organisms. EPS biofilms reinforce plant cell walls and alter their structures in such a way that it prevents the infection. Allard et al. (2016) reported that a PGPR *Bacillus subtilis* can detect young plant roots through chemotaxis and establish a thin biofilm on the root surface for long-term rhizosphere colonization.

23.4 Microbial exopolysaccharides in insect pest control

Microbial EPS hold significant potential for insect pest management in agriculture, primarily through their role in biopesticides, and to enhance plant defense mechanisms. While EPS themselves are not primarily used as pesticides, they can be utilized in the formulation of biopesticides. For example, some microbial biopesticides rely on the efficacy of specific microbes like *Bt*. The EPS produced by these beneficial microbes can contribute to the stability and delivery of these biopesticides, or even possess secondary effects that enhance their performance. EPS-producing microbes frequently coexist with parasitoids, other microbial products, predators, parasitoids, and other biocontrol agents. It offers a habitat for biocontrol agents and improves the effectiveness of biocontrol **for beneficial microorganisms.** Microbial EPS are vital in biopesticide formulations as they offer protection to microorganisms, enhance their adhesion to plant surfaces, and facilitate the establishment of MBCA.

23.4.1 Enhancing adhesion and colonization of microbial bio-control agents

Microbial (EPS) primarily contribute to insect pest control by enhancing the adhesion, colonization, and survival of beneficial microbes in the host plant, thereby indirectly aiding in biocontrol. EPS themselves are not directly toxic to insects. EPS-producing microorganisms are frequently combined with biocontrol agents like parasitoids, predators, or other microbial products. It can form a viscous gelatinous substance on the surface of host plants, thereby creating a more favorable environment for the growth of beneficial microbes, such as entomopathogenic bacteria or fungi, which adhere to and colonize the target insect's environment. This increases the likelihood of the microbes coming into contact with the pest and initiating their infection process.

23.4.2 Exopolysaccharides as insect antifeedant

An antifeedant is a substance that prevents an herbivore from feeding on the plant. It works by interfering with the pest's feeding behavior, potentially by affecting their taste receptors, disrupting their feeding cues, or making the plant unpalatable, or by disrupting the gut microbiome. EPS are not typically studied for their antifeedant activity against insects. Certain bacteria in biofilms have the ability to create secondary compounds that are poisonous or discouraging to herbivores. Some EPS may contain **bitter or toxic moieties**, making them unpalatable. EPS can act as an insect antifeedant through several mechanisms, as follows:

23.4.2.1 *Biofilm formation/forming a physical barrier*

The microbial control agents benefit from the protective habitat these biofilms offer, which increases their effectiveness and increases their survivability in harsh weather conditions. This property of microbial EPSs is utilized in developing formulations of microbial control agents. Biofilms can create a physical barrier on plant surfaces, making it more difficult for herbivores to access and consume the plant tissue. This approach is particularly relevant in agriculture, where plants are susceptible to herbivore damage, and it might be possible to reduce herbivore feeding by encouraging biofilm formation on plant surfaces and thereby improve crop yields. Through both direct and indirect means, biolfilm-forming rhizobacteria can defend plant roots against nematodes that feed on them. These biofilms can deter or kill plant parasitic nematodes by production of lytic enzymes, toxic nematicidal proteins, or volatile compounds, or by exertion of competition for nutrients and space in the rhizosphere, limiting their access to resources needed for survival and reproduction. Biofilm produced by some strains of *Pseudomonas,* which contain lipopeptides, pyrrolnitrin, and phenazines, exhibits insecticidal properties. These compounds can kill or inhibit the growth of insects that come into contact with the biofilm or the plant. Saha et al. (2020) reported that bacterial strains, such as *Pseudomonas aeruginosa, Azotobacter chroococcum, Sphingomonas paucimobilis, Paenibacillus spp., Klebsiella spp.,* and *Bacillus spp.* synthesize EPSs that are essential for plants for their normal physiology and development under different stress. *Bt* is an entomopathogenic bacterium that is commonly used in agriculture and forestry as a microbial insecticide. *Bt* can produce biofilm with the help of plant polysaccharide particularly pectin, which acts as an environmental cue to initiate biofilm formation. These polysaccharides trigger a signaling pathway involving kinases and the master regulator Spo0A, finally resulting in the transcription of matrix genes and the subsequent development of the biofilm. The synthesized matrix proteins and polysaccharides then mediate the adhesion of *Bt* cells to surfaces, forming a biofilm. Proteinaceous crystalline inclusions, or crystals, are produced by *Bt* during sporulation and are poisonous to a range of insects (Sauka and Benintende, 2008). The active principle of *Bt* products is insecticidal crystal proteins (ICPs) and *Bt* spores. *Bt* can colonize the surface of peas, clover, cabbage, and cotton. Up to eight weeks may pass during the colonization period. After colonization, it shows insecticidal efficacy against pests such as *Spodoptera frugiperda* on cotton and other vegetables and the Diamond-backed moth (*Plutella xylostella*) in crucifers (Normander et al., 1998; Bizzarri and Bishop 2007; 2008; Bizzarri et al., 2008; Monnerat et al., 2009).

23.4.2.2 *Disruption of insect microbiome symbiosis in the gut*

EPS produced by certain bacteria, like *Bt*, can be synergistic with ICP, enhancing their effectiveness in pest control. This synergism is particularly notable with *Bt* toxins like crystalline proteins and vegetative insecticidal proteins) where EPS can increase the insecticidal activity against various pests. Wang et al. (2021) isolated and characterized the EPS from *Bt* strain 4D19, and he could find that EPS addition increased the toxicity of ICP against *Plutella xylostella*. Although it did not attach to activated toxins, EPS served as a mediator to encourage the binding of the Cry1Ac protoxin to the *P. xylostella* gut lining. These results indicated that EPS may bind to the protoxin C-terminal region so that the insecticidal property is enhanced. EPS produced by *Bt* strains can therefore be utilized to make ICP more toxic, which could work in concert to create a *Bt*-based biopesticide. Crude and purified EPS of *Bt* strain HD270, or EPS-HD270, increased the toxicity of Cry1-type toxins and vegetative insecticidal protein (Vip3Aa11) against lepidopteran pests, especially *Spodoptera frugiperda* and *Helicoverpa armigera*, by about two times, according to Xue et al. (2023). Vip3Aa11 exhibited robust binding to EPS-HD270, which also made it easier for the pests' brush border membrane vesicles (BBMVs) to adhere to it. Additionally, it safeguarded Vip3Aa11 from proteolytic digestion in the larval midgut, maximizing its efficacy. The findings demonstrate that *Bt* EPSs, particularly EPS-HD270, have synergistic action which is distinct from other *Bacillus* spp. This is essential for increasing the potency of *Bt* toxin. EPSs act as synergistic agents, increasing the toxicity of *Bt*, Cry1Ac protoxin toward resistant pests like *Plutella xylostella*. It is achieved by modifying the gut microbiota composition of the target pest. EPSs were also found to have a role in increasing bacterial richness in the pest's gut, promoting beneficial microbial interactions while reducing harmful genera, which not only strengthened the Cry1Ac protoxin's efficacy but also disrupted the pest's feeding behavior, further reducing crop damage (Yang et al., 2024).

23.4.3 Exopolysaccharides as nano carrier of biopesticide

Microbial EPSs, such as alginate, chitosan, cellulose, curdlan, dextran, pullulan, hyaluronic acid, gellan gum, levan, mauran, xanthan gum, and schizophyllanfurnished with exceptional drug carrier properties. EPSs offer advantages like enhanced adherence and dispersion of pesticides in formulations, and their application can boost the effectiveness of bioactive components in the management of notorious pests like the Mediterranean fruit fly, *Cerratitis capitata* Wiedemann (Diptera: Tephritidae) (Falsini et al., 2023). EPS based nano-carriers were designed for capsaicin, which is a strong alkaloid from hot chili pepper, capsicum annum, and for hydroxytyrosol—a phenolic compound obtained form byproducts of extra-virgin olive oil. It was combined with egg phosphatidylcholine as a lipid component for target-specific release and EPS from the cyanobacteria *Neocyano spiracapsulata* to develop uniquely designed ad hoc nanocarriers. Because of its affinity for the chitin found in insect exoskeletons, a specific EPS was selected, and lipids were added to modify the carrier. When tested against *Ceratitis capitata*, it was found that the nano-carriers could enhance hydroxyl tyrosol's ovicidal action. Furthermore, a lower proportion of female moths landing on treated apricots demonstrated that the encapsulated formulation of either capsaicin or hydroxytyrosol had oviposition deterrence activity.

23.5 Use of microbial exopolysaccharides in weed management

By vying for vital resources like sunlight, water, and nutrients, weeds can seriously hinder agricultural growth, resulting in lower crop quality and yields. They can also act as alternate hosts for pests and diseases that further damage the crops. Weeds affect both the amount and quality of crops, causing an average loss of about 37% of global agricultural production. Weed control is therefore essential in all agricultural production systems. Numerous bacteria that live in soil have the properties of microbial herbicides. Examples include *Xanthomonas campestris* pv. *poannua*, *Pseudomonassyringae* pv. *tagetis* and *P. syringae* pv. *phaseolicola* strains, *Pseudomonas aeruginosa*, *P. fluorescens*, *Erwinia herbicola*, *Alcaligenes* sp., and fungi such as *Colletotrichumgloeosporioides*, *Aeschynomene virginica*, *Phomo chenopodicola*, and *Exserohilum monoceras*. The synthesis of basic chemicals such as cyanide, organic acids, secondary metabolites (antibiotic 2, 4-diacetyl-phloroglucinol), and plant growth hormones like auxins (indole acetic acid and δ-aminolevulinic acid) are only a few examples of the various ways that biocontrol agents function. Numerous phytotoxins that impede weeds' metabolism and result in its mortality are synthesized by microorganisms. As a result, the potential of creating microbial bioherbicides that could reduce the need for chemical herbicides is enormous. Several microorganisms, particularly strains of *Pseudomonas fluorescens* and *Xanthomonas campestris*, have been identified as producing EPS and other metabolites that contribute to weed suppression. While *P. fluorescens* strains D7 and ACK55 have shown promise in inhibiting specific weeds like downy brome, their effectiveness in field conditions can be inconsistent and dependent on environmental factors and mental stresses like drought (Roca et al., 2015), metals (Mishra and Jha 2013), salt (Shultana et al., 2020), and temperature (Shaik et al., 2011). Microbial EPS provides resilience to abiotic stresses in plants by several means, such as improving soil properties like soil structure, texture, porosity, bulk density, and water holding capacity, etc. Another way is directly safeguarding the plant cells by aiding in developing structural features to resist against abiotic stresses. For instance, EPSs secreted by beneficial microbes, PGPR, improve soil structure, moisture retention, and nutrient availability while also reducing the harmful effects of drought, salinity, etc. EPS contribute to soil aggregation and stability, increasing water retention and nutrient uptake by plant roots, creating a more favorable environment for plant growth, especially under adverse conditions (Bashan et al., 2014; Vardharajula et al., 2015). Nowadays, PGPR are commonly used to mitigate the harmful effects of chemical pollutants and to maintain crop growth by improving soil health and fertility (Upadhyay et al., 2012; Vejan et al. 2016; Basu et al., 2021). Diverse bacterial species like *Pseudomonas aeruginosa, Azotobacter vinelandii, Pseudomonas* spp., *Bacillus, Sphingomonas paucimobilis, Paenibacillus,* and *Klebsiella* produce EPS and play a significant role in optimizing plant growth under stressed environmental conditions (Saha et al., 2020). EPS-producing PGPR strains that are isolated from a stressed environment are more competent in giving resilience to abiotic stresses compared to the PGPR strains isolated from the nonstress ecosystem. Presence of these EPS strains influences plant growth, and thus such bacteria are promising candidate bio-formulation. Its herbicidal property is connected with extracellular proteins and lipopolysaccharides, which work in synergy to exert herbicidal effects (Lazaras et al., 2020).

23.5.1 Exopolysaccharides as bioemusifier

EPS is a natural emulsifying agent in herbicide formulations, contributing environment safe tactic to synthetic choices. EPS derived from *Rhodotorula mucilaginosa* YMM19 were demonstrated to have significant potential as a natural emulsifying substance in pesticide formulations. It also emphasizes that the bio-emulsifying properties of EPS, which marked a high emulsification activity of 71% and an emulsifying index of 60%. The durability and effectiveness of formulations were improved by the EPS's exceptional capacity to produce oil-in-water emulsions, a desired trait for pesticide applications (Mohammed et al., 2021). The application of EPS dramatically increased the herbicidal property of farnesol against *Melilotus indicus*, which is a naturally occurring sesquiterpene, resulting in total weed death in four days. These findings suggest that EPS could replace synthetic surfactants in weedicide formulations, offering an eco-friendly, biodegradable, and nontoxic alternative, while ensuring efficient delivery and enhanced bioavailability of the active ingredients. The EPS produced by *Fusarium fujikuroi* was found to reduce surface tension, which enhanced EPS spread and absorption on leaf surfaces, leading to increased herbicidal efficiency (Todero et al., 2019). *Fusarium, Alternaria,* and *Trichoderma* are the most diverse three genera of fungi that are extensively utilized in bioherbicide production and use (Todero et al., 2019). Bioactive principles, like cellulases produced by these fungi, efficiently disintegrate the plant cell wall, thus affecting membrane system integrity, distracting cellular metabolic pathways, and finally inducing systemic chlorosis and necrosis.

23.5.2 Exopolysaccharides as a weedicide by disrupting plant physiology

Some microbes produce allelopathic compounds along with EPS. The EPS matrix can stabilize and retain these allelochemicals in the rhizosphere that affect seed germination, root development, slow down their growth, and reduce their competitive ability for resources like water, nutrients, and sunlight or overall plant growth. Reliance on synthetic herbicides could be decreased by adopting this approach as a reliable and environmentally acceptable weed control strategy. Some studies suggest that allelopathic compounds, potentially including those derived from microbial EPS, can lessen leaf chlorophyll, impairing photosynthesis and leading to reduced rates of plant physiological functions and vigor. Microbial EPS and the microorganisms producing them can engage in complex biochemical interactions with plants and soil microorganisms. These interactions have the potential to influence weed growth and development as well as the establishment of the soil microbial community, which can enhance crop health and suppress weeds.

23.5.2.1 *Interference with water and nutrient uptake*

EPS are highly hydrophilic and can form a gel-like matrix around plant roots (rhizosphere) in some weed species. This gel may **block root hairs** from effectively absorbing water and nutrients. It alters **soil porosity**, making it harder for roots to breathe (reduced oxygen uptake). The stress induced by **nutrient deficiency** can weaken or kill weeds over time. EPS synthesized by *Pseudomonas spp.* were proved to be an efficient bioagent for weed control due to its ability to disrupt plant physiology and water conduction. EPSs facilitate bacterial colonization within vascular systems, which in turn results in wilting of the plant by obstructing water movement. Microbes like *Pseudomonas* and *Bacillus* sp. struggle with weeds for absorbing nutrients such as Fe through siderophore synthesis. EPS synthesis is a well-known characteristic of several *Bacillus sp.* and has been found to contribute to weed control, often by inhibiting seed germination or affecting seedling growth. Some species, like *Bacillus mucilaginosus,* enhance potassium solubilization, potentially impacting nutrient availability for weeds.

23.5.2.2 *Formation of biofilms and phyto-toxic effect*

EPS-producing microbes may colonize the weed surface or roots and create a biofilm barrier, physically impeding gas exchange or water movement. Release secondary metabolites or toxins along with EPS that can kill or stunt weed growth. Some EPS also chelate essential ions (e.g., iron, magnesium), starving the weed of key nutrients. *Xanthomonas campestris* bacterium has received significant attention as a bioherbicide candidate, producing various extracellular metabolites with phytotoxic effects. *Erwinia herbicola, Alcaligenes* sp., *Serratia plymuthica,* and *S. marcescens* have also been characterized as potential bioherbicides, with some producing metabolites like cyanide, organic acids, and auxins that inhibit weed growth. An EPS-producing bacteria, *Xanthomonas campestris* pv. *poae,* exhibit selective action, effectively targeting specific weeds such as annual bluegrass, and remain nontoxic to other species such as turfgrass. Additionally, EPS often work synergistically with other bioactive compounds, including hydrogen cyanide (HCN) and phytotoxins, to enhance their weed-suppressive activity (Phukan et al., 2021). Neto et al. (2021) found that the fungus *Phoma dimorpha* produces biologically activesecondary metabolite it can be used to suppress weed plants biologically. Fermented and concentrated broth of *P. dimorpha* with vegetable oils and adjuvants was found to be toxic to different weeds (*Echinochloa* sp., *Amaranthus cruentus, Senna obtusifolia*, and *Bidens pilosa*).

23.5.2.3 *Oxidative Stress Induction*

Plants naturally produce **reactive oxygen species** (ROS), such as **superoxide anions (O_2^-), hydrogen peroxide (H_2O_2), and hydroxyl radicals (OH),** but can normally detoxify them via antioxidants (e.g., SOD, CAT, APX, GSH). EPS can promote the accumulation of **ROS** to toxic levels in tissues, especially when coupled with microbial colonization. This leads to **cell damage**, lipid peroxidation, and ultimately cell death. Photosynthetic efficiency is reduced by the physiological stress exerted by the microbial EPSs. *Phoma sp.* generated EPS by submerged fermentation that had notable cidal property against *Cucumis* sp., which was directly proportional to the quantity of EPS produced. In order to enhance its phytotoxicity, hollow fiber member filtration and adsorption methods are successful in concentrating the EPSs from fermentation broth. This finding suggests the scope of developing broth liquid and solid and solid bioherbicide formulations, utilizing EPS's structural properties like hydroxyl and carboxyl groups for targeted agricultural applications (Luft et al., 2019). Similarly, EPSs with high molecular weight were generated by *Fusarium fujikuroi* mediated fermentation process, which had herbicidal properties. It synthesized EPS at 5.94 g/L in its crude broth. Among the concentration techniques studied lyophilization method gave the highest yield among the other methods. Concentrated EPS was demonstrated for its phytotoxicity indicators such as necrosis, chlorosis, and total chlorophyll loss (i.e., 100% depigmentation) on *Cucumis sativus* leaves, with lyophilized samples delivering the most potent herbicidal activity (Todero et al., 2020). Thus, microbial EPS is an ideal candidate as a bioherbicide, minimizing harmful effects on soil, plants, and the water ecosystem.

23.6 Microbial exopolysaccharides for abiotic stress tolerance

Abiotic stress in agriculture means the deleterious effect of nonliving environmental components on crop growth, establishment, and productivity. These include water stress, soil salinity, extreme weather conditions, especially temperature, heavy metal toxicity in soil, and nutrient deficiencies that can significantly affect crop production and quality (Raza et al., 2019). There are numerous ways that improve plants' resilience to abiotic stress through EPSs. EPS impart defense against a wide range of environmental stimuli for stress management strategies for plant growth and development (Schmid et al., 2015). A number of environmental challenges, like water stress (Roca et al., 2015), heavy metal toxicity (Mishra and Jha, 2013), salinity (Shultana et al., 2020), and temperature (Shaik et al., 2011), can be mitigated by EPS. Microbial EPS enhances abiotic stress tolerance in plants by influencing soil properties and directly protecting plant cells, offering a potential pathway for eco-friendly farming. PGPR and other helpful microorganisms secrete EPSs, which enhance soil structure, moisture retention, and nutrient availability. At the same time, it also minimizes the harmful effects of abiotic stresses. In order to preserve crop development and reduce the negative effects of chemical pollutants, PGPR are now frequently utilized to restore soil productivity (Upadhyay et al., 2012; Basu et al., 2021). According to

Vejan et al. (2016), PGPR improves the growth of roots and shoots, which enhances the uptake of water and nutrients from the soil substrate. Furthermore, EPS promote soil stability and aggregation, which improves water retention and nutrient uptake by plant roots and makes the environment more conducive to plant growth, particularly in unfavorable conditions (Bashan et al., 2014; Vardharajula et al., 2015). In order to maximize plant growth in stressful environments, a variety of bacterial species, including *Pseudomonas aeruginosa, Azotobacter vinelandii, Pseudomonas* spp., *Bacillus, Sphingomonas paucimobilis, Paenibacillus*, and *Klebsiella*, create extracellular polymeric substances (Saha et al., 2020). It is found that microorganisms that produce EPS have more potential in giving resilience to abiotic stresses compared to PGPR isolated from the nonstress ecosystem. Such bacteria are promising candidates as bio-stimulants for stress management strategies for plant growth and development (Schmid et al., 2015).

23.6.1 Role of exopolysaccharides in mitigating water stress

Drought is one of the major constraints in agriculture productivity. It indicates nonavailability or insufficient supply of moisture that is needed for normal crop growth and development for quiet long period (Lipiec et al., 2013, Vandana et al., 2020). It significantly affects general crop health and yield. Water stress has drastically escalated and intensified over the past few decades, mainly due to changes in climatic factors (Gupta and Huang 2014). Microbial EPS can help to mitigate soil moisture stress in plants by improving soil parameters such as structure, texture, bulk density, and porosity that improve water holding capacity of the soil. Microbial EPSs are essential for the formation of soil clods, and they aid in the conduction of water and nutrients through plant roots by forming biofilms (Vejan et al., 2016; Khan et al., 2019).

Bacillus cereus strain has been found to produce significant amounts of EPS and contributes significantly to soil moisture conservation and improving plant growth in arid regions, potentially through its role in enhancing soil aggregation and rhizo-sheath formation. *B. cereus* strain-inoculated maize seedlings showed enhanced soil moisture retention and optimal plant growth during drought. Similarly, maize seedlings that are primed with EPS producing PGPR strains of *Proteus penneri* (Pp1), *Pseudomonas aeruginosa* (Pa2), and *Alcaligenes faecalis* (AF3) showed water stress tolerance in plants and improved water-holding capacity of the soil. It promoted plant growth, development, biomass production, and overall health of the seedlings. It is also observed that, relative moisture content, protein, and sugar content in the plants under drought stress were higher, but the proline content and antioxidant enzyme activities were low. Among the PGPRs tested, the performance of Pa 2 strain (an isolate from semi arid region) was found to be more efficient. Naseem and Bano (2014) stated that a combination of PGPR strains and their relevant EPSs has a greater potential for drought tolerance than the PGPR inocula used alone. In another study, it is comprehended that, when EPS producing soil micro-organisms, viz., *Enterobacter hormaechei, Pseudomonas migulae*, and *P. fluorescens*, were used to cultivate drought-tolerant crops like foxtail millet (*S. italica* L.), they released EPS in direct proportion to the strain's population level (Mathur et al., 2021). It is documented that the application of EPSs synthesized by *Pseudomonas chlororaphis O6* on leaves of *Arabidopsis thaliana* L minimized wilt symptoms by reducing stomatal opening (Vardharajula and Ali, 2014). Similarly, Sandhya et al. (2009) documented that EPS synthesized by PGPR showed improved crop resilience to moisture stress. In the same way, when *Paenibacillus polymyxa* and *Rhizopus tropici*, which produce trehalose, were co-inoculated into bean plants (*Phaseolus vulgaris* L.), the plants showed notable improvements in crop growth, nitrogen content, and nodulation during drought. Likewise, when sunflower rhizosphere soil was co-inoculated with the EPS-yielding rhizobium-strain YAS34, a considerable rise in the ratio of root adhering soil (RAS) to root tissue was observed. Additionally, by maintaining a higher water potential surrounding the roots and enhancing soil aggregation, bacterial generation of EPS increased the permeability of RAS. Due to EPS production, the seedlings were shielded from drought by the bacteria (Alami et al., 2000). Dar et al. (2021) noticed a drastic improvement in soil properties when EPS producing strains of *Rhizobium phaseoli* and *Pseudomonas bathysetes* were treated either alone or in combination, thereby showing a drastic improvement in maize plant growth. *Rhizobium* KYGT207 strain produces EPSs, which are enriched with galactose (Gal), mannuronic acid (ManA), and glucose (Glc). These EPSs reported to increase RAS, which assists in the absorption of soil water and nutrients by wheat plants (*Triticum aestivum* L.) considerably (Kaci et al., 2005). *Azospirillum brasilense* Sp245 capsule is made up of compounds having a higher molecular weight, i.e., complexes between lipids, polysaccharides, and proteins, which could safeguard the plant from abiotic extremities like drought. Even after decapsulation, the *A. brasilense* Sp245 strain was able to flourish in dry circumstances (Svetlana et al., 2001). By the application of above mentioned EPS producing microorganisms, the soil aggregate stabilization was enhanced, which was evident as improved soil structure and texture. It results in microbes-assisted nutrient absorption by the root system, improved secretion of phytohormones, phosphate solubilization activity, plant protection by production of antibiotics, and general enhancement of plant growth and yield. Hartel and Alexander (1986) noticed that the amount of EPSs yielded by *Brady rhizobium* isolate and drought resistance in cowpea (*Vigna unguiculata* L.) is positively correlated. It is also determined that by boosting the population of EPS-producing rhizobacteria in the rhizosphere, drought stress in crops can be lessened. In a similar manner, the EPS-synthesizing *Pseudomonas* strain GAP-P45 reduces the impact of moisture stress in sunflower (*Helianthus indicus* L.) crops and acts as a PGPR. Additionally, it is essential for producing elements that promote plant growth and for enhancing the texture and structure of the soil. EPS's essential moisture sorption and colloidal stability qualities may be used in sophisticated agro-production techniques. EPS producing micro organisms, which impart drought tolerance in plants, are listed in Table 23.1.

23.6.2 Role of exopolysaccharides in mitigating salinity stress

Excessive salt buildup in farmed soil is referred to as soil salinity, and it negatively impacts crop production and plant physiology. This condition, also referred to as soil salinization, is one of the most limiting abiotic factors in agricultural production. In cultivable lands, it changes the environmental balance and improves water holding capacity and osmotic pressure while changing the physico-chemical characteristics of the soil, such as texture, osmotic potential, porosity, aeration, CEC, and EC (Upadhyay et al., 2011). Worldwide, around twenty per cent of cultivable land is affected by soil salinity (Gamalero et al., 2022), and this percentage is likely to be increasing and attain about fifty per cent by 2050. Arid or semi-arid climates account for the lion's share of the world's saline land (Haj-Amor et al., 2022). It occurs when anions like Cl- (chloride) and NO_3 (nitrate) and cations like Na+ (sodium), Ca+2 (calcium), and K+ (potassium) reduce the soil's water potential (Stavi et al., 2021). This makes the plants strenuous to absorb water and nutrients which in turn results in osmotic stress. Besides, plants may absorb certain saline ions more and deposit them in plant cells at high concentration, which becomes detrimental and disrupts the normal physiology. Particularly, high salt content can alter the pattern of nutrient assimilation and create a nutritional imbalance. It can cause early senescence, necrosis, and

TABLE 23.1 EPS-producing microorganisms for drought tolerance in crop plants.

Microbe	Plants	Uses	References
Rhizobium sp. Strain YAS34	*Helianthus annus* L.	Enhancement in RS/RT ratio improved dry biomass, Nfixation and absorption.	Alami et al. (2000)
Pseudomonas mendocina	*Lactuca sativa* L	Improve soil aggregate steadiness and total carbohydrates.	Kohler et al. (2009)
Proteus penneri *Pseudomonas aeruginosa* *Alcaligenes faccalis*	*Zea mays* L.	Increase the soil's ability to retain water and promote plant growth.	Naseem et al. (2014)
Bacillus sp. (B. amyloliquefaciens, B. licheniformis, B. thuringiensis, B. subtilis, Paenibacillus favisporus)	*Zea mays* L.	Boost plant growth andbiomass-production, Root adhereing soil to root ratio and soil aggregate consistency.	Khan et al. (2019)
Rhizobium leguminosarum *Mesorhizobium ciceri* *Rhizobium phaseoli*	*Triticum aestivum* L.	Enhance growth, biomass, and moisture stressresistance index.	Hussain et al. (2014)
Rhizobium strain KYGT207	*Triticum aestivum* L.	Improve soil aggregation.	Kaci et al. (2005)
Bacillus sp. (*B. licheniformis, B. megaterium, B. pumilus*)	*Glycine max* L.	Encourage the growth and development of plants.	Susilowati et al. (2018)
Klebsiella sp. IG3	*Triticum aestivum* L.	Boost the development of soil aggregation, RAS, and soil water-holding capacity.	Gontia-Mishra et al. (2016)
P. aeruginosa PM389 *P. aeruginosa* ZNP1 *B. endophyticus* J13 *B. tequilensis* J12		Encourage the synthesis of phytohormones, and EPS enhances the physiology, growth, and development of plants.	Ghosh et al. (2019)
Pseudomonas chlororaphis O6	*Arabidopsis thaliana*	Minimize the opening of stomata and minimize wilting.	Cho et al. (2018)
B. amyloliquefaciens 54	*Lycopersiconesculentam* L.	Assist plants in surviving stressful situations, lower malondialdehyde production. Increase the expression of stress-reactive genes (lea, tdi65, and Itpg2) and the activity of antioxidant enzymes.	Wang et al. (2020)
B. amyloliquefaciens FZB42	*Arabidopsis thaliana*	Improved proline synthesis, increased expression of drought defense genes including RD29, RD17, ERD1, and LEA14, increased fresh and dry shoot weights, increased survival rate, and increased activities of superoxide dismutase and peroxidase, among other factors.	Lu et al. (2018)

(*continued on next page*)

TABLE 23.1 EPS-producing microorganisms for drought tolerance in crop plants—cont'd

Microbe	Plants	Uses	References
Pseudomonas fluorescens DR7 *P. fluorescens* DR11 *Pseudomonasmigulae* DR35 *Enterobacterhormaechei* DR16	*Setaria italica* L.	Increase the RAS/RT, improve seed germination.	Niu et al. (2018)
P. fluorescens FAP2 *B. licheniformis* B642	*Triticum aestivaum* L.	Stimulate vegetative development. Enhance general physiology, including photosynthetic net rate (PN), internal CO_2 concentration (Ci), gs, transpiration rate, stomatal conductance, chlorophyll content, and leaf water potential (LWP).	Ansari et al. (2019)

EPS, exopolysaccharides; RS, Root to Shoot; RT, Root to Toal .

premature death of leaves (Zhu, 2001; Julkowska and Testerink, 2015). Microbial EPS contribute significantly to salinity tolerance in plants by forming protective layers around roots, binding sodium ions, enhancing soil aggregation, and promoting nutrient and water uptake. EPS also aids in biofilm formation, which provides a protective environment for microbes in saline conditions, improving plant development and resistance to stress. In order to lessen the harmful consequences of elevated sodium concentrations, EPS can attach to sodium ions in the soil and stop plant roots from absorbing them. EPS produced by bacteria binds with the Na^+ ion in the root, thereby Na^+ buildup in the plant root get reduced. Thus, bacterial EPS is said to minimize the stress due to soil salinity in crop plants. As per Kasotia et al. (2016), EPS synthesized by *Pseudomonas* sp. The AK-1 strain was capable of binding with free toxic cations in soil, such as Na^+ and thus restricting its availability to soybean plants and maintaining K^+/Na^+ balance. This strain decreased soil electrical conductivity (EC) from 1.1 to 0.9 dS/m at a concentration of 200 mM of sodium chloride. Hence, it can be concluded that the harmful effect of salinity stress can be reduced in faba bean by the treatment with EPS producing *Pseudomonas* sp. strain AK-1, which is attributed to improved plant growth, number of lateral roots, and lowered Na^+/K^+ ratio. Previously isolated PGPR strains from the root samples of different crops were assessed in vitro under a range of NaCl concentrations (0.0, 50, 100, and 150 mM) in order to determine the synthesis of biofilm and EPS by 20 PGPR strains under varying salt concentrations. Twelve of these strains of PGPR were observed to generate biofilms at every tested dosage. The PGPR strains that produced the most EPS and biofilm synthesis were chosen for additional research on faba bean plants under salt stress. When compared to uninoculated controls, plant growth was enhanced by inoculation with these PGPR strains at greater salt concentrations. Across all tested NaCl concentrations, the strain *Pseudomonas anguilliseptica* SAW 24 produced the most EPS and formed the most biofilms, which led to the highest plant height and fresh and dry weight per plant of faba beans. According to the results, PGPR's biofilm formation and EPS generation increase faba bean plants' resistance to different salt concentrations (Alaa, 2018). EPS, secreted by the endophyte *Pantoea alhagi* NX-11, play a crucial role in mitigating salinity stress in rice seedlings. These high-molecular-weight polymers enhance plant salt tolerance by improving physiological responses. Under salt stress, inoculation with EPS-producing *P. alhagi* markedly improved plant growth metrics like fresh weight, chlorophyll content, and root and shoot length. The EPS boosted proline content, a crucial osmo-protectant, decreased oxidative damage by lowering malondialdehyde (MDA) levels, and raised the K+/Na+ ratio. Furthermore, EPS reduced oxidative stress brought on by salt by boosting the activity of antioxidant enzymes such as peroxidase (POD), catalase (CAT), and superoxide dismutase (SOD). The crucial function of EPS in salt stress tolerance was highlighted by the failure of an EPS-deficient mutant strain (NX-11eps−) to display the aforementioned advantageous traits. According to Sunita et al. (2020), EPS's polyanionic composition probably helps bind excess sodium ions, lessening their harmful effects and enhancing plant resilience in general.

Tewari and Arora (2014) selected an isolate of *Pseudomonas aeruginosa*, PF23, from 110 fluorescent *Pseudomonads* for its salinity tolerance and ability to produce EPS even at NaCl concentrations as high as 2000 mM. An EPS-defective mutant of this isolate, PF23EPS-, exhibited 86% decline in EPS synthesis, which in turn resulted in loss of salt tolerance capacity. The purified EPS from PF23 demonstrated multifunctional roles: At lower concentrations, it acted as a biocontrol agent by effectively suppressing the phytopathogen *Macrophomina phaseolina* in sunflower plant and at high concentrations, it served as an osmo-protectant and stress ameliorator. Additionally, when introduced into saline soils, the EPS promoted plant growth and functioned as a seed biopriming agent. The studies conducted in sunflower plants showed that PF23 significantly enhanced plant growth and effectively suppressed the phytopathogen *Macrophomina phaseolina* under saline conditions, whereas the PF23EPS-mutant failed to promote growth or suppress disease in saline environments. As a promising bioagent for enhancing sunflower yield in semi-arid regions and lowering the induction of charcoal rot disease, *Pseudomonas aeruginosa* PF23 is highlighted in this study.

23.6.3 Exopolysaccharides for heavy metal tolerance

Microbial EPS play an important role in heavy metal tolerance and remediation by acting as protective barriers and effective biosorbents, with their production increasing under heavy metal stress. These high-molecular-weight biopolymers offer numerous functional groups that facilitate electrostatic interactions and complexation with heavy metal ions, thus sequestering them and preventing cellular damage. EPS producing PGPR have been suggested by various authors as a useful phyto-remediating agent against heavy metal accumulation in plants. In the presence of salinity and arsenic content, halophilic PGPR from the species *Halomonas* demonstrated increased EPS synthesis, which in turn led to resistance to these abiotic stresses through mechanisms for salt and arsenic sequestration. *Halomonas* species. Exo1 was also observed to oxidize more toxic (As (III)) to less toxic (As(V)), resulting in arsenic detoxification. Plant growth promotion activities of EPS secreted by *Halomonas* sp. Exo1 was also investigated by growing rice plants in pot culture under abiotic stresses, and it was observed that EPS treated salt sensitive plants were able to carry out normal cellular functions under salt stress. This was evident from the development of basic plant hormones like indole-3-acetic acid (IAA), cytokinins (CK), gibberellins (GA), and abscisic acid (ABA), as well as the restoration of their hormonal balance. A low rate of Arsenic translocation was noted in *Halomonas* sp. Exo1-treated plants outperformed the control plants, demonstrating the potential of bacteria that accumulate arsenic to lessen pollution of the food chain. Siderophore production was also observed to be enhanced in proportion with EPS production, thus proving itself as a successful biocontrol metabolite against phytopathogenic fungi and bacteria (Mukherjee et al., 2019).

23.7 Constraints

Microbial EPSs, while offering sustainable biotic and abiotic stress management, face several constraints that hold back their universal adoption. These include the time taken by the microbial strains to establish under diverse environmental conditions, the cost and human resources involved in their isolation, purification, and mass production. Moreover, it demands specialized knowledge and training for proper handling without losing its virulence and viability. AS of now, several microorganisms have been registered in India as MBCA, which are effective against a diverse group of pests and diseases. But the success of this technology for biotic/abiotic stress management is highly dependent on its biosafety and environmental impacts. Sometimes, on continuous use, there are chances to develop resistance against the target pathogen. However, long term impact of EPSs producing microorganisms in soil, water, and plants is yet to be studied. The high cost of producing EPS is a significant barrier, largely driven by the price of substrates and the complex downstream processing required to extract and purify the EPS. The amount of EPS produced can vary significantly, from 0.1 to 3 g/g of cell weight, and is heavily influenced by factors like fermentation conditions, medium composition, and aeration. During fermentation, the highly viscous broth can impede crucial processes like oxygen transfer, heat exchange, and the blending of nutrients, affecting overall production efficiency. To improve commercial viability, cheaper and more sustainable carbon sources, such as agricultural and industrial byproducts, are needed to reduce substrate costs. Some EPS-producing strains can be pathogenic, which limits their widespread use in agriculture. Safe, "generally regarded as safe" (GRAS) probiotic strains are preferred but must also meet production criteria. Moreover, registration of microbial strains faces several challenges, including high costs, complex regulatory procedures, and the need for specialized safety assessments that differ from chemical pesticides. These factors can hinder the development and adoption of biopesticides, despite their potential benefits for sustainable agriculture.

23.8 Conclusion

Microbial EPSs offer noteworthy contributions in farming by promoting soil health and improving other soil parameters, encouraging plant growth and development, minimizing the pernicious effects of biotic stresses, and providing resilience to abiotic stress. EPSs play crucial roles in nutrient entrapment, soil aggregation, and water retention, contributing to improved crop growth and yield. This makes EPS producing microorganism as a promising tool in sustainable agricultural practices. Moreover, these naturally derived substances are safe to nontarget organisms, including humans, birds, animals, fish, and beneficial insects, and are biodegradable. It plays a remarkable role in microbial biofilm synthesis. The ability of biofilm to shield the microorganisms from adverse environmental circumstances is one of its most crucial characteristics. Microbial EPSs play a key role in diverse rhizospheric changes like nutrient recycling, C-sequestrations and improving rhizospheric soil structureand are dependent on plant-microbe interaction. Microbial EPS is essential for biotic stress management because it helps plants develop systemic resistance (ISR), which prepares them to fight off pests and diseases. It is widely used as synergistic agents in formulations of microbial biological control agentsand as stable emulsifying agent in pesticide formulations. Dextran, alginate, hyaluronic acid, pullulan, xanthan gum, gellan gum, levan, curdlan, mauran, and schizophyllan are examples of EPS that have exceptional drug carrier qualities because they are biocompatible, biodegradable, and can be altered for controlled release and targeted delivery. Moreover, its use as an edible coating to prevent postharvest loss in fruits and vegetables is highly significant. EPSs in EC for fruits and vegetables aid in extending shelf life, reducing spoilage, and preserving quality, offering a protective barrier against microbial growth and moisture loss. These well-accepted, safe compounds, which are frequently generated from bacteria such as *Pseudomonas protegens* or LAB, can greatly improve the nutritional and functional qualities of vegetables.

Credit statement

This chapter draws on the work of researchers, farmers, and field experts whose contributions have enriched the content.

Conflict of interest

The authors have declared that they have no conflict of interest.

Ethical statement

This chapter does not contain any studies with human participants or animals performed by any of the authors.

Acknowledgments

The authors gratefully acknowledges the valuable contributions of colleagues, field staff of ICAR-Central Plantation Crops Research Institute (CPCRI), whose insights and support were instrumental in preparing this chapter.

References

Agriopoulou, S., Stamatelopoulou, E., Sachadyn-Krol, M., Varzakas, T., 2020. Lactic acid bacteria as antibacterial agents to extend the shelf life of fresh and minimally processed fruits and vegetables: Quality and safety aspects. Microorganisms, 8 (6), 1–23.

Agusti, L., Bonaterra, A., Moragrega, C., Camps, J., Montesinos, E., 2011. Biocontrol of root rot of strawberry caused by *Phytophthora cactorum* with a combination of two *Pseudomonas fluorescens* strains. J. Plant Pathol., 93 (2), 363–372.

Akram, W., Mahboob, A., Javed, A.A, 2013. *Bacillusthuringiensis* strain 199 can induce systemic resistance in tomato against *Fusarium* wilt. Eur. J. Microbiol. Immunol., 3 (4), 275–280.

Alaa, F.M., 2018. Effectiveness of exopolysaccharides and biofilm forming plant growth promoting rhizobacteria on salinity tolerance of faba bean *Vicia faba* L. Afr. J. Microbiol. Res., 12 (17), 399–404.

Alami, Y., Achouak, W., Marol, C., Heulin, T., 2000. Rhizosphere soil aggregation and plant growth promotion of sunflowers by exopolysaccharide producing *Rhizobium* sp. strain isolated from sunflower roots. Appl. Env. Microbiol., 66 (33933398), 108.

Allard-Massicotte, R., Tessier, L., Lecuyer, F., Lakshmanan, V., Lucier, J.F., Garneau, D. 2016. *Bacillus subtilis* early colonization of *Arabidopsisthaliana* roots involves multiple chemotaxis receptors mBio. e01664–16. doi: 10.1128/mBio.01664-16.

Alvarez, A., Manjarres, J.J., Ramirez, C., Bolivar, G., 2021. Use of an exopolysaccharide-based edible coating and lactic acid bacteria with antifungal activity to preserve the postharvest quality of cherry tomato. Food Sci. Technol., 151, 112225.

Ansari, F.A., Ahmad, I., 2019. Fluorescent pseudomonas-FAP2 and *Bacillus licheniformis* interact positively in biofilm mode enhancing plant growth and photosynthetic attributes. Sci. Rep., 9, 4547.

Arrebola, E., Tienda, S., Vida, C., de Vicente, A., Cazorla, F.M., 2019. Fitness features involved in the biocontrol interaction of *Pseudomonas chlororaphis* with host plants: The case study of PcPCL1606. Front. Microbiol., 10, 719.

Bais, H.P., Fall, R., Vivanco, J.M., 2004. Biocontrol of *Bacillussubtilis* against infection of *Arabidopsis* roots by *Pseudomonassyringae* is facilitated by biofilm formation and surfactin production. Plant Physiol., 134, 307–319. doi:10.1104/pp.103.028712.

Balko, O.I., Kiprianova, E.A., Kovalenko, A.G., SKohlhepelevych, V.V., Avdeeva, L.V., 2010. Antiphytoviral activity of Gaupsin biopreparation. Microb. Biotechnol., 2, 51–58.

Barna, B., 2022. Manipulation of senescence of plants to improve biotic stress resistance. Life (Basel) 26 12 (10), 1496. doi:10.3390/life12101496.

Bashan, Y., de-Bashan, L.E, Prabhu, S.R, Hernandez, J.-P., 2014. Advances in plant growth-promoting bacterial inoculant technology: Formulations and practical perspectives (1998–2013). Plant Soil 378, 1–33.

Basu, A., Prasad, P., Das, S.N, Kalam, S., Sayyed, R.Z, Reddy, M.S, et al., 2021. Plant growth promoting rhizobacteria (PGPR) as green bioinoculants: Recent developments, constraints, and prospects. Sustainability., 13, 1140.

Bizzarri, M.F., Bishop, A.H., 2007. Recovery of *Bacillus thuringiensis* in vegetative form from the phylloplane of clover (Trifolium hybridum) during a growing season. J. Invertebr. Pathol 94, 38–47.

Bizzarri, M.F., Bishop, A.H., 2008. The ecology of *Bacillus thuringiensis* on the phylloplane: Colonization from soil, plasmid transfer, and interaction with larvae of *Pieris brassicae*. Microb. Ecol 56, 133–139.

Bizzarri, M.F., Prabhakar, A., Bishop, A.H., 2008. Multiple-locus sequence typing analysis of *Bacillus thuringiensis* recovered from the phylloplane of clover (*Trifolium hybridum*) in vegetative form. Microb. Ecol., 55, 619–625.

Blainski, J.M.L., da Rocha Neto, A.C., Schimidt, E.C., Voltolini, J.A., Rossi, M.J., Di Piero, R.M., 2018. Exopolysaccharides from *Lactobacillus plantarum* induce biochemical and physiological alterations in tomato plant against bacterial spot. Appl. Microbiol. Biotechnol., 102, 4741–4753.

Bulyhina, T.V., Kyrychenko, A.M., Kharchuk, M.S., Varbanets, L.D., (2021) Anti-TMV activities of *Pantoeaagglomerans* lipopolysaccharides in vitro. Mikrobiol. Z., doi: https://doi.org/10.15407/microbiolj83.02.064

Chen, Y., Yan, F., Chai, Y.R., Liu, H.X., Kolter, R., Losick, R., Guo, J.H., 2013. Biocontrol of tomato wilt disease by *Bacillus subtilis* isolates from natural environments depends on conserved genes mediating biofilm formation. Env. Microbiol., 15, 848–864.

Cho, S.M., Anderson, A.J., Kim, Y.C, 2018. Extracellular polymeric substances of *Pseudomonas chlororaphis* 06 induce systemic drought tolerance in plants. Res. Plant Dis., 24, 242–247.

Choudhary, D.K., Kasotia, A., Jain, S., Vaishnav, A., Kumari, S., Sharma, K.P., Varma, A., 2016. Bacterial-mediated tolerance and resistance to plants under abiotic andbiotic stresses. J. Plant Growth Regul. 35276–35300 https://doi.org.

Dar, A., Zahir, Z.A., Iqbal, M., Mehmood, A., Javed, A., Hussain, A., Bushra, A.M., Ahmad, M., 2021. Efficacy of rhizobacterial EPSs in improving plant growth, physiology and soil properties. Environ. Monit. 193 (8). doi:10.1007/s10661-021-09286-6.

Drogue, B., Combes-Meynet, E., Moënne-Loccoz, Y., Wisniewski-Dyé, F., Prigent-Combaret, C., 2013. Control of the cooperation between plant growth-promoting rhizobacteria and crops by rhizosphere signals. In: Bruijn, F.J. (Ed.), Molecular Microbial Ecology of the Rhizosphere. John Wiley & Sons, Inc., NJ, USA, pp. 281–294.

Falsini, S., Rosi, M.C., Ravegnini, E., Schiff, S., Gonnelli, C., Papini, A., Adessi, A., Urciuoli, S., Ristori, S., 2023. Nanoformulations with exopolysaccharides from cyanobacteria: enhancing the efficacy of bioactive molecules in the Mediterranean fruit fly control. Env. Sci. Pollut. Res, 30 (35), 83760–83770.

Gamalero, E., Glick, B.R., 2022. Recent advances in bacterial amelioration of plant drought and salt stress. Biology, 11, 437.

Garmasheva, I., Tomila, T., Kharkhota, M., Oleschenko, L., 2024. Exopolysaccharides of lactic acid bacteria as protective agents against bacterial and viral plant pathogens. Int. J. Biol. Macromol., 276, 133851.

Ghosh, D., Gupta, A., Mohapatra, S., 2019. A comparative analysis of exopolysaccharide and phytohormone secretions by four drought-tolerant rhizobacterial strains and their impact on osmotic-stress mitigation in *Arabidopsis thaliana*. World J. Microbiol. Biotechnol. 35, 90.

Gontia-Mishra, I., Sapre, S., Sharma, A., Tiwari, S., 2016. Amelioration of drought tolerance in wheat by the interaction of plantgrowthpromoting rhizobacteria. Plant Biol., 18, 992–1000.

Gupta, B., Huang, B., 2014. Mechanism of salinity tolerance in plants: Physiological, biochemical, and molecular characterization. Int. J. Genom., 2014 (2014), 701596.

Haj-Amor, Z., Araya, T., Kim, D.G., Bouri, S., Lee, J., Ghiloufi, W., Yang, Y., Kang, H., Jhariya, M.K., Banerjee, A., Lal, R., 2022. Soil salinity and its associated effects on soil microorganisms, greenhouse gas emissions, crop yield, biodiversity and desertification: A review. Sci. Total Env. 843, 156946.

Hartel, P.G., Alexander, M., 1986. Role of extracellular polysaccharide production and clays in the desiccation tolerance of cowpea *Bradyrhizobia*. Soil Sci. Soc. Am. J 50, 1193–1198.

Hussain, M.B, Zahir, Z.A, Asghar, H.N, Asghar, M., 2014. Can catalase and exopolysaccharides producing rhizobia ameliorate drought stress in wheat? Int. J. Agric. Biol 16, 3–13.

Hyakumachi, M., Nishimura, M., Arakawa, T., Asano, S., Yoshida, S., Tsushima, S., Takahashi, H., 2013. *Bacillus thuringiensis* suppresses bacterial wilt disease caused by *Ralstonia solanacearum* with systemic induction of defense-related gene expression in tomato. Microbes Env., 28, 128–134.

Ipper, N.S., Cho, S., Lee, S.H., Cho, J.M., Hur, J.H., Lim, C.K., 2008. Antiviral activity of the exopolysaccharide produced by *Serratia sp.* strain Gsm01 against cucumber mosaic virus. J. Microbiol. Biotechnol., 18 (1), 67–73.

Jiang, C.H., Fan, Z.H., Xie, P., Guo, J.H., 2016. *Bacillus cereus* AR156 extracellular polysaccharides served as a novel micro-associated molecular pattern to induced systemic immunity to Pst DC3000 in *Arabidopsis*. Front. Microbiol., 7, 664.

Julkowska, M.M., Testerink, C., 2015. Tuning plant signaling and growth to survive salt. Trends Plant Sci 20, 586–594.

Kaci, Y., Heyraud, A., Barakat, M., Heulin, T., 2005. Isolation and identification of an EPS-producing *Rhizobium* strain from arid soil (Algeria): Characterization of its EPS and the effect of inoculation on wheat rhizosphere soil structure. Res. Microbiol., 156, 522–531.

Kasotia, A., Ajit, V., Tuteja, N., Choudhary, D.K, 2016. Amelioration of soybean plant from saline-induced condition by exopolysaccharide producing *Pseudomonas*-mediated expression of high affinity K+-transporter (HKT1) gene. Curr. Sci. 111 (12), 1961–1967 25.

Kelly, S.J., Muszynski, A., Kawaharada, Y., Hubber, A.M., Sullivan, J.T., Sandal, N., Carlson, R.W., Stougaard, J., Ronson, C.W., 2013. Conditional requirement for exopolysaccharide in the *Mesorhizobium-Lotus* symbiosis. Mol. Plant Microbe Interact., 26, 319–329.

Khan, N., Bano, A., Rahman, M.A, Guo, J., 2019. Comparative physiological and metabolic analysis reveals a complex mechanism involved in drought tolerance in chickpea *(Cicer arietinum L.)* induced by PGPR and PGRs. Sci. Rep., 9, 2097.

Kim, P.I., Bai, H., Bai, D., Chae, H., Chung, S., Kim, Y., Park, R., Chi, Y.T., 2004. Purification and characterization of a lipopeptide produced by *Bacillus thuringiensis* CMB26. J. Appl. Microbiol., 97, 942–949.

Kohler, J., Fuensanta, C., Roldan, A., 2009. Effect of drought on the stability of rhizosphere soil aggregates of *Lactuca sativa* grown in a degraded soil inoculated with PGPR and AM fungi. Appl. Soil Ecol., 42, 160–165.

Kumar, K.K, Sridharb, J., Murali-Baskaranb, R.K, Senthil-Nathanc, S., Kaushalb, P., Darad, S.K., Arthurs, S., 2019. Microbial biopesticides for insect pest management in India: Current status and future prospects. J Invertebr. Pathol.,.

Kyungseok, P., Joseph, W.K., Choong, M.R., 2008. Rhizobacterial exopolysaccharides elicit induced resistance on cucumber. J. Microbiol. Biotechnol., 18 (6), 1095–1100.

Lahlali, R., Ezrari, S., Radouane, N., Kenfaoui, J., Esmaeel, Q., El Hamss, H., Belabess, Z., Barka, E.A., 2022. Biological control of plant pathogens: A global perspective. Microorganisms, 10, 596.

Laspidou, C.S, Rittmann, B.E., 2002. A unified theory for extracellular polymeric substances, soluble microbial products, and active and inert biomass. Water Res 36, 2711–2720.

Lazarus Brynne, E., Feris, K., Germino, M.J., 2020. Weed-suppressive bacteria effects differ in culture compared to in soils and with or without microbial competition and separation of active ingredient. Biol. Control, 12 (10), 1496.

Lipiec, J., Doussan, C., Nosalewicz, A., Kondracka, K., 2013. Effect of drought and heat stresses on plant growth and yield: A review. Int. Agrophys 27, 463–477.

Lu, X., Liu, S.F, Yue, L., Zhao, X., Zhang, Y.-B., Xie, Z.-K., et al., 2018. Epsc involved in the encoding of exopolysaccharides produced by *Bacillus amyloliquefaciens* FZB42 act to boost the drought tolerance of *Arabidopsis thaliana*. Int. J. Mol. Sci 19, 3795.

Luft, L., Confortin, T.C., Todero, I., Neto, J.R.C., Tonato, D., Felimberti, P.Z., Zabot, G.L., Mazutti, M.A., 2019. Different techniques for concentration of extracellular biopolymers with herbicidal activity produced by *Phoma sp*. Env. Technol., 1–27. https://doi.org/10.1080/09593330.2019.1669720.

Luiz, C., Schauffler, G.P., Lemos-Blainski, J.M., Rosa, D.J., Di Piero, R.M., 2016. Mechanisms of action of aloe polysaccharides and xanthan gum for control of black rot in cauliflower. Sci. Hortic, 200, 170–177.

Manjarres Melo, J.J., Alvarez, A., Ramirez, C., Bolivar, G., 2021. Antagonistic activity of lactic acid bacteria against phytopathogenic fungi isolated from cherry tomato (*Solanum lycopersicum* var. cerasiforme). Curr. Microbiol.,, 0123456789. https://doi.org/10.1007/s00284-021-02416-w.

Marin, A., Plotto, A., Atar´es, L., Chiralt, A., 2019. Lactic acid bacteria incorporated into edible coatings to control fungal growth and maintain postharvest quality of grapes. Hort Sci., 54 (2), 337–343. https://doi.org/10.21273/HORTSCI13661-18.

Mathur, P., Roy, S., 2021. Insights into the plant responses to drought and decoding the potential of root associated microbiome for inducing drought tolerance. Physiol. Plant., 172, 1016–1029.

Meneses, C.H.S.G., Rouws, L.F.M., Simoes-Araujo, J.L., Vidal, M.S., Baldani, J.I., 2011. Exopolysaccharide production is required for biofilm formation and plant colonization by the nitrogen-fixing endophyte *Gluconacetobacter diazotrophicus*. Mol. Plant Microbe Interact., 24, 1448–1458.

Mishra, A., Jha, B., 2013. Microbial exopolysaccharides. In: Rosenberg, E, DeLong, EF, Lory, S, Stackebrandt, E, Thompson, F (Eds.), The Prokaryotes. Springer, Berlin, Heidelberg.

Mohammed, Y.M., Saad, M.M., Abdelgaleil, S.A., 2021. Production, characterization and bio-emulsifying application of exopolysaccharides from *Rhodotorula mucilaginosa* YMM19. 3 Biotech, 11 (7), 349. https://doi.org/10.1007/s13205-021-02898-2.

Monnerat, R.G., Soares, C.M., Capdeville, G., Jones, G., Martins, E.S., Praca, L., 2009. Translocation and insecticidal activity of *Bacillus thuringiensis* living inside of plants. Microb. Biotechnol., 2, 512–520.

More, T.T, Yadav, J.SS., Yan, S., Tyagi, R.D, Surampalli, R.Y., 2014. Extracellular polymeric substances of bacteria and their potential environmental applications. J. Env. Manage 144, 1–25.

Mukherjee, P., Mitra, A., Roy, M., 2019. *Halomonas* rhizobacteria of *Avicennia marina* of Indian Sundarbans promote rice growth under saline and heavy metal stresses through exopolysaccharide production. Front. Microbiol. 10. https://doi.org/10.3389/fmicb.2019.01207.

Naseem, H., Bano, A., 2014. Role of plant growth-promoting rhizobacteria and their exopolysaccharide in drought tolerance of maize. J. Plant Interact., 9, 689–701.

Neto, J.R.C., Santos, M.S.N., Mazutti, M.A., Zabot, G.L., Tres, M.V., 2021. Phoma dimorpha phytotoxic activity potentialization for bioherbicide production. Biocatal. Agric. Biotechnol 33 (2021), 1019.

Niu, X., Song, L., Xiao, Y., Ge, W., 2018. Drought-tolerant plant growth-promoting rhizobacteria associated with foxtail millet in a semiarid agroecosystem and their potential in alleviating drought stress. Front. Microbiol 8, 2580.

Nookongbut, P., Kantachote, D., Khuong, N.Q., Tantirungkij, M., 2020. The biocontrol potential of acid-resistant *Rhodopseudomonas palustris* KTSSR54 and its exopolymeric substances against rice fungal pathogens to enhance rice growth and yield. Biol. Control 150,, 104354.

Normander, B., Christensen, B.B., Molin, S., Kroer, N., 1998. Effect of bacterial distribution and activity on conjugal gene transfer on the phylloplane of the bush bean (*Phaseolus vulgaris*). Appl. Env. Microbiol., 64, 1902–1909.

Phukan, J., Deka, J., Kurmi, K., Kalita, S., 2021. Deleterious rhizobacteria as a potential bioherbicide- A Review. Int. J. Agric. Environ. Sci. 8 (2), 1–5.

Poveda, J., 2021. *Trichoderma* as biocontrol agent against pests: new uses for a mycoparasite. Biol. Control 159, 104634.

Raza, A., Razzaq, A., Mehmood, S.S, Zou, X., Zhang, X., Lv, Y., et al., 2019. Impact of climate change on crops adaptation and strategies to tackle its outcome: A review. Plants (Basel) 30 (8), 34.

Roca, C., Alves, V.D, Freitas, F., Reis, M.A., 2015. Exopolysaccharides enriched in rare sugars: Bacterial sources, production, and applications. Front. Microbiol 6, 288.

Saha, I., Datta, S., Biswas, D., 2020. Exploring the role of bacterial extracellular polymeric substances for sustainable development in agriculture. Curr. Microbiol 77, 3224–3239.

Sandhya, V., Ali, S.K, Minakshi, G., Reddy, G., Venkateswarlu, B., 2009. Alleviation of drought stress effects in sunflower seedlings by the exopolysaccharides producing *Pseudomonas putida* strain GAPP45. Biol. Fertil. Soils 46, 17–26.

Sauka, D.H., Benintende, G.B., 2008. *Bacillus thuringiensis*: General aspects. An approach to its use in the biological control of lepidopteran insects behaving as agricultural pests. Rev. Argent. Microbiol 40, 124–140.

Schmid, J., Sieber, V., Rehm, B., 2015. Bacterial exopolysaccharides: Biosynthesis pathways and engineering strategies. Front. Microbiol 6, 496.

Shaik, Z.A, Sandhya, V., Grover, M., Linga, V.R, Bandi, V., 2011. Effect of inoculation with a thermotolerant plant growth promoting *Pseudomonas putida* strain AKMP7 on growth of wheat (*Triticum spp.*) under heat stress. J. Plant Interact, 4, 239–246.

Shultana, R., Tan Kee Zuan, A., Yusop, M.R, Mohd Saud, H., Ayanda, A.F, 2020. Effect of Salt-tolerant bacterial inoculations on rice seedlings differing in salt-tolerance under saline soil conditions. Agronomy 10, 1030.

Stavi, I., Thevs, N., Priori, S., 2021. Soil salinity and sodicity in drylands: A review of causes, effects, monitoring, and restoration measures. Front. Env. Sci., 9, 712831.

Sunita, K., Mishra, I., JaiPrakash, M.J., Arora, N.K., 2020. Secondary metabolites from halotolerant plant growth promoting rhizobacteria for ameliorating salinity stress in plants. Front. Microbiol. 11. https://doi.org/10.3389/fmicb.2020.567768.

Susilowati, A., Puspita, A.A, Yunus, A., 2018. Drought resistant of bacteria producing exopolysaccharide and IAA in rhizosphere of soybean plant (Glycine max) in Wonogiri Regency Central Java Indonesia. IOP Conf. Ser.: Earth Environ. Sci., 142 (2018), 012058.

Svetlana, K., Brykova, O., Sachkova, O., Egorenkova, I., Ignatov, V., 2001. Protective role of the polysaccharide-containing capsular components of *Azospirillum brasilense*. Microbiology 70, 436–440.

Tewari, S., Arora, N.K., 2014. Multifunctional exopolysaccharides from *Pseudomonas aeruginosa* PF23 involved in plant growth stimulation, biocontrol and stress amelioration in sunflower under saline conditions. Curr. Microbiol 69 (4), 484–494.

Todero, A., Confortin, T.C., Luft, L., Selbel, J., Tres, M.V., Zabot, G.L., Mazutti, M., 2019. Concentration of EPS produced by *Fussarium fujikuroi* and application of bioproduct as an effective bioherbicide. Env. Technol 41 (00271), 21,. doi:10.1080/09593330.2019.1580775.

Todero, Z., Tassia, C.C., Luft, L., Jeferson, S., Raquel, C., Kuhnc, M.V., Tres, G.L., Zabot, G.L, Marcio, A.M., 2020. Concentration of exopolysaccharides produced by *Fusarium fujikuroi* and application of bioproduct as an effective bioherbicide. Env. Technol., 41 (21), 2742–2749.

Trias, R., Baneras, L., Montesinos, E., Badosa, E., 2008. Lactic acid bacteria from fresh fruit and vegetables as biocontrol agents of phytopathogenic bacteria and fungi. Int. Microbiol. 11,, 231–236.

Upadhyay, S.K., Singh, J.S., Saxena, A.K., Singh, D.P., 2012. Impact of PGPR inoculation on growth and antioxidant status of wheat under saline conditions. Plant Biol., 14, 605–611.

Upadhyay, S.K., Singh, J.S., Singh, D.P., 2011. Exopolysaccharide-producing plant growth-promoting rhizobacteria under salinity condition. Pedosphere, 21, 214–222.

Vandana, U.K., Singha, B., Gulzar, A.B.M., Mazumder, P.B., 2020. Molecular mechanisms in plant growth promoting bacteria (PGPR) to resist environmental stress in plants. In: Sharma, V., Salwan, R., Al-Ani, L.K.T. (Eds.), Molecular Aspects of Plant Beneficial Microbes in Agriculture. Academic Press, Cambridge, MA, USA, pp. 221–233.

Vardharajula, S., Ali, S.Z., 2015. The production of exopolysaccharide by *Pseudomonas putida* GAP-P45 under various abiotic stress conditions and its role in soil aggregation. Microbiology, 84, 512–519.

Vardharajula, S., Ali Sk, Z., 2014. Exopolysaccharide production by drought tolerant *Bacillus spp.* and effect on soil aggregation under drought stress. J. Microbiol. Biotechnol. Food Sci., 4, 51–57.

Vejan, P., Abdullah, R., Khadiran, T., Ismail, S., Nasrulhaq Boyce, A., 2016. Role of plant growth promoting rhizobacteria in agricultural sustainability-a review. J. Microbiol. Biotechnol., 18 (1), 67–73. doi:10.3390/molecules21050573virus.

Velmourougane, K., Prasanna, R., Saxena, A.K., 2017. Agriculturally important microbial biofilms: Present status and prospects. J. Basic Microbiol., 57, 548–573.

Wang, J., Song, L., Gong, X., Xu, J., Li, M., 2020. Functions of jasmonic acid in plant regulation and response to abiotic stress. Int. J. Mol. Sci., 21, 1446.

Wang, M., Lili, G., Bai, X., Zeyu, W., Wenyue, X.u, Changlong, S., Jie, Z., 2021. Structure characteristics and function of a novel extracellular polysaccharide from *Bacillusthuringiensis* strain 4D19. Int. J. Biol. Macromol 189 (2021), 956–964.

Wang, M., LiliGeng, S.J, KuiWang, W.Xu, ChanglongShu, J.Z, 2023. *Bacillus thuringiensis* exopolysaccharides induced systemic resistance against *Sclerotinia sclerotiorum* in *Brassica campestris* L. Biol. Control 183, 1–9.

Wu, L., Huang, Z., Li, X., Ma, L., Gu, Q., Wu, H., Liu, J., Borriss, R., Wu, Z., Gao, X., 2018. Stomatal closure and SA-, JA/ET-signaling pathways are essential for *Bacillusamyloliquefaciens* FZB42 to restrict leaf disease caused by *Phytophthoranicotianae* in *Nicotiana benthamiana*. Front. Microbiol 9, 847.

Xue, B., Wang, M., Wang, Z., Shu, C., Geng, L., Zhang, J., 2023. Analysis of synergism between extracellular polysaccharide from *Bacillus thuringensis* subsp. *kurstaki* HD270 and insecticidal proteins. Toxins (Basel) 15 (10), 590.

Yang, J., Xu, X., Wu, J., Champer, J., Xie, M., 2024. Involvement of miR-8510a-3p in response to Cry1Ac protoxin by regulating PxABCG3 in *Plutella xylostella*. Int. J. Biol. Macromol. 263 (Pt 1), 130271.

Zheng, M., Shi, J., Shi, J., Wang, Q., Li, Y., 2013. Antimicrobial effects of volatiles produced by two antagonistic *Bacillus* strains on the anthracnose pathogen in postharvest mangos. Biol. Control, 65, 200–206.

Zhu, J.K., 2001. Plant salt tolerance. Trends Plant Sci. 6, 66–71.

CHAPTER

24

Recent developments, deep insights into opportunities, failures, and the challenges of exopolysaccharides in biotechnological applications

Neeraj Kumar Ravi[a], *Arpita Avinash Prabhu*[b], *Vaibhav Soni*[a], *Partha Palit*[c], *Prashant Tiwari*[d], *Santosh Kumar Guru*[e] *and Pankaj Kumar Singh*[a]

[a]Department of Pharmaceutics, National Institute of Pharmaceutical Education and Research, Hyderabad, Telangana, India [b]Department of Biotechnology, National Institute of Pharmaceutical Education and Research, Hyderabad, Telangana, India [c]Department of Pharmaceutical Sciences, Assam University (A Central University), Silchar, Assam, India [d]College of Pharmaceutical Sciences, Dayanand Sagar University, Ramanagara, Karnataka, India [e]Department of Pharmacology, National Institute of Pharmaceutical Education and Research, Hyderabad, Telangana, India

24.1 Introduction

Exopolysaccharides (EPS) are complex carbohydrate compounds produced by a variety of microorganisms, including microalgae, yeasts, fungi, and bacteria. These may be produced intra- or extracellularly and then essentially functionalized extracellularly. It has a high molecular mass and is found attached to the cell surface of microorganisms in the form of a capsule or freely released into the surrounding environment as slimy substances. EPS combines with other biological substances, such as uronic acid, lipids, and proteins, and forms a key part of the extracellular matrix that microbes produce. EPS constitutes 40%–95% of the total extracellular polymeric substances produced by microbes (Rana and Upadhyay, 2020). EPS

DOI: https://doi.org/10.1016/B978-0-443-36619-2.00024-3

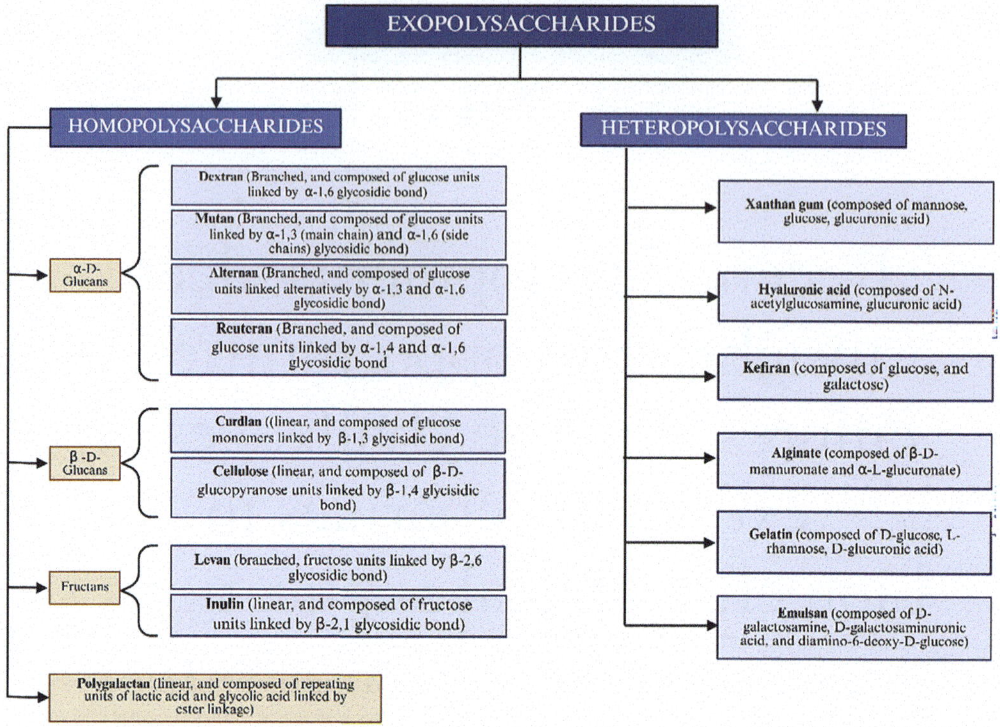

FIGURE 24.1 Classification of EPS. *EPS*, Exopolysaccharides.

consists of monosaccharide units, which form a chain and are linked by glycosidic bonds (Netrusov et al., 2023). EPS are majorly categorised into two classes, homopolysaccharides and heteropolysaccharides (Fig. 24.1). Homopolysaccharides are made up of single sugars like fructose or glucose, whereas heteropolysaccharides are composed of multiple units of different sugars, which include rhamnose, galactose, and glucose, and even include nonsugar groups like acetyl, amine, or phosphate (Xu et al., 2019a). These heteropolysaccharide EPS molecules are differentiated based on their sizes, particularly ranging from a million Daltons, and are found in various forms, such as linear, branched, and oligomeric patterns (Nguyen et al., 2024).

EPS are produced by a variety of microorganisms. These microorganisms include lactic acid bacteria, in which the primary strains responsible for EPS production are *Lactobacillus* and *Bifidobacterium*. These strains are comparatively safer in use and are mainly used in the health and food industries. Other natural sources include cyanobacteria, microalgae, yeasts, and fungi. EPS synthesis occurs via a particular metabolic pathway within microbial cells. Structurally, EPS are found outside the cell wall, either attached to the cells, forming a capsule-like structure, or sometimes present in the extravascular space where they are freely secreted and form a slimy layer. Apart from various bacterial strains, EPS are also produced by archaea, algae, and fungi. These microorganisms largely come into the thermophilic, halophilic, and mesophilic classes. Within these groups, fungi, bacteria, microalgae, and yeasts are the most prominent EPS producers (Nguyen et al., 2024). Future research could be done to understand deeply the biosynthetic pathways of EPS of various microbes, further exploring enzymes and genes that can enhance the structural properties of EPS. This would lead to a better efficient production process for biomedical applications.

24.1.1 Importance and applications of exopolysaccharides

Over the many years, research into EPS has explored starting from industrial level utilization to finding its major role in biological functions. Mainly, EPS is being used as stabilizers (Rana and Upadhyay, 2020) and thickening agents (Torino et al., 2015) in food processing industries. Later on, studies showed that various microbial EPSs have biomedical advantages, which include boosting immune responses (Fanning et al., 2012), acting as antioxidants (Xu et al., 2019b), and possessing anticancer (Giavasis, 2013) and antimicrobial (Abdalla et al., 2021) effects. Therefore, they are identified as probable therapeutic agents for various medical conditions. EPS also possess characteristics like biocompatibility, ability to degrade naturally and their good safety margin, portraying them as a promising candidate for their use in the formulation of drugs, vaccine development, and other biomedical fields, such as targeting at resisting cancers (Wahab et al., 2024), infections, and immune deficiencies (Fan et al., 2014). The majority of these applications require biotechnological aids to produce EPS in quantities required for mass production at an industrial scale (Nadzir et al., 2021).

EPS that are obtained from bacterial sources are largely used in medicine and healthcare because of their various benefits. It is mainly used as drugs delivery vehicle, as tissue rebuilding support, material for wound dressings, and dental repair (Darilmaz and Beyatli, 2012). Blandon et. al. prepared kefiran-alginate microspheres to be used as a release-modifying formulation that releases antibiotics in a sustained manner, hence providing protection to the drug, such as ciprofloxacin, in harsh gastrointestinal environments (Blandón et al., 2018). Similarly, succinate-modified xanthan gum delays the release of drugs like gentamicin, while cellulose from bacterial sources is being used as an effective material for dental care and wound treatment (Sirin and Aslim, 2020) because of its strength and compatibility with biological tissues (Finore et al., 2014). Other EPSs, like levan and xanthan gum, are used in nano-formulation to improve intracellular drug transport and are used as a basic foundation in tissue engineering (Nadzir et al., 2021). Exploration of various EPS sources and their formulation into nano delivery systems holds significant promise for cancer and other infectious diseases. These can lead to the revolutionization of current approaches of medical science by leveraging EPS's properties like biodegradability, biocompatibility, and other bioactivities.

24.1.2 Advantages of exopolysaccharides and other considerations

EPS shows several advantages as compared to synthetic materials. It is biocompatible, has lesser toxicity than synthetic polymers available for use, is biodegradable, and hence reduces environmental damage. Another important benefit is its ability to be used for specific functions, that includes regulating the drug release profiles thereby improving drug profile stability and effectiveness in vivo (Ba-akdah and Satheesh, 2021). As EPS are produced by microbial fermentation processes via biotechnological aids and principles, they are renewable and support sustainable development, making them more environmentally friendly than any conventional synthetic material available (Najafi et al., 2025).

The rising global demand for EPS from bacterial sources like xanthan gum and cellulose describes their commercial potential and economic utility. EPS production usually involves scalable fermentation methods by using renewable biological resources, hence avoiding harsh chemical processes, and is more cost-effective (Sørensen et al., 2022). Due to their natural biodegradability, there is a significant reduction in the generation of industrial as well as medical waste, thus decreasing environmental pollution. Combining EPS with other polymers can further improve their properties, potentially replacing nonbiodegradable and synthetic materials, promoting eco-friendly biomedical transformation (Wang et al., 2024). Furthermore, future research could focus on optimizing microbial fermentation techniques to enhance the EPS yield and polish their various properties for specific biomedical applications, enhancing their value economically and scalability. Conventional synthetic polymers can be replaced by EPS based polymer composites for sustainable development and environmental conservation (Sørensen et al., 2022).

24.2 Opportunities for exopolysaccharides in biotechnological applications

EPS with the aid of biotechnology is opening a whole new area of various disciplines as an economical, environmentally friendly, relatively cheaper, and better option. These disciplines include, but are not limited to, pharmaceutical, biomedical, environmental, as well as food and agricultural biotechnology.

24.2.1 Medical and pharmaceutical biotechnology

24.2.1.1 *Drug delivery systems*

EPS obtained from various microbial sources like bacteria, fungi, etc., offer a great opportunity for formulating drug delivery systems for various applications like anticancer, antibacterial, and antiviral therapy (El Awady et al., 2019). It is hydrophilic in nature and biocompatible, biodegradable, nontoxic, and extremely stable, making it a suitable candidate for biomedical and pharmaceutical applications (Moscovici, 2015). Most of the EPS are polyanionic in nature. They can also be modified easily because of the presence of carboxyl and hydroxyl groups. EPS can be used as a carrier to hold or encapsulate a drug to protect it from degradation in the body for targeted delivery purposes and to control the release rate of the drug as well. So, selection of a suitable carrier is an important step for formulating a drug delivery system (Tabernero and Cardea, 2020; Asgher et al., 2021).

Alginate, which is obtained from *Azotobacter vinelandii* bacteria, is used as an encapsulating agent. In this, alginate gelation is coupled with the electrospinning process to produce 25–30 μm fibres with chlorhexidine gluconate drug encapsulated in it (Tabernero and Cardea, 2020). Silver nanoparticles, when integrated with pullulan polysaccharide, showed higher antibacterial properties (Salleh et al., 2021). When carboxymethyl pullulan was copolymerized with polyetheramine, it showed sustained action due to sol-gel transformation in the body. Pullulan/chitosan nanostructure also shows sustained action when prepared by the electrospinning method. It completely dissolves within 60 seconds in aqueous media (Asgher et al., 2021). Novel carriers of amphiphilic α-tocopherol/Pullulan have been prepared for anticancer drugs (Fard et al., 2022). They can easily self-assemble into nano-micelles. It has a high entrapment efficiency of up to 98.3% (Asgher et al., 2021). β-oligomers of β-amyloid hydrophobically bonded with cholesterol-pullulan nanostructures can be used for Alzheimer's disease (Mishra et al., 2011). Epirubicin encapsulated in deoxycholic acid copolymerized with β-glucan increases the drug uptake in tumor cells of mice. EPS can also be used in gene delivery. siRNA was loaded in a cationic dextran nano-carrier for targeted delivery (Asgher et al., 2021). *Pseudomonas otitidis*, an EPS producing bacteria (endophytic) obtained from the *Tribulus terrestris L* plant, is used for formulating nanoparticles loaded with quercetin, with glutaraldehyde as a cross-linking agent. These nanoparticles are used for the treatment of breast cancer (Kalimuthu et al., 2023). New studies are going on EPS nanostructures (iron oxide, quantum dots) for theragnostic applications (Asgher et al., 2021).

EPS based films and patches are recently being used in transdermal drug delivery systems because they provide superior adhesiveness and skin contact compared to other polymers used. Levan obtained from the bacterium *Halomonas smyrnensis* forms a multilayer film (Sima et al., 2011). EPS can also be combined with chitosan-polyethylene oxide to form a film by the solvent casting method. It had shown better biocompatibility compared to chitosan-PEO films (Bostan et al., 2014). Pullulan obtained from *Aureobasidium pullulans* is also used to produce a film using the casting method. Fulvestrant (anticancer drug) was loaded in a multilayer film made up of alginate/chitosan/alginate layered by silica nanocapsules (Xu et al., 2019b). Mauran/Chitosan nanoparticles of fluorouracil were also prepared to give sustained release (Laubach et al., 2021). Continuous exploration of EPS based delivery systems represents a promising aspect of personalized medication needs in the future. This will enhance the utility of EPS based formulation with a reduction in the side effects by improving the drug targeting, opening new markets in the pharmaceuticals (Tabernero and Cardea, 2020).

24.2.1.2 *Tissue engineering and regeneration*

Due to the various inherent properties of EPS, it has also been portrayed to play an important role in tissue engineering purposes and regenerative medicine in the changing landscape of pharmaceuticals toward environment-friendly options. Hyaluronic acid (HA) plays an important role in bone regeneration due to its ability to interact with growth factors and cells involved in bone formation. Acrylate-HA is responsible for improving cytocompatibility and enhancing osteogenic factor expression in mature bone formation in vivo with vascular markers. Similarly, with aldehyde-modification to HA and integrin-binding domain (FN III9*-10, fibronectin)-conjugated HA, the stem cell attachment and spreading are improved and show better bone formation and collagen fibre organization in vivo (Wang et al., 2018). Hydrazide-modified HA is nontoxic and releases BMP-2 in a controlled manner and improves bone regeneration in vivo (Cuthbertson et al., 2010). Trans-cyclooctene-conjugated HA and tetrazine-conjugated hyaluronic shows great biocompatibility and in vitro improve osteogenic activities (Tabernero and Cardea, 2020). Bacterial cellulose (BC) can also be combined with hexagonal boron nitride and polyvinyl alcohol (PVA) to prepare a 3-D scaffold using 3-D printing technology for bone tissue engineering (Ahuja et al., 2023). Alginate hydrogel and alginate-starch aerogels can also be used for tissue engineering. It is also used as an osteoinductive carrier for tissue regeneration (Tabernero and Cardea, 2020).

Gellan gum obtained from *Sphingomonas elodea* is anionic, water-soluble, and acts as a strong gelling agent. It is combined with gelatin for skin regeneration purposes. Gellan EPS is used for tissue engineering and promoting platforms for proliferating cells, bone, skin, and cartilage tissues owing to its cell differentiation properties (Ahuja et al., 2023). Gellan-based hydrogel injectables are also used to treat bone defects and are widely used in composites as it has less mechanical resistance and high polyelectrolytic behavior (Das and Giri, 2020). Kefiran delivers diclofenac via a cryogel scaffold, antibiotic doxycycline loaded in a nanofiber, to be used in wound dressing and also as a mucosal adjuvant. From BC, a wound healing membrane was created filled with vaccarin, an angiogenesis-promoting drug (Nadzir et al., 2021). Xanthan gum, obtained from *Xanthomonas campestris,* is used for producing hydrogels with citric acid as a cross-linking agent. It can be used for wound healing purposes, as xanthan gum swells in the presence of water and releases the drug. Curdlan is an EPS that has a special type of conformation depending on the manufacturing process. It forms reversible and irreversible, both types of gels, depending on the final temperature, and can be combined with chitosan to produce a transparent hydrogel which has wound healing application (Tabernero and Cardea, 2020). Novel EPS composites, when combined with 3-D printing advancements, hold significance in developing personalized, biocompatible implants with improved wound-healing abilities. Addressing the current challenges, like mass manufacturing and regulatory approval, will further unlock the therapeutic and commercial potential of EPS in biomedical applications.

24.2.1.3 *Immunomodulation and therapeutics*

In the pharmaceutical and medical application, EPS play a key role as immunobiotics due to their prebiotic, anticancer, antioxidant, and immunomodulation effects toward intestinal infections (Nácher-Vázquez et al., 2015). EPSs have a beneficial effect on gut microbiota due to their capability to destroy microbial pathogens and prebiotic effects, thus helping to regulate gut microbiota. Fermentation of EPS by the gut microbiome produces short-chain fatty acids that reduce pH levels and initiate the growth and diversity of gut flora. EPS from *Bifidobacterium breve* depicts antagonistic effects and protects the host from pathogens, typically by their probiotic strains and by modulating the immune system. *Lactobacilli* and *Bifidobacteria* produce an EPS levan, having anticancer activity in human kidney (Nácher-Vázquez et al., 2015), breast (Abdelnasser and Abu-Shahba, 2024), liver, and colon (Li et al., 2022). Similarly, EPS produced by *L. plantarum* also shows anticancer effects toward HT-29, BGC-823, and HepG-2 cancer cell lines (Sheng et al., 2022). Also, EPS produced by *L. plantarum* C88 exhibits antioxidant activity, which inhibits apoptosis induced by hydrogen peroxide (Nguyen et al., 2024).

Immunoadjuvants are the components that increase the strength of the immune response. β-glucans have been exploited as an immunomodulator and as an adjuvant. It impacts the production of cytokines like IL-6, IL-8, and TNF-α. It has been discovered that β-glucans attract the macrophages and effector T-cells, improving the Th-1 immune response to help in the clearance of Hepatitis-B virus DNA (Singh et al., 2018; Abbasi et al., 2022). EPS obtained from *Lactobacillus casei* with *S. aureus* antigen MntC (rMntC-EPS 30) upregulates INF and IL-17 in CD4+ T-cells, which provide a high level of protection against *S. aureus* infection (Zhang et al., 2021; Banerjee et al., 2021). Recently, various EPS products have been gaining approval from regulatory markets and establishing their own foothold in the global market, especially in pharmaceutical and medical biotechnology (Table 24.1). Advancements in the structural modification and change in delivery strategies can enhance efficacy as anticancer and immunoadjuvants, increasing the novel therapeutics development with reduced side effects and improved specificity. Furthermore, expanding clinical studies to validate the prebiotic, antioxidant, and immune-boosting functions of EPS will be crucial for their translation into mainstream pharmaceutical and nutraceutical products (Nadzir et al., 2021).

TABLE 24.1 Marketed EPS products.

Product name	EPS	Source	Marketed by	Application	Reference
TIMOPTIC-XE	Gellan gum	*Shingomonas elodea*	Merck	Glaucoma	TIMOPTIC-XE® (1993)
Macrodex	Dextran	*Leuconostoc mesenteroides*	Pharmacosmos	Burns	Johnston et al. (1953)
Hyalgan	Hyaluronic acid	*Streptococcus zooepidemicus*	Fidi pharma	osteoarthritis	Hyalgan (n.d.)
Restylane	Hyaluronic acid	*Streptococcus zooepidemicus*	Galderma pharma	Cosmetic filler	Rohrich et al. (2007)
Keltrol	Xanthan gum	*Xanthomonas camprestis*	CP Kelco	Cosmetics	Graham (1971)
Hyalobarrier	Hyaluronic acid	*Streptococcus zooepidemicus*	Anika Therapeutics	Anti adhesion for surgical procedures	Anika (n.d.)
RoC Capsules	Hyaluronic acid	*Streptococcus zooepidemicus*	Johnson and Johnson	Antiaging and skin hydration	RoC® (n.d.)
Artz	Hyaluronic acid	*Streptococcus zooepidemicus*	Seikagaku corporation	osteoarthritis	Kaken Pharmaceutical Co. Ltd. (n.d.)

EPS, exopolysaccharides.

24.2.2 Environmental biotechnology

Various aspects of environmental remediation and agriculture have long been in search of green approaches involving different strains of microbes and their products. EPS produced with the aid of newer biotechnological tools has simply opened up a new domain in this area owing to the versatile application of EPS in environmental biotechnology.

24.2.2.1 *Bioremediation and wastewater treatment*

EPS contribute to efficient, rapid, and economical application in environmental biotechnology, such as decontaminating agents in textile and mining industries, and in wastewater treatment, because of their emulsifying, biofilm-forming, coagulating, and also flocculation of ionic wastes, textile dyes, metal ions, hydrocarbons, and heavy metals bioremediation (Costa et al., 2018). Bacterial EPS from extremophiles is rich in anionic charged ligands, such as sulfates and uronic acid residues (galacturonic and glucuronic acid). Both of these groups provide EPS with a negative charge, thus helping in scavenging toxic heavy metals from wastes, and have shown massive ability to restore the environment (Banerjee et al., 2021). Also, bacterial EPS consisting of reducing functional groups helps in detoxification and reduction of selenite into elemental selenium nanoparticles (Zhang et al., 2020). *Klebsiella oxytoca* EPS (KO-EPS) possesses iron chelation property and is being used in bioremediation (Costa et al., 2018). *Psychrophilic* microorganisms produce EPS, which removes heavy metals like cadmium up to of 48% (López-Ortega et al., 2021). Furthermore, bacteria like *Deinococcus radiodurans* R1 and *Lactobacillus plantarum*-605 have shown efficient sorption of various heavy metals in NO_3^- salt such as Cd^{2+},

Co^{2+}, Ni^{2+}, Pb^{2+} and Cu^{2+}. Halophilic EPSs are employed for in situ degradation of petroleum hydrocarbons and n-alkanes with long chains (Nguyen et al., 2024). Future research can focus on optimizing EPS production from the versatile microbial sources, exploring combinations with cheap substrates to enhance efficiency. Understanding the molecular mechanisms of metal ion binding and detoxification by EPS can drive the development of tailored biopolymers with improved specificity and capacity, thereby contributing substantially to pollution control and environmental restoration efforts.

24.2.2.2 *Sustainable agriculture*

EPS are important for successfully controlling both the environment and agronomy in sustainable agriculture (Costa et al., 2018). EPS have been studied for their ability to bind soil particles intact, and function like a glue due to its viscous nature and ionic charges. EPS improves soil characteristics, helps to create a moist environment, accumulates soil particles, and provides protection to plant cells from environmental predators and stress (Young et al., 2021). In agriculture, EPS are also used as biocontrol agents and biofertilizers. Levan acts as a biocontrol agent against the virulence of plant pathogens (Nguyen et al., 2024).

In agriculture, to improve crop yield and soil health, EPS act as biofertilizers and biocontrol agents. EPSs also show the capability to reduce the stress of salt on growing plants in saline or toxic environments. Thus, the introduction of purified EPS or the EPS produced by microbes may help to improve the quality of biofertilizers and improve the tolerance of plants to salinity. In the future, microbial EPS can be optimized for characterization, manufacturing, and application to develop biocontrol formulations which enhances health of soil, promote sustainable agriculture, and improve plant tolerances to environmental stresses like drought and salinity, etc (Banerjee et al., 2021).

24.2.2.3 *Enhanced oil recovery*

In the petroleum industry, xanthan EPS is widely recognized and considered a primary biopolymer and is commonly employed in enhancing oil recovery. It plays a role in maintaining viscosity during oil drilling methods, thus fastening oil recovery efforts and reducing oil spread (Ahmed and Hutton-Prager, 2024). Furthermore, in high-temperature and high-salinity reservoirs, another EPS, Welan gum, shows great potential in petroleum engineering, particularly in polymer flooding for enhanced oil recovery (Nguyen et al., 2024). This is due to its strong viscosity-modifying properties and ability of thickening aqueous solutions. In extreme environments, the oil recovery process can be enhanced by performing chemical modifications to this biopolymer, thus improving its solubility, resistance to bacterial degradation, and thermoviscosifying performance (Mouro et al., 2024). Continuous advancements in the chemical modification and gene editing of the xanthan and welan EPS can make them into effective biopolymers for oil recovery by improving their solubility, thermal stability, and resistance to microbial degradation. This kind of improvement has the potential to significantly enhance oil recovery rates, reduce impact on the environment as compared to synthetic polymers, and also extend the viability of oil extraction from challenges such as high temperature and high salinity reservoirs.

24.2.3 Food biotechnology and livestock management

Xanthomonas campestris produces the highest level of EPS Xanthan gum, which is mainly used in the food industry as a rheology modifier in sauces (Abid et al., 2021). Another, EPS Kefiran produced by *Lactobacillus* spp. primarily *Lentilactobacillus kefiri* and *Lactobacillus kefiranofaciens* have techno-functional properties such as prebiotics, texturing, and gelling agents in functional foods. These EPS are also used in bakery products to help in improving volume, softness, and crumb texture (Korcz and Varga, 2021).

EPS provides an alternative to antibiotics as feed additives for protecting animal health and enhancing livestock production efficiency. Experimental studies showed that EPS reduces the occurrence of pathogenic bacteria, including *Salmonella, Enterococcus,* and *E. coli* species in poultry intestines, indicating their potential to be used for inhibiting the intestinal pathogens' growth in poultry feed (Wang et al., 2019). Furthermore, in mice, EPS from *L. buchneri* TCP016 ameliorated liver damage induced by D- galactosamine/lipopolysaccharide and intestinal dysbiosis by inhibiting the *Lachnospiraceae, Helicobacteraceae*, and *Enterobacteriaceae* growth, and enhanced the growth of *Lactobacillus* and *Bacteroidaceae* (Mitra et al., 2021). Similarly, in aging mice, EPS secreted by *L. plantarum* YW11 was able to reduce the affluence of *Flexispira* and helped the growth of *Butyricicoccus* and *Blautia,* thus significantly enhancing gut flora and improving animal health by inhibiting pathogenic bacteria in the intestine (Zhang et al., 2017). EPS produced by some LAB strains, such as *L. plantarum* and *B. bifidum,* against enteric bacteria shows strong broad-spectrum antibacterial properties (Xie et al., 2020). Therefore, EPS has high potential as an excellent replacement for antibiotics in the feed industry. Apart from inhibitory activities of EPS, many in vivo and in vitro studies have demonstrated immunomodulatory effects of LAB-EPS. The existing literature suggested adding LAB-EPS as an immunomodulatory and anti-inflammatory agent to animal diets as supplements to improve overall animal health (Banerjee et al., 2021). Future research could be focused on the EPS, such as xanthan gum and kefiran, for improving functional properties, as a natural alternative to antibiotics in animal feed, enhanced utilization in food formulations to improve gut health and overall livestock quality.

24.2.4 Industrial biotechnology

24.2.4.1 *Bioplastics and packaging*

BC is an exopolysaccharide composed of repeating cellobiose units, employed in various industrial applications, including paper, cardboard, and composite materials. For packaging development, optical properties such as transparency, as compared to other biodegradable films like chitosan-PVA, BC films impart good transparency. Despite imparting transparency, BC films have demonstrated good UV protection and prevent lipid oxidation in food. In addition to these properties, it also exhibits excellent thermal properties, making it suitable for packaging (O'Toole et al., 2000). Interactions between antimicrobial compounds and BC, such as nanoparticles, organic acids, and essential oil extracts, have shown great potential in the development of bioactive food packaging materials, and when incorporated into BC, these bioactive compounds have demonstrated antimicrobial activity, thus helping in extending the shelf life of food products. BC loaded with cerium oxide nanoparticles inhibits the growth of *Staphylococcus aureus* and *Escherichia coli* by more than 90%, with no cytotoxicity (Harutoshi, 2013). Similarly, BC nanowhiskers reinforced by incorporating oregano oil into an edible whey protein film showed significant antimicrobial properties (Papadaki et al., 2024). This suggests that EPS can generate safe and effective antimicrobial packaging solutions. In industries such as cosmetics, food, and pharmaceuticals, Pullulan is an essential polymer. Pullulan applications are closely related to its particular properties, including biocompatibility, biodegradability, nontoxicity, and water solubility. Its molecular weight and structure display particular physical attributes, such as its ability to form fibres and thin films. It has excellent adhesive capacity and forms thin films that are impermeable to oxygen and are transparent (Santos et al., 2024). BC based food packaging holds the substantial promise of large-scale production to overcome the economic and technical challenges due to its antimicrobial, mechanical, and thermal properties. Film-forming abilities and biocompatibility of pullulan offers sustainable packaging solution, contributing to eco-friendly materials that enhance food preservation and environmental sustainability in the packaging industry.

24.2.4.2 *Textile and cosmetic industries*

In the textile industry, EPS, especially xanthan produced by *Xanthomonas campes*tris, helps in improving the adhesive properties of textiles and in cosmetics products, functions as oil control and acts as an antibacterial agent (Nguyen et al., 2024). Currently, HA, due to its viscoelastic and hygroscopic properties, is widely used in the cosmetics industry. In moisturizing products, potassium and sodium hyaluronate are widely used as active ingredients intended for antiwrinkle cosmetics, facial and body care, as well as in eye-area treatment. HA is commonly used in creams, gels, serums, masks, and lotions, and is widely used in aesthetic medicine, cosmetology, and clinical medicine. Numerous studies have confirmed the effectiveness and safety of HA in reducing visible signs of aging and improving skin hydration. HA is also an ingredient in tissue fillers, commonly used for restoring lost volume of tissues over the course of aging (Chylińska and Maciejczyk, 2025). Future research could focus on the applications of the xanthan gum and hyaluronic acid, such as viscoelastic, moisturizing, and antimicrobial properties, to formulate advanced products that improve skin hydration, elasticity, and antimicrobial protection, as well as improve adhesion performance in the textile industry, driving to sustainable innovation with improved consumer benefits.

24.3 Failures and challenges in exopolysaccharides production

EPS can be obtained through biotechnological processes in which the culture of bacteria/fungi is grown. On smaller or laboratory scales, it can be done in culture plates (Fig. 24.2), whereas industrial-scale production of EPS requires large-scale fermenters and suitable downstream processing methods (Fig. 24.3). It can be produced from various microbes like bacteria, yeast, fungi, molds, etc. Xanthan gum by *Xanthomonas campetris* is the most produced EPS with a production of 50,000 tonnes per year and market valuation of 600–800 million dollars per year. EPS have their own biotechnological pathways, and they are obtained after growth of a particular microbe in a fermentation tank, and this process can be divided into four phases: (1) Incorporation of carbon source, (2) oligosaccharide synthesis, (3) arrangement of polysaccharides from repeating units, and (4) EPS extraction (Barcelos et al., 2020).

24.3.1 Scalability, production, and technological issues

Production and quality of EPS are directly related to media composition. The presence of carbon and nitrogen sources in the media has opposite effects on EPS yield, such that the greater the carbon source better the EPS yield, and vice versa. It is because a higher amount of nitrogen source leads to the enzymatic degradation of EPS. Glycosyl hydrolase enzyme can degrade EPS

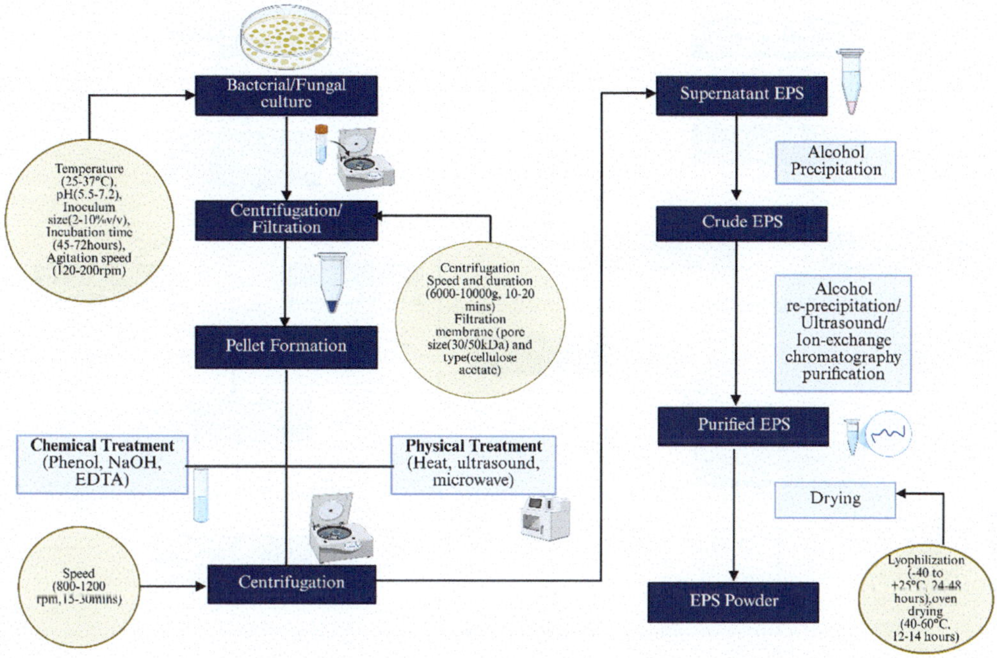

FIGURE 24.2 Small scale production of EPS with a mention of relevant critical parameters. *EPS*, Exopolysaccharides.

due to the long fermentation time. Use of industrial residues provides an alternative to reduce the cost of the fermentation process (Barcelos et al., 2020). Production of EPS from *Bacillus subtilis* increases when sucrose is substituted with cane molasses and rice bran, and it also reduces the cost (Asgher et al., 2021). For extraction and purification, the chemical method is used in which EDTA and ion exchange resins are used, but they can contaminate the extracted product (Comte et al., 2006). Physical methods are also used, like ultrasonication, centrifugation, heating, etc., but they are less efficient (Pavlou et al., 2025).

Extraction of EPS is not simple, and it does not guarantee the extraction of all polysaccharide units. Dextran EPS yield varies with different constituents, like a mixture of molasses and 10% cheese whey gives the highest yield, the lowest yield when 2% mixture is used, and no yield when only whey is used (Özcan and Öner, 2015). Xanthan gum is the most expensive EPS despite being the most produced, as glucose and sucrose can only be used as carbon source and involve costly downstream process (Barcelos et al., 2020). Levan EPS have different molecular weights depending on which organism is being used for fermentation (*Brenneria goodwinii, Pseudomonas syringa, Bacillus subtilis*). Its scale-up is an issue due to difficulty in handling a large amount of alcohol for production, an additional step of enzyme inhibition at the end of the fermentation, and unavailability of ultrafiltration at that scale (González-Garcinuño et al., 2019). In the pullulan fermentation process, there is an issue of melanin pigmentation, which turns the broth into a dark green/black color. This requires the use of a special solvent for precipitation to circumvent the issue. EPS produced from *Colwellia psychrerythraea* 34H has a higher yield in

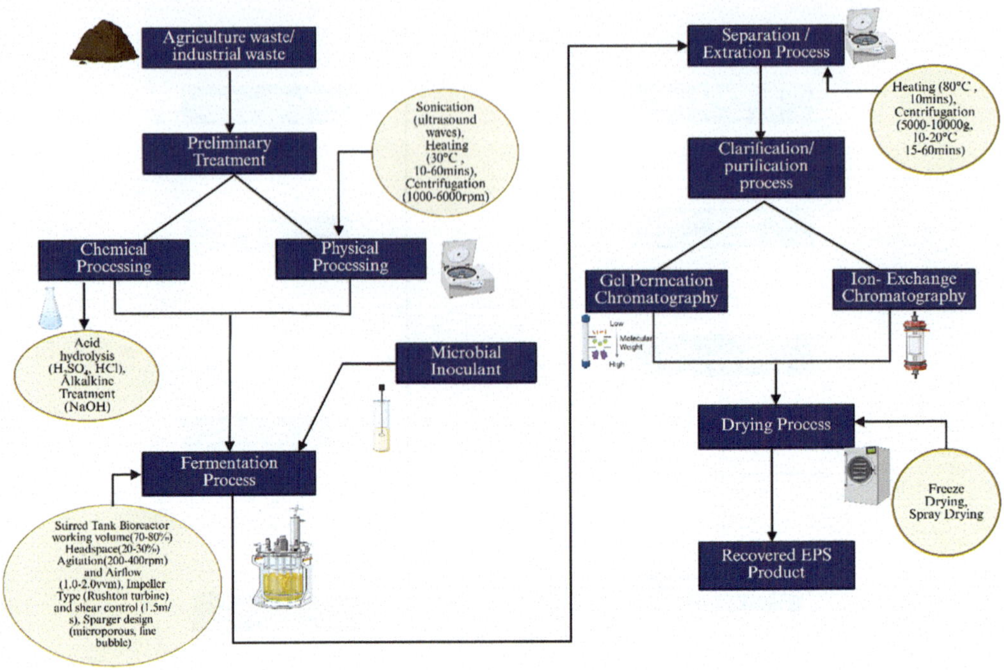

FIGURE 24.3 Industrial production of EPS with listing of critical process parameters. *EPS*, Exopolysaccharides.

the temperature range of -8 °C to -14 °C, especially under higher salinity (Barcelos et al., 2020; Pavlou et al., 2025).

When oxygen is sparged into a bioreactor, bubbles are produced of different sizes, which depend on gas flow rate, media density, viscosity, etc., which can complicate the process for extremophiles due to high viscosity. To prevent this, the use of specialized bioreactors is needed (Pavlou et al., 2025). Scleroglucan obtained from *Sclerotium glucanicum* is a fungal β-glucan known for its anticancer activity. Its method of processing can affect the stability, as below 28 °C, an undesirable byproduct, i.e., oxalic acid, is observed (Barcelos et al., 2020). Use of different nitrogen sources can alter the structure of EPS. In *R. meliloti*, the use of sodium glutamate favors glucose rich polymer, while the use of other nitrogen sources alters the composition of monosaccharides. That's why the selection of nitrogen source and knowledge of its effect on stability and production of EPS is important to know. Carbon to nitrogen (C/N) ratio also influences the structure of EPS. An imbalanced C/N ratio can modify the structure, impact the quality of polymers, and can impact the efficacy of EPS. EPS are prone to degradation, and they can endure only for a few hours to a few weeks, mostly depending on the nature and environmental conditions of the composite, and hence require specialised processing and storage environment tailored to their specifications on a case-to-case basis (Ahuja et al., 2023). Future research should prioritize the comprehensive optimization of media components, particularly the carbon-to-nitrogen ratio and the use of cost-effective industrial residues, to maximize EPS yield and quality while minimizing enzymatic degradation; advancing bioreactor design and extraction methodologies will be critical to overcoming current scale-up and purity challenges, ultimately enabling sustainable, economically viable large-scale production of diverse EPS with tailored structural and functional properties.

24.3.2 Regulatory hurdles and complexities in characterization

For the characterization of EPS, various techniques have been developed. Scanning and transmission electron microscopy can be used for revealing details about the arrangement of sugar groups attached to the surface of the cells. They have an issue of sample preparation because in EPS higher content of water is present, sample must be in a dry form for analysis. To circumvent this, atomic force microscopy can be used. This technique is expensive, has low efficiency, samples can get easily damaged, and molecules of EPS can get aggregated (Nadda and Sajna, 2021).

Turbidometric and colorimetric methods of EPS characterization require only pure samples, as the presence of other substances will interfere with the measurement and analysis of EPS. HPLC is also used for EPS characterization, but EPS that are isolated by acid hydrolysis cannot be used as derivatization leads to a loss in the yield, and there is also a loss in reproducibility. There are high incidences of byproducts such as furfurals or improper cleavage of glycosidic bonds leading to loss of resolution in peaks or multiple peaks that interfere with the analysis (Hang et al., 2020) nuclear magnetic resonance is also widely used for characterization but it has limitation of low sensitivity compared to other optical methods, and also it requires labeled substrates which can change the physiology of EPS (Nadda and Sajna, 2021).

Implementation of EPS as drug delivery systems can be considered as a long way to go because a molecule has to pass through a very strict regulatory guidelines to be considered as a drug and a drug should be of high purity, high efficacy, and safe unlike EPS because their medical efficacy are not so much high compared to chemically synthesized drugs (Nadda and Sajna, 2021). Future advancements in EPS characterization should focus on developing integrated, cost-effective methodologies that can accurately analyze hydrated samples without extensive preparation, improve sensitivity and reproducibility, and reduce contamination risks during extraction; such progress will be vital to better understanding EPS structure-function relationships, thereby facilitating their optimized application across biomedical, food, and industrial fields.

24.3.3 Market acceptance and competition

EPS provides many pharmaceutical and cosmetic applications, but due to their high production costs and low production yield, they are not much being widely accepted in the market. Currently, cheaper substrates are being investigated to reduce the cost of production of EPS, but they have their own limitations. EPS biosynthesis is a complex process, and its extraction and purification costs are around 50% of the global cost compared to synthetic polymers, which are a lot cheaper to obtain. There is a very restrictive regulatory framework for EPS, which limits the use of just any cheaper substrate as starting materials for EPS production impacts the quality of EPS produced (Benhadda et al., 2023). In large-scale production of EPS, strain selection is one of the major challenges due to differences in quality and yield. Due to these variations and the presence of cheaper, synthetic alternatives in the market, there is severe competition among the various EPS available in the market. The higher costs incurred by the manufacturer in using EPS to prepare a product, leading to higher product costs makes the consumer market somewhat hesitant to accept EPS based biotechnological products despite the advantages coupled with the product (Ahuja et al., 2023). Future efforts in the production of EPS for pharmaceutical and cosmetic applications should focus on overcoming high production costs and low yields by optimizing fermentation processes and exploring cost-effective, high-quality substrates within regulatory limits. Addressing challenges in strain selection, extraction, and purification will be essential to ensure product consistency and efficacy, enabling EPS-based biotechnological products to become competitive with synthetic alternatives and gain wider market acceptance for their unique benefits. This progress will support the sustainable growth of EPS use in high-value industries despite existing market barriers.

24.4 Future perspectives and emerging trends

Even though the polysaccharides of various sources have been explored extensively, microbial EPS only account for a smaller market share on a global scale. Recently, microbial EPS are gaining momentum in different prospects with the aid of biotechnology and advancements in it. The new era of EPS is aimed at its efficient production using emerging technologies and the incorporation of EPS in personalized therapeutics.

24.4.1 Integration of artificial intelligence and machine learning

Despite EPS being produced in large quantities by microbial cells, intricate control over media components and culture environment is necessary for an optimum yield of EPS. These parameter needs to be optimized. Use of AI and ML to optimize process and environmental parameters for EPS production promises unprecedented growth in the EPS yields with ease in culture sustenance. The tools of AI and ML utilized for these purposes can be support vector machine (SVM), particle swarm optimization (PSO), as well as teaching learning based optimization (TLBO). Rabiya and Sen (2022) demonstrated that integration of these AI and ML tools yielded far superior results to that of EPS cultivation procedures optimized with the use of response surface methodology (RSM).

Other than these AI tools, several other machine learning approaches can be used to create several models to optimize the otherwise uncertain EPS production models to guarantee more predictive results of exopolysaccharide production through biotechnological means. Some of these models are multiple regression models, a genetic algorithm optimized artificial neural network (ANN). These ANN models can be single as well as multiple objectives optimized to develop predictable EPS production models by experimenting with controllable variables within the established design space (Al Rawahi et al., 2025). Future research should aim to integrate advanced modeling and optimization strategies with traditional bioprocess approaches, enabling more precise and sustainable enhancement of EPS production and thereby supporting a wide range of industrial and biomedical applications.

24.4.2 Personalized medicine applications

Functionalized EPS prepared through advancements of biotechnology, especially those isolated from bacterial strains, have great potential in the area of personalized therapeutics. They have been shown to be compatible with live human cells, such as osteoblast cells. This indicates their capability to be used as constructs to prepare artificial blood vessels, implants (both transdermal as well as for orthopaedic purposes). Based on the conjugating component, EPS have different properties with potential in different aspects. Biotechnology plays the key role in generating various conjugates of EPS, such as EPS-PVA-hexagonal boron nitride (for personalized therapeutics concerned with cell proliferation). When this hexagonal boron nitride is replaced with dextran, it further enhances the conjugates' biodegradability and healing prowess, which has a promising future in terms of internal injuries, concerned with which critical care is concerned with (Ahuja et al., 2023).

Furthermore, biotechnology-driven EPS generation has also shown immense growth in intelligent therapeutics. EPS can be utilized to formulate various drug delivery systems that have capabilities to modify the release profile of the drug depending on the patient's homeostasis and metabolism. These can also be utilized as a raw material for 3D printing of pharmaceuticals for personalized therapeutics (Asgher et al., 2021). Future prospects for biotechnologically enhanced functionalized EPS include expanding their use as biocompatible constructs for regenerative medicine, such as artificial blood vessels and orthopedic implants, benefiting from their compatibility with human cells and tunable properties via conjugation. Additionally, EPS-based drug delivery systems hold promise for developing personalized therapeutics with controlled release tailored to patient-specific metabolic needs, and their utility as raw materials for 3D printed pharmaceuticals offers transformative potential for individualized medicine. Continued advancements in biotechnology will further unlock and optimize these diverse therapeutic applications, supporting improved healing, targeted treatment, and innovation in personalized healthcare.

24.4.3 Sustainable and ecofriendly approaches toward microbial exopolysaccharides

Microbial EPS is now an emerging option for the substitution of polymers used in therapeutics and the development of pharmaceuticals. Due to this prospect of EPS, nowadays, efforts are directed toward the sustainable development of EPS in an eco-friendly manner. These approaches include the preparation and isolation of EPS from waste materials with the help of biotechnological tools and processes. These tools include various developments in fermentation, such as submerged fermentation. The most prominent microbial EPS to serve as an example for these approaches are xanthan gum and pullulan, using kitchen waste, fruit pomace, and husk, etc., by the process of submerged fermentation. This is a classic example of biotechnological waste management that generates useful products to aid the treatment and therapy of diseases (Tiwari et al., 2024; Wang et al., 2025). The future scope of microbial EPS lies in advancing eco-friendly and sustainable production methods, particularly through the use of waste materials and biotechnological fermentation processes like submerged fermentation. These approaches not only promote waste valorization but also enable the scalable production of important EPS such as xanthan gum and pullulan, supporting their wider application as biodegradable and biocompatible polymers in pharmaceuticals and therapeutics. Continued development in these areas promises to enhance resource efficiency and contribute to greener pharmaceutical manufacturing while expanding EPS utility in disease treatment and therapy (Nguyen et al., 2024).

24.5 Conclusion

EPS is a versatile and promising biotechnological material in this emerging market with a plethora of applications spanning across a variety of industries such as pharmaceutical, biomedical, environmental, etc. This biomaterial has unique properties such as biocompatibility, biodegradability, and functionality, which portray it as a lucrative alternative to synthetic polymers. With the aid of biotechnological principles, EPS has shown great potential in the pharmaceutical and biomedical fields for the purpose of drug delivery, regenerative medicine, and immunomodulation. It also has exemplary uses in environmental remediation, agriculture, livestock maintenance, food, and cosmetic industries.

However, scaling up and commercializing EPS still remains a challenge. Various challenges associated with the cultivation of EPS source, extraction, characterization, and regulatory hurdles need to be addressed in detail for the hope of a better alternative than the synthetic polymers. The higher cost of EPS in comparison to its synthetic counterparts creates a barrier to its market acceptance. Ongoing research and advancements are aimed at offering promising solutions in these aspects, such as the incorporation of waste material in production procedures to reduce the cost of production. Integration of AI and ML is aimed at developing reproducible and scaled-up EPS production procedures with increased yields. Personalized medicine is yet another domain of EPS conjugates open for exploration to aid in therapeutics and theragnostics.

As the biotechnological field is progressing, it is moving toward fully harnessing the infinite potential of microbial EPS with efforts aimed at decoding the structure-function relationship amongst EPS. With the inherent renewable nature and unique properties of EPS, they are set to act as a key component in the development of sustainable biomaterials and technologies that will rule the coming decades, spanning industries of different disciplines. Realization of this vast potential of EPS needs targeted research and development focused on EPS.

Credit author statement

Neeraj Kumar Ravi: Conceptualization, investigation, writing—original draft. Arpita Avinash Prabhu: Writing—original draft, visualization. Vaibhav Soni: Writing—original draft, visualization. Partha Palit: Writing—review and editing, validation. Prashant Tiwari: Writing—review and editing, validation. Santosh Kumar Guru: Writing—review and editing, validation. Pankaj Kumar Singh: Conceptualization, supervision, writing—review and editing.

Conflict of interest

The authors declare no conflict of interest.

Ethical statement

This work has not been published nor is it under consideration for publication elsewhere, and appropriate attribution and citation are given for any material reproduced from any other source. The material in the manuscript has been acquired according to modern ethical standards. Generative AI has not been used to generate text or images included in this manuscript.

Data availability

The references below are not mentioned in the content. Please check these references.
Created in BioRender. N, S. (2025) https://BioRender.com/hojphku
Created in BioRender. N, S. (2025) https://BioRender.com/gl9nr9y
EPS large
Created in BioRender. N, S. (2025) https://BioRender.com/r11rrrx

Acknowledgments

The authors gratefully acknowledge the support and guidance provided by their respective institutions during the preparation of this book chapter. The authors also thank the colleagues for their valuable insights and feedback that helped improve the quality of this work.

References

Abbasi, A., Saadat, T.R, Saadat, Y.R, 2022. Microbial exopolysaccharides–β-glucans–as promising postbiotic candidates in vaccine adjuvants. Int. J. Biol. Macromol 223 (Pt A), 346–361. doi:10.1016/j.ijbiomac.2022.11.003.

Abdalla, A.K., Ayyash, M.M., Olaimat, A.N., Osaili, T.M., Al-Nabulsi, A.A., Shah, N.P., Holley, R., 2021. Exopolysaccharides as antimicrobial agents: Mechanism and spectrum of activity. Front. Microbiol 12. doi:10.3389/fmicb.2021.664395.

Abdelnasser, S.M., Abu-Shahba, N., 2024. *Bacillus sonorinses* derived exopolysaccharide enhances cell cycle arrest, apoptosis, necrosis, autophagy and COX-2 down regulation in liver cancer cells. Biotechnol. Rep. 43, e00848. doi:10.1016/j.btre.2024.e00848.

Abid, Y., Azabou, S., Blecker, C., Gharsallaoui, A., Corsaro, M.M, Besbes, S., Attia, H., 2021. Rheological and emulsifying properties of an exopolysaccharide produced by potential probiotic *Leuconostoc citreum*-BMS strain. Carbohydr. Polym 256, 117523. doi:10.1016/j.carbpol.2020.117523.

Ahmed, F., Hutton-Prager, B., 2024. Xanthan gum modification to surface and interfacial properties between soil-based matrixes and petroleum oils to minimize soil pollution. Appl. Res., 3 (6). doi:10.1002/appl.202400096.

Ahuja, V., Bhatt, A.K., Banu, J.R., Kumar, V., Kumar, G., Yang, Y.H., Bhatia, S.K., 2023. Microbial exopolysaccharide composites in biomedicine and healthcare: Trends and advances. Polymers 15 (7). doi:10.3390/polym15071801.

Al, A.M., Zafar, M., Khan, T.A., Al Araimi, S., Mahanty, B., Behera, S.K., 2025. Genetic algorithm-optimized artificial neural network for multi-objective optimization of biomass and exopolysaccharide production by *Haloferax mediterranei*. In: Rawahi, Al., Zafar, M., Khan, TA, Araimi, S.Al., Mahanty, B., Behera, S.K. (Eds.). In: Bioprocess and Biosystems Engineering, 48. SpringerAM, pp. 785–798. doi:10.1007/S00449-025-03143-3 2025.

Anika, Post-surgical adhesion barrier - hyalobarrier - hyaloglide, n.d. Available from: https://anika.com/medical/products/surgical-solutions/ (Accessed 7 October 2025).

Asgher, M., Qamar, S.A., Iqbal, H.M.N., 2021. Microbial exopolysaccharide-based nano-carriers with unique multi-functionalities for biomedical sectors. Biologia 76 (2), 673–685. doi:10.2478/s11756-020-00588-7.

Asgher, M., Rani, A., Khalid, N., Qamar, S.A, Bilal, M., 2021. Bioconversion of sugarcane molasses waste to high-value exopolysaccharides by engineered *Bacillus licheniformis*. Case Stud. Chem. Env. Eng 3, 100084. doi:10.1016/j.cscee.2021.100084.

Ba-akdah, M.A., Satheesh, S., 2021. Characterization and antifouling activity analysis of extracellular polymeric substances produced by an epibiotic bacterial strain *Kocuria flava* associated with the green macroalga *Ulva lactuca*. Acta Oceanol. Sin 40 (4), 107–115. doi:10.1007/s13131-020-1694-x.

Banerjee, A., Sarkar, S., Govil, T., González-Faune, P., Cabrera-Barjas, G., Bandopadhyay, R., Salem, D.R., Sani, R.K., 2021. Extremophilic exopolysaccharides: Biotechnologies and wastewater remediation. Front. Microbiol 12. doi:10.3389/fmicb.2021.721365.

Barcelos, M.C.S., Vespermann, K.A.C., Pelissari, F.M., Molina, G., 2020. Current status of biotechnological production and applications of microbial exopolysaccharides. Crit. Rev. Food Sci. Nutr 60 (9), 1475–1495. doi:10.1080/10408398.2019.1575791.

Benhadda, F., Zykwinska, A., Colliec-Jouault, S., Sinquin, C., Thollas, B., Courtois, A., Fuzzati, N., Toribio, A., Delbarre-Ladrat, C., 2023. Marine versus non-marine bacterial exopolysaccharides and their skincare applications. Mar. Drugs 21 (11), 582. doi:10.3390/md21110582.

Blandón, L.M., Noseda, M.D., Islan, G.A., Castro, G.R., de Melo Pereira, G.V., Thomaz-Soccol, V., Soccol, C.R., 2018. Optimization of culture conditions for kefiran production in whey: The structural and biocidal properties of the resulting polysaccharide. Bioac. Carbohydr. Diet. Fibre 16, 14–21. doi:10.1016/j.bcdf.2018.02.001.

Bostan, M.S., Mutlu, E.C., Kazak, H., Sinan Keskin, S., Oner, E.T., Eroglu, M.S., 2014. Comprehensive characterization of chitosan/PEO/levan ternary blend films. Carbohydr. Polym 102 (1), 993–1000. doi:10.1016/j.carbpol.2013.09.096.

Chylińska, N., Maciejczyk, M., 2025. Hyaluronic acid and skin: Its role in aging and wound-healing processes. Gels 11 (4), 281. doi:10.3390/gels11040281.

Comte, S., Guibaud, G., Baudu, M., 2006. Relations between extraction protocols for activated sludge extracellular polymeric substances (EPS) and EPS complexation properties: Part I. Comparison of the efficiency of eight EPS extraction methods. Enzyme Microb. Technol 38 (1-2), 237–245. doi:10.1016/j.enzmictec.2005.06.016.

Costa, O.Y.A., Raaijmakers, J.M., Kuramae, E.E., 2018. Microbial extracellular polymeric substances: Ecological function and impact on soil aggregation. Front. Microbiol 9. doi:10.3389/fmicb.2018.01636.

Cuthbertson, L., Kos, V., Whitfield, C., 2010. ABC Transporters involved in export of cell surface glycoconjugates. Microbiol. Mol. Biol. Rev 74 (3), 341–362. doi:10.1128/mmbr.00009-10.

Darilmaz, D.O., Beyatli, Y., 2012. Investigating hydrophobicity and the effect of exopolysaccharide on aggregation properties of dairy propionibacteria isolated from Turkish homemade cheeses. J. Food Prot 75 (2), 359–365. doi:10.4315/0362-028X.JFP-11-225Turkey.

Das, M., Giri, T.K, 2020. Hydrogels based on gellan gum in cell delivery and drug delivery. J. Drug Deliv. Sci. Technol 56, 101586. doi:10.1016/j.jddst.2020.101586.

El Awady, M.E., Eldin, M.A.N., Ibrahim, H.M., Al Bahnasy, M.E., Abdel Aziz, S.H., 2019. In vitro evaluation of antioxidant, anticancer, and antiviral activities of exopolysaccharide from *Streptomyces hirsutus* NRC2018. J. Appl. Pharm. Sci., 9 (11), 10–18. doi:10.7324/JAPS.2019.91102.

Fan, H., Wu, X., Yu, F., Bai, Y., Long, B., 2014. Oral Immunization with recombinant *Lactobacillus acidophilus* expressing the adhesin Hp0410 of *Helicobacter pylori* induces mucosal and systemic immune responses. Clin. Vaccine Immunol 21 (2), 126–132. doi:10.1128/CVI.00434-13China.

Fanning, S., Hall, L.J., Cronin, M., Zomer, A., MacSharry, J., Goulding, D., Motherway, M.O.C., Shanahan, F., Nally, K., Dougan, G., Van Sinderen, D., 2012. Bifidobacterial surface-exopolysaccharide facilitates commensal-host interaction through immune modulation and pathogen protection. Proc. Nat. Acad. Sci. U.S.A 109 (6), 2108–2113. doi:10.1073/pnas.1115621109Ireland.

Fard, G.H., Moinipoor, Z., Anastasova-Ivanova, S., Iqbal, H.M.N., Dwek, M.V., Getting, S.J., Keshavarz, T., 2022. Development of chitosan, pullulan, and alginate based drug-loaded nano-emulsions as a potential malignant melanoma delivery platform. Carbohydr. Polym. Technol. Appl., 4. doi:10.1016/j.carpta.2022.100250.

Finore, I., Di Donato, P., Mastascusa, V., Nicolaus, B., Poli, A., 2014. Fermentation technologies for the optimization of marine microbial exopolysaccharide production. Mar. Drugs 12 (5), 3005–3024. doi:10.3390/md12053005.

Giavasis, I., 2013. Production of Microbial Polysaccharides for Use in Food Microbial Production of Food Ingredients, Enzymes and Nutraceuticals. Elsevier Ltd., Greece, pp. 413–468. doi:10.1533/9780857093547.2.413.

González-Garcinuño, Á, Ruiz, S., Sánchez-Muñoz, Á, Tabernero, A., Valle, E.Mín, 2019. Biotechnological strategies to produce levan: Mass transfer and techno-economical evaluation. Chem. Eng. Process. - Process Intensif., 141, 107529. doi:10.1016/j.cep.2019.107529.

Graham, H.D., 1971. Microdetermination of keltrol (xanthan gum). J. Dairy Sci 54 (11), 1622–1628. doi:10.3168/jds.S0022-0302(71)86081-2.

Hang, L.T.T., Chau, H.B., Nam, T.V.H., Phu, D.H., Hiep, D.M., Thang, N.T., 2020. A survey on acid hydrolysis in analyzing the monosaccharide composition of exopolysaccharide from Ophiocordyceps. Int. J. Agric. Technol. 16 (3), 619–628.

Hyalgan, Pain relief for osteoarthritis (OA) of the knee., n.d. Available from: https://www.hyalgan.com/ (Accessed 7 September 2025).

Harutoshi, T., 2013. Exopolysaccharides of lactic acid bacteria for food and colon health applications. In *Lactic acid bacteria-R & D for food, health and livestock purposes*. IntechOpen. https://www.intechopen.com/chapters/42337.

Johnston, E.V., Lundy, J.S., Bennett, W.A., Janes, J.M., 1953. Use of dextran (macrodex) in burns. II. Clinical evaluation in eight cases. Am. J. Surg 85 (6), 720–728. doi:10.1016/0002-9610(53)90558-3.

Kaken Pharmaceutical Co., Ltd., ARTZ Dispo intra-articular injection 25mg, n.d. Available from: https://www.rad-ar.or.jp/siori/english/search/result?n=35012 (Accessed 15 July 2025).

Kalimuthu, A.K., Pandian, S.R.K., Pavadai, P., Panneerselvam, T., Kabilan, S.J., Sankaranarayanan, M., Ala, C., Kunjiappan, S., 2023. Drug delivery applications of exopolysaccharides from endophytic bacteria *Pseudomonas otitidis* from *Tribulus terrestris* L. J. Polym. Env., 31 (8), 3632–3649. doi:10.1007/s10924-023-02848-4.

Korcz, E., Varga, L.áó, 2021. Exopolysaccharides from lactic acid bacteria: Techno-functional application in the food industry. Trends Food Sci. Technol 110, 375–384. doi:10.1016/j.tifs.2021.02.014.

Laubach, J., Joseph, M., Brenza, T., Gadhamshetty, V., Sani, R.K., 2021. Exopolysaccharide and biopolymer-derived films as tools for transdermal drug delivery. J. Control. Release 329, 971–987. doi:10.1016/j.jconrel.2020.10.027.

Li, F., Jiao, X.u, Zhao, J., Liao, X., Wei, Y., Li, Q., 2022. Antitumor mechanisms of an exopolysaccharide from *Lactobacillus fermentum* on HT-29 cells and HT-29 tumor-bearing mice. Int. J. Biol. Macromol 209 (Pt A), 552–562. doi:10.1016/j.ijbiomac.2022.04.023.

López-Ortega, M.A., Chavarría-Hernández, N., López-Cuellar, M.d.R., Rodríguez-Hernández, A.I., 2021. A review of extracellular polysaccharides from extreme niches: An emerging natural source for the biotechnology. From the adverse to diverse!. Int. J. Biol. Macromol 177, 559–577. doi:10.1016/j.ijbiomac.2021.02.101.

Mishra, B., Suneetha, V., Rath, K., 2011. The role of microbial pullulan, a biopolymer in pharmaceutical approaches: A review. J. Appl. Pharm. Sci. 1 (6), 45–50. Available from http://www.japsonline.com/admin/php/uploads/122_pdf.pdf .

Mitra, A., Chatterjee, S., Kataki, S., Rastogi, R.P., Gupta, D.K., 2021. Bacterial tolerance strategies against lead toxicity and their relevance in bioremediation application. Env. Sci. Pollut. Res 28 (12), 14271–14284. doi:10.1007/s11356 021 12583-9.

Moscovici, M., 2015. Present and future medical applications of microbial exopolysaccharides. Front. Microbiol 6. doi:10.3389/fmicb.2015.01012.

Mouro, C., Gomes, A.P., Gouveia, I.C., 2024. Microbial exopolysaccharides: Structure, diversity, applications, and future frontiers in sustainable functional materials. Polysaccharides, 5 (3), 241–287. doi:10.3390/polysaccharides5030018.

Nácher-Vázquez, M., Ballesteros, N., Canales, Á., Saint-Jean, S.R.í, Pérez-Prieto, S.I., Prieto, A., Aznar, R., López, P., 2015. Dextrans produced by lactic acid bacteria exhibit antiviral and immunomodulatory activity against salmonid viruses. Carbohydr. Polym, 124, 292–301. doi:10.1016/j.carbpol.2015.02.020.

Nadda, A.K., Sajna, 2021. K.V. In: Sharma, S. (Ed.), *Microbial exopolysaccharides as novel and significant biomaterials*. Springer International Publishing, Berlin/Heidelberg, Germany, pp. 2–9.

Nadzir, M.M., Nurhayati, R.W., Idris, F.N., Nguyen, M.H., 2021. Biomedical applications of bacterial exopolysaccharides: A review. Polymers 13 (4), 1–25. doi:10.3390/polym13040530.

Najafi, A., Heidary, M., Martinez, R.M., Baby, A.R., Morowvat, M.H., 2025. Microalgae-based sunscreens as green and sustainable cosmetic products. Int. J. Cosmet. Sci 47 (2), 213–222. doi:10.1111/ics.13019.

Netrusov, A.I., Liyaskina, E.V., Kurgaeva, I.V., Liyaskina, A.U., Yang, G., Revin, V.V., 2023. Exopolysaccharides producing bacteria: A review. Microorganisms, 11 (6). doi:10.3390/microorganisms11061541.

Nguyen, H.T., Pham, T.T., Nguyen, P.T., Le-Buanec, H., Rabetafika, H.N., Razafindralambo, H.L., 2024. Advances in microbial exopolysaccharides: Present and future applications. Biomolecules, 14 (9). doi:10.3390/biom14091162.

O'Toole, G., Kaplan, H.B., Kolter, R., 2000. Biofilm formation as microbial development. Annu. Rev. Microbiol 54, 49–79. doi:10.1146/annurev.micro.54.1.49.

Özcan, E., Öner, E.T., 2015. Microbial Production Of Extracellular Polysaccharides From Biomass Sources Polysaccharides: Bioactivity and Biotechnology. Springer Nature, Srpinger International Publishsing Switzerland, pp. 161–184. doi:10.1007/978-3-319-16298-0_51.

Papadaki, A., Lappa, I.K., Manikas, A.C., Pastore Carbone, M.G., Natsia, A., Kachrimanidou, V., Kopsahelis, N., 2024. Grafting bacterial cellulose nanowhiskers into whey protein/essential oil film composites: Effect on structure, essential oil release and antibacterial properties of films. Food Hydrocoll, 147. doi:10.1016/j.foodhyd.2023.109374.

Pavlou, A., Radchenkova, N., Yaşar, S., Yıldız, Y., Angeloff, S., 2025. Advanced optimization of bioprocess parameters for exopolysaccharides synthesis in extremophiles. Process 13, 822. https://doi.org/10.3390/PR13030822.

Rabiya, R., Sen, R., 2022. Artificial intelligence driven advanced optimization strategy vis-à-vis response surface optimization of production medium: Bacterial exopolysaccharide production as a case-study. Biochem. Eng. J 178. doi:10.1016/j.bej.2021.108271.

Rana, S., Upadhyay, L.S.B., 2020. Microbial exopolysaccharides: Synthesis pathways, types and their commercial applications. Int. J. Biol. Macromol 157, 577–583. doi:10.1016/j.ijbiomac.2020.04.084.

RoC®, Hydration+ serum capsules, n.d. Available from: https://www.rocskincare.com/products/hydration-serum-capsules?srsltid=AfmBOop1oJxhhfa6Tw52GpkHIFW1tOuyhwTuwtvU8_PK6YczI91lAsNZ (Accessed 7 October 2025).

Rohrich, R.J., Ghavami, A., Crosby, M.A., 2007. The role of hyaluronic acid fillers (Restylane) in facial cosmetic surgery: Review and technical considerations. Plast. Reconstr. Surg 120 (6). doi:10.1097/01.prs.0000248794.63898.0f.

Salleh, M.S.N., Ali, R.R., Shameli, K., Hamzah, M.Y., Kasmani, R.M., Nasef, M.M., 2021. Interaction insight of pullulan-mediated gamma-irradiated silver nanoparticle synthesis and its antibacterial activity. Polymers 13 (20). doi:10.3390/polym13203578.

Santos, M.R.d., Durval, I.J.B., Medeiros, A.D.M.d., Silva Júnior, C.J.G.D., Converti, A., Costa, A.F.d.S., Sarubbo, L.A., 2024. Biotechnology in food packaging using bacterial cellulose. Foods, 13 (20). doi:10.3390/foods13203327.

Sheng, S., Fu, Y., Pan, N.a, Zhang, H., Xiu, L., Liang, Y., Liu, Y., Liu, B., Ma, C., Du, R., Wang, X., 2022. Novel exopolysaccharide derived from probiotic *Lactobacillus pantheris* TCP102 strain with immune-enhancing and anticancer activities. Front. Microbiol 13. doi:10.3389/fmicb.2022.1015270.

Sima, F., Mutlu, E.C., Eroglu, M.S., Sima, L.E., Serban, N., Ristoscu, C., Petrescu, S.M., Oner, E.T., Mihailescu, I.N., 2011. Levan nanostructured thin films by MAPLE assembling. Biomacromolecules 12 (6), 2251–2256. doi:10.1021/bm200340b.

Singh, P.K., Srivastava, A.K., Dev, A., Kaundal, B., Choudhury, S.R., Karmakar, S., 2018. 1, 3β-Glucan anchored, paclitaxel loaded chitosan nanocarrier endows enhanced hemocompatibility with efficient anti-glioblastoma stem cells therapy. Carbohydr. Polym 180, 365–375. doi:10.1016/j.carbpol.2017.10.030.

Sirin, S., Aslim, B., 2020. Characterization of lactic acid bacteria derived exopolysaccharides for use as a defined neuroprotective agent against amyloid beta1–42-induced apoptosis in SH-SY5Y cells. Sci. Rep. 10 (1). doi:10.1038/s41598-020-65147-1.

Sørensen, H.M., Rochfort, K.D., Maye, S., MacLeod, G., Brabazon, D., Loscher, C., Freeland, B., 2022. Exopolysaccharides of lactic acid bacteria: Production, purification and health benefits towards functional food. Nutrients 14 (14). doi:10.3390/nu14142938.

Tabernero, A., Cardea, S., 2020. Microbial exopolysaccharides as drug carriers. Polymers 12 (9), 2142. doi:10.3390/polym12092142.

TIMOPTIC-XE® 0.25% and 0.5% (Timolol Maleate Ophthalmic Gel Forming Solution), 1993.

Tiwari, O.N., Bobby, M.N., Kondi, V., Halder, G., Kargarzadeh, H., Ikbal, A.M.A., Bhunia, B., Thomas, S., Efferth, T., Chattopadhyay, D., Palit, P., 2024. Comprehensive review on recent trends and perspectives of natural exopolysaccharides: pioneering nano-biotechnological tools. International journal of biological macromolecules 265, 130747.

Torino, M.I., de Valdez, G.F., Mozzi, F., 2015. Biopolymers from lactic acid bacteria. Novel applications in foods and beverages. Front. Microbiol 6. doi:10.3389/fmicb.2015.00834.

Wahab, W.A.A., Shafey, H.I., Mahrous, K.F., Esawy, M.A., Saleh, S.A.A., 2024. Coculture of bacterial levans and evaluation of its anti-cancer activity against hepatocellular carcinoma cell lines. Sci. Rep 14 (1). doi:10.1038/s41598-024-52699-9.

Wang, J., Salem, D.R., Sani, R.K., 2019. Extremophilic exopolysaccharides: A review and new perspectives on engineering strategies and applications. Carbohydr. Polym 205, 8–26. doi:10.1016/j.carbpol.2018.10.011.

Wang, S.S., Han, Y.H., Chen, J.L., Zhang, D.C., Shi, X.X., Ye, Y.X., Chen, D.L., Li, M., 2018. Insights into bacterial cellulose biosynthesis from different carbon sources and the associated biochemical transformation pathways in *Komagataeibacter sp.* W1. Polymers 10 (9). doi:10.3390/polym10090963.

Wang, Y., Mehmood, S., Henry Maina, N., Katina, K., Coda, R., 2024. Synthesis in situ of heteropolysaccharide by *Levilactobacillus brevis* AM7 during fermentation of oat and hemp and its effect on the techno-functional properties of oat yogurt type model. Food Hydrocoll, 147, 109416. doi:10.1016/j.foodhyd.2023.109416.

Wang, Z., Zheng, Y., Guo, J., Lai, Z., Liu, J., Li, N., Li, Z., Gao, M., Qiao, X., Yang, Y., Zhang, H., An, L., Xu, K., 2025. Recent advance on the production of microbial exopolysaccharide from waste materials. Bioprocess. Biosyst. Eng 48 (8), 1239–1254. doi:10.1007/s00449-025-03169-7.

Xie, J., Feng, N., Wang, R., Guo, Z., Dong, H., Cui, H., Wu, H., Qiu, G., Liu, X., 2020. A Reusable biosorbent using Ca-alginate immobilized *Providencia vermicola* for Pd(II) recovery from acidic solution. Water Air Soil Pollut 231 (2). doi:10.1007/s11270-020-4399-z.

Xu, L., Chu, Z., Wang, H., Cai, L., Tu, Z., Liu, H., Zhu, C., Shi, H., Pan, D., Pan, J., Fei, X., 2019a. Electrostatically assembled multilayered films of biopolymer enhanced nanocapsules for on-demand drug release. ACS Appl. Bio Mater 2 (8), 3429–3438. doi:10.1021/acsabm.9b00381.

Xu, Y., Cui, Y., Yue, F., Liu, L., Shan, Y., Liu, B., Zhou, Y., Lü, X., 2019b. Exopolysaccharides produced by lactic acid bacteria and *Bifidobacteria*: Structures, physiochemical functions and applications in the food industry. Food Hydrocoll, 94, 475–499. doi:10.1016/j.foodhyd.2019.03.032.

Young, I.D., Latousakis, D., Juge, N., 2021. The immunomodulatory properties of β-2,6 fructans: A comprehensive review. Nutrients 13 (4). doi:10.3390/nu13041309.

Zhang, H., Pan, N.a, Ma, C., Liu, B., Xiu, L., Tong, H.e, Sheng, S., Liang, Y., Li, H., Ma, F., Bao, X., Hu, W., Wang, X., 2021. Vaccine composition formulated with a novel *Lactobacillus*-derived exopolysaccharides adjuvant provided high protection against *Staphylococcus aureus*. Vaccines 9 (7), 775. doi:10.3390/vaccines9070775.

Zhang, X., Fan, W.Y., Yao, M.C., Yang, C.W., Sheng, G.P., 2020. Redox state of microbial extracellular polymeric substances regulates reduction of selenite to elemental selenium accompanying with enhancing microbial detoxification in aquatic environments. Water Res 172. doi:10.1016/j.watres.2020.115538.

Zhang, Z., Cai, R., Zhang, W., Fu, Y., Jiao, N., 2017. A novel exopolysaccharide with metal adsorption capacity produced by a marine *Bacterium Alteromonas sp.* JL2810. Mar. Drugs 15 (6), 175. doi:10.3390/md15060175.

CHAPTER

25

Current regulations governing extracellular polysaccharide (EPS): Application and commercialization in food, pharmaceutical, and medical sectors

Tanmay Jit[a], *Abu Md Ashif Ikbal*[a], *Kesavan Bhaskar Reddy*[b], *Indira Mikkili*[c], *Onkar Nath Tiwari*[d] *and Partha Palit*[a]

[a]Department of Pharmaceutical Sciences, Assam University (A Central University), Silchar, Assam, India [b]Apollo Institute of Pharmaceutical Sciences, The Apollo University, Chittoor, Andhra Pradesh, India [c]Department of Biotechnology, Vignan's Foundation for Science, Technology and Research, Vignan University, Vadlamudi, Guntur, Andhra Pradesh, India [d]Centre for Conservation and Utilization of Blue Green Algae (CCUBGA), Division of Microbiology, ICAR-Indian Agricultural Research Institute (PUSA), New Delhi, India

25.1 Introduction

The structural, functional, and useful polysaccharides produced by eukaryotic and prokaryotic microbial cells can be homopolymers or heteropolymers in composition. These polymeric substances are known as extracellular polysaccharides (EPS). Homopolymer EPS molecules are self-possessed of repeating units of a single monosaccharide, usually fructose or glucose, while heteropolysaccharides are usually branched and calm of repeating units of multiple monosaccharides, primarily galactose, glucose, and fructose, along with other noncarbohydrate groups. In lower amounts, intracellular intermediates are converted into heteropolysaccharides,

DOI: https://doi.org/10.1016/B978-0-443-36619-2.00025-5

which are primarily associated with immunological control. "EPS" stands for "extracellular polysaccharides" as well as "extracellular polymeric substances." Only carbohydrates make up extracellular polysaccharides, whereas repeating units of proteins, nucleic acids, and monosaccharides make up extracellular polymeric substances. EPSs are extracellular polysaccharides formed by a variety of bacterial species. Among the various shapes they can take are capsular polysaccharides (CPS), also known as "cell-bound extracellular polymeric substances"(Costa et al., 2018). Numerous bacterial species have them on their outermost surface. Furthermore, the EPS can be purchased as "free EPS," or unbound. It suggested that the EPS matrix shielded the cells from the harsh environmental circumstances. Moreover, EPS is the main component of bacterial biofilms, which are multifunctional components of microbial communities that offer physiology, adhesion, structure, protection, and recognition (Dittrich and Sibler, 2013). EPS are associated with the proliferation of microbial cells on solid surfaces, which is a common phenomenon in nature. According to their structural and functional relationships with the cell, different kinds of microbial polysaccharides found outside of their cell walls can either demonstrate loose attachment or strong binding. Sutherland coined the term "EPS" in 1972 to describe these polysaccharides (Cerning, 1995). Due to the poly-anionic properties, it is possible to view EPS molecules with a light microscope either through harmful staining or special stains. Scanning electron microscopy (SEM) methods are commonly employed in order to identify exopolysaccharides, but SEM cannot differentiate the chemical properties of EPS. Transmission electron microscopy (TEM), on the other hand, can give important information about surface features. These properties that promote cell adherence to surfaces serve a number of functions, including protection against antimicrobial substances, desiccation protection in terrestrial bacteria, or protection against protozoan or phagocytic predation (Hibbing et al., 2010). Its industrial uses have been altered by its biochemical properties made it widely used in the food industry as thickening agents, gelling agents, adhesives, stabilizer and suspending agents (Moscovici, 2015). Diversity of chemical structure and distinctive properties of EPS has led to their application in biomedical, bioremediation, pharmaceutical, and culinary uses. EPS is an alternative stabilizer, viscous thickening, and gelling agent because it has a high viscosity in aqueous solutions (Meybodi and Mohammadifar, 2015). EPS molecules can be used as a catalyst due to their less toxicity and more sustainable production processes than chemically synthesized polymers (Silva et al., 2019). This is because of the capability of EPS molecules to establish flexible and durable films, which can be used to develop films or coating, which can help the packaging sector to increase the shelf life of fruits and other foods. Crucially, substituting chemical additions in food production processes can be accomplished with the least amount of capital cost.

Future investigations could explore comparative regulatory policies across regions to identify pathways for unified global standards in EPS approval. In addition, mechanistic studies linking EPS structural features to their regulatory classifications can improve safety evaluation and risk assessment.

25.1.1 Emphasize the need for regulations controlling the use of extracellular polysaccharides

25.1.1.1 Food sector

There are numerous microbial EPS that are employed as thickeners, stabilizers, or prebiotic additives, such as dextran, pullulan, gellan, and xanthan. Purified EPS must undergo a safety evaluation prior to approval because they frequently have no history of ingestion, making them subject to the EU's "novel food" law (Patel et al., 2022).For instance, the United States Food and Drug Administration (FDA) and European Food Safety Authority (EFSA) only issue generally recognized as safe (GRAS)/ qualified presumption of safety (QPS) certification to some strains and pure EPS (such as β-glucans and xanthan gum), not all of them. Regulations guard against the dangers of microbiological contamination, toxicity, and allergies (Lynch et al., 2018).

25.1.1.2 *Pharmaceutical sector*

EPS are being investigated extensively for usage as vaccine adjuvants, medication delivery, and nanoparticle coating due to their biocompatibility and bioactivity. Microbial fermentation may, however, result in the contamination of undesirable biomolecules, endotoxins, or poisons (Rana and Upadhyay, 2020). EPS is guaranteed to meet stringent requirements for purity, repeatability, and therapeutic safety by regulations including good manufacturing practices (GMP) and ICH (International Council for Harmonisation of Technical Requirements for Pharmaceuticals for Human Use) guidelines (Freitas et al., 2019). Without supervision, poor EPS formulations could jeopardize medication safety or diminish effectiveness.

25.1.1.3 *Medical sector*

Antimicrobial coatings, tissue engineering scaffold and wound healing hydrogels are under development as EPS. The reason behind the need to have regulatory systems is that interactions between EPS and immunological pathways can lead to inflammatory or immunogenic responses (Ladrat et al., 2014). Moreover, biofilms are produced with EPS and help to protect infections against host defense and drugs. This may result in the emergence of antimicrobial resistance when it is not managed properly (Hall-Stoodley et al., 2004).

In future research, EPS safety, metabolism, and immunological outcome would be evaluated using in-vitro and clinical models, and standardized genetically modified organism (GMO) validation and harmonized regulations should be improved to provide reproducible, labeling-transparent, biosafety, and controlled therapeutic application in food, pharmaceutical, and medical industries.

25.1.2 Rationality behind commercialization of extracellular polysaccharides

25.1.2.1 *Food sector*

Functional role in the food industry: EPS, like levan, kefiran, and curdlan, enhance the stability, viscosity, and rheology of dairy, baked goods, and beverages (Cerning, 1995).

Effects on nutrition and health: Some EPS have antioxidant, immunomodulatory, and cholesterol-lowering properties that support their usage in nutraceuticals (Welman and Maddox, 2003).

Market demand: There is a significant commercial push for EPS over synthetic stabilizers due to growing customer interest in "natural and safe additives."

25.1.2.2 *Pharmaceutical sector*

EPS as biopharmaceutical agents: Because of their biocompatibility and biodegradability, EPS, like pullulan, dextran, and alginate, are being investigated as drug delivery and gene therapy carriers (Rehm, 2010).

Immunological applications: Certain microbial EPS improve the efficacy of treatment by acting as immunostimulants and vaccine adjuvants.

Industrial rationale: EPS is economical for pharmaceutical research due to its scalable fermentation techniques and renewable microbial origin (Sutherland, 1998).

25.1.2.3 *Medical sector*

Biomaterials in regenerative medicine: Extracellular matrix is mimicked by EPS, which are utilized in tissue engineering and include bacterial cellulose, hyaluronic acid, and dextran derivatives (Jain et al., 2013).

Wound healing and antimicrobial use: EPS-based hydrogels have the potential to be antibacterial wound dressings that speed up tissue regeneration.

Clinical rationality: Their compatibility, biodegradability, and low toxicity support their commercialization in innovative therapeutic materials and medical devices (Nwodo et al., 2012).

Future research should enhance EPS production efficiency, consumer acceptance, and structural versatility while advancing sustainable synthesis for pharmaceutical and biomedical uses, enabling multifunctional drug delivery, regenerative scaffolds, and innovative nanocomposite-based medical materials for scalable, safe, and eco-friendly commercialization.

25.1.3 Extracellular polysaccharide composition and structure

To conduct a more in-depth study of the physicochemical properties of EPS molecules, it is crucial to understand their structure. Some of the various carbohydrates that comprise various exopolysaccharides are glucose, galactose, and rhamnose, among others, in varying proportions (Wingender and Flemming, 1999). It may contain organic and inorganic substances, including uranic acid, methyl esters, sulfates, and pyruvates, among other carbohydrates (Sutherland, 2001). The EPS of bacteria either are covalently bound to the microbial cell surface or loosely bound. Functional groups such as carboxyl and hydroxyl attach their heteropolysaccharides, which are normally acidic, making them highly attracted to metal ions such as Ca^{2+} and Zn^{2+}. Numerous studies indicated that EPS molecules were discharged into the environment during the microbes' developmental stage without exhibiting any long-term adherence to the microbial cell membrane. This characteristic sets them apart from structurally related polysaccharides known as CPS, which originate on the surfaces of microbial cells. The two classifications for EPS are homopolysaccharides, which comprise of a single type of monosaccharide and heteropolysaccharides which contain repeating units of many monosaccharides. Differences in branching areas, glycosidic connections, and chain length lead to different properties even if homopolysaccharides polymers have similar repeating monosaccharide units. The majority of the sugars in microbial exopolysaccharides are those commonly found in plants and animals, but their carbohydrate composition is known to vary widely. The production circumstances and substitute decoration are directly related to the composition of EPS and can either increase or decrease its endurance (Gupta and Diwan, 2017). Heteropolysaccharides are made up of multiple monosaccharide sugar residues and are produced by different glycosyltransferases working together (Årsköld et al., 2007). It is hard to accurately regulate the chemical composition of carbs through biochemical analysis and to choose downstream processes, like the extraction and separation of EPS molecules, because of the presence of several carbohydrates with low individual concentrations and their interactions. Furthermore, minimal or negligible EPS disturbance or alteration is required by the extraction process. For the correct usage of EPS, chemical moieties must be characterized using contemporary spectroscopic techniques. Levansucrase and dextransucrase, two extracellular sucrose enzymes, catalyze the formation of homopolysaccharides, including levan and dextran, while also releasing fructose monosaccharide into the culture (Zannini et al., 2016). Besides the fact that sugar nucleotide precursors are needed in the synthesis and formation of heteropolysaccharides such as xanthan and alginate, the associated enzymes play vital roles in internal housekeeping activities (Welman and Maddox, 2003).

Future structural clarification with the help of the highly sophisticated spectroscopic and computational techniques will help to learn more about the EPS-metal ion interactions and bioactivity. This knowledge will be used to customize the design of EPS to specific functionality in the operations of pharmaceuticals, food, and biomedical applications.

25.2 Types and sources of extracellular polysaccharides

25.2.1 Types of extracellular polysaccharide

EPS are classified based on their monomer composition, structure and origin. Many microbes synthesise these polymers and they are used in the development of biofilms, cell adhesion, protection and retention of nutrition.

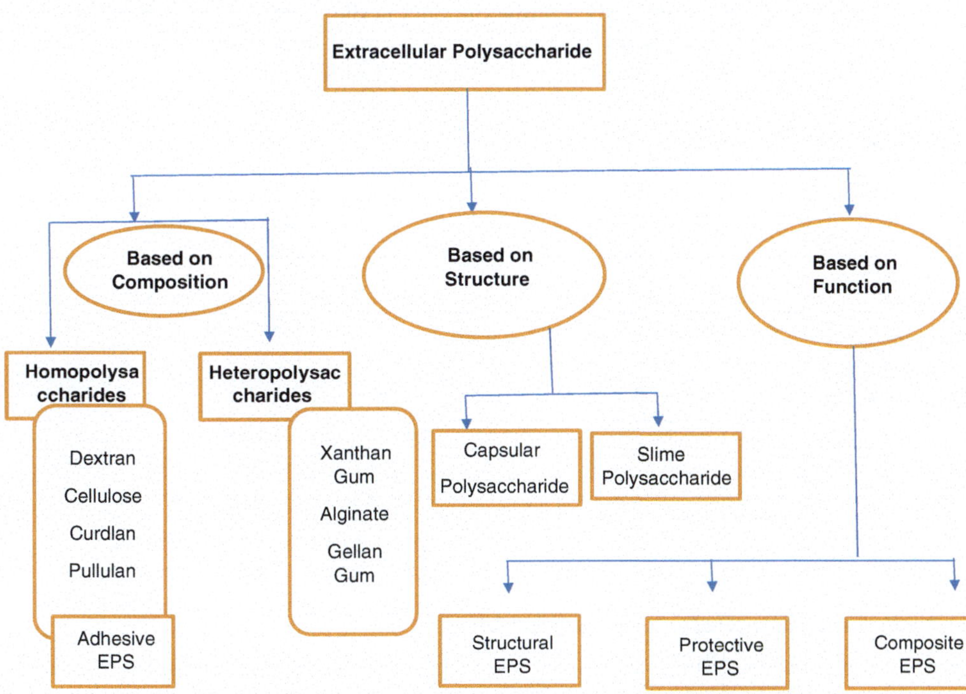

25.2.1.1 *Based on composition*

Homopolysaccharides—homopolysaccharides are polysaccharides consisting solely of a single type of monosaccharide unit repeated along the polymer chain. Although including just one kind of sugar, they can exhibit considerable variation in bond types, branching, and biological characteristics (Freitas et al., 2011).

25.2.1.2 *Heteropolysaccharides*

Heteropolysaccharides are intricate polysaccharides consisting of two or more distinct monosaccharide units organized in repetitive sequences. They are often branched, include diverse functional groups, and demonstrate a broad spectrum of biological and commercial applications (Sutherland, 2001).

Example: Xanthan gum, alginate, gellan gum

Xanthan gum: It is the most researched and well-known commercial microorganism and the first biopolymer to be made scientifically (Rosalam and England, 2006). This heteropolysaccharides is made up of a backbone of glucose, and trisaccharide side chains with acetyl and glucuronic acid residues, which is produced by bacteria in the genus *Xanthomonas* (Rottava et al., 2009).

Gellan gum: The genus *Sphingomonas* produces gellan, a heteropolysaccharides that is categorized by a common tetrasaccharide backbone made up of glucuronic acid and glucose. The side chain alignments and connections of gellan, welan, rhamsan, and diutan vary. (Coleman et al., 2008).

Alginate: Bacteria belonging to the genera *Azotobacterial* and *Pseudomonas* produce a linear polysaccharide made up of guuronic and mannuronic acids that create poly-guuronic acid and poly-mannuronic acid block structures and mixed sequences. The main difference between bacterial and algal alginates is that the latter is acetylated (Rehm, 2009).

25.2.2 Sources of extracellular polysaccharide

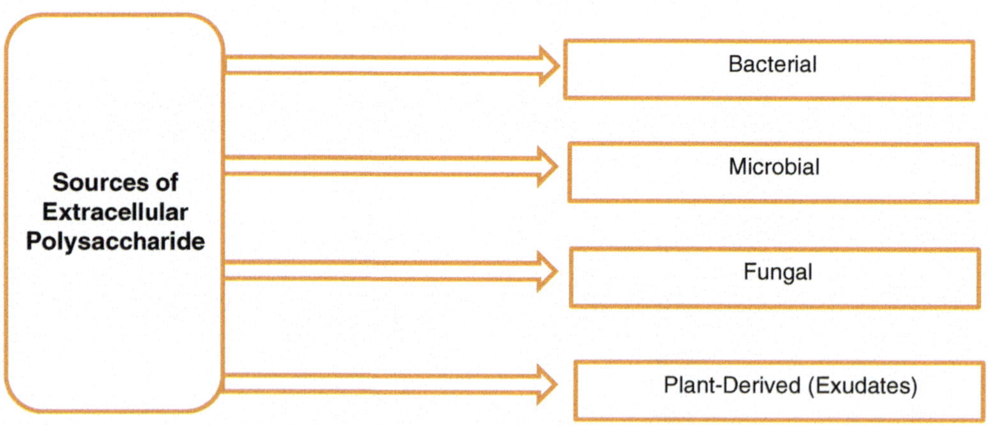

25.2.2.1 *Bacterial extracellular polysaccharides*

A diverse range of extracellular polymeric substances comprising extracellular polymeric materials, such as nucleic acids, proteins, and polysaccharides, with varying molecular weights and structural properties, are expelled by bacteria (Jayaratne et al., 1993).

The often-cited biological roles of bacterial extracellular polysaccharide encompass confrontation with and protection against both generic and particular host immunity and adhesion (Roberson and Firestone, 1992). For a number of harmful bacteria, CPS are important components of pathogenicity because they decrease human serum's complement-mediated bactericidal action. They may impede antibody opsonization and phagocytosis, facilitate tissue and surface colonization, and provoke inflammation and abnormal complement activation that might harm host tissues. The capsule's presence may substantially enhance a pathogen's environmental survival by serving as a permeability barrier that promotes selective nutrient transport while simultaneously offering protection against harmful substances, including antimicrobial agents (Beveridge and Fyfe, 1985).

25.2.2.1.1 Bacterial extracellular polysaccharides chemistry and their structures

Repeated monosaccharide units joined by glycosidic linkages make up bacterial polysaccharides. They can be homo- or heteropolymers, and heteropolysaccharides frequently have repeated units of oligosaccharides.

The high structural complexity, importance in microbial adhesion and biofilm formation, close relationship to bacterial pathogenicity, and role in host-bacteria interactions of bacterial cell external polysaccharides, such as capsular and lipopolysaccharides, drive efforts to recognize their production and export. (Woodward et al., 2010). The distribution and properties of bacterial polysaccharides greatly influenced by variables such as the food source, solution chemistry, and stage of cell development (Omoike and Chorover, 2004). Typically, linear, bacterial CPS consist of one to six regularly repeating monosaccharide subunits. They range in molecular weight from 100 to 1000 kDa.

The O-antigen is a polysaccharide with repeating units that range from two to more than one hundred and a core oligosaccharide that is embedded in the bacterial membrane (Strauss et al., 2009). Lipid A's chemical structure is highly conserved, consisting of a fatty acid-based conventional backbone associated with a 2-amino-2-deoxy-D-glucose β-1′, 6-linked disaccharide (D-glucosamine, D-GlcN). Ester or amide linkages are used to conjugate acids, particularly 3-hydroxyalkanoic acids.

The internal portion of the core oligosaccharide is more flexible than the lipid moiety and is frequently composed of L-glycero-D-manno-heptose (LD-Hep) and 3-deoxy-D-manno-oct-2-ulosonic acid (Kdo). The anionic properties of lipid A and the inner core region are impacted by KDO, phosphate, and, less frequently, hexuronic and other sugar acid residues. The phosphate groups frequently act as a link between other amino sugar moieties, 4-amino-4-deoxy-L-arabinose (L-Ara4N), or an amino alcohol (ethanolamine, Et_3N). The polymerized oligosaccharide units that make up the O-antigenic side chains vary greatly in both their chemical makeup and composition. Furthermore, the common has a great deal of inherent complexity and variation potential (Caroff and Karibian, 2003). In addition to hexoses alone, less frequent enantiomers and other stereoisomers can occasionally be added to the variety of EPS. Additional structural changes, such as esterification, positional and cumulative permutations of deoxy, amino, and carboxyl functions, further expand the diversity of these polysaccharides. Along with noncarbohydrate residues such as polyols, carboxylic acids, and amino acids, there is etherification and amidation. Furthermore, because every monosaccharide has a significant number of hydroxyl groups that can create glycosidic bonds, a wide variety of configurations can be formed between any two structural units. The O-antigen's structure is flexible and dependent on the bacterium's immediate surroundings. The O-antigen may be situated on the nonpolar areas of the cell surface, horizontally across the saturated fats and phospholipids of lipid A in the preferred configuration of lipopolysaccharides. (Flemming, 1995).

Future studies should elucidate bacterial EPS biosynthetic pathways and environmental adaptability using omics and molecular modeling tools. The knowledge of the structure-function dynamics would inform engineered polysaccharides to be used in antimicrobial resistance management, vaccine development, or specific biotechnological or pharmaceutical purposes.

25.2.2.2 Microbial extracellular polysaccharides

Microbial EPS finds wide applications in the culinary, pharmaceutical, medicinal, and bioremediation industries due to the variety of chemical structures and unique characteristics. Extracellular is very viscous in aqueous solutions, and thus it is commonly used as a stabilizing, thickening, as well as gelling factor. Gum's high viscosity and solubility make it a popular stabilizer and emulsifier in the food business. Additionally, certain varieties of EPS, like pullulan and levan, have recently been developed for use as food coatings. The recommended materials may enhance food shelf life owing to their antibacterial characteristics (Widyaningrum and Meindrawan, 2020).

25.2.2.2.1 Microbial extracellular polysaccharide structure and function

Microbial EPS are classified as homopolysaccharides and heteropolysaccharides according to their sugar composition. Homopolysaccharides consist of a single type of monosaccharide, whereas heteropolysaccharides are composed of 3–7 different monosaccharides. Conventional classification of EPS is also based on their microbial origin, chemical composition, molecular structure, and functional properties.

Different microbial species have different EPS compositions and structures. For example, complex heteropolysaccharides made up of six to ten monosaccharides make up the bulk of cyanobacterial EPS. Dextran, the lactic acid bacteria's extrecellular polysaccharide (EPS), is made entirely of glucose (Khalikova et al., 2005). Moreover, less than four monosaccharides make up the bulk of bacterial extracellular polymeric compounds (Czaczyk and Myszka, 2005). Additionally, these factors affect exopolysaccharide physical properties. The hexose homopolymers are neutral. An anionic charge is imparted by the elevated uronic acid content. EPS is hydrophobic due to the ester-linked acetyl groups and deoxy-sugar. Physical factors affect how EPS interacts with the cell surface, whereas the medium's organic macromolecules further determine the unique functional characteristics.

Future research should focus on engineering microbial EPS with tunable rheological and antimicrobial properties through genetic and process optimization. Elucidating structure–function relationships may expand applications in sustainable food packaging, targeted drug delivery, and eco-friendly bioremediation systems.

25.2.2.3 Fungal extracellular polysaccharides

Fungal EPS are high-molecular-weight biopolymers excreted by diverse fungi into their ambient environment. These chemicals are crucial for biofilm formation, cellular protection, adhesion, and regulation of host immunity, making them very relevant for biomedical, food, and pharmaceutical applications (Tsurkan et al., 2021).

25.2.2.3.1 Composition and structure of fungal extracellular polysaccharides

Fungal EPS are typically heteropolysaccharides, composed of different types of monosaccharides such as glucose, mannose, galactose, fucose, rhamnose, xylose, and uronic acids. These EPS may also contain noncarbohydrate substituents such as sulfate, phosphate, or proteins, contributing to their biological activity and structural diversity. The composition of fungal EPS differs according to species and environmental factors. They may be linear or branched, exhibiting various glycosidic linkages (e.g., β-1, 3; β-1, 6; α-1, 4; α-1, 6), which considerably affect their functional characteristics (Herrera, 1991).

Future studies should explore fungal EPS biosynthesis pathways and structural modifications influencing bioactivity. Advanced omics and material science approaches can optimize their biomedical, nutraceutical, and pharmaceutical applications while promoting sustainable large-scale production and therapeutic biocompatibility.

25.3 Extracellular polysaccharides regulations in the food industry

EPS have a number of commercial applications in a variety of industries, such as the food, pharmaceutical and agricultural biotechnology sectors, because of their distinct topologies and compatibility. Since their inception in the 1960s, microbial polymers have seen a substantial increase in commercial use (Kleber et al., 2015). The food, beverage, detergent, textile, pharmaceutical, cancer treatment, drug, and fermentation media formulation sectors have all made use of them. Petroleum, papermaking, coatings, biotechnology, and agriculture have all made use of them (Fenibo et al., 2019). Firmness and creaminess are crucial sensory attributes influencing customer preferences for dairy products (Mende et al., 2016). In order to strengthen the stiffness of the casein matrix, EPS may serve as texturizers and stabilizers by making the finished product more viscous and then coalescing water hydration with new milk components, such as proteins and micelles (Tabibloghmany and Ehsandoost, 2014). As a result, EPS can improve product quality and reduce the production of dangerous byproducts. Furthermore, it has been demonstrated that EPS significantly affects gut health. *S. thermophiles* that produce EPS can recover the functional qualities of mozzarella cheese, but they are not phage-resistant (Broadbent., 2003) (Table 25.1).

25.3.1 Applications of different extracellular polysaccharides in the food processing sectors

25.3.1.1 Dextran's

The alpha-linked D-glucopyranosyl repeating units that make up dextrans are microbial homopolysaccharides with a linear backbone. α-D-glucans, or dextrans, are distinguished by their key chain of α-(1-6) glycosidic bonds and varying concentrations of branching connections with α-(1-2), α-(1-3), or α-(1-4) (Fig. 25.1) (Nouvel et al., 2003). Its biodegradable and biocompatible qualities have won it praise from EPSs all over the world. Dextran was studied as a dietary supplement before the 1950s. Dextran was recently designated as a GRAS food additive by the FDA (Patel et al., 2010). Dextran is frequently used in a variety of food products for emulsification, texture alteration, gel formation, and viscosity increase. LAB dextran is used in the pharmaceutical and food manufacturing sectors (Leemhuis et al., 2013) (Table 25.2).

TABLE 25.1 Catalogue of primary extracellular polysaccharides together with their uses and roles in the food processing sector.

Type	Application	Role	References
Dextran	Freezing food as a potential prebiotic agent.	The inclusion of dextran does not compromise the product's viscosity and viscoelastic qualities.	Lopez et al. (2015)
Xanthan	Frozen goods drinks.	-Enhances the dependability of freeze-thaw cycles. -Alongside fibers, xanthan exhibited reduced degradation reactions with an effect of protection.	Palaniraj and Jayaraman (2011)
Pullulan	Improvement of the functionality of various polysaccharide-based edible films.	The increased viscosity and decreased drift activity index (n) of gel solutions were caused by the higher pullulan concentration. In addition to reducing weight loss, carotenoids, fruit, ascorbic acid degradation, and softening, the coatings also prevented the formation of mold. The 10% pullulan film is said to be high-quality and effective.	Eroglu et al. (2014)
Gellan	Emulsion-based gel beverages.	Elastic behavior was mostly displayed by gellan emulsions at high concentrations, producing flexible gels. Alkali, enzymes, or high temperatures can all partially diacylate native gellan during fermentation or other treatments.	Dreher (1987)
Curdlan	As a deterrent in deep oil frying.	Curdlan exhibits a linear influence on moisture transfer, and a reduction in oil was seen by the incorporation of curdlan. The impact of curdlan was probably attributable to its thermal gelling properties, which, during frying, may have functioned as a barrier to oil and moisture due to heat-induced gelation. The se of curdlan boosted the noodle's chewiness, adaptability, and durability.	Funami et al. (1999)

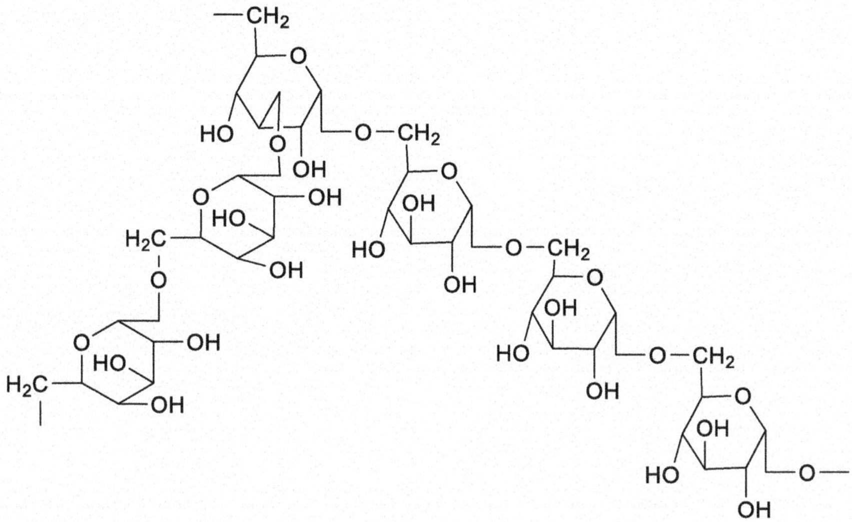

FIGURE 25.1 Chemical structure of dextran.

TABLE 25.2 Dextran uses and functions in the food industries.

Application	Function	References
Bakery	Improves freshness, texture, tenderness, crumb quality, and loaf volume.	Katina et al. (2009)
Ice cream	Function as a cryoprotectant.	Naessens et al. (2005)
Confectionary	Moisture preservation while inhibiting the formation of jelly candies, gelling agents in gum, and sugar crystallization.	Martins et al. (2000)
Frozen foods	Safeguarding against chemical modifications and oxidation, while preserving texture and taste.	Bhavani and Nisha (2010)
Fermented dairy products	Increases diminish syneresis, creaminess, and viscosity.	Mende et al. (2013)
Reduced-fat cheese	Improves water compulsory and enhances moisture content in the components without fat.	Awad et al. (2005)
Protein–dextran conjugates	Improves the Maillard reaction's ability to emulsify, froth, gel, and solubilize proteins.	Zhang et al. (2012)

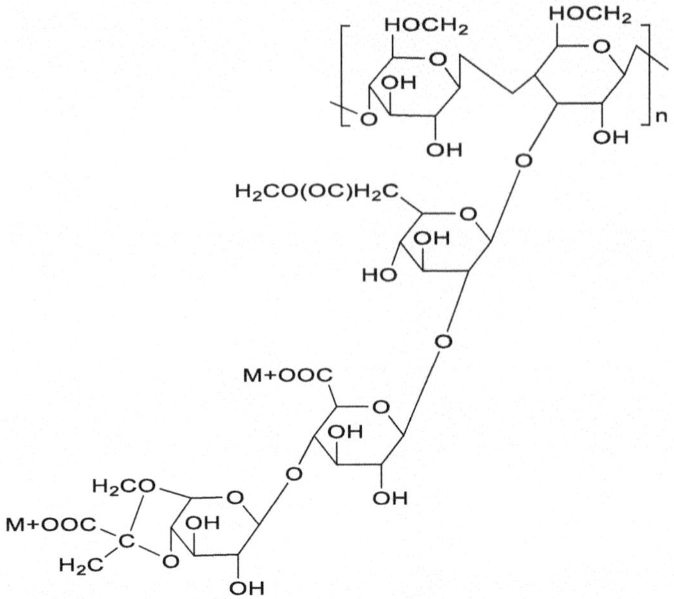

FIGURE 25.2 Chemical structure of xanthan gum.

25.3.1.2 *Xanthan*

Xanthan is a major industrial biopolymer and the second most marketable natural microbial polysaccharide, behind dextran. D-glucuronic acid, D-mannose, and D-glucose are the building components of this long-chain polymer (where acetylation occurs in one sugar residue, whereas the other sugar residue has pyruvate) with a high number of trisaccharide side chains at a molecular ratio of 3:3:2 (Fig. 25.2) (Bylaite et al., 2005). *Xanthomonas campestris* NRRL B-1459 produces xanthan gum, which has been extensively studied for its potential to complement other well-known natural and manufactured water-soluble gums. The significant commercial production of xanthan gum began in the first part of 1964. The safety and toxicity of xanthan gum in food and medicine have been extensively investigated (Jit et al., 2024). Xanthan, of course, is nontoxic and never stops developing. Because of its nonsensitizing properties, xanthan would not irritate the skin or eyes. In both acidic and alkaline environments, this polymer remains stable and is highly soluble in water. It is resistant to enzymes and incredibly resilient in salty environments. These qualities make xanthan gum one of the most significant polysaccharides used in the food processing industry today. (Casas et al., 2000) (Table 25.3).

25.3.1.3 *Pullulan*

In cultures of sugar and starch, *Aureobasidium* pullulans generates pullulan, a neutral EPS. Mostly composed of maltotriose repeating entities connected by α-(1, 6) bonds, it is a linear linked linkage (Fig. 25.3) (Karim et al., 2009).

The α-(1, 4) and α-(1, 6) bonds' frequent repetition promotes improved solubility and structural stability. For technical purposes, such as stabilizing emulsions with process aids or providing them with the inherent structure needed for supply or packing, EPS are commonly employed in food items (Gniewosz and Reinhard, 2008). The "consuming consistency" of the product is often enhanced by their use. Pullulan is used as dietary fiber in rats and humans

TABLE 25.3 Xanthan gum applications and functions in food processing industries.

Application	Function
Water-based gels	Gels for dessert, aspics
Dairy products	Yogurt, ice cream, milkshakes
Confectionery	Jelly candies, marshmallows
Beverages	Fruit juice, beverages made from plant proteins, alcoholic beverages, and beverages made from milk
Elderly food	Sweet, An elderly-friendly mousse with the ideal texture
Encapsulation	Tablets, glaze

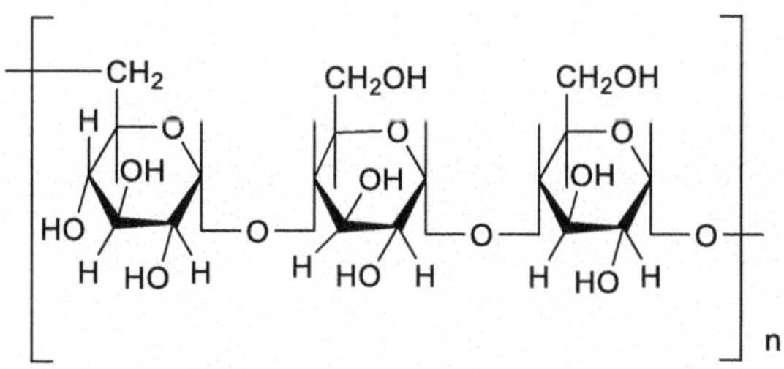

FIGURE 25.3 Chemical structure of pullulan.

and has a low-calorie content. Its resistance to mammalian amylases is the cause of it (Tiwari et al., 2021). According to the studies, nutritional pullulan can be used as a prebiotic to promote healthy Bifidobacterial.

Similar to Arabic gum, pullulan solutions are low viscosity. Pullulan can also be employed in place of starch in both liquid and solid foods, giving them the usual qualities of starch, such as consistency, density, and moisture retention (Kycia et al., 2020). Pullulan can also be used to partially substitute starch in pasta dishes or baked products. Because pullulan is not an easily digestible carbon source for the bacteria, molds, and fungi that cause food deterioration, it prolongs the shelf life of food. Dehydrating food delays deterioration because pullulan is better than starch at preserving water.

25.3.1.4 Gellan

The food processing sector makes extensive use of gellan, the newest biopolymer on the market, as a gelling agent. It follows a linear path. Two β-D-glucose residues, one of β-D-glucuronic acid, and one of α-L-rhamnose make up the repeating tetrasaccharide units of a 2:1:1 anionic polymer (Fig. 25.4) (Srisuk et al., 2018).

Natural polysaccharides have L-glycerides at O-2 from the third residue of the glucose moiety connected to the tetrasaccharide unit and one acetyl group at O-6 from the same residue in at least multiple glucose units (Okiror and Jones, 2012). Gellan gum's distinctive and creative

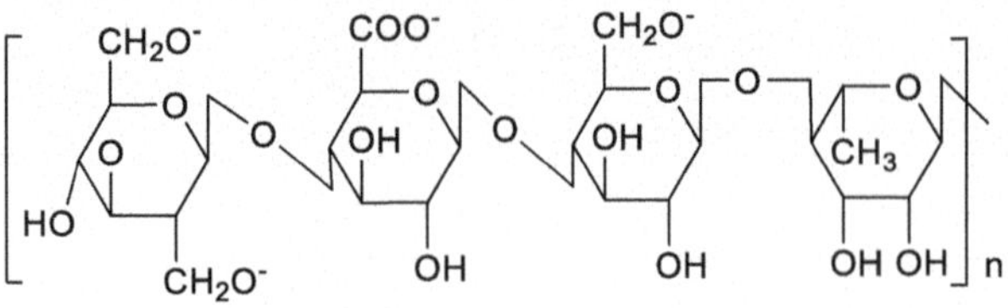

FIGURE 25.4 Chemical structure of gellan gum.

TABLE 25.4 Food applications and function of gellan gum.

Application	Function
Water-based gels	Gels for dessert, aspics
Dairy products	Cheese, flavored milk, milkshakes, ice cream, and yogurt
Confectionery	Jelly chocolates, marshmallows, and high stringy candies
Beverages	Fruit juice, beverages made from plant proteins, alcoholic beverages, and beverages made from milk
Jams, jellies	Agent for glazing, bakery fillings, and jellies

qualities make it useful in a wide range of industrial culinary applications. A few particular applications for gellan gum are covered in Table 25.4 under these general headings.

Future research should aim to harmonize international food safety standards for EPS use, integrating toxicological data and consumer risk assessments. Developing eco-friendly, traceable production systems can strengthen regulatory acceptance and ensure sustainable application of EPS in global food industries

25.4 Extracellular polysaccharides regulations in pharmaceutical applications

25.4.1 Drug/hydrogel controlled release systems

25.4.1.1 Nano-microsystems for long-term administration

Polysaccharides, including EPS, have emerged as extremely promising materials for "intelligent therapeutics" or "intelligent drug delivery systems." Owing to their ability to create hydrogels, these hydrophilic polymers are very potent drug carriers, which are cross-linked three-dimensional network structures that retain a significant amount of water while staying insoluble (Caldorera and Peppas, 2009). Since these gels are macroparticles, particularly nanoparticles, they can trick medications or biomolecules inside their internal structures, adsorb the therapeutic molecules onto their exterior surfaces, and enter target organs through cells and tissue gaps. Once they have arrived, the drug delivery systems' diverse properties—such as pH, ion, and temperature sensitivity, bioadhesion, and biodegradability—can exhibit varying drug rate-release characteristics. They can increase a drug's bioavailability and allow for lower doses with a correspondingly less harmful effect by prolonging its residence and, consequently, its therapeutically useful fraction. Due to their numerous reactive groups, varying molecular weights, and varied chemical composition, natural polysaccharides exhibit a wide variety of shapes and characteristics; they are also safe, nontoxic, and biodegradable (Liu et al., 2008).

25.4.1.2 *Amphiphilic extracellular polysaccharides as controlled delivery systems*

Amphiphilic polymers, which have an inner hydrophobic core and self-associate to form thermodynamically stable nanogel formations, are created when hydrophobic segments are grafted onto the hydrophilic polymeric backbone. These polymeric micelles are identified as prospective drug carriers due to their stable size over time and ability to capture hydrophobic medicines due to their hydrophobic core-shell structure (Popa, 2011). The most referenced study in this area was conducted in 1993 and laid the groundwork for possible future drug-release complexes. It examined how cholesteryl-bearing pullulan self-assembled to form stable hydrogel nanoparticles. Consequently, mitoxantrone was added to self-aggregating nanoparticles of pullulan modified with cholesterol and succinyl bonds (which showed a continuous release of the model anticancer) (Yang et al., 2010).

25.4.1.3 *Interpenetrating polymer networks*

Polysaccharide hydrogels have the potential to fulfil intriguing medical uses. These "smart" systems, which are made up of molecularly penetrating polymers, promise to be potential narcotics traffickers. It has been demonstrated that interpenetrating polymer network microspheres in the form of alginate and synthetic or natural polymers entrap and release weakly soluble antihypertensive pindolol, anticancer 5-fluorouracil, anti-inflammatory indomethacin, antibiotic Gemifloxacin, anticoagulant heparin, and NSAID sulindac over time through the intestinal route. (Matricardi et al., 2013).

Gellan gum and pectin beads ionotropically gelated with Al^{3+} as a crosslinker have demonstrated appropriate mucoadhesive and trapping qualities for enteric coated in-vitro controlled release of the anti-inflammatory drug. Ketoprofen. The antibiotic amoxicillin was captured and showed sustained release in-vitro by poly(ε-caprolactone)-grafted dextran nanoparticles (Prezotti et al., 2014). Recent EPS research largely ignores steroid hormones, which are significant pharmacological ingredients whose administration may be enhanced by prolonged delivery. These compounds' micro- and nanoparticles can achieve clinically significant concentrations at lower dosages when employed in transdermal drug delivery system (TDDS), avoiding a lot of the negative effects associated with first-pass metabolism, gastrointestinal absorption, and increased plasma levels following oral dosing. (Saldías et al., 2015).

25.4.2 Drug-targeting, macromolecule, and cell carriers

25.4.2.1 *Drug-targeting*

Polysaccharide conjugates with different levels of hydrophilization, especially the amphiphilic ones, showed unique binding and piercing properties to cellular receptors, showing promise as drug-targeting carriers. By hydrophobically interacting with beta-amyloid's beta oligomer forms, cholesteryl-pullulan nanoparticles can significantly reduce their toxicity. According to the findings, they might be utilized in combination with other therapies for neurologic conditions like Alzheimer's disease, where soluble hazardous aggregates develop (Boridy et at., 2009).

25.4.2.2 Anticancer drug targeting

Anticancer epirubicin-physically loaded self-assembled nanoparticles enhanced medication uptake in tumors while decreasing it in tumor-bearing mice's kidneys and hearts. They demonstrated a 70% decrease in tumor volume and a sustained release pattern. One identified (1—→3)-β-glucan that is hydrophobicized by deoxycholic residues is carboxymethyl curdlan (Gao et al., 2010). In-vitro and in-vivo study of human lung epithelial adenocarcinoma (A549), human breast adenocarcinoma (MCF-7) and SCC7 carcinoma were shown to respond to receptor-mediated, sustained, and targeted administration of the anticancer drug vorinostat via nanoparticles. (Choi and colleagues, 2009).

To accomplish target delivery of such anticancer and anti-inflammatory medications, a unique gellan nanohydrogel system was recently created that can simultaneously carry two poorly water-soluble pharmaceuticals (chemically linked prednisolone and physically entrapped paclitaxel). Their synergistic action on three different types of cancer cells produced encouraging in-vitro results (Tran et al., 2014).

In skin conditions, transdermal drug administration is a significant pharmaceutical option for drug targeting. Hyaluronic acid nanoemulsions demonstrated a successful selective proclamation of methylene-dioxycamptothecin, an inhibitor of keloid skin tumors, both in-vitro and in-vivo (D'Arrigo et al., 2014).

25.4.2.3 Recombinant macromolecular biopharmaceuticals

A quickly growing category of modern therapies, such as peptides, proteins, and small interfering RNA (siRNA), vaccines and hormones, is more effective than traditional small medications for treating serious and fatal illnesses as well as for diagnostic purposes. However, because of their propensity to swiftly be removed in the liver or other human tissues and to lose their native structure, they have trouble passing through biological membranes and mucosal surfaces. In addition to the commonly administered routes of administration—pulmonary, oral, and nasal polymeric hydrogels laboring as carriers could reduce their unsteadiness and enhance bioavailability. Insulin, a medication used to treat diabetes mellitus over the long term, is given particular consideration in this regard. Insulin was protected from enzymatic breakdown by cholesterol-bearing pullulan nanoparticles, and its action remained unaltered following intravenous infusion. In a simulated intestinal media, insulin was electrostatically linked to dextran sulfate/chitosan nanoparticles 85% of the time, indicating a potential oral administration method. After oral administration, dextran nanoparticles coated with vitamin B12 derivatives contained over 65% insulin and showed a sustained hypoglycaemic effect in diabetic mice and rats.

25.4.2.4 Gene delivery

To enhance transport and prevent lysosomal enzyme degradation, nonviral vectors—such as polymers—are favored for delivering nucleic acid molecules. Pullulan encapsulation of plasmids. The nanoparticles showed effective internalization (Gupta et al., 2004).

For delivering nucleic acid molecules, nonviral vectors such as polymers are favored because they enhance transport and prevent lysosomal enzyme degradation. The nanoparticles showed effective internalization when plasmids were encased in pullulan (Khan et al., 2012). Hepatocellular carcinoma Huh-7 cells were successfully transfected using cationic dextran derivative nanoparticles filled with gene-silencing siRNA and a photosensitizer. Additionally, PEG-ylated dextran was used to transfer the same genetic material to Huh 7 and A 431 human epithelial carcinoma cells (Raemdonck et al., 2010).

Future studies should standardize EPS-based drug delivery and targeting systems through GMP-aligned protocols and clinical validation. Integrating nanotechnology, biopolymer chemistry, and regulatory harmonization will advance safe, effective, and sustainable EPS therapeutics for controlled release, gene delivery, and personalized medicine applications.

25.5 Extracellular polysaccharides in medical applications

25.5.1 Bacterial cellulose

Depending on the strains or growing conditions, bacterial fermentation yields bacterial cellulose with a controlled three-dimensional structure. The resulting network of water-insoluble nanofibrils is highly similar to collagen and demonstrates remarkable biocompatibility (Lin et al., 2013). BC is used in wound healing and skin repair since it is a porous material and an exudate absorbent. Compared to plant cellulose, BC is more crystalline, has a greater water-absorbing capacity, and is more resilient to microbial or enzymatic degradation (Torres et al., 2012). To achieve certain objectives, its potential as a scaffold for wound healing and tissue engineering might be modified and enhanced by chemically changing it or combining it with other synthetic or mineral materials, biopolymers, or cell-growth agents. In clinical investigations, BC dressings fared better on both acute and chronic wounds than similar plant-cellulose dressings. BC Glycerine For dryness caused by skin disorders, including psoriasis and atopic dermatitis, membranes with a clinically proven moisturizing effect may be advantageous.

25.5.2 Vaccines

EPS may be useful as antigens or antigen carriers in vaccination formulations. Tetanus toxoid, or anatoxin, was successfully released from its entrapment on carboxymethyl pullulan microparticles hydrophobized using Poloxamer (Mocanu et al., 2011). Rabbits given alginate-tetanus toxoid microparticles intranasally experienced a robust immunoreaction. The intravenous administration of influenza hemagglutinin-containing hyaluronic acid esters H1N1 to mice, rabbits, and micropigs also proved effective. (Sharma et al., 2009) (Table 25.5 and 25.6).

Future research should advance biocompatible EPS scaffolds, vaccine adjuvants, and biomedical materials through genetic engineering and nanotechnology integration. Regulatory collaboration and clinical validation can enhance safety, scalability, and therapeutic efficiency of EPS-based wound dressings, implants, and drug-loaded medical devices.

25.6 Challenges and gaps in regulation

25.6.1 Different rules in different countries

EPS regulations differ greatly between nations and areas, which makes global commerce and commercialization challenging. In a similar vein, China and Japan have different systems for evaluating and labeling food additives. Particularly for businesses that export EPS-based goods internationally, these regulatory disparities can cause delays in market entry and raise compliance expenses (Freitas et al., 2011).

TABLE 25.5 Extracellular polysaccharides with recognized viable medical uses.

EPS	Monomer composition	Main producing microorganism	Applications	References
Dextran	Glucose	*Leuconostoc mesenteroides*	Expander for blood plasma volume (regulating wound shock).	Amspacher and Curreri (1953)
Xanthan	Mannose, glucuronic acid, acetate, pyruvate	*Xanthomonas campestris*	Pharmaceutical creams and suspensions use thickeners and suspension stabilizers.	Morris and Harding (2009)
Alginate	Mannuronic acid, acetate	*Azotobacterial vineland*, *Pseudomonas aeruginosa*	Tablets' disintegrating agent, pharmaceutical suspensions' thickening and stabilizer; and Dental impressions, emulsions, stomach protectors that prevent reflux, drug delivery microspheres, and fibers in bandages and hemostatic dressings for wounds.	McHugh (1987)
Gellan	Rhamnose, glycerate	*Sphingomonas paucimobilis* (formerly *Pseudomonas elodea*)	Excipient for sustained/controlled tablet disintegration in oral, ophthalmic, and nasal medication formulations release.	Dumitriu (1994)
Hyaluronic acid/hyaluronan	N-acetyl-glucosamine	*Streptococcus equisimilis/ zooepidemicus; Bacillus subtilis* (recomb.Str. *equisimilis*)	Treatment of osteoarthritis (intraarticular injection); eye surgery; persistent, challenging wound healing (vitreous replacement or substitution).	Kogan et al. (2007)

25.6.2 Confusion arises when extracellular polysaccharides is used across multiple sectors

Pharmaceuticals (as excipients or drug delivery vehicles), cosmetics (as moisturizing agents), medical devices (as biomaterials), and food (as stabilizers and thickeners) all use EPS. Nonetheless, distinct regulatory procedures and product classification standards apply to each sector. For example, the same EPS might be used as a medical excipient in one product and as a food additive in another, necessitating different certifications and compliance records. Manufacturers

TABLE 25.6 Patented and commercialized EPS-derived products with food, pharmaceutical and medical sector.

EPS (source/type)	Category	Brand/Patent	Application
Xanthan gum (microbial, *Xanthomonas campestris*)	Food	Keltrol (Tate and Lyle/CP Kelco	Superior thickening and stabilizing agent for dressings, sauces, and drinks.
Gellan gum (microbial, *Sphingomonas elodea*)	Food	KELCOGEL (Tate and Lyle/CP Kelco)	Low-acyl gelling agent that gives suspension/clarity in beverages and desserts.
Pullulan (microbial, *Aureobasidium pullulans*)	Pharma / Nutraceutical delivery	Plantcaps (Lonza/Capsugel)	Pullulan-based hard vegetarian capsules without starch (high O_2 barrier).
Dextran 70 (microbial, *Leuconostoc* spp.)	Pharmaceutical (OTC/Rx)	Tears Naturale II (Alcon)	Dextran 70 plus Hypromellose are artificial tears used to lubricate dry eyes.
Hyaluronan (fermentation-derived HA derivatives)	Medical device (viscosupplement)	Synvisc/Synvisc-One (hylan G-F 20)	An intra-articular injection of a crosslinked HA derivative is authorized to treat knee OA discomfort.
Hyaluronan (NA-based HA gels)	Medical device (dermal fillers)	Restylane family	FDA-approved stable HA gel implants for lip augmentation and wrinkle repair.
Hyaluronan (NA-based HA gels)	Medical device (dermal fillers)	Juvéderm family	Nonanimal HA fillers (like NLFs) that are FDA-approved for mid-to-deep dermal injection.

EPS, extracellular polysaccharides.

and regulators become confused by this multisectoral use, which frequently results in clearance delays and unclear labeling and marketing claims (Wang et al., 2020).

25.6.3 Purity and stability problems with natural extracellular polysaccharides

Because of variations in microbial strains, fermentation techniques, and downstream purification, naturally produced EPS frequently show diversity. Impurities such as endotoxins, leftover proteins, and fermentation by-products might arise from such variability, compromising the safety and quality of the final product. Additionally, EPS's sensitivity and structural complexity make them vulnerable to deterioration during processing and storage, which makes ensuring their stability and shelf life more difficult. Naturally manufactured EPS finds it difficult to comply with regulatory bodies' stringent requirements for purity, stability, and reproducibility (Nair et al., 2020).

25.6.4 The need for common global rules

One major obstacle to innovation and knowledge transfer is the lack of unified international regulations for EPS. Although the Codex Alimentarius offers general guidelines for food safety, there are no consistent international standards that particularly address the use of EPS in the food, pharmaceutical, and medical industries. Companies are forced to apply for several approvals for the same product in several regions due to this lack of harmonization, which raises costs and delays commercialization. Harmonized international frameworks could speed up EPS-based inventions, assist small businesses, and simplify approvals (Rezoagli et al., 2023).

25.6.5 Difficulty for start-ups due to strict regulations

Significant financial and technical resources, such as thorough safety evaluations, stability studies, and manufacturing process documentation, are frequently needed for regulatory compliance. Small businesses and startups usually lack the funding and regulatory know-how needed to handle these intricate frameworks, especially when their products are intended for several foreign markets. Long approval processes and the expense of regulations deter entrepreneurship, which hinders innovation and the creation of new EPS-based technologies (Rehm and Moradali, 2021).

25.7 Conclusion

The regulatory environment surrounding EPS in the food, pharmaceutical, and medical industries is changing, although it is still disjointed. Frameworks like the FDA's GRAS system and the EFSA's additive approval procedures are advantageous for food applications, but because they directly affect human health, pharmaceutical and medical applications of EPS are subject to more stringent scrutiny. Businesses that use the same EPS in several applications face complications because of variations in sector-specific categories, such as food additive, excipient, or medical device component. Furthermore, there are issues with consistency, stability, and purity with organically generated EPS, which frequently makes it more difficult to comply with safety and quality regulations. The lack of internationally standardized standards makes worldwide commercialization much more difficult, especially for creative businesses with limited funding to handle various complex laws. A foundation is provided by initiatives like the Codex Alimentarius, although there is currently little harmonization across sectors. The full promise of EPS in biomedical materials, enhanced medicine delivery, and sustainable agricultural products requires a uniform, risk-based global regulatory framework. Along with expediting the clearance process, such an organization would promote international trade, innovate, and guarantee consumer safety across all applications.

Credit authorship contribution statement

Tanmay Jit: Writing, formal analysis; Abu Md Ashif Ikbal: Formal analysis, software; Bhaskar Reddy: Investigation, visualization; M. Indira: Validation; Onkar Nath Tiwari: Write-up-review and editing, data curation, formal analysis, investigation; Partha Palit: Visualization, validation, conceptualization, supervision, write-up-review, editing, methodology, project administration, resources.

Declaration of generative AI and AI-assisted technologies in the writing process of chapter

During the end process after drafting of the chapter, the author used Grammarly, an AI-assisted editing tool for language correction, specifically to improve spelling, grammar, word selection, readability, and clarity. After using this tool, the authors re-examined and revised the text as required.

Conflict of interest

The authors declare no conflict of interest.

Ethical statement

Not applicable

Acknowledgments

Dr. Partha Palit and Dr. O.N. Tiwari are highly grateful to the authorities Assam University, Silchar and ICAR-IARI, New Delhi, for their continuous encouragement, support, and facilities for preparing the chapters.

References

Amspacher, W.H., Curreri, A.R., 1953. Use of dextran in control of shock resulting from war wounds. AMA Arch. Surg. 66 (6), 730–740. https://doi.org/10.1001/archsurg.1953.01260030742009.

Awad, S., Hassan, A.N., Halaweish, F., 2005. Application of exopolysaccharide-producing cultures in reduced-fat cheddar cheese: Composition and proteolysis. J. Dairy Sci. 88 (12), 4195–4420. https://doi.org/10.3168/jds.S0022-0302(05)73105-2.

Årsköld, E., Svensson, M., Grage, H., Roos, S., Rådström, P., van Niel, E.W., 2007. Environmental influences on exopolysaccharide formation in *Lactobacillus reuteri* ATCC 55730. Int. J. Food Microbiol. 116 (1), 159–167. https://doi.org/10.1016/j.ijfoodmicro.2006.12.031.

Beveridge, T.J., Fyfe, W.S., 1985. Metal fixation by bacterial cell walls. Can. J. Earth Sci. 22 (12), 1893–1898. https://doi.org/10.1139/e85-199.

Bhavani, A.L., Nisha, J., 2010. Dextran—The polysaccharide with versatile uses. Int. J. Pharm. Biol. Sci. 1 (4), 569–573.

Boridy, S., Takahashi, H., Akiyoshi, K., Maysinger, D., 2009. The binding of pullulan-modified cholesteryl nanogels to Aβ oligomers and their suppression of cytotoxicity. Biomaterials 30 (29), 5583–5591. https://doi.org/10.1016/j.biomaterials.2009.06.055.

Broadbent, J.R., McMahon, D.J., Welker, D.L., Oberg, C.J., Moineau, S., 2003. Biochemistry, genetics, and applications of exopolysaccharide production in *Streptococcus thermophilus*: A review. J. Dairy Sci. 86 (2), 407–423. https://doi.org/10.3168/jds.S0022-0302(03)73619-4.

Bylaite, E., Adler-Nissen, J., Meyer, A.S., 2005. Effect of xanthan on flavor release from thickened viscous food model systems. J. Agric. Food Chem. 53 (9), 3577–3583. https://doi.org/10.1021/jf048098i.

Caldorera-Moore, M., Peppas, N.A., 2009. Micro- and nanotechnologies for intelligent and responsive biomaterial-based medical systems. Adv. Drug. Deliv. Rev. 61 (15), 1391–1401. https://doi.org/10.1016/j.addr.2009.09.004.

Caroff, M., Karibian, D., 2003. Structure of bacterial lipopolysaccharides. Carbohydr. Res. 338 (23), 2431–2447. https://doi.org/10.1016/j.carres.2003.07.010.

Casas, J.A., Santos, V.E., García-Ochoa, F., 2000. Xanthan gum production under several operational conditions: Molecular structure and rheological properties. Enzyme Microb. Technol. 26 (2–4), 282–291. https://doi.org/10.1016/S0141-0229(99)00159-9.

Cerning, J., 1995. Production of exopolysaccharides by lactic acid bacteria and dairy propionibacteria. Le Lait 75 (4–5), 463–472. https://doi.org/10.1051/lait:1995546.

Choi, K.Y., Min, K.H., Na, J.H., Choi, K., Kim, K., Park, J.H., Kwon, I.C., Jeong, S.Y., 2009. Self-assembled hyaluronic acid nanoparticles as a potential drug carrier for cancer therapy. J. Mater. Chem. 19 (24), 4102–4107. https://doi.org/10.1039/b901213f.

Coleman, R.J., Patel, Y.N., Harding, N.E., 2008. Identification and organization of genes for diutan polysaccharide synthesis from *Sphingomonas* sp. ATCC 53159. J. Ind. Microbiol. Biotechnol. 35 (4), 263–274. https://doi.org/10.1007/s10295-008-0302-8.

Costa, O.Y., Raaijmakers, J.M., Kuramae, E.E., 2018. Microbial extracellular polymeric substances: Ecological function and impact on soil aggregation. Front. Microbiol. 9, 1636. https://doi.org/10.3389/fmicb.2018.01636.

Czaczyk, K., Myszka, K., 2007. Biosynthesis of extracellular polymeric substances (EPS) and its role in microbial biofilm formation. Pol. J. Environ. Stud. 16 (6), 799–806.

D'Arrigo, G., Navarro, G., Di Meo, C., Matricardi, P., Torchilin, V., 2014. Gellan gum nanohydrogel containing anti-inflammatory and anti-cancer drugs. Eur. J. Pharm. Biopharm. 87 (1), 208–216. https://doi.org/10.1016/j.ejpb.2013.12.017.

Delbarre-Ladrat, C., Sinquin, C., Lebellenger, L., Zykwinska, A., Colliec-Jouault, S., 2014. Exopolysaccharides produced by marine bacteria and their applications as novel biomaterials. Mar. Drugs 12 (1), 233–253. https://doi.org/10.3390/md12010233.

Dittrich, M., Sibler, S., 2010. Calcium carbonate precipitation by cyanobacterial polysaccharides. Geomicrobiol. J. 27 (6–7), 491–502. https://doi.org/10.1080/01490451003703012.

Dreher, M.L., 1987. Handbook of Dietary Fiber: An Applied Approach. Marcel Dekker, New York, NY, USA.

Dumitriu, S., 1994. Polymeric Biomaterials. Marcel Dekker, New York, NY, USA.

Eroglu, E., Torun, M., Dincer, C., Topuz, A., 2014. Influence of pullulan-based edible coating on some quality properties of strawberry during cold storage. Packag. Technol. Sci. 27 (10), 831–838. https://doi.org/10.1002/pts.2069.

Fenibo, E.O., Ijoma, G.N., Selvarajan, R., Chikere, C.B., 2019. Microbial surfactants: The next generation multifunctional biomolecules. Microorganisms 7 (11), 581. https://doi.org/10.3390/microorganisms7110581.

Flemming, H.C., 1995. Biofouling und Biokorrosion—die Folgen unerwünschter Biofilme. Chemie Ingenieur Technik 67 (11), 1425–1430. https://doi.org/10.1002/cite.330671104.

Freitas, F., Alves, V.D., Reis, M.A., 2011. Advances in bacterial exopolysaccharides: From production to biotechnological applications. Trends Biotechnol. 29 (8), 388–398. https://doi.org/10.1016/j.tibtech.2011.03.008.

Freitas, F., Alves, V.D., Reis, M.A.M, 2019. Advances in bacterial exopolysaccharides: From production to biotechnological applications. Trends Biotechnol. 37 (1), 18–34. https://doi.org/10.1016/j.tibtech.2018.09.002.

Funami, T., Funami, M., Tawada, T., Nakao, Y., 1999. Decreasing oil uptake of doughnuts during deep-fat frying using curdlan. J. Food Sci. 64 (5), 883–888. https://doi.org/10.1111/j.1365-2621.1999.tb15930.x.

Gao, F., Li, L., Zhang, H., Yang, W., Chen, H., Zhou, J., Zhou, Z., Wang, Y., Cai, Y., Li, X., Liu, L., 2010. Deoxycholic acid modified-carboxymethyl curdlan conjugate as a novel carrier of epirubicin: In vitro and in vivo studies. Int. J. Pharm. 392 (1–2), 254–260. https://doi.org/10.1016/j.ijpharm.2010.03.032.

Gniewosz, M., Duszkiewicz-Reinhard, W., 2008. Comparative studies on pullulan synthesis, melanin synthesis and morphology of white mutant *Aureobasidium pullulans* B-1 and parent strain Ap-3. Carbohydr. Polym. 72 (3), 431–438. https://doi.org/10.1016/j.carbpol.2007.09.004.

Gupta, M., Gupta, A.K., 2004. Hydrogel pullulan nanoparticles encapsulating pBUDLacZ plasmid as an efficient gene delivery carrier. J. Controlled Release 99 (1), 157–166. https://doi.org/10.1016/j.jconrel.2004.06.003.

Gupta, P., Diwan, B., 2017. Bacterial exopolysaccharide mediated heavy metal removal: A review on biosynthesis, mechanism and remediation strategies. Biotechnology Reports 13, 58–71. https://doi.org/10.1016/j.btre.2016.12.006.

Hibbing, M.E., Fuqua, C., Parsek, M.R., Peterson, S.B., 2010. Bacterial competition: Surviving and thriving in the microbial jungle. Nat. Rev. Microbiol. 8 (1), 15–25. https://doi.org/10.1038/nrmicro2259.

Hall-Stoodley, L., Costerton, J.W., Stoodley, P., 2004. Bacterial biofilms: From the natural environment to infectious diseases. Nat. Rev. Microbiol. 2 (2), 95–108. https://doi.org/10.1038/nrmicro821.

Jain, R., Raghukumar, S., Tharanathan, R., Bhosle, N.B., 2013. Extracellular polysaccharide production by marine bacteria. BioMed. Res. Int. 2013, 1–13. https://doi.org/10.1155/2013/762975.

Jayaratne, P.A.D.M., Keenleyside, W.J., MacLachlan, P.R., Dodgson, C., Whitfield, C., 1993. Characterization of rcsB and rcsC from *Escherichia coli* O9:K30:H12 and examination of the role of the rcs regulatory system in expression of group I capsular polysaccharides. J. Bacteriol. 175 (17), 5384–5394. https://doi.org/10.1128/jb.175.17.5384-5394.1993.

Jit, T., Roy, S.D., Shil, D., Chakraborty, J., Paul, A., Roy, S., De, D., 2024. The potential of tannins from medicinal plants as anti-cancer agents. J. Cancer Res. Ther.,.

Karim, M.R., Lee, H.W., Kim, R., Ji, B.C., Cho, J.W., Son, T.W., Oh, W., Yeum, J.H., 2009. Preparation and characterization of electrospun pullulan/montmorillonite nanofiber mats. Carbohydr. Polym. 78 (2), 336–342. https://doi.org/10.1016/j.carbpol.2009.04.019.

Katina, K., Maina, N.H., Juvonen, R., Flander, L., Johansson, L., Virkki, L., Tenkanen, M., Laitila, A., 2009. In situ production and analysis of *Weissella confusa* dextran in wheat sourdough. Food Microbiol. 26 (7), 734–743. https://doi.org/10.1016/j.fm.2009.07.007.

Khalikova, E., Susi, P., Korpela, T., 2005. Microbial dextran-hydrolyzing enzymes: Fundamentals and applications. Microbiol. Mol. Biol. Rev. 69 (2), 306–325. https://doi.org/10.1128/MMBR.69.2.306-325.2005.

Khan, W., Hosseinkhani, H., Ickowicz, D., Hong, P.D., Yu, D.S., Domb, A.J., 2012. Polysaccharide gene transfection agents. Acta Biomater. 8 (12), 4224–4232. https://doi.org/10.1016/j.actbio.2012.07.014.

Kleber, M., Eusterhues, K., Keiluweit, M., Mikutta, C., Mikutta, R., Nico, P.S., 2015. Mineral–organic associations: Formation, properties, and relevance in soil environments. Adv.n Agron. 130, 1–140. https://doi.org/10.1016/bs.agron.2014.10.005.

Kogan, G., Šoltés, L., Stern, R., Gemeiner, P., 2007. Hyaluronic acid: A natural biopolymer with a broad range of biomedical and industrial applications. Biotechnol. Lett. 29 (1), 17–25. https://doi.org/10.1007/s10529-006-9219-z.

Kycia, K., Chlebowska-Śmigiel, A., Szydłowska, A., Sokół, E., Ziarno, M., Gniewosz, M., 2020. Pullulan as a potential enhancer of *Lactobacillus* and *Bifidobacterium* viability in synbiotic low fat yoghurt and its sensory quality. LWT 128, 109414. https://doi.org/10.1016/j.lwt.2020.109414.

Leemhuis, H., Pijning, T., Dobruchowska, J.M., van Leeuwen, S.S., Kralj, S., Dijkstra, B.W., Dijkhuizen, L., 2013. Glucansucrases: Three-dimensional structures, reactions, mechanism, α-glucan analysis and implications in biotechnology. J. Biotechnol. 163 (2), 250–272. https://doi.org/10.1016/j.jbiotec.2012.06.037.

Lynch, K.M., Zannini, E., Arendt, E.K., 2018. Microbial exopolysaccharides in foods and beverages: Current applications and future trends. Crit. Rev. Food Sci. Nutr. 58 (3), 334–366. https://doi.org/10.1080/10408398.2016.1176477.

Lin, S.P., Loira Calvar, I., Catchmark, J.M., Liu, J.R., Demirci, A., Cheng, K.C., 2013. Biosynthesis, production and applications of bacterial cellulose. Cellulose 20 (5), 2191–2219. https://doi.org/10.1007/s10570-013-9994-3.

Liu, Z., Jiao, Y., Wang, Y., Zhou, C., Zhang, Z., 2008. Polysaccharides-based nanoparticles as drug delivery systems. Adv. Drug. Deliv. Rev. 60 (15), 1650–1662. https://doi.org/10.1016/j.addr.2008.09.001.

Lopez, E.C., Champion, D., Blond, G., Le Meste, M., 2005. Influence of dextran, pullulan and gum arabic on frozen sucrose solution properties. Carbohydr. Polym. 59 (1), 83–91. https://doi.org/10.1016/j.carbpol.2004.08.020.

Martins, S.I., Jongen, W.M., Van Boekel, M.A., 2000. A review of Maillard reaction in food and implications to kinetic modelling. Trends Food Sci. Technol. 11 (9–10), 364–373. https://doi.org/10.1016/S0924-2244(01)00022-X.

Matricardi, P., Di Meo, C., Coviello, T., Hennink, W.E., Alhaique, F., 2013. Interpenetrating polymer network polysaccharide hydrogels. Adv. Drug. Deliv. Rev. 65 (9), 1172–1187. https://doi.org/10.1016/j.addr.2013.04.002.

McHugh, D.J., 1987. Production and Utilization of Products From Commercial Seaweeds. Food and Agriculture Organization of the United Nations, Rome, Italy.

Mende, S., Peter, M., Bartels, K., Dong, T., Rohm, H., Jaros, D., 2013. Concentration dependent effects of dextran on acid milk gels. Carbohydr. Polym. 98 (2), 1389–1396. https://doi.org/10.1016/j.carbpol.2013.07.022.

Mende, S., Rohm, H., Jaros, D., 2016. Influence of exopolysaccharides on yoghurt texture and stability. Int. Dairy J. 52, 57–71. https://doi.org/10.1016/j.idairyj.2015.08.002.

Meybodi, N.M., Mohammadifar, M.A., 2015. Effect and application of bacterial exopolysaccharide. J. of Food Qual. Hazards Control 2, 112–117.

Mocanu, G., Mihai, D., Dulong, V., Picton, L., Le Cerf, D., Moscovici, M., 2011. Anionic polysaccharide hydrogels with thermosensitive properties. Carbohydr. Polym. 83 (1), 52–59. https://doi.org/10.1016/j.carbpol.2010.07.010.

Morris, G., Harding, S., 2009. Polysaccharides, microbial. In: Schaechter, M. (Ed.), Encyclopedia of Microbiology. Elsevier, pp. 482–494. https://doi.org/10.1016/B978-012373944-5.00288-9.

Moscovici, M., 2015. Present and future medical applications of microbial exopolysaccharides. Front. Microbiol. 6, 1012. https://doi.org/10.3389/fmicb.2015.01012.

Naessens, M., Cerdobbel, A.N., Soetaert, W., Vandamme, E.J., 2005. *Leuconostoc dextransucrase* and dextran: Production, properties and applications. J. Chem. Technol. Biotechnol. 80 (8), 845–860. https://doi.org/10.1002/jctb.1285.

Nair, M.S., Nair, D.V., Johny, A.K., Venkitanarayanan, K., 2020. Use of food preservatives and additives in meat and their detection techniques. In: Al-Marzooq, A.A., Hossain, M.W., Barros-Velázquez, D.C. (Eds.), Meat Quality Analysis. Academic Press, Cambridge, MA, USA, pp. 187–213. https://doi.org/10.1016/B978-0-12-819233-7.00013-6.

Nouvel, C., Dubois, P., Dellacherie, E., Six, J.L., 2003. Silylation reaction of dextran: Effects of conditions. Biomacromolecules 4 (5), 1443–1450. https://doi.org/10.1021/bm0340608.

Nwodo, U.U., Green, E., Okoh, A.I., 2012. Bacterial exopolysaccharides: Functionality and prospects. Int. J. Mol. Sci. 13 (11), 14002–14015. https://doi.org/10.3390/ijms131114002.

Okiror, G.P., Jones, C.L., 2012. Effect of temperature on dielectric properties of low acyl gellan gel. J. Food Eng. 113 (1), 151–155. https://doi.org/10.1016/j.jfoodeng.2012.05.020.

Omoike, A., Chorover, J., 2004. Spectroscopic study of extracellular polymeric substances from *Bacillus subtilis*. Biomacromolecules 5 (4), 1219–1230. https://doi.org/10.1021/bm0345160.

Palaniraj, A., Jayaraman, V., 2011. Production, recovery and applications of xanthan gum. J. Food Eng. 106 (1), 1–12. https://doi.org/10.1016/j.jfoodeng.2011.03.035.

Patel, S., Kasoju, N., Bora, U., Goyal, A., 2010. Structural analysis and biomedical applications of dextran. Bioresour. Technol. 101 (17), 6852–6855. https://doi.org/10.1016/j.biortech.2010.03.125.

Patel, S., Majumder, A., Goyal, A., 2022. Potentials of exopolysaccharides from lactic acid bacteria. Biotechnol. Rep. 33, e00714. https://doi.org/10.1016/j.btre.2021.e00714.

Popa, V., 2011. Polysaccharides in medicinal and pharmaceutical applications. Med. Mater. Sci.

Prezotti, F.G., Cury, B.S.F., Evangelista, R.C., 2014. Mucoadhesive beads of gellan gum/pectin. Carbohydr. Polym. 113, 286–295. https://doi.org/10.1016/j.carbpol.2014.06.085.

Raemdonck, K., Naeye, B., Høgset, A., Demeester, J., De Smedt, S.C., 2010. Prolonged gene silencing by siRNA nanogels. J. Controlled Release 145 (3), 281–288. https://doi.org/10.1016/j.jconrel.2010.04.027.

Rana, S., Upadhyay, L.S.B., 2020. Microbial exopolysaccharides: Synthesis pathways, types and their commercial applications. Int. J. Biol. Macromol. 157, 577–595. https://doi.org/10.1016/j.ijbiomac.2020.04.084.

Rehm, B.H. (Ed.), 2009. Alginates: Biology and Applications. Springer, The Netherlands. https://doi.org/10.1007/b103816.

Rehm, B.H.A., 2010. Bacterial polymers: Biosynthesis, modifications and applications. Nat. Rev. Microbiol. 8 (8), 578–592. https://doi.org/10.1038/nrmicro2354.

Rehm, B.H., Moradali, M.F. (Eds.), 2021. Biopolymers for Biomedical and Biotechnological Applications. John Wiley & Sons, Hoboken, NJ, USA.

Rezoagli, E., Major, I., Buckley, C., Zainol Abidin, I.I., Waltzer Fehrenbach, G., Montgomery, T., Pogue, R., Murphy, D.E.J., Murray, P., 2023. Polysaccharides—Naturally occurring immune modulators. Front. Immunol. 14, 1154285. https://doi.org/10.3389/fimmu.2023.1154285.

Roberson, E.B., Firestone, M.K., 1992. Relationship between desiccation and exopolysaccharide production in a soil *Pseudomonas* sp. Appl. Environ. Microbiol. 58 (4), 1284–1291. https://doi.org/10.1128/aem.58.4.1284-1291.1992.

Rosalam, S., England, R., 2006. Review of xanthan gum production from unmodified starches. Enzyme Microb. Technol. 39 (2), 197–207. https://doi.org/10.1016/j.enzmictec.2005.10.019.

Rottava, I., Batesini, G., Silva, M.F., Lerin, L., de Oliveira, D., Padilha, F.F., Toniazzo, G., Mossi, A., Cansian, R.L., Di Luccio, M., Treichel, H., 2009. Xanthan gum production and rheological behavior. Carbohydr. Polym. 77 (1), 65–71. https://doi.org/10.1016/j.carbpol.2008.12.008.

Ruiz-Herrera, J., 1991. Fungal Cell Wall: Structure, Synthesis, and Assembly. CRC Press, Boca Raton, FL.

Saldías, C., Velásquez, L., Quezada, C., Leiva, A., 2015. Physicochemical assessment of Dextran-g-PCL micellar nanoaggregates. Carbohydr. Polym. 117, 458–467. https://doi.org/10.1016/j.carbpol.2014.09.095.

Sharma, S., Mukkur, T.K.S., Benson, H.A., Chen, Y., 2009. Pharmaceutical aspects of intranasal delivery. J. Pharm. Sci. 98 (3), 812–843. https://doi.org/10.1002/jps.21473.

Silva, L.A., Lopes Neto, J.H.P., Cardarelli, H.R., 2019. Exopolysaccharides produced by *Lactobacillus plantarum*. Ann. Microbiol. 69 (4), 321–328. https://doi.org/10.1007/s13213-019-01432-9.

Srisuk, P., Berti, F.V., da Silva, L.P., Marques, A.P., Reis, R.L., Correlo, V.M., 2018. Electroactive gellan gum/polyaniline spongy-like hydrogels. ACS Biomater. Sci. Eng. 4 (5), 1779–1787. https://doi.org/10.1021/acsbiomaterials.8b00067.

Strauss, J., Burnham, N.A., Camesano, T.A., 2009. Role of LPS O-antigen on adhesion of *E. coli*. J. Mol. Recognit. 22 (5), 347–355. https://doi.org/10.1002/jmr.940.

Sutherland, I.W., 2001. Biofilm exopolysaccharides: A strong and sticky framework. Microbiology 147 (1), 3–9. https://doi.org/10.1099/00221287-147-1-3.

Sutherland, I.W., 1998. Novel and established applications of microbial polysaccharides. Trends Biotechnol. 16 (1), 41–46. https://doi.org/10.1016/S0167-7799(97)01139-6.

Tabibloghmany, F.S., Ehsandoost, E., 2014. Exopolysaccharides produced by lactic acid bacteria in dairy industry. Eur. J. Food Res. Rev. 4 (4), 150–168.

Tiwari, S., Kavitake, D., Devi, P.B., Shetty, P.H., 2021. Bacterial exopolysaccharides for yoghurt quality. Int. J. Biol. Macromol. 183, 1585–1595. https://doi.org/10.1016/j.ijbiomac.2021.04.092.

Torres, F.G., Commeaux, S., Troncoso, O.P., 2012. Biocompatibility of bacterial cellulose based biomaterials. J. Funct. Biomater. 3 (4), 864–878. https://doi.org/10.3390/jfb3040864.

Tran, T.H., Choi, J.Y., Ramasamy, T., Truong, D.H., Nguyen, C.N., Choi, H.G., Yong, C.S., Kim, J.O., 2014. Hyaluronic acid-coated solid lipid nanoparticles for cancer therapy. Carbohydr. Polym. 114, 407–415. https://doi.org/10.1016/j.carbpol.2014.07.041.

Tsurkan, M.V., Voronkina, A., Khrunyk, Y., Wysokowski, M., Petrenko, I., Ehrlich, H., 2021. Progress in chitin analytics. Carbohydr. Polym. 252, 117204. https://doi.org/10.1016/j.carbpol.2020.117204.

Wang, K., Liu, M., Mo, R., 2020. Polysaccharide-based biomaterials for protein delivery. Med. Drug Disc. 7, 100031. https://doi.org/10.1016/j.medidd.2020.100031.

Welman, A.D., Maddox, I.S., 2003. Exopolysaccharides from lactic acid bacteria. Trends Biotechnol. 21 (6), 269–274. https://doi.org/10.1016/S0167-7799(03)00106-3.

Widyaningrum, D., Meindrawan, B., 2020. Application of microbial extracellular polymeric substances in food industry. IOP Conf. Ser.: Earth and Environ. Sci. 426 (1), 012181. https://doi.org/10.1088/1755-1315/426/1/012181.

Wingender, J., Neu, T.R., Flemming, H.C., 1999. What are bacterial extracellular polymeric substances? In: Wingender, J., Neu, T.R., Flemming, H.C. (Eds.), Microbial Extracellular Polymeric Substances: Characterization, Structure and Function. Springer, Berlin, Germany, pp. 1–19. https://doi.org/10.1007/978-3-642-60147-7_1.

Woodward, R., Yi, W., Li, L., Zhao, G., Eguchi, H., Sridhar, P.R., Guo, H., Song, J.K., Motari, E., Cai, L., Kelleher, P., 2010. Bacterial polysaccharide biosynthesis: Functions of Wzy and Wzz. Nat. Chem. Biol. 6 (6), 418–423. https://doi.org/10.1038/nchembio.364.

Yang, W.Z., Chen, H.L., Gao, F.P., Chen, M.M., Li, X.M., Zhang, M.M., Zhang, Q.Q., Liu, L.R., Jiang, Q., Wang, Y.S., 2010. Cholesterol-modified pullulan nanoparticles. Curr. Nanosci. 6 (3), 298–306. https://doi.org/10.2174/157341310791162703.

Zannini, E., Waters, D.M., Coffey, A., Arendt, E.K., 2016. Production, properties, and industrial food application of lactic acid bacteria-derived exopolysaccharides. Appl. Microbiol. Biotechnol. 100 (3), 1121–1135. https://doi.org/10.1007/s00253-015-7173-2.

Zhang, D., Chen, L., Cai, J., Dong, Q., Din, Z.U., Hu, Z.Z., Wang, G.Z., Ding, W.P., He, J.R., Cheng, S.Y., 2023. Pullulan-based nano-drug delivery systems. J. Drug Delivery Sci. Technol. 79, 104021. https://doi.org/10.1016/j.jddst.2022.104021.

CHAPTER

26

The emerging technology of biofuel production from exopolysaccharides

Hrishav Das Purkayastha[a], *Debasmita Deb*[a], *Ananya Bhattacharjee*[b], *Nirupam Das*[a] and *Abhishek Bhattacharjee*[a]

[a]Department of Pharmaceutical Sciences, Assam University (A Central University), Silchar, Assam, India [b]Symbiosis Institute of Technology, Symbiosis International (Deemed University), Lavale, Pune, Maharashtra, India

26.1 Introduction

The content under this heading explains biofuels as renewable energy sources derived from biomass through various processes. It highlights the environmental and economic benefits of biofuels over fossil fuels. The introduction emphasizes the growing global energy demands and environmental impact from fossil fuels, underscoring biofuels' potential as affordable, clean energy alternatives. It also notes the need for continued research to optimize production methods and further develop biofuels as dependable energy sources.

26.1.1 Background of biofuel

Biofuels can be solid, liquid, and gaseous fuels formed from renewable resources either through the biochemical process, like fermentation, anaerobic digestion, or thermochemical process like pyrolysis, gasification, or transesterification. Scientists are paying more attention to biofuels since they have a number of benefits over fossil fuels (Ambaye et al., 2021). Biofuel is referred to as a type of fuel produced from biomass that comprises plant, algae material, or animal waste. Unlike fossil fuels like coal, petroleum, and natural gas, biofuel is a sustainable energy source since its feedstock material is readily regenerated. Biofuel is marketed as an affordable and environmentally friendly substitute for petroleum and other fossil fuels,

DOI: https://doi.org/10.1016/B978-0-443-36619-2.00026-7

especially in light of growing petroleum prices and global warming (Ruan et al., 2019). There are severe issues in the modern era due to rising global energy consumption and carbon dioxide emissions from the burning of fossil fuels. Furthermore, biofuels have unique properties like nontoxicity, biodegradability, cost-effectiveness, and availability (Malik et al., 2024). Biofuels have been used as sustainable and environmentally friendly energy sources by considering the biomass's composition and ideal production characteristics, such as temperature, catalytic reaction time, chemicals, and catalyst type. A dire need is there to find sustainable, dependable, and consistent energy sources that may be used effectively in place of conventional fuels. Since the term was first used in the early 1970s authors have well-defined "biofuel" as follows: (1) A fuel made up of or by living, micro or macroorganisms; (2) a fuel made directly or indirectly from biomass; (3) a liquid fuel made from biomass, such as biodiesel made from fats and oils, biogas made from animal waste, etc.; (4) a bio-based fuel that is produced chemically from biomass to charcoal, biodiesel, bioethanol, and biomethane or which is naturally derived from wood and wood chips or agricultural residues (Awogbemi et al., 2021). Based on such criteria, it can be concluded that biofuel can be transformed into another variety of biofuel and is produced from solid, liquid, or gaseous materials such as plants, animal waste, manure, sludge, etc. Most biofuels are made using safe manufacturing methods and locally accessible resources (Navas et al., 2020). The production and use of biofuels boost domestic agricultural investment and development. Human health and living situations are also improved by biofuels. Biofuels lower energy imports, improve local livelihoods, and generate jobs (Pugazhendhi et al., 2020). At the macroeconomic level, biofuel contributes to job creation, foreign exchange conservation, and energy price stability.

26.1.2 Definition of exopolysaccharide

Exopolysaccharides (EPS) are special polymers produced by bacteria, fungi, and algae to shield them from external influences. These polymers are recovered from the medium culture using a fermentative process (Li et al., 2023). EPS have been investigated for their immunomodulatory, antiviral, antibacterial, and antitumor properties. Their essential qualities, such as biocompatibility, biodegradability, and absence of irritation, have drawn a lot of interest in innovative drug delivery techniques (Mouro et al., 2024). Dextran, hyaluronic acid, pullulan, xanthan gum, alginate, gellan gum, levan, curdlan, cellulose, chitosan, mauran, and schizophyllan are the exopolysaccharide that exhibit excellent drug carrier properties.

Bacteria create most EPS, which are either expelled from the cell or, depending on the environment, may stay firmly or loosely attached to the bacterial cell surface. In the food, dairy, pharmaceutical, cosmetics, bioremediation, and agricultural sectors, EPS has shown promise as a water-soluble, biocompatible, and renewable substance (Ruijgrok et al., 2024). In a variety of industries, such as biomedicine, food, cosmetics, petroleum, pharmaceuticals, and environmental remediation, bacterial EPS act as vital natural biopolymers. Their unique abilities, such as biocompatibility, biodegradability, greater purity, hydrophilia, anti-inflammatory, antioxidant, anticancer, antibacterial, immune-modulating, and prebiotic activities, are the main reasons for their interest (Netrusov et al., 2023). The EPSs made from bacteria are superior than those that are isolated from algae (agar, fucoidan, carrageenans, and alginates), plants (cellulose, starch, and pectin), and animals (glycogen and chitin). On an industrial scale, bacterial EPSs can be produced in every season or under any kind of weather (Mahmoud et al., 2021).

26.2 Classification of biofuels

(The content classifies biofuels based on physical state into solid, liquid, and gaseous forms, and further categorizes them by technology maturity into conventional and advanced biofuels.)

26.2.1 Based on physical state

26.2.1.1 Solid biofuel

The type of solid feedstock that can be transformed into biofuel is basically considered solid biomass. Solid biofuel can be defined as any solid biomass substance (Chua et al., 2020). Examples include lignocellulosic biomass and various types of solid waste.

26.2.1.2 Liquid biofuel

Any type of renewable fuel in a liquid state is specified as a liquid biofuel. Their primary function is as fuel for transportation. Biodiesel, butanol, biomethanol, bioethanol, propanol, bio-oil, jet fuel, and so forth are noteworthy liquid biofuels (Sandesh and Ujwal, 2021).

26.2.1.3 Gaseous biofuel

The most dominant types of gaseous biofuels are biogas/biomethane, biohydrogen, and biosyngas. Among its many uses, thermal, transportation, and heat applications as well as the production of electricity and power, stood first (Krishnan et al., 2022).

26.2.2 Based on technology maturity

26.2.2.1 Conventional biofuel

Conventional biofuels are produced using technology that is already available on the market. Due to their profitability and economic feasibility, these technologies are still being developed. Bioethanol, biodiesel, and biogas are examples of traditional biofuels that are made by fermentation, transesterification, and anaerobic digestion, respectively (Priya et al., 2022). These technologies have been created, brought to market, and modified for usage in homes or communities.

26.2.2.2 Advanced biofuel

The conversion methods for the majority of advanced biofuels are still primarily in the early phases of development, testing, or commercialization (Vickram et al., 2023). Microalgae biodiesel, biomass-generated esters, lignocellulosic bioethanol, hydrotreated vegetable oil, and biohydrogen are some prominent examples. Furthermore, another classification of biofuel based on the basis of generation of products has been tabulated below in Table 26.1.

26.3 Conversion technologies for bioenergy generation

(The content explains that biomass stores chemical energy via photosynthesis, which can be converted to bioenergy through various technologies influenced by feedstock availability and environmental concerns. Biofuel production technologies are classified into thermochemical and

TABLE 26.1 Classification of biofuels on the basis of the generation of products.

Types	Advantages	Disadvantages
Primary biofuel	• Cost-effective, • Easily accessible, • Do not need technologies for energy conversion.	• Insufficient applications and uses, • Combustion inefficiency (Singh et al., 2019).
First generation Biofuel (1GB)	• Safe to handle, biodegradable, • Beneficial to the environment and readily available, • Assure energy supply for extensive use.	• Compete with food, • It takes time and a lot of arable land to increase feedstock, • Compared to previous generations of biofuels, they have the largest carbon footprint (Saini et al., 2020).
Second generation biofuel (2GB)	• Low cost of feedstock and lower requirement of land, • Every part of the grain, seed, and residue can be used as feedstock, • Readily available in large stock, • Do not compete with the food chain.	• Conversion technologies are still in their infancy. There is a dearth of commercially viable technology, • Complex methods are involved, • Less effective conversion than fuels derived from fossil sources (Guerra et al., 2020).
Third generation biofuel (3GB)	• The energy density per acre is higher than that of traditional crops, • Renewability of bioengineered algae is possible, • The feedstock has a higher nutritional variety, • Very compatible with traditional gasoline-powered engines, • Very low cost of feedstock.	• Expensive system infrastructure, • Compared to diesel, their combustion generates more NO_2 and NO, • The development of production technology is still in its early phases (Molinuevo-Salces et al., 2019).
Fourth generation biofuel (4GB)	• Improved carbon entrapment ability, • Adaptable to genetic and metabolic engineering, • Better methods for fermentation, harvesting, and cultivation.	• Technology is still under development stage, • High cost, • High energy inputs are needed for the process (Bhan et al., 2020).

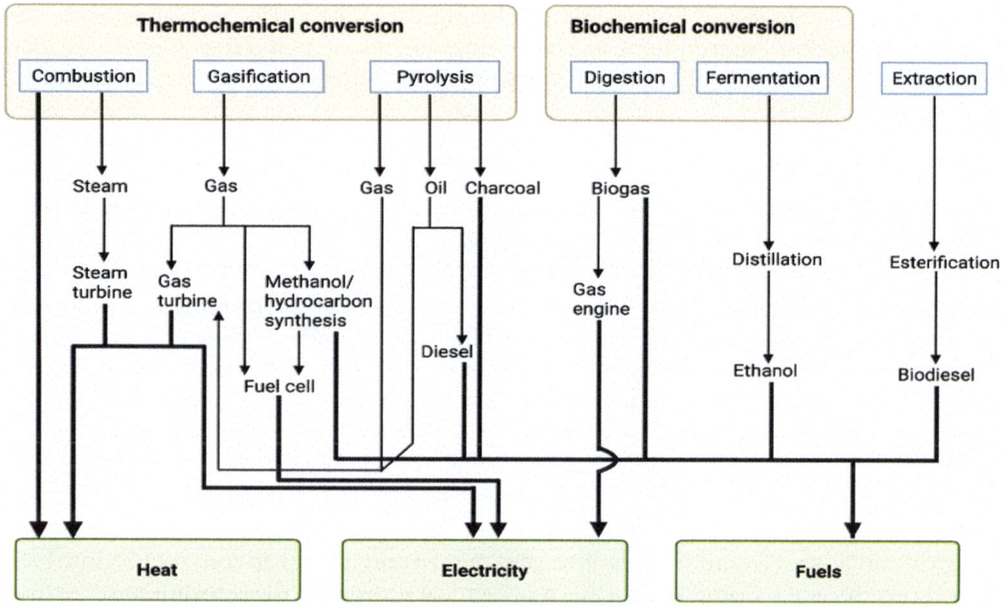

FIGURE 26.1 Pictorial illustration of conversion technique for bioenergy generation (Garba, 2020).

biochemical methods, with recent focus on thermochemical processes for their ability to convert diverse biomass sources into solid, liquid, and gaseous biofuels, supporting sustainable energy solutions.)

During photosynthesis, biomass uses a mix of carbon dioxide and solar energy to store chemical energy in the form of carbohydrates. Because the carbon dioxide absorbed during photosynthesis may be released when it burns, for this reason, it might be used as a source of energy (Sansaniwal et al., 2017). Furthermore, in Fig. 26.1, a streamlined diagram has been given below to make you more aware about all the conversion technologies available for bioenergy generation. It is inexpensive and accessible in a number of formats, including wood, biological material byproducts, organic materials from municipal and sludge wastes, and forest and agricultural residues. Biomass can be transformed into several beneficial products for energy generation and chemicals (Fig. 26.1). The choice of the conversion method to be used on the biomass is influenced by a few aspects. These elements include the biomass feedstock's availability, quality, and quantity, end-product selection, process economics, and environmental concerns (Garba, 2020). Technologies for producing biofuels come in a variety of forms and can be divided into two categories: Thermochemical and biological/biochemical technologies. Biochemical conversion techniques, specifically anaerobic digestion and fermentation, use microorganisms, bacteria, and enzymes to break down biomass into gaseous and liquid biofuels, such as biogas and bioethanol, respectively (Ara Begum et al., 2024). In recent years, research and technical attention have focused on thermochemical biofuels and bioenergy generation technologies due to their ability to process a wide range of biomass feedstocks to produce a number of bioenergy sources, such as solid, liquid, and gaseous biofuels like syngas, biochar, and bio-oil, as well as direct heat energy (Wang and Wu, 2023). These conversion methods present viable options to lessen dependency on fossil fuels while producing useful products and sustainable energy. Their advantages for the environment, as well as ongoing research and development to increase their effectiveness and financial feasibility, motivate us to help create a more sustainable future (Chaudhary et al., 2025).

26.4 Different classified methodology for exopolysaccharides to biofuel production

(The document outlines biological conversion methods for transforming EPS into biofuels using enzymes or microorganisms, emphasizing their eco-friendly nature and mild operating conditions. Key pathways include microbial fermentation producing bioethanol and biohydrogen, anaerobic digestion generating methane-rich biogas, and aerobic degradation/composting for waste-to-energy processes, each influenced by specific environmental factors.)

26.4.1 Biological conversion methods

Biological conversion techniques use enzymes or microorganisms to convert EPS into biofuels via several precise steps. Compared to thermochemical processes, these techniques are thought to be more ecologically friendly and work in milder environments (Mahapatra et al., 2021). The main biological pathways for converting EPS into biofuel are listed below.

26.4.1.1 Microbial fermentation

This is the main biological method in which monomeric sugars produced by EPS hydrolysis are converted into biofuels like bioethanol, biobutanol, or biohydrogen by microorganisms (bacteria, fungi, and yeast) (Keasling et al., 2021). For ethanolic fermentation, microorganisms such as *Saccharomyces cerevisiae, Pichia stipitis*, and *Zymomonas mobilis* are frequently used. High biocatalytic selectivity, low operating temperatures (usually between 30°C and 37°C), and the avoidance of metal toxicity problems that are typical of certain chemical processes are all advantages of this process (El-Fawal et al., 2025). Additionally, we tried to simplify down most of the well-known methodologies by illustration given below in Fig. 26.2.

26.4.1.2 Anaerobic digestion

Contains several microbially driven processes that transform complex organic structures, such as hydrolyzed EPS into biogas, primarily methane and CO_2, in the absence of oxygen. Macromolecules undergo hydrolysis, methanogenesis, the process of producing methane from acids, and acidogenesis/acetogenesis, the conversion to organic acids, hydrogen, and CO_2. Temperature, pH, redox potential, and the balance of the microbial community all affect anaerobic digestion (Osman et al., 2023).

26.4.1.3 Aerobic degradation/composting

In the presence of oxygen, microorganisms break down biomass to produce compost, heat, CO_2, and water. Aerobic composting is a crucial biological process for waste-to-energy cycles and can increase biogas generation in integrated systems, even though it is not a direct biofuel (El-Fawal et al., 2025).

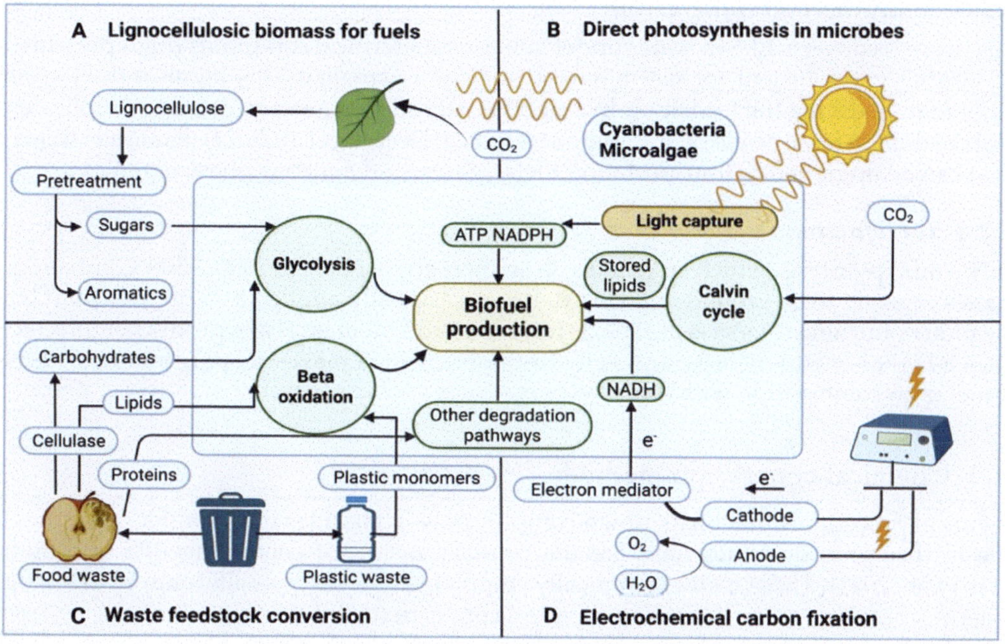

FIGURE 26.2 Pictorial representation of EPS to biofuel production. *EPS*, Exopolysaccharides.

26.4.2 Thermochemical conversion methods

In order to convert complicated polysaccharides into fuels without the need for direct microbial or enzymatic action, thermochemical conversion methods for turning EPS into biofuels use high temperatures, frequently in conjunction with regulated atmospheres or certain catalysts (Canabarro et al., 2013). These techniques may process a variety of feedstocks, are usually quick, and produce solid fuel products, syngas, charcoal, or bio-oil (Chen et al., 2015). The main thermochemical conversion techniques applicable to EPS and similar biomass substrates are:

26.4.2.1 *Pyrolysis*

In this process, EPS breaks down without oxygen, usually at temperatures between 300°C and 700°C. While slow pyrolysis produces more charcoal, fast pyrolysis is preferred for the generation of liquid biofuel. This method works well with materials high in polysaccharides, such as microalgal biomass and EPS (Jerzak et al., 2024). The resulting products include syngas, biochar, and bio-oil.

26.4.2.2 *Gasification*

In gasification, partial oxidation of EPS occurs at high temperatures (700°C–1200°C) in the presence of controlled amounts of oxygen, steam, or air (Tezer et al., 2022). EPS's carbon content is effectively converted by gasification into a clean, adaptable fuel gas (Gao et al., 2023). Syngas is the product that can be further processed into liquid fuels.

26.4.2.3 *Hydrothermal liquefaction*

This process converts EPS in water under sub- or supercritical conditions (high pressure and 250°C–374°C) very efficient for wet biomass, such as microalgal EPS, because the technique does not require drying the feedstock. In a significantly shorter amount of time, it replicates the natural creation of fossil crude oil (Ocampo et al., 2023; Deepika et al., 2024). Example: Bio-crude oil (that can be upgraded to transportation fuels), gases, and aqueous phase organics.

26.4.2.4 *Torrefaction*

Only a mild pyrolysis, which takes place in an inert atmosphere at 200°C–300°C enhances the biomass's capacity to be ground and stored; it is occasionally applied as a pretreatment before gasification or burning (Piersa et al., 2022). Torrefied solid biomass is a byproduct of torrefaction that has a higher energy density and is hydrophobic, which makes it appropriate for use as a solid fuel or in combination with coal.

26.4.3 Chemical conversion methods

Chemical procedures that break down complex polysaccharides into simpler molecules that can be further processed into fuels are the primary means of converting EPS into biofuels (Osman et al., 2021). These methods typically complement or serve as alternatives to biological and thermochemical conversions. Key chemical conversion methodologies include:

26.4.3.1 *Chemical hydrolysis*

The enzyme process of fermentable monosaccharide releases divides the monosaccharide polymers within EPS. High temperatures depolymerize the polysaccharides by concentrated or diluted acids, including hydrochloric or sulfuric acid (Lee et al., 2019). This is necessary to prepare carbohydrates to ferment them by microorganisms into biofuels such as butanol or ethanol (Swiatek et al., 2020). The negative factors are that it needs neutralization procedures, harsh conditions of the reactions, and the potential damage of sugars.

26.4.3.2 *Chemical pretreatment*

Chemical emissions can be used to break down EPS structures by using bases such as sodium hydroxide or acids such as diluted sulfuric acid, which needs further fermentation or hydrolysis. Solvent extraction has the potential to improve the quality of feedstock or help with the extraction of components to convert (Yap et al., 2022).

26.4.3.3 *Catalytic conversion*

One can also transform bio-oil into a higher-quality liquid via catalytic treatment of bio-oil obtained through hydrolysis of the EPS or thermal liquefaction (Sant'Ana Junior et al., 2025). The heterogeneous catalysts, such as metal oxides and zeolites, help in transforming the bioproducts intermediate into liquid fuels that are of good quality and whose quality is enhanced.

26.4.4 Hybrid and emerging methods

Hybrid and emerging methods for biofuel production combine aspects of biological converting biomass, such as EPS, into biofuels.

26.4.4.1 *Emerging consolidated bioprocessing*

To provide the CBP functional organism the necessary capabilities, a high rate of hydrolysis, tolerance to a pretreatment-derived chemical, and appropriate fermentation, two distinct methods have been used. This entirely biological process employs powerful bacteria that can hydrolyze polysaccharides and ferment the resultant sugars in a single step, either naturally occurring or through genetic engineering. Clostridium thermocellum and other thermophilic organisms validate this by combining fermentation and enzyme synthesis, which lowers process stages and expenses (Olguin-Maciel et al., 2020; Periyasamy et al., 2023).

26.4.4.2 *Pulsed electric field pretreatment*

PEF is the emerging pretreatment technology that improves biomass cell disruption, increasing hydrolysis rates and enzyme accessibility, when paired with fermentation, these increase the yield of biofuel (Ranjha et al., 2021; Bocker and Silva, 2024). PEF is not thermal and involves a less energy-intensive comparison to conventional thermal processes that offer a more effective energy use of the plant cell wall components, and is thus potentially a more efficient AD process. The corresponding influence of PEF on producing biogas has yet to be extensively studied. Primarily, up until today, the investigation of PEF has been by means of liquid and semi-liquid substrates, wastewaters, or sludges, among others. The impact of the PEF on solid, LC-type materials degradation has not been researched to a much extent. Nevertheless, the findings of the available studies lead to the conclusion that PEF increases cell permeability and thus releases soluble components and breaks down complex organic molecules into simpler ones, they become more responsive to being the substrates of the biological reactions.

26.4.4.3 *Integrated biorefinery approaches*

In biorefineries, hybrid techniques are frequently used to maximize all biomass fractions through the combinatorial use of thermochemical, chemical, and biological processes to create a variety of fuels and chemicals, enhancing environmental and economic sustainability (Barik et al., 2024).

26.5 Origin of exopolysaccharides assisted biofuel production

(EPS serve as both biofuel feedstocks and biotechnological enhancers, improving efficiency, sustainability, and cost-effectiveness in biofuel production. Initially known for their industrial uses as thickening agents, EPS are now recognized as multifunctional biopolymers with significant roles in microbial ecology, biofilm formation, and emerging applications in bioenergy and environmental engineering.)

EPS are both possible feedstocks and bio-technological enhancers of biofuel production efficiency, sustainability of biofuel production, and cost-effectiveness of biofuel production. The following is an in-depth description of the roles EPS play in the manufacture of biofuels, not only as a source of carbon materials, but also as biofuels production materials (Liu et al., 2025). With the growing pressure on the necessity of environmentally friendly and sustainable alternatives to fuel, research on biologically based systems that can support the production of biofuels has been stirred up. Microbial EPS, high-molecular-weight polymers secreted into the external milieu, have become a subject of not only biotechnology applications, but also biotechnological

TABLE 26.2 List of EPS from various microorganism sources assisting biofuel production.

Microorganism	EPS Type	Assisting Role	Biofuel Type	References
Xanthomonas spp.	Xanthan gum	Hydrolyzed to glucose	Bioethanol	Bhat et al. (2022)
Bacillus spp.	Levan	Biofilm matrix	Biobutanol	Hamada et al. (2022)
Anabaena spp. (cyanobacteria)	Heteropolysaccharide	Bio flocculant	Biodiesel (from algae)	Irawan et al. (2023)
Rhizobium spp.	Gel-like EPS	Methanogenesis enhancer	Biogas	Kim et al. (2024)

EPS, exopolysaccharides.

enhancement of biofuel production (Kaur and Dey, 2022). Although currently, EPS have been known to produce material with good rheological characteristics in the food and pharmaceutical industries, the use of EPS in bioenergy is a new addition (Nguyen et al., 2024). This chapter dates back the comprehension of EPS microbial ecology and biotechnology to its new direction in energy sciences. Further, in Table 26.2, a few lists of EPS and their origin sources have been summarized, which assisted in biofuel production.

It is in the aftermath of the Second World War that, in 1952–1953, vehicles producing EPS such as *Leuconostoc mesenteroides* in a type of slime fall squarely into this category and were found to secrete extracellular materials which altered the viscosity of the medium in which they lived (Ates, 2015). In older studies, the driving interest was industrial in thickening agents. Nonetheless, microbiologists in the 1970s and 1980s started to realize the importance of EPS as key players of biofilms, facilitating the adhesion, protection, and communication of microbes (Mirghani et al., 2022). The introduction of microbial fermentation technologies and the chemistry of carbohydrates made the structure-function relationship of EPS less ambiguous, as a result of which new applications like bioenergy and environmental engineering were developed (Waoo et al., 2023). As microbial fermentation technologies rose to prominence in the late 20th century, and with a synthesis of the relationship between carbohydrate chemistry and functionality, the structural complexity and functional plasticity of EPS have become progressively clearer (Sawant et al., 2025). This knowledge displacement changed the attitude toward EPS as rheological agents only into a perception of EPS as multifunctional biopolymers with possible applications in environmental engineering and renewable energy installations. The combination of the physicochemical properties of their components, i.e., molecular weight distribution, monosaccharide composition, their branching, and functional group chemistry, and their biological functions allowed global researchers to imagine EPS both as biofuel feedstock materials and process supplements (Das et al., 2022).

Within the bioenergy scenario, EPS may be used as a direct carbon source, especially after conversion of degradable fermentable sugars like glucose, galactose, or mannose in hydrolysis (Ojo, 2023). The sugars can now be used by biofuel-producing microorganisms such as the *Saccharomyces cerevisiae* to produce bioethanol or *Clostridium acetobutylicum* to produce biobutanol (Ponsetto et al., 2024). Indicatively, cyanobacteria and micro-algal-derived EPS contain large amounts of hexoses and pentoses that become accessible following weak acid or enzyme-catalyzed hydrolysis, which makes them easily compatible with current fermentation processes without having to significantly invest in lignocellulosic pretreatment. This adaptability decreases the energy and chemical intensity requirements in comparison to traditional biomass processing and enhances the life-cycle energy balance of biofuel production (Chowdhury et al., 2025). EPS can be used as a feedstock, but because of their unique rheological and binding characteristics, they can also boost the efficiency of biofuel production. The EPS excreted by the granule-forming bacteria like *Methanosaeta* and *Methanosarcina* in anaerobic digestion systems adds to the stability of microbial aggregates as well, which enhances the sludge settling behavior as well as the mass transfer (Kazimierowicz and Dębowski, 2022). This leads to more stable methane yields and to increased process resilience to variable operational conditions. In the algal biofuel system, EPS also mediates flocculation of microalgal cells, which leads to energy-efficient harvesting of microalgal biomass, an essential but very challenging step in large-scale microalgal biodiesel production (Diao et al., 2024). An example is the opportunity to aggregate *Chlorella vulgaris* cells using bioflocculants based on *Bacillus* subtilis EPS, which was successfully used, and their primary utilization aim is to aggregate algal cells, so that they are highly concentrated through centrifugation (Wang et al., 2015).

Also, EPS contributes to biofuel-producing cell immobilization, which is a method used to increase productivity, keeping high cell numbers involved and preventing microbial communities against shear-stress or harmful byproducts (Patel et al., 2020). The use of immobilized *Zymomonas mobilis* in EPS-based carrier matrices has already shown the ability to enhance stability of ethanol production over long periods of time when compared to free cell systems, especially in cases where high gravity fermentation is used (Todhanakasem et al., 2019). Also, the use of EPS in biofilm reactors has the capacity to enhance the substrate utilization rates and make biofuel processes continuous, which are in tandem with industrial requirements of affordable and scalable biofuel facilities (Ennaceri et al., 2023). Environmentally-engineeringly speaking, the EPS-mediated bioremediation can go hand in hand with biofuel production in a conceptual biorefinery platform. Treatment of wastewater high in organic material can be achieved with EPS-generating microbial consortia, and concurrently, the loads of pollutants can be reduced, and bioenergy-rich homeostatic carbohydrate-rich biomass can be made (Han et al., 2022). The microalgae biorefinery with the combination of two functions has been proven by pilot-scale works with the activated sludge limit, where the technology of microalgal EPS recovery and further laying of the product has been used as an asset of water treatment efficiency, as well as the obtainment of renewable energy (Mouro et al., 2024). The newly arising role of EPS in bioenergy, therefore, is not restricted to one-dimensional applications but to a scale ranging between principal feedstock use to secondary process intensification and environmental co-gains. Consequently, the EPS-based strategies have potential in meeting various sustainability objectives with regard to valorizing of waste, intensity of processes, as well as a decrease in reliance on fossil fuels. Nevertheless, to unleash their full potential, the following aspects need to be addressed depending on the optimal yield optimization, structural variability, and cost-effective downstream recovery (Jyoti et al., 2024).

26.6 Challenges of bioenergy generation from exopolysaccharides

(This section highlights the challenges in bioenergy generation from EPS, focusing on feedstock limitations such as low natural EPS yields, strain-dependent variability, and costly substrates. It also discusses production difficulties like energy-intensive fermentation, contamination risks, and costly downstream processing, alongside bioconversion hurdles due to EPS's complex structure, all complicating large-scale, cost-effective biofuel production.)

26.6.1 Feedstock-related challenges

As far as EPS-based bioenergy generation is concerned, the nature and the availability of the feedstock pose a major constraint. The natural production of these polymers, under normal conditions of growth, is generally of relatively low concentrations by most of the EPS-producing microorganisms, which makes economically feasible large-scale production challenging (Ahorsu et al., 2018). Moreover, the yield and the makeup of EPS is sensitive to strain, and the amount of a product generated and the physical-chemical characteristics differ significantly in microbial species and strains. This variation requires a specific strain choice or improved genetic engineering techniques to boost productivity (Dirbeba et al., 2022). Price and accessibility to appropriate substrates is also a significant limitation since most EPS-producing microbes demand costly carbon sources like sucrose or glucose in order to obtain optimal yields. The inconsistencies in the supply are further complicated by seasonal changes in the biomass quantity and quality to yield an agricultural residue or algal biomass supply used in production of EPS, hence reducing the efficiency of the process and making it not always economically feasible (Pulighe et al., 2019).

26.6.2 Production process limitations

There are a number of operational challenges with the industrial-scale growth of EPS-producing microorganisms. Fermentation process usually needs high amounts of aeration, agitation and regulation of the temperature, which raise the energy consumption and operation costs. The generation of EPS is also relatively slow, which leads to a longer fermentation process leading to a diminished overall productivity (An et al., 2021). Also, when working with systems of large-scale, they are extremely vulnerable to contamination by undesirable microbial organisms that can quickly affect the quality and yield. The next-step processing can cause even more issues, because EPS recovery off-culture medium can be very viscous, and thus, specialized apparatus and large amounts of energy are usually needed (Nair et al., 2022). A combination of these variables adversely influences scalability and cost-effectiveness of the production of EPS with the purpose of bioenergy use. There can be high initial and continued cost of bioenergy conversion plants particularly a thermochemical in the long-term perspective. Dedicated machinery and excessive temperatures eliminate costs (Nair et al., 2022).

26.6.3 Bioconversion challenges

EPS has a complicated structure that poses a great challenge in converting it to bioenergy. EPS are polysaccharides of high molecular weight, branched, and violently unbranched monosaccharide structure, which is resistant to straightforward fermentation by microbes (Liu et al., 2022). Pretreatment and enzymatic hydrolysis can also be used to get fermentable sugars, which are therefore susceptible to effective conversion, but these processes require effortlessly and the expensive enzymes, including glycosidases (Mignogna et al., 2024). Poor hydrolysis may lead to residual unfermented substances thus decreasing the overall bioenergy. Besides, some pretreatment approaches may lead to the formation of inhibitory products, including furfural and phenolics, that negatively impact downstream fermentation activities and reduce the efficiency of the conversion even further (AlBahnasawi et al., 2024).

26.6.4 Bioenergy yield and efficiency

The energy output of EPS even when successfully converted is usually lower than when using other biomass feeds such as lipids or lignocellulosic feeds. During fermentation, loss of carbon, especially carbon dioxide, is also another factor (Welfle et al., 2023) that decreases net energy recovery. The transition of laboratory to industrial level is usually full of inefficiencies as the perfect conditions in the controlled setting cannot be reproduced in high scale processes. This gap between theory and practice is still a major drawback to making EPS-based bioenergy systems commercially viable (Simon and Hulsbergen, 2021).

26.6.5 Economic and environmental concerns

The commodity price of the production of bioenergy using the EPS method is still high in addition to the combined cost of fermentation, pretreatment, enzymatic hydrolysis, and downstream process. What makes it even more difficult is that EPS have already cleared high-value markets within their pharmaceutical, food, and cosmetic fields, which pose a competition with their usage (Nie et al., 2021). Environmentally, spent biomass and by-products have to be disposed in a manner that does not cause secondary pollution. Also, it is found that, when optimized processes are not carried out, or incorporated into biorefinery systems, the total energy input to the process of EPS production and conversion can offset the net energy output, compromising its sustainability targets (Wu et al., 2023). There is the danger to societal perception and acceptance of bioenergy when bioenergy is perceived to be disruptive on land and food security or the quality of water. The economic viability of the bio energy projects can be affected by the policy and the government incentives as well as the availability of markets where to sell to the bio energy products (Edrisi and Abhilash, 2015).

26.6.6 Technological gaps

Although interest has grown, industrial-scale demonstration plants of bioenergy based on the use of EPS are currently rare, and the standardization of the process is not as established. Recent developments on strain engineering, including CRISPR/Cas-mediated metabolic changes, can potentially increase the productivity of EPS, and their implementation in biofuel-related production is currently at the initial phase (Rehman Zia et al., 2020). In addition, little has been done in terms of linking the production of EPS bioenergy with the co-generation of high-value bio-products that, in turn, might enhance the process economics and sustainability substantially (Tshikovhi and Motaung, 2023).

26.6.7 Regulatory and market challenges

The commercial uptake of EPS-based bioenergy is further constrained by a lack of targeted policy support, subsidies, and regulatory frameworks tailored to this niche sector. Market acceptance is hindered by the absence of proven large-scale economic advantages compared to established biofuels (Rodionova et al., 2017). Additionally, intellectual property restrictions on microbial strains, enzyme technologies, and process designs can limit accessibility and increase production costs for new entrants in the field (M and Rajesh, 2025). Currently, the policy, the government is subsidizing the domestic fuel price; subsequently, causing the production cost of electricity using conventional sources to be cheaper than that of renewable fuels. No special regulations govern the working of utilization of biomass resources, and there are no specific punishments on the nonuse behavior, which should be utilized in a comprehensive manner (Onn et al., 2025). Additionally, the development of the biomass resources industry has no special mechanism, nor does the implementation of corresponding standards and policies have a specialized department (Alsaleh et al., 2021).

26.7 Socioeconomic future scope of exopolysaccharides in bioenergy production

(Future prospects include multiproduct biorefineries co-producing biofuels and value-added products, supported by policy frameworks and innovation, positioning EPS bioenergy as a strategic contributor to energy diversification and a sustainable bioeconomy.)

The EPS-based bioenergy has a socioeconomic potential because it can incorporate the sustainability of microbial biotechnology with renewable energy systems, which has both environmental and economic prospects (M and Rajesh, 2025). With the ever-dropping cost of production with the ongoing strain engineering endeavors, through fermentation technology and the optimization of bioprocess, it is not inconceivable that EPS-derived biofuels may appear as a viable competitor to fossil fuels, resulting in a reduction of greenhouse gases, and an addition to national renewable energy goals. Socioeconomically, the decentralized manufacturing of EPS-based bioenergy through locally sourced bio-substrates like agricultural waste, food processing byproducts, and wastewater streams has the potential to generate employment opportunities in the rural communities and increase energy independence in remotely situated locations (Garg et al., 2023). The technology could also be used to diversify economies of rural areas through the creation of smaller biorefineries, therefore cutting to shorten reliance on centralized supply chains of fossil fuels. Moreover, the valorization of waste biomass to high-value EPS and bioenergy is a concept consistent with the concept of the circular economy that promotes resource-efficiency and minimizes the responsibility of the environment (Varela Perez et al., 2022).

Co-producing biofuels and bioplastics, pharmaceuticals, and functional food ingredients can give full profitability to integrating EPS bioenergy systems into multiproduct biorefineries, in the long run (Igbokwe et al., 2022). In this way, it is possible to attract the interests of private investments, initiate the development of innovations, and create industrial symbiosis among the energy, agricultural, and biotechnology industries. Nevertheless, its socioeconomic gains will be hinged on friendly policy frameworks, economic incentives, government-to-business relationships, and knowledge transfer between the research institutions and the industry stakeholders (Casady et al., 2024). Globally, as climate change mitigation strategies gain momentum, niche bioenergy technologies like EPS-based fuels could gain strategic importance in specialized applications such as aviation biofuels, marine transport, or integration with microbial fuel cells. Thus, while currently at a developmental stage, EPS bioenergy production holds promise for contributing to energy diversification, rural development, and a more sustainable bioeconomy in the coming decades (Li et al., 2024). The fossil energy has the persistent weaknesses despite of the considerable danger to environment even as the energy demand in the global level continues to rise. The global fuel consumption combined with the demand is bound to increase swiftly, and the fossil energy use leads to a serious issue and adverse effect on the environment on earth. The ongoing energy crisis in the world today has attracted major interest in the world. One of the elements needed to resolve the energy problem in the world is renewable energy. The best alternative to address such problems is biofuels (Aslan et al., 2021; Raina et al., 2024).

26.8 Conclusion

The origin of exopolysaccharide-assisted biofuel production is rooted in decades of microbiological and biochemical research. Initially valued for their physical properties, EPS are now being redefined as enablers of sustainable bioenergy. As research continues, EPS stands to play a critical role in advancing the efficiency and sustainability of future biofuel technologies. Although EPS have potential as either a substrate or facilitator to generate bioenergy, notably through their renewability and ability to interact bio-safely, one may keep in mind that various technical, economic, and biological factors prevent the large-scale use of EPS. Future studies in the area should be biased toward gene engineering, streamlining of their processes, minimization of their costs, and life-cycle analysis so as to entirely justify their prospective in sustainable bioenergy systems. In spite of these opportunities, there are enormous scientific, technological, and economic roadblocks still. Limiting factors currently exist, such as low natural yields, strain-specific variability, high substrate prices, complicated downstream processing factors, and inefficient bioconversion efficiencies. Also, the lack of large-scale industrial scale models, and the lack of policy back-ups and competing higher-value EPS applications in the food, pharmaceutical, and cosmetic sectors still limit their market implementation as bioenergy applications. Here, EPS-generated bio energy can become a niche yet significant source of renewable energy portfolio, especially in decentralized, resource-intensive, and rural energy generation systems. With the help of interconnecting biotechnology, energy engineering, and socioeconomic structures, EPS-based bioenergy may become a part of a new sustainable energy projection.

Credit statement

Dr. A. Bhattacharjee: Conceptualization. H. D. Purkayastha, D. Deb: Writing—original draft, visualization. A. Battacharjee: Methodology, writing—review and editing, supervision. Dr. N. Das: Data curation, investigation.

Conflict of interest

All authors declare no competing interest with the work presented in this chapter.

Acknowledgment

Bio render software is acknowledged for the illustrations in this chapter.

References

Ahorsu, R., Medina, F., Constantí, M., 2018. Significance and challenges of biomass as a suitable feedstock for bioenergy and biochemical production: a review. Energies, 11, 3366. https://doi.org/10.3390/en11123366.

Albahnasawi, A., Eyvaz, M., Alazaiza, M.Y.D., Özdoğan, N., Gurbulak, E., Alhout, S., Yuksel, E., 2024. Biomass waste and bioenergy production: challenges and alternatives. In: Srivastav, A.L., Bhardwaj, A.K., Kumar, M. (Eds.), Valorization of Biomass Wastes for Environmental Sustainability: Green Practices for the Rural Circular Economy. Springer Nature Switzerland, Cham, pp. 51–67. https://doi.org/10.1007/978-3-031-52485-1_3.

Alsaleh, M., Abdul-Rahim, A.S., Abdulwakil, M.M., 2021. The importance of worldwide governance indicators for transitions toward sustainable bioenergy industry. J. Env. Manage. 294, 112960. https://doi.org/10.1016/j.jenvman.2021.112960.

Ambaye, T.G., Vaccari, M., Bonilla-Petriciolet, A., Prasad, S., van Hullebusch, E.D., Rtimi, S., 2021. Emerging technologies for biofuel production: a critical review on recent progress, challenges and perspectives. J. Env. Manage. 290, 112627. https://doi.org/10.1016/j.jenvman.2021.112627.

An, Y., Liu, Yu, Liu, Yijing, Lu, M., Kang, X., Mansfield, S.D., Zeng, W., Zhang, J., 2021. Opportunities and barriers for biofuel and bioenergy production from poplar. GCB Bioenergy 13, 905–913. https://doi.org/10.1111/gcbb.12829.

Ara Begum, Y., Kumari, S., Kumar Jain, S., Chandra Garg, M., 2024. A review on waste biomass-to-energy: integrated thermochemical and biochemical conversion for resource recovery. Environ. Sci.: Adv. 3, 1197–1216. https://doi.org/10.1039/D4VA00109E.

Aslan, S., Ceylan, Z.D., Yolcular Karaoğlu, S., Şahin, B., 2021. Sustainability and socioeconomic impacts of bioenergy. bes, 1, 37–42. https://doi.org/10.51606/bes.2021.5.

Ates, O., 2015. Systems biology of microbial exopolysaccharides production. Front. Bioeng. Biotechnol. 3. https://doi.org/10.3389/fbioe.2015.00200.

Awogbemi, O., Kallon, D.V.V., Onuh, E.I., Aigbodion, V.S., 2021. An overview of the classification, production and utilization of biofuels for internal combustion engine applications. Energies 14, 5687. https://doi.org/10.3390/en14185687.

Barik, S.K., Kumar, P., Pal, U.J., Aikat, K., 2024. Integrated biorefinery approach for sustainable biofuel production from algal biomass. Clean Techn Env. Policy. https://doi.org/10.1007/s10098-024-02851-w.

Bhan, C., Verma, L., Singh, Dr.J., 2020. Alternative fuels for sustainable development. In Environmental Concerns and Sustainable Development. Springer, Singapore. 317–331. https://doi.org/10.1007/978-981-13-5889-0_16

Bhat, I.M., Wani, S.M., Mir, S.A., Masoodi, F.A., 2022. Advances in xanthan gum production, modifications and its applications. Biocatal. Agric. Biotechnol. 42, 102328. https://doi.org/10.1016/j.bcab.2022.102328.

Bocker, R., Silva, E.K., 2024. Pulsed electric field technology as a promising pre-treatment for enhancing orange agro-industrial waste biorefinery. RSC Adv. 14, 2116–2133. https://doi.org/10.1039/D3RA07848E.

Canabarro, N., Soares, J.F., Anchieta, C.G., Kelling, C.S., Mazutti, M.A., 2013. Thermochemical processes for biofuels production from biomass. Sustain. Chem. Process 1, 22. https://doi.org/10.1186/2043-7129-1-22.

Casady, C.B., Cepparulo, A., Giuriato, L., 2024. Public-private partnerships for low-carbon, climate-resilient infrastructure: insights from the literature. J. Clean. Prod. 470, 143338. https://doi.org/10.1016/j.jclepro.2024.143338.

Chaudhary, A., Rathour, R.K., Solanki, P., Mehta Kakkar, P., Pathania, S., Walia, A., Baadhe, R.R., Bhatia, R.K., 2025. Recent technological advancements in biomass conversion to biofuels and bioenergy for circular economy roadmap. Renew. Energy 244, 122714. https://doi.org/10.1016/j.renene.2025.122714.

Chowdhury, P., Mahi, N.A., Yeassin, R., Chowdhury, N.U.R., Farrok, O., 2025. Biomass to biofuel: impacts and mitigation of environmental, health, and socioeconomic challenges. Energy Convers. Manag. X 25, 100889. https://doi.org/10.1016/j.ecmx.2025.100889.

Chen, W.-H., Lin, B.-J., Huang, M.-Y., Chang, J.-S., 2015. Thermochemical conversion of microalgal biomass into biofuels: a review. Bioresour. Technol. 184, 314–327. https://doi.org/10.1016/j.biortech.2014.11.050.

Chua, S.Y., Periasamy, L.A., Goh, C.M.H., Tan, Y.H., Mubarak, N.M., Kansedo, J., Khalid, M., Walvekar, R., Abdullah, E.C., 2020. Biodiesel synthesis using natural solid catalyst derived from biomass waste — a review. J. Ind. Eng. Chem. 81, 41–60. https://doi.org/10.1016/j.jiec.2019.09.022.

Das, A., Ringu, T., Ghosh, S., Pramanik, N., 2022. A comprehensive review on recent advances in preparation, physicochemical characterization, and bioengineering applications of biopolymers. Polym. Bull. 80 (7), 7247–7312. https://doi.org/10.1007/s00289-022-04443-4.

Deepika, C., Mrinal, Pohrmen, C.B., Jaiswal, K.S., Sangmesh, B., Jaiswal, K.K., Ramasamy, A.P., Jaiswal, A.K., 2024. Hydrothermal liquefaction of wet microalgal biomass for biofuels and platform chemicals: advances and future prospects. Discov. Appl. Sci. 6, 245. https://doi.org/10.1007/s42452-024-05911-4.

Diao, Y., Gong, X., Xu, D., Duan, P., Wang, S., Guo, Y., 2024. From culture, harvest to pretreatment of microalgae and its high-value utilization. Algal Res. 78, 103405. https://doi.org/10.1016/j.algal.2024.103405.

Dirbeba, M.J., Werkelin, J., Dirbeba, M.J., Werkelin, J., 2022. Challenging biomass feedstocks for energy and chemicals. Biomass, Biorefineries and Bioeconomy. IntechOpen. https://doi.org/10.5772/intechopen.103936.

Edrisi, S.A., Abhilash, P.C., 2015. Book review: socio-economic impacts of bioenergy production. Front. Bioeng. Biotechnol. 3, 174. https://doi.org/10.3389/fbioe.2015.00174.

El-Fawal, E.M., El Naggar, A.M.A., El-Zahhar, A.A., Alghandi, M.M., Morshedy, A.S., El Sayed, H.A., Mohammed, A., elshifa, M.E., 2025. Biofuel production from waste residuals: Comprehensive insights into biomass conversion technologies and engineered biochar applications. RSC Adv., 15, 11942–11974. https://doi.org/10.1039/d5ra00857c.

Ennaceri, H., Ishika, T., Mkpuma, V.O., Moheimani, N.R., 2023. Microalgal biofilms: towards a sustainable biomass production. Algal Res. 72, 103124. https://doi.org/10.1016/j.algal.2023.103124.

Gao, Y., Wang, M., Raheem, A., Wang, F., Wei, J., Xu, D., Song, X., Bao, W., Huang, A., Zhang, S., Zhang, H., 2023. Syngas production from biomass gasification: influences of feedstock properties, reactor type, and reaction parameters. ACS Omega 8, 31620–31631. https://doi.org/10.1021/acsomega.3c03050.

Garba, A., 2020. Biomass conversion technologies for bioenergy generation: an introduction. Biotechnological Applications of Biomass. IntechOpen. https://doi.org/10.5772/intechopen.93669.

Garg, J., Barnwal, P.P., Suman, J., Rakshit, A., 2023. Bioenergy crops. Handbook of Energy Management in Agriculture. Springer, Singapore, pp. 435–450. https://doi.org/10.1007/978-981-19-7736-7_23-1.

Guerra, S., Denadai, M., Saad, A., Spadim, E., Costa, M., 2020. Sugarcane: Biorefinery, Technology, and Perspectives. Academic Press, 49–65. https://doi.org/10.1016/B978-0-12-814236-3.00003-2

Hamada, M.A., Hassan, R.A., Abdou, A.M., Elsaba, Y.M., Aloufi, A.S., Sonbol, H., Korany, S.M., 2022. Bio_fabricated levan polymer from *Bacillus* subtilis MZ292983.1 with antibacterial, antibiofilm, and burn healing properties. Appl. Sci. 12, 6413. https://doi.org/10.3390/app12136413.

Han, M., Zhang, C., Ho, S.-H., 2022. Immobilized microalgal system: an achievable idea for upgrading current microalgal wastewater treatment. Environ. Sci. Ecotechnology 14, 100227. https://doi.org/10.1016/j.ese.2022.100227.

Igbokwe, V.C., Ezugworie, F.N., Onwosi, C.O., Aliyu, G.O., Obi, C.J., 2022. Biochemical biorefinery: a low-cost and non-waste concept for promoting sustainable circular bioeconomy. J. Env. Manage. 305, 114333. https://doi.org/10.1016/j.jenvman.2021.114333.

Irawan, A.P., Rahmawati, A., Fahmi, U.A., Budiman, A., Maghfiroh, K.Q., Erfianti, T., Andeska, D.P., Putri, R.A.E., Nurafifah, I., Sadewo, B.R., Suyono, E.A., 2023. Studies on bioflocculant exopolysaccharides (EPS) produced by Anabaena sp. and its application as bioflocculant for low cost harvesting of Chlorella sp. Asian J Agric Biol doi:10.35495/ajab.2022.150.

Jerzak, W., Acha, E., Li, B., 2024. Comprehensive review of biomass pyrolysis: conventional and advanced technologies, reactor designs, product compositions and yields, and techno-economic analysis. Energies 17, 5082. https://doi.org/10.3390/en17205082.

Jyoti, K., Soni, K., Chandra, R., 2024. Optimization of the production of exopolysaccharide (EPS) from biofilm-forming bacterial consortium using different parameters. Microbe 4, 100117. https://doi.org/10.1016/j.microb.2024.100117.

Kaur, N., Dey, P., 2022. Bacterial exopolysaccharides as emerging bioactive macromolecules: From fundamentals to applications. Res. Microbiol. 174 (4), 104024. https://doi.org/10.1016/j.resmic.2022.104024.

Kazimierowicz, J., Dębowski, M., 2022. Aerobic granular sludge as a substrate in anaerobic digestion—current status and perspectives. Sustainability 14 (17), 10904. https://doi.org/10.3390/su141710904.

Keasling, J., Garcia Martin, H., Lee, T.S., Mukhopadhyay, A., Singer, S.W., Sundstrom, E., 2021. Microbial production of advanced biofuels. Nat. Rev. Micro. 19, 701–715. https://doi.org/10.1038/s41579-021-00577-w.

Kim, K., Oh, E., Park, S., Jeong, J., Jeon, S., Lee, S., Shin, Y., Jung, S., 2024. Enhanced rheological and structural properties of the exopolysaccharide from *Rhizobium leguminosarum* VF39 through NTG Mutagenesis. Polymers 16, 3179. https://doi.org/10.3390/polym16223179.

Krishnan, R.Y., Manikandan, S., Subbaiya, R., Kim, W., Karmegam, N., Govarthanan, M., 2022. Advanced thermochemical conversion of algal biomass to liquid and gaseous biofuels: a comprehensive review of recent advances. Sustain. Energy Technol. Assess. 52, 102211. https://doi.org/10.1016/j.seta.2022.102211.

Lee, S.Y., Sankaran, R., Chew, K.W., Tan, C.H., Krishnamoorthy, R., Chu, D.-T., Show, P.-L., 2019. Waste to bioenergy: a review on the recent conversion technologies. BMC Energy 1 (4). https://doi.org/10.1186/s42500-019-0004-7.

Li, B., Shi, Q., Miao, B., Liu, D., Jin, S., Wang, Z., 2023. Application of exopolysaccharide directionally synthesized by *Xanthomonas campestris* as the green selective depressant for the clean flotation of talc: statistical optimization and mechanism analysis. J. Clean. Prod. 383, 135381. https://doi.org/10.1016/j.jclepro.2022.135381.

Li, Y., Sohail, M.T., Zhang, Y., Ullah, S., 2024. Bioenergy for sustainable rural development: elevating government governance with environmental policy in China. Land 13, 2147. https://doi.org/10.3390/land13122147.

Liu, L., Zhang, X., Yin, L., Zhang, H., Li, J., Ma, Y., 2025. Advances and challenges in bioproduction of microbial exopolysaccharides: synthesis mechanisms, engineering strategies, and future perspectives. Carbohydr. Polym., 124010. https://doi.org/10.1016/j.carbpol.2025.124010.

Liu, T., Miao, P., Shi, Y., Tang, K.H.D., Yap, P.-S., 2022. Recent advances, current issues and future prospects of bioenergy production: a review. Sci. Total Env. 810, 152181. https://doi.org/10.1016/j.scitotenv.2021.152181.

M, L.H., Rajesh, P., 2025. Bioenergy in India: challenges, potential and policy perspectives. Int. J. Innov. Sci. Eng. Manag. 148–154. https://doi.org/10.69968/ijisem.2025v4i1148-154.

Mahapatra, S., Kumar, D., Singh, B., Sachan, P.K., 2021. Biofuels and their sources of production: a review on cleaner sustainable alternative against conventional fuel, in the framework of the food and energy nexus. Energy Nexus 4, 100036. https://doi.org/10.1016/j.nexus.2021.100036.

Mahmoud, Y.A.-G., El-Naggar, M.E., Abdel-Megeed, A., El-Newehy, M., 2021. Recent advancements in microbial polysaccharides: synthesis and applications. Polymers 13, 4136. https://doi.org/10.3390/polym13234136.

Malik, K., Capareda, S.C., Kamboj, B.R., Malik, S., Singh, K., Arya, S., Bishnoi, D.K., 2024. Biofuels production: a review on sustainable alternatives to traditional fuels and energy sources. Fuels 5, 157–175. https://doi.org/10.3390/fuels5020010.

Mignogna, D., Szabó, M., Ceci, P., Avino, P., 2024. Biomass energy and biofuels: perspective, potentials, and challenges in the energy transition. Sustainability 16, 7036. https://doi.org/10.3390/su16167036.

Mirghani, R., Saba, T., Khaliq, H., Mitchell, J., Do, L., Chambi, L., Diaz, K., Kennedy, T., Alkassab, K., Huynh, T., Elmi, M., 2022. Biofilms: formation, drug resistance and alternatives to conventional approaches. AIMS Microbiol. 8 (3), 239–277. https://doi.org/10.3934/microbiol.2022019.

Molinuevo-Salces, B., Riaño, B., Hernández, D., García-González, M., 2019. Microalgae and Wastewater Treatment: Advantages and Disadvantages. Springer Nature, Singapore. 505–534. https://doi.org/10.1007/978-981-13-2264-8_20

Mouro, C., Gomes, A.P., Gouveia, I.C., 2024. Microbial exopolysaccharides: structure, diversity, applications, and future frontiers in sustainable functional materials. Polysaccharides 5, 241–287. https://doi.org/10.3390/polysaccharides5030018.

Nair, L.G., Agrawal, K., Verma, P., 2022. An overview of sustainable approaches for bioenergy production from agro-industrial wastes. Energy Nexus 6, 100086. https://doi.org/10.1016/j.nexus.2022.100086.

Navas, M.B., Ruggera, J.F., Lick, I.D., Casella, M.L., 2020. A sustainable process for biodiesel production using Zn/Mg oxidic species as active, selective and reusable heterogeneous catalysts. Bioresour. Bioprocess 7 (4). https://doi.org/10.1186/s40643-019-0291-3.

Netrusov, A.I., Liyaskina, E.V., Kurgaeva, I.V., Liyaskina, A.U., Yang, G., Revin, V.V., 2023. Exopolysaccharides producing bacteria: a review. Microorganisms 11, 1541. https://doi.org/10.3390/microorganisms11061541.

Nguyen, H.-T., Pham, T.T., Nguyen, P.T., Le-Buanec, H., Rabetafika, H.N., Razafindralambo, H.L., 2024. Advances in Microbial exopolysaccharides: present and future applications. Biomolecules 14 (9), 1162. https://doi.org/10.3390/biom14091162.

Nie, A., Kung, S.-S., Li, H., Zhang, L., He, X., Kung, C.-C., 2021. An environmental and economic assessment from bioenergy production and biochar application. J. Saudi Chem. Soc. 25, 101173. https://doi.org/10.1016/j.jscs.2020.11.006.

Ocampo, E., Beltrán, V., Ocampo, D., Gómez, E., Ríos, L., 2023. Hydrothermal liquefaction process: review and trends. Curr. Res. Green Sustain. Chem. 7, 100382. https://doi.org/10.1016/j.crgsc.2023.100382.

Ojo, A., 2023. An overview of lignocellulose and its biotechnological importance in high-value product production. Fermentation 9 (11), 990. https://doi.org/10.3390/fermentation9110990.

Olguin-Maciel, E., Singh, A., Chable-Villacis, R., Tapia-Tussell, R., Ruiz, H.A., 2020. Consolidated bioprocessing, an innovative strategy towards sustainability for biofuels production from crop residues: an overview. Agronomy 10, 1834. https://doi.org/10.3390/agronomy10111834.

Onn, S.M., Koh, G.J., Yap, W.H., Teoh, M.-L., Low, C.-F., Goh, B.-H., 2025. Recent advances in genetic engineering of microalgae: bioengineering strategies, regulatory challenges and future perspectives. J. Appl. Phycol. 37, 247–264. https://doi.org/10.1007/s10811-024-03367-y.

Osman, A.I., Lai, Z.Y., Farghali, M., Yiin, C.L., Elgarahy, A.M., Hammad, A., Ihara, I., Al-Fatesh, A.S., Rooney, D.W., Yap, P.-S., 2023. Optimizing biomass pathways to bioenergy and biochar application in electricity generation, biodiesel production, and biohydrogen production. Env. Chem. Lett. 21, 2639–2705. https://doi.org/10.1007/s10311-023-01613-2.

Osman, A.I., Mehta, N., Elgarahy, A.M., Al-Hinai, A., Al-Muhtaseb, A.H., Rooney, D.W., 2021. Conversion of biomass to biofuels and life cycle assessment: a review. Env. Chem. Lett. 19, 4075–4118. https://doi.org/10.1007/s10311-021-01273-0.

Patel, A., Karageorgou, D., Rova, E., Katapodis, P., Rova, U., Christakopoulos, P., Matsakas, L., 2020. An overview of potential oleaginous microorganisms and their role in biodiesel and omega-3 fatty acid-based industries. Microorganisms 8 (3), 434. https://doi.org/10.3390/microorganisms8030434.

Periyasamy, S., Beula Isabel, J., Kavitha, S., Karthik, V., Mohamed, B.A., Gizaw, D.G., Sivashanmugam, P., Aminabhavi, T.M., 2023. Recent advances in consolidated bioprocessing for conversion of lignocellulosic biomass into bioethanol – a review. Chem. Eng. J. 453, 139783. https://doi.org/10.1016/j.cej.2022.139783.

Piersa, P., Unyay, H., Szufa, S., Lewandowska, W., Modrzewski, R., Slezak, R., Ledakowicz, S., 2022. An extensive review and comparison of modern biomass torrefaction reactors vs. biomass pyrolysis—1. Energies 15, 2227. https://doi.org/10.3390/en15062227.

Ponsetto, P., Sasal, E.M., Mazzoli, R., Valetti, F., Gilardi, G., 2024. The potential of native and engineered Clostridia for biomass biorefining. Front. Bioeng. Biotechnol. 12. https://doi.org/10.3389/fbioe.2024.1423935.

Priya, Deora, P.S., Verma, Y., Muhal, R.A., Goswami, C., Singh, T., 2022. Biofuels: an alternative to conventional fuel and energy source. Mater. Today: Proc. SCPINM-2021, 48, 1178–1184. https://doi.org/10.1016/j.matpr.2021.08.227.

Pugazhendhi, A., Alagumalai, A., Mathimani, T., Atabani, A.E., 2020. Optimization, kinetic and thermodynamic studies on sustainable biodiesel production from waste cooking oil: an Indian perspective. Fuel 273, 117725. https://doi.org/10.1016/j.fuel.2020.117725.

Pulighe, G., Bonati, G., Colangeli, M., Morese, M.M., Traverso, L., Lupia, F., Khawaja, C., Janssen, R., Fava, F., 2019. Ongoing and emerging issues for sustainable bioenergy production on marginal lands in the Mediterranean regions. Renew. Sustain. Energy Rev. 103, 58–70. https://doi.org/10.1016/j.rser.2018.12.043.

Raina, N., Chuetor, S., Elalami, D., Tayibi, S., Barakat, A., 2024. Biomass valorization for bioenergy production: current techniques, challenges, and pathways to solutions for sustainable bioeconomy. Bioenerg. Res. 17, 1999–2028. https://doi.org/10.1007/s12155-024-10792-x.

Ranjha, M.M.A.N., Kanwal, R., Shafique, B., Arshad, R.N., Irfan, S., Kieliszek, M., Kowalczewski, P.Ł., Irfan, M., Khalid, M.Z., Roobab, U., Aadil, R.M., 2021. A critical review on pulsed electric field: a novel technology for the extraction of phytoconstituents. Molecules 26, 4893. https://doi.org/10.3390/molecules26164893.

Rehman Zia, U.U., Rashid, T., Awan, W.N., Hussain, A., Ali, M., 2020. Quantification and technological assessment of bioenergy generation through agricultural residues in Punjab (Pakistan). Biomass Bioenergy 139, 105612. https://doi.org/10.1016/j.biombioe.2020.105612.

Rodionova, M.V., Poudyal, R.S., Tiwari, I., Voloshin, R.A., Zharmukhamedov, S.K., Nam, H.G., Zayadan, B.K., Bruce, B.D., Hou, H.J.M., Allakhverdiev, S.I., 2017. Biofuel production: challenges and opportunities. Int. J. Hydrog. Energy 42, 8450–8461. https://doi.org/10.1016/j.ijhydene.2016.11.125.

Ruan, R., Zhang, Y., Chen, P., Liu, S., Fan, L., Zhou, N., Ding, K., Peng, P., Addy, M., Cheng, Y., Anderson, E., Wang, Y., Liu, Y., Lei, H., Li, B., 2019. Biofuels: introduction. In: Pandey, A., Larroche, C., Dussap, C.-G., Gnansounou, E., Khanal, S.K., Ricke, S. (Eds.), Biofuels: Alternative Feedstocks and Conversion Processes for the Production of Liquid and Gaseous Biofuels. Academic Press, pp. 3–43. https://doi.org/10.1016/B978-0-12-816856-1.00001-4.

Ruijgrok, G., Wu, D.-Y., Overkleeft, H.S., Codée, J.D.C., 2024. Synthesis and application of bacterial exopolysaccharides. Curr. Opin. Chem. Biol. 78, 102418. https://doi.org/10.1016/j.cbpa.2023.102418.

Saini, S., Chandel, A., Sharma, K., 2020. Past practices and current trends in recovery and purification of first generation ethanol: a learning curve for lignocellulosic ethanol. J. Clean. Prod. 268, 122357. https://doi.org/10.1016/j.jclepro.2020.122357.

Sandesh, K., Ujwal, P., 2021. Trends and perspectives of liquid biofuel – Process and industrial viability. Energy Convers. Manag.: X 10, 100075. https://doi.org/10.1016/j.ecmx.2020.100075.

Sansaniwal, S.K., Pal, K., Rosen, M.A., Tyagi, S.K., 2017. Recent advances in the development of biomass gasification technology: a comprehensive review. Renew. Sustain. Energy Rev. 72, 363–384. https://doi.org/10.1016/j.rser.2017.01.038.

Sant'Ana Júnior, D.B., Kelbert, M., Hermes de Araújo, P.H., de Andrade, C.J., 2025. Physical pretreatments of lignocellulosic biomass for fermentable sugar production. Sustain. Chem. 6, 13. https://doi.org/10.3390/suschem6020013.

Sawant, S.S., Park, H.Y., Sim, E.Y., Kim, H.S., Choi, H.S., 2025. Microbial fermentation in food: impact on functional properties and nutritional enhancement—a review of recent developments. Fermentation 11 (1), 15. https://doi.org/10.3390/fermentation11010015.

Simon, R.O., Hülsbergen, K.-J., 2021. Energy balance and energy use efficiency of annual bioenergy crops in field experiments in Southern Germany. Agronomy 11, 1835. https://doi.org/10.3390/agronomy11091835.

Singh, D., Sharma, D., Soni, S., Sharma, S., Sharma, P., Jhalani, A., 2019. A review on feedstocks, production processes, and yield for different generations of biodiesel. Fuel 262, 116553. https://doi.org/10.1016/j.fuel.2019.116553.

Świątek, K., Gaag, S., Klier, A., Kruse, A., Sauer, J., Steinbach, D., 2020. Acid hydrolysis of lignocellulosic biomass: sugars and furfurals formation. Catalysts 10, 437. https://doi.org/10.3390/catal10040437.

Tezer, Ö., Karabağ, N., Öngen, A., Çolpan, C.Ö., Ayol, A., 2022. Biomass gasification for sustainable energy production: a review. Int. J. Hydrog. Energy Green Hydrog. Glob. Decarbonization 47, 15419–15433. https://doi.org/10.1016/j.ijhydene.2022.02.158.

Todhanakasem, T., Salangsing, O.L., Koomphongse, P., Kaewket, S., Kanokratana, P., Champreda, V., 2019. *Zymomonas mobilis* biofilm reactor for ethanol production using rice straw hydrolysate under continuous and repeated batch processes. Front. Microbiol. 10. https://doi.org/10.3389/fmicb.2019.01777.

Tshikovhi, A., Motaung, T.E., 2023. Technologies and innovations for biomass energy production. Sustainability 15, 12121. https://doi.org/10.3390/su151612121.

Varela Pérez, P., Greiner, B.E., von Cossel, M., 2022. Socio-economic and environmental implications of bioenergy crop cultivation on marginal African drylands and key principles for a sustainable development. Earth 3, 652–682. https://doi.org/10.3390/earth3020038.

Vickram, S., Manikandan, S., Deena, S.R., Mundike, J., Subbaiya, R., Karmegam, N., Jones, S., Kumar Yadav, K., Chang, S., woong, Ravindran, B., Kumar Awasthi, M., 2023. Advanced biofuel production, policy and technological implementation of nano-additives for sustainable environmental management – A critical review. Bioresour. Technol. 387, 129660. https://doi.org/10.1016/j.biortech.2023.129660.

Wang, Y., Wu, J.J., 2023. Thermochemical conversion of biomass: potential future prospects. Renew. Sustain. Energy Rev. 187, 113754. https://doi.org/10.1016/j.rser.2023.113754.

Wang, Y., et al., 2015. Optimization of *Chlorella vulgaris* and bioflocculant-producing bacteria co-culture: enhancing microalgae harvesting and lipid content. Lett. Appl. Microbiol. 60 (5), 497–503. https://doi.org/10.1111/lam.12403.

Waoo, A.A., et al., 2023. Microbial exopolysaccharides in the biomedical and pharmaceutical industries. Heliyon 9 (8), e18613. https://doi.org/10.1016/j.heliyon.2023.e18613.

Welfle, A.J., Almena, A., Arshad, M.N., Banks, S.W., Butnar, I., Chong, K.J., Cooper, SamuelJ.G., Daly, H., Garcia Freites, S., Güleç, F., Hardacre, C., Holland, R., Lan, L., Lee, C.S., Robertson, P., Rowe, R., Shepherd, A., Skillen, N., Tedesco, S., Thornley, P., Verdía Barbará, P., Watson, I., Williams, O.S.A., Röder, M., 2023. Sustainability of bioenergy – mapping the risks & benefits to inform future bioenergy systems. Biomass Bioenergy 177, 106919. https://doi.org/10.1016/j.biombioe.2023.106919.

Wu, B., Lin, R., Bose, A., Huerta, J.D., Kang, X., Deng, C., Murphy, J.D., 2023. Economic and environmental viability of biofuel production from organic wastes: a pathway towards competitive carbon neutrality. Energy 285, 129322. https://doi.org/10.1016/j.energy.2023.129322.

Yap, B.J.T., Heng, G.C., Ng, C.A., 2022. Chemical extraction methods for activated sludge extracellular polymeric substances (EPS): a review. E3S Web Conf., 347, 04015. https://doi.org/10.1051/e3sconf/202234704015.

Index

Page numbers followed by "*f*" and "*t*" indicate, figures and tables respectively.

F

Printed and bound by CPI Group (UK) Ltd, Croydon, CR0 4YY
25/09/2026
02222749-0014